GEOTECHNICAL AND GEOENVIRONMENTAL ENGINEERING HANDBOOK

GEOTECHNICAL AND GEOENVIRONMENTAL ENGINEERING HANDBOOK

Edited by

R. Kerry Rowe

QUEEN'S UNIVERSITY,
KINGSTON, ONTARIO, CANADA

Springer Science+Business Media, LLC

 Electronic Services <http://www.wkap.nl>

Library of Congress Cataloging-in-Publication Data
Geotechnical and geoenvironmental engineering handbook/edited by
  R. Kerry Rowe
      p.   cm.
  Includes bibliographical references.
  ISBN 978-1-4613-5699-8   ISBN 978-1-4615-1729-0 (eBook)
  DOI 10.1007/978-1-4615-1729-0
  1. Engineering geology handbooks, manuals, etc.   I. Rowe, R. K.
TA705.G427   2000
624. 1'51—dc21   99-37319
                                                CIP

Copyright © 2001 by Springer Science+Business Media New York
Originally published by Kluwer Academic Publishers in 2001

Printed on acid-free paper.

Dedicated in memory of
J. R. Booker
E. H. Davis
M. Novak
R. M. Quigley

CONTENTS

11. Foundations on Rock   305

K. Y. Lo and A. M. Hefny

12. Dynamics of Foundations   337

M. H. El Naggar

III. SLOPE, EMBANKMENT AND WALL STABILITY, AND SOIL IMPROVEMENT   395

14. Slopes and Mass Movements   397

S. Leroueil, J. Locat, G. Sève, L. Picarelli and R. M. Faure

16. Embankments Over Soft Soil and Peat   463

S. Leroueil and R. K. Rowe

21. Earthquake Engineering   615

W. D. L. Finn

V. GEOENVIRONMENTAL ENGINEERING   661

22. Geoenvironmental Problem Identification and Risk Management   663
M. Whittaker, J. G. Sprenger and D. D. DuBois

27. Covers for Waste 825

R. Bonaparte and E. K. Yanful

LIST OF FIGURES

LIST OF TABLES

CONTRIBUTING AUTHORS

Dr S. L. Barbour
Department of Civil Engineering
University of Saskatchewan
57 Campus Drive
Saskatoon
Saskatchewan, Canada S7N 5A9
TEL: 306-966-5369
FAX: 306-966-5427
e-mail: S_L_Barbour@engr.usask.ca

Dr R. J. Bathurst
Department of Civil Engineering
Royal Military College of Canada
Kingston, Ontario, Canada K7K 7B4
TEL: 613-541-6000 x6479
FAX: 613-545-8336
e-mail: bathurst-r@rmc.ca

Dr D. Becker
Golder Associates Ltd
2180 Meadowvale Boulevard
Mississauga, Ontario, Canada L5N 5S3
TEL: 905-567-4444
FAX: 905-567-6561
e-mail: dbecker@golder.com

Dr D. T. Bergado
School of Civil Engineering
Asian Institute of Technology
P.O. Box 4
Klong Luang, Pathumthani, Thailand
TEL: 66-2-516-0110
FAX: 66-2-524-6050
e-mail: bergado@ait.ac.th

Dr D. Blowes
Department of Earth Sciences
University of Waterloo
Waterloo, Ontario, Canada N2L 3G1
TEL: 519-888-4878
FAX: 519-746-3882
e-mail: blowes@sciborg.uwaterloo.ca

Dr R. Bonaparte
GeoSyntec Consultants
1100 Lake Hearn Drive N.E.
Atlanta, Georgia 30342-1523 USA
TEL: 404-705-9500
FAX: 404-705-9400
e-mail: rbonaparte@geosyntec.com

Dr J. A. Cherry
Department of Earth Sciences
University of Waterloo
Waterloo, Ontario, Canada N2L 3G1
TEL: 519-888-4516
FAX: 519-883-0220
e-mail: cherryja@sciborg.uwaterloo.ca

Dr D. DuBois
Golder Associates Ltd
2180 Meadowvale Boulevard
Mississauga, Ontario, Canada L5N 5S3
TEL: 905-567-4444
FAX: 905-567-6561
e-mail: DDuBois@golder.com

Mr W. Dyck
Conestoga-Rovers & Associates Ltd
651 Colby Drive
Waterloo, Ontario, Canada N2V 1C2
TEL: 519-725-3313
FAX: 519-725-1394
e-mail: wdyck@craworld.com

Dr M. H. El Naggar
Department of Civil & Environmental Engi-
    neering
University of Western Ontario
London, Ontario, Canada N6A 5B9
TEL: 519-661-4219
FAX: 519-661-3942
e-mail: naggar@uwo.ca

R. M. Faure
Etudes et recherches en Gènie Civil

Centre d'Etude des Tunnels
25 Av. F. Mitterrand
69500 Bron, France
TEL: 33-472-143481
FAX: 33-472-143490
e-mail: rene-
michel.faure@cetu.equipement.gouv.fr

Dr W. D. L. Finn
Anabuki Chair of Foundation Geodynamics
Kagawa University
2217-20 Shinmachi
Hayashi-cho
Takamatsu City
761-0396 Japan
TEL: 81-87-864-2170
FAX: 81-87-864-2031
e-mail: finn@eng.kagawa-u.ac.jp

Mr G. Ford
Presidio Trust
P.O. Box 29052
San Francisco, CA 94129-0052 USA
TEL: 415-561-4292
FAX: 415-561-4180
e-mail: gford@presidiotrust.gov

Dr D. G. Fredlund
Department of Civil Engineering
University of Saskatchewan
57 Campus Drive
Saskatoon
Saskatchewan, Canada S7N 5A9
TEL: 306-966-5342
FAX: 306-966-5427
e-mail: Del.Fredlund@skyfox.usask.ca

Dr R. Gillham
Department of Earth Sciences
University of Waterloo
Waterloo, Ontario, Canada N2L 3G1
TEL: 519-888-4658
FAX: 519-746-1829
e-mail: rwgillha@sciborg.uwaterloo.ca

Dr J.-P. Giroud
GeoSyntec Consultants
621 N.W. 53rd St., Suite 650
Boca Raton, FL 33487 USA
TEL: 561-995-0900 x213
FAX: 561-995-0925
e-mail: jpgiroud@geosyntec.com

Dr Ralph Haas
Department of Civil Engineering
University of Waterloo
Waterloo, Ontario, Canada N2L 3G1
TEL: 888-4567 x2176
FAX: 519-888-6197
e-mail: haas@uwaterloo.ca

Dr A. M. Hefny
School of Civil & Structural Engineering
Division of Geotechnics & Surveying
Nanyang Technological University
Nanyang, Singapore 639798
TEL: 65-790-6936
FAX: 65-792-1650
e-mail: camhefny@ntu.edu.sg

Dr R. D. Holtz
Department of Civil & Environmental Engi-
neering
University of Washington
Wilcox Hall 260, Box 352700
Seattle, WA 98195-2700 USA
TEL: 206-543-7614
FAX: 206-685-3836
e-mail: holtz@u.washington.edu

Dr Y. G. Hsuan
Geosynthetics Institute
475 Kedron Avenue
Folsom, PA 19033-1208 USA
TEL: 610-522-8440
FAX: 610-522-8441
e-mail: ghsuan@coe.drexel.edu

Dr C. J. F. P. Jones
Department of Civil Engineering
The University
Newcastle Upon Tyne NE1 7RU UK
TEL: 44-191-222-7117
FAX: 44-191-222-6613
e-mail: c.j.f.p.jones@newcastle.ac.uk

Dr R. M. Koerner
Geosynthetics Institute
475 Kedron Avenue
Folsom, PA 19033-1208 USA
TEL: 610-522-8440
FAX: 610-522-8441
e-mail: Robert.Koerner@cbis.ece.drexel.edu

Dr J.-M. Konrad
Département de génie civil
Fac. des Sciences et de Génie
Université Laval
Sainte-Foy, Quebec, Canada G1K 7P4
TEL: 418-656-2131 x3878
FAX: 418-656-2928
e-mail: Jean-Marie.Konrad@gci.ulaval.ca

Dr P. V. Lade
Department of Civil Engineering
Aalborg University
Sohngaardsholmsvej 57
9000 Aalborg
Denmark
TEL: 45-96-358452
FAX: 45-98-142555
e-mail: Lade@civil.auc.dk

Dr S. Leroueil
Département de Génie civil
Université Laval
Ste-Foy, Qué, Canada G1K 7P4
TEL: 418-656-2601
FAX: 418-656-2928
e-mail: Serge.Leroueil@gci.ulaval.ca

Dr K. Y. Lo
Department of Civil & Environmental
    Engineering
University of Western Ontario
London, Ontario, Canada N6A 5B9
TEL: 519-661-2125
FAX: 519-661-3942
e-mail: jlemon@eng.uwo.ca

Dr Jacques Locat
Département de Géologie et génie
géologique
Université Laval
Ste-Foy, Québec, Canada G1K 7P4
TEL: 418-656-2179
FAX: 418-656-7339
e-mail: Jacques.Locat@ggl.ulaval.ca

Mr R. Loughney
Loughney Associates, Inc.
25 Arrow Point Road
New Preston, CT 06777 USA
TEL: 860-868-9995
FAX: 860-868-1025
e-mail: richard.loughney@snet.net

Dr P. Lucia
GeoSyntec Consultants
1500 Newell Avenue, Suite 800
Walnut Creek, CA 94596 USA
TEL: 925-943-3034
FAX: 925-943-2366

Dr E. McBean
Conestoga-Rovers & Associates Ltd
651 Colby Drive
Waterloo, Ontario, Canada N2V 1C2
TEL: 519-725-3313
FAX: 519-725-1736
e-mail: emcbean@craworld.com

Dr G. Milligan
Geotechnical Consulting Group
9 Lathbury Road
Oxford OX2 7AT UK
FAX: 44-1865-516407
e-mail: g.w.e.milligan@gcg.co.uk

Dr J. K. Mitchell
Via Department of Civil & Environmental
    Engineering
Virginia Tech
Blacksburg, VA 24061-0105 USA
TEL: 540-231-7351
FAX: 540-231-7532
e-mail: jkm@vt.edu

Dr I. D. Moore
Department of Civil & Environmental
    Engineering
University of Western Ontario
London, Ontario, Canada N6A 5B9
TEL: 519-661-3997
FAX: 519-661-3942
e-mail: ian@hazzstanl.engga.uwo.ca

Mr Luciano Picarelli
Seconda Università di Napoli
Dipartimento di Ingegneria Civile
Via Roma 21
80125 Aversa, Italy
TEL: 39-81-5010213
FAX: 39-81-5037370
e-mail: picarell@cds.unina.it

Dr H. G. Poulos
Coffey Geosciences Pty. Ltd
142 Wicks Road

North Ryde, NSW 2113 Australia
TEL: 61-2-9888-7444
FAX: 61-2-9888-9977
e-mail: harry_poulos@coffey.com.au

Mr B. L. Rodway
Bruce Rodway and Associates Pty. Ltd
4/15 Mosman Street
Mosman 2088, Australia
TEL: 61-2-9969-6295
FAX: 61-2-9969-6295
e-mail: bruce.rodway@big pond.com

Dr C. D. F. Rogers
School of Civil Engineering
University of Birmingham
Edgbaston
Birmingham B15 2TT UK
TEL: 44-121-414-5066
FAX: 44-121-414-3675
e-mail: c.d.f.rogers@bham.ac.uk

Mr F. Rovers
Conestoga-Rovers & Associates Ltd
651 Colby Drive
Waterloo, Ontario, Canada N2V 1C2
TEL: 519-884-0510
FAX: 519-725-1158

Dr R. K. Rowe
Vice-Principal (Research)
Queen's University
Kingston, Ontario, K7L 3N6
TEL: 613-533-3113
FAX: 613-533-2128
e-mail: kerry@civil.queensu.ca

Dr R. A. Schincariol
Department of Earth Sciences
The University of Western Ontario
London, Ontario, Canada N6A 5B7
TEL: 519-661-3732
FAX: 519-661-3198
e-mail: schincar@julian.uwo.ca

Dr K. Schmidtke
Conestoga-Rovers & Associates Ltd
651 Colby Drive
Waterloo, Ontario, Canada N2V 1C2
TEL: 519-884-0510
FAX: 519-884-0111

Dr Gilles Sève
CETE Mediterraneé
Laboratoire de Nice
56 Bd. Stalingrad
F-06300 Nice
France
TEL: 33-4-92-00-8182
FAX: 33-4-92-00-8199
e-mail: gilles.seve@equipement.gouv.fr

Dr J. Q. Shang
Department of Civil & Environmental
    Engineering
University of Western Ontario
London, Ontario, Canada N6A 5B9
TEL: 519-661-4218
FAX: 519-661-3942
e-mail: jshang@uwo.ca

Dr J. C. Small
School of Civil Engineering
The University of Sydney
Sydney 2006, NSW Australia
TEL: 61-2-9351-2128
FAX: 61-2-9351-3343
e-mail: j.small@civil.usyd.edu.au

Mr D. Smyth
Department of Earth Sciences
University of Waterloo
Waterloo, Ontario, Canada N2L 3G1
TEL: 519-888-4567 (Ext. 2899)
FAX: 519-746-3882
e-mail: dsmyth@sciborg.uwaterloo.ca

Mr J. Sprenger
GE Canada
2300 Meadowvale Blvd.
Mississauga, Ontario, Canada L5N 5P9
TEL: 905-858-5708
FAX: 905-858-5276
e-mail: james.sprenger@corporate.ge.com

Mr H. A. Tuchfeld
GeoSyntec Consultants
1500 Newell Avenue, Suite 800
Walnut Creek, CA 94596 USA
TEL: 925-943-3034
FAX: 925-943-2366
e-mail: HalT@geosyntec.com

Dr M. Whittaker
Consultant
161 Westminster Avenue
Toronto, Ontario, Canada M6R 1N8
FAX: 905-707-9084
e-mail: mwhittaker@home.com

Dr G. W. Wilson
Department of Mining and Mineral
    Processing
The University of British Columbia
Forward Building
517-6350 Stores Rd.
Vancouver, BC Canada V6T 1Z4
TEL: 604-822-6781
FAX: 604-822-5599
e-mail: gww@mining.ubc.ca

Dr E. K. Yanful
Department of Civil & Environmental
    Engineering
University of Western Ontario
London, Ontario, Canada N6A 5B9
TEL: 519-661-4069
FAX: 519-661-3942
e-mail: eyanful@eng.uwo.ca

PREFACE

This handbook aims to discuss, in one volume, a wide array of topics that have entered the mainstream of geopractice (i.e. geotechnical and geoenvironmental engineering) over the past two decades, while at the same time not losing sight of the more conventional aspects of the discipline that remain a core part of the work of geoprofessionals. These topics range from conventional saturated soil mechanics, to unsaturated soil behavior, rock mechanics, hydrogeology and geosynthetics. The book deals with pavements, shallow and deep foundations, embankments, slopes, retaining walls, buried structures, dynamics and earthquakes, risk assessment and management, contaminant transport, groundwater monitoring, and containment, treatment and remediation of contaminated sites.

The 50 contributors to the 30 chapters of the book are all recognized experts in their field—from industry and the academic world. Each chapter was edited and, in order to keep this handbook as one volume, many were reduced to half their original length. Each chapter has been reviewed by at least two experts in the field. Many of these reviewers were contributors to other chapters and are listed in the accompanying list of contributors. In addition, I would like to thank the following individuals who also served as reviewers of one or more chapters: Dr F. S. Barone, Dr S. F. Brown, Dr J. P. Carter, Mr J. M. A. Costa, Dr R. D. Holtz, Dr H. P. Hong, Dr F. Kulhawy, Dr B. Ladanyi, Mr C. Lake, Dr J. Mlynarek, Dr B. Ruth, Dr F. W. Schwartz, Mr G. Skinner, Dr D. Smith, Mr J. Thompson and Dr P. Ullidtz.

Notwithstanding the effort that was made to ensure that each chapter is as correct as practicable, there is little doubt that some errors (typographical or otherwise) will creep into a document of this size, especially in its first printing. Readers are requested to advise the undersigned of any errors (typographical or otherwise) or significant omissions they identify—to the extent possible they will be corrected in subsequent printings.

I am indebted to the numerous contributors to this book who have labored, in their "spare time" to produce a manuscript, respond to editorial and reviewer comments, and helped provide what I hope will be useful contributions to the profession.

Finally, I owe a deep debt of gratitude to my family (Kathy, Katrina, Kieron and Kendall) for their patience and understanding over the past four years as this book has taken shape; to the families of all the authors who will have contributed in a similar fashion to the compilation of each chapter; and to the typists and draftspeople who have assisted in the preparation of the manuscript—especially to Joanne Lemon and Kathy Rowe who have been an invaluable help in getting this book to the publisher.

R. Kerry Rowe
Department of Civil Engineering
Queen's University
Kingston, Ontario, Canada, K7L 3N6
kerry@civil.queensu.ca

IV. SPECIAL TOPICS

18. BURIED PIPES AND CULVERTS

I. D. Moore

18.1 Introduction

18.1.1 THE BURIED PIPE SYSTEM

A variety of buried pipe infrastructure is needed to service the needs of our communities and resource industries for electrical, water, gas and oil supply, power development, storm-water and sanitary sewer systems, and highway and railway culverts. The conduit surrounded by soil is both loaded and supported by the earth and porewater: and geotechnical engineers are frequently called upon to provide advice regarding various design and construction issues. This chapter describes issues associated with pipe loading and load capacity, and the construction of buried pipeline infrastructure. Other chapters contain related material associated with trenchless pipe-laying technologies (Chapter 19), slope stability (Chapter 14) and frost and thermal effects (Chapter 20).

Figure 18.1a shows a gravity flow pipe or culvert buried under an earth embankment. Generally, the zone of "structural" backfill soil placed directly adjacent to the pipe is specially selected to provide a stable supporting environment, and often consists of granular soil. The "bedding" soil placed beneath the pipe is also specially prepared. Beyond these materials lie lower quality soils: undisturbed native material and lower quality (generally native) soil backfill.

Design and selection of the soil materials is important, since the pipe does not act as an isolated structural element subjected to clearly defined loads, but as one component in the complete pipe–soil system. The pipe carries part of the earth pressures, and part is carried around the pipe through the backfill. The stiffness and uniformity of the backfill soil influences both the proportion of loads reaching the pipe, and the pipe's load carrying ability.

Figure 18.1b defines key locations around the pipe circumference. The crown, springlines and invert are generally those positions where the pipe reaches critical limit states, though shoulders and haunches can also be critical as a result of non-uniform surface loading or narrow bedding at the invert.

18.1.2 PIPE TYPES AND LIMIT STATES

Table 18.1 contains a summary of six different pipe categories, together with a classification as rigid or flexible, and details of key performance limits. The wide variety of materials used in pipe manufacture (ceramics, metals and polymers) have stiffness, strength, ductility and durability characteristics that can vary by orders of magnitude. There are also substantial variations in pipe wall geometry (plain, corrugated or profiled, uniform or composite systems) that lend further complexity. However, in all cases, the stiffness and configuration of the soil around the pipe can affect pipe performance. Performance limits significantly influenced by the surrounding soil are marked in the table. This table also includes the categories of pipe stiffness relative to the surrounding ground. Relative stiffness is influenced by two different

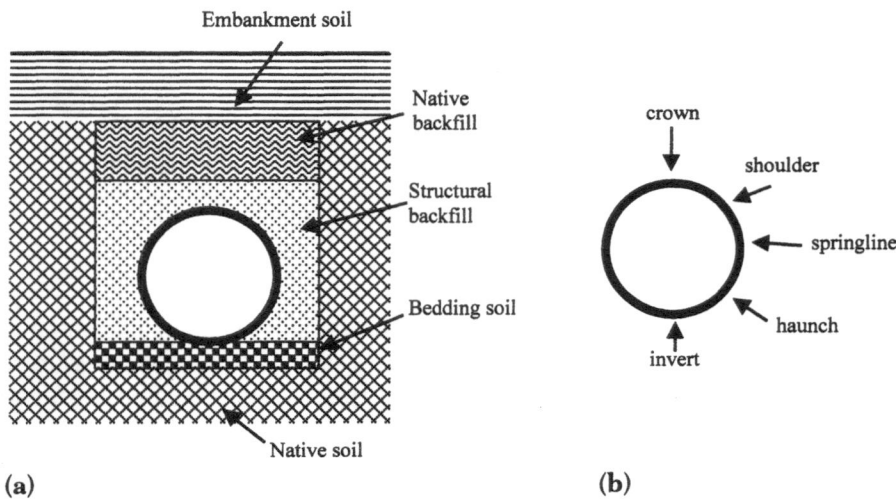

FIGURE 18.1. Definition of terms for the pipe and the surrounding soil: (a) buried pipe and surrounding soil zones; and (b) location of crown, invert, springline, shoulder and haunch.

types of pipe and soil deformation: bending (associated with non-uniform external pressures leading to deformations from circular to oval or other non-circular shape); and hoop compression (uniform and non-uniform pressures reducing pipe circumference). Pipe stiffness categories are:

- Rigid: where the pipe stiffness in bending and ring compression is very large relative to the soil.
- Semi-flexible: where the pipe stiffness in bending is of similar magnitude to the soil, but where stiffness in ring compression is very large.
- Flexible: where the pipe is flexible in bending relative to the soil, but where stiffness in ring compression is very large.
- Compressible: where the pipe is flexible in bending relative to the soil, and stiffness in ring compression is similar to that of the soil.

These stiffness categories will be defined more precisely in Section 18.3.4, as will the manner in which the stiffness influences load sharing between the soil and pipe, and the resistance to deformation.

18.1.3 ANALYSIS OF BURIED PIPES

There are many different pipe types, and many different empirical and semi-empirical procedures have been developed for pipe analysis and design. Product development history and geography have led to design practice that varies considerably from region to region. A major breakthrough occurred with the development of closed-form solutions by Burns & Richard (1964) and Hoeg (1968), examining the pipe response based on elastic response of a thin circular shell and the solid material surrounding it. This permits rational predictions of behavior for many different pipes, based on the elastic properties and the geometry of each particular product. A unified method of analysis that can predict the different responses of rigid, flexible and compressible pipes is then possible, independent of the empirical and semi-empirical procedures.

Another key development has been the formulation of computer analyses based on the finite element method. These developments are summarized in the following paragraphs.

Two-dimensional procedures can model

TABLE 18.1. Pipe stiffness categories and typical performance limits

Pipe materials	Stiffness[a]	Material failure	Deflection	Buckling	Durability
Clay	Rigid	Cracking[a]	No	No	Abrasion
Reinforced concrete	Rigid	Cracking[a], steel yield[a]	No	No	Abrasion, corrosion[a]
Cast iron	Rigid	Cracking	No	No	Corrosion[a]
Ductile iron (DI), corrugated steel (CS) aluminum (Al)	Semi-flexible, flexible	Yield or crushing	Ovaling[a]	Global[a]	Abrasion corrosion[a]
Thermoplastics (PVC, HDPE, PP)	Flexible, compressible	Short-term yield[a], long-term cracking[a]	Ovaling[a] and hoop compression[a]	Global[a] and local[a]	UV degradation, solvents
Glass-reinforced plastic (GRP)	Flexible	Cracking[a]	Ovaling[a]	Global[a]	Solvents

[a] Performance limit influenced by backfill soil.

TABLE 18.2. Pipe design standards[a]

Pipe materials	ASTM	AS/NZS	BS and EN	CSA	ISO
Concrete (C)	C14 (ASTM)	3725 (AS/NZ)	BS 8010	A5 (CSA)	
Corrugated steel (CS)	A796 (ASTM)	2566 (AS/NZ)	EN 1295-1	C140 (CSA)	
Ductile iron (DI)					
Culvert	A716 (ASTM)		BS 8010		TR 7186 (ISO/TR)
Sewer	A746 (ASTM)				
Glass-reinforced plastic (GRP)	D3262 (ASTM)	2566 (AS/NZ)	EN 1295-1		TR 7370 (ISO)
High density polyethylene (HDPE)					
Smooth wall	F810 (ASTM)	2566 (AS/NZ)	EN 1295-1		TR 7074 (ISO/TR)
Corrugated	F892 (ASTM)				
Profile wall	F894 (ASTM)		EN 12666-1		TR 8772 (ISO/TR)
Polyvinylchloride (PVC)					
Sewers	D3034 (ASTM)	2566 (AS/NZ)	EN 1295-1	B182.1 (CSA)	TR 7074 (ISO/TR)
Large diameter	F679 (ASTM)				
Corrugated	F949 (ASTM)		EN 1401-1	B182.4 (CSA)	DIS 4435 (ISO/DIS)
Reinforced concrete (RC)	C76 (ASTM)	3725 (AS/NZ)	EN 1295-1	A257.2 (CSA)	
Vitrified clay pipe (VC)	C301 (ASTM)	4060 (AS/NZ)	EN 1295-1	A60.1 (CSA)	

[a] ASTM, American Society for Testing and Materials; AS/NZS, Australian and New Zealand Standard; BS, British Standard; EN, European Standard; CSA, Canadian Standards Association; ISO, International Standards Organization (TR, technical report).

the specific geometry and material response of the pipe structure, and the zones of soil below, beside and above the pipe (e.g. Katona 1978; Duncan 1979). The influence of soil placement, which can be significant for flexible and long-span structures, can also be modeled. The response to live load can be estimated, provided vehicles can be represented as line loads extending along the culvert axis. Ultimate limit states can be assessed once suitable procedures are available to deal with plastic mechanisms and buckling response (e.g. Moore 1987; Seed & Raines 1988).

Three-dimensional procedures can more accurately capture the influence of live load effects, including three-dimensional load distribution through the fill covering the structure (e.g. Moore & Brachman 1994; Fernando & Carter 1998).

New constitutive models have been incorporated into finite element procedures, allowing consideration of more complex material response on the pipe–soil interaction, e.g. the reinforced concrete pipe wall model of Katona 1978, and the visco-plastic models for HDPE pipe of Zhang & Moore 1998.

This chapter presents a unified approach to analysis of buried pipes and culvert structures based on the elastic continuum analysis of Hoeg (1968). Key empirical equations used for flexible and rigid pipe design are also described and explained.

18.1.4 DESIGN STANDARDS
Table 18.2 shows some of the relevant standards used in pipe design and specification. Design examples are included in this chapter to illustrate use of a unified design approach, as well as traditional empirical design methods.

18.2 Loads

Loads on buried pipes arise in a number of ways, and the following subsections describe the influence of the geostatic stresses (loads associated with earth weight), surface live

loads, fluid loads, loads induced by ground movements and dynamic events, and other load sources. Some can be characterized effectively using simple equations. Others are more difficult to quantify, as a result of the complexity of the mechanical response of the system or because of the vagaries of the load source.

Loading must be distinguished as either acting on the pipe–soil system, Fig. 18.2a, or directly on the pipe itself, Fig. 18.2b. The former will generally be used here, since it permits unified assessment of pipes of different material, wall type and stiffness categories. Empirical design procedures generally focus on the vertical load acting across the pipe at the level of the pipe crown, W_v, Fig. 18.2c, rather than the more complete loading pattern. This may be augmented by approximate lateral loading conditions for some of the empirical design methods, Fig. 18.2d.

Buried pipe analysis is generally conducted based on considerations of the total soil stresses acting on the system. In some circumstances, the loads are best subdivided into components associated with the soil skeleton (the effective stresses) and the external water pressures (the pore pressures). Most of the discussion in this chapter is based on the use of total stresses, though fluid pressure loading is explicitly described in Section 18.2.5.

18.2.1 STATIC LOADS: EMBANKMENTS
Vertical, σ_v, and horizontal, σ_h, earth pressures applied to the pipe–soil system, Fig. 18.2a, can be calculated in the conventional way where geostatic conditions apply:

$$\sigma_v = \gamma H \tag{18.1}$$

$$\sigma_h = K\sigma_v \tag{18.2}$$

for soil unit weight γ, depth to the crown, H, and coefficient of lateral earth pressure, K. The quantity $W_p = \sigma_v D$ is often referred to as the "prism load", meaning the weight of

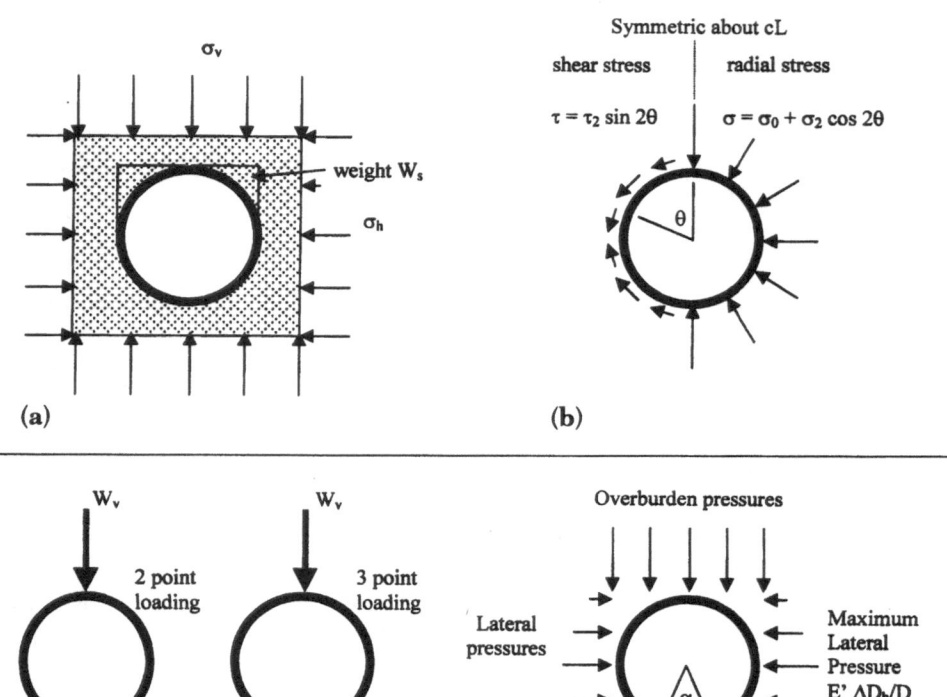

FIGURE 18.2. Load conditions for circular pipes: (a) vertical and horizontal earth pressures on pipe–soil system, (b) external loads acting directly on the pipe, (c) vertical earth loads at the pipe crown and invert, and (d) approximate vertical and horizontal earth pressure distributions on the pipe (modified from Spangler 1956).

the column of soil extending across the pipe diameter, D (from springline to springline), and up to the ground surface.

These equations ignore the weight:

$$W_s = 0.25\gamma D^2(1 - \pi/4) = \gamma 0.054 D^2 \quad (18.3)$$

of backfill soil in each of the two zones of soil directly over the shoulders, Fig. 18.2a. This weight can be important for long span (large diameter) culverts at shallow burial depth. The equations are also based on the assumption that the influence of groundwater can be neglected (the water table is below the pipe)

or can be implicitly included in the soil parameters (for example, the unit weight, γ, and lateral earth pressure coefficient, K).

18.2.2 STATIC LOADS: TRENCH INSTALLATION

Is has long been recognized that backfill soil placed into a trench moves downward relative to the native soil at the sides, and that frictional shear stresses develop between the backfill and the trench walls. The shear stresses reduce the vertical force, W_t, transmitted through the trench backfill to the pipe crown at depth, H. Marston & Anderson

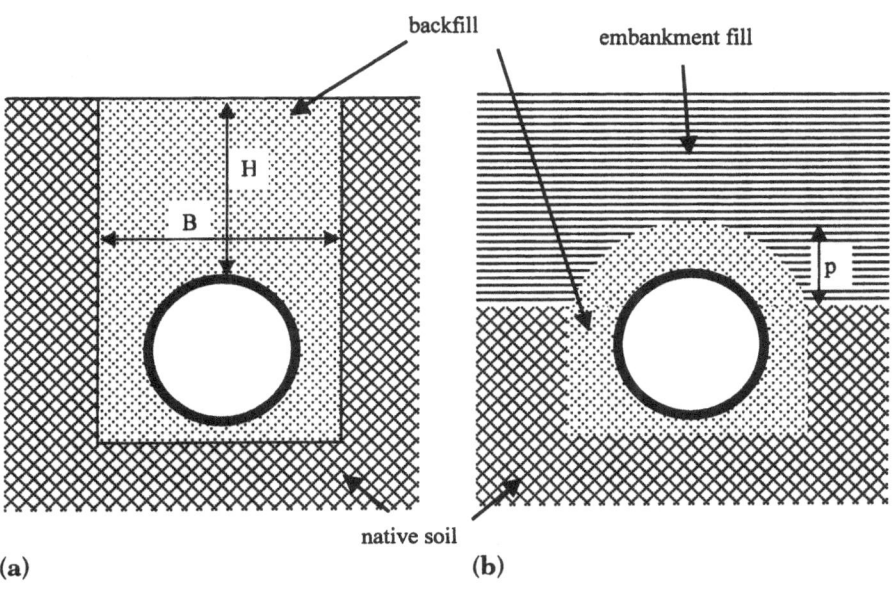

FIGURE 18.3. Pipe burial conditions: (a) conventional trench burial; and (b)pipe under embankment in positive projection condition (pipe crown projects above native ground level).

(1913) adapted the silo arching theory of Janssen (1895) to express vertical load, W_t, as a function of soil unit weight, γ, trench width, B, lateral earth pressure coefficient, K, and friction coefficient, μ, between backfill and native soil at the sides of the trench, Fig. 18.3a. Re-expressing the trench arching solution in terms of average vertical stress in the backfill, σ_v:

$$\sigma_v = W_t/B = C_t\gamma B \qquad (18.4)$$

$$C_t = (1 - e^{-2K\mu H/B})/(2K\mu). \qquad (18.5)$$

For very shallow or very wide trenches where H/B is small, differentiation of C_t reveals that C_t approaches H/B, so that σ_v is given by γH and the geostatic stress solution, Eq. 18.1, is recovered.

As trench depth becomes large, the exponential term in Eq. 18.5 becomes small and C_t approaches $1/(2K\mu)$. The vertical stress, σ_v, then approaches the limit $\gamma B/(2K\mu)$. This occurs because the shear stresses at the sides

of the trench are then large enough to balance the extra weight of each additional layer of backfill. Table 18.3 contains a standard set of $K\mu$ values often used in pipe design. "Arching" coefficient C_t is sometimes presented in graphical form in codes of practice and design guides.

Many other burial configurations have been defined in terms of the trench arching theory, where the distance the pipe crown protrudes above the native ground surface (the pipe "projection", p) is varied, Fig. 18.3b. These empirical procedures have been extended to include discussions of pipe under

TABLE 18.3. Soil parameters for trench load analysis

Soil type	$K\mu$
Granular, no cohesion	0.19
Maximum for sand and gravel	0.165
Maximum for saturated top soil	0.150
Maximum for most soils	0.130
Maximum for saturated clay	0.110

embankment loading, with the intention of explaining measurements of pipe load in the field that exceed overburden pressures (the phenomenon referred to as "negative arching"). These procedures all rely on empirical "settlement" coefficients to capture the effects of geometry and the movement of the prism of soil over the pipe relative to the surrounding soil. Negative arching will be discussed further in Section 18.3.3, where techniques are introduced that permit load calculation without the need for empirical "trench arching" theory. Spangler & Handy (1973) provide details of the empirical load calculation procedures.

18.2.3 CONSTRUCTION LOADING

One important source of applied loading relates to temporary construction loads during placement of soil at the sides and over the top of a buried pipe. These are associated with transient loads from construction vehicles, compaction equipment and the influence of unbalanced earth loads (when backfill is placed unevenly on each side of the pipe). While earth loads generally dominate the design and performance of deeply buried pipelines, construction loads can be critical for shallow buried structures. Recent studies by Webb et al. (1995) provide guidelines for soil placement and compaction around metal, concrete and polymer pipes.

18.2.4 SURFACE LIVE LOADS

The effect of live loads acting at the ground surface on underlying buried pipes has often been considered in terms of the corresponding vertical stress, σ_v, at the crown of the pipe. It is common to simply add this to the static vertical earth pressures prior to calculations of thrust and moment, though such procedures ignore the effect of pipe–soil interaction.

One commonly used procedure estimates σ_v using the Boussinesq solution, examining the stress that develops at a depth H and lateral distance x from the surface point load P:

$$\sigma_v = 3P/(2\pi H^2)(1 + x^2/H^2)^{-5/2}. \quad (18.6)$$

Another alternative is to use simple prism models to estimate load attenuation with depth:

$$\sigma_v = P/A_H \quad (18.7)$$

where the vertical surface load, P, is spread over horizontal area, A_H, at depth H. If the surface load is applied over a surface rectangle of length l_0 and width b_0, the area at depth H is:

$$A_H = (l_0 + mH)(b_0 + mH). \quad (18.8)$$

Parameter m = the extent to which the area increases with depth (it defines the side-slopes of the prism). Recommended values of m include values of 1 or 1.15 from AASHTO (1994) and 1.45 from AS/NZS 2566.

18.2.5 FLUID LOADING

Pressure from gas or water within the pipe, σ_i, leads to tensile hoop stresses in the walls of the pipe and a wall force, N, per unit length:

$$N = -\sigma_i r. \quad (18.9)$$

Submarine structures can experience large external fluid pressures, σ_e, producing a compressive wall force per unit length of:

$$N = \sigma_e r. \quad (18.10)$$

18.2.6 OTHER LOADING SOURCES

Ground movements in the vicinity of a pipeline due to seismic ground motion, landslides and frost action may all affect the pipeline within it, Selvadurai (1985). Both lateral and axial ground movements can cause structural distress. Offshore pipeline structures are particularly susceptible to seabed instability or

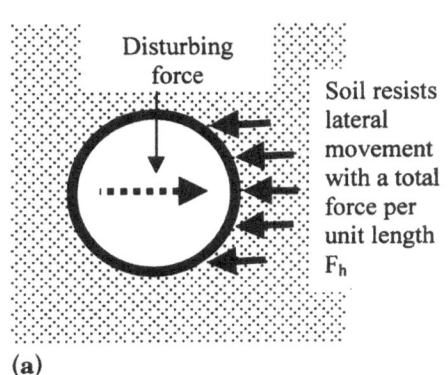
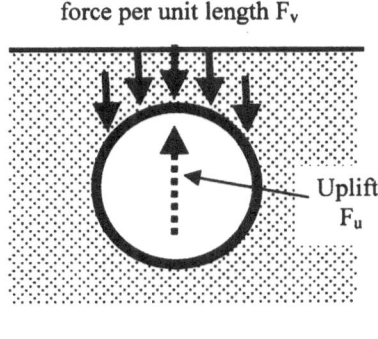

FIGURE 18.4. Action of the pipe as a ground anchor, limiting forces applied by the ground to the pipe perpendicular to the pipe axis: (a) horizontal force applied by the soil to resist the lateral movement of the pipe, and (b) vertical force applied by soil to the pipe to resist uplift.

creep over longer periods of time. Pipeline location should be chosen to avoid or minimize the risk of such ground movements.

Lateral earth movements past a pipeline will generate a lateral force per unit length, F_h, which must be resisted by the pipeline (see Fig. 18.4a). Analysis of the pipeline performance will require an estimate of the lateral force, F_h.

Trautmann & O'Rourke (1985) have investigated the lateral response of pipelines in sand, using soil box tests and reviewing available solutions. They conclude that the anchor force coefficient, $N_{hs}(\phi, H/D)$, of Rowe & Davis (1982b) provides reasonable estimates of peak lateral force per unit length, F_h, accounting for friction angle of the sand, ϕ', unit weight, γ' (bulk unit weight if above the water table, submerged unit weight if below), and normalized burial depth, H/D:

$$F_h = N_{hs}\gamma'HD. \qquad (18.11)$$

Trautmann & O'Rourke also provide guidelines on methods of calculating force–displacement relationships, to obtain lateral force, F_h, as a function of lateral soil movement (where F_h has not fully mobilized soil strength and reached a peak value).

Pipe response in clay has been investigated by Paulin et al. (1998), who confirm that the anchor force coefficient, $N_{hc}(H/D)$, of Rowe & Davis (1982a) also provides reasonable, though somewhat large solutions for clays of undrained shear strength, s_u:

$$F_h = N_{hc}s_uD. \qquad (18.12)$$

Buoyancy force, F_u, per unit length:

$$F_u = \gamma_w\pi D^2/4 \qquad (18.13)$$

can be significant when pipelines are placed underwater at shallow covers, and burial needs to be sufficient to generate a resisting force, F_v (see Fig. 18.4b) greater than F_u in the circumstances where the pipeline can be filled (or partially filled) with air.

Trautmann et al. (1985) have examined resistance to uplift for pipelines in sands, and have concluded that the anchor force coefficient, $N_{vs}(\phi', H/D)$ of Rowe & Davis (1982b) provides reasonable solutions for force per unit length resisting uplift, F_v, provided by:

$$F_v = N_{vs}\gamma'HD \qquad (18.14)$$

though forces are overestimated at depths greater than $4D$.

Thermally induced loads can be substantial when the pipe is restrained axially and large temperature changes develop. This is particularly true when the pipe is manufactured from a material like HDPE, which has a high coefficient of thermal expansion. Careful design of thrust end restraints is then required, or provision made to accommodate the large axial movements that develop when the ends are not restrained. Assessment of the influence of the compressive or tensile axial forces may also be needed during design.

Seismic analysis of long span metal culverts by Byrne *et al.* (1996) suggests that the effect of vertical acceleration exceeds that of horizontal acceleration. They suggest that seismically induced thrust can be estimated from static thrust using F_v, the ratio of vertical acceleration to acceleration due to gravity, g, CHBDC (1998). Earthquake loading of pipes is discussed further in Section 21.3.3.

Finally, significant forces can arise in curved pipes filled with fluid moving at high velocity. Thrust blocks and other forms of additional restraint are needed where forces are large and ground support is inadequate to resist those forces.

18.3 Response

18.3.1 PIPE AND SOIL PROPERTIES

18.3.1.1 Pipe modulus and wall properties

Pipe structural properties in the circumferential direction generally control the structural design of pipelines and culverts. These include the modulus, E_P, the cross-sectional area per unit length, A_P, and the second moment of area per unit length, I_P.

For a long plain pipe of wall thickness t:

$$A_P = t \qquad (18.15)$$

$$I_P = t^3/12. \qquad (18.16)$$

Pipe modulus should be reduced by the factor $(1 - v_P^2)$ when the plain pipe is very long (and axial restraint is sufficient to result in plane strain conditions).

For profiled wall structures the section properties A_P and I_P are a function of the wall geometry. Modulus should not be reduced by $(1 - v_P^2)$ for such structures, since the axial restraint is typically low.

Modulus E_P represents the effective elastic modulus of the pipe wall. This requires averaging in the case of composite structures (e.g. reinforced concrete, glass-reinforced plastic) and a "secant" modulus parameter where stress–strain response is either non-linear or time dependent.

18.3.1.2 Pipe Stiffness

Historically, the results of laboratory pipe load tests have been used to classify pipe, and develop empirical and semi-empirical design methods. For example, "pipe stiffness" (PS) is measured for thermoplastic pipes by compressing the pipe between two rigid steel plates and measuring the deflection as a function of the vertical load (e.g. ASTM D2412):

PS = vertical load per unit length/

vertical deflection.   (18.17)

PS for thermoplastic pipes varies with the magnitude of the vertical deflection, and a deflection corresponding to 5% decrease in vertical diameter is generally used for evaluations of standard pipe stiffness.

PS for a short pipe segment can be expressed as a function of flexural stiffness, $E_P I_P$, and radius, r:

$$PS = E_P I_P/[r^3(\pi/4 - 2/\pi)]. \qquad (18.18)$$

Ring stiffness constant is another index measure of pipe wall rigidity obtained using the parallel plate load test:

RSC = vertical load per unit length/0.03   (18.19)

$$RSC = 6.44 E_P I_P/D^2. \qquad (18.20)$$

In this case, the pipe is deflected to 3% vertical diameter decrease.

In practice, the parallel plate loading test produces a complex three-dimensional pipe response so the theoretical relationships given by Eqs 18.18 and 18.20 can substantially overestimate stiffness. Furthermore, the time dependent response of polymer materials is such that measurements made at different rates of deformation and at different deflection levels will have different effective pipe moduli, E_P. Three-dimensional analysis capable of capturing the rate dependent stiffness of the material is needed to establish true relationships between PS, RSC, A_P and I_P, Moore (1994).

18.3.1.3 Soil Properties

Backfill properties have a great influence on pipe performance. While it is unusual to have precise information regarding the constitutive behavior of the backfill and native soils, Selig (1990) tested a series of backfill materials and developed non-linear soil parameters for three materials: well-graded granular soil (SW), silt (ML) and lean clay (CL). McGrath (1998) used these parameters to develop design values of soil modulus, which were successfully used to interpret a field testing program for reinforced concrete, corru-

gated steel and profiled polyethylene pipe. McGrath's modulus values are for three different densities relative to peak values obtained using the Standard Proctor test (85, 90 and 95%). They are presented in Table 18.4 for six different levels of effective vertical earth pressure (7, 35, 70, 140, 275 and 410 kPa).

18.3.2 SOIL–PIPE INTERACTION SOLUTIONS

Various workers have published theoretical solutions to estimate the response of a pipe of radius r, modulus E_P and section properties I_P and A_P when it is buried in a soil of modulus E_s and Poisson's ratio v_s. Solutions have been developed for two idealized interface conditions: (a) perfect adhesion of the soil to the structure (the perfectly rough, no-slip or bonded interface condition), and (b) zero adhesion (the full-slip or smooth interface condition). Intermediate response can be evaluated using numerical solutions that account for the finite shear strength of the pipe–soil interface (e.g. Katona 1978).

Elastic solutions for pipe–soil interaction include those of:

- Burns & Richard (1964), who examined the response of a pipe installed with no ap-

TABLE 18.4. Design values of soil modulus in MPa (adapted from McGrath 1998)

Soil type[a]	RD[b]	Vertical stress level[c]/(kPa)					
		7	35	70	140	275	410
Granular materials (SW, SP, GW, GP)	85	3.2	3.6	3.9	4.5	5.7	6.9
	90	8.8	10.3	11.2	12.4	14.5	17.2
	95	13.8	17.9	20.7	23.8	29.3	34.5
Silty backfill (ML)	85	2.5	2.7	2.8	3.0	3.5	4.1
	90	4.6	5.1	5.2	5.4	6.2	7.1
	95	9.8	11.5	12.2	13.0	14.4	15.9
Clay soils (CL)	85	0.9	1.2	1.4	1.6	2.0	2.4
	90	1.8	2.2	2.4	2.7	3.2	3.6
	95	3.7	4.3	4.8	5.1	5.6	6.2

[a] Unified Classification System.
[b] Density relative to maximum from Standard Proctor Test, e.g. ASTM D-698.
[c] Level of vertical effective stress at end of burial.

plied stress and then subjected to vertical overburden pressure, σ_v, and horizontal pressure, $\sigma_h = [v_s/(1 - v_s)] \sigma_v$; and Hoeg (1968), who examined the more general case where $\sigma_h = K\sigma_v$ (for arbitrary K). These solutions approximate loading conditions such as pipe burial where ground stresses develop after the pipe is in place.

- Einstein & Schwartz (1979), who considered the response of a thin elastic pipe inserted in prestressed elastic ground, $\sigma_h = K\sigma_v$; this solution is useful for examining problems such as tunneling and pipe jacking where the ground stresses are active before the pipe is installed.
- Moore (1988), who examined the behavior of thin elliptical pipes for both the initially unstressed and prestressed ground cases.
- Moore (1990), who developed three-dimensional solutions for thin or thick elastic pipes in non-uniform elastic ground.

The solution of Hoeg (1968) is adapted here for use in buried circular pipe or culvert analysis. Use will be made of the techniques developed by Gumbel (1981) and Moore (1988) to relate solutions for the initially unstressed ground case to those for prestressed ground (allowing both pipe burial and tunneling/jacking problems to be examined using the same solution).

The following equations can be used to calculate the distribution of radial stress, σ, and shear stress, τ, on the external boundary of the pipe at angle θ from the crown, Fig. 18.2b:

$$\sigma = \sigma_0 + \sigma_2 \cos 2\theta \qquad (18.21)$$

$$\tau = \tau_2 \sin 2\theta. \qquad (18.22)$$

Once these are established, the stress resultants and deformations of the pipe can be calculated (expressions are provided in later sections). For elastic soil responding under onedimensional conditions, coefficient of lateral pressure at rest, K_0, is related to Poisson's ratio of the soil:

$$K_0 = [v_s/(1 - v_s)]. \qquad (18.23)$$

More realistic ground response features higher lateral earth pressures as given in Section 3.4, which provides values for normally and overconsolidated soils.

For pipes buried then loaded through placement of backfill, mean σ_m and deviatoric σ_d stress components are defined:

$$\sigma_m = (\sigma_v + \sigma_h)/2 \qquad (18.24)$$

$$\sigma_d = (\sigma_v - \sigma_h)/2 \qquad (18.25)$$

For pipes inserted within prestressed ground:

$$\sigma_m = 0.25(\sigma_v + \sigma_h)/(1 - v_s) \qquad (18.26)$$

$$\sigma_d =$$
$$0.125(\sigma_v - \sigma_h)/[(1 - v_s)/(3 - 4v_s)]. \qquad (18.27)$$

Factors A_m, $A_{d\sigma}$ and $A_{d\tau}$ are defined to give stresses on the pipe in terms of the mean and deviatoric field stresses. The uniform component of radial stress is given by:

$$\sigma_0 = A_m\sigma_m \qquad (18.28)$$

$$A_m = 2(1 - v_s)/[1 + C(1 - 2v_s)]. \qquad (18.29)$$

The non-uniform component of radial stress is given by:

$$\sigma_2 = A_{d\sigma}\sigma_d \qquad (18.30)$$

$$A_{d\sigma} = 4(1 - v_s)[4 + 3C(1 - 2v_s) - 2F]/\Delta$$
$$\text{for a bonded interface} \qquad (18.31)$$

$$A_{d\sigma} = 12(1 - v_s)/(2F + 5 - 6v_s)$$
$$\text{for a smooth interface.} \qquad (18.32)$$

The non-uniform component of shear stress is given by:

$$\tau_2 = A_{d\tau}\sigma_d \qquad (18.33)$$

$$A_{d\tau} = 16(1 - v_s)(F + 1)/\Delta$$
$$\text{for a bonded interface} \qquad (18.34)$$

$$A_{d\tau} = 0 \quad \text{for a smooth interface.} \qquad (18.35)$$

The denominator in Eqs 18.31 and 18.34 is given by:

$$\Delta = C(1 - 2v_s)(5 - 6v_s + 2F)$$
$$+ 2F(3 - 2v_s) + 4(3 - 4v_s) \qquad (18.36)$$

where the two stiffness parameters C and F are defined by Hoeg (1968) as:

$$C = E_s D / [2(1 + v_s)(1 - 2v_s)E_p A_p] \qquad (18.37)$$

$$F = E_s D^3 / [48(1 + v_s)E_p I_p]. \qquad (18.38)$$

18.3.3 POSITIVE AND NEGATIVE ARCHING

"Arching" is the term generally used to express load transfer from low stiffness to high stiffness components of a multicomponent load-support system. Factors A_m, $A_{d\sigma}$ and $A_{d\tau}$ are used to define the extent of arching between the backfill and the pipe.

Figure 18.5a shows a block of soil subjected to earth pressures. Radial and shear stresses around the disk of soil are given by: $\sigma_0 = \sigma_m$, $\sigma_2 = \sigma_d$, $\tau_2 = \sigma_d$. If a pipe is inserted into the ground at this location prior to application of the earth pressures, then if the pipe stiffness:

- is equal to the disk of soil it "displaces" (Fig. 18.5a), then a "neutral" condition arises, where the pipe carries the "soil prism" directly over it and:

$$A_m = 1 \qquad (18.39)$$

- is less than the disk of soil it displaces, then "positive arching" develops, where the arching factor A_m is less than one (Fig. 18.5b);
- is greater than the disk of soil it displaces, then "negative arching" occurs, where arching factor A_m is greater than one (Fig. 18.5b).

For pipe inserted into prestressed ground, the "neutral condition" occurs if the pipe is rigid and permits no displacements at its external boundary and soil stresses are maintained equal to the geostatic stress values, Eqs 18.24 and 18.25. Pipes of any lower stiffness permit deformations that reduce the stresses acting on the pipe below the pre-existing ground stresses.

18.3.4 CLASSIFICATION FOR FLEXIBLE AND RIGID PIPES

Pipe rigidity or flexibility can be defined using Hoeg's relative stiffness parameters C

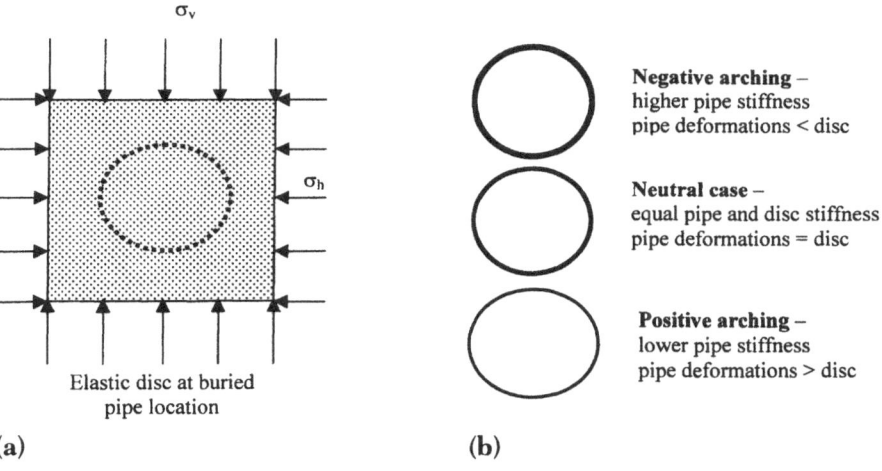

(a)   **(b)**

FIGURE 18.5. Explanation of arching mechanism: (a) the reference case: a block of elastic soil subjected to earth pressures; (b) positive and negative arching stiffness less than or greater than the disk.

and F (Eqs 18.37 and 18.38). Structures with low C or F experience negative arching, while those with large values of C or F experience positive arching. Elastic predictions of pipe response can be inspected to classify the soil–pipe system using C and F, as shown in Table 18.5. Also shown are typical examples of these pipe classes.

Tables 18.6 and 18.7 give examples of C and F for six specific reinforced concrete, corrugated metal and high density polyethylene pipes. Also shown are values of arching factors A_m and $A_{d\sigma}$ calculated using Eqs 18.29 and 18.32. Results are given for two different soil moduli, E_s, to reveal how soil modulus influences the soil structure interaction.

An inspection of Tables 18.6 and 18.7 indicates that:

- Rigid pipes have arching factors that are largely independent of soil modulus: the role of the soil around rigid pipes is to spread the earth pressures around the circumference rather than change the magnitude of those earth pressures (this is discussed further in Section 18.5.1).
- Flexible pipes are stiff in hoop compression and attract a uniform component of earth pressure from the soil ($A_m > 1$ indicating negative arching); the non-uniform components of earth pressure are reduced by the backfill soil (positive arching); soil modulus has a significant effect on the magnitude of $A_{d\sigma}$ (i.e. the deflections and bending moments).
- Compressible structures like profiled HDPE pipes have negligible value of $A_{d\sigma}$; and positive arching means that $0 < A_m < 1$, which is controlled by soil modulus.

TABLE 18.5. Pipe stiffness classes

Pipe class	C values	F values	Examples
Rigid pipe	$0.1 > C$	$0.1 > F$	Cast iron, reinforced concrete and clay
Semi-rigid pipe	$0.1 > C$	$10 > F > 0.1$	Long span reinforced concrete, ductile iron, rib stiffened corrugated steel
Flexible pipe	$0.1 > C$	$F > 10$	Corrugated steel and aluminum, PVC, glass reinforced plastic, polypropylene, plain polyethylene
Compressible pipe	$100 > C > 0.1$	$F > 10$	Profiled polyethylene
Empty cavity	$C > 100$	$F > 10$	

TABLE 18.6. C values for a variety of pipes and two backfill moduli shown

Product	Diameter (mm)	Modulus (GPa)	A_p (mm² mm⁻¹)	$E_s = 40$ MPa C	$E_s = 40$ MPa $A_m{}^c$	$E_s = 10$ MPa C	$E_s = 10$ MPa $A_m{}^c$
Reinforced concrete pipe[a,b]	450	30	90	0.006	1.39	0.002	1.40
	2100	30	220	0.012	1.39	0.003	1.40
Corrugated steel pipe	450	200	1.3	0.066	1.33	0.017	1.38
	2100	200	5.8	0.070	1.33	0.017	1.38
HDPE							
Plain[b]	200	0.6	18	0.71	0.92	0.18	1.24
Profiled	450	0.6	6.4	4.51	0.32	1.13	0.76

[a] Reinforced concrete pipe properties adjusted to include area of steel with modulus of concrete.
[b] Moduli of plain polyethylene and reinforced concrete pipes adjusted by $(1 - v_p^2)^{-1}$ to characterize long pipes.
[c] Arching factor.

TABLE 18.7. F values for a variety of pipes and two backfill moduli

Product	Diameter (mm)	Modulus (GPa)	I_p (mm^4 mm^{-1})	$E_s = 40$ MPa		$E_s = 10$ MPa	
				F	$A_{d\sigma}^c$	F	$A_{d\sigma}^c$
Reinforced concrete pipe[a,b]	450	30	60 000	0.03	2.57	0.01	2.61
	2100	30	887 333	0.22	2.30	0.06	2.54
Corrugated steel pipe	450	200	25	12	0.31	3	0.92
	2100	200	1770	17	0.23	4	0.73
HDPE							
Plain[b]	200	0.6	501	17	0.23	4	0.72
Profiled	450	0.6	127	767	0.01	192	0.02

[a] Reinforced concrete pipe properties adjusted to include area of steel with modulus of concrete.
[b] Moduli of plain polyethylene and reinforced concrete pipes adjusted by $(1 - v_p^2)^{-1}$ to characterize long pipes.
[c] Arching factor. Calculations are based on a smooth pipe–soil interaction.

18.3.5 THRUST AND MOMENT

Three approaches can be used to evaluate stress resultants thrust N and bending moment M: (a) general expressions in terms of interface stresses σ_0, σ_2 and τ_2; (b) expressions derived from elastic continuum solutions for rigid or flexible conditions; and (c) empirical procedures developed for specific pipe classes. The first of these is used below for the general design procedure; the latter two are used to provide limiting values for rigid and flexible pipes.

18.3.5.1 General Expressions

Thrust and moment can be estimated once interface pressure components σ_0, σ_2 and τ_2 have been evaluated (Eqs 18.28, 18.30 and 18.33). For harmonic interface stresses defined by Eqs 18.21 and 18.22, thrusts at the crown, N_{cr}, and springline, N_{sp}, are given by:

$$N_{cr} = \sigma_0 r - (\sigma_2/3 + 2\tau_2/3)r \quad (18.40)$$

$$N_{sp} = \sigma_0 r + (\sigma_2/3 + 2\tau_2/3)r \quad (18.41)$$

while bending moments at the crown, M_{cr}, and the springline, M_{sp}, are given by:

$$M_{cr} = (\sigma_2/3 + \tau_2/6)r^2 \quad (18.42)$$

$$M_{sp} = -(\sigma_2/3 + \tau_2/6)r^2. \quad (18.43)$$

18.3.5.2 Limiting Values

For buried flexible pipes, moments are assumed negligible and thrust limits can be calculated assuming C is small, and F is large:

$$N_{sp} = \sigma_v r[(1 - v_s)/(3 - 2v_s)][5 - 2v_s + K(1 - v_s)]$$
$$\text{bonded interface} \quad (18.44)$$

$$N_{sp} = \sigma_v r(1 - v_s)(1 + K)$$
$$\text{smooth interface.} \quad (18.45)$$

For buried rigid pipes, C and F are small, yielding:

$$N_{sp} = \sigma_v r(1 - v_s)[1 + K + 2(1 - K)/(3 - 4v_s)]$$
$$\text{bonded interface} \quad (18.46)$$

$$N_{sp} = \sigma_v r(1 - v_s)[1 + K + 2(1 - K)/(5 - 6v_s)]$$
$$\text{smooth interface} \quad (18.47)$$

$$M_{sp} = -\sigma_v r^2(1 - K)(1 - v_s)/(3 - 4v_s)$$
$$\text{bonded interface} \quad (18.48)$$

$$M_{sp} = -\sigma_v r^2 (1 - K)(1 - v_s)2/(5 - 6v_s)$$
$$\text{smooth interface.} \quad (18.49)$$

Analogous expressions exist for pipes inserted within prestressed ground.

Each of these expressions is dependent on Poisson's ratio, v_s, but is independent of other elastic soil or pipe properties. This occurs because soil and pipe moduli affect the interaction stresses only for semi-rigid pipes ($10 > F > 0.1$) and compressible pipes ($100 > C > 0.1$).

18.3.5.3 Non-Circular Pipes and Culverts

Parametric studies examining the behavior of non-circular pipes and culverts indicate that springline thrust is a function of the ratio of rise R (crown to invert distance) to span S (springline to springline distance), (e.g. see Moore 1988, CHBDC 1998). Circular culvert solutions Eqs 18.21–18.38 ($S/R = 1$) will overpredict thrust (i.e. are conservative) for $S/R > 1$, but may underpredict thrust (i.e. may be unconservative) for $S/R < 1$. Moments are also influenced by pipe shape; this is particularly important for rigid pipes.

18.3.5.4 Empirical Limits

Marston & Anderson (1913) developed empirical guidelines for rigid pipes buried in trenches. They assumed such pipes carry all of the vertical force acting in the column of backfill soil (Fig. 18.2a):

$$W_t = 2N_{sp} = B\sigma_v = C_t\gamma B^2. \quad (18.50)$$

Empirical limits were also developed for the loads, W_e, that develop on rigid pipes buried under embankments. These had:

$$W_e = 2N_{sp} = (1.2 \text{ to } 1.8)D\sigma_v \quad (18.51)$$

depending on the backfill soil and the location of the pipe relative to the original ground surface. A review of Eqs 18.46 and 18.47 indicates that for $K = 0.5$ and $v_s = 0.1$ to 0.4, $W_e = 2.N_{sp} = (1.1 \text{ to } 1.7)D\sigma_v$, agrees well with the empirical earth load calculations like those presented by Spangler & Handy (1973).

For "flexible" pipes like corrugated metal culverts, White & Layer (1960) suggested that the "prism load", W_p (the weight of the column of soil directly above the pipe), generates thrusts at the pipe sides:

$$N_{sp} = r\sigma_v. \quad (18.52)$$

This equation forms the basis of some design procedures (e.g. AISI, 1990).

Evaluation of Eq. 18.44 gives $N_{sp} = 1.2$ to $1.7r\sigma_v$ for bonded interface and $v_s = 0.1$ to 0.4. These higher values match field measurements, (e.g. Webb et al. 1999) where conventional "ring compression theory" based on the soil prism load has been found to be unconservative. Negative arching is considered in some flexible metal culvert design standards, (e.g. CHBDC, 1998).

18.3.6 DEFLECTION

18.3.6.1 Deflection Limits

Pipe deflection results from both permanent and transient loads. While negligible for rigid pipes that are correctly designed and installed, deflections for non-rigid pipes can be significant and are an important design consideration. The deflection of these flexible and compressible pipes is largely controlled by the stiffness of the backfill supporting the buried structure.

While pipe response to static earth pressures can be estimated using rational procedures, response to construction loads is difficult or impossible to predict, since it is subject to the particular installation practices, soil placement, compaction procedure and loading history. Deflection of very flexible and compressible structures must be carefully controlled during construction. Contractors are obliged to maintain deflection within limits specified by a code or the pipe designer.

Typical deflection limits are:

- 5% for corrugated metal structures, e.g. AS/NZS 2566.1.
- 5–6% for GRP pipes; e.g. ASTM D3262-96, AS/NZS 2566.1.
- 7.5% for profile wall HDPE pipe; e.g. ASTM F894-95, AS/NZS 2566.1.
- 3.3–11% for plain HDPE pipe; e.g. ASTM F714-97, AS/NZS 2566.1.
- 6–7.5% for PVC pipes; e.g. ASTM F679-95, 949-96, D3034-97, AS/NZS 2566.1, CSA B182.1-87.

18.3.6.2 Deflection Resulting From Static Earth Pressures: Elastic Theory

Pipe deflections occur as a result of the uniform stress, σ_0, acting on the pipe, as well as non-uniform stresses, $\sigma_2 \cos 2\theta$ and $\tau_2 \sin 2\theta$ (see Fig. 18.2b). Using the estimate of uniform component of radial stress, σ_0, and the non-uniform components, σ_2 and τ_2, from Eqs. 18.28, 18.30 and 18.33 the radial deflection of the pipe, w, is given by:

$$w = w_0 + w_2 \cos 2\theta \qquad (18.53)$$

$$w_0 = \sigma_0 r^2 / E_p A_p \qquad (18.54)$$

$$w_2 = (2\sigma_0 + \tau_2) r^4 / (18 E_p I_p). \qquad (18.55)$$

The deflections vary around the pipe circumference as a function of θ in a manner similar to the applied stresses (Fig. 18.6a). Dependence of displacement components w_0 and

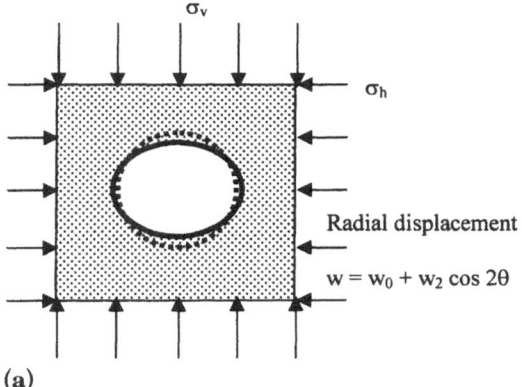

(a)

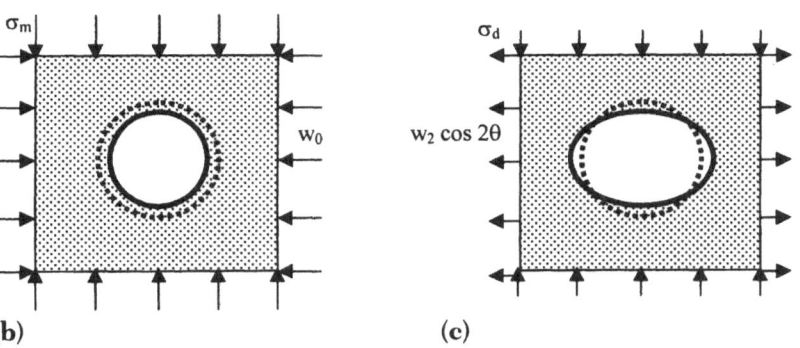

(b)   (c)

FIGURE 18.6. Idealized deformation components around a buried circular pipe: (a) radial pipe deformations under geostatic earth pressures; (b) deformation w_0 for uniform earth pressure component, σ_m; and (c) deformation w_2 for non-uniform earth pressure component, σ_d.

w_2 on σ_m and σ_d, respectively, is illustrated in Fig. 18.6b and c.

The changes in the vertical and horizontal pipe diameters, ΔD_v and ΔD_h, can be found from w:

$$\Delta D_v = -2(w_0 + w_2) \qquad (18.56)$$

$$\Delta D_h = -2(w_0 - w_2). \qquad (18.57)$$

18.3.6.3 Deflection Resulting from Static Earth Pressures: Limiting Values

The decrease in vertical pipe diameter can be evaluated for limits of soil to pipe stiffness parameters C and F. For an empty cavity (F and C very large):

$$\Delta D_v =$$
$$-2\sigma_v r(3 - K)(1 - v_s)(1 + v_s)/E_s. \qquad (18.58)$$

For a flexible pipe that is stiff in hoop compression (F large, C negligible):

$$\Delta D_v = -8\sigma_v r$$
$$(1 - K)(1 - v_s)(1 + v_s)/[(3 - 2v_s)E_s]. \qquad (18.59)$$

For a rigid pipe (F and C negligible):

$$\Delta D_v = 2\sigma_v r(1 - K)r^4(1 - v_s)/[3E_p I_p(3 - 4v_s)]$$
$$- 2\sigma_v r^2(1 + K)(1 - v_s)/(E_p A_p). \qquad (18.60)$$

18.3.6.4 Deflection Resulting from Static Earth Pressures: Spangler's Equation

Spangler (1956) developed a procedure for predicting the changes in horizontal diameter of flexible metal pipes. The modified Iowa deflection equation, as the revised form of Spangler's semi-empirical equation is known, provides an estimate of horizontal pipe diameter increase:

$$\Delta D_h =$$
$$D_L K_b(2\sigma_v r)/(E_p I_p/r^3 + 0.061E'). \qquad (18.61)$$

Spangler employed an earth pressure distribution like that shown in Fig. 18.2d, assum-

ing: (a) a finite width bedding support at the invert characterized using K_b, which is chosen based on the angle of bedding support at the invert α, Fig. 18.2d (e.g. $K_b = 0.105$ for $\alpha = 45°$); (b) a time-dependent response quantified using an empirical "deflection lag factor" D_L (typically between one and 1.5); (c) a parabolic distribution of side pressures with maximum magnitude occurring at the springline and equal to the product of soil stiffness parameter, E', and normalized lateral pipe deflection equal to $\Delta D_h/D$:

$$\sigma_h = E'\Delta D_h/D. \qquad (18.62)$$

It is also generally assumed that the change in horizontal pipe diameter is equal and opposite to the change in the vertical diameter. While this is not unreasonable for corrugated metal pipes where the uniform component of radial deflection is negligible, it is not valid for compressible pipes. Furthermore, the soil parameter E' is an empirical model-dependent quantity that depends on pipe size and pipe material, as well as soil type. Finally, the procedure Eq. 18.62 for calculating lateral earth pressures implies zero pressure at zero lateral deflection. Instead, non-zero lateral earth pressures such as $\sigma_h = K\sigma_v$ should be recovered.

Despite these shortcomings, studies have been conducted to provide guidance on selection of E', Howard (1977), and various design procedures employ this procedure (e.g. glass-reinforced plastic pipe design, AWWA C905-88).

18.3.6.5 Pipe Stiffness and Flexibility

Pipe deflection is controlled for some products by specifying limits for pipe stiffness (PS), ring stiffness constant (RSC), or flexibility factor (FF) defined by:

$$FF = S^2/E_p I_p \qquad (18.63)$$

where S is the culvert span (springline to springline). These limits effectively control the flexural rigidity of the pipe, $E_P I_P$, the objective being to limit the bending deformations that occur, particularly during construction (when soil resistance to pipe bending is low). Control of FF is used in design of long span metal culverts (e.g. ASTM A796-98).

18.3.7 STRESS AND STRAIN

Once thrust and moment have been estimated, the maximum compressive circumferential wall stress, σ_c (compression positive), can be calculated from:

$$\sigma_c = N/A_P + |M|y_c/I_P \qquad (18.64)$$

and the maximum tensile circumferential pipe wall stress, σ_t, from:

$$\sigma_t = N/A_P - |M|y_t/I_P. \qquad (18.65)$$

Estimates are best made at the crown and springline using the thrust and moment given in Eqs 18.40–18.49. The y_c and y_t values used here are the distance from the neutral axis of the pipe section to the extreme fibers, respectively, in compression and tension. Stability against material failure can be determined after comparing σ_c and σ_t with the stress limits in compression and tension.

Circumferential (tension positive) strain, ε_θ, can be estimated from circumferential wall stress, σ:

$$\varepsilon_\theta = -\sigma/E_p$$

for uniaxial stress (e.g. profiled pipe) (18.66)

$$\varepsilon_\theta = -\sigma(1 - v_p^2)/E_p$$

for plane strain (e.g. plain pipe). (18.67)

One alternative for calculating strain is to express maximum circumferential strain as a function of pipe deflection, ΔD, distance to the extreme fibre, y, pipe diameter, D, and an empirical "strain factor" D_f (sometimes referred to as a shape factor):

$$\varepsilon_{max} = D_f(t/D)(\Delta D_v/D) \qquad (18.68)$$

where t may be replaced by $2y$ for profiled or corrugated pipes. Empirical values of D_f ranging 4.5–8 are employed, depending on pipe stiffness (e.g. ASTM D3262-96). This semi-empirical approach seeks to account for strain amplification in the haunches due to localized bedding support at the invert, Fig. 18.2d. The effect of construction-induced deformations on strain is sometimes estimated by setting $\Delta D_v/D$ in Eq. 18.68 to the levels of pipe deflection allowed during installation (see Section 18.3.6).

18.3.8 BUCKLING

Since the earth and groundwater pressures induce compressive hoop thrusts in the buried pipe, the potential exists for flexible pipe systems to buckle circumferentially. Two methods have generally been used to estimate the compressive thrusts that induce buckling, N_b.

Early workers characterized the backfill soil as a series of elastic springs. The soil stiffness for this "Winkler" or "subgrade reaction" model is given by E'. The critical thrust can then be represented as:

$$N_b = 2(E_P I_P E'/r)^{1/2} \qquad (18.69)$$

where E' = an empirical value of soil stiffness. It is not necessarily the same as the value of E' used in the modified Iowa deflection equation. Both should be back-calculated from field data; one for deflection, the other for pipe buckling. This approach has been used in many codes of practice, with empirical correction factors added to account for the effect of shallow burial and groundwater pressure (e.g. AWWA C905).

Alternatively, the soil can be represented as an elastic solid (e.g. AS2566.1 and the ASTM standard for polyethylene manhole design). In that case, critical hoop thrust is given by

$$N_b = 1.2p_f(E_P I_P)^{1/3}\{E_s/(1 - 2v_s)\}^{2/3}R_h \qquad (18.70)$$

where p_f = a performance factor; E_s and v_s = the modulus and Poisson's ratio for the soil, respectively; and R_h = a factor that accounts for the influence of burial depth and extent of backfill soil (as discussed below).

Comparisons of Eqs 18.69 and 18.70 with measurements of maximum thrust at buckling indicate that the performance of Eq. 18.70 is superior (Moore 1989). Performance factor p_f was estimated by Moore (1989) as 0.55 for granular backfill soils.

The elastic continuum model also permits rational assessment of backfill geometry through correction factor R_h (Moore *et al.* 1994). R_h is a function of H/D, normalized bending stiffness, F, minimum width of backfill soil, W, adjacent to the pipe and the ratio of moduli for native and backfill soils, E_0/E_s. Lower bounds for R_h are given by:

$$R_h = 11.4/(11 + D/H) \leq 1$$
$$\text{for } W/D \text{ large} \quad (18.71)$$

$$R_h = 20/(56 + D/H)$$
$$\text{for } W/D = 0.1 \text{ and } E_0/E_s = 0.1. \quad 18.72)$$

More detailed values of R_h are provided in Moore *et al.* (1994).

Some pipe products also have the potential for local buckling. Corrugated and profile walled structures composed of thin structural elements can develop local short wavelength buckles within individual components of the wall. Soil support is important in these circumstances since it controls the level of bending and hoop strain that develops in the pipe, and also acts to provide local lateral resistance to buckling deformations (Moore & Laidlaw, 1997).

18.4 Backfill Selection and Pipe Installation

Guidelines on pipe installation practice are available (e.g. ASTM A798-97 and D2321-89, AS/NZS 2566.2, BS 8010, ISO/TR 7073: 1998 and 10465-1:1993). Issues to be considered include:

- Care is needed when the native foundation soil beneath the pipe is expansive, soft or unstable; special treatment may be required in such conditions to stabilize the material and prevent excessive deformations.
- Over-excavation under the pipeline must be controlled since longitudinal bending will likely occur where variable bedding settlements develop along the pipeline.
- Restrictions are placed on the types of backfill materials, with preference to granular materials; clay backfills are generally avoided around flexible structures; fines in granular backfills are usually limited to a maximum of 5%; pipe design should include consideration of stone impingement where large size particles are employed for drainage applications.
- Minimum trench widths are specified to permit effective placement and compaction of backfill materials at the sides of the pipe; the placement and compaction of that backfill is far more important than efforts to minimize loads by keeping trench width narrow (see the discussion in the following paragraph); a space of width 500 mm between the pipe and the trench wall is generally needed (even larger for pipe diameters greater than 1 m); the space should be wider than the compaction equipment in use.
- Maximum trench widths are specified, since construction of trench width greater than that used in design, leads to increases in earth loads expected to reach the crown of the pipe.
- Uniform bedding soil is placed in the bottom of the trench below the pipe; compaction of the bedding directly under the pipe can lead to line loading at the invert and possibly distress at invert and haunches; bending moments are reduced if the bedding is not compacted directly under the pipe, though the outer bedding should still be compacted.
- Placement of soil under the haunches and compaction of that material reduces local bending in the pipe; material can be sliced in this area with shovels, or tampers may be used.

- Rammers generally provide the best method of soil compaction in the vicinity of the pipe and are considerably more effective than vibratory plates (McGrath, 1998).
- Granular bedding is permeable, and may serve as a path for groundwater; trench plugs or lateral drains may be needed to control groundwater flow.
- Minimum spacing between parallel pipelines of one diameter or 600 mm (whichever is larger), is needed to permit placement of sidefill.
- Restrictions should be placed on the use of large size compaction equipment in the vicinity of the pipe (the mass of static compaction equipment, and the energy of dynamic equipment).
- Backfill soils placed more than 300 mm or one pipe diameter above the pipe may be native materials, but care is needed to keep large particles away from the pipe, and to compact the backfill beneath roads, water crossings and other pipelines; native soil should also be compacted in populated areas and under cultivated land.
- Native and backfill soils must have compatible gradations to prevent fines migration; alternatively, suitable geotextiles should be used to separate the different soil zones;

care is needed in selecting a suitable geotextile and ensuring adequate construction controls (Rowe & Seychuk 1995).
- Care is needed when removing trench boxes or other temporary trench-side support (such as sheet piling); the objective is to prevent loss of side support to the pipe; trench walls must have adequate strength, particularly when flexible pipe is being installed.
- Care is needed with jointing, to prevent inflow of groundwater into the pipe and subsequent loss of backfill.
- Adequate supervision of pipe laying work is required unless large variations in pipe placement, trench excavation and backfill selection, placement and compaction can be tolerated (this is not generally the case).

18.5 Design Examples

Three example problems are included to demonstrate pipe design using the procedures outlined in this chapter: (a) rigid pipe: a reinforced concrete storm sewer (Fig. 18.7a) (b) flexible pipe: a corrugated steel culvert (Fig. 18.7b) (c) compressible pipe: a plain high density polyethylene landfill drainage pipe (Fig. 18.7c).

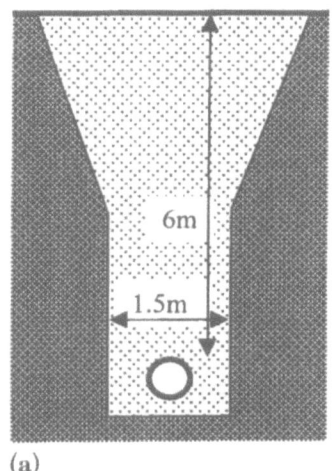

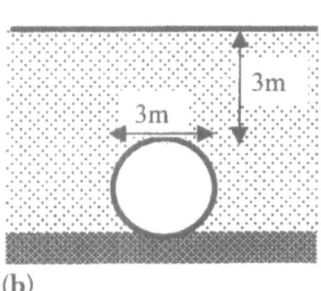

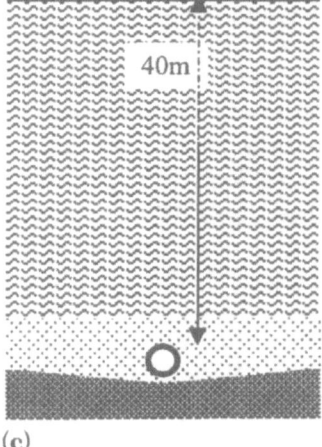

(a)   (b)   (c)

FIGURE 18.7 Example problems: (a) reinforced concrete storm sewer, 0.5-m diameter; (b) corrugated steel culvert, 3-m diameter; and (c) plain HDPE leachate collection pipe, 0.3-m diameter.

18.5.1 CONCRETE STORM SEWER

A reinforced concrete pipe with external diameter of 500 mm is examined at a burial depth of 6 m within a trench of 1.5 m width. Table 18.8 contains all dimensions, material properties and estimates of pipe response. Pipe response is assumed to be unrestrained in the axial direction, due to the flexible joints that would be used in this application. Comparison of values calculated for C and F with Table 18.5 verifies that the pipe is effectively rigid compared with the soil surrounding it.

18.5.1.1 Structural Design to Limit Crack Width

Structural capacity of reinforced concrete pipe is quantified by measuring crack development in the laboratory under diametrically opposed line loads. A limit to crack width is generally specified in the pipe design standard being used (e.g. 0.25 mm, ASTM C655M-95). This serves as a serviceability limit, designed to limit the ingress of moisture and prevent damage to the reinforcing steel. A two- or three-point load test is used to measure W_{cr}/SF, Fig. 18.2c, the factored

TABLE 18.8. Design example I: 500-mm diameter reinforced concrete pipe

Description	Value	Source
Cover, H (m)	6	
Backfill classification Well-graded sand (% of max. Proctor)	SW90	
Trench width, B (m)	1.5	
Trench friction coefficients, $K\mu$	0.165	Table 18.3
Bedding factor, BF	2	Table 18.9
Soil modulus, E_s (MPa)	12	Table 18.4
Poisson's ratio, ν_s	0.3	
Unit weight, γ (kNm^{-3})	20	
Coefficient of lateral pressure, K	0.4	Table 18.9
Radius, r (mm)	250	
Thickness, t (mm)	40	
Modulus, E_p (MPa)	33 000	
Area, A_P (mm^2 mm^{-1})	40	Eq. 18.15
Second moment of area, I_P (mm^4 mm^{-1})	5333	Eq. 18.16
Relative hoop stiffness, C	0.0044	Eq. 18.37[b]
Relative bending stiffness, F	0.137	Eq. 18.38[c]
Trench arching, C_t	2.22	Eq. 18.5
Trench load, $W_v = C_t\gamma B^2$ (kNm^{-1})	100	Eq. 18.50
Embankment load, $W_v = 2N_{sp}$ (kNm^{-1})	87[a]	Eq. 18.51
Cracking load, W_{cr} (kNm^{-1})	43	Eq. 18.75
Vertical stress		
$\quad\sigma_v = \gamma H$ (kPa)	120[a]	Eq. 18.1
$\quad\sigma_v = C_t\gamma B^2/D$ (kPa)	200	Eq. 18.4
Cracking moment, M_{sp} (kN m m^{-1})	-1.75	Assume rigid Eq. 18.48
Cracking load, W_{cr} (kN m^{-1})	39	Eq. 18.76

[a] Embankment case governs.
[b] Stiff in hoop compression.
[c] Stiff in circumferential bending.

vertical load per unit length that generates cracks of this size. The safety factor (SF) is used to adjust "ultimate" to "design" load, e.g. values of 1.25–1.5 are specified in ASTM C655M-95 depending on pipe class. The three-point load test has a pair of closely spaced supports at the invert, and produces bending moments and their associated cracks, which are virtually identical to those resulting from two-point loading.

Pipe performance in the field is then evaluated, using either "direct" or "indirect" design procedures to relate W_{cr} to the vertical earth load, W_v, supported by the pipe. This earth load is calculated based on the embankment or trench burial condition:

$$W_v = W_e \quad \text{for embankment burial} \quad (18.73)$$

$$W_v = W_t \quad \text{for trench burial.} \quad (18.74)$$

18.5.1.2 Indirect Design
In indirect design, the vertical earth load, W_v, is related to the factored vertical cracking load, W_{cr}/SF, using the empirical bedding factor (BF):

$$W_{cr}/SF = W_v/BF. \quad (18.75)$$

The bedding factor relates earth load, W_v, to the vertical force that produces the same peak bending moment (and therefore cracking) under parallel plate loading, W_{cr}/SF (Fig. 18.2c). The bedding factor depends on the extent to which backfill distributes earth loads around the pipe. Table 18.9 specifies BF values for different types of burial condi-

tion in a manner which is typical of codified design practice. Increases in backfill stiffness and uniformity, i.e. backfill quality, lead to increases in BF.

18.5.1.3 Direct Design
An alternative procedure is to work directly using bending moment (so-called "direct design"). Springline moment associated with the limiting cracks can be estimated from the vertical cracking load, W_{cr}, assuming the pipe sample responds as an elastic ring (e.g. Young & Trott, 1984):

$$M_{sp} = W_{cr}r(\pi^{-1} - 0.5)/SF \quad (18.76)$$

This is then equated to the springline bending moment, M_{sp}, calculated to occur in the field. Soil–structure interaction analysis is used to estimate M_{sp} based on burial condition (in particular, the lateral earth pressure ratio, K, since this controls bending moment and is influenced by the quality of backfill soil and its placement).

Rational "direct" design methods using soil–structure interaction analysis can be developed for any standard bedding conditions provided standard definitions of K(BF) are established. Table 18.9 gives approximate values of K(BF) for the four bedding conditions. More comprehensive earth pressure distributions and their associated values of K(BF) have also been proposed and implemented for reinforced concrete pipe design, McGrath & Kurdziel (1991).

TABLE 18.9. Bedding factors and equivalent lateral earth pressures for use in direct design

Class	Bedding & haunch[a]	Cover	BF Trench	BF Embankment	K
A	15-MPa concrete	Dense backfill	3–5	3–5	0.6–0.75
B	SW90	Compacted granular	1.9	2–3	0.4–0.6
C	SW80 or ML85	Lightly compacted	1.5	1.8–2.5	0.2–0.5
D	Loose	Loose	1.1	1.3	0–0.1

[a] Compaction to percentage of maximum modified Proctor density.

18.5.1.4 Trench versus Embankment Loading

This particular design example demonstrates that the trench burial condition may not lead to effective decreases in vertical earth loads. In this case, the trench is too wide to induce effective arching. Both embankment, W_e, and trench, W_t, loads are estimated in Table 18.8. The embankment load is found to be less than the trench load. Conventional design assumes the whole weight of the trench backfill rests on the rigid pipe, rather than just the column of soil over the pipe, and the weight of this wide column exceeds that predicted for embankment loading. Clearly, shear stresses at the sides of the trench can reduce, not increase, earth loads on the pipe; so the use of all the weight of the backfill in this wide trench leads to overconservative estimates of pipe load.

18.5.2 LARGE DIAMETER CORRUGATED STEEL CULVERT

A 3-m diameter corrugated metal culvert is examined in the second design example. Table 18.10 contains details of all calculations and results. The culvert is buried 3-m within a road embankment. Corrugated steel plate of 152-mm wavelength and depth 52-mm is used, where thickness is 2-mm. Section properties I_p and A_p are values obtained from a product handbook, AISI (1990).

Design calculations using flexible pipe lim-

TABLE 18.10. Design example II: 3-m diameter corrugated steel pipe

Description	Value	Source
Cover, H(m)		
Backfill classification Well-graded sand	SW85	
Soil modulus, E_s (MPa)	10	Table 18.4
Poisson's ratio, soil, ν_s	0.3	
Unit weight, γ (kN m^{-3})	20	
Coefficient of lateral pressure, K	0.35	
Radius, r (mm)	1500	
Thickness, t (mm)	2	
Area, A_p (mm^2 mm^{-1})	2	Tabulated data for 152 × 52 plate
Second moment of area, I_p (mm^4 mm^{-1})	12500	Tabulated data for 152 × 52 plate
Pipe modulus, E_p (MPa)	200000	
Poisson's ratio, pipe, ν_p	0	Assume zero for corrugated material
Yield stress, σ_y (MPa)	250	
Relative hoop stiffness, C	0.072	Eq. 18.37
Relative bending stiffness, F	1.7	Eq. 18.38
Vertical stress, σ_v (kPa)	60	Eq. 18.1
Springline thrust, N_{sp} (kN m^{-1})	119.2	Assume flexible Eq. 18.44
Factor of safety against crushing, SF$_y$	4	Eq. 18.77
Correction factor, for extent of backfill, R_h	1	Assume deeply buried in wide zone of backfill
Critical thrust (for buckling), N_b(kN m^{-1})	2199	Eq. 18.70
Factor of safety against buckling SF$_b$	18	Eq. 18.81
Deflection Earth load only, ΔD_v (mm)	−17.7	Assume perfectly flexible Eq. 18.59
Per cent deflection %ΔD_v	−0.6	

its for thrust and deflection are illustrated. The relative flexibility, F, for this structure is calculated to be 1.7, which would be assessed as being in the "semi-rigid" category in Table 18.5. This results from use of the large size corrugation pattern for a structure of relatively small diameter (3 m). Consideration of bending effects is therefore included, by calculating the bending stresses that result at springlines (the location of maximum wall thrust).

The performance limits are now discussed in turn:

- Wall crushing at the springline. This results from springline thrust; a factor of safety (SF_y) is introduced to ensure that wall stress at working loads is less than the yield stress:

$$SF_y = \sigma_y A_P / N_{sp}. \qquad (18.77)$$

The strength of bolted wall seams should also be designed for $SF_y \, N_{sp}$.

- Deflection (decrease in vertical diameter) as obtained from Eq. 18.59 must be less than the deflection limit (the conventional value is 5%); a value of 0.6% is calculated (very low due to the high bending stiffness of the structure).
- Bending stresses at the springlines associated with pipe ovaling leading to change in diameter, ΔD_v, and no change in pipe circumference, $\Delta D_h = -\Delta D_v$, are given by:

$$\sigma_t = 6E_P \,(y_t/D)(\Delta D_v/D) \qquad (18.78)$$

$$\sigma_c = -6E_P \,(y_c/D)(\Delta D_v/D). \qquad (18.79)$$

For the deflection associated with static earth load (0.6%), the bending stress is 60 MPa, leaving a factor of safety against first yield of:

$$F_y =$$

$$\sigma_y/[(N_{sp}/A_P) + 6E(y_c/D)(-\Delta D_v/D)] \qquad (18.80)$$

TABLE 18.11. Design example III: 300-mm diameter HDPE pipe

Description	Value	Source
Cover, H(m)	40	
Backfill classification, uniform stone	SW95	
Soil modulus, E_s (MPa)	29	Table 18.4 for 275-kPa stress
Poisson's ratio, soil, v_s	0.25	
Unit weight of waste, γ (kN m^{-3})	12.5	
Coefficient of lateral pressure, K	0.35	
Radius, r (mm)	160	
Dimension ratio, DR	23	
Thickness, t (mm)	13.9	
Area, A_p (mm^2 mm$^{-1} = t$)	13.9	
Second moment of area, I_p (mm^4 mm$^{-1} = t^3/12$)	224	
Pipe modulus, E_p (MPa)	200	50 year modulus for HDPE
Poisson's ratio, pipe, v_p	0	Assume zero axial restraint
Yield stress, σ_y (MPa)	9	
Relative hoop stiffness, C	2.67	Eq. 18.37
Relative bending stiffness, F	353	Eq. 18.38
Vertical stress, σ_v (kPa)	500	Eq. 18.1
Correction factor, R_h	1	Deeply buried, wide backfill
Critical thrust, N_b (kN m^{-1})	120	Eq. 18.70

TABLE 18.12. Arching solutions for HDPE pipe example

	Bonded interface condition		Smooth interface condition	
	Value	Source	Value	Source
Arching factor				
A_m	0.64	Eq. 18.29	0.64	Eq. 18.29
$A_{d\sigma}$	0.77	Eq. 18.31	0.01	Eq. 18.32
A_{dt}	1.56	Eq. 18.34	0.00	Eq. 18.35
Springline thrust, N_{sp} (kN m^{-1})	55.1	Eq. 18.41	34.8	Eq. 18.41
Crown thrust, N_{cr} (kN m^{-1})	13.3	Eq. 18.40	34.6	Eq. 18.40
Springline moment, M_{sp}, (kN mm^{-1})	0.015	Eq. 18.43	0.018	Eq. 18.43
Maximum hoop stress, σ_{max} (MPa)	4.4	Eq. 18.64	3.0	Eq. 18.64
Factor of safety				
Yield, SF_y	2.0		3.0	
Buckling, SF_b	2.0	Eq. 18.81	3.5	Eq. 18.81
Uniform displacement, w_0 (mm)	2.0	Eq. 18.54	2.0	Eq. 18.54
Non-uniform displacement, w_2 (mm)	2.9	Eq. 18.55	3.3	Eq. 18.55
Diameter change, ΔD_v (mm)	−9.8	Eq. 18.56	−10.7	Eq. 18.56

equal to 2.1. Construction induced deformation will also produce local bending stresses. The conventional deflection limit associated with flexible metal pipe (5%), gives an elastic bending stress of 500 MPa. This structure, therefore, should be limited to, say, 1% during construction to limit local bending stresses to 100 MPa (giving a final factor of safety of 1.4 against first yield). This should not be difficult, since the structure is very stiff.

- Springline thrust leading to elastic buckling with factor of safety (SF_b) of 18 being obtained from:

$$SF_b = N_b/N_{sp}. \qquad (18.81)$$

18.5.3 HIGH DENSITY POLYETHYLENE DRAINAGE PIPE

The last example problem features a HDPE drainage pipe, Table 18.11. The ends of this structure have been assessed where there is zero axial constraint. Relative stiffness values calculated are $C = 0.8$ and $F = 353$, placing the pipe in the compressible pipe range (Table 18.5). Soil–structure interaction has therefore been evaluated using arching factors A_m, $A_{d\sigma}$ and A_{dt} (Table 18.12). The calculated value of $A_m = 0.64$ indicates that there will be positive arching and soil stresses around the pipe, which are reduced below those in the free field. Therefore, soil modulus in the vicinity of the pipe is chosen from

TABLE 18.13. HDPE pipe results for bonded, smooth and frictional interface

Quantity	Bonded	$\mu_i = 0.18$	Smooth
N_{sp}, kN m^{-1}	55.1	39.9	34.8
N_{cr}, kN m^{-1}	14.3	29.9	34.6
M_{sp}, kN m m^{-1}	0.015	0.016	0.018
σ_{max}, MPa	4.9	4.3	3.0
$\Delta D_v/D$, %	−3.0	−3.3	−3.3

TABLE 18.14. Flexible pipe design using Iowa deflection equation

Coefficient of soil reaction, E' (MPa)	20	
Bedding factor, K_b	0.105	Assumes bedding angle $\alpha = 45°$
Deflection, ΔD_v (mm)	-9	Eq. 18.61
Percent deflection, $\%\Delta D_v$	-4.7	

Table 18.4 for the 275-kPa stress level, rather than the total vertical stress 500 kPa.

The interface condition between soil and pipe must be chosen, and the example calculation shows values estimated for both "smooth" and "bonded" interface cases. More advanced analysis using a frictional interface (tan $\delta = 0.18$) is presented in Table 18.13. These results show that the bonded interface approximation produces conservative pipe wall stresses. The smooth interface solution produces a conservative prediction of pipe deformation.

Pipe design considers factor of safety against yield:

$$SF_y = \sigma_y/\sigma_c \qquad (18.82)$$

as well as buckling and pipe deflection. Pipe design should also consider the influence of stress concentrations resulting from stone contact forces and perforations (Brachman 1999). Since the pipe wall stresses in this particular example are already at maximum working limits, this pipe design is inadequate if it is to be perforated and placed within large-sized gravel.

Also shown is a calculation for pipe deflection made using the Iowa deflection equation (Table 18.14). This produces a higher estimate of vertical pipe deflection.

19. TRENCHLESS TECHNOLOGY

G. W. E. Milligan
C. D. F. Rogers

19.1 Introduction

Trenchless technologies cover a wide range of methods of installing new cables, pipes, ducts and small-diameter tunnels in the ground without open excavation between the start point and end point. For the smaller sizes, the installed structure is of circular cross-section, but for larger sizes in suitable ground conditions non-circular sections may be possible. The larger-size installations at substantial cover depth provide an alternative to conventional tunneling methods, while at shallow depths trenchless technologies usually compete with installation in open trench. For small-diameter pipes, techniques such as microtunneling and directional drilling provide the only economic methods of installation at depth. Trenchless technologies also include various methods of on-line replacement and renovation of existing pipes and ducts. A useful summary of the methods available is provided in *Trenchless Technology Guidelines*, produced by the International Society for Trenchless Technology (ISTT).

Once in the ground, pipes and ducts installed by trenchless methods are subject to similar load regimes to those installed in trenches; design for these conditions is covered in Chapter 18. However, the installation procedures often impose more onerous loading on the pipes than do the long-term in-service conditions. This chapter is therefore mainly concerned with the geotechnical aspects of the installation procedures and their influence on the success of the work and the design of the pipes or culverts. One by-product of this is that the installed structures may need to be of a higher quality than would normally be the case for similar structures installed in open trench.

Trenchless technologies are summarized in Table 19.1. However, the boundaries between different techniques are not always distinct, for instance between auger boring and some methods of microtunneling using auger systems. Variations in terminology may also occur between different countries. In Europe, "microtunneling" is used to describe installation of pipes smaller than man-entry size (less than 900 or 1000 mm internal diameter, ID) by pipe jacking methods, while "pipe jacking" refers to installation of pipes up to about 3000 mm ID; in the USA "microtunneling" covers installation of pipes of all diameters up to about 3000 mm ID using a remotely controlled tunneling machine.

Trenchless technologies are mainly used for the installation of new service pipes and ducts, but there are also important techniques for replacement of existing services at either the same or larger diameter. Techniques may be divided generally into those involving excavation of the ground, covered in Section 19.2, and those involving displacement of the ground, covered in Section 19.3. Techniques for rehabilitation of existing pipelines are introduced briefly in Section 19.4, while some of the many uses of trenchless technologies are discussed in Section 19.5.

TABLE 19.1. Summary of trenchless technology methods

Method	Pipe internal diameter (mm)	Drive length (m)	Ground conditions	Directional control
Pipe jacking	900–3000+	Typically 100–500	All soils	Yes; accuracy ± 50–100 mm
Microtunneling	250–1000	Up to about 120	All soils; problems with obstructions, e.g. boulders	Yes; accuracy ± 25–50 mm
Auger boring	150–900	Up to about 120	All except hard rock; problems in unstable soils, below water table, also with obstructions, e.g. boulders	Some control; accuracy ≈50 mm
Horizontal directional drilling	500–1500	Up to about 2000	All soils; problems with obstructions, e.g. boulders	Steerable both horizontally and vertically; accuracy ≈1% of length
Guided drilling	50–600	Up to about 600	All soils; problems with obstructions, e.g. boulders	Steerable both horizontally and vertically; accuracy ≈150 mm
Moling	50–200	Typically 10–20; up to 75 possible	Not hard rock or very soft soils; beware obstructions	Only if steerable moles are used; accuracy ±50–100 mm
Pipe bursting	100–500	Up to about 120	Most soils	Follows line of existing pipe
Pipe ramming	50–1200	Up to about 75	Not through rock	None

19.2 Techniques Involving Excavation

19.2.1 PRINCIPLES OF OPERATION

The techniques involving excavation all depend on the same basic principle, of excavating the ground to the external dimensions of the pipe to be inserted, or a slightly larger diameter, and pushing or pulling a lining pipe into the excavated hole. With pipe jacking, microtunneling and auger boring, the pipe is of relatively large diameter in relation to the drive length, and the lining pipe is pushed forward from the rear end. Directional and guided drilling usually involves pipes with much greater length/diameter ratio; in this case a pilot hole is drilled, then the hole enlarged if necessary by pulling back a reaming tool. One or two stages of reaming may be required to reach the required diameter, and the lining pipe is then pulled behind the final reamer to complete the installation.

Some details of the different methods are covered in the next three sections, and the geotechnical engineering considerations relevant to them in Section 19.2.5.

19.2.2 PIPE JACKING AND MICROTUNNELING

These techniques usually involve pushing a "string" of pipes through the ground, excavating at the front end using a suitable steerable tunneling shield, and adding additional pipes at the rear end as the drive advances. Work starts from a jacking pit or shaft in which the hydraulic jacks to push the pipeline are located, reacting against a thrust wall at the back of the pit or a previously installed length of pipeline. Spoil from the tunnel excavation is removed at the jacking shaft, and new pipes inserted.

The method of excavation in stable ground conditions may vary from hand excavation using pneumatic tools from within a simple shield, to mechanical excavation using backacters, cutter booms or tunnel-boring machines with rotating cutting heads. For all except the last, excavated sections of noncircular cross-section are possible. In unstable ground, various traditional methods such as breasting boards may allow hand excavation, but fully enclosed systems using slurry or earth-pressure-balance types of tunneling machine are becoming increasingly popular. In very soft clays, simple shields in which the soil is allowed to extrude through controlled ports in the closed face of the shield can be very effective. Shields are usually driven forward by the pipeline, but some incorporate their own jacks so that the cutting head may be driven forward independently, reacting against the leading pipe, and the pipes then closed up behind by the main rams. All shields are steerable, using two or more short-stroke jacks. A drive normally ends at a reception pit, from which the shield may be recovered or relaunched for the next drive. However, "blind" drives, in which the pipe terminates underground, are possible, although the shield may have to be sacrificed unless it is specially designed to be recovered through the installed pipeline.

To help reduce frictional forces between the pipeline and the ground, the ground is often excavated (whether by hand or machine) to slightly larger dimensions than the extrados of the pipes. The presence of a small open annulus around the pipes allows the introduction of a lubricating fluid, such as bentonite slurry, the effects of which are discussed further below.

Excavated spoil is usually removed in wheeled skips through the pipeline, though other methods such as conveyors or clay pumps may be used. In slurry systems the spoil is transported by the slurry, which is pumped back to the surface, where separation facilities are needed to remove the soil from the slurry.

Pipes for pipe jacking are made in a wide variety of materials: clay, glass-reinforced plastic, steel, ductile iron and, most commonly, plain or reinforced concrete. The pipes have to be strong enough to resist the high longitudinal forces imposed on them during jacking, and in particular the stress concentrations caused at pipe joints due to small angular misalignments between one pipe and the next. To maintain a smooth outer surface profile along the pipeline, joints must be formed within the thickness of the pipe wall. The effects of stress concentrations are usually reduced by the incorporation of a compressible packing material in the joint. Research has shown that chipboard and medium density fiberboard are more effective packing materials than plywood or solid timber of similar thickness (Milligan & Ripley 1989; Boot & Husein 1991). Joints are normally made watertight by the incorporation of a rubber sealing ring, often supplemented by internal caulking of the joint gap. Details of pipe dimensions, joint details, allowable loads, etc. should be obtained from the manufacturers.

The main jacking system typically involves two, three or four hydraulic jacks mounted in a frame in the jacking pit. The forces applied to the rearmost pipe are distributed using a thrust ring between the jacks and the pipe. The jacking system may involve long-stroke jacks, or short-stroke jacks and a set of spacer units; the system must be tailored to the arrangements for removing spoil and inserting pipes, and to the expected rate of progress. A thrust wall to provide reaction for the jacks is usually built into the rear wall of the jacking pit, or, if space is limited, into a short "back-shunt" tunnel. The jacking forces are ultimately transmitted into the ground by passive pressures against the thrust wall. Alternatively, reaction may be provided by a length of pipeline installed in a previous drive, through the frictional resistance between the pipes and the soil. Pipe jacks at shallow depth or through embankments may require piles or ground anchors to provide the necessary jacking reaction.

In long drives, the jacking force may increase to the extent that it exceeds the capacity of the jacks, thrust wall or pipes. In these cases, before failure occurs, an interjack station may be inserted. This consists of a telescopic combination of steel sleeve and special pipe incorporating subsidiary jacks. The interjack station allows the pipes ahead of it to be advanced using these jacks, then the pipes behind it to be brought up, closing the interjack again, using the main rams. In principle, any number of interjack stations may be incorporated. The maximum length of drive may then be greatly extended, especially in combination with lubrication; the practical limit is usually controlled by other requirements such as access, safety for workers, or the economics of spoil removal over increasing lengths.

Most pipe jacks are designed to be straight, though it is possible to jack around curves of reasonably large radius; special techniques are being developed to allow jacking around tighter curves. Straight drives are usually controlled by a laser, mounted in the thrust pit and illuminating a target in the shield. Checks are made by conventional surveying methods. Curved pipelines are more difficult and may require sophisticated laser systems or gyrocompass controls. Tolerances for line and level are usually specified, typically in the range ±25–100 mm depending on end-use and pipe diameter. Level control is generally more important, particularly for gravity sewers. However, it should be realized that tight line and level tolerances may not be needed, and are not in themselves sufficient to ensure that angular misalignments between successive pipes are kept small enough. Steering corrections within these limits must be kept gradual, and it may be preferable to allow line tolerances to be exceeded rather than correct a deviation too rapidly. Allowable jacking loads on pipes reduce rapidly with joint misalignment angle, due to stress concentrations in the joints. There is a strong case for specifications to include an allowable jacking load limitation based on joint angles. Well-controlled pipe jacks should be able to keep joint angles in the range 0–0.3°, or 0.5° in difficult ground conditions (such as ground of rapidly varying density, containing boulders or other obstructions, or very soft soils that cannot provide the reaction necessary to facilitate steering).

Microtunneling is required for smaller diameter pipelines. This is identical in principle to pipe jacking, but since man-entry is not possible, the excavation and removal of spoil must be done by remote control. Two main systems have been developed: the first uses an auger along the pipeline to remove the spoil, with a cutting head to suit the ground conditions, driven from the thrust pit; the second is a miniaturized version of a slurry-tunneling machine, with a full-face cutting head and slurry spoil-transport system. Microtunneling drives are usually relatively short, less than 100 m in length between shafts; interjacks are not feasible, provision of lubrication is more difficult, and the load capacity of the small-diameter pipes is limited. While machines are available to cope with all soil types from soft clay to hard rock, mixed or unstable ground conditions may cause problems. For instance, an obstruction such as a large boulder will halt the machine and retrieval may be very expensive and disruptive, involving excavation from the surface or a rescue tunnel from the far end. Some systems allow the machine to be retrieved through the pipeline in the event of it becoming stuck.

Various specialist systems have been developed: one, for instance, allows the installation of a flexible plastic pipe by driving a steel pipe containing the plastic pipe, and then withdrawing the steel pipe for re-use; in another, an existing pipeline is "eaten" by the tunneling machine and a new pipeline of larger diameter jacked in to replace it.

For further general information on pipe jacking and microtunneling see Thomson (1993). Much practical information and advice is contained in the publications of the UK Pipe Jacking Association, in particular the *Guide to Best Practice* (1995).

19.2.3 AUGER BORING

Auger boring predates microtunneling and is a simple method commonly used for driving a steel pipe through an existing embankment, to provide a casing through which a service pipe, duct or cable is subsequently installed. Precise directional tolerance is therefore less important. Most systems have very limited ability to steer the pipe, which usually consists of lengths of steel pipe welded together and pushed through the embankment while soil is excavated and removed from its front end. The cutting head is powered by flight augers, which are located in the steel casing and transport the muck back to the launch pit. Auger microtunneling differs mainly in that the pipe is made up of short discrete units so that it can be undertaken at depth from a shaft of limited diameter. The cutting head may be independently driven and the augers are located in temporary steel casings inside the permanent pipe. A space is deliberately left to allow a laser beam to focus on a target at the head, and an adjustable shield allows close tolerance steering.

Auger boring construction takes place at ground level or between shallow launch and reception pits. The boring equipment is driven by hydraulic jacks along a track in the drive pit.

19.2.4 DIRECTIONAL DRILLING

Directional drilling includes two different techniques, horizontal directional drilling (HDD) and guided drilling or boring (see Table 19.1). They are both normally operated by surface-mounted machines and both install a pilot bore, which is reamed out for larger sizes. In the past there have been major differences in technology and application between the two, but there is now an increasing overlap in methods and capabilities.

HDD uses large, powerful and expensive equipment for installing relatively long lengths of pipe for crossings under major waterways or similar obstructions. The pilot drill string is not rotated, but a mud motor turns the cutting head, creating the bore. The drill is usually mounted on a wheeled or tracked vehicle, with pickets or ground anchors used to provide additional reaction when necessary. Various drill bits are available for different ground conditions. The drilling fluid is also used to transport the drill returns and may be plain water in fine-grained soils, but is often a sophisticated drilling mud designed to transport material, lubricate and provide hole stability. Curvatures in both horizontal and vertical directions are possible, steering being by means of a "bent sub" in the drill string behind the cutting head, which is turned to produce a change in orientation.

Guided drills are small, relatively low-cost machines much used in urban and suburban situations for the installation at shallow depth of pressure pipes and cable ducts. The majority of the work is for pipes up to 200 mm in diameter, but recently more powerful "midi" machines capable of installing pipes up to 600-mm diameter have been introduced and find a use for mid-length crossings. Guided drilling machines use a rotating drill rod, with wedge-shaped "chisel" heads to allow steering. When continually rotated, the drill follows a straight line through the ground, but when advanced without rotation the wedge causes it to deviate in the direction required. The location and orientation of the head is often detected by signals from a sonde housed in the drill head. Small-diameter pipes are installed by drilling a hole of slightly greater diameter than the pipe, then pulling the pipe through the excavated hole as the drill is retracted. For larger-diameter pipes, a small-diameter pilot hole is drilled, then enlarged using a reaming tool, which is pulled back with the service pipe attached behind it. There are now several thousand guided drilling machines in operation and in terms of lengths of installation, this is the most important of the trenchless technology methods. In the past, HDD and guided drilling have mainly been used for pressure pipes, cable ducts, etc., for which precise control of gradient was not essential, while pipe jacking and microtunneling were necessary for gravity sewer pipes. However, improved control of alignment is likely to make these techniques increasingly viable for gravity pipelines.

19.2.5 GEOTECHNICAL ENGINEERING CONSIDERATIONS

19.2.5.1 Introduction

There are four main elements of any trenchless technique involving excavation: (a) a method of excavation; (b) a means of removing the spoil; (c) a lining pipe; and (d) a system for pushing or pulling the pipe. Many systems have been developed. However, similar geotechnical considerations apply in most cases. These relate to:

• the stability of the excavated face;
• the choice of excavation method and resulting shield thrust force;
• stability of the excavated hole, and "frictional" forces between pipe and ground;
• closure of ground onto the pipe and resulting ground movements; and
• design of the thrust wall or other reaction system.

Geotechnical aspects are considered here mainly in relation to pipe jacking and microtunneling, which have been the subject of detailed studies. These aspects, covered briefly

in the following sections, are discussed in greater detail in Thomson (1993). Some of the theoretical concepts apply equally to other techniques, but in general their success owes mainly to practical experience in the development and use of equipment and procedures; particular problems associated with these are discussed briefly in Section 19.2.5.7

19.2.5.2 Face Stability, Excavation Method and Shield Thrust Force

The method of excavation adopted is dependent on the stability of the excavated face. Hand excavation is only safe if the face is stable and serious soil collapse or water inflow will not occur. On the other hand, for cohesionless soils below the water table, a tunneling machine (probably slurry-type) that can control soil and water pressures will be essential. Stability of tunnel faces in cohesive soils has been studied theoretically and by model tests, and the results summarized by Atkinson & Mair (1981). The pressure, σ_T, required to maintain stability of the tunnel face is given by:

$$\sigma_T \geq \gamma(H + D/2) - T_c s_u \qquad (19.1)$$

where γ = unit weight of soil, H = depth of cover, D = diameter of tunnel, s_u = undrained strength of the soil, and T_c = the stability number.

The value of T_c may be obtained from Fig. 19.1, in which P is an unsupported length of tunnel behind the face. In pipe jacking, microtunneling and auger boring, P is usually small or zero.

Obviously, if σ_T is zero or negative, no pressure is needed and the face is stable unsupported; this is commonly the case in trenchless technologies, unless the drive is very deep or the soil very soft. At the other extreme, an excessively pressurized tunnel face will yield outwards, causing heave of the ground. To prevent this occurring:

$$\sigma_T \leq \gamma(H + D/2) + T_c s_u \qquad (19.2)$$

Within these limits for collapse or "blowout", a factor of safety of 1.5–2.0 on s_u is

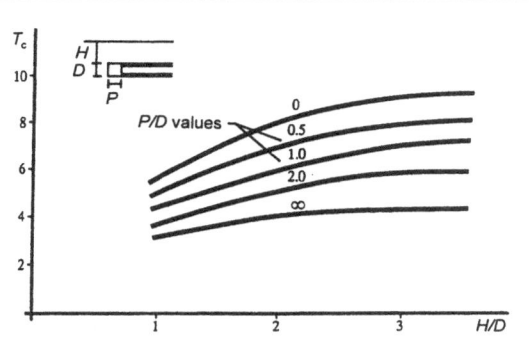

FIGURE 19.1. Stability ratios for tunnel in cohesive soil (modified from Atkinson & Mair 1981).

needed to prevent excessive ground settlement or heave (Mair 1987).

More generally, Anagnostou & Kovari (1996) have presented an expression for the effective support pressure, σ'_T, for soil with both cohesion and friction:

$$\sigma'_T = F_0\gamma'D - F_1 c' + F_2\gamma'\Delta h - F_3 c'\frac{\Delta h}{D} \quad (19.3)$$

where γ' = submerged unit weight of soil; D = tunnel diameter; c' = soil cohesion; $\Delta h = h_0 - h$, where h_0 = external water head and h = water head at face; and F_0–F_3 are dimensionless coefficients obtained from the plots in Fig. 19.2. To obtain the total support

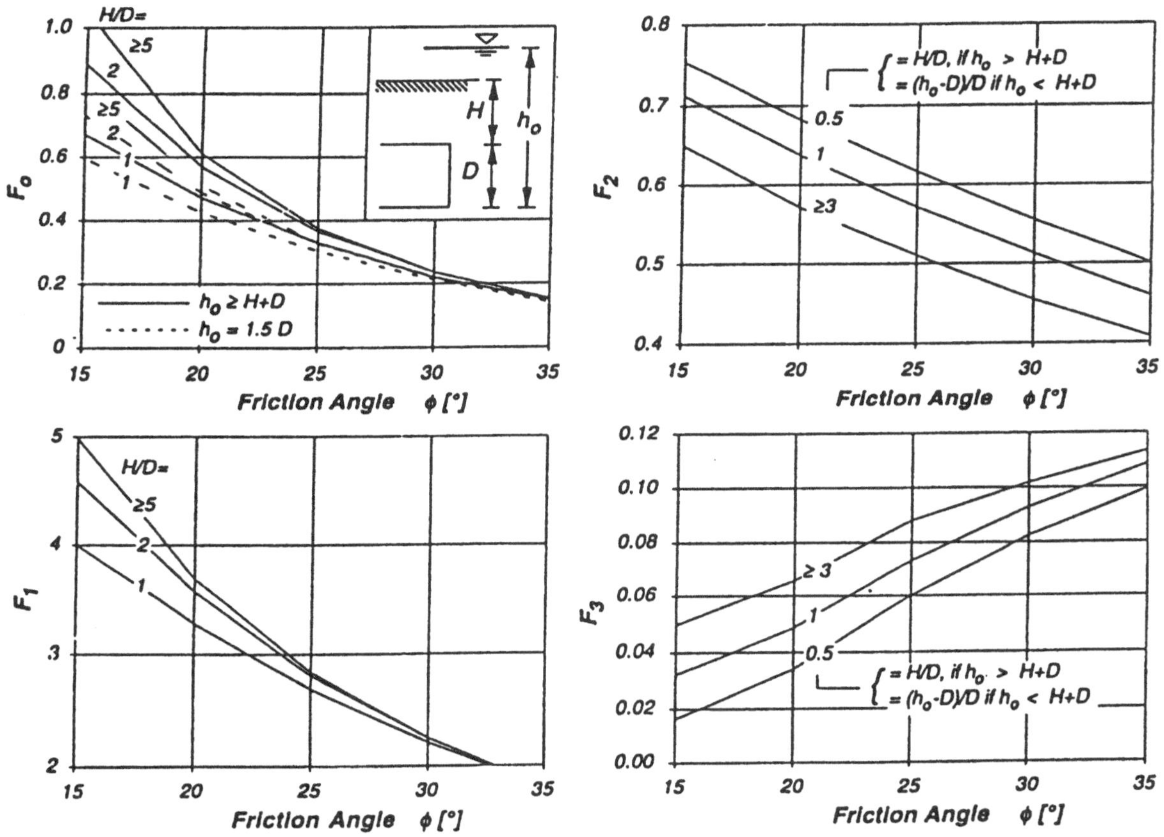

FIGURE 19.2. Dimensionless coefficients F_0–F_3 (reprinted from Tunnelling and Underground Space Technology, 11(2), 165-173, Anagnostou & Kovari, "Face stability conditions with earth-pressure-balance shields," 1996, with permission from Elsevier Science Ltd, The Boulevard, Langford Lane, Kidlington OX5 1GB, UK).

pressure the water pressure at the face must be added to the effective support pressure.

The total face resistance transmitted to the pipeline is the sum of the force due to the support pressure over the face area and the cutting force, P_s, from the edge of the shield being driven into the soil. The latter is given by:

$$P_s = \pi D_s t_s p_s \qquad (19.4)$$

where D_s = external diameter of shield, t_s = thickness of cutting edge of shield, p_s = tip resistance pressure. Typical values of p_s, after Herzog (1985), are given in Table 19.2. However, the actual value of tip resistance is very sensitive to the details of face excavation. In rock and very stiff cohesive soils it is usual to excavate, whether by machine, pneumatic tools or drill-and-blast, to a diameter slightly greater than the external diameter of the shield, in which case the tip resistance may be reduced to zero. Damp, fine sands and silts above the water table will usually also provide a stable face in the short term due to capillary tensions, allowing similar overexcavation.

The final component of resistance to the forward movement of the shield is the friction between the sides of the shield and the ground. This may be assessed using similar methods to those for estimating pipe–soil friction discussed in the next section.

19.2.5.3 Tunnel Stability and Pipeline Friction

The resistance to forward sliding of the pipeline behind the shield depends very much on

TABLE 19.2. Shield tip resistance pressures, p_s

Soil type	p_s (kPa)
Soft rock, cemented soil	12000
Gravel	7000
Dense sand	6000
Medium-dense sand	4000
Loose sand	2000
Stiff–hard clay	3000
Soft–firm clay	1000
Silt, alluvium	400

the stability of the excavated tunnel bore. Stability in cohesive ground may be assessed using Eq. 19.1 and Fig. 19.1, with the value of T_c for an infinite length P. Again, for most trenchless bores the hole will be stable except in very soft clays, at large depth, or for large-diameter pipe jacks. The internal pressure, σ_T, to maintain stability of tunnel bores in cohesionless soil, may be assessed from the following equation, after Atkinson & Mair (1981):

$$\sigma_T = \sigma_s T_s + \gamma D T_\gamma \qquad (19.5)$$

where σ_s = surface surcharge loading, T_s = stability number for surface surcharge, γ = unit weight of soil, D = diameter of tunnel bore, and T_γ = stability number for soil weight. The values of the stability numbers are given in Fig. 19.3. For a tunnel below the

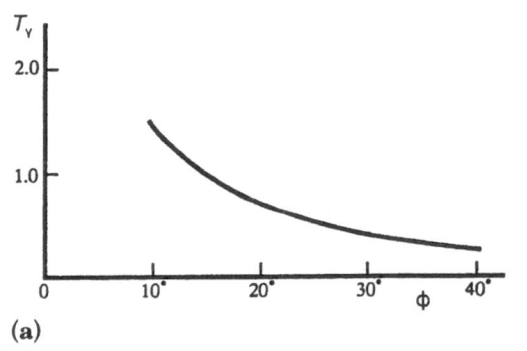

(a)

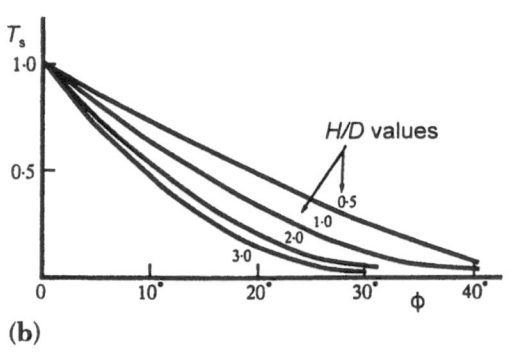

(b)

FIGURE 19.3. Stability ratios for tunnel in cohesionless soil (modified from Atkinson & Mair 1981): (a) stability number for soil weight; and (b) stability number for surface surcharge.

water table, a buoyant unit weight, γ', must be used instead of γ in Eq. 19.5, and the water pressure added to obtain the total pressure.

Traditionally, the resistance to sliding of the pipeline has been estimated on the basis of empirical data such as tabulated in Craig (1983) and Stein *et al.* (1989). These are given as apparent frictional forces per unit external area of the pipe; however, they show a very wide scatter of possible values even in a single soil type, and better estimates may be obtained by considering the basic mechanics of pipe–soil interaction under different conditions.

In most cases, the shield used for pipe jacking or microtunneling is of slightly larger diameter than the pipes; this "overcut", combined with any overexcavation at the face, ensures that in stable ground the pipes are sliding along the base of an open bore. The overcut is generally kept to the minimum required to ensure that the ground does not close on to the pipes, as a large overcut may lead to unacceptable ground settlements above the tunnel. Estimation of likely ground

closure and settlement is discussed further below. At the end of a drive, any overcut remaining open may be grouted up to minimize further settlements in the long term.

Two simple theoretical models are appropriate for assessing sliding resistance of pipes in an open bore. In rock, stiff clays, silts and sands, the interaction between pipe and ground is frictional, and the resistance, F, per unit length of pipe is given theoretically by $F = W_p \tan \delta$, where W_p = the weight of unit length of pipe and δ = the interface friction angle between soil and pipe (see Fig. 19.4). Values of δ may be obtained from interface tests in a shear box, or from published values. Some values quoted in Herzog (1985) and Norris & Milligan (1992) are given in Table 19.3.

Norris & Milligan (1992) also found from field measurements that the theoretical resistance should typically be increased by about 25% to allow for the effects of misalignments in nominally straight drives. O'Reilly & Rogers (1987) have suggested that for drives through rock, an additional factor to account for non-vertical reactions between pipe and

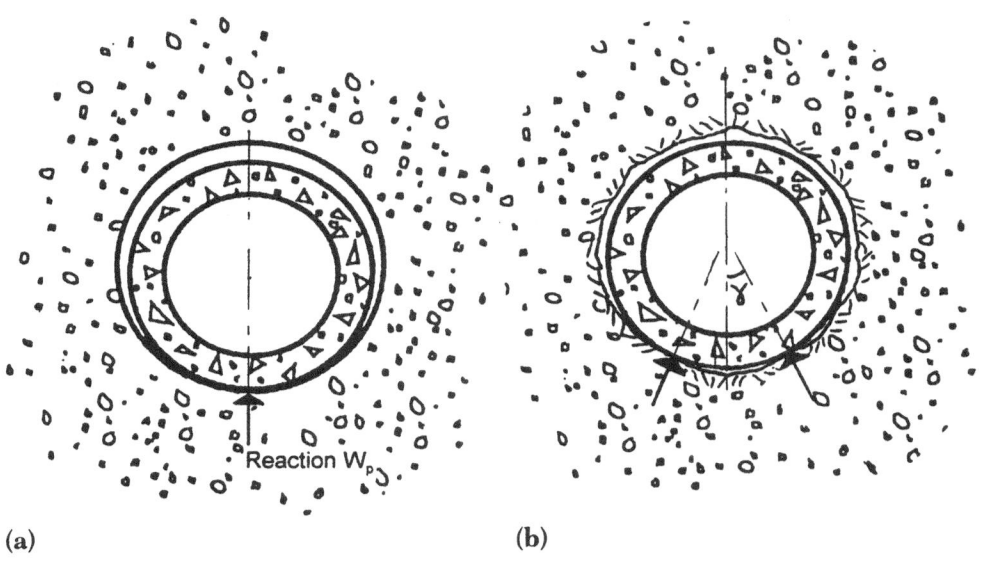

(a)   (b)

FIGURE 19.4. Pipes sliding in a stable bore: (a) pipe in soil bore; and (b) pipe in rock bore.

TABLE 19.3. Pipe–soil interface friction coefficients, tan δ

For steel and concrete pipes (Herzog 1985)			For concrete pipes (Norris & Milligan 1992)		
Soil type	Steel	Concrete	Soil type	δ(°)	tan δ
Gravel	0.55	0.88	Dense silty sand	38	0.78
Sand	0.45	0.65	Loose sand and gravel	32	0.62
Loam, marl	0.35	0.40	Glacial till (clay)	19	0.34
Low-grade clay	0.30	0.35	Weathered mudstone	17	0.31
Clay	0.20	0.25	London Clay	13	0.23

rock should be included, so that $F = W_p \tan \delta / \cos \gamma$ (see Fig. 19.4b). In softer cohesive soils, the interaction between pipe and ground may be adhesive in nature, and the resistance to forward sliding per unit length, F, is then given by $F = \alpha s_u b$, where α = an adhesion factor as used in pile design (see Chapter 10), s_u = the undrained soil strength, and b = the width of contact between pipe and ground. Haslem (1986) has suggested that b may be calculated using elastic theory for contact between two curved surfaces, but a common-sense estimate is probably sufficient given the other uncertainties involved.

If the ground is not stable, it will collapse onto the pipes, making contact with the full periphery of the pipes. In soft clay, the resistance to forward sliding will then be given by the undrained "adhesion", probably equal to the remolded shear strength of the soil, acting over the full surface area of the pipes. For cohesionless soils, a number of models of the soil–pipe interaction have been proposed, mostly based on or similar to Terzaghi's (1943) "trap door" analysis. This is shown in Fig. 19.5. The frictional resistance, F, to forward sliding due to the earth pressures on the pipes is then given by:

$$F = \frac{\pi D_e}{2}(\sigma_v' + \sigma_h')\tan \delta \qquad (19.6)$$

where

$$\sigma_h' = K(\sigma_v' + 0.5\gamma' D_e) \qquad (19.7)$$

$$\sigma_v' = \sigma_{v_1}' \exp\left[\frac{-K \tan \phi'(H - H_1)}{B}\right]$$

$$+ \frac{\gamma' B}{K \tan \phi'}\left\{1 - \exp\left[\frac{-K \tan \phi'(H - H_1)}{B}\right]\right\} \qquad (19.8)$$

$$\sigma_{v_1}' = \frac{\gamma B}{K \tan \phi'}\left[1 - \exp\left(\frac{-K \tan \phi' H_1}{B}\right)\right] \qquad (19.9)$$

and

$$B = \frac{D_e \tan(45° - \phi'/2)}{2} \qquad (19.10)$$

$$+ \frac{D_e}{2 \sin(45° + \phi'/2)}.$$

Here γ = the bulk unit weight above the water table, γ' = the submerged unit weight below the water table, H_1 = the depth to the

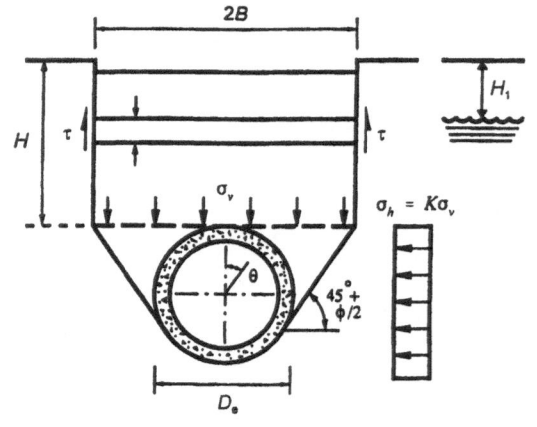

FIGURE 19.5. Ground loading in a cohesionless soil.

water table, and D_e = the external diameter of the pipes. Prior to tunneling, hydrostatic conditions are assumed in the groundwater. The parameter K is the earth pressure coefficient relating horizontal and vertical effective stresses; it must have a value between the active and passive coefficients, and may reasonably be taken as the coefficient at rest, K_0, and equal to $(1 - \sin \phi')$ in sands and gravels (see Chapter 3). The frictional resistance due to the self-weight of the pipes must then be added to give the total sliding resistance; the end result is a resistance very much greater than for pipes sliding in an open tunnel bore.

Even when the tunnel bore is stable, some ground closure will occur due to the unloading of the ground around the opening; if the overcut is insufficient the ground may make contact with the pipes and increase the sliding resistance over that for the simple "open-bore" model. By treating the ground as being elastic, standard solutions from elasticity theory may be applied (Poulos & Davis 1974). For initial total stresses in the ground at depth z of $\sigma_v(= \gamma z)$ and σ_h, the reduction in horizontal and vertical radius, δ_h and δ_v, respectively, are:

$$\delta_h = \frac{(1 - \nu)}{4G}D(3\sigma_h - \sigma_v),$$

$$\delta_v = \frac{(1 - \nu)}{4G}D(3\sigma_v - \sigma_h). \qquad (19.11)$$

While, if internal supporting pressure, p, is applied, the radius in all directions is increased by:

$$\delta_p = \frac{pD}{4G} \qquad (19.12)$$

where G and ν = respectively, the elastic shear modulus and Poisson's ratio for the soil; and D = the diameter of the tunnel bore.

Stress concentrations will also occur around the tunnel, and the maximum circumferential stress, σ_c, at the surface of the tunnel will be given by $(3\sigma_h - \sigma_v)$ or $(3\sigma_v - \sigma_h)$,

whichever is the greater (the relative values of σ_v and σ_h will depend on the degree of overconsolidation of a clay deposit, as discussed in Chapter 3). Plastic yielding of clay will occur, and tunnel deformations start to increase more rapidly, when $(\sigma_c - 2p)$ exceeds $2s_u$. There is then no simple general solution for ground deformations, but for axisymmetric conditions with σ_h and σ_v both equal to σ_0:

$$\frac{\delta}{a} = \frac{s_u}{2G}\left(\frac{a}{r}\right)\exp(N^\circ - 1) \qquad (19.13)$$

where δ = radical inward movement at radius, r; a = radius of tunnel bore $\approx D/2$; $N^\circ = \sigma_0/s_u$ and is called the stability ratio.

Initial undrained deformations will set up changes in porewater pressure, which will dissipate with time, allowing further movements to occur. The complete analysis of the time-dependent closure of ground onto pipes or tunnels, and the resulting ground–pipe interaction pressure, is complex and would have to be solved by finite element analysis. Such sophistication is unlikely to be justified for most applications, and the simple equations given above are sufficient to assess whether problems are likely to occur and make a reasonable decision about the overcut needed.

Even for pipes sliding in an open bore, time effects can be important in clays of high plasticity. The low value of interface friction angle for London Clay given in Table 19.2 is applicable while the pipeline is moving. Once the pipe has stopped, jacking resistance to restart movement increases in only a few minutes, and may increase by 50% or more after a few hours. Increases in low plasticity glacial clay have been found to be smaller, with a maximum of about 35% over a three-day break.

19.2.5.4 Calculation of Ground Movements
Here again, methods developed for tunnels may be applied to trenchless techniques. For

small pipe diameters, movements are likely to be very small provided ground collapse and loss of soil into the excavated bore is prevented. However, for pipe jacking and larger microtunneling work, movements may be sufficient to affect surface structures, or services or foundations beneath the surface.

The most commonly used method for predicting a surface settlement profile is an empirical one, based on a Gaussian distribution curve. The surface settlement, S_0, is described by:

$$S_0 = S_{max} \exp^{(-x^2/2i^2)} \qquad (19.14)$$

$$S_{max} = \frac{V_s}{i(2\pi)^{1/2}} \qquad (19.15)$$

where S_{max} = the settlement above the centerline of the tunnel, x = the horizontal distance from the centerline, i = the value of x at the point of inflexion of the curve (see Fig. 19.6), and V_s = the total volume of the settlement trough. Provided the tunnel face is kept stable, settlements due to movements there are very small; the total settlement volume may then be obtained as the volume loss along the pipeline, which is equal to the volume of the overcut around the pipe. O'Reilly & New (1982) published the following expressions:

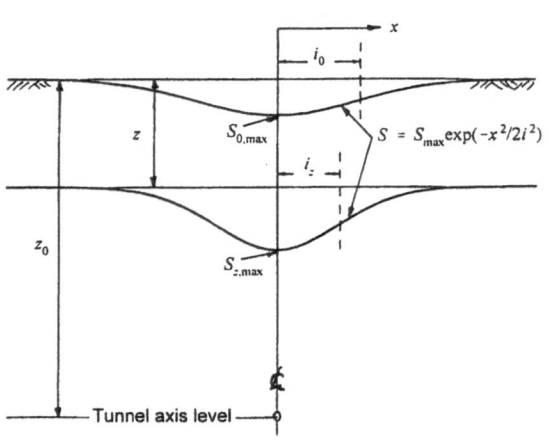

FIGURE 19.6. Surface and subsurface settlement curves.

$$i = 0.28z_0 - 0.12 \quad \text{in granular soils} \quad (19.16)$$

$$i = 0.43z_0 + 1.1 \quad \text{in cohesive soils} \quad (19.17)$$

with z_0 = the depth to tunnel axis and dimensions of meters.

A method of estimating subsurface settlement in cohesive soil has been presented by Mair *et al.* (1993). The volume of the settlement trough is taken to be the same at all depths, with the width at depth z varying in accordance with:

$$\frac{i}{z_0} = 0.175 + 0.325\left(1 - \frac{z}{z_0}\right). \quad (19.18)$$

When $z = 0$, this expression gives a close approximation to the O'Reilly & New expression for normal values of z_0.

An alternative method of calculating ground movements is to use the flow method of Sagaseta (1987), discussed below in Section 19.3.5.3 in relation to pipe bursting.

Large settlements (or heaves) may occur if inadequate or excessive pressures are applied at the tunnel face in soft ground (see Section 19.2.5.2). These pressures may be applied by the slurry in the head of a slurry-tunneling machine, or by the soil in the pressure chamber of an earth-pressure-balance machine (EPBM). The latter results from the balance between the rate of excavation at the cutter head and the rate of extraction of spoil by the screw conveyor from the pressure chamber. A skilled operator is needed to maintain the pressure within the required limits, though increasingly assistance is provided by pressure sensors at various point within the head of the machine.

In addition to these short-term ground movements, long-term settlements may occur due to consolidation of clay soils, particularly when seepage towards the tunnel causes increased effective stresses in the ground around it.

19.2.5.5 Use of Lubrication

The resistance to forward movement of the pipes may be reduced by the use of a lubricating layer between the pipes and the soil. Although simply coating the pipes with some lubricating material may have some effect, full lubrication requires that the tunnel bore is stable and the overcut remains open and is completely filled with lubricating fluid. In unstable ground, the primary action of the fluid is to provide an internal support pressure to prevent collapse of the tunnel bore. A bentonite slurry is usually used, designed to form a filter cake in cohesionless ground and allow sufficient pressure to be applied to counteract groundwater pressure and support the soil (Eq. 19.5). Various polymer materials are also available as an alternative to bentonite slurries.

Provided the overcut is kept open, the pipes are able to become partially or completely buoyant in the fluid, depending on their diameter and weight. The effective weight causing friction forces with the soil will then be reduced; for fully buoyant pipes (larger diameter pipe-jacked pipes), the pipes will in fact press against the roof of the tunnel as the effective weight becomes negative. Since the shear strength of the slurry itself is minimal, the jacking resistance can be reduced to very low values, allowing drive lengths of several hundred meters with moderate jacking forces. As an example, a 1.2-m external diameter pipe was jacked through fine sands below the water table. With fully effective lubrication, the jacking resistance was only $2.6 \, \text{kN m}^{-1}$ (equivalent to about 0.7 kPa averaged over the external surface area of the pipe), and this matched quite well the calculated frictional resistance for a buoyant pipe in a stable bore. At this rate, a drive of over 1.5-km length would be possible with a jacking force of 4000 kN. In contrast, the jacking resistance in this job increased to $41.6 \, \text{kN m}^{-1}$ when the lubrication system was not operational.

19.2.5.6 Jacking Shafts

Drive and reception shafts may be formed in a variety of ways, such as sheet piling with timber framing, concrete segments, or caissons. The thrust wall may be constructed as part of the shaft, or as a separate structure to distribute jacking forces onto the shaft wall. Resistance will usually be provided by the earth pressure at the back of the wall. In calculating this resistance, it must be recognized that intimate contact between wall and ground be achieved, probably by grouting, and that some movement is needed to mobilize passive pressures in the ground. In difficult ground conditions the design and construction of the shafts for pipe jacking or microtunneling may constitute a major part of the work, but cannot be covered in detail in this chapter.

19.2.5.7 Geotechnical Considerations for Other Techniques

Geotechnical problems with auger boring generally relate to the lack of precise control over the stability of the face. It should not be used in cohesionless soils below the water table, as the inflow of soil cannot readily be prevented. Similarly, there is a danger that voids may be created in any loose soil. Steering is fairly primitive, so that if an obstruction such as a boulder is encountered, which cannot be removed or forced out of the way, the bore will have to be abandoned.

With guided boring, directional control is good, so that if an obstruction is encountered it is possible to withdraw the drill a suitable distance and then continue again along a different line. The relatively shallow depth of most bores allows the drill head to be tracked quite accurately from the surface. Drills can operate in most ground conditions, and it is possible to retract the drill if necessary to change bits to cope with a change in ground conditions. However, it may not be possible to penetrate coarse gravel and cobbles. The water or "mud" used to assist drilling also helps to stabilize the drill hole and to lubricate the pipe as it is pulled through the ground.

Horizontal directional drilling often suffers

from the lack of good information on ground conditions, due to its use under major obstructions such as a river or airport runways, where boreholes are very expensive or even impossible. Likely ground conditions should be assessed carefully in the light of the geological history of the site; for instance, the sediments beneath a river crossing may be related to the past regimes of the river. Relatively high mud pressures in the drill hole are required to return the excavated material over the full length of the drill, and zones of high permeability, including fractures and fissures in rock, may lead to unacceptable losses, while unexpectedly low soil strength with insufficient overburden may lead to blow-out failures. Correct design of the drilling mud for the ground conditions is essential, as it must be able to perform its functions of lubrication, spoil transport and powering of the mud motors at the drill head, without excessive loss into the ground.

19.3 Techniques Involving Ground Displacement

19.3.1 PRINCIPLES OF OPERATION

The techniques involving ground displacement all work on the same fundamental principle of soil displacement to create a void into which a pipe is pulled or pushed. In the case of moling, the displacement tool creates an opening in an undisturbed soil mass, which is marginally larger in diameter than the external diameter of the pipe or cable that is pulled into place. Where an existing pipeline needs replacing, the replacement pipeline can be installed along the same route by pipe bursting or pipe removal. Pipe bursting involves a tool which is generally pulled through the existing pipe to burst or split it and expand the void to a diameter that is larger than the new pipe. The replacement pipe can have a similar or larger internal diameter than the original and can either be pulled or jacked into position, depending upon its type. Pipe removal refers to the process of pulling the existing pipe into a shaft, where it is shattered, prior to insertion of the new pipeline. Pipe ramming involves the installation of a relatively thick-walled steel pipeline, with an annular cutting shoe fixed to its leading edge, by hydraulic or pneumatic impacts. The soil that enters the pipe during this process is subsequently removed and the pipe(s) or cable(s) are fed through the permanent casing.

These techniques are described in more detail in the next three sections. Sufficient detail is provided to allow the geotechnical engineer to understand the important features of the operations, although for more information on the construction aspects, reference should be made to Kramer *et al.* (1992). The important geotechnical engineering considerations for practical application are discussed in Section 19.3.5.

19.3.2 MOLING AND GUIDED DRILLING

There are several techniques that work on the principle of void creation in an undisturbed soil mass, including rod pushing (sometimes termed thrust boring), impact moling, steerable moling and jet drilling. As with the excavation techniques, there can be confusion caused by different usage of terminology and similarity of methods. All techniques require an adequate site investigation to ensure that nothing lies in the path or within the zone of influence of the operation.

In rod pushing, a string of solid steel rods is pushed into the soil by hydraulic rams, with the rams being retracted and a new rod added to the string once it is sufficiently advanced. The leading rod is usually fitted with a tapered head having a greater base diameter than the rods. Since there is no facility for steering, a pit usually needs to be excavated and the jacking frame needs to be installed, anchored to enable both pushing and pulling operations, and accurately set to line and level. Once the leading rod has reached its target, often an excavated reception pit, either the pipe can be pulled in directly (if sufficiently small) or an expander can be drawn through in progressive stages prior to pipe installation. This technique is usually restricted to pipes of up to 150-mm diameter and lengths of 30 m.

Impact moles were introduced to improve both drive length and efficiency relative to rod pushing. These consist of a pneumatically or hydraulically powered hammer contained within a torpedo, or mole, of typically 1–2-m length. The mole has a tapered leading edge and is advanced progressively by the hammer action of the drive mechanism. Once the hole has been formed, the pipe is attached to the head of the mole and pulled into position as the tool is drawn back to the starting point. There is no facility for steering a standard impact mole. To minimize the potential for the mole to rise to the surface, the technique is normally restricted to a depth of at least ten times the diameter of the mole. The mole also can deviate from its desired route by encountering the interface between strata of different consistencies or by meeting obstacles (cobbles, rubble, roots, etc.). Thus it is necessary to track the mole as it progresses. This can be done by fitting a sonde, or signal generator, to the head of the mole and monitoring the strength of the signal transmitted to the ground surface to allow the operator to detect any significant changes in direction. Since there is no facility to steer an impact mole, however, the only recourse available to the operator is to withdraw the mole and try again at a different location.

Steerable moles, which have a head that is angled as though sliced obliquely through a cylinder, and a facility for mole rotation, have been developed but are not in widespread use. Jet drilling operates according to the same principle as steerable moling except that a water jet is forced from the leading edge of the mole. Impact and steerable moling are generally used for smaller-diameter pipelines (typically up to 200 mm) and, al-though the length of drives is increasing, the practical limit on length concerns not only the robustness of the pipe being drawn in, but also the power losses down the hoses that feed the hammer.

19.3.3 PIPE BURSTING

Pipe bursting or splitting is used where an existing pipe needs to be replaced due to structural or operational inadequacy. The bursting unit is usually pulled along the existing pipe and aims to cause outward displacement by multiple fragmentation of a rigid pipe or longitudinal splitting of a flexible pipe. In its simplest form, the burster consists of a tapered mandrel that is steadily pulled through the pipe. Where a steel, lead or ductile plastic pipe needs replacement, the burster will have a single, prominent, sharpened fin that splits the pipe prior to enlargement by increasing the diameter, i.e. the gap at the point of splitting is increased accordingly. The pulling forces required to break out the existing pipe can prove to be very large, however, and thus either static hydraulic expansion or dynamic thrusting is usually employed. In the case of hydraulic expansion, the cone of the burster is made up of segments. The burster is advanced with the segments in the closed position, stopped when further advancement is resisted and the segments are expanded outwards from the axis of the burster to create a void that is greater in diameter than the body of the burster (which is in turn greater in diameter than the following replacement pipes). The segments are then closed and the burster is advanced once more. In the case of dynamic pipe bursting, an impact hammer is built into the burster such that it operates in a similar manner to a guided impact mole. Repeated dynamic thrusts advance the burster in small increments. This generally fragments the pipe more thoroughly than hydraulic expansion.

The choice of burster for any particular site depends upon the type of existing pipe, the surrounding soil and the likely zone of influence of the burster. The primary problem faced by a bursting unit occurs in cases where the joints in the existing pipe are very strong (e.g. bolted collar joints around cast iron pipes), where the existing pipe is surrounded by concrete (perhaps only at isolated points along its length), where lengths of existing pipe are reinforced or where the existing pipe passes close by or through an existing structure. In such cases, local excavation may prove necessary to effect the break-out of the existing pipe.

The replacement pipe is either pulled (continuous flexible pipe) or jacked (pipe in discrete lengths) into position. Damage to the external surface of the replacement pipe can occur due to scoring by fragments of the burst pipe. In addition, some countries do not allow shards of broken pipe to remain in the ground. One means of avoiding these problems is to use the pipe removal technique.

19.3.4 PIPE RAMMING

Pipe ramming is generally used for relatively short crossings where good access is available. The technique essentially involves a strong, relatively thick-walled steel pipe being driven into the ground by repeated blows of a pneumatic or hydraulic hammer. The leading edge of the steel pipe is generally fitted with a cutting shoe, which has a marginally greater external diameter than the pipe. The process is thus not dissimilar to that of driving a tube for soil sampling. The pipe is advanced into the ground, the hammer is withdrawn, a new length of steel pipe is welded on and the pipe is advanced once again. The resistance to penetration, which together with the structural capacity of the pipe determines the length that can be driven, derives from a combination of face resistance on the cutter, and frictional resistance on the external and internal surfaces of the pipe. However, in some soil conditions a plug may form in the end of the pipe and the

resistance to driving will rise to that for a closed-ended tube (*cf.* a cylindrical driven pile). The ground displacements in this case would increase considerably (Rogers & Chapman 1995a).

19.3.5 GEOTECHNICAL ENGINEERING CONSIDERATIONS

The geotechnical engineer must consider the means of and factors controlling void creation, the stability of the displaced soil around the void and the effects of outward soil displacement away from the void. These factors not only influence the choice of equipment and the way in which it should be used, but also whether damage to the adjacent buried infrastructure or overlying structures could occur.

19.3.5.1 Factors Influencing Void Creation

The two primary influences on void creation are the means by which the soil is displaced and the geotechnical properties of the soil. The means of void creation described above can be categorized as dynamic displacement, or static displacement either with or without forward displacement. The dynamic displacement techniques cause associated vibrations in the surrounding soil, which are dependent upon the frequency of the dynamic action used to displace the soil. Slow hammer action and high frequency vibrations will thus affect the soil in different ways, depending upon the soil type and state. Consideration of the type of construction operation will consequently be addressed for each category of soil type in turn. In addition, the effects of the operation will be influenced by the stress state (essentially the effective confinement, p') in the surrounding soil.

In an unsaturated, essentially cohesionless soil, there is the facility for immediate shearing and compression to take place. In a very loose soil, such as a sand dune, compression is likely to occur readily (i.e. with relatively little effort) in the immediate vicinity of the operation. However, disturbance of the soil structure is likely also to induce collapse settlements in the soil overlying the operation, since volume reduction associated with shearing results in no arching in the overlying soil. This can increase the resistance to outward expansion considerably above that expected for a compressible soil. Such collapse is more likely to occur if dynamic techniques are used, high frequency vibrations being highly effective at causing collapse in such soils. If the soil is relatively dense, however, the soil will not readily compress and there will be larger magnitudes of outward displacement that extend over a greater distance from the operation. The process of outward displacement necessarily causes shearing within the soil and in dense soils this is associated with dilation, or volume increase. The difficulty in achieving outward displacement is consequently wholly dependent upon the mean normal effective stress. Vibration induced by dynamic techniques in this case is likely to restrict the extent of displacements and ease the passage of the void creation tool.

In a saturated, relatively impermeable soil, neither compression nor dilation can occur immediately and thus void creation is associated with undrained shearing. If the soil is heavily overconsolidated, undrained shearing will be associated with negative porewater pressure generation (the soil wishes to dilate) and the mean normal effective stress will increase. This will make void creation harder. Conversely, if the soil is normally consolidated or lightly overconsolidated, undrained shearing will be associated with positive porewater pressure generation and void creation will be easier since the mean effective stress will decrease. This in turn has implications for directional control of, for example, steerable moling, which relies upon the resistance along the length of the mole to effect directional change.

The second influence on porewater pressures concerns the dynamic nature of the operation. Vibration, which is in effect repeated small shearing, will result in a progressive increase in porewater pressures, which will reduce the resistance to void creation. Vibration will also facilitate local densification of soil and, if applied at an appropriate frequency, can fluidize the soil.

If the soil does not fall into one of the above categories, then engineering judgment has to be applied to predict the resistance to void creation as the tool attempts to shear/compress the soil. In cases of cemented soil, weak rock and intact rock, the resistance to soil displacement increases greatly, and can result in the expansive technique becoming impractical or impossible. In such a case, some form of "down-the-hole" hammer drilling is necessary, a technique that needs to be employed with care. Where boulders, cobbles, buried timber, objects in fill or tree roots lie in the path or close to the operation, these will similarly increase the resistance to void expansion and in some cases cause progress to be halted. As mentioned earlier, local enhancement of the pipeline structure (e.g. bolted collars, concrete surrounds) or adjacent structures can similarly inhibit progress of a pipe bursting operation and must be dealt with carefully. If progress is greatly impeded or halted for whatever reason, then either the tool must be retracted or local excavation will be required. In order to avoid these undesirable actions in all but the most exceptional of circumstances, it is important that the expansion equipment is (considerably) more than sufficient for the length and anticipated ground conditions, accounting also for any power loss to the head that may occur when working at the farthest end of the drive.

19.3.5.2 Void Stability and Pipeline Friction

When assessing void stability, similar considerations to those presented for excavation techniques (Section 19.2.5.2) are relevant. However, the techniques involving expansion often occur at much shallower depths than those involving excavation and the surrounding ground has been altered due to shearing/compression. A void can remain stable due to either negative porewater pressure (sometimes termed suction, or cohesion), and/or the creation of a stable surrounding structure in which arching can develop. The former is a time-dependent effect, void instability being more likely if delays occur in the operation.

The pipeline friction depends on the coefficient of friction between the pipe material and the surrounding soil/fragment(s) of original pipe and the mean effective stress acting normal to the pipe wall. The ease of pipe insertion is thus greatly dependent on whether the void collapses or remains stable and at what point(s) along the length of the pipeline this occurs. In addition, a single continuous length of pipe with a smooth external surface will be easier to install than a pipe made up of discrete lengths that is jacked into place. This is because there is the facility for soil or shards to be trapped in the joints (depending on their configuration), thereby increasing the resistance to forward movement.

The process of insertion of the pipeline can also have a considerable influence on the pipe itself. The pipe will tend to undergo scoring when advanced against sharp soil particles and (particularly) shards of existing pipe, the degree of scoring being dependent upon the hardness of the pipe. This can be allowed for by thickening the wall of the pipe (i.e. including a sacrificial thickness), by coating the pipe with a harder material (such as a thin steel coating), installing a liner and subsequently inserting the service pipe, and/or providing protective sleeves over (any) pipe joints. It should be noted that severe scoring could weaken the pipe structure, especially if it is subsequently subjected to cyclic loading, and in very severe cases could lead to isolated puncturing of the pipe wall. A further important influence concerns pipes that are pulled into the line, particularly if the pipe material is relatively extensible. The tensile forces involved in insertion will result in residual stresses and/or elongation that can alter the structural performance of the installed pipe as the material undergoes relaxation. It should be noted that the tensile stress application will be differential along the length of the pipeline and there will be highly concentrated stresses around the connection with the void creation tool. Jacked pipes, which are installed under compression, are less likely to suffer from residual stresses, but must be capable of withstanding the likely maximum jacking loads without structural distress (both to the pipe barrel and, where concentrated stresses occur, to the area around the joints).

19.3.5.3 Calculation of Ground Movements

The trenchless pipelaying techniques described in Section 19.3 all result in either forward soil displacement as the void creation tool approaches any point under consideration, (differential) radial outward displacement, possibly some degree of backward displacement as the tool passes and possibly some further forward displacement as the pipeline is inserted. In addition, long-term heave (upward) or settlement (downward) displacements can occur as a result of excess porewater pressure dissipation. The degree of forward displacement is likely to be most pronounced in the case of pipe ramming, but significant forward displacements will also occur with moling and dynamic pipe bursting. This is of concern where buried services cross the line of the drive within the zone of influence of the construction technique, since the service will undergo additional distress due to radial displacement. Prediction of the forward movements is not straightforward, but can be attempted on the basis of published research papers, e.g. Rogers & Chapman (1995a,c). The radial displacements are usually greater and thus more important to predict.

The prediction of radial displacements usually involves calculating the short-term displacements caused by static outward expansion, adjusting the pattern in the case of dynamic expansion and then anticipating the long-term displacements on the basis of the excess negative or positive porewater pressures generated. The radial displacements can be predicted using finite element analyses, via charts based upon empirical measurements (Leach & Reed, 1985) or by theoretical equations. Chart analyses usually rely only upon surface observations and thus give no clear or accurate indication of the likely subsurface displacement pattern, which is often of most importance. Simple theoretical analysis calibrated using site measurements or laboratory test data (e.g. Rogers & Chapman 1995b) thus provides an attractive option in cases where the subsurface

pattern is predicted from the operation outwards (rather than from the surface downwards). As with all such predictions, the findings must be tempered by carefully applied engineering judgment.

Sagaseta (1987) proposed an analytical technique based on a fluid flow model that could be applied to tunnel displacements. According to this model, the ground is simulated by a fluid. The ground displacement pattern generated by a volume loss at a point (a sink) at a specified depth, z, can be estimated by assuming that undrained, constant volume soil displacements occur as the result of an equal and opposite source located vertically above it at a distance z above the ground surface (Fig. 19.7). An adjustment to the dis-

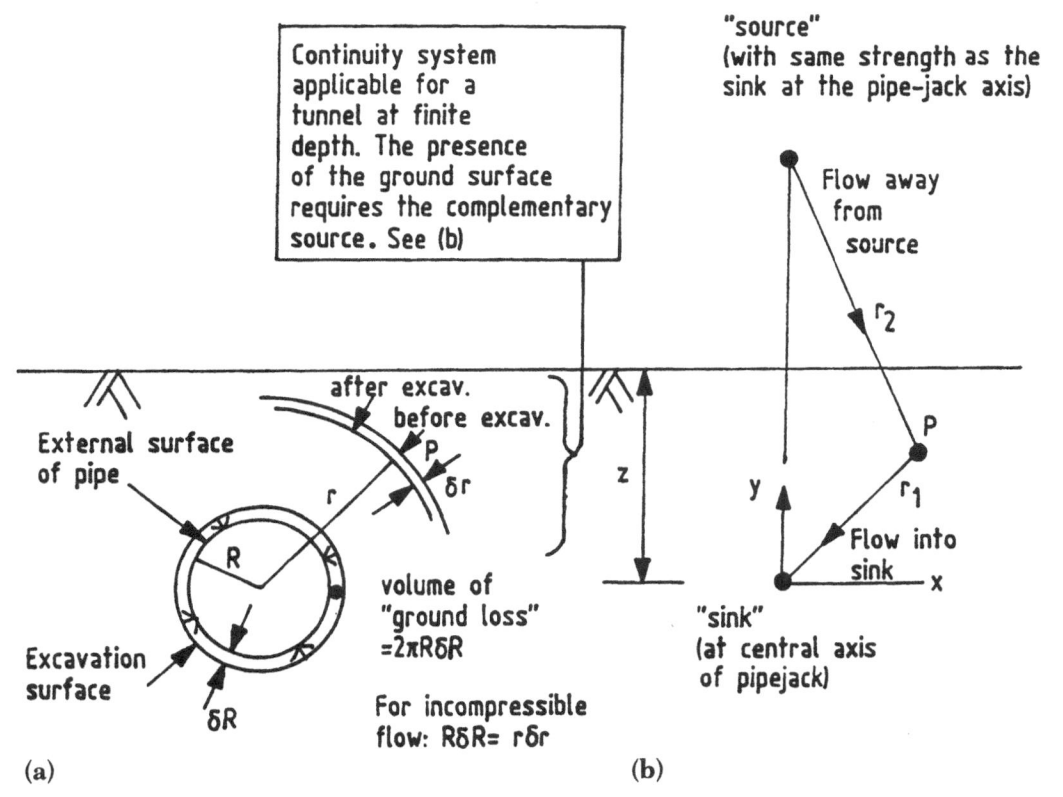

FIGURE 19.7. Application of the modified fluid flow analysis to trenchless techniques involving excavation (similar for expansion, but with "source" and "sink" interchanged): (a) schematic section through pipe jack, with definitions; and (b) analytical framework.

placement pattern close to the ground surface is made to account for the surface shear stresses generated as a result of the assumptions made. This technique can therefore be used to predict the immediate displacement pattern in saturated, fine-grained soils. The long-term displacements that result from equilibration of porewater pressures can subsequently be estimated on the basis of the initial state of the soil.

This technique has been reversed in order to predict the displacements resulting from an expansive operation such as pipe bursting (Rogers & Chapman 1998). In addition, the technique has been adjusted to allow for contraction and dilation of the soil in order that a truer picture of the displacements in a wide variety of soils can be determined directly. The principle is similar to that shown in Fig. 19.7, but with the "sink" and "sources" reversed. Where dynamic pipe bursting is used, the residual displacements are likely to be limited to a smaller zone of influence due to more effective local densification of the soil (Chapman *et al.* 1996).

19.3.5.4 Influence on the Buried Infrastructure and Adjacent Structures and Services

The patterns of ground displacement must first be predicted, as discussed in Section 19.3. The response of adjacent buried and/or overlying services or structures to the displacement pattern then needs to be assessed based on the relative stiffnesses and the form of the service/structure. This is done by tracing it onto the displacement pattern and hence estimating the maximum displacement that it would undergo (i.e. if its stiffness were less than or equal to the soil). The relative displacement of the service/structure can then be calculated on the basis that no movement will occur outside the zone of influence of the operation. If the stiffness of the service/structure is significantly greater than the soil and it can either move relative to the soil or is strong enough to restrain the soil

and hence limit the soil displacements, then the relative displacement (e.g. maximum curvature) can be reduced.

Important in this analysis is a consideration of the strength and continuity of the service/structure concerned. For example, if the expansive operation were to take place close below the surface of a continuous road pavement structure that is strong enough to resist cracking, the surface displacements will be lower than those of a free ground surface. The displacement pattern in such a case will be more radial in nature and there will be a greater degree of compression within the soil. If the surface consists of a thin, weak pavement structure, then a single crack is likely to occur directly above the line of the operation. If, however, the surface were to consist of small, discrete blocks then a more extensive pattern of smaller cracks might be expected over the full width of the surface zone of influence.

19.4 Pipeline Rehabilitation Techniques

If an existing pipeline becomes, or is likely soon to become, structurally or operationally inadequate, then there are five fundamental options available:

1. The pipeline can be abandoned and a new pipeline can be constructed in parallel.
2. The pipeline can be replaced on-line using a "pipe eating" type of microtunneling, drilling or auger boring equipment.
3. The pipeline can be replaced following pipe removal or bursting/splitting and diametrical expansion.
4. The pipeline can be replaced using trench excavation.
5. The existing pipeline can be rehabilitated.

Methods by which the first three options can be effected have been discussed earlier in this chapter, with all three being suitable for some degree of up-sizing and size-for-size replacement. The fourth, traditional option is always to be considered, although it will result in considerable surface disruption that might be or might not prove acceptable depending upon where it is located. However, it should be noted that where the pipeline has lateral connections, some means of remotely remaking the connections will be needed if surface disruption is to be avoided totally using trenchless techniques. If local excavations are required to make lateral connections, however, then the benefits over trenching can become significantly reduced, although small local excavations will nevertheless be much less disruptive to pavement structures and surface operations than full trench excavation. The fifth option, discussed below, is only viable where some size reduction of the pipeline is acceptable.

The choice of rehabilitation over other techniques depends partially upon the reason for renovation/replacement. If the existing pipeline has become operationally inadequate due to poor hydraulic performance or leakage (either egress of fluid, in which case beware of local surrounding soil erosion, or ingress of water), but otherwise remains structurally sound, the case for simple restoration of the hydraulic condition or watertightness is strong. If the existing pipeline has deteriorated in a structural sense also, the rehabilitation technique will (usually) need to provide structural enhancement, with the new liner/lining acting in combination with the existing pipe structure. It is thus important to determine the condition of the existing pipeline and this is usually done using a closed-circuit television camera survey.

The geotechnical considerations of the structural performance of a new pipeline constructed using trenchless techniques (i.e. covering the first three options listed above) have been considered earlier in this chapter and will not be discussed further here. When compared with pipelines installed by trenching, the advantages of rehabilitation are that

the surrounding soil remains undisturbed and will usually therefore be relatively stiff. When installing a pipe in a trench, it is usually impossible to reinstate the natural soil to provide the same degree of support as when it is in its undisturbed state. For this reason, a high quality fill is commonly introduced to surround the pipes and backfill the trench (depending upon its location), creating additional cost and environmental damage. Thus it is the pipe–soil system, rather than the pipe in isolation, that needs to be considered in determining the structural performance of rehabilitated pipes and pipes installed by trenching.

One further important consideration here is the intimacy of the association between an existing pipe and a new liner or coating. Many rehabilitation processes will produce a lining that has insufficient structural capacity alone to resist applied loading. Such processes are prohibited in general in current design unless there is full bonding and hence interaction between the existing pipe and the liner. The lining needs to act in combination with the existing pipe to counteract soil loading and any dynamic or static imposed loading. This type of structural calculation has been discussed in Chapter 18. However, if the existing pipeline is no longer watertight, or if it might lose its watertightness as a result of the rehabilitation process (for example, during intensive cleaning operations prior to treatment), then the hydrostatic effects of water pressure acting on the lining alone must be considered. If the pipeline is deeply buried beneath the groundwater table, the hydrostatic pressure can become considerable and the liner, which is often thin, can be susceptible to buckling, if flexible, or cracking, if rigid. In such a case the liner must be designed to be sufficiently strong to resist buckling/cracking, or some means of grouting behind the liner should be used to prevent the hydrostatic pressure acting directly on it.

The traditional means of providing a full

lining is to use sliplining techniques. The pipeline to be treated is taken out of service, cleaned and a new (generally thin-walled) liner is pulled and/or pushed into place. Any lateral connections then need to be made, either by hand in local surface excavations or remotely using robotic techniques, prior to testing the system for water/airtightness and recommissioning. High density polyethylene liners, for example, are often installed in this manner since they can often be bent sufficiently to be passed into the pipeline in one continuous length without additional excavation (e.g. via a manhole). An alternative system is to use welded lengths of a less deformable material, such as thin-wall stainless steel.

In order to make this process more efficient, techniques of reducing the diameter of the liner during insertion, and then expanding the diameter once it is in place, have been developed. This greatly eases the insertion process and reduces the risk of scoring of the external pipe surface, as well as ensuring a tight fit to the existing pipe structure and minimizing the loss of cross-sectional area of the pipeline. Such techniques are applied to continuous lengths of medium- or high-density polyethylene pipe. One technique is known as *rolldown*, in which the liner is diametrically compressed by passing it through a series of rollers such that it is elastically, but not plastically, deformed (i.e. the yield point of the plastic is not reached). The liner is then fed, in one continuous operation, into the host pipeline, the ends of the liner thereafter being sealed and the original diameter being restored by pumping in water under pressure. It should be noted that rerounding would occur automatically within weeks, although pressure is usually applied as an expediency.

A second technique is known as *die drawing* or *swage lining*. In this case, the liner is pulled by a winch through a die. For smaller diameters, the process is carried out at ambi-

ent temperatures, whereas for larger pipes, the die and even the liner may be heated prior to diametrical reduction. The liner is thereafter expanded when in place and any lateral connections remade, as before.

A further variation of this idea consists of the insertion of collapsed, or folded, sections of thermoplastic pipe, which are subsequently rerounded using hot air, hot water or steam. The liner in this case is folded into a C or U shape using a thermomechanical process, delivered to site in its deformed shape and then heated to make it flexible prior to insertion. One advantage of the heating process used to reround the liner is that it will deform readily around bends and into any surface irregularities in the pipe wall, thus ensuring a close fit. In addition, indentations in the liner will occur where lateral connections occur in the host pipe, making it easier to determine their location for making remote controlled connections.

Cured-in-place resin-impregnated soft liners can be installed to produce a similar end result, although via a different process. A flexible, permeable tube made from a geotextile is fully impregnated with a polyester or epoxy resin and a catalyst, and is cut to the required length. This tube is then either drawn into place and filled with cold water to force the liner against the host pipe wall or installed by inversion. For installation by inversion, the tube is fixed to a ring at the insertion point (for example, a manhole for sewer renovation) and is progressively installed using cold water under pressure such that the tube is turned inside out as it passes through the ring and the host pipe. Once it has reached the reception point (often a second manhole), the cold water is replaced by recirculated hot water for a period of a few hours to cure (heat cured) or an ultraviolet light is passed down the inflated tube (light cured). Whichever technique is used, and several companies offer such a service, the resins and/or geotextile thickness can be designed for the particular job being undertaken. In general, epoxy resins produce superior adhesion with the host pipe, whereas polyester resins have a superior chemical resistance. Liner design is commonly offered by the contracting companies.

A further form of liner that can be installed in small spaces, such as from the base of a manhole, is a spirally wound liner. This type of liner is formed from a continuous strip of plastic, which is ribbed on one side and smooth on the other, and has interlocking edges. The strip is fed into a winding device at the base of the manhole, which produces a pipe by interlocking the edges. The pipe, which is typically 25-mm smaller in diameter than the host pipe (assuming that a reasonably true, intact bore exists), is progressively fed into the host pipe and the annular gap is subsequently grouted using a cementitious grout or polyurethane foam. The winding process produces a ribbed external surface, which adds to the liner strength, and a smooth internal surface. An alternative process was developed that allowed a spirally wound pipe to be formed and inserted as above, but then the liner was wound in reverse allowing the diameter to increase, as the spiral joints slip, with the aid of an expander tool. This allows the liner to come into contact with the host pipe wall and obviates the need for grouting, although grout can be added during the winding process to aid sealing, bonding with the host pipe and resistance to hydrostatic pressure.

The final technique involves a continuous sprayed liner. This can be formed from cementitious mortar or resin. For non-man-entry pipelines a polymer mix can be applied as a thin coating to improve the hydraulic properties of the host pipe, using robotic devices to spray the material. However, this technique is still undergoing commercial development. This type of liner rarely provides significant structural strength. In the case of man-entry pipes, reinforcement can be placed prior to spraying shotcrete or Ferrocement. The reinforced lining in this case is designed for full internal and external load.

19.5 Practical Application

In some cases, the range of trenchless techniques now available will prove highly attractive since access cannot be readily gained to the surface. In the majority of cases, however, the engineer is faced with a choice between surface excavation or trenchless construction. In these cases, the engineer must make an appropriate engineering decision based upon both hydraulic and structural considerations and on cost. The hydraulic considerations focus on both the size of the new/replacement pipe and its hydraulic performance in relation to pipes laid by trenching and, where relevant, to an existing pipeline that is being replaced. The structural considerations of the two broad alternatives have been isolated in this chapter and Chapter 18 with respect to the pipeline itself and its immediate surround. In addition the influence of trenchless technology operations on adjacent services and buried structures has been highlighted, as would be the case for traditional tunneling. Similar considerations are necessary for trenching operations, however, although the effects of trenching are often overlooked. For example, trenching or utility cut patching can cause a significant reduction in the life of road pavement structures (Shahin & Crovetti 1987).

The principal considerations for geotechnical engineering in connection with trenchless methods have been introduced in the sections above. It is important to emphasize the need for appropriate site investigation if these techniques are to be used successfully. Problems arise mainly where ground conditions vary significantly along a drive, where "mixed-face" conditions can occur, from unexpected groundwater pressures, and from the presence of obstructions, whether natural or man-made, or other services already present. The relative importance of these will depend on the type of project and the ground conditions. For example, obstructions may be dealt with relatively easily in a stable face in a pipe-jacked tunnel, but may prevent completion of a microtunneling, moling or ramming operation. However, the same pipe jack encountering a lens of water-bearing sand will suffer face collapse, which may endanger miners and lead to large surface settlements. Thus although trenchless techniques can provide considerable advantages, serious problems such as obstruction or deviation of a drive, or large settlements, can require expensive remedial measures.

Trenchless techniques are often used for crossings beneath rivers and canals, major roads and railways. Conventional ground investigation in such cases may be very expensiveor unacceptable, but the consequences of problems due to unexpected ground conditions may be very serious. Use of directional drilling as a ground exploration technique along the line of a proposed pipe jack might be considered in these circumstances. In other cases, consideration should be given to the use of geophysical methods to interpolate information between boreholes, for instance, on the depth to a sand–clay or soil–rock interface. In urban areas,

the ground will often be highly congested with buried services. Underground mapping techniques, such as ground-probing radar and infrared thermography, should be added to the ground investigation program, and the desk study should be as extensive as possible. The site investigation process needs especially careful design since techniques that, for example, recognize cables may not react to plastic water-filled pipelines. In addition, the observational approach involving careful site monitoring should be adopted while construction takes place so that any changes in conditions or unexpected events can be perceived and appropriate action taken at the earliest opportunity.

Trenchless technology is a relatively new subject that is increasingly being introduced into general engineering practice. New trenchless techniques are being developed, the currently available techniques are being refined and confidence in the use of equipment is increasing. The direct costs of trenchless technology are consequently falling and the risk associated with the use of the techniques is reducing. The instinctive decision to dig a trench for even a small-scale construction operation must be reconsidered. The pipeline engineer must therefore weigh carefully the disadvantages and advantages of both trenching and trenchless techniques.

20. COLD REGION ENGINEERING

J.-M. Konrad

20.1 Introduction

Permafrost underlies about one-quarter of the world's land surface and most of the engineering aspect has been well-covered in Andersland & Anderson (1978), Johnston (1981) and Andersland & Ladanyi (1994). This chapter considers more specifically the problems associated with seasonal freezing and thawing of soils in which there is no perennially frozen ground. Frost action may cause extensive damage to buildings, infrastructure such as roads and utility lines and other civil-engineering structures. All such facilities should be designed and constructed to avoid serious functional problems, costly maintenance and unduly short service life. The behavior of soils is strongly influenced by temperature and, therefore, an appreciation of heat transfer in soils is of utmost importance to cold region engineering.

Ground temperatures can be predicted, in principle, if the boundary conditions and the thermal properties of each soil layer are known, especially with available finite element models. It is beyond the scope of this chapter to give details of numerical methods since it is intended to provide basic information about heat transfer in soils and guidance for simplified predictions of frost depth, frost heave and freezing-induced water content increase, and some design mitigations against frost action. Frost heave mechanics is covered extensively to provide the necessary background for the design engineer.

20.2 Thermal Considerations

20.2.1 HEAT FLOW THROUGH SOILS

20.2.1.1 Steady State

Steady state (i.e. time independent) heat flow through homogeneous soils by conduction is given by the Fourier equation:

$$Q_x = -\lambda \frac{dT}{dx} A$$

$$\text{or} \quad q_x = \frac{Q_x}{A} = -\lambda \frac{dT}{dx} = -\lambda T_G \quad (20.1)$$

where Q_x = heat flow in the x-direction per unit time (W), q_x = heat flux (W m^{-2}), λ = thermal conductivity (W m^{-1} K^{-1}), T = temperature (K); x = distance (m), A = area (m^2).

Heat flux is thus proportional to the temperature gradient, T_G, and the proportionality constant is referred to as the thermal conductivity. The negative sign in Eq. 20.1 indicates that heat flow is in the direction of decreasing temperature, i.e., warm to cold.

In homogeneous materials not undergoing phase change, the distribution of temperature at steady state is linear for planar one-dimensional conduction. In layered soils, each with its own constant thermal conductivity, the heat flux normal to the bedding can be computed from:

$$q_x = T_1 - T_2 \bigg/ \sum_{i=1}^{n} \frac{d_i}{\lambda_i} \quad (20.2)$$

or from

$$q_x = \bar{\lambda} \frac{T_1 - T_2}{D}$$

where $\bar{\lambda} = D/\sum_{i=1}^n d_i/\lambda_i$ = the equivalent thermal conductivity. D = the total thickness of the layered system, d_i is the thickness of layer i, T_1 and T_2 = imposed temperature boundary conditions. The temperature distribution is piecewise linear and obeys the following relationship at each interface:

$$\lambda_{i+1} T_{G_{i+1}} = \lambda_i T_{G_i} . \qquad (20.3)$$

The heat flux per unit length for one-dimensional radial heat flow in homogeneous material without phase change is:

$$q_r = 2\pi\lambda(T_0 - T_1)/\ln\frac{r_1}{r_0} \qquad (20.4)$$

where r_0 is the inner radius maintained at a fixed temperature, T_0; and r_1 is the outer radius at a temperature T_1.

The temperature distribution given by Eq. 20.5 is non-linear as heat flows through elements with increasing areas if $T_0 > T_1$ or with decreasing areas if $T_0 < T_1$:

$$T(r) = T_0 - (T_0 - T_1)\ln\frac{r}{r_0}\bigg/\ln\frac{r_1}{r_0}. \qquad (20.5)$$

20.2.1.2 Transient State

Transient conduction processes, in which the temperature at a given point varies with time, are commonly encountered in engineering design. They generally fall into two classes: the process that ultimately reaches steady state conditions, and the process in which the thermal boundary conditions change with time.

For a homogeneous material, without phase change, the governing equation of one-dimensional heat transfer is:

$$\frac{\partial T}{\partial t} = \alpha\frac{\partial^2 T}{\partial x^2} \qquad (20.6)$$

where t = time; α = thermal diffusivity = λ/C (m^2 s^{-1}), where C = volumetric heat capacity (J m^{-3} K^{-1}).

Equation 20.6 is analogous to the well-known consolidation equation in which temperature is replaced by pore pressure and thermal diffusivity by the coefficient of consolidation. As for the consolidation problem, solutions for unsteady state heat transfer can be presented in terms of temperature–time charts (Fig. 20.1) using dimensionless

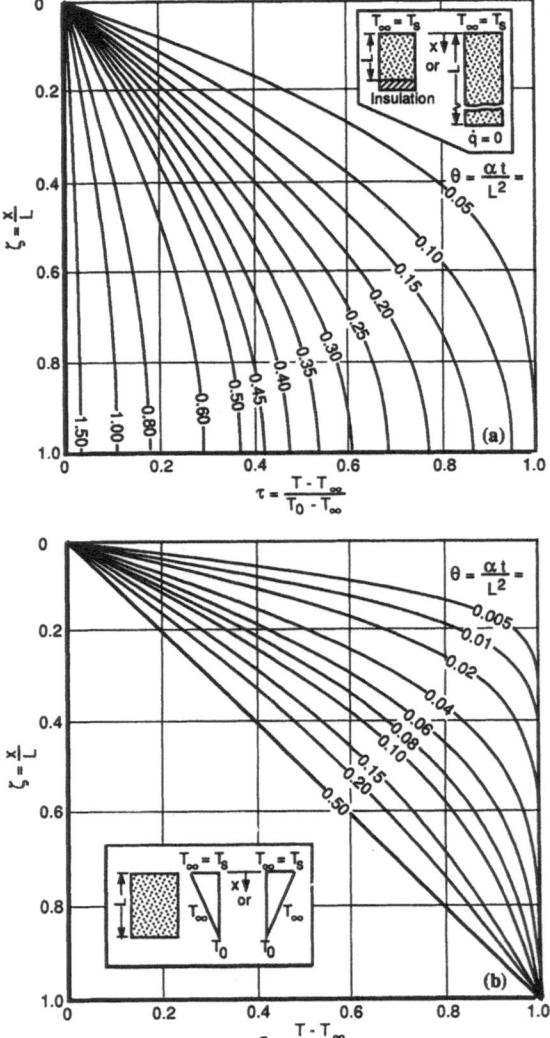

FIGURE 20.1. Temperature–time charts for unsteady heat flow by conduction.

ratios θ, relative time $= \alpha t/L^2$; ζ, relative position $= x/L$; and τ, relative temperature $= (T - T_\infty)/(T - T_0 - T_\infty)$, where T_∞ represents the temperature at steady state. These charts may be used to evaluate temperature profiles for cases involving conductive heat flow if the following conditions apply: (a) constant thermal diffusivity and no internal heat source; (b) the conducting medium has a uniform initial temperature, T_0; and (c) the temperature at the boundary is changed to a new value, T_∞, for $t > 0$.

In many time-dependent conduction processes actual initial and/or boundary conditions may be more complex than those corresponding to the analytical solutions, geometry may be irregular, thermal properties may be temperature dependent and phase change may be present. Numerical techniques may be used to compute temperature–time distributions using the many available computing codes. This aspect will not be discussed further in this chapter.

20.2.2 THERMAL PROPERTIES OF SOILS AND OTHER MATERIALS

The thermal properties needed for the analysis of thermal problems in soils are the thermal conductivity, the volumetric heat capacity and the latent heat of fusion associated with phase change of interstitial water. In this section, sufficient information on the thermal properties of different materials is provided to enable the reader to apply various analytical methods for calculation of temperature–time distributions.

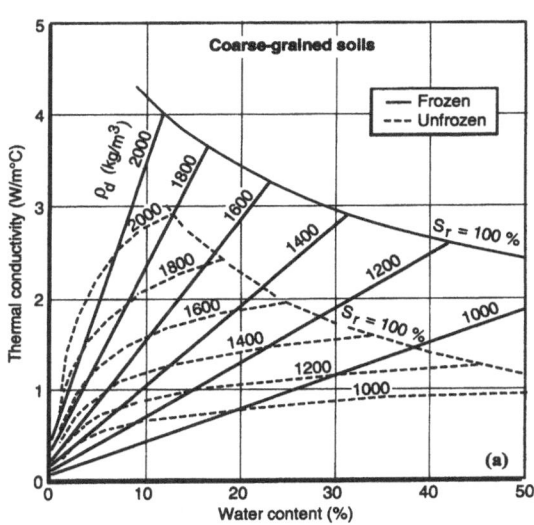

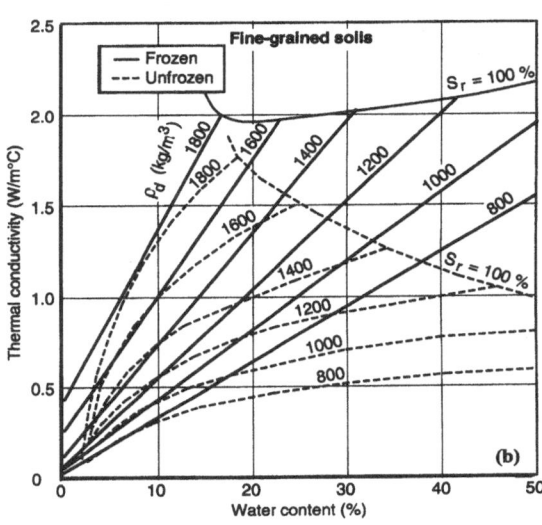

FIGURE 20.2. Thermal conductivity of (a) coarse-grained, and (b) fine-grained soils (modified from Kersten 1949). — frozen, (---) unfrozen.

20.2.2.1 Thermal Conductivity

The thermal conductivity, λ, of soil depends on its density, water content, mineralogy, grain size and temperature. Kersten (1949) has measured the thermal conductivity for a wide range of both frozen and unfrozen soils at different water contents and dry densities, as shown in Fig. 20.2.

20.2.2.2 Heat Capacity

The specific heat is defined as the quantity of heat required to raise the temperature of a given material by 1 °C. When expressed on a per unit mass basis, this quantity of heat is referred to as the specific heat capacity, c, and when expressed on a per unit volume basis, it is known as the volumetric heat capacity, C.

For unfrozen soils, the volumetric heat capacity can be estimated from:

$$C_u = c_s \rho_d + \frac{w}{100} c_w \rho_d \qquad (20.7)$$

where c_s = specific heat capacity of soil grains = 800 J kg^{-1} K^{-1} (average value), c_w = specific heat capacity of water = 4200 J kg^{-1} K^{-1}, w = water content (% dry weight), ρ_d = dry density (kg m^{-3}).

For frozen soils, the volumetric heat capacity is estimated from:

$$C_f = c_s \rho_d + \frac{w_u}{100} c_w \rho_d + \frac{w - w_u}{100} c_i \rho_d \quad (20.8)$$

where c_i = specific heat capacity of ice = 2050 J kg^{-1} K^{-1}, w_u = unfrozen water content (see section below).

The thermal diffusivity, α, which is defined as the ratio of thermal conductivity to volumetric heat capacity, is readily calculated ($\alpha = \lambda/C$). It represents the rate at which a material will undergo a temperature change in response to boundary condition changes.

20.2.2.3 Latent Heat

As depicted on Fig. 20.3, the latent heat of pure water, L, is associated with phase change from solid to liquid. When ice is progressively heated, thermal energy is absorbed by the ice–water system under constant phase change temperature until all ice is melted. Conversely, when water is progressively cooled, thermal energy is released by the ice–water system under constant phase change temperature until all water is frozen. In both cases, the latent heat for water, L, is 334 kJ kg^{-1} or 334 MJ m^{-3}.

For a soil (saturated or unsaturated), the latent heat of freezing or thawing is related to water content as:

$$L_s = \frac{w}{100} \frac{\rho_d}{\rho_w} L. \qquad (20.9)$$

20.2.2.4 Thermal Resistance

The concept of thermal resistance is widely used for characterizing the thermal properties of construction materials such as walls, polystyrene insulation and bricks. The thermal resistance, R, per unit area is defined as the difference in temperature between two surfaces divided by the heat flux between them. It can be viewed as another form of Eq. 20.1 since:

$$R = \frac{\Delta T}{q} = \frac{\Delta x}{\lambda}. \qquad (20.10)$$

R is thus expressed as m^2 KW^{-1} and manufacturers usually also specify Δx.

For instance, extruded polystyrene has a thermal resistance of 0.92 m^2 K^{-1} W for a 25.4-mm thick panel, which yields a thermal conductivity of 0.0254/0.92 = 0.028 W m^{-1} K^{-1}. Table 20.1 presents a summary compiled from various sources (Dysli 1990; Johnston 1981) of thermal properties for materials used by the geopractitioner.

20.2.3 THERMAL BOUNDARY CONDITIONS

20.2.3.1 Generality

Several thermal boundary conditions can be used with the heat conduction equation:

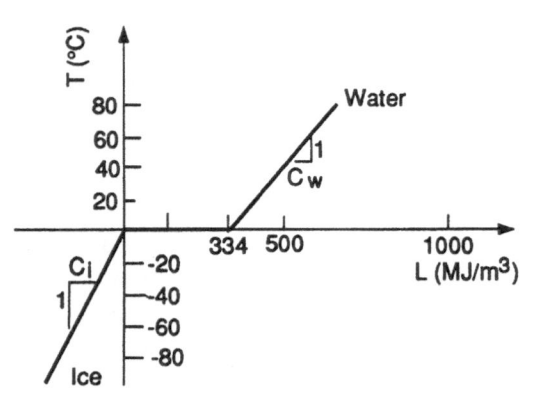

FIGURE 20.3. Thermal energy of water.

TABLE 20.1. Thermal properties of various materials[a]

Material	λ(W m^{-1} K^{-1})	c(kJ kg^{-1} K^{-1})	C(kJ m^{-3} K^{-1})
Air (20 °C)	0.026	1.00	1.2
Water			
10 °C	0.55	4.22	4217
20 °C	0.60	4.18	4174
Ice			
0 °C	2.24	2.11	1935
−10 °C	2.32	2.03	1865
−20 °C	2.43	1.96	1800
Snow			
$\rho = 100$ kg m^{-3}	0.06	0.21	210
$\rho = 500$ kg m^{-3}	0.59	1.05	1050
Minerals			
Quartz	4.8–7.7	~0.75	~2000
Feldspar	1.9		
Biotite	2.0		
Rocks			
Granite	2.9–4.1	~0.80	~2100
Gneiss	2.7–4.6	~0.85	~2200
Quartzite	5.4–8.1	~0.85	~2100
Limestone	1.5–3.3	0.85–2.20	2000–5200
Sandstone	2.3–6.5	0.90–2.00	2200–5000
Mudstone	1.5–3.5	1.00–2.20	2100–4600
Soils (unfrozen, saturated)			
Clays	0.9–1.8	1.4–2.1	2600–3400
Silts	1.2–2.4	1.2–1.8	2500–3100
Gravelly sands	1.2–3.0	1.1–1.8	2400–3000
Peat	~0.6	~3.0	~4000
Common materials			
Steel	70	0.48	3800
Aluminum	220	0.95	2600
Copper	400	0.39	3500
Glass	0.8	0.9	2400
Concrete	1.8	1.10	2600
PVC	0.22	1.50	2000
Bituminous concrete	~1.5	1.0	2300
Insulation			
Polystyrene			
Extruded	0.03–0.04	1.40	40
Expanded	0.05–0.06		
Saw dust			
Dry	0.16		
Unfrozen, saturated	0.30–0.35		
Frozen, saturated	0.60–0.70		
Wood chips			
Unfrozen	0.70–0.85		
in situ, dry	0.18		
Wood bark	0.5		
Straw fibers	0.04–0.05		
Recycled plastic chips			
Laboratory	0.3–0.4		
In situ	0.5		
Tire chips (dry)	0.40		

[a] Moisture pick-up in the field must be considered.

1. Prescribed temperature: this temperature may be constant, or a function of time, or position, or both.
2. No flux across the boundary: according to Eq. 20.1, this condition means that $dT/dx = 0$, or that the temperature profile is normal to the boundary.
3. Prescribed flux across the surface: $dT/dx \neq 0$ and is equal to q/λ.
4. Linear heat transfer at the surface: if the flux across the surface is proportional to the temperature difference between the surface and the surrounding medium: $\kappa(T - T_M)$, where T_M = the temperature of the medium and κ is a constant.
5. Other boundary conditions: such as forced convection, natural convection, non-linear heat transfer and radiation may be used, but will not be discussed herein. Further details may be found in Carslaw & Jaeger (1959).

20.2.3.2 Freezing and Thawing Indices at Ground Surface

The actual thermal boundary condition at the ground surface is best described in terms of a prescribed net heat flux that varies with time both seasonally and daily. In practice, however, the time-dependent net heat flux is difficult to estimate since it results from solar radiation (dependent upon surface reflectivity and orientation, clouds, latitude, season and day), long-wave radiation (dependent upon temperature of surface and surrounding medium, clouds, temperature and humidity gradients in the atmosphere), and air flow over the surface (dependent upon temperature, wind speed, surface roughness and flow regime, evaporation, condensation, evapotranspiration).

Historically, the depth of frost penetration was first estimated by Stefan (1891) for a large water body with the assumptions that the latent heat is the only heat to be removed during freezing and that thermal energy stored as volumetric heat in the frozen zone, i.e. the ice sheet, is neglected. The initial tem-

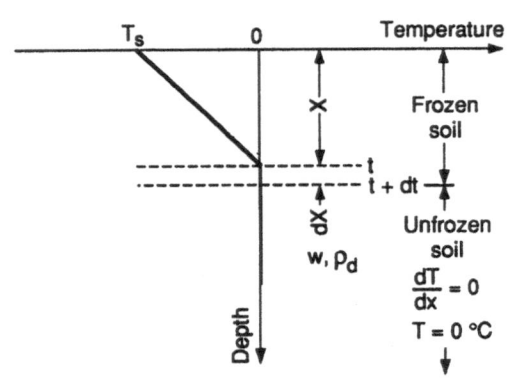

FIGURE 20.4. Simplified temperature profile in Stefan's equation.

perature in the water is assumed to be 0°C and constant with depth. Adapting Stefan's solution to a saturated soil at given water content and dry density subjected to a step-freezing temperature, T_s, the equation of continuity at the frozen–unfrozen interface is (Fig. 20.4):

$$L_s \frac{dX}{dt} = \lambda_f \frac{T_s}{X} \qquad (20.11)$$

which, after integration, yields the relationship between frost depth and duration of freezing:

$$X = \left(\frac{2\lambda_f T_s t}{L_s} \right)^{1/2} \quad \text{with} \quad L_s = \frac{w}{100} \frac{\rho_d}{\rho_w} L \qquad (20.12)$$

where X = the depth of frost penetration and t = the duration of freezing.

If the surface temperature varies with time, integration of Eq. 20.11 gives:

$$X = \left[\frac{2\lambda_f \int_0^t T_s(t)dt}{L_s} \right]^{1/2} = \left(\frac{2\lambda_f I_{sf}}{L_s} \right)^{1/2} \qquad (20.13)$$

where I_{sf} = the integral of the surface temperature with time. It is a measure of freezing intensity, referred to as the ground surface freezing index, and is usually expressed in °C-days.

Stefan's simplified approach can also be

used to estimate the thaw depth in permafrost as:

$$X = \left[\frac{2\lambda_t \int_0^t T_s(t)\mathrm{d}t}{L_s} \right]^{1/2} = \left(\frac{2\lambda_t I_{st}}{L_s} \right)^{1/2} \quad (20.14)$$

where I_{st} = the integral of the surface temperature with time and is a measure of thawing intensity, referred to as the ground surface thawing index. It is also expressed in degrees C days. λ_t is the thermal conductivity of the thawed soil. L_s corresponds to the latent heat of the frozen soil and accounts for the actual ice content, which can be significant if segregated ice is present. In areas of seasonal freezing, Eq. 20.14 can be used to estimate the length of the thawing period, given a maximum frost depth computed from Eq. 20.13 as discussed below.

Since air-temperature data are readily available from meteorological stations and ground surface temperatures are not, engineers have established correlations between these two parameters. Furthermore, in practice, the actual surface temperature variations are replaced by the surface freezing and thawing indices. Thus, thermal boundary conditions are usually related to the air-freezing, I_{af}, and air-thawing, I_{st}, indices:

$$I_{sf} = n_f I_{af} \quad \text{and} \quad I_{st} = n_t I_{at} \quad (20.15)$$

where n_f and n_t = empirical coefficients called the "n-factor" for freezing and thawing, respectively. The magnitude of the n-factors depends on the same factors as those affecting the net heat flux at ground surface. Table 20.2 compiles available data from (US Army 1966; Smith 1986; Dysli 1990).

The extreme range of n-factor values may reflect the range of geographical location, but also of climatic characteristics such as atmospheric relative humidity, solar radiation, precipitation, wind speed, etc. For these rea-

TABLE 20.2. Values of n-factors for different surfaces

Surface type	n_f	n_t
Snow	1.0	—
Sand and gravel (probable range for northern conditions)	0.6–1.0	1.3–2.0
Asphalt pavement (free of snow or ice)	0.4–1.0	1.4–2.3
Concrete pavement (free of snow or ice)	0.25–0.95	1.3–2.1

sons, Dysli (1990) proposed to use the following relationship to estimate surface-freezing index from air-freezing index:

$$I_{sf} = I_{af} - \mathrm{RI} \quad (20.16)$$

where RI = the radiation index, which can be related to the net global radiation over the freezing period measured by a pyranometer. RI can also be estimated directly from the average length of exposure to sun over the whole freezing period or more crudely from an estimate of exposure to sun (nil, low, medium, high). The relationship presented in Fig. 20.5 is valid for Switzerland and is useful for the design of highways.

Applying the Swiss relationship to North American conditions, say to Quebec City characterized by an average air-freezing in-

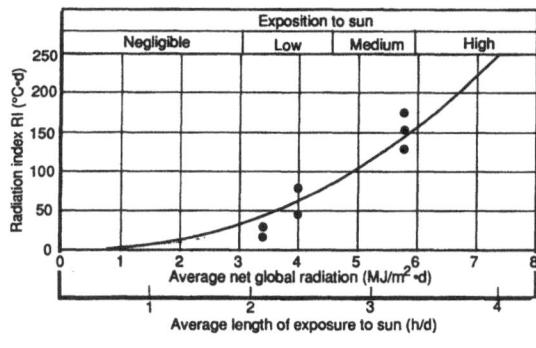

FIGURE 20.5. Radiation index (modified from Dysli 1990).

dex of 1000 °C-days and by about 3 h per day of sun exposure from November to April yields a radiation index of 120 °C-days and a surface-freezing index of 880 °C-days. It is usual to consider that $n_f = 0.8$–0.9 for an asphalt pavement, i.e. I_{sf} is 800–900 °C-days. Both approaches suggest similar values of I_{sf}, but more data on radiation are needed for North America.

20.3 Ice Formation in Freezing Soils

When a saturated coarse-grained soil freezes with unimpeded drainage of the excess porewater away from the freezing front, porewater will be changed into pore ice without change in the soil's porosity. However, if the same soil freezes in a closed system, the pore space expands by 9% as water changes into ice. Usually, these soils are referred to as non frost-susceptible.

When a fine-grained-soil freezes, water from the unfrozen soil is attracted to the freezing front where it forms an ice lens causing frost heave well in excess of the 9% original unfrozen pore space increase. These soils are referred to as frost-susceptible.

20.3.1 UNFROZEN WATER CONTENT

When a fine-grained soil is frozen not all the water within the soil pores freezes at 0 °C (Bouyoucos 1916; Lovell 1957). In air-free and solute-free soils, water coexisting with ice is presumed to exist as thin films of adsorbed water and as capillary water that lies outside the range of adsorption forces but fails to freeze because it occupies spaces too narrow to be penetrated by a curved ice–water interface. Several factors govern the phase composition of a frozen soil: specific surface area, temperature, applied pressure, osmotic pressure of the soil solution, pore-size distribution, particle packing geometry, and exchangeable adsorbed ions (Anderson & Tice 1972; Nersesova & Tsytovitch 1963; Ander-

son & Morgenstern 1973). For a given soil, most of these factors are fixed and the unfrozen water content, w_u, is the amount of liquid water in a soil at a given sub-zero temperature, usually expressed as a percentage of the dry weight of the soil.

Typical unfrozen water content curves are shown in Fig. 20.6 for various soils. In general, significant amounts of unfrozen water exist in clays at temperatures lower than 10 °C, while most of the porewater freezes within a few tenths degree Celsius of zero in silty soils. As for unsaturated soils in which air–water menisci in the pores produce negative porewater pressures (or suctions), the ice–unfrozen water menisci in partially frozen soils also cause negative pressures in the water films.

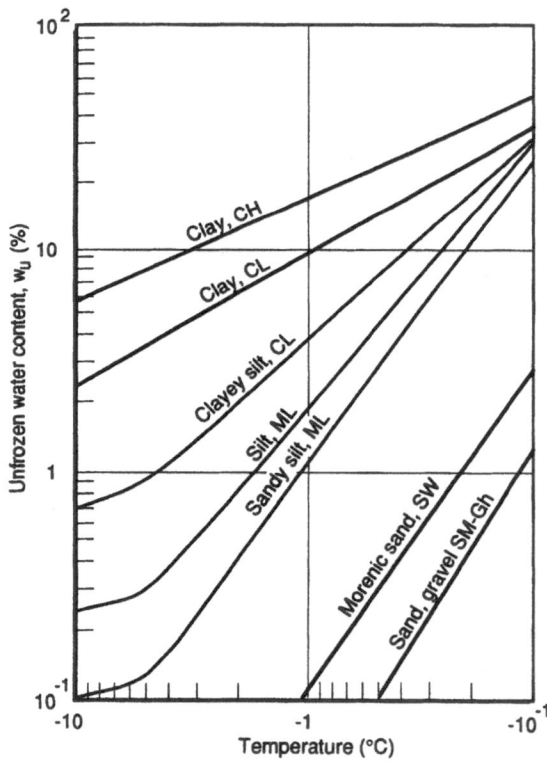

FIGURE 20.6. Typical unfrozen water content–temperature relationships for various soils (modified from Nordal & Refsdal 1994).

20.3.2 WATER MIGRATION AND ICE LENS FORMATION

The fact that not all the water in a fine-grained soil freezes at a unique temperature, but rather over a given range and that it exists in a state of suction, has a significant impact on water migration when a saturated porous medium is frozen under a temperature gradient. Once nucleated, the ice crystals propagate into pore space through the maximum pore-size openings. The highest temperature at which ice can grow in the soil pores is termed the pore freezing temperature, T_p. It is mainly a function of soil type, pressure and solute concentration. In general, the more finer grained a soil, the lower the pore freezing temperature. The pore freezing front is defined herein as the T_p isotherm.

Experimental observations on specimens frozen under a temperature gradient suggest that even though much of the porewater is frozen, water transport still occurs in the frozen soil past the pore freezing front in re-sponse to temperature-induced unfrozen water content gradients and suction gradients. The migratory water freezes at the segregation freezing temperature, T_s, which is lower than T_p (Freden 1965; Hoekstra 1969; Penner & Goodrich 1980; Konrad & Morgenstern 1982b). The zone between the pore freezing front, T_p, and the segregation freezing front, T_s, has been referred to as the frozen fringe (Miller 1972).

The hydraulic conductivity in the frozen fringe is related to the amount of unfrozen water and decreases continuously from the pore freezing front to the segregation freezing front. Furthermore, because water flows through the frozen fringe, unfrozen water content is a function of both temperature and suction as outlined in Konrad & Duquennoi (1993). This has serious impacts on the hydraulic conductivity of the frozen fringe, which cannot be considered as a simple soil property.

Figure 20.7 illustrates schematically the hydrodynamic and thermal conditions for a

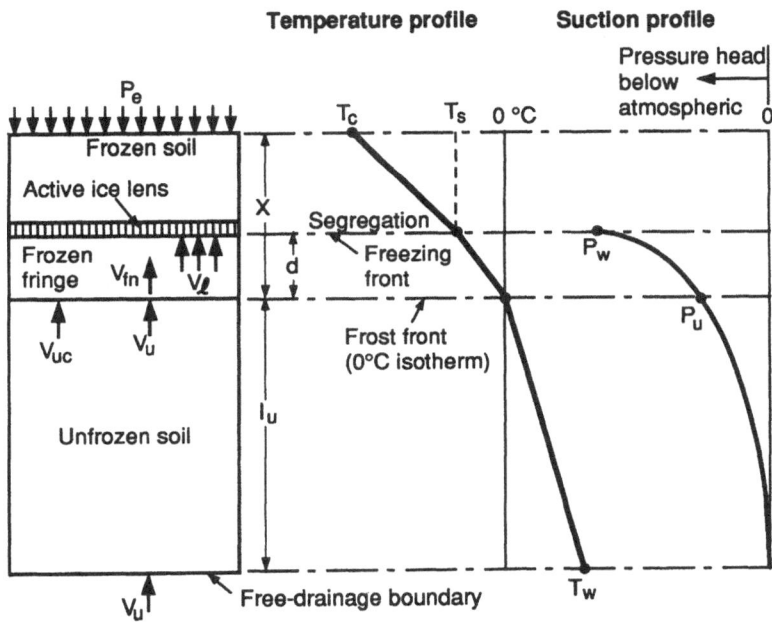

FIGURE 20.7. Schematic of hydrodynamic and thermal processes in freezing soils (modified from Konrad & Seto 1994).

soil freezing under open-system conditions, i.e. with free access to water. Water flowing to the active ice lens may be drawn from three distinct sources:

1. water generated within the frozen fringe, where freezing occurs at invariant porosity and accumulates at the segregation freezing front at a rate v_{fn};
2. water leaving the unfrozen soil as a result of suction-induced consolidation and entering the frozen fringe at a rate v_{uc}; and
3. water from an external supply entering at a rate v_u.

The overall rate of water migration to the active ice lens, v_l is thus the sum of these three distinct fluxes:

$$v_l = v_{fn} + v_{uc} + v_u. \qquad (20.17)$$

At the segregation freezing front, the migrating water freezes and the rate of ice lens growth, dh/dt, is readily obtained considering an increase in volume of 9%.

$$dh/dt = 1.09v_l. \qquad (20.18)$$

Pore freezing in the frozen fringe without changing the pore volume requires that a given quantity of porewater must be removed as ice forms and is flowing to the ice lens where it freezes. The flux v_{fn} is associated with transient freezing and can be approximated by:

$$v_{fn} = 0.09\varepsilon n dX/dt \qquad (20.19)$$

where n = the porosity, ε = a factor taking into account the amount of unfrozen water remaining in the frozen fringe, dX/dt = the rate of frost penetration. The rate of ice lens growth corresponding to this mechanism is:

$$(dh/dt)_{fn} = 1.09v_{fn} = (0.1)\varepsilon n dX/dt. \qquad (20.20)$$

This has been known as the heave rate associated with the freezing of porewater of the unfrozen soil solely.

20.3.3 SEGREGATION POTENTIAL OF FREEZING SOILS FOR LABORATORY CONDITIONS

From a phenomenological point of view, frost heave mechanics can be regarded as a problem of impeded drainage in a layered medium to an ice–water interface that exists in the frozen soil at the segregation freezing front. The soil freezing characteristics should therefore be related to the characteristics of the frozen fringe, i.e. the relationship between hydraulic conductivity and unfrozen water content. In view of the complexity of one-dimensional frost heave modeling (Konrad 1994), it appears that precise point measurements of hydraulic conductivity, temperature and suction within the frozen fringe would not ultimately be of direct value in a predictive frost heave model. Instead, representative overall frozen fringe characteristics should be deducible from controlled laboratory freezing tests or from field observations and should constitute adequate input parameters to a general formulation of simultaneous heat and mass transfer in freezing soils.

If the suction-induced effective stresses in the unfrozen soil do not exceed the preconsolidation pressure, it is justified to set v_{uc} to zero in Eq. 20.17 in overconsolidated soils. Furthermore, if freezing is studied near thermal steady state, i.e. at the formation of the final ice lens in laboratory step-freezing tests, v_{fn} is equal to zero and Eq. 20.17 reduces to:

$$v_l = v_u. \qquad (20.21)$$

Konrad & Morgenstern (1980) demonstrated that overall frozen fringe characteristics in incompressible soils (high OCR value) can be obtained in the laboratory without detailed measurements at the scale of the frozen fringe. A simple analysis of seepage in the frozen fringe at the onset of formation of the final ice lens in step-freezing tests, i.e. at a rate of cooling close to zero, showed that the porewater velocity entering into the unfrozen soil, v_u, is proportional to temperature gradient in the frozen fringe provided that the suction at the pore-freezing front is constant:

$$v_u = \frac{P_w - P_u}{d} \bar{k}_f \qquad (20.22)$$

$$= \left(\frac{P_w - P_u}{T_s} \bar{k}_f\right) T_G = SP(T_G)$$

where P_w = suction head at the ice lens (m); P_u = suction head at the frost front (m); $\bar{k}_f$ = overall hydraulic conductivity of the frozen fringe (m s^{-1}).

The constant of proportionality, i.e. the slope of the linear relationship between v_u and the temperature gradient in the frozen fringe, is defined as the segregation potential parameter, SP. Since SP is a parameter that partly reflects the overall hydraulic conductivity of the frozen fringe, which, in turn, is dependent upon the overall unfrozen water content, it is expected that factors affecting w_u will also affect SP.

The segregation potential extended to transient freezing is at least dependent upon the following variables (Konrad 1989a):

$$SP = f(\text{soil type, porosity, OCR,}$$
$$\text{pore fluid, } dT_f/dt, P_u, P_e, N, \ldots). \qquad (20.23)$$

The first four factors relate basically to the porous medium subjected to freezing. Soil type includes all the physical properties such as gradation, mineralogy of the fine fraction, specific surface and surface charge density. The porosity, n, reflects the degree of packing of the soil. The overconsolidation ratio, OCR, controls freezing-induced consolidation of the unfrozen soil. Finally, the pore fluid relates to porewater chemistry.

The other factors given in Eq. 20.23 relate to the freezing conditions. The rate of cooling of the frozen fringe, dT_f/dt, is defined herein as the change of temperature per unit time at the frost front and can be calculated (Konrad 1987) as:

$$\frac{dT_f}{dt} = \frac{dT_f}{dX}\frac{dX}{dt} \qquad (20.24)$$

where dX/dt = the rate of frost front advance and dT_f/dX = the temperature gradient in the frozen fringe. P_u = the suction head at the frost and P_e = the overburden pressure at the segregation freezing front. Finally, the segregation potential is also affected by the number of freeze–thaw cycles, N (Konrad 1989b).

20.3.4 SEGREGATION POTENTIAL FOR FIELD CONDITIONS

To successfully predict water migration rates in the field using Eq. 20.23, SP must be obtained from laboratory freezing tests on representative samples using representative values of dT_f/dt, P_u and P_e. In the field, small frost penetration rates combined with small temperature gradients in the frozen soil near the frost front result in extremely small rates of cooling of the frozen fringe. The suctions at the frost front are generally very small owing to small water migration rates consistent with relatively small temperature gradients in the field. The overburden pressure at the freezing front can be estimated as a function of frost depth, X, as:

$$P_e = \gamma_f X \qquad (20.25)$$

where γ_f = the unit weight of the frozen soil.

These field conditions can be satisfactorily approximated by step-freezing tests on representative samples, i.e. undisturbed, subjected to the range of field overburden pressures. Conditions with small rates of cooling are obtained at the formation of the final ice lens, i.e. when the rate of frost penetration approaches zero. Low values of P_u would be achieved by choosing a warm plate temperature close to 0 °C (about 0.5–1 °C) in order to force the final ice lens to form close to the external water source, hence a relatively small flow path.

It is emphasized that in the field where frost penetration rates may be different from zero during most of the freezing period, water for ice lens growth comes first from the excess water in the pores, $(v_{fn})_{field}$, and if required from the surrounding medium, $(v_u)_{field}$, such that:

$$(v_{fn} + v_u)_{field} = SP_{field}(T_{G_{field}}) \qquad (20.26)$$

with $SP_{field} = (v_u/T_G)_{lab}$.

20.3.4.1 SP_{field} from Laboratory Freezing Tests

It is recommended to use undisturbed samples wherever possible so that the factors characterizing the *in situ* conditions (soil structure, void ratio, degree of saturation and pore fluid chemistry) are preserved. If, for economic reasons, it is not possible to justify tests on undisturbed samples, reconstituted soil specimens, preferably overconsolidated, reproducing as close as possible soil struc-

ture, water content and porewater chemistry can be used. Figure 20.8 is a simplified flow chart that specifies the type of testing that may be used by the designer to estimate the frost heave parameter operative in the field depending on the stress and thermal history of a natural soil deposit.

Frost heave data are shown on Fig. 20.9 for conditions representative of field freezing. Most of the data were obtained from step-freezing tests. The segregation potential is sensitive to overburden pressure and can be expressed for many soils ranging from sandy silts to clays as:

$$SP_{field} = SP_0 \exp(-aP_e) \qquad (20.27)$$

where SP_0 = the segregation potential at zero overburden and a is a soil constant. Since the segregation potential considers the actual temperature boundary conditions, it is not too critical to define a standard test, provided

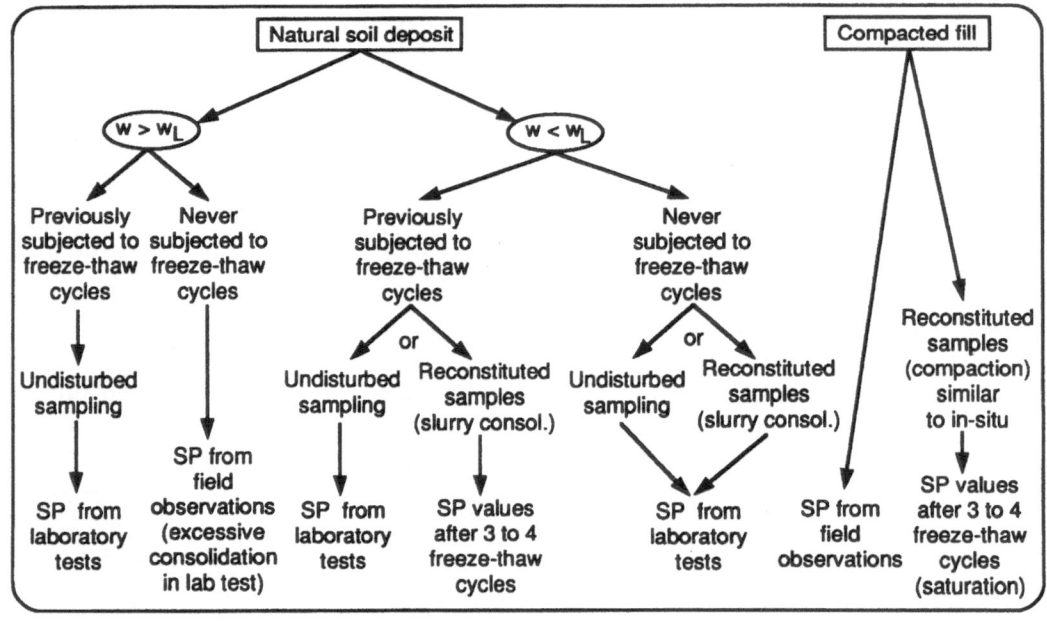

FIGURE 20.8. Flow chart for obtaining representative values of SP field (w/w_L = ratio of water content, w, to liquid limit, w_L).

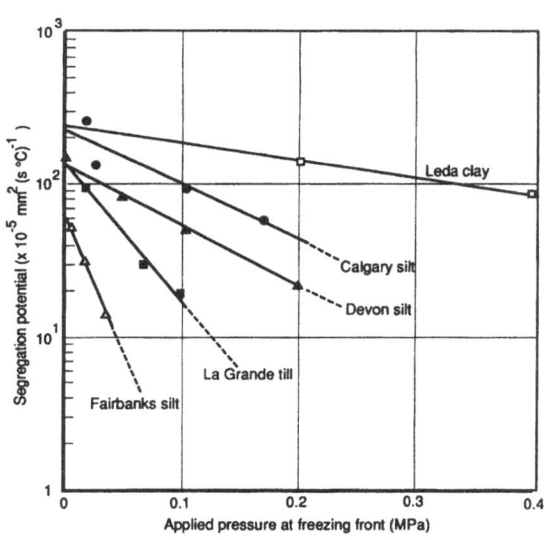

FIGURE 20.9. Segregation potential versus overburden pressure for various soils.

that the test results consider the factors discussed above.

20.3.4.2 SP$_{field}$ from Field Observations

The segregation potential can also be obtained directly from field observations where surface heave and temperature data are available from:

$$SP_{field} = \frac{dh/dt}{1.09 T_{G_{field}}}. \qquad (20.28)$$

Figure 20.10 illustrates the procedure for determining SP of a subgrade soil. Frost penetrates through the base material without significant surface heave and enters subgrade soil in early December. During freezing of the subgrade, until reaching a depth of 1.3 m, frost heave rate is relatively constant (0.68 mm per day) and the average temperature gradient in the frozen soil near 0°C is 9.2 °C

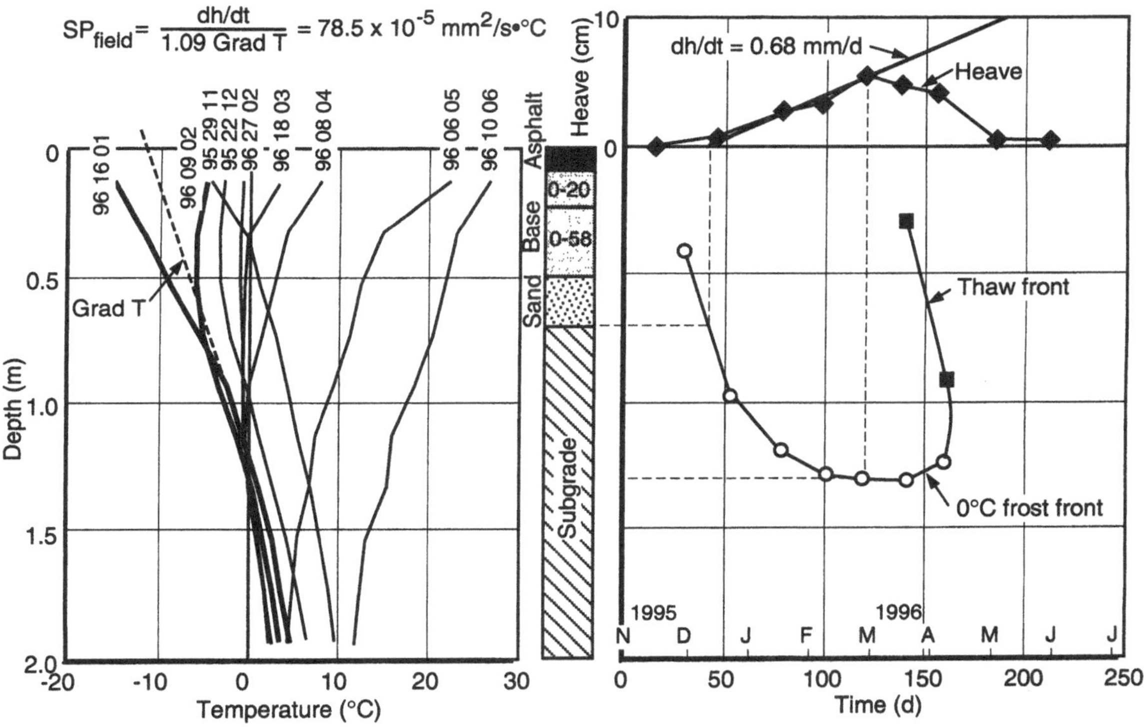

FIGURE 20.10. Segregation potential inferred from field data.

m^{-1}. SP_{field} is thus $0.68/(1.09)(0.0092) = 68$ $mm^2\ °C^{-1}$ per day $= 79.10^{-5}\ mm^2\ °C^{-1}\ s^{-1}$. This approach is especially recommended in compressible soils such as lightly overconsolidated soft clays and when undisturbed samples are difficult or too expensive to obtain.

20.3.4.3 SP_{field} From Empirical Correlations

An estimate of the SP value for a given soil can also be obtained from empirical correlations. Rieke et al. (1983) showed that the frost heave susceptibility increased with increasing percentage of fines, decreasing activity of the fine fraction, and for a specific fine fraction mineralogy, increasing liquid limit of the fine fraction. Recently, Konrad (1998) has shown that SP is best related to d_{50} of the fines fraction ($<74\ \mu m$), w/w_L (ratio of water content to liquid limit) and mineralogy.

The segregation potential for zero applied pressure can be estimated using the empirical relationship presented in Fig. 20.11. For a ratio w/w_L equal to 0.7, the empirical relationship is best fit by:

$$SP_0(S_s) = [116 - 75\ \log d_{50}(FF)] \times 10^3\ mm^4\ °C^{-1}\ s^{-1}\ g^{-1}. \quad (20.29)$$

where $d_{50}(FF)$ is expressed in μm.

Knowing the value of the specific surface area of the fines fraction (FF) SP_0 readily follows. If the soil is at a different water content, it is suggested to apply a correction to SP_0 using the following relationship:

$$SP_0(w) = SP_0 + 9 \times 10^{-3} \log(w/0.7w_L) \quad mm^2\ °C^{-1}\ s^{-1}. \quad (20.30)$$

20.3.5 PREDICTION OF FROST DEPTH IN FROST-SUSCEPTIBLE SOILS

While frost depth in non-frost-susceptible soils is readily estimated with Eq. 20.13, the calculation of frost depth in frost-susceptible soils must account for the release of latent heat associated with the formation of ice lenses. An extension to Stefan's approach yields enough accuracy for practical considerations. Using the segregation potential to quantify the rate of ice formation with Stefan's assumptions gives the modified Stefan equation:

$$X = \left[\frac{2(\lambda_f - SP\ L)I_{sf}}{L_s}\right]^{1/2} \quad (20.31)$$

where SP = the value of the segregation potential in $m^2\ s^{-1}\ °C^{-1}$; L = the latent heat of water, i.e. 334 MJ m^{-3}; λ_f = the thermal conductivity of the frozen soil from Kersten's relationship given in Fig. 20.2, L_s is given by Eq. 20.12 and I_{sf} is obtained as discussed in Section 20.2.3.2.

Worked Example

Calculate the frost depth for a saturated clayey silt characterized with an SP value of $250 \times 10^{-5}\ mm^2\ s^{-1}\ °C^{-1}$ and a water content of 25% in an area where the freezing index at the ground surface I_{sf}, is 1000 °C-days. (NB: For a non-frost-susceptible soil SP = 0.)

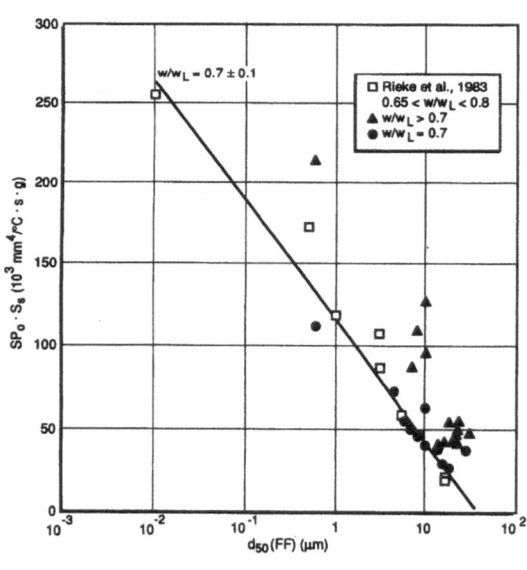

FIGURE 20.11. Empirical relationship between segregation potential and soil index properties (modified from Konrad 1998): $\blacktriangle$, $w/w_L > 0.7$; $\bullet$, $w/w_L = 0.7$; $\square$, $0.65 < w/w_L < 0.8$ (Rieke et al. 1983).

Solution: Figure 20.2 gives $\lambda_f = 1.9$ W m^{-1} °C^{-1} $= 1.9$ J(s m °C)$^{-1}$.

SP $L = (250 \times 10^{-11})(334 \times 10^6) = 0.835$ J/(s m °C)

$L_s = (0.25)(1600/1000)(334 \times 10^6)$

$\quad = 134 \times 10^6$ J m^{-3} (from Eq. 20.12)

$I_{sf} = 1000 \times 24 \times 3600 = 86.4 \times 10^6$ °C s

and hence from Eq. 20.31:

$$X = \left[\frac{2(1.9 - 0.835)86.4 \times 10^6}{134 \times 10^6}\right]^{1/2} = 1.17 \text{ m}.$$

It is noted here that X represents in fact the thickness of the frozen soil rather than the frost depth because it includes the surface heave due to ice lensing.

Neglecting the latent heat from ice lensing, i.e. using Eq. 20.13, would result in a prediction of frost depth of 1.56 m, i.e. an overprediction of about 30%.

Equation 20.31 always overpredicts the depth of freezing because the volumetric heat component from both the frozen and unfrozen soil is neglected. Furthermore, the heat flow from the unfrozen soil is also neglected. One might consider a dimensionless correction factor, λ°, dependent on the thermal ratio β and the fusion parameter μ to improve the accuracy of the prediction (Aldrich 1956):

$$X_{corr} = \lambda^\circ X \qquad (20.32)$$

For soils with high water content, λ° is close to one and Eq. 20.31 is reasonable, especially because it considers ice lensing. For northern climates with high air-freezing indices, β is relatively small and λ° is larger than 0.9. Equation 20.31 is then also satisfactory. The correction becomes more important, however, in more temperate climates and in relatively dry or well-drained soils. Values of λ° are given in the *Canadian Foundation Engineering Manual* (CGS 1992).

This simplified approach can also be used efficiently in layered soils, such as highways.

Worked Example

Calculate the frost depth in layered system consisting of a 1-m thick silty sand at a water content of 8% underlain by a deposit of a saturated clayey silt characterized with an SP value of 250×10^{-5} mm^2 s^{-1} °C and a water content of 25% in an area where the freezing index at the ground surface, I_{sf}, is 1000 °C-days with a freezing period, t_f, of 130 days. The surface layer is unsaturated and does not result in ice lensing.

Solution: Figure 20.2 gives $\lambda_f = 1.0$ W(m °C)$^{-1}$ $= 1.0$ J(s m °C)$^{-1}$.

$L_s = (0.08)(1800/1000)(334 \times 10^6)$

$\quad = 48 \times 10^6$ J m^{-3}.

Equation 20.13 can be used to calculate the time required to completely freeze the upper layer:

$$I_{s1} = \frac{D^2 L_s}{2\lambda_f}$$

where I_{s1} is the freezing index for a frost penetration of depth D.

$$I_{s1} = \frac{(1)^2(48 \times 10^6)}{(2)(1)} = 24\ 106 \text{ °C s} = 277 \text{ C days}.$$

To transform I_{s1} into a period, it is assumed that the temperature variation is sinusoidal, i.e.:

$$t_1 = \frac{\cos^{-1}(I_{sf} - 2I_{s1})/I_{sf}}{\pi} t_f$$

$$= \frac{\cos^{-1}[1000 - (2 \times 277)/1000]\ 130}{\pi} \qquad (20.33)$$

$$= 41 \text{ days}.$$

To calculate the maximum frost depth, use Eq. 20.2 to calculate the equivalent thermal conductivity of the system as:

$$\bar{\lambda} = \frac{X}{(D_1/\lambda_{f_1}) + \{(X - D_1)/[\lambda_{f_2} - SP(L)]\}} \qquad (20.34)$$

and the weighted value of the latent heat of the system as:

$$\bar{L}_s = \frac{L_{s1}D_1 + L_{s2}(X - D_1)}{X}. \qquad (20.35)$$

The maximum frost depth is then obtained from Eq. 20.13 with the weighted parameters:

$$X = \left(\frac{2\bar{\lambda}I_{sf}}{\bar{L}_s}\right)^{1/2}. \qquad (20.36)$$

Since X is the unknown, it must be obtained iteratively from Eqs 20.34–20.36. The example gives the following results:

$$X = \left(\frac{2\bar{\lambda}I_{\text{sf}}}{\bar{L}_s}\right)^{1/2}$$

$$= \left[\frac{2(1.02)1000 \times (86.4)10^3}{7710^6}\right]^{1/2} = 1.51 \text{ m.}$$

The same approach can also be used to estimate the required thickness of a given insulation to prevent undue frost penetration into a frost-susceptible horizon.

Worked Example

Given the previous example, calculate the styrofoam thickness to be placed above the frost-susceptible layer in order to have a frost depth of only 1.15 m. The thermal conductivity of the styrofoam is 0.03 W m^{-1} °C^{-1}:

$$\bar{\lambda} = X/(D_1/\lambda_{f_1}) + \{(X - D_1)/[\lambda_{f_2} - SP(L)]\} + \frac{e}{\lambda_i}$$

and

$$\bar{L}_s = \frac{L_{s1}D_1 + L_{s2}(X - e - D_1)}{X}$$

where X is known.

Solving iteratively gives $e = 0.075$ m.

20.3.6 FROST HEAVE PREDICTION

To predict frost heave in the field, the engineer must first obtain a representative value of the segregation potential at the active ice lens, SP_{field}, as discussed in Section 20.3.4. The frost heave, h, is then calculated from:

$$h = 1.09\int SP_{\text{field}}T_{G_{\text{field}}}dt \qquad (20.37)$$

where $T_{G_{\text{field}}}$ = the temperature gradient in the frozen zone near the frost front and is obtained by solving the coupled heat- and mass-transfer equations for the specific field problem. Figure 20.12 outlines the procedure for the prediction of one-dimensional frost heave versus time for given thermal conditions. The temperature distribution in the field is obtained either from a numerical ap-

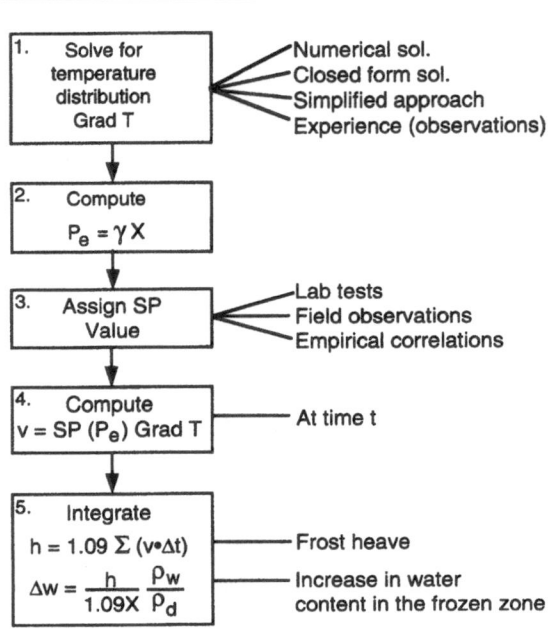

FIGURE 20.12. Methodology for frost heave prediction.

proach, from a closed-form solution, from a simplified method as discussed above, or directly from observations at an instrumented site. The temperature gradient in the frozen soil adjacent to the frost front (it approximates the temperature gradient in the frozen fringe) is calculated as a function of time. Knowing the relationship between SP and overburden pressure (laboratory freezing tests, field data or empirical correlations), the rate of water migration to the freezing front is readily computed from Eq. 20.26. Integration over time provides total heave. Often, a simpler calculation by assuming a constant value of $T_{G_{\text{field}}}$ during the freezing yields a good approximation of frost heave. It is noted that heave from expansion of *in situ* pore-water is already included in SP_{field} as discussed earlier.

Worked Example

Calculate the amount of frost heave for the layered system examined in the second worked example.

Solution: The average surface temperature is $1000/130 = -7.7$ °C, and the average temperature gradient in the frost-susceptible layer is $-7.7/(1 + 0.25) = -6.15$ °C m^{-1} since the frost depth is about 1.5 m. Surface heave is thus computed using Eq. 20.37 as:

$$h = 1.09 SP_{\text{field}} T_{G_{\text{field}}}(t_{\text{f}} - t_1)$$

$$= (1.09)(250 \times 10^{-11})(6.15)(89 \times 24 \times 3600)$$

$$= 0.13 \text{ m}.$$

The increase in water content in the frost-susceptible layer can also be calculated as:

$$\Delta w = \frac{1}{1.09} \frac{h}{(X - D_1)} \frac{\rho_w}{\rho_d}$$

$$= \frac{(0.13)(1000)}{(1.09)(1.5 - 1.0)(1600)} \qquad (20.38)$$

$$= 0.16 = 16\%.$$

For non-frost-susceptible soils, i.e. no water migration towards the freezing front, frost heave depends on whether excess porewater can be expelled during freezing as would be the case for a thick layer of clean sand or gravel or when excess porewater cannot freely escape as ice enters into the soil pores because of the presence of a clay layer below the sand layer. In the first case, frost heave is negligible, while in the latter frost heave in saturated soils during closed-system is obtained from:

$$h = 0.09(\rho_d/\rho_w)wX \qquad (20.39)$$

where X = the thickness of frozen non-frost-susceptible soil.

20.3.7 THAW WEAKENING

With thawing the moisture accumulated during the freezing process is released and generally causes a decrease in suction (porewater pressures may even become positive). As a result, the resilient modulus is reduced and considerable deformation or failure may occur. In highway engineering, considerable work has been done by the US Army Corps of Engineers and the reader may find key references in Cole *et al.* (1986), Johnson *et*

al. (1978) and Chamberlain (1973). While the significance of moisture tension in modeling resilient response of soils is well recognized (Bergen & Monismith 1973; Fredlund *et al.* 1975; Brown & Pappin 1981), more attention needs to be devoted to the prediction of moisture content with time under the actual environmental conditions.

Thawing of frozen soil is also associated with volume change leading to thaw settlement and thaw consolidation. Considerable work was done on these aspects by Morgenstern & Nixon (1971) and the reader is referred to the textbooks edited by Johnston (1981) and Andersland & Anderson (1978). A special mention is made for sensitive clays that are highly frost susceptible when subjected to freezing for the first time (excavation, cuts, etc.) since they display significant strength loss during thaw. The undrained shear strength, the thaw consolidation volume change and the strength after consolidation may be related to the initial liquidity index (Leroueil *et al.* 1991).

20.4 *Frost Susceptibility of Soils*

20.4.1 LIMITATIONS OF EXISTING FROST-SUSCEPTIBILITY CRITERIA

From a survey of more than 100 frost-susceptibility test methods (Chamberlain 1981), it is possible to identify, in general, three levels of sophistication in estimating the frost susceptibility of soils. The first level, type I, methods, are primarily based on the per cent of soil finer than a specified particle size, commonly 0.074 or 0.02 mm. Since frost susceptibility is directly related to unfrozen water content and applied stress and other factors discussed in Section 20.3.3, type I methods cannot successfully yield adequate and reliable decisions.

The second level, type II, methods, are generally based on soil type or classification of particle-size curves, additional soil properties such as Atterberg limits, pore-size distribution, hydraulic conductivity of unfrozen soil, capillary rise, etc. Type II methods, although more time-consuming and costly than type I methods, have limitations as well, since most of the parameters are only partially related to the characteristics of the frozen fringe.

The third level, type III, methods require a laboratory freezing test and/or observations of frost heave and/or thaw weakening and eventually some predictions. Freezing test procedures differ widely and generally provide heave rate values specific to each procedure. Substantial differences in temperature gradient (not always reported in the literature) in the various tests result then, according to Sections 20.3.2 and 20.3.3, to significant variations in heave rate and consequently in apparent degree of frost susceptibility.

20.4.2 RATIONAL EVALUATION OF FROST SUSCEPTIBILITY

The frost-susceptibility criteria described above lead only to the acceptance or rejection of soils without specifying the degree of damage resulting from frost action. Konrad & Morgenstern (1983) have proposed a more rational approach for assessing the relative frost susceptibility of saturated soils. The method is illustrated in Fig. 20.13 and essentially uses the concept of the segregation potential to predict frost heave and/or water (ice) content after freezing for given thermal and geological conditions, but also includes project requirements and acceptable risk. Non-frost-susceptible backfill is thus defined according to the acceptable response during freezing. It should not be specified solely based upon grain-size characteristics, as discussed in Section 20.3.4.3. Water availability is essential for severe frost action. For instance, compaction on the dry side of optimum in backfill material reduces frost heave.

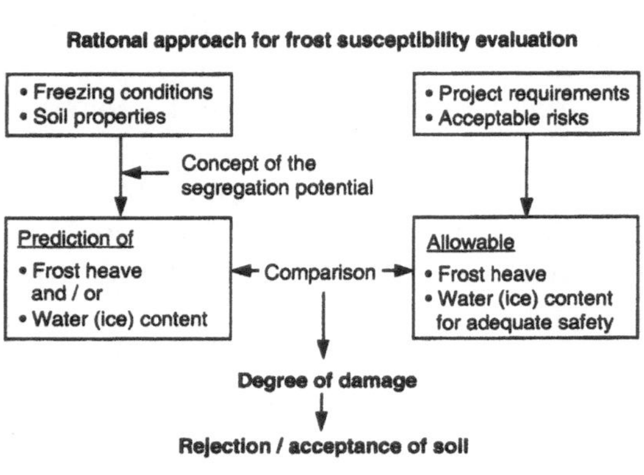

20.5 Frost Action in Civil-Engineering Works

20.5.1 DESIGN CONSIDERATIONS

In cold regions, the design engineer evaluates the following:

1. The thermal boundary conditions, usually associated with the temperature variation in the air, at the structure–air or soil–air interface, as discussed in Section 20.2.3. For design purposes, as recommended by the *Canadian Foundation Engineering Manual* (CGS 1992), it is common practice to use a design air-freezing index, I_{ad}, defined with respect to the mean air-freezing index, I_{am}, as:

$$I_{ad} = 100 + 1.29 I_{am}. \qquad (20.40)$$

When historical climatic data are available, the design air-freezing index can also be taken as that of the coldest winter over the last ten year period.

The actual design freezing index at the air–structure or air–soil interface is obtained through the empirical "n"-factors outlined in Section 20.2.3.2. The presence of snow must also be considered, using a layered system approach with thermal properties given in Table 20.1.

In artificial freezing, the thermal boundary conditions are usually tempera-

Rational approach for frost susceptibility evaluation

FIGURE 20.13. Rational approach for frost-susceptibility evaluation.

ture conditions imposed by the freezing system and must specify the duration of freezing. In some cases, artificial thaw may also be considered.

2. The geological conditions used to assess the type of soils, the position of the groundwater table and the proximity to other water sources, the drainage paths, stress history, water content and degree of saturation of each soil unit.

3. The geotechnical soil properties, including both physical and thermal properties.

4. The project requirements in terms of allowable deformations (heave and thaw), allowable strength and allowable freezing-induced expansion pressures.

5. The effect of frost action in terms of adfreezing, differential responses under unheated or partially heated appurtenances to a primary structure, temporary excavations in winter subjected to fairly rapid thaw and the formation of frost-induced shrinkage cracks.

20.5.2 FOUNDATIONS

The conventional approach for protection of building foundations against frost action is to locate shallow foundations at a depth greater than the design depth of frost penetration (Eqs 20.13, 20.31 and 20.40). In some cases, thin soil cover may be supplemented with insulating material. A design methodology for insulated foundations has been presented by Robinsky and Bespfug (1973) and given in the *Canadian Foundation Engineering Manual* (CGS 1992).

Moist soil in contact with shallow foundations or piles can freeze to their sides, developing a substantial adfreeze bond. Backfill that is frost susceptible can heave and transmit uplift forces to the foundation. Average adfreeze strengths typically range from 65 kPa for fine-grained soils frozen to wood or concrete to 100 kPa for steel structures (Penner 1974). According to the *Canadian Foundation Engineering Manual* (CGS 1992), it is good practice to backfill against foundations

with non-frost-susceptible soil (see Section 20.4). Furthermore, provision should be made for drainage around the foundation perimeter, below the maximum frost depth.

20.5.3 EARTH RETAINING STRUCTURES AND EXCAVATIONS

Earth retaining structures are subjected to many pressure distributions depending on the direction of relative wall movement, drainage and frost action. Horizontally and vertically acting frost heave forces can greatly exceed any earth pressure and may cause cracking and tilting. Experimental data is scarce! Sui *et al.* (1988) have measured horizontal frost heave forces on full-scale retaining walls backfilled with relatively frost-susceptible soils. For a wall height of 1.6 m, the horizontal pressure distribution was trapezoidal with a maximum value of 0.15–0.2 MPa at a depth of about two-thirds of the wall height. Further details are given in Konrad & Shen (1994).

Traditionally, frost protection has been accomplished by using non-frost-susceptible backfill and by placing the foundation below the frost depth. Nordal & Refsdal (1994) have proposed alternative solutions using thermal insulating materials.

Morgenstern & Sego (1981) found a strong correlation between tie-back loads and average air temperature, which can be accounted for by the alternate freezing and thawing of the ground adjacent to the sheet pile wall. Freezing of fine-grained soils around a sheeted excavation may result in an increase in water content in a zone close to the excavation due to ice lensing and frost heave forces. In unsupported excavations subjected to unplanned freezing (late winter–early spring), stability problems may be encountered during thaw, especially in highly frost-susceptible clays.

20.5.4 BURIED STRUCTURES

Utility systems (water, sewage) in cold regions must be operative at all time despite severe climatic conditions. The reader is referred to the *Cold Climate Utilities Manual* (1986) for details and design methods.

20.5.5 HIGHWAYS

Frost action in highways and urban roads is responsible on the one hand for differential surface movements both in the longitudinal direction and in cross-sections owing to the formation of ice lenses in frost-susceptible soils and on the other hand for the seasonal variations in bearing capacity due to ice melting during thaw, which causes a variation in moisture content with time in the different soil layers. The rational understanding of these frost-induced phenomena requires the prediction of: (a) the amount and spatial distribution of frost heave with time, which is also related to the amount of excess ice in each soil layer; and (b) water content distribution with time as thaw progresses.

The thermal regime of cold regions with freezing indices ranging from 800 to 2000 °C-days generally produces the following features:

1. A frost penetration period of about 3–4 months, in which the frost advance rate is higher in the granular base materials and slower in the frost-susceptible subgrade.
2. Frost depths larger than 1.2 m (depending essentially on moisture content of materials and on frost susceptibility of subgrade soil).
3. A relatively rapid thaw with duration of about 2–3 weeks.
4. Thaw fronts progressing from pavement surface downward and from the base of the frozen subgrade upwards.
5. Occurrence of one or several thaw cycles during the freezing period resulting in shallow thaw beneath the pavement.

20.5.5.1 Frost Heave

The key of a successful frost heave prediction is the use of relevant soil properties. Assuming that both base and subbase granular material are non-frost-susceptible, it can be argued that surface heave depends mainly on the freezing characteristics of subgrade soil. Close examination of the conditions in the subgrade reveals that three distinct zones may be found in the subgrade:

- Zone 1, which is an unsaturated zone if the water table is fairly low.
- Zone 2, which is saturated by capillary action and subjected to negative pore pressures.
- Zone 3, which is a saturated subgrade with hydrostatic pore pressures.

In zone 3, laboratory freezing tests have shown that SP varies with overburden pressure. In zone 2, SP decreases with increasing suction (Konrad & Seto 1994). Field tests confirm the dependence of SP on suction, as shown by Fig. 20.14 for a sandy silt obtained at two sections of the same road with Shelby tubes. At Section A, frost heave rate was 0.3 mm per day with a water table at a depth of 4.0–5.0 m, and at Section B it was 0.48 mm per day with a water table at a depth of 3.0 m. In both cases, water content measurements confirmed that the subgrade was saturated by capillary action. In zone 1, SP varies with degree of saturation. In general, it is acceptable to assume a value of SP close to 0.

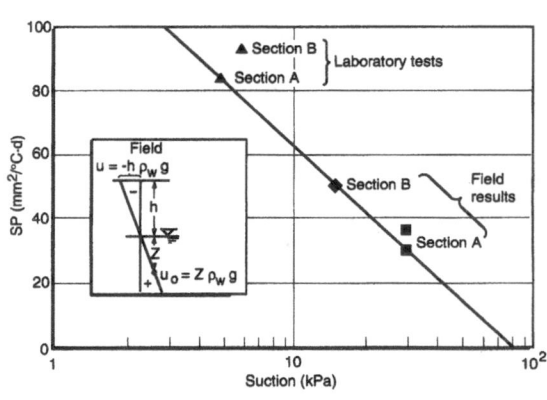

FIGURE 20.14. Influence of suction in a saturated sandy silt.

20.5.5.2 Thaw-Weakening

The existence, in many cases, of two thaw fronts renders the problem more complex than that treated by Morgenstern & Nixon in the early 1970s (1991) for one-dimensional consolidation of thawing permafrost. During and after the thaw period, the pavement design engineer must assess the rate of pore pressure dissipation in the frost-susceptible subgrade and the variation of water content in the subbase and base material with time.

The latter is of utmost importance as the deformation and strength properties are strongly related to the degree of saturation of the granular pavement foundation materials. Soil–water characteristic curves of these materials are required to assess the evolution of stress state and deformation properties with time. Needless to say that pore pressure development during traffic loading depends upon the drainage capacity of these pavement foundation layers; which, in turn, is directly related to the hydraulic conductivity and thus to gradation and type of material as well.

As the lower thaw front retreats upwards and water from the ice lenses is liberated and cannot readily escape owing to the presence of an impervious frozen subgrade barrier, it will carry some of the overburden pressure. Under this pressure gradient, water redistribution occurs into subgrade soil that stayed unfrozen during the whole winter. This water redistribution mechanism leads to an increase in water content of the subgrade that never froze and geotechnical engineers referred to it as swelling. Once, the subgrade is completely thawed, water then migrates in an upward direction and may contribute to further water content increase in the foundation materials before it can drain laterally.

20.5.5.3 Trends in Pavement Design

Pavement design in cold regions is a complex task. Clearly, it is the water content in each layer, including natural subgrade soils, that controls structural deterioration and ultimately failure of a flexible pavement. Since the water content varies with time, pavement design must take into account these variations. In frost-free areas, it is acceptable that pavements are designed as discussed in Chapter 13. However, this may not be sufficient in regions with significant variations in layer properties throughout the year.

A mechanistic pavement design procedure developed by the US Army Corps of Engineers considers critical strains and pavement damage that are calculated daily and assume that damage results are additive over the life of a pavement. To calculate strains at different locations in the pavement structure, it is, however, necessary to consider the actual layers with their respective thickness, water content and properties.

While this pavement design procedure appears quite complex, not necessarily in principles but because many soil properties may be required as input, it is clear that pavement damage will be a function of climate, type of subgrade and its ability to create excess ice in forms of ice lenses, rate of thaw and the ability of foundation material to evacuate excess water liberated at the thaw front and finally of the swelling characteristics of the subgrade. For all these reasons, it is unlikely that standard pavement design may be an optimal solution; rather it is suggested that more geotechnical investigations and engineering be done on each site. Adequate design may be more expensive that standardized design, but it will lead to economy from a long-term perspective since the actual life of the pavement structure is increased.

21. EARTHQUAKE ENGINEERING

W. D. L. Finn

21.1 Introduction

This chapter presents the concepts and procedures used by geotechnical earthquake engineers to design safe structures in a seismic environment. Geotechnical engineers tackle a wide range of problems; establishing design ground motions, the seismic design of foundations, analysis of soil–structure interaction, estimating seismic pressures against retaining and basement walls, evaluation of liquefaction potential, seismic response analysis of earth structures, post-liquefaction behavior of soil structures, evaluating the design of remediation measures and seismic risk analysis for critical facilities such as embankment dams with seismic liquefaction hazard potential.

It is impossible to deal in detail with most of these topics within the confines of a single chapter. In some areas, such as evaluating liquefaction potential and estimating seismic pressures, the key concepts and procedures can be presented compactly. In others, such as evaluating the seismic response of embankments, this is not possible. In these cases, the concepts are explained but for the most part the reader is directed to references for details of the procedures.

There is a limited review of seismological concepts. Seismological reports are the basis for seismic design of any critical project. It is imperative that the geotechnical engineer who will participate in the development of seismic design criteria should be able to understand the basis for the recommendations of a seismological report and be able to appreciate the uncertainties associated with them.

21.2 Seismological Aspects

21.2.1 EARTHQUAKE OCCURRENCE

Earthquakes are generated by the interaction between tectonic plates. These plates move with velocities that are considered constant over geological time. There are two main classes of earthquakes: interplate earthquakes, which occur mainly at the boundaries of the tectonic plates; and intraplate earthquakes, which occur in the interior of the plates far from the boundaries.

Interaction at the plate boundaries determines the characteristics of the interplate earthquake sources. If the plates slide past each other, strike-slip earthquakes occur at fairly shallow depths less than 70 km. An example is the San Andreas strike-slip fault in California, which is the boundary between the North American plate and the Pacific plate. If one plate overrides the other and forces it to bend down into the asthenosphere, a subduction zone develops, which can give rise to earthquakes at a variety of depths. Some of these earthquakes occur in the subducting plate, others in the overriding slab. Large earthquakes are associated with slip between the locked segments of the subducting and overriding plates. On the west coast of Chile, subduction earthquakes have occurred at depths up to 650 km. Finally,

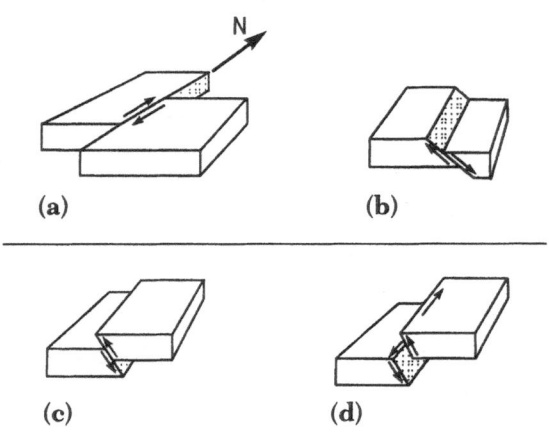

FIGURE 21.1. Various types of faulting: (a) right-lateral strike-slip; (b) dip-slip (normal); (c) dip-slip (reverse); and (d) oblique-slip (left-lateral reverse).

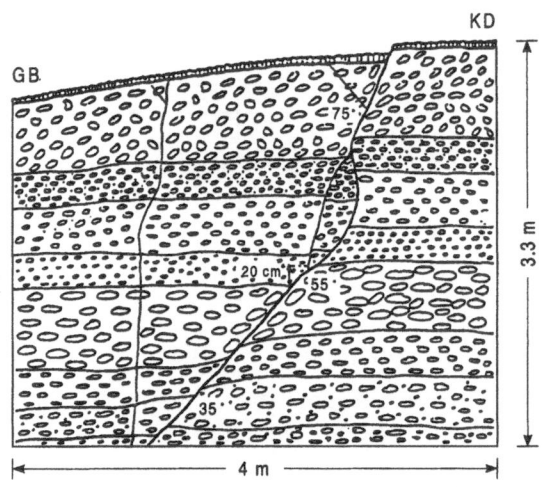

FIGURE 21.2. Offsets in strata caused by past earthquakes (modified from Demirtas *et al.* 1996).

when plates collide head on, very large mountains are created and the potential for very large earthquakes exists. The collision between the Indo-Australian plate and the Eurasian plate created the Himalayas. In this region, design earthquakes with magnitudes up to $M = 8.5$ have been proposed.

Earthquakes are associated with faults (Fig. 21.1), and are described with reference to movement of the fault: strike-slip, normal, reverse and oblique. Each of these types of slip generates its own specific radiation pattern of seismic ground motions (Reiter 1990).

In addition to natural seismicity, earthquakes are also induced by the impoundment of large reservoirs behind dams. Loading the ground surface in this way appears to activate local faults and can generate earthquakes of substantial magnitude. In some instances, earthquakes have also been generated by the pumping of fluids under high pressures back into the geological structures. This type of seismicity is classified as induced seismicity.

There are three sources of seismicity data: instrumental, historical and geological. Because of the limited time frame of instrumental data (about 100 years), historical and geological data play a major part in extending the

earthquake record. Historical data are derived from diverse sources, such as inscriptions on stone columns (e.g. in the eastern Mediterranean) and from chronicles and letters (Ambraseys 1988). The use of geological data (called paleoseismic data) is increasing and is becoming routine for major projects involving critical structures. The data on past events are derived from digging trenches across faults and inferring both the magnitude and the date of the events from dislocations and offsets in the sedimentary sequences (Sieh & Jahns 1984). A typical example is shown in Fig. 21.2.

21.2.2 WAVE TYPE

Four principal wave types generated by an earthquake are important in engineering applications. The deformation pattern of the ground during passage of these waves is illustrated in Fig. 21.3 (Bolt 1976). The compression waves, also called P-waves, travel fastest, at about 5.5 km s^{-1} in hard rock. The particle displacements occur along the direction of propagation and create zones of compression and extension as shown in Fig. 21.3a. The shear or S-waves travel at about 3.0 km s^{-1}.

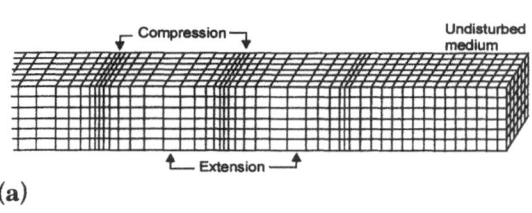

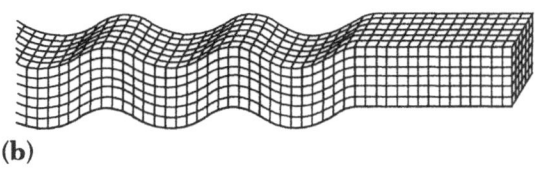

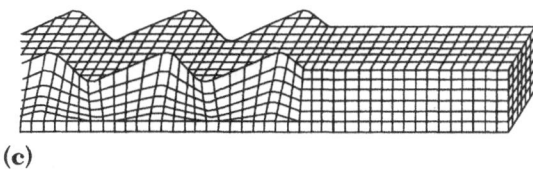

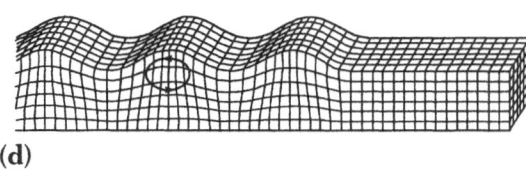

FIGURE 21.3. Deformation patterns of the major wave types (modified from Bolt 1976): (a) P-wave, (b) S-wave; (c) Love wave; and (d) Rayleigh wave.

The particles move normal to the direction of wave propagation, as shown in Fig. 21.3b. The Rayleigh waves travel at about 92% of the speed of the S-waves. The particles describe the retrograde motion shown in Fig. 21.3d. The Love waves (Fig. 21.3c) are surface shear waves with no vertical component, and are slower than Rayleigh waves. The shear waves (S), cause most damage in earthquakes, followed by Rayleigh waves, which tend to have the longest periods and therefore affect long period structures the most.

21.2.3 EARTHQUAKE MAGNITUDE

The magnitude frequently quoted in newspaper accounts of earthquakes is the Richter or local magnitude, M_L. It is defined as "the logarithm to the base 10 of the maximum seismic wave amplitude in thousands of a millimeter recorded on the (standard) Wood–Anderson seismograph at a distance of 100 km from the epicenter," (Richter 1958).

The most commonly used magnitude scales are the surface wave magnitude, M_s, the body wave magnitude, m_b, and the moment magnitude, M_w. M_s is based on the amplitude of (Rayleigh) waves with a period of about 20 s. It is limited to relatively shallow earthquakes <70-km deep and is usually measured at distances of >1000 km from events $M_s \geq 5$. During earthquakes at great depths, Rayleigh waves are not generated well. The body wave magnitude, m_b, is a world-wide scale based on the maximum amplitude in the first few cycles of P-wave motion observed on vertical seismograms. The waves have a period of about 1 s. These waves are best observed at distances greater than 1000 km. The body wave magnitude does not adequately describe the strength of large earthquakes that generate very large fault ruptures because it is based on a relatively short period motion. In eastern North America, a magnitude, m_{Lg}, is used. This scale uses Love waves with a period of about 1 s. In Japan, a different local magnitude, M_{JMA}, is used by the Japanese Meteorological Agency. This is a long period measurement. Since Japanese earthquakes are often used in Western databases, it is important to keep the differences in the definitions of magnitude in mind.

Moment magnitude was developed to achieve a single unified magnitude scale applicable to earthquakes of all sizes, depths and locations. This scale is based on seismic moment, M_0, defined by:

$$M_0 = GAD \tag{21.1}$$

where G = the shear modulus, A = the area of the fault rupture, and D = the average slip between opposite sides of the fault.

The moment magnitude, M_w, is defined as:

$$M_w = (\log M_0 - 16.05)/1.5. \qquad (21.2)$$

This magnitude is consistent with other magnitudes over a wide range of magnitudes. It does not saturate as the other magnitudes do that are period dependent. For example, magnitudes based on a 1-s period can reflect properly only the energy release or deformation from faults whose dimensions are of the order of tens of kilometers or less, because their wavelength is about 10 km. M_s, which is based on a period of 20 s and a wavelength of 80 km, cannot fully represent the energy release or deformations from faults whose rupture length is many hundreds of kilometers.

21.2.4 EARTHQUAKE STATISTICS

Gutenberg & Richter (1944), on the basis of world-wide seismicity, developed the linear relationship (G–R line) between frequency of occurrence and magnitude:

$$\log_{10} N(M) = a - bM \qquad (21.3)$$

where $N(M)$ = number of earthquakes per year with magnitude greater than or equal to M, and a and b = constants for a given seismic zone. The number N is associated with a given area and time. Alternatively, $N(M)$ can be given by:

$$N(\geq M) = \exp^{\alpha - \beta M} = N_0 \exp^{-\beta M}. \qquad (21.4)$$

There is a threshold magnitude, M, below which the occurrence of earthquakes are not recorded completely for a given seismographic layout, population distribution, etc. Fitting the Gutenberg–Richter (G–R) equation to incomplete data (at low end) will flatten the slope of the line and indicate higher occurrence rates for larger earthquakes. Therefore, it is imperative to ensure that any data used to establish the rates of occurrence of earthquakes are complete for the time period under consideration.

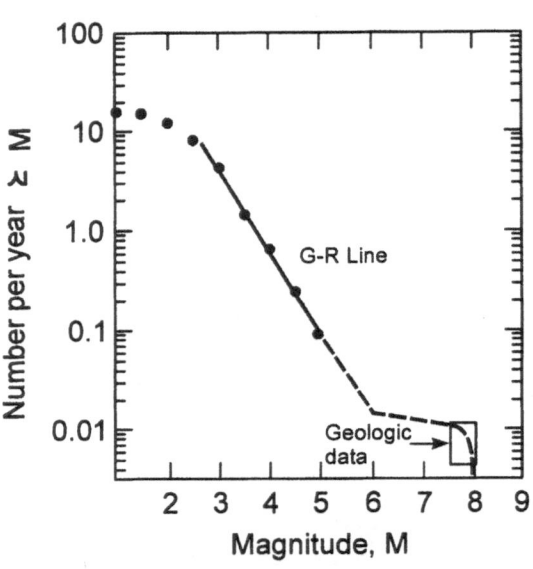

FIGURE 21.4. Earthquake recurrence rates showing G–R line and characteristic earthquakes (modified from Schwartz & Coppersmith 1984).

The G–R equation does not always hold. Some fault segments tend to have many occurrences of an earthquake of a particular size or of earthquakes in a narrow range of magnitudes. These earthquakes are often called characteristic earthquakes, and tend not to lie on the G–R line based on smaller earthquakes. An example is shown in Fig. 21.4.

21.2.5 DISCRETE FREQUENCY OCCURRENCE

It is necessary to know how often an earthquake of a particular magnitude may occur so that its contribution to seismic hazard or seismic risk, such as liquefaction, can be independently assessed. However, it is not possible to deal directly with discrete magnitudes, so the earthquakes are put into bins with magnitude labels, M, but including earthquakes with $M \pm dM$ where dM is arbitrary. The frequency occurrence of the bin magnitudes is given by a discrete frequency relation.

The number of earthquakes in each bin can be calculated as follows. Assuming that the range in magnitude is $dM = \pm\frac{1}{4}$, the number of earthquakes in magnitude bin M, $n(M)$, is given by:

$$n(M) = N_0\{\exp^{[-\beta(M-1/4)]} - \exp^{[-\beta(M+1/4)]}\}$$
$$= N_0[\exp^{(\beta/4)} - \exp^{(-\beta/4)}] \exp^{-\beta M} \qquad (21.5)$$

yielding the discrete frequency relation:

$$n(M) = N_{0d} \times \exp^{(-\beta M)}. \qquad (21.6)$$

21.2.6 EARTHQUAKE SOURCE ZONES

Seismicity of a region is defined by finite source zones each with uniform seismicity, (i.e. with a specified rate of occurrence defined by values of the parameters a and b or α and β in the G–R equation and a specified maximum magnitude). The source zones may be areal sources, line (vertical fault) or inclined plane (dipping fault) sources. Probabilistic seismic hazard analysis is based on these homogeneous source zones.

21.2.7 MAXIMUM MAGNITUDE, M_x

There is an upper limit to the maximum magnitude expected in a given source zone based on historical, geological or tectonic reasons. This entails a curved portion in the frequency of occurrence curves so that the equation can have the form:

$$N(\geq M) =$$
$$N_0 \exp(-\beta M)\{1 - \exp[-\beta(M_x - M)]\}. \qquad (21.7)$$

The factor $\{1 - \exp[-\beta(M_x - M)]\}$ produces a curve giving a smooth approach to zero rate of occurrence at $M = M_x$ in the plotted cumulative distribution (Fig. 21.5).

21.2.8 ESTIMATION OF MAXIMUM MAGNITUDE

The following procedures are among those used to estimate the maximum magnitude:

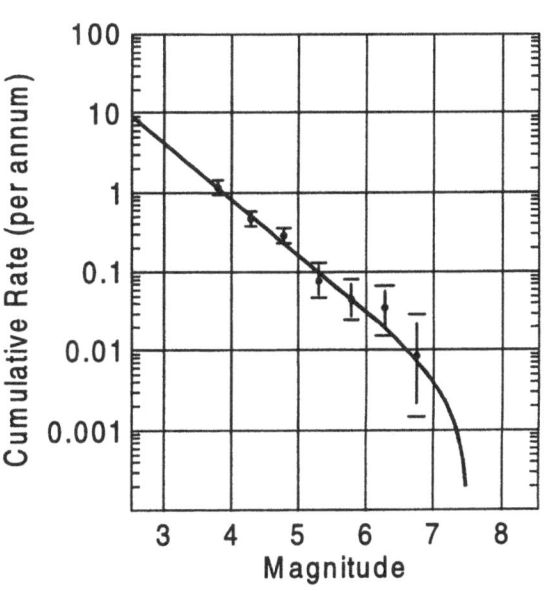

FIGURE 21.5. Magnitude recurrence relation with a maximum earthquake magnitude.

1. Use the largest previous earthquake if the observation period is significantly longer than the return period of the maximum earthquake.
2. Use correlations between maximum fault break and magnitude (Bonilla *et al.* 1984).
3. Estimate maximum magnitudes from plate tectonic models, slip rates, etc.
4. Base magnitude on geological studies, such as trenching to detect past events and strata displacements, or height of fault scarps, etc.
5. In some cases, arbitrary estimates are used: 0.5-magnitude greater than largest known historical event. This approach is typically used when the type of information required for accurate estimates is not available or the cost of acquiring the information, if retrievable, is not justified.

For zones of high seismicity, such as California, the contribution to significant risk at moderate probabilities comes from earthquakes near the estimated maximum magnitude. For zones of low seismicity, such as eastern North America, the probability of occurrence of earthquakes near the maximum can be much less than the probability being considered in the risk estimate. In these cases, the choice of maximum magnitude may be less important for projects with a time horizon that is short relative to the return period of the maximum earthquake.

21.2.9 MAGNITUDE FROM FAULT BREAK AND FAULT AREA

Various attempts have been made to relate the magnitude of an earthquake to the characteristics of the fault break, such as rupture length and rupture area. Bonilla *et al.* (1984) gave an expression for the surface magnitude:

$$M_s = 6.04 + 0.708 \log L \qquad (21.8)$$

with a standard error, $s = 0.306$. The length of rupture, L, is in kilometers. This equation is based on all fault types. Some authors, such as Slemmons *et al.* (1989), give equations for each fault type.

Other researchers have correlated magnitude with fault rupture area, e.g. Wyss (1979):

$$M_s = 4.15 + \log A \qquad (21.9)$$

where A = the area of the fault rupture surface in square kilometers.

21.2.10 PROBABILITY OF OCCURRENCE

The following basic assumptions are made in estimating the probability of occurrence:

- Earthquake occurrence is independent of space and time.
- The probability density function is given by the Poisson distribution:

$$p(n) = \frac{(Nt)^n \exp(-Nt)}{n!} \quad n = 0, 1, 2, \text{etc.} \quad (21.10)$$

where N = the rate of occurrence per year for the source, t = the time period under consideration, and $p(n)$ = the probability of n earthquakes occurring in time t. The

probability of at least one earthquake occurring is p_e:

$$p_e = 1 - \exp(-Nt). \qquad (21.11)$$

Thus, for example, the probability that an earthquake with a return period of 100 years will occur during 100 years is obtained from Eq. 21.11 using $t = 100$ years and $N = 0.01$, giving $p_e = 63\%$. More generally, the probability of an earthquake occurring within its return period is 63%.

21.3 Wave Propagation

21.3.1 ELEMENTS OF WAVE PROPAGATION

Assume the largest wave amplitude, a_m, is associated with a harmonic wave of circular frequency, ω (wave period, $T = 2\pi/\omega$). Since the wave is propagating in space and time, both types of variables must be present in the wave form:

$$a = a_m \sin(\kappa x - \omega t). \qquad (21.12)$$

For a constant wave amplitude, $\sin(\kappa x - \omega t)$ must remain constant, thus:

$$\kappa x - \omega t = C_1$$

$$\kappa \frac{dx}{dt} - \omega = 0 \qquad (21.13)$$

$$\frac{dx}{dt} = V_w = \frac{\omega}{\kappa}$$

where V_w = wave speed and x = displacement in direction of propagation. A similar analysis for spatial period shows that the wave number, $\kappa = 2\pi/L$, where L = wavelength.

21.3.2 STRAINS IN THE GROUND

The strains in the ground caused by the propagating wave may be readily calculated for any wave type using the following procedure. Assume that for a compressive harmonic wave, the particle displacement, y_p, is in the direction of propagation, x, and is given by:

$$y_p = A \sin(\omega t - \kappa x). \qquad (21.14)$$

The particle velocity, v_p, is:

$$v_p = \frac{\partial y_p}{\partial t} = \omega A \cos(\omega t - \kappa x) \qquad (21.15)$$

therefore:

$$\therefore A \cos(\omega t - \kappa x) = v_p/\omega. \qquad (21.16)$$

The strain in the ground, ε_x, is given by:

$$\varepsilon_x = \frac{\partial y_p}{\partial x} = -\kappa A \cos(\omega t - \kappa x). \qquad (21.17)$$

Therefore, combining Eqs 21.16 and 21.17 gives:

$$\varepsilon_x = \left(-\frac{\kappa}{\omega}\right)v_p = -\frac{v_p}{V_p} \qquad (21.18)$$

where V_p = the compression wave velocity.

The result may be expressed as:

$$\text{Strain} = \frac{\text{particle velocity}}{\text{wave velocity}} \qquad (21.19)$$

where particle and wave velocities are in the same direction.

21.3.3 STRAINS IN BURIED PIPES

Strains in buried pipes caused by earthquake shaking are readily calculated if it is conservatively assumed that there is no slip between the pipe and the ground. Thus, for compressional waves along the pipe:

$$\varepsilon_{\text{pipe}} = \varepsilon_{\text{ground}} = \frac{v_p}{V_p}. \qquad (21.20)$$

When the wave is propagating at an angle, θ, to a pipeline, the same formula for strain can be employed provided the particle velocity is resolved parallel to the pipe and the apparent wave velocity along the pipe is substituted for the wave propagation velocity. The procedure is explained by means of an example.

Consider compressional wave fronts 1-s

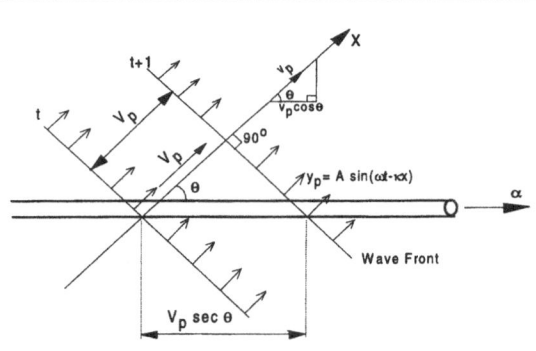

FIGURE 21.6. Compressional wave propagating at an angle θ to the pipeline.

apart as shown in Fig. 21.6. From Fig. 21.6, the apparent propagation velocity along the pipe is:

$$V_\alpha = V_p (\sec \theta) \qquad (21.21)$$

Particle velocity in the direction of the pipe is:

$$v_\alpha = v_p \cos \theta. \qquad (21.22)$$

Therefore, the maximum strain in the pipe is:

$$\varepsilon_\alpha = \frac{\text{particle velocity}}{\text{apparent wave velocity}} = \frac{v_p \cos^2 \theta}{V_p}. \qquad (21.23)$$

The axial strain is a maximum at $\theta = 0°$.

When the propagating wave is a shear wave (Fig. 21.7), then the particle velocity is nor-

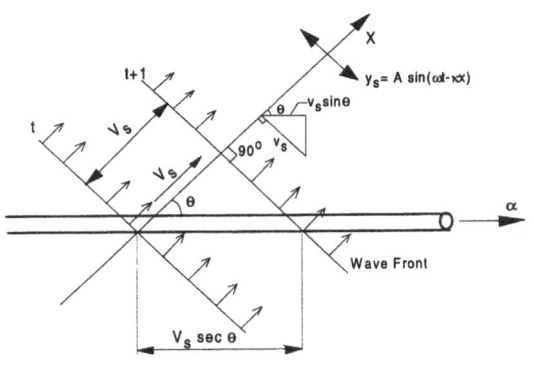

FIGURE 21.7. Shear wave propagating at an angle θ to the pipe.

mal to the direction of propagation and the strain in the pipe, ε_α, is given by:

$$\varepsilon_\alpha = \frac{\text{particle velocity along pipe}}{\text{apparent wave velocity along pipe}} \quad (21.24)$$

$$= \frac{v_s}{V_s} \sin\theta \cos\theta.$$

The axial strain is a maximum at $\theta = 45°$.

21.4 Design Ground Motions

21.4.1 BASIC ELEMENTS

The simple one-degree-of-freedom system in Fig. 21.8a contains most of the significant elements controlling dynamic response; mass = m, spring stiffness = k, viscous damping coefficient = c, and excitation by ground acceleration = $\ddot{u}_g(t) = d^2u/dt^2$.

The force system for dynamic equilibrium is shown in Fig. 21.8b, where F_D = damping force = $c\dot{u}$, F_S = spring force = ku, F_I = inertia force = $m\ddot{u}(t)$.

Note that the stiffness and damping depend on the relative movement between the structure and the ground. The total displacement of the structural mass $u_t(t)$ = the sum of the ground displacement, $u_g(t)$, and the displacement of the structural mass relative to the ground, $u(t)$:

$$u_t(t) = u_g(t) + u(t). \quad (21.25)$$

The equilibrium equation for the structure is:

$$F_I + F_D + F_S = 0 \quad (21.26)$$
$$m(\ddot{u}_g + \ddot{u}) + c\dot{u} + ku = 0 \quad (21.27)$$
$$\ddot{u} + (c/m)\dot{u} + (k/m)u = -\ddot{u}_g(t) \quad (21.28)$$
$$\ddot{u} + 2\lambda\omega\dot{u} + \omega^2 u = -\ddot{u}_g(t) \quad (21.29)$$

where $\ddot{u}_g(t)$ and $\ddot{u}(t)$ = the accelerations of the ground and of the mass relative to the ground, respectively; $\dot{u}(t)$ = velocity of the mass relative to the ground; $\omega = (k/m)^{1/2}$ natural (circular) frequency; and $\lambda = 0.5c/(km)^{1/2}$ fraction of critical damping or damping ratio. The critical damping above which no free vibration occurs is:

$$c_{cr} = 2(km)^{1/2} \quad \text{or} \quad 2m\omega. \quad (21.30)$$

21.4.2 RESPONSE SPECTRA

The response spectrum of a ground motion parameter such as acceleration, velocity or displacement for a specified level of damping is a plot of the maximum value of that quantity as a function of the frequency or period of a one-dimensional structure under a specific seismic excitation. A good introduction to response spectra is given by Chopra (1981).

Spectra can be computed for peak displacement, peak velocity and peak acceleration. However, peak velocity and peak acceleration spectra are not usually used in design. In practice, spectra are drawn for pseudo-relative velocity (S_v or PSV) and pseudo-acceleration, S_a or PSA. S_v and S_a are defined in terms of S_d, the ordinates of the peak displacement spectrum:

$$S_d = U_{max} \quad (21.31)$$

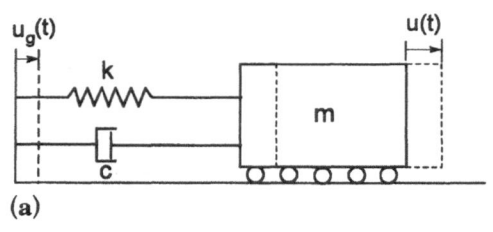

(a)

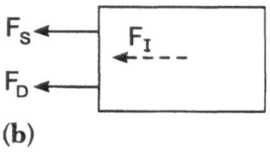

(b)

FIGURE 21.8. Simple one-degree-of-freedom system: (a) mass–spring damper system; and (b) free-body diagram.

where U_{max} = the maximum displacement at each period for a constant level of damping:

$$S_v = \omega S_d \qquad (21.32)$$

$$S_a = \omega^2 S_d. \qquad (21.33)$$

Since S_d, S_v and S_a are all related, they may be plotted on the same chart as a function of frequency or period. Such a plot, for the El Centro earthquake of 1940, known as a tripartite logarithmic plot, is shown in Fig. 21.9. The motivation for defining S_v and S_a, as described above, is as follows. During an earthquake, the maximum elastic energy, E_{max}, is stored in a structure of weight W at peak displacement U_{max} and can be shown to be related to the pseudo-velocity, S_v, not the peak velocity, by $E = (\frac{1}{2})(W/g)S_v^2$. The elastic base shear in the structure, V_{max} is related to S_a by $V_{max} = (W/g)S_a$.

Different parts of the response spectrum control the seismic behavior of different kinds of structures. Stiff structures with low period (high natural frequency) will be affected most by the acceleration parts of the spectrum. Structures of intermediate frequency will be controlled by the pseudo-velocity part of the spectrum and long period structures will be controlled by the displacement region of the spectrum.

21.4.3 SIMPLE DESIGN SPECTRA

Design is generally carried out using smooth spectral shapes. The peaks and valleys in spectra from real records (Fig. 21.9) cannot be used effectively in design because of the major uncertainties associated with the frequency content of any future motions at the site and the uncertainties associated with the dynamic periods of the structure itself.

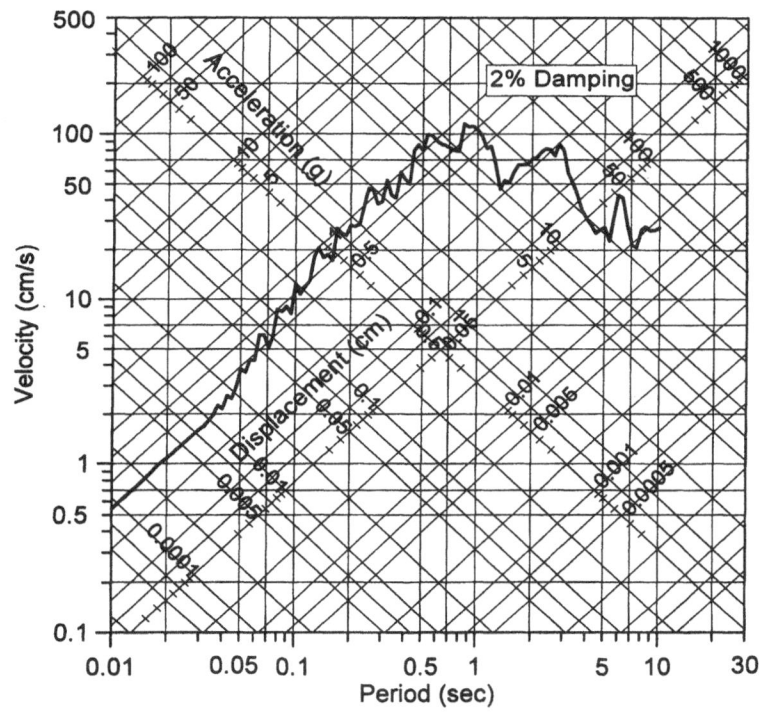

FIGURE 21.9. Logarithmic plot of response spectrum of El Centro corrected accelerogram, 270°, Caltech IIA001.

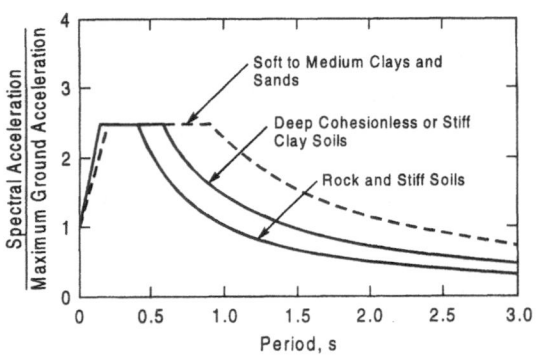

FIGURE 21.10. Normalized spectral curves suitable for use in building codes (modified from Seed & Idriss 1982).

Spectral shapes for use in design are found in the *National Building Code of Canada* (NRCCa,b 1995), and in the US *Uniform Building Code* (ICBO 1997). Some industry organizations such as the American Petroleum Institute have recommended design spectra for use by their members (API 1993).

As an example of design spectra, the spectra proposed by Seed and Idriss (1982) for use in building codes, which are normalized to the maximum ground acceleration, are shown in Fig. 21.10. Spectral shapes can be anchored to acceleration levels established by probabilistic analysis to provide spectra for design of particular projects. These idealized Seed and Idriss spectra are derived from the spectra of ground motions recorded on different soil types.

21.4.4 SITE-SPECIFIC RESPONSE SPECTRA

For important and critical structures, it is usual to develop site-specific design spectra, following procedures described below. Spectra are usually developed for two different levels of motions: operating level events (OLE) and design level events (DLE). The names of these levels of design motions vary with area of application.

Operating level events are usually moder-ate earthquakes with a high probability of occurring at least once in the life of the structure. The structure is designed to resist these events without any significant damage so that the structure or facility can continue to operate. Design level events are rare large earthquakes. Because of the low probability of occurrence, significant damage is acceptable, but collapse and loss of life must be avoided.

The ground motions for these events can be characterized by response spectra and/or time histories of motion. For critical structures, design is often done on the basis of response spectra and confirmatory dynamic analyses are conducted using ground motions that cover the frequency range of the design spectra adequately.

The development of site-specific spectra involves:

1. Seismotectonic characterization, i.e. seismic source evaluation, source to site attenuation (see Section 21.5) and site evaluation.
2. Seismic exposure assessment: for OLE events probabilistic analysis may be used to help rank the significance of the earthquake sources and to select events that are likely to contribute most to ground motions at a site for a range of spectral periods. In the most general case, near-field, mid-field and far-field events are considered. The selection of DLE events involves a more deterministic approach. Events are often classified as maximum credible events.
3. Ground motion characterization: select and suitably scale ground motion records from earthquakes similar to the design events in magnitude, epicentral distance, types of faulting, site conditions, etc. Estimate the site specific motions by site response analysis using rock input motions if possible. Non-linear analysis is recommended (Schnabel *et al.* 1972; Lee & Finn 1978). The result of these steps is a set of tuned ground motion records for the site and associated response spectra.

4. Design ground motion specification: specify smooth response spectra and associated sets of representative records for dynamic analysis. The smoothed spectra are usually based on the mean or 84% levels obtained from a statistical analysis of the representative ground motion spectra.

Design spectra should not be based solely on a statistical analysis of the response spectra from representative records, though such analyses play a major part. Consideration should be given to the following items before selecting a design spectrum: (a) acceptable level of damage; (b) consequences of failure; (c) method of analysis and associated level of conservatism; (d) uncertainties in material properties; (e) brittleness or ductility of structure; (f) level of damping; and (g) extent and quality of seismic data base, also how well is local seismicity understood and associated with known geologic features.

21.4.5 CONFIRMATORY DYNAMIC ANALYSES

Select a sufficient number of ground motion records representing the range of design ground motions associated with OLE and DLE events to cover the desired frequency range. Most records have significant energy over a fairly narrow band. Thus, no single record in the original group from which the design response spectra were developed contains sufficient energy to be representative over the broad frequency band represented by the design spectrum. For this reason, confirmatory analysis should be conducted using records from the near-, mid- and far-field to match the frequency content of the design spectra.

21.4.6 UNIFORM HAZARD SPECTRA

The design spectra described earlier were derived from earthquake records scaled to peak ground accelerations only. The peak ground acceleration may be selected at a specified probability of exceedance based on a probabilistic seismic hazard analysis.

The probability of exceedance for periods other than that of the peak ground acceleration is unknown. To ensure that the probability of exceedance at any period is the same, a uniform hazard spectrum is used to define the design spectrum. This is developed by conducting a seismic hazard analysis for a specified probability of exceedance of the spectral acceleration (or velocity) over the full range of periods relevant to the design. Recorded earthquake motions are then scaled to match the design spectrum at all periods. This scaling is conducted using computer software such as RASCAL (Silva & Lee 1987). A very lucid explanation of the process of developing spectrum-compatible motions may be found in Clough & Penzien (1993).

The uniform hazard response spectrum leads to a consistent treatment of hazard in the design of structures of different periods. However, no natural earthquake motions would have the spectral amplitude characteristics of the spectrum-compatible motion over the entire spectrum. Using spectrum-compatible motions for predicting non-linear analysis can lead to inflated estimates of seismic displacements (Naeim & Lew 1995).

21.5 Earthquake Ground Motions

21.5.1 GROUND MOTION ATTENUATION RELATIONS FOR CRUSTAL EARTHQUAKES

The basic input to seismic hazard calculations is an estimate of the expected ground motion at a site at a given distance from an earthquake of a given magnitude, usually expressed as moment magnitude. This motion may be characterized by peak ground motion parameters such as acceleration, velocity and displacement, or ground motion spectra. Ground motion attenuation relationships, which define how ground motion parameters decay with distance from the source, are used

TABLE 21.1. Comparison of site classification for attenuation
models for shallow crustal events in active regions

Boore et al. (1993)	Abrahamson & Silva (1997); Sadigh et al. (1997)	Campbell (1997)	Idriss (1990)
Average $V_s = 620$ m s^{-1}	Rock, shallow soil	Hard rock, soft rock	Rock and stiff soil
Average $V_s = 310$ m s^{-1}	Deep soil	Firm soil	Deep soil
$V_s < 180$ m s^{-1}	Soft soil	Soft soil, shallow soil	Soft soil

to provide ground motion estimates for a site. There are many such relationships and as more data become available, existing attenuation relationships are updated. A state-of-the-art assessment of the main attenuation relations in use in North America may be found in a special issue of *Seismological Research Letters* (SSA 1997).

Attenuation relations tend to be regionally specific. For example, in North America, very different attenuation relations are used in the west and the east because of radically different geological and tectonic structures. Relations also tend to be specific with respect to the type of faulting (see Section 21.2.1 for definition of types of faults). Attenuation from subduction sources is different than from strike-slip sources. There are also differences between attenuation from strike-slip sources and reverse or thrust faults.

Attenuation relations may be site-specific in the sense that the relation may be established for a particular soil condition such as rock, soft soil, deep stiff soil, shallow stiff soil, etc. Some attenuation relations define these conditions by means of descriptive adjectives. Boore et al. (1993) characterized the site conditions by means of the time averaged shear wave velocity in the top 30 m of the site. Site

classifications used in five different attenuation models are shown in Table 21.1.

The characteristics of attenuation relations will be illustrated using the attenuation relation for crustal sources by Boore et al. (1997):

$$\ln Y = b_1 + b_2(M - 6) + b_3(M - 6)^2 + b_5 \ln r + b_v \ln \frac{V_S}{V_A} \quad (21.34a)$$

where

$$r = (r_{jb}^2 + h^2)^{1/2} \quad (21.34b)$$

$b_1 =$

$$\begin{cases} b_{1SS} & \text{for strike-slip earthquakes} \\ b_{1RS} & \text{for reverse-slip earthquakes} \\ b_{1ALL} & \text{if mechanism is not specified.} \end{cases} \quad (21.34c)$$

and examples of other parameters are given in Table 21.2.

In this equation, Y = the ground-motion parameter (peak horizontal acceleration or pseudo-acceleration response, g); the predictor variables are moment magnitude, M, distance, r_{jb}(km), and average shear-wave velocity to 30 m, V_s (m^{-1} s). Note that h = a regression parameter.

TABLE 21.2. Smoothed coefficients used to estimate pseudo-acceleration
response spectra (g) for the random horizontal component at 5% damping

Period	b_{1SS}	b_{1RS}	b_{1ALL}	b_2	b_3	b_5	b_v	V_A	h	$\sigma_{\ln Y}$
0.000	−0.313	−0.117	−0.242	0.527	0.000	−0.778	−0.371	1396	5.57	0.520
0.200	0.999	1.170	1.089	0.711	−0.207	−0.924	−0.292	2118	7.02	0.502
1.000	−1.133	−1.009	−1.080	1.036	−0.032	−0.798	−0.698	1406	2.90	0.613

The many different definitions of distance, r, from the earthquake source used in attenuation relations are illustrated in Abrahamson & Shedlock (1997). The Boore *et al.* (1997) distance, r_{jb}, is the closest horizontal distance to the vertical projection of the rupture surface. In the case of a vertical fault, this is the same as the distance to the fault break. For dipping faults, the distance can be as low as zero when the source is within the vertical projection of the rupture surface.

The attenuation relation (Eq. 21.34) is valid for earthquake magnitudes ranging from $M = 5.5$ to 7.5 and for distances $D \leq$ 80 km. Values of the coefficients in Eq. 21.34 are given for the attenuation of spectral accelerations at periods from 0 to 2 s. The complete table of coefficients is given in Boore *et al.* (1997). A few lines from the table of parameters are given in Table 21.2 to illustrate the structure of such tables. The entries for zero period are the coefficients for peak horizontal acceleration. In Table 21.2, $\sigma_{\ln Y}$ is the square root of the overall variances of the regression, so that the mean plus one σ value of the natural logarithm of the ground motion parameter Y in Eq. 21.34 is $\ln Y + \sigma_{\ln Y}$.

Abrahamson & Silva (1997) have developed attenuation relations valid for periods up to 5 s. These may be particularly useful in developing spectra for tall buildings.

21.5.2 GROUND MOTION ATTENUATION RELATIONS FOR SUBDUCTION ZONE EARTHQUAKES

Youngs *et al.* (1997) provide attenuation relations for horizontal response spectral acceleration (5% damping) for subduction earthquakes applicable to rock and soil sites. These attenuation relationships are considered appropriate for earthquakes with $M = 5$ and greater, and for distances to the rupture surface from 10 to 500 km.

Most attenuation models for subduction zone events are based on recordings from Japan and South America. Most of these events were recorded at large distances. However, recordings were made during the 1985 Michoacan earthquake at distances as small as 13 km, although recordings within 30 km were sparse. Youngs *et al.* (1997) show that peak ground motions from subduction zone earthquakes attenuate more slowly than those from shallow crustal earthquakes and that intraslab earthquakes produce larger peak ground accelerations than interface earthquakes for the same magnitude and distance. However, the database contains a very limited number of intraslab recordings.

21.6 Effects of Soil Conditions on Ground Motion

A number of factors contribute to the amplification of incoming rock motions at soil sites. The first important factor is the relative stiffness between the soil and the rock. This is usually characterized by the impedance ratio, ζ. The impedance of a soil layer is the mass density multiplied by the shear-wave velocity of the material. Therefore, the impedance ratio $\zeta = \rho_s V_{ss}/\rho_r V_{sr}$, where ρ = the mass density, V_s = the shear-wave velocity, and S and R refer to the surface layer and underlying rock, respectively.

For example, consider a uniform elastic undamped surface layer. If the incoming wave motion in the bedrock from the source is a harmonic wave and has the same period as that of the elastic surface layer, then the amplification of the motion in the surface of the soil layer according to Okamoto (1973) is:

$$A = 2/\zeta. \qquad (21.35)$$

This amplification includes the effect of impedance, resonance and the doubling of the motion at the free surface. If there is no surface layer of soil, then the amplification of the incoming motion at the surface of the layer is $A = 2$.

In the above example, the incoming mo-

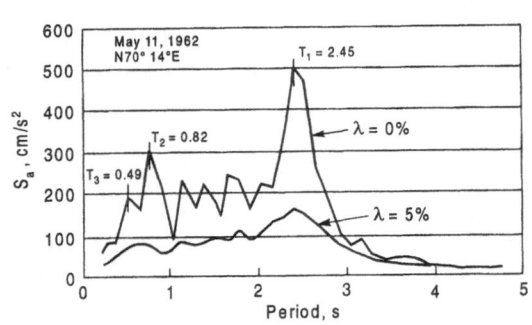

FIGURE 21.11. Pseudo-acceleration spectrum, Alameda Park, May 11, 1962.

tions were assumed to have the same period as the surface layer of soil. Now consider how the periods of the surface layer are manifested in recorded ground motions. The natural periods, T_n, of a uniform elastic layer of thickness H and a shear-wave velocity of V_s are:

$$T_n = (4H/V_s)/(2n - 1) \quad n = 1, 2, \text{ etc.} \quad (21.36)$$

The ratios between the individual periods of the layer are:

$$T_1 = (4H/V_s), T_2$$
$$= (\tfrac{1}{3})T_1, T_3 = (\tfrac{1}{5})T_1, \text{ etc.} \quad (21.37)$$

This relationship may be seen clearly in the ground response at Alameda Park in Mexico City during an earthquake in 1962. The ground motions are characterized in Fig. 21.11 by the response spectrum for pseudo-acceleration. The zero damping spectrum is used to clearly define the peaks. Significant peaks in the spectrum are associated with the natural periods of the ground. Note the location of T_1, T_2, T_3, etc., and that they have the relation to each other close to that described above.

In Fig. 21.11, the response spectra are shown for different damping ratios. Even a modest 5% damping, which would not be untypical of structures undergoing reasonable

strong motion without damage, reduces very significantly the response of the structure. Therefore, there are three important factors controlling the effects of surface layers on ground motion: (a) the impedance ratio, ζ; (b) the relationship between the high energy periods of the bedrock motions and the periods of the surface layer; and (c) the amount of damping in the soil layer.

21.6.1 NON-LINEAR SOIL BEHAVIOR

The amplification of the softer sites during the Loma Prieta earthquake was studied by Idriss (1990), and he provided the amplification chart shown in Fig. 21.12, which gives the amplification of the ground motions at the surface with respect to rock or firm ground. The greatest amplifications occur at the lower accelerations. At accelerations higher than 0.4g, the motions are attenuated at the site. This is due to the fact that the stiffness and damping of the soft soils are strain-dependent. Under very strong shaking, the stiffness is reduced substantially and the damping is increased. The net effect for strong shaking is to reduce the amplitude of the high–medium frequency ground motions. The strengths of the weak soils, of course, impose a definite limit on the maximum acceleration that can be transmitted to the surface. Accelerations cannot generate

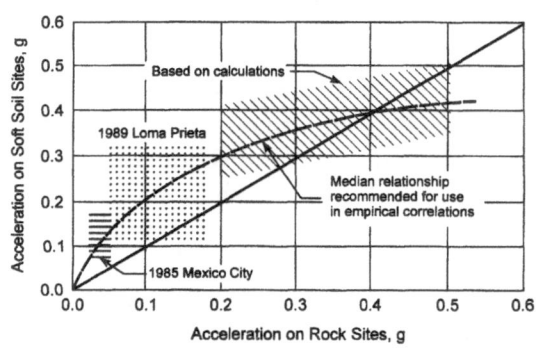

FIGURE 21.12. Amplification of ground motions in soft soils (modified from Idriss 1990).

shearing stresses in excess of the strength of the soil under dynamic loading conditions.

21.6.2 SITE CLASSIFICATION-BASED AMPLIFICATION FACTORS

For any given uniform site, the impedance ratio is fixed and the level of response will depend heavily on the correspondence between the site periods and the periods of the incoming motion. The fundamental site period is given by $T = 4H/V_s$. Therefore, the period, and hence, the response to a given input motion, depends on the thickness of the layer and the shear-wave velocity. For layers of equal thickness, the shear-wave velocity accurately reflects the period. Furthermore, if the differences in mass densities between rock and surface layer are neglected, the shear-wave velocity is also an index of relative impedance.

These ideas have led to a new way of characterizing sites for estimating the potential amplification of earthquake motions. The time-averaged shear-wave velocity in a constant thickness layer comprised of the materials in the top 30 m of a site is taken as an index of potential site amplification. This procedure has been adopted by the US National Hazards Reduction Program (NEHRP) for characterizing sites for code purposes.

The classification of sites based on V_s is shown in Table 21.3 (FEMA 1997). Because the shear-wave velocity may not be known in a given area, or may not be measured at a particular site, alternative indices of amplification that correspond to the shear-wave velocity criteria are also used. These are the average standard penetration resistance, $\bar{N}$, for cohesionless soils, and the average undrained shear strength, $\bar{s}_u$, for clays in the top 30 m of the geological profile at the site (see Section 4.4.6 for a discussion of how N and s_u are obtained in the field).

TABLE 21.3. Site classification[a]

Site class	Site-class name/generic description	Site-class definition
A	Hard rock	$\bar{V}_s > 1500$ m s^{-1}
B	Rock	$760 < \bar{V}_s \leq 1500$ m s^{-1}
C	Very dense soil and soft rock	$360 < \bar{V}_s \leq 760$ m s^{-1}, or with either $\bar{N} > 50$; or $\bar{s}_u \geq 100$ kPa
D	Stiff soil	$180 < \bar{V}_s \leq 360$ m s^{-1}, or with either $15 \leq \bar{N} \leq 50$; $50 \leq \bar{s}_u \leq 100$ kPa
E	A soil profile with	$\bar{V}_s < 180$ m s^{-1}, or with either $\bar{N} < 15$; or $\bar{s}_u < 50$ kPa
	Or	
	Any soil profile with	More than 3 m of soft clay defined as soil with PI > 20, $w \geq 40\%$, and $s_u < 25$ kPa
F	F_1: soils vulnerable to potential failure or collapse under seismic loading such as liquefiable soils, quick and highly sensitive clays, collapsible weakly cemented soils	
	F_2: peats and/or highly organic clays ($H > 3$ m of peat and/or highly organic clay where H = thickness of soil)	
	F_3: very high plasticity clays ($H > 8$ m with PI > 75)	
	F_4: very thick "soft/medium stiff clays" ($H > 36$ m)	

[a] Exception: when the soil properties are not known in sufficient detail to determine the site class, class D shall be used. Site classes E or F need not be assumed unless the authority having jurisdiction determines that site classes E or F could be present at the site or in the event that site classes E or F are established by geotechnical data.

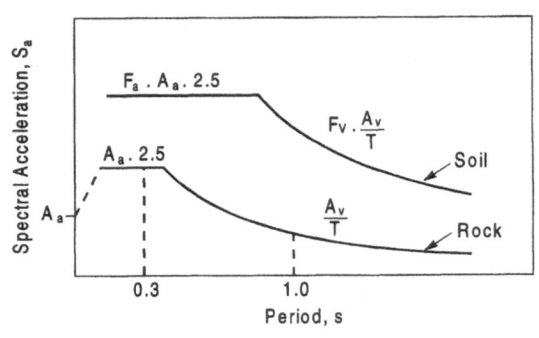

FIGURE 21.13. Development of design spectra using period-dependent site amplification factors (modified from Martin & Dobry 1994).

21.6.3 NEHRP PROVISIONS

NEHRP adopted the spectral shape for site class B (Table 21.3), shown in Fig. 21.13 as the basis for design. The spectrum is constructed using spectral values at periods of $T = 0.3$ and 1.0 s. At $T = 0.3$ s, the spectral value is $2.5A$, where A is the effective site acceleration at the appropriate probability of exceedance. Similarly, the spectral value at the period $T = 1.0$ s is the value with the appropriate probability of exceedance. The curved part of the spectrum varies as $1/T$.

NEHRP adopted two site coefficients for constructing free-field acceleration response spectra for soil sites from the given rock spec-

trum as shown in Fig. 21.13. The coefficient F_a is used for the short period motion and the coefficient F_v for the longer period motion. Together the factors are intended to cover the period range 0.2–3.0 s.

These site coefficients are a function of site class and intensity of shaking, specified by spectral acceleration. The five different general site classes, A–F, are given in Table 21.3. Site specific geotechnical investigations and dynamic site response analyses are recommended for soils falling into class F.

The values of the site coefficients, F_a and F_v, appropriate for the different site classes and the different levels of shaking were drawn from the results of hundreds of site response analyses using both equivalent linear and non-linear methods. The analyses were all calibrated first on data from the 1989 Loma Prieta earthquake. Variations in the values of the site coefficients, F_a and F_v, reflect the non-linear response of soil to strong shaking. Hence, they become smaller as the acceleration increases. Values of F_a are given in Table 21.4, and of F_v in Table 21.5, for different levels of peak ground accelerations. The values for F_a are mean values. The values of F_v were highly variable depending on site conditions and input motions. Therefore, F_v values are given at the mean plus one standard deviation level. For more detailed discussion of foundation factors see Finn (1995).

TABLE 21.4. F_a as a function of site class and earthquake spectral acceleration, S_a, at 0.3 s

| Site class | Spectral response acceleration at 0.3 s[a] | | | | |
	$S_s \leq 0.25$	$S_s = 0.5$	$S_s = 0.75$	$S_s = 1.00$	$S_s \leq 1.25$
A	0.8	0.8	0.8	0.8	0.8
B	1.0	1.0	1.0	1.0	1.0
C	1.2	1.2	1.1	1.0	1.0
D	1.6	1.4	1.2	1.1	1.0
E	2.5	1.7	1.2	0.9	[b]
F	[b]	[b]	[b]	[b]	[b]

[a] Use straight-line interpolation for intermediate values of S_s.
[b] Site-specific geotechnical investigation and dynamic site response analyses are recommended.

TABLE 21.5. F_v as a function of site class and spectral acceleration, S_a, at 1 s period

	Spectral response acceleration at 1 s period[a]				
Site class	$S_s \leq 0.25$	$S_s = 0.5$	$S_s = 0.75$	$S_s = 1.00$	$S_s \leq 1.25$
A	0.8	0.8	0.8	0.8	0.8
B	1.0	1.0	1.0	1.0	1.0
C	1.7	1.6	1.5	1.4	1.3
D	2.4	2.0	1.8	1.6	1.5
E	3.5	3.2	2.8	2.4	[b]
F	[b]	[b]	[b]	[b]	[b]

[a] Use straight-line interpolation for intermediate values of S_s.
[b] Site-specific geotechnical investigation and dynamic site response analyses are recommended.

21.6.4 EFFECTS OF TOPOGRAPHY

Aki (1988) used the simple structure of a triangular wedge (Fig. 21.14a) to illustrate the effects of topography. This structure may be used to model approximately ridge–valley topography as shown in Fig. 21.14b by Faccioli (1991). An exact solution exists for the wedge for SH waves propagating normal to the ridge and polarized parallel to the ridge axis. Displacement amplification at the vertex is $2/v$ where the ridge angle is $v\pi$ ($0 < v < 2$). In

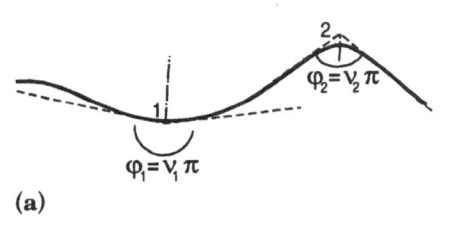

(a)

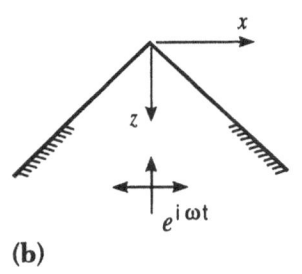

(b)

FIGURE 21.14. (a) Approximating a ridge formation by a triangular wedge, and (b) infinite wedge excited by plane SH waves (modified from Faccioli 1991).

Fig. 21.14a, the amplification of the crest relative to the base is v_1/v_2. Thus the simple solution provides a rough estimate of the relative amplification at the crest of the ridge or de-amplification in a valley.

Amplification of motions at the crest of a ridge relative to the base is also supported by damage patterns during the 1980 Fruili earthquakes in Italy (Brambati et al. 1980), the Chilean earthquake of 1985 (Celebi 1987, 1991), the 1985 Loma Prieta earthquake and the 1994 Northridge earthquake.

21.6.5 MOTIONS IN SEDIMENTARY BASINS

Surface waves have been recorded by strong-motion instruments at sites in deep sedimentary basins. These waves arrive later than the shear body waves and have periods in the general range of 3–10 s. At some sites the largest amplitudes at long periods may be due to surface waves.

Sediment filled basins are usually the locations of greatest residential and industrial development. Ground motions on these sites generated by shear waves propagating vertically are usually estimated by one-dimensional shear beam models, using either equivalent linear methods (Schnabel et al. 1972) or nonlinear models (Finn 1988; Lee & Finn 1978). Body waves trapped in the basin may amplify the motion and increase the duration over that predicted by one-dimensional analysis. These effects were very pronounced in the 1985 Mexico City and 1994 Northridge earthquakes.

Ground motion estimations near the edges of sediment-filled valleys require two- and three-dimensional analysis in order to ensure that the surface waves appear in the computed responses (e.g. Graves *et al.* 1998).

The effects of surface topography and sediment-filled valleys on site response have been summarized by Silva (1988).

21.6.6 NEAR-FAULT EFFECTS

Large velocity pulses have been noted in ground motion recordings near faults where the motions have propagated towards the recording station. The conditions that give rise to this directivity effect have been reviewed by Somerville & Graves (1996).

Somerville (1998) suggested that the directivity effects in the near-fault region of strike-slip faults may not be adequately represented in empirical ground motion attenuation relations. This is especially true of relations that use an average of the two horizontal ground motion components, since the fault normal motions may be up to 1.5 times the average motion. Data from the Kobe earthquake (Finn *et al.* 1996) confirm that potential directivity effects should be taken into consideration in estimating design loads in the near field.

21.7 Seismic Soil–Structure Interaction

Seismic soil structure interaction has the following beneficial effects: (a) the period and damping of the vibrating structure are increased; (b) energy can flow from the vibrating building back into the soil; (c) uplift of foundation slab reduces forces transmitted to building; (d) large foundation slabs can reduce the high frequency motions and hence reduce the input motions to the structure; but (e) rocking increases the relative displacements and hence can increase overturning moments (called the P–δ effect).

The NEHRP *Recommended Provisions for Seismic Regulations for New Buildings and Other Structures* proposed in March 1997 (FEMA 1997) give approximate methods for calculating the damping and period of a building on a flexible foundation soil. Based on these parameters, a reduction in the seismic base shears for the rigid base structure is allowed. The procedures are somewhat complicated and cannot be conveniently summarized here.

21.7.1 KINEMATIC AND INERTIAL SOIL–STRUCTURE INTERACTION

Soil–structure interaction may be classified into two types: kinematic and inertial interaction. Kinematic interaction describes the modification of the ground motions by the structure due to its stiffness only. The additional modifications of ground motions introduced by the accelerating mass of the structure is attributed to inertial interaction.

The effects of kinematic interaction are produced by the scattering and reflection of incident waves from the foundation. These secondary motions modify the free-field ground motions to produce resultant ground motions at foundation levels that may be significantly different from the free-field motions. The significance of kinematic interaction depends on the characteristics of the incident waves and the size, embedment and flexibility of the foundation. Kinematic interaction is generally most pronounced for the high frequency components of the motion because the associated wavelengths are comparable to the dimensions of large foundations such as rafts or slabs.

Inertial interaction describes the secondary motions generated in the foundation soils by the inertia forces generated in the structure and foundation. The inertial forces are usually more significant at the fundamental period of the structure, so inertial interaction is more pronounced in a narrow period range around the fundamental period.

Therefore, kinematic and inertial interaction modify different parts of the spectrum. Kinematic effects are pronounced at the higher frequencies and inertial interaction at the fundamental frequency with little effect on the higher frequency responses.

21.7.2 TAU METHOD FOR ASSESSING INTERACTION EFFECTS

Clough & Penzien (1993) proposed a method for estimating the effect of kinematic interaction on the input motions to the rigid basemat. The ratio of the amplitude of a harmonic component of the basemat in translation to the same harmonic of the free-field motion is given by τ, where:

$$\tau = \frac{1}{\alpha} [2(1 - \cos \alpha)]^{1/2} \qquad (21.38a)$$

where

$$\alpha = \frac{\omega D}{V_a} = \frac{2\pi D}{L} \qquad (21.38b)$$

and ω = the circular frequency of the harmonic under consideration, D = the dimension of the structure in the direction of wave propagation, V_a = the apparent wave speed in the ground, and L = the wavelength corresponding to ω. The ratio τ is plotted in Fig. 21.15.

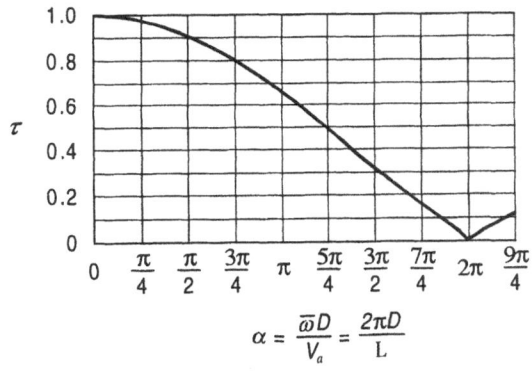

$$\alpha = \frac{\overline{\omega} D}{V_a} = \frac{2\pi D}{L}$$

FIGURE 21.15. Amplitude reduction ratio, τ, as a function of frequency and wave velocity (modified from Clough & Penzien 1993).

21.8 Seismic Design of Retaining Structures

The static design of retaining structures was discussed in Chapter 17 and reference should be made to that chapter for the basic concept of wall design. Seismic design may be based on two different approaches. The walls can be designed to resist the inertial earthquake forces or to limit displacements of the structure to tolerable levels. A basic element in both approaches is the calculation of the active or passive seismic pressures against the retaining structure during seismic excitation. These pressures are usually calculated using the Mononobe–Okabe method (Okabe 1926; Mononobe & Matsuo 1929).

In this method, the effects of earthquake shaking are simulated by applying vertical and horizontal inertia forces, $k_h W$ and $k_v W$, to the active and passive failure wedges of weight, W, behind the wall as shown in Fig. 21.16. The parameters k_h and k_v are called the seismic coefficients and are defined by $a_h = k_h g$ and $a_v = k_v g$, where a_h and a_v are horizontal and vertical accelerations. The values of k_h and k_v may be defined by codes or selected on the basis of experience. When not defined by codes, k_h is usually selected between 0.5 and 0.8 times the peak ground acceleration, a_p; k_v is often taken as 0.5–0.7 k_h. Studies of the failure of quay walls in Japan during earthquakes suggest that $k_h = 0.33$ $(a_p/g)^{1/3}$.

The Mononobe–Okabe method assumes that the wall movements are sufficient to fully mobilize the shearing resistance along the base of the failure wedge, as is the case for Coulomb's active and passive earth pressure theories (see Section 17.2). Dynamic tests on model retaining walls indicate that the movements required to develop the dynamic active earth pressure force are on the order of those movements required to develop the static active earth pressure force (Sherif *et al.* 1982).

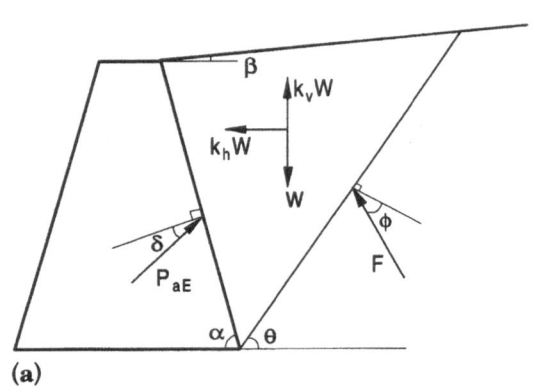

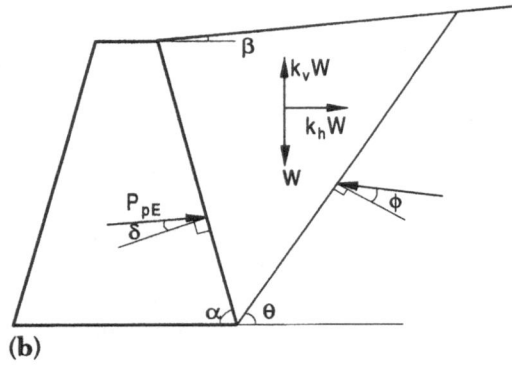

FIGURE 21.16. Forces acting on (a) active, and (b) passive failure wedges in Mononobe–Okabe analysis.

21.8.1 DYNAMIC ACTIVE AND PASSIVE EARTH PRESSURE FORCE

The Mononobe–Okabe relationship for P_{AE} for dry backfills with unit weight γ is given by:

$$P_{AE} = \frac{1}{2}\gamma H^2[K_{aE}(1 - k_v)] \quad (21.39)$$

and acts at the angle of wall friction, to the normal to the back of the wall of height H (Whitman & Christian 1990).

The dynamic active earth pressure coefficient, K_{aE}, is equal to:

$$K_{aE} = \sin^2(\phi + \alpha - \psi)/\cos\psi\sin^2\alpha\sin(\alpha - \delta - \psi)$$

$$\times\left\{1 + \left[\frac{\sin(\phi + \delta)\sin(\phi - \psi - \beta)}{\sin(\alpha - \delta - \psi)\sin(\alpha + \beta)}\right]^{1/2}\right\}^2. \quad (21.40)$$

The seismic inertia angle, ψ, represents the angle that the resultant of the gravity force and the inertial forces make with the vertical direction and is given by:

$$\psi = \tan^{-1}\left[\frac{k_h}{(1 - k_v)}\right]. \quad (21.41)$$

If the wall and soil wedge are rotated through the angle ψ, the resulting force vector representing the vectorial sums of W, k_hW and k_vW becomes vertical and the dynamic problem becomes equivalent to the static Coulomb earth pressure problem. Existing tables of lateral pressure coefficients for static loading can then be used to determine the dynamic lateral pressures. Then the active lateral dynamic force is given by:

$$P_{aE} = [K_a(\beta^\circ, \alpha^\circ) \cdot F_{aE}] \cdot \frac{1}{2}[\gamma(1 - k_v)]H^2 \quad (21.42)$$

where H = actual height of the wall, $\beta^\circ = \beta + \psi$, $\alpha^\circ = \alpha - \psi$, and ψ is computed using Eq. 21.41 and:

$$F_{aE} = \frac{\sin^2(\alpha - \psi)}{\cos\psi\sin^2\alpha}. \quad (21.43)$$

The dynamic passive earth pressure coefficient, K_{pE}, is equal to:

$$K_{pE} = \sin^2(\alpha - \phi + \psi)/\cos\psi\sin^2\alpha\sin(\alpha + \delta + \psi)$$

$$\times\left\{1 - \left[\frac{\sin(\phi + \delta)\sin(\phi - \psi + \beta)}{\sin(\alpha + \delta + \psi)\sin(\alpha + \beta)}\right]^{1/2}\right\}^2 \quad (21.44)$$

where $\psi = \tan^{-1}[k_h/(1 - k_v)]$.

The radical in the equation for K_{aE} and K_{pE} must remain positive for real solutions. Therefore, for active conditions $k_h \leq (1 - k_v)\tan(\phi + \beta)$, and for passive conditions $k_h \leq (1 - k_v)\tan(\phi + \beta)$. The vertical seismic coefficient does not have a significant effect on the seismic lateral forces for the usual ratios of k_v/k_h ranging from 0.3 to 0.7 for values of k_h up to 0.4. The error in neglecting k_v is generally less than 10%, except when $k_h > 0.5$ and $k_v/k_h > 0.5$.

21.8.2 SATURATED BACKFILL

The presence of water in the backfill behind a retaining wall has three effects: (1) it increases the total unit weight and hence the inertial forces within the backfill, (2) seismic pore-water pressures may be generated by the earthquake shaking, and (3) hydrodynamic pressures may develop independently against the wall.

The inertial forces in saturated soils depend on the relative movement between the backfill soil particles and the porewater that surrounds them. If the permeability of the soil is small enough, the porewater moves with the soil during earthquake shaking. For this restrained water condition, the inertial forces will be proportional to the total unit weight of the soil. In the case of clean gravels, where the permeability is very high, however, the porewater may act independently while the soil skeleton moves back and forth. Under this free-water condition, inertial forces from the saturated soil will be proportional to the buoyant unit weight of the soil.

In the free-water condition, hydrodynamic pressures can develop against the wall. The resultant hydrodynamic force is calculated using the equation (Westergaard 1931):

$$P_w = (7/12) \cdot k_h \gamma_w h^2 \qquad (21.45)$$

where γ_w = the unit weight of water and h = the depth of water. The force is assumed to act at a height $0.4h$. The static water force must be added to the hydrodynamic thrust to give the total water force against the wall.

The Mononobe–Okabe method can be modified for the restrained water conditions to account for the presence of porewater within the backfill (Matsuzawa *et al.* 1985). If no significant porewater pressures are developed, the soil thrust from partially submerged backfills may be computed using an average unit weight based on the relative vol-

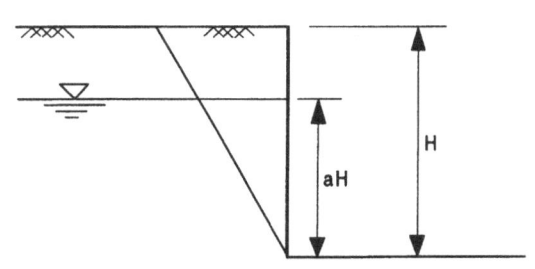

FIGURE 21.17. Geometry and notation for partially submerged backfill.

umes of soil within the active wedge that are above and below the phreatic surface:

$$\bar{\gamma} = a^2\gamma_{sat} + (1 - a^2)\gamma_m \qquad (21.46)$$

where γ_m = the unit weight of moist soil, γ_{sat} = the saturated unit weight, and a is defined in Fig. 21.17. The static water pressure must be added to the total soil thrust.

If seismic porewater pressures develop during shaking, represented on the average by the ratio, r_u, of the porewater pressure, u, to the effective overburden pressure, σ', the active soil thrust acting on a yielding wall can be computed using Eq. 21.39 with:

$$\gamma = \gamma_b(1 - r_u) \qquad (21.47)$$

where γ_b = buoyant unit weight. The seismic inertia angle, ψ, is now given by:

$$\psi = \tan^{-1}\left[\frac{\gamma_{sat}k_h}{\gamma_b(1 - r_u)(1 - k_v)}\right]. \qquad (21.48)$$

An equivalent hydrostatic thrust based on a fluid of unit weight $\gamma_{eq} = \gamma_w + r_u\gamma_b$ must be added to the soil thrust (Ebeling & Morrison 1992).

21.8.3 ANCHORED BULKHEADS

Ebeling & Morrison (1992) have proposed a seismic design procedure for anchored bulkheads based on the free earth support method and Rowe's moment reduction procedure.

The bulkhead is first designed for static

loading conditions. The active seismic force on the back of the wall and the seismic passive force in front of the wall are calculated using the Mononobe–Okabe method for the different drainage conditions described in the previous section. The required depth of wall penetration is obtained by summing moments about the tie–rod connection to the wall. The tie–rod resistance is obtained by summing all horizontal forces acting on the wall. The distribution of bending moments are then calculated and the design bending moments determined using Rowe's moment reduction factors, based on wall flexibility.

The design tie-rod force is conventionally set at a level 30% greater than the anchor resistance computed above. The anchor block is located a sufficient distance from the wall so that the active wedge developing behind the wall does not intersect the passive wedge in front of the anchor block. The active failure plane is assumed to originate at the point of rotation of the wall. The size of the block is based on horizontal force equilibrium.

Kitajima & Uwabe (1979) summarized the performance of 110 quay walls during earthquakes in Japan since 1960. Most of the walls were anchored bulkheads. They categorized the damage to anchor bulkheads as a function of the permanent horizontal displacement at the top of the sheet pile. Their study shows that the level of damage increased in proportion to the displacements greater than 10 cm. Details of this study are given by Kramer (1996).

21.8.4 REINFORCED SOIL WALLS

Seismic design and analysis of reinforced soil walls are, in many aspects, very similar to that for gravity-type retaining walls. The primary difference is that an internal seismic stability analysis should also be performed as part of the design for these walls. The internal stability analysis incorporates the effects of the inertial force generated by the reinforced soil volume on individual reinforcing elements

during an earthquake as a pseudo-static horizontal load. The FHWA has recently published design guidelines for seismic design and analysis for reinforced soil walls (Elias & Christopher 1996).

21.8.5 SEISMIC WALL DISPLACEMENTS

Richards & Elms (1979) proposed a method for the seismic design of gravity walls based on displacement criteria. The displacements are estimated using the Newmark (1965) sliding-block procedure. The forces on the wall are shown in Fig. 21.18.

The wall will begin to slide when the sum of the horizontal forces, $k_h W_w$ and $(P_{AE})_h$ exceeds the shearing resistance, T. If the coefficient of sliding friction is ϕ_s, then the resistance to sliding horizontally is given by the equation:

$$T = [W_w + P_{aE} \cos(\alpha - \delta)]\tan \phi_s. \quad (21.49)$$

The horizontal driving forces are $k_h W_w$ and $P_{aE} \sin(\alpha - \delta)$. Equating driving and resisting forces gives:

$$k_h =$$

$$\times \left[\tan \phi_s - \frac{P_{aE} \sin(\alpha - \delta) - P_{aE} \cos(\alpha - \delta) \tan \phi_s}{W_w} \right]$$

$$(21.50)$$

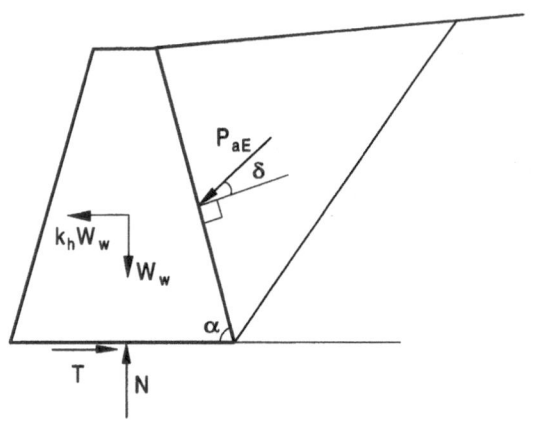

FIGURE 21.18. Static and seismic forces acting on a gravity retaining wall.

This seismic coefficient, k_h, is just sufficient to initiate sliding of the wall. The associated acceleration $a_y = k_h g$ is called the yield acceleration. Richards & Elms (1979) proposed the following equation for evaluating the permanent displacement of the wall during an earthquake with peak ground velocity v_p and peak ground acceleration a_p. The permanent displacement is given by:

$$d_{perm} = 0.087 \frac{v_p^2 a_p^3}{a_y^4}. \qquad (21.51)$$

The Newmark (1965) and Richards & Elms (1979) equations tend to provide an upper bound for the wide range of displacements computed by Franklin & Chang (1977). Kramer (1996) gives an excellent description of displacement analysis.

21.8.6 SEISMIC DESIGN FOR TOLERABLE WALL DISPLACEMENT

The first step is to calculate the yield acceleration, a_y, that will allow the tolerable deformation, d_t, to occur under ground motions with a peak velocity v_p and a peak acceleration a_p using Eq. 21.51. This gives:

$$a_y = \left(0.087 \frac{v_p^2 a_p^3}{d_t}\right)^{1/4}. \qquad (21.52)$$

The weight of wall, W_w, required to limit the permanent displacement to the tolerable value, from Eq. 21.50 is:

$$W_w =$$

$$\frac{P_{aE} \sin(\alpha - \delta) - P_{aE} \cos(\alpha - \delta) \tan \phi_s}{\tan \phi_s - k_h}. \qquad (21.53)$$

The dynamic force P_{aE} in Eq. 21.53 is computed using Eq. 21.40 with $k_h = a_y/g$. Whitman & Liao (1985) have shown that a factor of safety of 1.1–1.2 reduces the probability of exceeding the allowable tolerable displacement to less than 5%.

21.8.7 UNYIELDING WALLS (RIGID)

The Mononobe–Okabe method is not applicable to rigid immovable walls because a failure plane does not develop behind the wall. Several researchers have used elastic wave theory to derive expressions for seismic pressures against rigid walls (Matsuo & Ohara 1960; Wood 1973; Veletsos & Younan 1994). Others have used a modified shear beam model, which incorporates horizontal normal stresses (Arias et al. 1981; Finn et al. 1994; Veletsos et al. 1995; Wu & Finn 1996, 1999). For many years, Wood's solution has been the one used most in practice.

Wood provided non-dimensional static elastic solutions for lateral forces, F_{sr}, and overturning moments, M_{sr} for different L/H ratios and Poisson ratios, ν, for 1g horizontal acceleration (Fig. 21.19) (where $L = $ the distance between the wall and another boundary that is rigid relative to the backfill, and $H = $ the wall height). The point of application of F_{sr} above the base of the wall, Y_{sr}, is given by:

$$Y_{sr} = \frac{M_{sr}}{F_{sr}}. \qquad (21.54)$$

Forces and moments caused by a harmonic excitation with peak acceleration, $k_h g$, are obtained by multiplying the non-dimensional values by $\rho H^2 k_h g$ or $\gamma H^2 k_h$.

In the cases of a wide backfill and a peak horizontal acceleration, $A_{max} = k_h g$, the lateral seismic force against the wall, when $\Omega < 0.5$ is given approximately for $\nu = 0.4$ by:

$$F_{sr} = \gamma H^2 (\cdot k_h) = \rho H^2 (\cdot A_{max}) \qquad (21.55)$$

acting at a height of about $0.63H$ above the back of the wall. This is called Wood's static force (Fig. 21.21).

21.8.8 EARTH PRESSURES FOR DESIGN OF RIGID WALLS

Wu & Finn (1996) proposed an alternative analysis based on the modified shear beam model, which gives pressures within 5% of the exact modal solution proposed by Wood

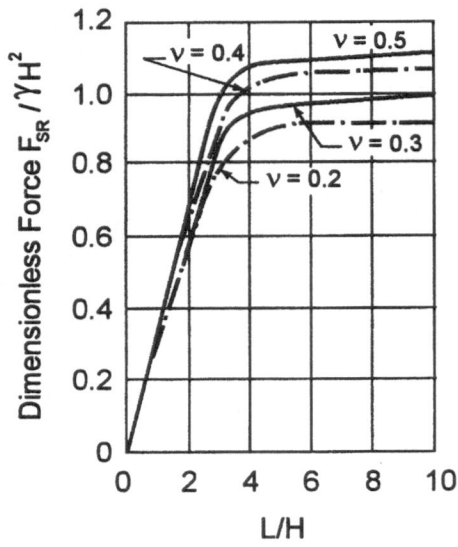 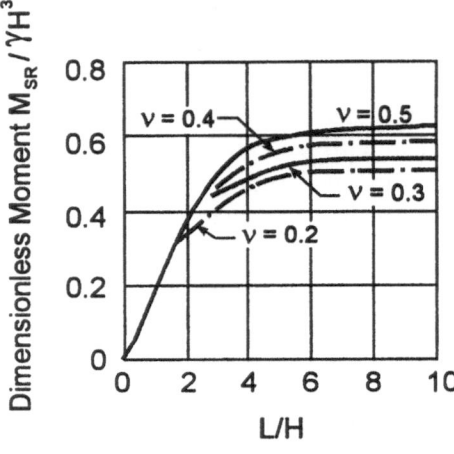

FIGURE 21.19. Non-dimensional lateral forces and moments on a rigid wall with unyielding backfill for 1-g static horizontal body forces.

(1973) for a wide range of all controlling parameters and which converges very rapidly.

Wu & Finn (1999) developed design charts for estimating the dynamic pressures against rigid walls for three types of soil profile, namely; uniform, linear and parabolic variations of shear modulus with depth. For the latter two cases, the shear moduli, G, are assumed to vary from zero at the ground surface to G_{soil} at the bottom of the soil profile. For most analyses, a Poisson's ratio of $\nu = 0.4$ and a damping ratio $\lambda = 10\%$ were used. The analyses were conducted assuming viscoelastic response. Simple corrections for nonlinear response are given later.

Peak dynamic thrusts are presented as functions of the ratio of the predominant frequency of the earthquake motions, ω, to the fundamental frequency of the wall–soil system, ω_{11}, which is established in the course of analysis and may initially be estimated from the maximum shear modulus in the backfill (Fig. 21.20).

In the presentation of the results, the computed seismic lateral forces, Q_{max}, are normalized by dividing by $\rho H^2 A_{max}$, where $\rho =$ the mass density, $H =$ the height of the wall, and $A_{max} =$ the peak-ground acceleration. Envelopes of lateral forces corresponding to upper bound and 84% levels for a parabolic variation in G are shown in Fig. 21.21 for $L/H = 5.0$ and Fig. 21.22 for $L/H = 1.5$. These charts cover the cases of relatively long backfill and short backfill, respectively. These charts give design forces at the upper bound and 84 percentile levels for the common case where the modulus varies parabolically with depth. Wu & Finn (1999) should be consulted for additional results, including the effects of variations in ν and damping ratio.

The principal findings are:

- Under seismic loads, the upper bound peak seismic thrusts, Q_{max}, against walls with semi-infinite backfills could be as high as $1.7\rho H^2 A_{max}$ for a uniform soil profile, $1.35\rho H^2 A_{max}$ for a parabolic soil profile, and $1.15\rho H^2 A_{max}$ for a linear soil profile.
- The more uniform the backfill is, the larger the seismic thrust against the wall.

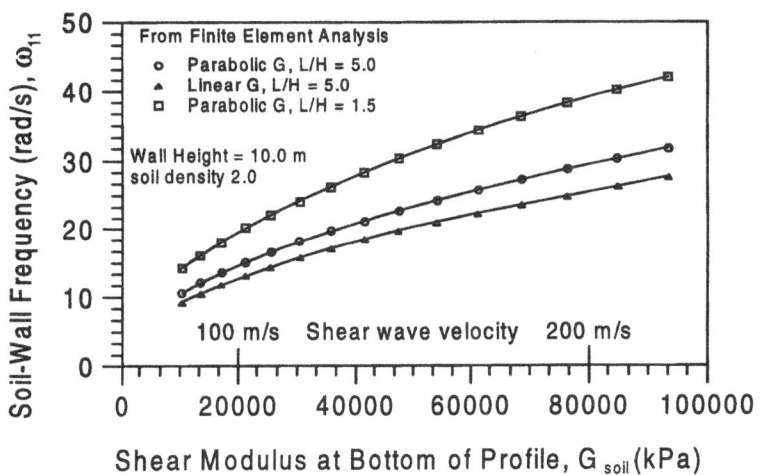

FIGURE 21.20. Chart for determining soil–wall system frequency on the basis of shear modulus: ○, parabolic G, $L/H = 5$; △, linear G, $L/H = 5$; □, parabolic G, $L/H = 1.5$. Wall height = 10 m, soil density = 2 (Mg m^{-3}).

- Amplification of seismic thrusts at resonance are much larger for backfills with $L/H = 1.5$ than for backfills with $L/H = 5.0$. The peak seismic thrust against walls with $L/H = 1.5$ reaches $1.7\rho H^2 A_{max}$ because of this amplification.
- The height of the resultant seismic thrust

above the base of the wall varies from $0.64H$ for the uniform G profile to approximately $0.5H$ for the linear and parabolic G profiles.

- The static lateral pressures, typically at rest pressures, must be added to the dynamic pressures for design.

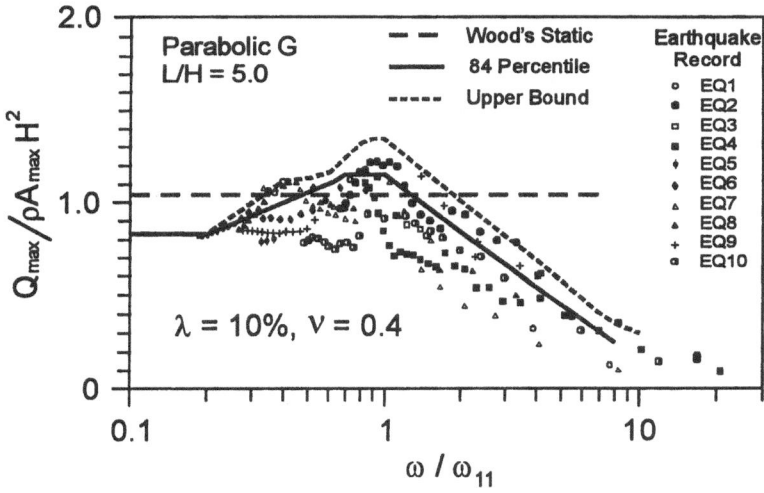

FIGURE 21.21. Peak seismic thrusts for soil profiles with parabolic variation in G ($L/H = 5.0$, $\lambda = 10\%$, $\nu = 0.40$) (modified from Wu & Finn 1999): – –, Wood's (1973) static; —, 84 percentile; ---, upper bound.

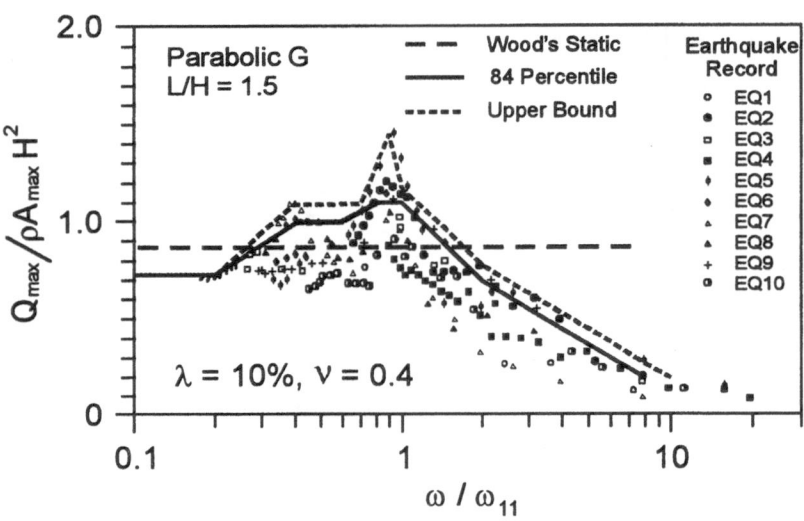

FIGURE 21.22. Peak seismic thrusts for soil profiles with parabolic variation in G ($L/H = 1.5$, $\lambda = 10\%$, $\nu = 0.40$) (modified from Wu & Finn 1999): – –, Wood's (1973) static; —, 84 percentile; ---, upper bound.

21.8.9 Practical Considerations

The shear modulus can be estimated from shear-wave velocity measurements made by seismic cone-penetration tests or from correlations with cone-bearing values, q_c, or standard penetration blow counts, N (see Chapter 4). During strong shaking, the modulus will be reduced because the modulus of soil is strain-dependent. The reduction in modulus for strong shaking can be estimated approximately on the basis of judgment or by a one-dimensional response analysis of the backfill using the program SHAKE (Schnabel *et al.* 1972). The modulus reduction table in NEHRP (FEMA 1997) is also useful (Table 21.6). In cases where the distribution of shear moduli with depth does not follow one of the profiles above, it is suggested that an average

TABLE 21.6. Values of G/G_o and V_s/V_{so}

	Ground acceleration coefficient, A_v			
	≤ 0.10	≤ 0.15	≤ 0.20	≥ 0.30
G/G_o	0.81	0.64	0.49	0.42
V_s/V_{so}	0.90	0.80	0.70	0.65

shear modulus weighted with respect to depth be used. In many cases involving basement walls, the elastic analysis is likely to be accurate enough except for very strong shaking. The advice of the geotechnical engineer in establishing the frequency of the soil–wall system is recommended.

21.9 Seismic Design of Foundations

21.9.1 ELEMENTS OF SEISMIC DESIGN OF FOUNDATIONS

Very few foundation failures have occurred during earthquakes except in cases where liquefaction occurred or sensitive clays lost their strength and stiffness under cyclic loading. The foundation failures in Mexico City during the 1985 Michoacàn–Guerrero earthquake provide some of the few examples of failure under earthquake loading that could not be attributed to liquefaction. In this earthquake, foundations with low static safety factors or significant load eccentricities behaved poorly. However, even in these cases,

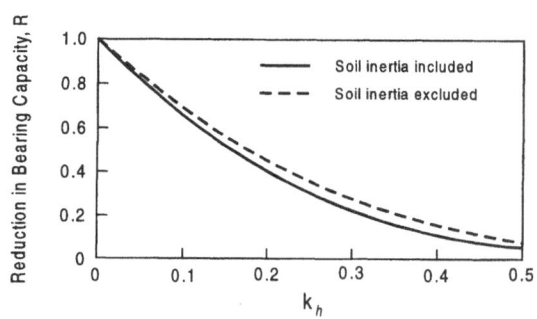

FIGURE 21.23. Reduction in bearing capacity, R, under seismic loading conditions with (---) and without (—) the inclusion of soil inertia effects as a function of the horizontal seismic coefficient, k_h (modified from Pecker 1996).

there is a possibility that the soil strength was also degraded (Girault 1986).

Eurocode 8, Part V (CEN 1994) deals explicitly with the effects of these seismic forces and states that in calculating the bearing capacity of a shallow foundation under earthquake loading, one should "include the load inclination and eccentricity arising from the inertial forces of the structure as well as the possible effects of the inertia in the soil".

Bearing capacity formulae are developed by analyzing the stability of a mass of soil moving under the footing load along slip surfaces. If the inertia forces acting on this soil mass are included, the effect is to generate additional shear stresses, which reduce the bearing capacity. Recent studies (Paolucci & Pecker 1997; Shi & Richards 1995) show that the reduction in bearing capacity due to soil inertia is not more than 15–20% for $k_h \leq 0.3$ (Fig. 21.23). The major reduction in bearing capacity comes from load inclination and the eccentricity of the resultant load on the foundation.

21.9.2 SEISMIC DESIGN REQUIREMENTS FOR FOUNDATIONS

A basic assumption of seismic design of foundations is that the foundation soils do not liquefy during the earthquake or undergo significant degradation of strength and stiffness. If these conditions are likely to occur, then a decision must be made to either: (a) improve the soil conditions; or (b) develop a design that can cope with these conditions.

There are two distinct requirements for the seismic design of foundations on stable soils. First, the overturning moment taken about the toe of the footing should be balanced by the dead load of the building with an adequate factor of safety. Second, one must ensure that the stresses generated in the foundation soils by the seismic moments do not exceed the specified design stresses based on the overall stability of the foundation nor the specified stresses to ensure that the settlements remain within tolerable limits.

21.9.3 DESIGN PRESSURE BASED ON BEARING CAPACITY

The bearing capacity of a footing loaded eccentrically by an inclined resultant force is considered to be the same as the bearing capacity of the reduced effective footing shown in Fig. 21.24, when loaded centrally by the same inclined force. The dimensions of the reduced area, B' and L', in terms of the plan dimensions of the actual footing (B and L) and the eccentricity, e, are width $B' = B - 2e$, and length $L' = L - 2e$. The bearing capacity of the effective footing may be calculated using a bearing capacity formula for

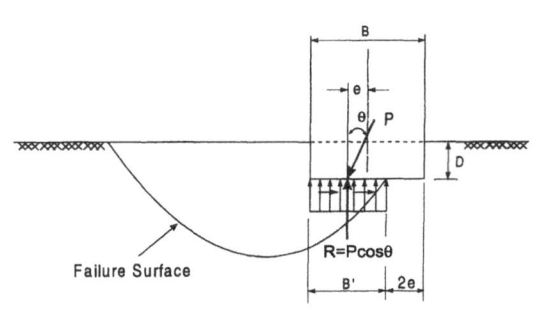

FIGURE 21.24. Equivalent footing width for footing subjected to eccentric and inclined loading at failure (modified from CGS 1992).

concentric inclined loading (see Section 9.3; CGS 1992; Meyerhof 1963; Bowles 1996). Because of the short-term application of the seismic loads, a smaller factor of safety can be adopted for seismic design of foundations.

21.9.4 DESIGN LOADS

The seismic design of structures according to the provisions in building codes are now based on capacity design in many jurisdictions, particularly North America, the EU and New Zealand. The design loads may be less than the loads that initiate yielding in the constructed structure for the following reasons. The yield moment used in design is taken as a fraction of the yield moment of the material. For steel, the fraction is typically 0.85–0.90. The yield stress is a specified stress, but the yield stress of the steel supplied by the manufacturer is usually greater than this to avoid problems of meeting the specifications with prescribed reliability. The yield stress supplied may be on the average 10% higher than the specified yield stress. Finally, the practical requirements of design and construction may require somewhat larger members and somewhat greater reinforcement than calculated for design. These effects result in a structure that yields at loads that may be in the range of 1.5–2.8 times the design loads, depending on the code used in design.

It is very important to take the overstrength factor into account when specifying the design loads for the foundation (CEN 1994; NRCC 1995a,b; ICBO 1997). Not only because the loads are greater, but because under seismic loading a significant increase is allowed in the bearing capacity of the footing, which reduces the factor of safety.

21.10 Liquefaction

21.10.1 KEY FACTORS

The primary factors controlling the liquefaction of saturated cohesionless soils are the in-

tensity and duration of earthquake shaking, and the density of the soil. In engineering practice, the potential for liquefaction is assessed usually by using liquefaction assessment charts where the *in situ* compactness of the soil is characterized by the standard penetration resistance (SPT), N, cone penetration resistance (CPT), q_c, the Becker penetration resistance or *in situ* shear-wave velocity, V_s (see Chapter 4 for details regarding these tests). The standard charts are based on a magnitude 7.5 earthquake. The intensity of earthquake shaking is usually represented by the effective average cyclic stress ratio (CSR).

Seed & Idriss (1971) formulated the following equation for calculation of CSR:

$$CSR = (\tau_{av}/\sigma'_{v0})$$
$$= 0.65(a_{max}/g)(\sigma_{v0}/\sigma'_{v0})r_d \qquad (21.56)$$

where a_{max} = the peak horizontal acceleration at ground surface generated by the earthquake; g = the acceleration of gravity; σ_{v0} and σ'_{v0} = total and effective vertical overburden stresses, respectively; and r_d = a stress reduction coefficient (see NCEER 1997 for the latest values for r_d). The latter coefficient provides an approximate correction for flexibility of the soil profile. An approximate expression for r_d used in Japan is:

$$r_d = 1.0 - 0.015H \qquad (21.57)$$

where H = the depth below the surface in meters of the location where the stress is being evaluated.

21.10.2 SPT(N) LIQUEFACTION ASSESSMENT CHART

Seed and his colleagues (Seed 1979a; Seed *et al.* 1985) developed correlations between the SPT(N)-value and the cyclic stress ratio to cause liquefaction (CSR_{crit}) during earthquakes of magnitude $M = 7.5$. The correlations, which are presented in Fig. 21.25, were based on the observed response of sites during earthquake loading. Sites were consid-

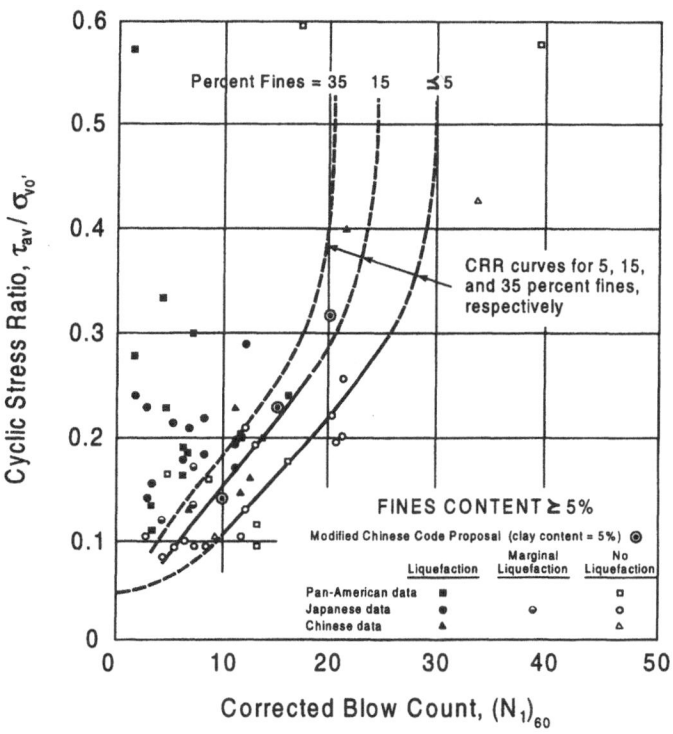

FIGURE 21.25. Simplified base curve recommended for calculation of CRR from SPT data, along with empirical liquefaction data (modified from Seed *et al.* 1985).

ered to have liquefied based on observed surface features, such as sand boils. Lower bound curves, separating liquefied from non-liquefied sites, are shown in Fig. 21.25, corresponding to various fine contents of the sand.

To compare the ground conditions at one site with those of another, it is necessary to standardize the measured penetration values to a standard driving energy and effective overburden pressure. Seed normalized the SPT to an energy level of 60% of the free-fall potential energy of the hammer and an effective overburden pressure of 100 kPa. Hence, the correlations presented in Fig. 21.25 show the normalized SPT(N) value, i.e. $(N_1)_{60}$.

The correction for energy level may be made using the data in Table 21.7 or by direct measurement of the energy imparted to the rods by the falling hammer.

A commonly used correction factor, C_N, for overburden pressure is that proposed by Liao and Whitman (1986):

$$C_N = (P_a/\sigma'_{v0})^{0.5} \qquad (21.58)$$

where σ'_{v0} is expressed in the same units as standard atmospheric pressure, P_a. $(N)_{60}$ at σ'_{v0} is corrected to $(N_1)_{60}$, i.e. to $(N)_{60}$ at 100 kPa nominal pressure by:

$$(N_1)_{60} = C_N(N)_{60}. \qquad (21.59)$$

Judgment is required in limiting the maximum value of the correction factor. Generally, the upper limit is taken in the range 1.5–2.0.

The chart in Fig. 21.25 is based on a magnitude $M = 7.5$, an effective overburden pressure of 100 kPa and level ground conditions. The last condition means that there are

TABLE 21.7. Summary of energy ratios for SPT procedures (adapted from Seed *et al.* 1985)

Country	Hammer type	Hammer release	Estimated rod energy (%)	Correction factor for 60% rod energy
Japan[a]	Donut	Free-fall	78	78/60 = 1.30
	Donut	Rope and pulley with special throw release	67	67/60 = 1.12
USA	Safety	Rope and pulley	60	60/60 = 1.00
	Donut[b]	Rope and pulley	45	45/60 = 0.75
Argentina	Donut	Rope and pulley	45	45/60 = 0.75
China	Donut	Free-fall[c]	60	60/60 = 1.00
	Donut	Rope and pulley	50	50/60 = 0.83

[a] Japanese SPT results have additional corrections for borehole diameter and frequency effects.
[b] Prevalent method in the USA today.
[c] Pilcon-type hammers develop an energy ratio of about 60%.

no static shear stresses on potential failure planes. The popularity of the Seed & Idriss (1982) method for assessing liquefaction resulted in it being applied outside the conditions on which the chart in Fig. 21.25 was based. Especially, it began to be applied to embankment dams and their foundations. In these situations, overburden pressures are much higher and large static shear stresses exist on potential failure planes. These applications challenged engineers and researchers to adapt the basic liquefaction resistance curve to these changed conditions and to other earthquake magnitudes. This adaptation resulted in corrections being applied to the critical cyclic resistance ratio (CCRR) read from the chart at a specified $(N_1)_{60}$ by means of correction factors K_m for magnitude, K_σ for pressure effects, and K_α for static shear effects. The corrected critical cyclic resistance ratio, CCRR (τ_{av}/σ'_0), corresponding to $(N_1)_{60}$, is then given by:

$$\text{CCRR}_{\text{CORR}} = \text{CCRR} \times K_m \times K_\sigma \times K_\alpha. \quad (21.60)$$

The development of these factors is reviewed here and values recommended by NCEER (1997) for use in practice, are given.

The state-of-the-art for evaluating liquefaction was first reviewed in 1985 by a committee of the US National Research Council (NRC

1985). It was reviewed again by a committee set up by the National Center for Earthquake Engineering Research (NCEER) in 1997 (NCEER 1997). The procedures described here reflect the current state of the document, although some further revisions may occur in 1999. This document defines the current state of knowledge and is likely to be the reference document for practice.

For some time, there has been concern about the fact that the liquefaction resistance curve is aligned through the origin. After reviewing the available data, the NCEER committee proposed a modification to the basic Seed curve in Fig. 21.25. The modification, shown by the dotted curve, is tangent to the old curve at $N = 10$ and cuts the vertical axis at a stress ratio of $\tau/\sigma'_0 = 0.05$.

21.10.3 MAGNITUDE SCALING FACTOR, K_m

A range of magnitude scaling factors are presented. These differ from the factors suggested by Seed & Idriss (1982), which are widely used in practice, in assigning higher scaling factors to magnitudes less than 7.5 to reflect the observed field data on liquefaction occurrence at the lower magnitudes. NCEER (1997) recommends the scaling factor K_m: developed by Idriss (NCEER 1997)

$$K_m = 10^{2.24}/M^{2.56}. \quad (21.61)$$

21.10.4 OVERBURDEN PRESSURE SCALING FACTOR, K_σ

A correction for overburden pressure must be made to the critical cyclic resistance ratio associated with the corrected $(N_1)_{60}$ value by the correction factor K_σ. K_σ is defined here as the cyclic resistance ratio of a soil at $\sigma'_{v0}(\sigma'_{v0} > 100$ kPa) divided by the ratio at $\sigma'_{v0} = 100$ kPa. The cyclic resistance is defined at 100% porewater pressure or 5% axial strain in cyclic triaxial tests. Values for K_σ used widely in practice are taken from the average curve proposed by Seed & Harder (1990). The data represent K_σ values from a wide variety of sand types and geological environments. Therefore, there is a wide scatter in the data about the average curve.

Since 1990, a lot of new data have become available on K_σ, up to confining pressures as high as 2500 kPa. Important sources of new data are given in Arango (1996), Vaid & Thomas (1994), and Pillai & Byrne (1994). After reviewing all these data, NCEER (1997) recommended the curve shown in Fig. 21.26 for taking into account the effects of effective overburden pressure on liquefaction potential. The committee are currently considering providing three correction curves, with the curve in Fig. 21.26 being applicable to medium-dense sand. A higher curve would be recommended for loose sand, a lower curve for dense sand.

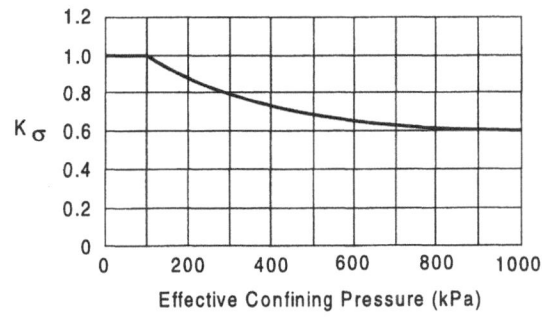

FIGURE 21.26. Minimum values for K_σ recommended for clean and silty sands and gravels (modified from NCEER 1997).

21.10.5 CORRECTION FACTORS FOR STATIC SHEAR, K_α

The Seed *et al.* (1985) liquefaction assessment chart is based on data from level or nearly level ground. There are no initial static shear stresses on potential failure planes. When the chart is used to evaluate the liquefaction potential of earth dams or their foundations, potential failure planes may carry large initial static shear stresses. Correction factors, K_α, have been developed to correct the critical cyclic resistance ratio for the effects of static shear. The K_α factors currently used in practice are those recommended by Seed & Harder (1990).

Harder & Boulanger (1997) have suggested a revised version of the Seed & Harder (1990) chart, which has a much narrower range of K_α values. For this chart, a triggering criterion of 3% shear strain was adopted (Harder & Boulanger 1997). Harder & Boulanger were influenced in selecting this criterion by the results of their tests involving static shear, both parallel and perpendicular, to the direction of shaking. They found that the influence of the direction of static shear was not very significant if a 3% triggering criterion was adopted. The correction factor recommended by Harder & Boulanger (1997) is shown in Fig. 21.27. The correction factor applies for effective overburden pressures less than 300 kPa.

The NCEER committee are still tentative about recommending values of K_α for use in practice. It would be better to incorporate the effect of static shear by conducting tests directly on high quality samples to obtain a combined factor $K_{\sigma\alpha}$, which will reflect the coupled effects of confining pressure and static shear on cyclic resistance.

21.10.6 CORE-PENETRATION TEST (CPT)

The liquefaction assessment chart based on the normalized CPT tip resistance, q_{c1}, recommended in NCEER (1997), is shown in Fig. 21.28. The chart is valid for an earth-

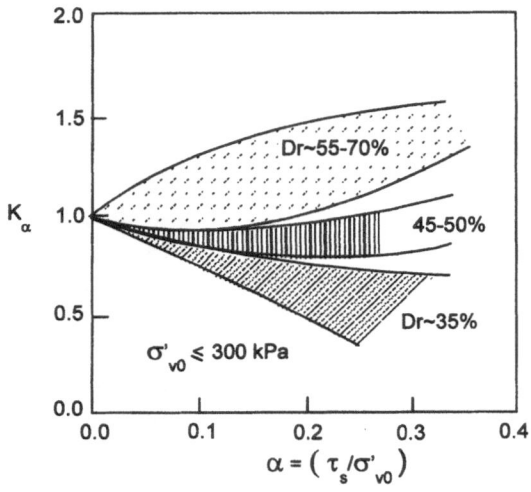

FIGURE 21.27. Correction factors to CRR as a function of the initial static shear stress ratio, τ_{st}/σ'_{v0}, and relative density (modified from Harder & Boulanger 1997).

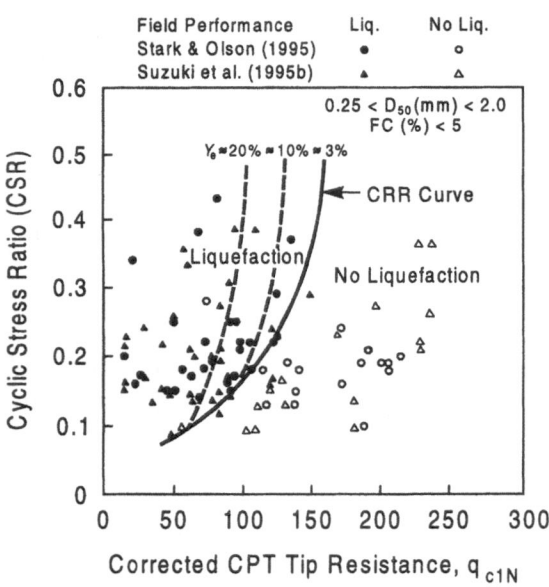

FIGURE 21.28. Curve recommended for calculation of CRR for earthquake magnitude $M = 7.5$ from CPT data along with empirical liquefaction data (modified from NCEER 1997). Liquefaction data: ● (Stark & Olson 1995), ▲ (Suzuki *et al.* 1995). No liquefaction: ○ (Stark & Olson 1995), △ (Suzuki *et al.* 1995).

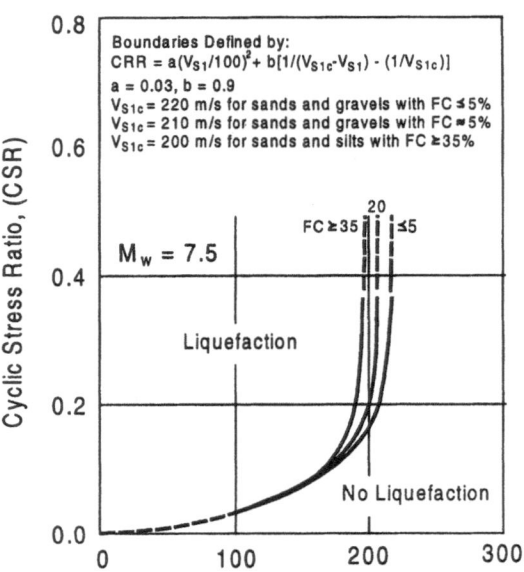

FIGURE 21.29. Curves recommended by workshop for calculation of CRR from corrected shear-wave velocity (modified from NCEER 1997). Boundaries defined by: $CRR = a \, (V_{sl}/100)^2 + b \, [1/(V_{slc} - V_{sl}) - (1/V_{slc})]$, $a = 0.03$ and $b = 0.9$; $V_{slc} = 220$, 210 and 200 ms^{-1} for sands and gravels with FC ≤ 5, ≈ 5 and $\geq 35\%$, respectively.

quake with a magnitude $M = 7.5$. The correction factors used on the SPT data are also used on the CPT data.

21.10.7 SHEAR-WAVE VELOCITY

The liquefaction assessment chart, based on average shear-wave velocity normalized to a vertical confining pressure of 100 kPa, is shown in Fig. 21.29 (NCEER 1997). The chart is valid for uncemented Holocene age soils and an earthquake magnitude, $M = 7.5$. The modified shear-wave velocity, V_{sl}, is defined as:

$$V_{sl} = V_s(P_a/\sigma'_{v0})^{0.25}. \qquad (21.62)$$

The other correction factors used with the SPT chart are also applied here.

21.10.8 BECKER PENETRATION TEST (BPT)

In the BPT, a double-walled casing is driven into the ground with a diesel pile hammer (Section 4.4.6.4). The test is standardized by reducing the blow counts to constant standard combustion conditions, following procedures established by Harder & Seed (1986). The standardized Becker blowcounts are then converted to SPT blowcounts, N. This is done on the basis of correlations developed by Harder & Seed (1986) on the basis of BPT and SPT tests run side by side to depths of less than 10 m. The Becker test has been used in evaluating gravelly soils in dam foundations where the penetration depths are far deeper than those for which correlations are available.

There is hardly any direct data linking the Becker blowcount directly to field liquefaction. Even if friction is accounted for properly (Sy & Campanella 1994), there is still concern whether correlations established in sand can be applied to the Becker casing when it is driven through coarse gravel. There is as yet no fully accepted solution to the friction problem associated with deep penetrations or driving through dense material after remediation.

21.10.9 SEISMIC SETTLEMENT

Both saturated and unsaturated sands tend to settle during earthquake shaking. Settlements in unsaturated sands are unlikely to be very significant unless the sand is loose. Large settlements are typical in loose saturated sands because of the development of high porewater pressures and even full liquefaction. Settlements in saturated sands may be estimated using the chart developed by Tokimatsu & Seed (1987) shown in Fig. 21.30. This figure gives the volumetric strains for an earthquake with reference magnitude $M_w = 7.5$ as a function of the normalized standard penetration resistance $(N_1)_{60}$ and cyclic stress ratio. The cyclic stress ratio is computed using Eq. 21.56. The volumetric strains corresponding to other earthquake magnitudes

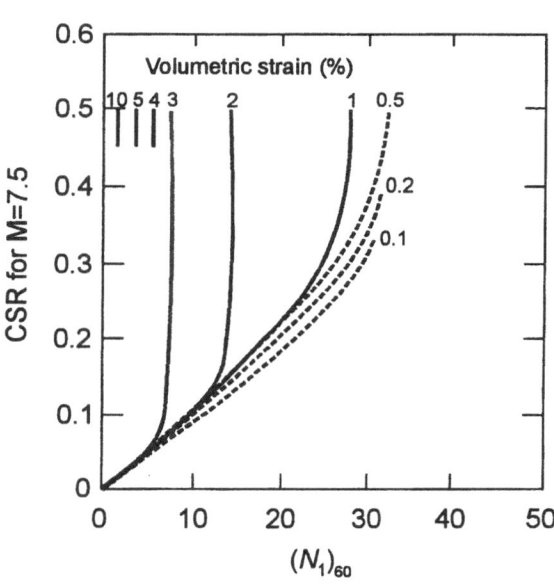

FIGURE 21.30. Post-liquefaction volumetric strains as a function of normalized standard penetration resistance and average cyclic stress ratio for $M_w = 7.5$ earthquakes (modified from Tokimatsu & Seed 1987).

may be calculated using the magnitude scaling factors for volumetric strain, given in Table 21.8. In order to calculate the seismic settlements, the site is divided into layers of roughly constant $(N_1)_{60}$ values. The seismic settlement of each layer is obtained by multiplying the layer thickness by the estimated volumetric strain for that layer. The layer set-

TABLE 21.8. Influence of earthquake magnitude on volumetric strain ratio for dry sands (adapted from Tokimatsu & Seed 1987)

Earthquake magnitude	Number of representative cycles at $0.65\tau_{max}$	Volumetric strain ratio, $\varepsilon_{C,N}/\varepsilon_{C,N=15}$
8.5	26	1.25
7.5	15	1.0
6.75	10	0.85
6	5	0.6
5.25	2–3	0.4

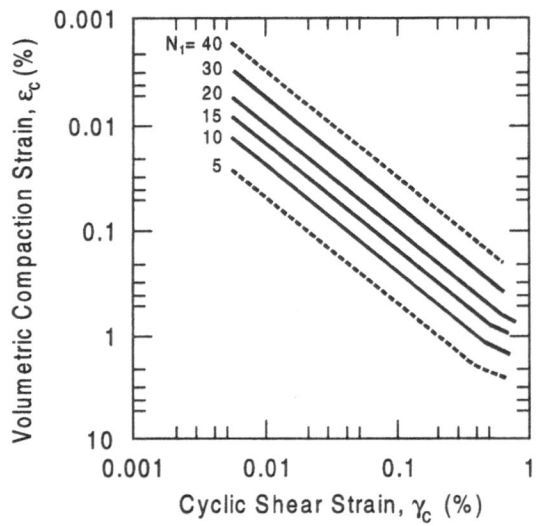

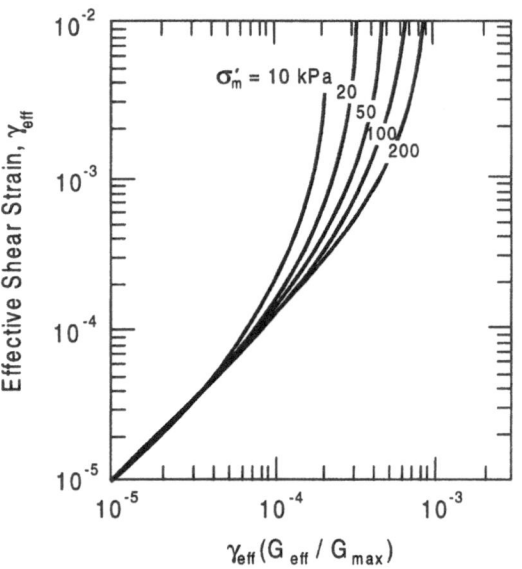

FIGURE 21.31. Relationship between volumetric strains in unsaturated sands as a function of cyclic shear strain and penetration resistance for an earthquake magnitude $M_w = 7.5$ (modified from Tokimatsu & Seed 1987).

FIGURE 21.32. Chart for determining the effective shear strain in a soil layer during earthquake shaking (modified from Tokimatsu & Seed 1987).

tlements are summed to obtain the total seismic settlement for the site.

The settlements in unsaturated sands are estimated using the chart by Tokimatsu & Seed (1987) shown in Fig. 21.31. In this chart (developed for $M = 7.5$), the volumetric strain is given as a function of the normalized standard penetration resistance, N_1, and the cyclic shear strain γ_c (%). The cyclic shear strain, γ_c, is taken to be the average effective shear strain during shaking, γ_{eff}. The effective strain, γ_{eff}, is obtained from Fig. 21.32. The parameter $\gamma_{eff}(G_{eff}/G_{max})$ in Fig. 21.32 is given by Eq. 21.63, which is a variation of Eq. 21.56:

$$\gamma_{eff}(G_{eff}/G_{max}) = (0.65 a_{max}\sigma_v r_d)/(g G_{max}) \quad (21.63)$$

where G_{max} = the shear modulus of the soil at small strain. The term $\gamma_{eff}(G_{eff}/G_{max})$ = the effective average cyclic shear stress in Eq. 21.56 divided by G_{max}. G_{max} (in kPa) can be

estimated from the correlation given by Seed & Idriss (1970).

$$G_{max} = 4400[(N_1)_{60}]^{1/3}(\sigma'_m)^{1/2} \quad (21.64)$$

where σ'_m = mean normal effective stress in kPa.

The volumetric strains at an earthquake magnitude other than $M_w = 7.5$ may be found using the strain ratio factors in Table 21.8.

The procedure described above takes only one horizontal component of motion into account. Multiply the volumetric strain due to compaction for each layer by two to correct for multidirectional shaking effects, as recommended by Tokimatsu & Seed (1987), to get the representative volumetric strain for each layer.

21.10.10 EFFECTS OF GROUND IMPROVEMENT

Yasuda et al. (1996) conducted post-earthquake settlement measurements on Port and

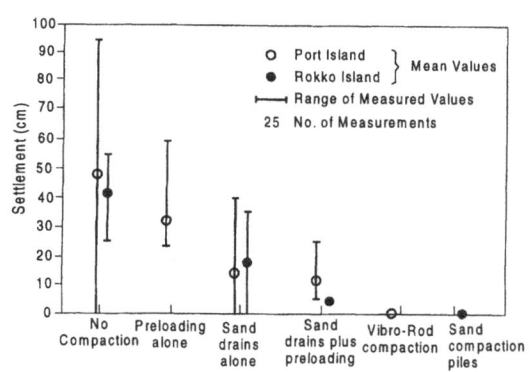

FIGURE 21.33. Effect of ground improvement techniques on settlement (modified from Yasuda *et al.* 1996). Clear values for: ○, Port Island: ●, Rokko Island. —, range of measured values. Total number of measurements = 25.

Rokko Islands to establish the effectiveness of the different remediation methods for improving the seismic performance of the site during the Kobe earthquake in 1995. The results of their survey are summarized in Fig. 21.33, which shows the variation in settlement for the different remediation techniques.

Sand drains were widely used in conjunction with preloading to accelerate the settlement of the alluvium clay stratum before construction on Port and Rokko Islands. The sand drains were installed by first driving closed pipes through the fill into the clay deposit. This produced some densification of the fill. The densification increased the resistance to liquefaction. At a number of sites, the fill was densified using a vibro-rod or sand compaction piles (see Section 15.2 for a description of foundation soil improvement techniques).

21.11 Residual Strength and Post-Liquefaction Deformations

The post-liquefaction behavior of earth structures such as dams or slopes should be as-

sessed using both limiting equilibrium analysis and deformation analysis. Limiting equilibrium is a useful screening tool for potential instability problems, but the extent and location of remediation should be assessed primarily on the basis of deformation criteria. For many dams, especially those with substantial freeboard, criteria based on factor of safety alone can result in unnecessary remediation costs.

Large displacement post-liquefaction analysis has been used in the evaluation of the consequences of liquefaction and the design of remediation measures since 1988. It has been the subject of several reviews (Finn 1993, 1999; Ledbetter & Finn 1993), which may be consulted for background theory and case histories. The post-liquefaction behavior of liquefied soils is controlled by the residual strength and the strain level required to reach it.

21.11.1 FACTORS CONTROLLING RESIDUAL STRENGTH

Recent research (to be summarized below) suggests that the residual strength measured within the strain capacity of laboratory equipment is a function of: (a) sample preparation technique, (b) stress path followed during loading, and (c) effective confining pressure.

21.11.1.1 Effect of Sample Preparation

Many laboratory studies of liquefaction use samples prepared by moist tamping, air pluviation or pluviation under water. Moist tamped samples can be formed at void ratios unrepresentative of field conditions and are less uniform than pluviated samples (Vaid & Negussey 1988). Vaid *et al.* (1999) have demonstrated that the residual strengths measured on samples prepared in different ways are quite different (Fig. 21.34). Vaid *et al.* (1999) have also shown that by using pluviation in water it was possible to reconstitute samples which gave residual strengths very

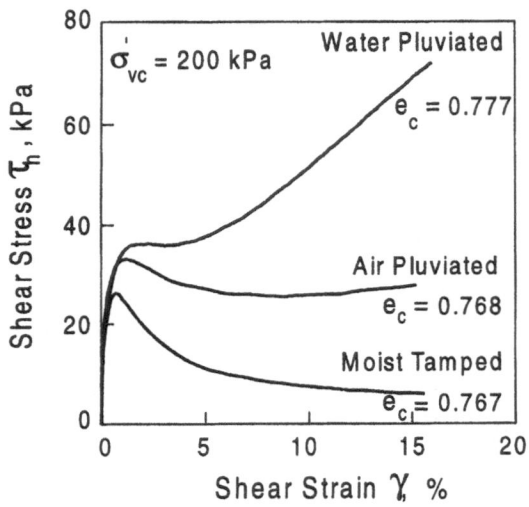

FIGURE 21.34. The effect of sample preparation on undrained simple shear response of Syncrude sand (modified from Vaid *et al.* 1999).

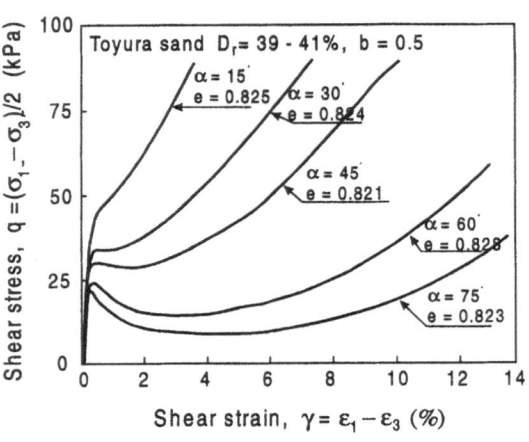

FIGURE 21.35. Effect of stress path on undrained behavior of Toyoura sand (modified from Yoshimine *et al.* 1998). $D_r = 39$–41%, $b = 0.5$.

similar to those obtained from the original undisturbed samples of recent sands.

These results provide a strong argument for using pluviation under water to form representative samples of soils that were originally deposited under water or were placed by hydraulic fill construction. The moist tamping method would seem to be more appropriate for unsaturated compacted soils.

21.11.1.2 Stress Path

Vaid & Chern (1985) were among the first to draw attention to the fact that the residual strength measured in extension was much smaller than the strength in compression and that sands in a given state were much more contractive in extension than in compression. These samples showed remarkable uniformity before and after testing and the test results should not be dismissed as being due to non-uniformity of the test specimens in extension tests and the development of necking at large strains.

Uthayakumar & Vaid (1998) and Yoshimine *et al.* (1998) have explored the effects

of stress path on residual strength over a wide range of stress paths defined by α, the inclination of the principal stress to the vertical axes of the sample, and the parameter $b = (\sigma_2 - \sigma_3)/(\sigma_1 - \sigma_3)$, which is a measure of the intermediate principal stress. The samples were tested using the hollow cylinder torsional shear device. Typical examples of this kind of data (Yoshimine *et al.* 1998) are shown in Fig. 21.35. These results suggest that different residual strengths should be assigned to different parts of the liquefied region depending on the predominant stress conditions. This selective use of shear strength for design is not new. Bearing capacity under offshore structures in the North Sea is evaluated using compression, simple shear and extension strength data to suit stress conditions at different locations along potential sliding surfaces.

At some loose densities, sands may not show contractive behavior in triaxial compression tests. This is especially true for angular sands. Therefore, design decisions based only on compressive undrained tests on loose sands may be potentially unconservative. Tests conducted by Vaid *et al.* (1998) suggest that these sands may not become compres-

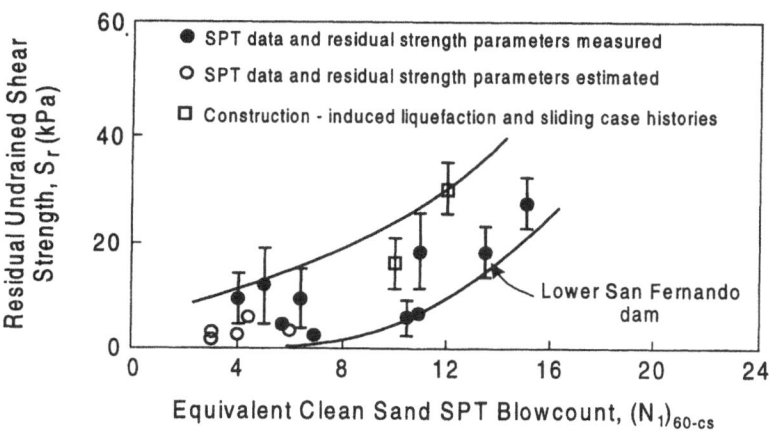

FIGURE 21.36. Correlation of residual strength, S_r, with $(N_1)_{60}$ (modified from Seed & Harder 1990; 100 kPa = 2000 psf). ●, measured SPT data; ○, estimated SPT data; □, construction-induced liquefaction and sliding case histories.

sive until very high confining pressures are used, approaching 1000 kPa.

21.11.1.3 Residual Strength as a Function of Effective Confining Pressure

The concept of expressing the residual strength as a fraction of the effective confining pressure has been used in engineering practice on several water-retaining and tailings dams, including the Sardis Dam in 1989 (Finn *et al.* 1991). For the most part, the ratio selected has been between 0.06 and 0.1. Similar results have been reported by Baziar & Dobry (1995) and by Ishihara (1993). A value of $S_r/p' = 0.23$ was used in the analysis of Duncan Dam, based on extensive testing of frozen samples (Byrne *et al.* 1994). Laboratory justification of this concept has been provided by Vaid & Sivathayalan (1996) and Vaid & Thomas (1994). The triaxial extension tests conducted by Vaid & Thomas (1994) show an almost linear dependence of the normalized residual strength on the void ratio, with the normalized residual strength varying from 0.05 to 0.20 as the void ratio changes from 0.95 to 0.82. This covers the range of normalized strengths found by other investigators.

Seed (1987) developed a correlation between $(N_1)_{60}$ and the undrained residual strength, S_r, from the back analysis of past flow slides. An updated version of this correlation was proposed by Seed & Harder (1990), which is shown in Fig. 21.36. Lower bound strengths from this correlation are very often used in practice, although occasionally values approaching the 33rd percentile have been used. In either case, these values are small at low $(N_1)_{60}$ values and frequently result in the prediction of instability and the need for substantial remediation.

21.11.2 POST-LIQUEFACTION DEFORMATIONS

Large deformation analysis to evaluate post-liquefaction deformations was first used in 1989 on the Sardis Dam (Finn 1990; Finn *et al.* 1991). The analysis, incorporated in the finite element computer program TARA-3FL (Finn & Yogendrakumar 1989) is based on the undrained stress–strain relations of the liquefied materials. The analysis considers the change in shear strength due to liquefaction and models the progressive deformation until equilibrium is reached.

Since the deformations may become large,

it is necessary to update progressively the finite element mesh. Each calculation of incremental deformation is based on the current shape of the dam, not the initial shape as in conventional finite element analysis. An independent assessment of the equilibrium of the final position should be conducted using a conventional static stability analysis. The factor of safety calculated in this way should be unity or greater, depending on whether the deformations occurred relatively slowly after the earthquake or during it when inertia forces were acting.

21.12 Embankment Dams

21.12.1 EQUIVALENT LINEAR ANALYSIS

The dynamic response of an earth dam is usually computed in engineering practice using an equivalent linear (EQL) method of two-dimensional analysis, such as that incorporated in the computer programs QUAD-4 (Idriss *et al.* 1973) or FLUSH (Lysmer *et al.* 1975). The results may be corrected approximately for three-dimensional effects (Mejia & Seed 1983). These corrections were used in the back analyses of Oroville Dam to the 1975 earthquake (Vrymoed 1975). The corrections are based on altering the shear modulus in the two-dimensional analysis so that the fundamental two-dimensional period matches the equivalent three-dimensional elastic period. Dakoulas & Gazetas (1986) have studied this problem further. Despite matching the fundamental period, the contributions of higher harmonics may be substantially underestimated. Therefore, assessing the seismic response of embankment dams in narrow valleys requires the exercise of engineering judgment, since the higher harmonics are likely to have their greatest effect at the crest of the dam.

The EQL analyses are conducted in terms of total stresses and so the effects of seismically induced porewater pressures are not reflected in the computed stresses and accelerations. Also, since the analyses are elastic,

they cannot predict the permanent deformations directly. Therefore, equivalent linear methods are used only to get the distribution of accelerations and shear stresses in the dam. Semi-empirical methods are used to estimate the permanent deformations and pore-water pressures using the acceleration and stress data from the equivalent linear analyses (Seed *et al.* 1975). Finn (1993) has presented a detailed review of these methods.

21.12.2 DEFORMATIONS FROM ACCELERATION DATA

Various potential sliding surfaces in the embankment are analyzed statically to find the inertia force $F_I = (W/g)a_y$ required to cause failure (Fig. 21.37). The average yield acceleration, a_y, is then deduced from this force. The average acceleration time-history of the sliding block is obtained usually from a QUAD-4 analysis. The yield acceleration is deducted from the average acceleration time-history and the net acceleration (the shaded area in Fig. 21.37), is available to generate permanent displacements (Newmark 1965). The analysis is conducted on the equivalent model of a horizontal sliding block on a plane with only one-way motions allowed (Fig. 21.37).

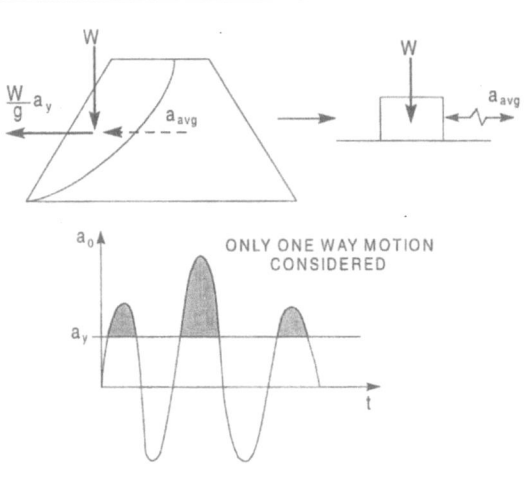

FIGURE 21.37. Elements of deformation analysis.

21.12.3 DEFORMATIONS FROM STRESS DATA

A more detailed picture of potential strains and deformations is obtained using Seed's semi-empirical method (Seed 1979b). The computed dynamic stresses in soil elements in the dam are converted to equivalent uniform stress cycles and are applied to laboratory specimens in consolidated states similar to corresponding elements in the dam. The resulting strains in the laboratory specimens are assigned to the corresponding elements in the dam. This procedure gives an incompatible set of strains which, however, are an indication of the potential for straining at selected locations within the dam.

A major motivation for the development of more general constitutive relations has been the need to model non-linear behavior in terms of effective stresses and to provide reliable estimates of porewater pressures and permanent deformations under seismic loading.

21.12.4 NON-LINEAR METHODS OF ANALYSIS

A hierarchy of constitutive models is available for the non-linear dynamic response of embankment dams to earthquake loading. The models range from the relatively simple hysteretic non-linear models to complex elastic-kinematic hardening plasticity models. Detailed critical assessments of these models may be found in Finn (1988) and Marcuson et al. (1992). This review presents the main true non-linear procedures used in current practice and outlines their advantages and limitations.

21.12.4.1 Elastic–Plastic Methods

The more comprehensive elastic–plastic models of soil behavior under cyclic loading are based on either a kinematic hardening theory of plasticity using multi-yield surfaces, or a bounding surface theory with a hardening law that defines the evolution of the plastic modulus. These constitutive models are complex and incorporate some parameters

not usually measured in field or laboratory testing. Soil is treated as a two-phase material using coupled equations for the soil and water phases. The coupled equations and the more complex constitutive models make heavy demands on computing time (Finn 1988).

Validation studies of the elastic–plastic models suggest that, despite their theoretical generality, the quality of response predictions is strongly path dependent (Saada and Bianchini 1987; Finn 1988). When loading paths are similar to the stress paths used in calibrating the models, the predictions are good. As the loading path deviates from the calibration path, the prediction becomes less reliable.

Typical elastic–plastic methods used in practice to evaluate the seismic response of embankment dams are DYNAFLOW (Prevost 1981), DIANA (Kawai 1985), DSAGE (Roth 1985), DYNARD (Moriwaki et al. 1988), and FLAC (Itasca 1996).

21.12.4.2 Direct Non-Linear Analysis

The direct non-linear approach is presented as incorporated in the program TARA-3 (Finn et al. 1986) because there is extensive experience in using this method in practice. In addition, the program has been validated in an extensive series of centrifuge tests conducted on behalf of the US Nuclear Regulatory Commission (Finn 1988).

In this model, the behavior of soil in shear is assumed to be non-linear and hysteretic. The response of the soil to uniform all-round pressure is assumed to be non-linearly elastic and dependent on the mean normal effective stress.

The objective during analysis is to follow the stress–strain curve of the soil in shear, during both loading and unloading. Checks are built into the program to establish whether or not a calculated stress–strain point is on the stress–strain curve and correction forces are applied to bring the point back on the curve if necessary. To simplify the computations, the stress–strain curve is assumed to be hyperbolic. This curve is defined by two parameters, which are fundamental soil properties, i.e. the strength, τ_{max}, and the *in situ* small strain shear modulus, G_{max}.

The response of the soil to an increment in load, either static or dynamic, is controlled by the tangent shear and tangent bulk moduli appropriate to the current state of the soil. The moduli are functions of the level of effective stress and therefore excess porewater pressures must be continually updated during analysis, and their effects on the moduli taken progressively into account.

During seismic shaking, two kinds of porewater pressures are generated in saturated soils: transient and residual. The residual porewater pressures are due to plastic deformations in the sand skeleton. These persist until dissipated by drainage or diffusion and therefore they exert a major influence on the strength and stiffness of the soil skeleton. These pressures are modeled in TARA-3 using the Martin *et al.* (1975) porewater pressure model.

21.12.5 RECOMMENDATIONS FOR ANALYSIS

The dynamic response analyses of embankment dams is still largely based on technology developed in the 1970s and which represents the very first attempts to carry out non-linear analysis by equivalent linear procedures. The stresses and accelerations calculated in this way are input into other procedures for assessing the performance of the dam. These procedures appear to work quite well provided the behavior of the dam is not strongly non-linear and significant pore pressures do not develop. Seed *et al.* (1989) note that this method can occasionally predict large displacements that do not occur when there is strong non-linearity or high porewater pressures. The Newmark (1965) procedure, based on sliding-block analysis, is particularly inappropriate when a large zone has liquefied in the embankment or foundation. The more comprehensive non-linear methods described above are especially appropriate for

evaluating the permanent displacements resulting from strong shaking with or without the presence of liquefaction.

21.13 Seismic Risk in Geotechnical Earthquake Engineering

21.13.1 SEISMIC RISK ASSESSMENT FOR DAMS

Seismic risk analysis provides a rational basis for the evaluation of dam safety during an earthquake. It is a process by which the consequences of exposure to a range of probabilistic seismic hazards are determined. The consequences are most often expressed in terms of both loss of life and economic loss at various probabilities of exceedance.

Risk analysis in the field of embankment dams is relatively new in engineering practice and most engineers do not have the knowledge or experience for making decisions on the basis of risk assessment. Risk assessment was adopted by the US Bureau of Reclamation (USBR) in 1995 and the US Army Corps of Engineers in 1997. The objective of USBR was "to ensure that structures do not create unacceptable risks to public safety and welfare, property, the environment and cultural resources". The Canadian Standards Association has developed the general framework for risk management shown in Fig. 21.38. There are two main structures in the framework, risk assessment and risk control, both of which form the basis for decisions on risk management.

21.13.2 APPROACHES TO RISK ASSESSMENT

The simplest approach to introducing probabilistic methods into the evaluation of dam safety is to formulate safety guidelines in which the hazard to the dam is defined probabilistically as a function of the potential consequences of failure expressed in general terms. The probability of the failure is not evaluated. The dam safety guidelines formu-

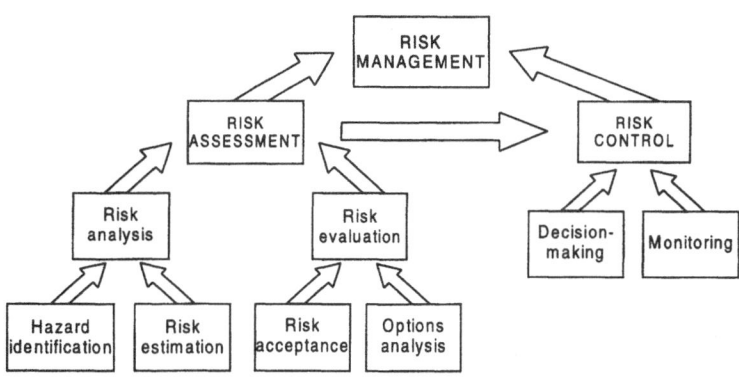

FIGURE 21.38. Risk management framework proposed by Canadian Standards Association (modified from International Journal on Hydropower and Dams 1998).

lated by the Canadian Dam Safety Association (CDSA) (1995), are a good example of this approach. These guidelines have been adopted nationally. The CDSA recommendations on hazards specifications are outlined in Table 21.9 as a function of three categories of consequences of failure. The categories are

TABLE 21.9. Usual minimum criteria for design earthquakes (adapted from CDSA 1995)

Consequence category	Minimum design earthquake (MDE)[a]	
	Deterministically derived	Probabilistically derived (Annual exceedance probability)
Very high	MCE[b]	10^{-4}
High[c]	50–100% MCE[b]	10^{-3}–10^{-4}
Low	—	10^{-2}–10^{-3}

[a] MDE (minimum design earthquake) firm ground accelerations and velocities can be taken as 50–100% of MCE[b] values. For design purposes, the magnitude should remain the same as the MCE.
[b] For a dam site, MCE (maximum credible earthquake) ground motions are the most severe ground motions capable of being produced at the site under the presently known or interpreted tectonic framework.
[c] In the "high" consequence category, the MDE is based on the consequences of failure. For example, if one incremental fatality would result from failure, an annual exceedance probability (AEP) of 10^{-3} could be acceptable, but for consequences approaching those of a "very high" consequence dam, design earthquakes approaching the MCE would be required.

defined in Table 21.10. The procedure for safety analysis of dams, once the seismic hazard is selected, follow conventional deterministic procedures such as stability calculations based on limit equilibrium and finite element methods. However, for "high consequence dams" in moderate to high seismic zones, seismic response and displacement analyses should be considered. Formal risk analysis is commonly used in many areas of engineering practice, but it is by no means widely embraced as yet for the seismic safety evaluation of embankment dams.

The Australian National Committee on Large Dams (ANCOLD) has been in the forefront in promoting formal risk assessment. In 1994, ANCOLD released its guidelines on risk assessment (McDonald 1997). Key elements of their guidelines, which are representative of emerging practice, are given below:

• For new dams, and the upgrading of existing dams, ensure that the average risk of death of particular members of the public from dam failure does not exceed 10^{-6} per exposed person per annum. Do not subject any person, being a member of the public, to a risk greater than 10^{-5} per annum.

• For existing dams, individual risks up to ten times those for new dams could be tolerable, subject to application of the ALARP

TABLE 21.10. Consequence classification of dams (adapted from CDSA 1995)

Consequence	Potential incremental consequences of failure[a]	
	Loss of life	Economic, social, environmental
Very high	Large increase expected[b]	Excessive increase in social, economic and/or environmental losses
High	Some increase expected[b]	Substantial increase in social, economic and/or environmental losses
Low	No increase expected	Low social, economic and/or environmental losses
Very low	No increase	Small dams with minimal social, economic and/ or environmental losses. Losses generally limited to the owner's property; damages to other property are acceptable to society

[a] Incremental to the impacts that would occur during an earthquake, but without failure of the dam. The type of consequence, e.g. loss of life or economic losses, with the highest rating, determines which category is assigned to the structure.
[b] The loss-of-life criteria, which separate the "high" and "very high" categories, may be based on risks that are acceptable or tolerable to society, taken to be 0.001 lives per year for each dam. Consistent with this tolerable societal risk, the minimum criteria for a "very high" consequence dam (MCE) should result in an annual probability of failure less than 1 : 100 000.

principle, i.e. keep the risk "as low as reasonably possible".

- Ensure that new dams, and dams being upgraded, satisfy the societal risk criterion given by the objective curves in Fig. 21.39.
- Ensure that existing dams satisfy the societal risk criterion given by the lower curve of Fig. 21.39, but carefully consider the ALARP principle.

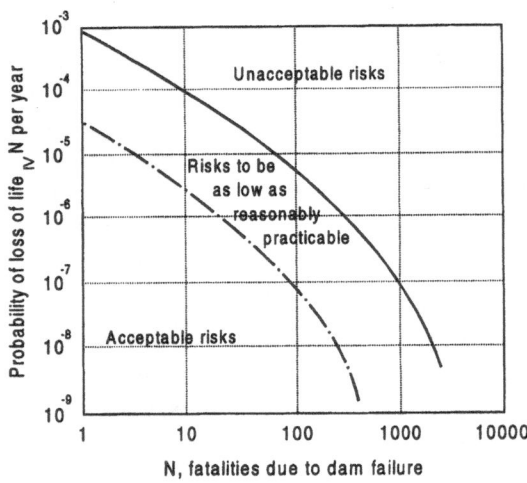

FIGURE 21.39. ANCOLD societal risk criteria (based on International Journal on Hydropower and Dams 1998).

- Ensure that a dam complies with both individual risk and societal risk criteria. In assessing compliance with individual and societal risk criteria, use a recognized methodology, such as the procedures set out by the US Bureau of Reclamation to estimate expected loss of life.

Note the crucial difference between formal risk assessment and the standards approach. The probabilities are now specified for the consequences, whereas in the standard approach, the emphasis is on the probability of the hazard.

21.13.3 SOME CONTENTIOUS ISSUES

Estimating the risk of loss of life is a contentious issue. In particular, there is debate about whether it should be expressed as expected losses per annum or losses associated with an actual event. Losses per event are more likely to arouse social and political opposition than expected values. In some cases, even economic losses probably should be considered on an event basis. Coping with losses from catastrophic events like earthquakes is not like dealing with losses from car accidents or on life insurance policies. Insurance companies when considering earthquake losses do not usually consider the expected losses. They are more concerned with the conditional probabilities of loss. Even the conditional probabilities are often ignored by moving the largest earthquake closest to the insured portfolio. The companies are very sensitive to the threat that, if and when the earthquake occurs, they may go broke because of having to meet all the losses at once. These features of potentially catastrophic events do not seem to be well-represented in conventional seismic risk procedures.

21.13.4 FRAMEWORK FOR RISK ASSESSMENT

The framework for formal risk assessment defines how to go from probabilistic specification of seismic hazard to the determination of the probabilities associated with different levels of consequences. The form of the framework depends on the potential failure modes of the dam. An example is presented from a paper by Lee *et al.* (1998). The paper describes how the probability of different levels of post-liquefaction damage and consequences were assessed for Keenleyside Dam in British Columbia by Hydro and Power Authority. BC Hydro has been actively promoting risk assessment for its projects (Hartford 1997). The emphasis here is on describing the framework of the risk assessment process.

21.13.4.1 Example: Risk Assessment for Keenleyside Dam

The framework used by Lee *et al.* (1998) is given in Fig. 21.40. There is a long sequence of complex steps between the specification of the seismic hazards and the calculation of the probabilities of different levels of damage and associated consequences. The sequence of steps is called an event tree. The development of an appropriate event tree is a complex task requiring a wide range of skills and considerable judgment. The event tree by it-

self, even without the probability assessments, leads to a much better understanding of the risk to the dam. It can be useful in planning remedial measures or risk management.

The first step is to specify the seismic hazard in level 1 (Fig. 21.40). A range of shaking intensities is specified by peak ground accelerations. Each level of acceleration has its probability of exceedance. The potential for liquefaction, all other things being equal, depends on the duration of strong shaking. Duration is represented by the earthquake magnitude in the techniques used in engineering practice for assessing liquefaction potential (Seed *et al.* 1985). Therefore, the magnitudes contributing most to the specified levels of shaking are obtained from a conventional seismic hazard assessment using a program such as EZ-FRISK, 4.0 (Risk Engineeering 1997), which allows de-aggregation of the magnitudes contributing most to each level of seismic risk. On this basis, the magnitudes in level 2 can be obtained. Level 3 in the event tree is the designation of the liquefaction model to be used in the analysis. The conventional models such as Seed *et al.* (1985) and Robertson & Campanella (1985), based on the standard penetration tests (SPT) and the cone penetration tests (CPT), respectively, are used deterministically. Liao *et al.* (1988) and Youd & Noble (1998), using the database for the deterministic models, have developed charts that allow the estimation of the probability of liquefaction for a given level of shaking and a given level of liquefaction resistance to cyclic loading (CRR) specified by data from standard penetration or cone penetration tests. In the Keenleyside study, only the Liao *et al.* (1988) probability model was actually used. If both models had been used, then weighting factors would have been assigned to each. Cyclic resistance ratios, CRR, with probabilities of liquefaction of 0.85, 0.50 and 0.15 are used to assess the potential damage to the dam in

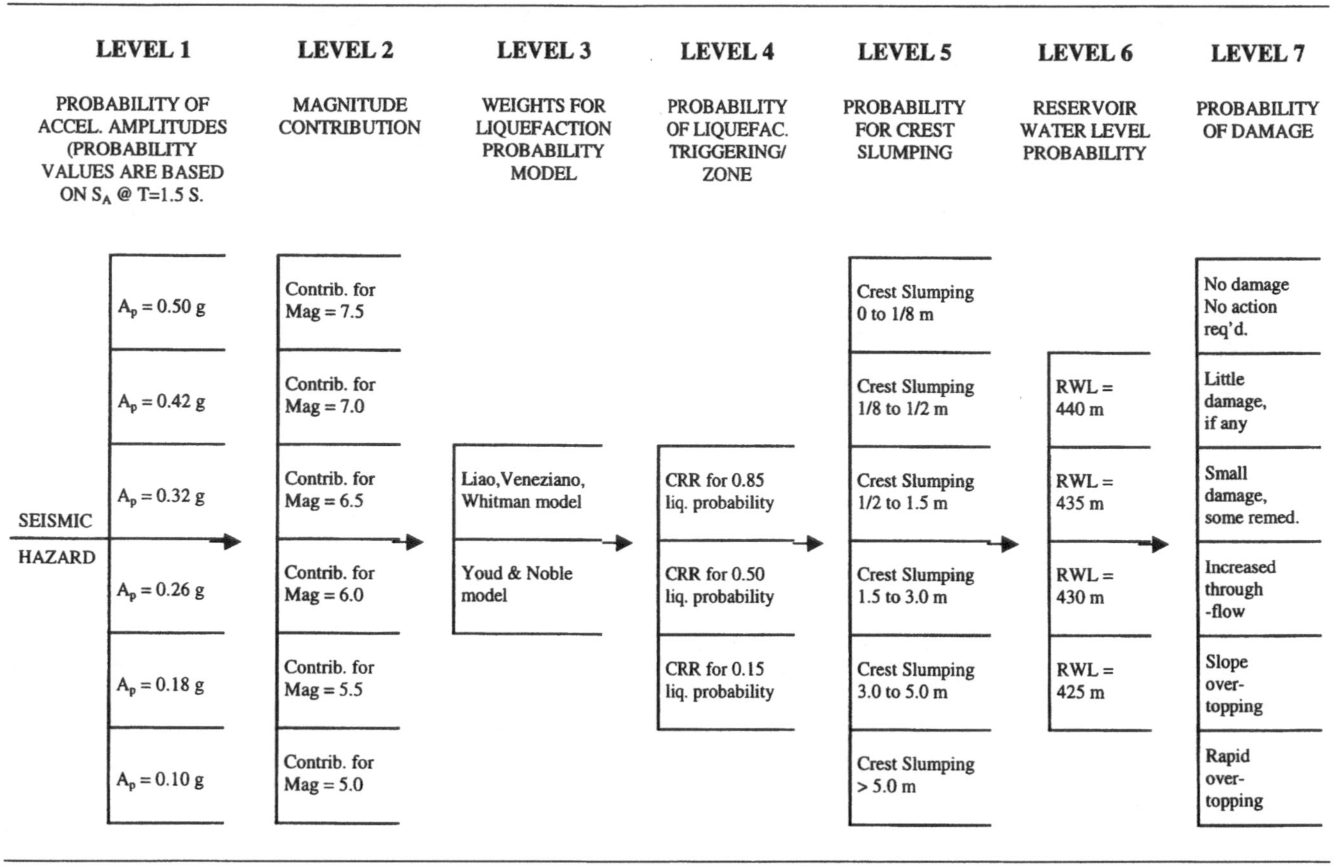

FIGURE 21.40. Event tree used for the probability of liquefaction-failure (modified from *Lee et al.* 1998).

terms of displacements such as loss of free-board (crest slumping). These analyses are carried out by finite element analyses using post-liquefaction stress–strain–strength properties in the liquefied soils.

Six different levels of crest slumping are used to characterize the potential performance of the dam. At this stage, the conditional probabilities of these different levels may be introduced, based on computed response data and taking into account potential variations in properties of the materials, uncertainties in the analysis procedure itself and the exercise of judgment. The definition of performance is a very critical part of the entire process. A simple criterion based on safety of factor against slope failure has been found to be unsatisfactory (Lee *et al.* 1998; Whitman 1984). The probabilities based on such a narrow criterion appear to be overly sensitive to some details of the analyses.

The next step is evaluating the consequences. The consequences of the loss of freeboard associated with crest slumping obviously depend on the level of the reservoir. The probability of the reservoir being at specific levels can be calculated from the long-term records of reservoir operation, and consequently, probabilities can be assigned to the reservoir level being at a number of representative stages, as shown in Fig. 21.40. Judgments must now be made of the damage potential for the different combinations of crest settlement and reservoir level. This is the difficult part and requires expert judgment to assign subjective probabilities.

Placing the range in reservoir levels, in level 6 of the event tree, after the displacement analyses in level 5, emphasizes the role that the reservoir level plays in assessing the consequences of failure. However, the level

of the reservoir also affects the stresses in the dam before failure and hence, needs to be incorporated into the finite element analyses so that strictly speaking the node for reservoir level should also be placed before the node for displacement analysis.

Whatever reservations one may have about the probability of the various consequences derived from a risk assessment analysis, it is clear that constructing the event tree forces one to think in a very detailed way about the process by which failure develops in the dam, and therefore, it contributes to a deep understanding of how the dam is likely to behave. This is a very positive benefit of any well-conducted risk assessment study.

If one now considers how the consequences might be mitigated, other events may need to be taken into account. The consequences may be dealt with in some situations by relying on evacuation of those who might be affected downstream. For such an evacuation to be effective, the consequences of the earthquake in the downstream area on transportation routes, on the communities at risk and on communications need to be taken into account. This would require an extension of the event tree to cope with these other events, which impact the mitigation of the consequences. If mitigation depends on lowering the reservoir level, then the impact of the earthquake on procedures for doing this needs to be taken into account. Clearly, the full range of risk assessment involves a very comprehensive investigation on all consequences of the earthquake, both directly on the dam itself, on the operating environment of the dam and on the region surrounding dam and reservoir. Full risk assessment is a very expensive process and requires a range of high-level skills that is not yet widely available.

V. GEOENVIRONMENTAL ENGINEERING

22. GEOENVIRONMENTAL PROBLEM IDENTIFICATION AND RISK MANAGEMENT

M. Whittaker

J. G. Sprenger

D. D. DuBois

22.1 Introduction

The issue of contaminated land carries with it significant environmental and economic implications. A key question is how best to fulfill the interests of all those with a stake in contaminated land (owners, potential vendors/purchasers, consultants, the public, the regulators and so on) while ensuring that the land is returned to a condition that is protective of the users and does not present a long-term environmental liability.

Increasingly, contaminated sites are being managed through consideration of the risks they pose to human and environmental health. The process of identifying and evaluating the significance of risks is known as risk assessment; the process through which identified risks are controlled or mitigated is known as risk management.

The objective of this chapter is to provide the reader with an overview of how geoenvironmental issues relating to the contaminated subsurface may be addressed by considering the risks to human and environmental health. In essence, risk assessment can be viewed as a framework for evaluating the likelihood of adverse effects on humans and the wider ecological environment resulting from exposure to a contaminant. Use of risk assessment at contaminated sites provides a customized, clearer, more comprehensive means of discriminating between, and identifying appropriate solutions for the potential issues of concern associated with a contaminated site than by direct comparison with generic non-health-based standards.

Typically, the risk-based approach has proven to be most valuable in situations where financial or technical constraints rule out a more conventional solution, e.g. where portions of the site underlie existing building structures or services, where landfill disposal is prohibitively expensive, or where the contaminated material is deep lying and cannot be removed. However, its application is by no means confined to such situations, and risk-based site management is now frequently being applied in a wider sense as a recognized clean-up framework in many jurisdictions across Canada, the USA and Europe.

This chapter is specifically intended to provide an overview of and commentary on the risk assessment/risk management framework as it applies to the management of contaminated sites. It is intended primarily as a commentary for risk assessment practitioners and not as a prescriptive manual or work sheet. The key technical elements of the risk assessment process are considered, with emphasis placed upon the underlying principles essential to understanding the issues and applying them in practice. Where appropriate, relevant non-technical issues such as risk communication and legal matters are also discussed.

22.1.1 FRAMEWORK FOR RISK-BASED SITE MANAGEMENT

Environmental risk assessment at contaminated sites entails characterization of the source of an environmental hazard, identification of exposure routes through which a contaminant may come into contact with a receptor, assessment of the relationship between the contaminants and the adverse effects produced (if any) and, finally, estimation of the effect (CCME 1996). The primary role of a risk assessment is a technical framework in which potentially contaminated land can be examined, target remediation levels derived, and the environmental issues caused by the presence of contamination managed to protect human and environmental health.

During risk assessment, a major requirement is to identify the key contaminants, pathways and receptors associated with a site, and, at the same time, rule out possible risk scenarios found to be of little or no significance. The three risk components are shown in Fig. 22.1. To achieve this, most site assessments are progressed to the appropriate level of investigation by means of an iterative process of refinement, which seeks to draw an accurate picture of the site through the gradual accumulation of information obtained using increasingly intrusive examination techniques. For risk-based site assessments, the information gathering should be done in an interactive (with all project stakeholders, including the public) and progressive, or "top-

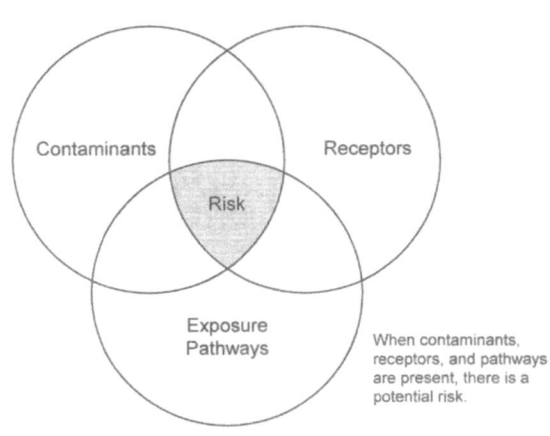

FIGURE 22.1. Risk components (modified from Health Canada 1994).

down refinement," manner. The top–down approach is illustrated in Fig. 22.2.

Beginning at the top level, with the most simplified approach, Table 22.1 summarizes the characteristics of the four main types of risk assessment frameworks in use at the present time. If conservative models based on gross assumptions indicate that no significant environmental problems are present, then more sophisticated examination is not warranted. The most widely used frameworks are the types III and IV, and these approaches will be examined in more detail throughout this chapter.

22.1.2 DEFINING GOALS

The principal goal of contaminated site remediation is to provide acceptable protection to the site users in the context of a specified land use. With the key word "acceptable" being somewhat subjective, the intensity and extent of restoration activities can vary significantly with differing interpretations of what is, or is not, acceptable. Regulators often strive to provide guidance on this issue by issuing generic soil and groundwater remediation criteria, and these are utilized in traditional ap-

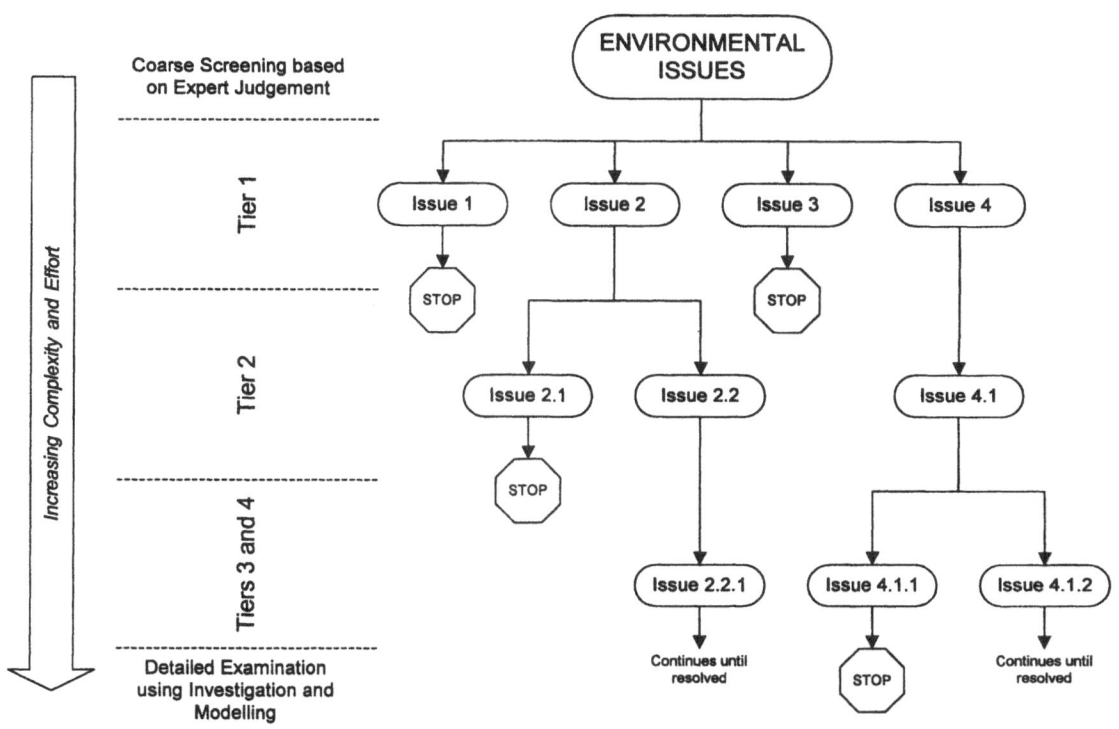

FIGURE 22.2. Top–down approach for risk assessment using step-wise refinement.

proaches to dealing with the definition of adverse effects of contamination by providing general "trigger" levels and remediation goals for the project team (e.g. MoE 1997). These generic criteria are often health-based risk management principles, but necessarily incorporate assumptions that are conservative, in order that they can be applied widely to many situations.

Assessment and management of contaminated land within a site-specific risk-based framework, as opposed to using generic criteria, addresses the issue of acceptability by allowing the environmental issues arising from the presence of contaminating substances to be prioritized according to their significance and in a way that considers site-specific factors such as land use, proximity to

environmentally sensitive areas, and the actual geologic and hydrogeologic conditions encountered at the site. In this way, insignificant or trivial environmental effects can be identified and quickly removed from the site management plan, allowing site management resources to be focused on the significant environmental issues; furthermore, the conservative assumptions (on, for example, soil porosity, the rate of contaminant degradation or the groundwater flow regime) used in the generic criteria can be identified and modified accordingly.

22.1.3 REGULATORY AND SOCIETAL ISSUES

The application of risk assessment in the management of contaminated land in North

TABLE 22.1. Types of risk assessment frameworks

Type I: non-analytical method
- Simplest method (advantage)
- Based on the comparison of observed environmental concentrations to tabulated criteria determined according to conservative, exposure assumptions
- Lack of site-specific realism
- Lack of quantitative confidence in results due to generalized nature of generic criteria

Type II: best professional judgment methods
- Risk assessment performed according to the best professional judgment of the environmental consultant/risk assessment team
- Very little procedural guidance
- Outcome highly dependent on the expertise of the assessor
- Can be highly site specific (advantage)
- Can be flexible (advantage)
- Results may not be independently reproducible (disadvantage)

Type III: RAGS-type method
- Consistent with framework developed in the US Environmental Protection Agency's method entitled "Risk Assessment Guidance for Superfund" (RAGS)
- High degree of procedural guidance available, which produces independently reproducible results (advantage)
- Widely accepted method
- Framework consists of four stages: problem formulation, toxicity assessment, exposure assessment and risk characterization
- Lack of site specificity, since RAGS method advocates the use of default factors (and conservatism) in various model parameters (can be disadvantage)

Type IV: enhanced RAGS-type method
- Method based on RAGS but with a significant evolutionary advancement because of the incorporation of scientifically based procedures (advantage)
- More realistic and able to quantify risk estimates and degree of uncertainty in estimates (advantage)
- Disadvantage to regulatory and regulated communities since unfamiliarity with such issues as probabilistic and statistical concepts
- Increased effort, time and cost associated with increased degree of site-specific realism (can be disadvantage)

America has gained widespread recognition by regulatory agencies in recent years (USEPA 1997; CCME 1996). In fact, the majority of Western industrialized countries have developed or are in the process of developing risk-based soil and groundwater quality criteria (CARACAS 1998). These are often based on the "Risk Assessment Guidance for Superfund" (RAGS)-type approach, but do differ in some aspects, and so the risk assessment team should become familiar with applicable regulations in the appropriate jurisdiction at the outset of the risk assessment.

A single universal approach has not yet been defined, although the 1996 American Society for Testing and Materials (ASTM) document (ASTM 1996) for risk-based corrective action (RBCA, pronounced "Rebecca") is the most widely practiced. The ASTM RBCA standard was developed to provide an overall process to assess risks associated with total petroleum hydrocarbon (TPH) sites. It incorporates methods developed under the USEPA superfund program for hazardous waste sites. In essence, it integrates site assessment, remedial action selection, and monitoring through the use of USEPA risk assessment methodologies.

22.2 Site Assessment for Risk-Based Site Management

22.2.1 SCOPE AND OBJECTIVES

Site assessment is considered to be the most critical stage of problem identification and risk management at contaminated sites. Information obtained during this phase of the site management program invariably provides the basis upon which the entire site management plan is founded and may be used to determine the nature and magnitude of any risks the site may pose, and the type and extent of risk management alternatives that might be required. The situation is often further complicated by a variety of social, financial and legal issues that must be factored into the assessment process. The primary objective of any site assessment is to provide accurate, reliable data on the environmental condition of the site.

Key site assessment goals may include any or all of the following items: identifying sources of contamination both on and off site; determining the extent, location and nature of contamination both on site and off site; assessing the geological and hydrogeological characteristics of the subsurface; identifying the type and locations of any underground structures, including service utilities, that may be present on site; obtaining information pertinent to geotechnical matters, e.g. soil stability, foundation design; identifying site users, both human and ecological. An account of conventional site characterization components and methods is provided in Chapter 4.

22.2.2 SITE ASSESSMENT PROCEDURAL COMPONENTS

An environmental site assessment (ESA) is usually carried out in two stages in a top–down manner:

22.2.2.1 Stage 1: Non-Intrusive Phase I Investigations

The first stage of site assessment is usually a conventional non-intrusive investigation and is intended to identify the potential for a site to have been impacted. Available information is assembled on the existing condition of the site, its past and present uses and any issues of potential environmental concern. At this stage, the focus is not only on the most immediate or obvious risk scenarios that could potentially arise at the site, but also those areas where a more detailed investigation could be required.

As described in Chapter 4, phase I assessments typically comprise one or more site visits and a background/historical survey. A limited, non-intrusive field monitoring program may also be undertaken. Knowledge of the site is maximized through the use of site photographs, personal interviews with individuals familiar with the historic site, site records, maps, aerial photographs, local and regional authority records and library sources. Information collected at this stage may relate to any or all of the components of the contaminant–exposure–receptor paradigm, including actual or potential site users (human and ecological), groundwater migration pathways, the presence of neighboring receptors (human and ecological) or sensitive water bodies, land usage at the site and the immediate vicinity.

In proceeding this way, an inventory is compiled of potential hazards associated with the site, the likely exposure routes and the potential human and/or ecological receptors. If, upon completion of the initial investigation, it appears that potential sources of contamination exist at or adjacent to the site, or that there is potential for off-site migration of contaminants, then an intrusive phase II investigation is usually warranted.

For an in-depth account of standard procedures in phase I environmental site assessments (ESA), the reader is directed toward the numerous technical manuals and standard practice documents that are currently available (e.g. CSA 1994; ASTM E1527).

22.2.2.2 Stage 2: Intrusive Site Assessment

Progression to the second stage of site assessment involves the accumulation of more detailed information on the subsurface environmental condition of the site. The objective of the phase II environmental assessment is to better characterize the nature, extent and distribution of contaminants identified during the initial stage as being of actual or potential concern, such that the actual risks posed by the site can be more accurately evaluated. This is accomplished by a combination of intrusive environmental sampling and subsequent chemical analysis.

Ideally, phase II ESAs should themselves comprise several stages: an initial planning stage, a site investigation and sample analysis stage, and a data evaluation stage. Allowance should be made to continually move information between these stages, since this often results in a more focused and insightful final product. It is not uncommon, for example, for the preliminary results of an analytical program to be used to direct further sampling and analysis.

The initial planning stage of the investigation will draw heavily upon the findings of the phase I ESA to determine the number, location and type of samples to be taken. At this stage, consideration is given to the nature of the contaminants under investigation, as different contaminants exhibit varying subsurface distribution characteristics. This point is elaborated upon in the following section. Sampling is usually accomplished through the groundwater or vapor use of boreholes/monitoring wells and/or test pits, which permit sampling of subsurface soil/groundwater and/or vapors at various depth intervals (see Chapter 4 for details). Figure 22.3 illustrates the phased approach to environmental site assessment (leading to remediation) and the interaction of ESA with the risk assessment process.

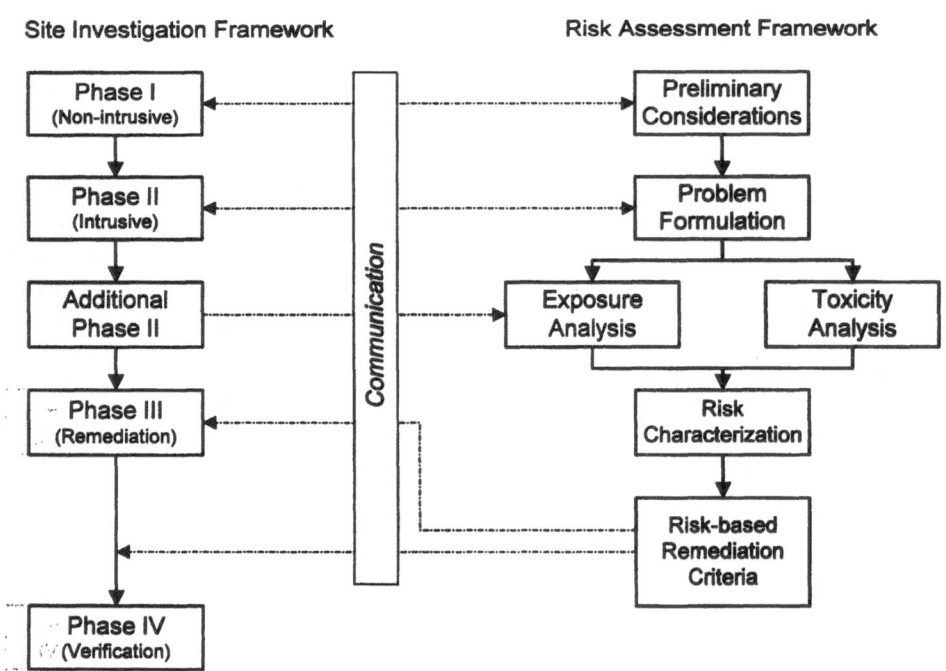

FIGURE 22.3. Environmental site management framework.

22.2.3 SAMPLING STRATEGIES AND METHODS

When designing a site sampling strategy, it is important to be mindful of the spatial and temporal variations in contaminant concentrations that may occur across a site, as well as the heterogeneity of the site subsurface.

There are two main approaches to sampling at contaminated sites: probabilistic (random) sampling; and deterministic sampling, in which decisions are made on where to sample based on some previous knowledge of the site and/or the investigation. Random sampling is favorable for any statistical treatment of data collected, which is often the case where quantitative risk assessment is applied or when the site is evenly or randomly contaminated.

Sampling should be carried out according to the recognized USEPA, ASTM (both US standards) or CCME (Canadian) protocols to ensure the necessary level of quality assurance (e.g. USEPA 1997; ASTM E1527;

TABLE 22.2. Summary of environmental sampling strategies[a]

Type of sampling design	Conditions when the sampling design is useful
Haphazard sampling	A very homogeneous population over time and space is essential if unbiased estimates of population parameters are needed. This method of selection is *not* recommended due to difficulty in verifying this assumption
Judgment sampling	The target population should be clearly defined, homogeneous, and completely assessable so that sample selection basis is not a problem. Specific environmental samples are selected for their unique value and interest rather than for making inferences to a wider population
Probability sampling, simple random sampling	The simplest random sampling design. Other designs below will frequently give more accurate estimates of means if the population contains trends or patterns of contamination
Stratified random sampling	Useful when a heterogeneous population can be broken down into parts that are internally homogeneous
Multistage sampling	Needed when measurements are made on subsamples or aliquots of the field sample
Cluster sampling	Useful when population units cluster together areas of specific historic use, and every unit in each randomly selected cluster can be measured
Systematic sampling	Usually the method of choice when estimating trends or patterns of contamination over space. Also useful for estimating the mean when trends and patterns in concentrations are not present or they are known *a priori* or when strictly random methods are impractical
Double sampling	Useful when there is a strong linear relationship between the variable of interest and a less expensive or more easily measured variable
Search sampling	Useful when historical information, site knowledge, or prior samples indicate where the object of the search may be found

[a] Modified from Gilbert (1987).

CCME 1996; MoEE 1996a). Gilbert (1987) provides an insightful discussion of different sampling strategies (see Table 22.2).

22.2.4 COMMON DIFFICULTIES ENCOUNTERED DURING THE SITE ASSESSMENT PROCESS

As stated in the opening paragraph of this section, information obtained during site assessment often underpins the entire problem identification and risk management process. It is crucial, therefore, to ensure that data of sufficient quality and volume are collected throughout the site assessment phase. The number and location of data points should reflect not just the size of the contaminated site, but also the number of environmental compartments into which the contaminant(s) may have partitioned and the type of sampling regime selected (deterministic versus random). Where a number of subsurface phases are suspected of having been impacted by a particular contaminant, then this should be reflected in the variety of samples taken. Petroleum products, for example, can potentially give rise to soil vapors, dissolved phase contamination, a residual phase adsorbed to soil natural organic matter and free-phase contaminant. It may be necessary to sample each of these subsurface phases in order to gain an accurate picture of the extent of contamination. It follows that composite samples, taken from more than one location are only used for contaminant screening but are of little use as input data for risk assessments.

22.3 Contaminant Identification

22.3.1 REQUIREMENTS AND OBJECTIVES

Contaminant identification within geoenvironmental problem identification and risk management is concerned principally with establishing the source, abundance, distribution and direction, and speed of movement of contaminants in the subsurface media. Invariably,

a combination of complex contaminants together with their degradation products, heterogeneous soil subsurfaces and confounding site-specific factors, e.g. site location, topography, give rise to a variety of technical challenges that must be addressed to fully characterize the contaminants present at a site.

22.3.2 TYPICAL CONTAMINANTS AND ASSOCIATED HAZARDS

The inherent diversity in materials and processes used in industrial activities has given rise to a multitude of potentially contaminating substances. Certain classes of contaminant are routinely associated with certain industries, and this assists the risk assessor in selecting which contaminants could be of concern at a particular site. In general, contaminants are classified according to their chemical characteristics. There are two main families of chemical contaminant: inorganic and organic. Examples of each, together with typical sources are provided in Table 22.3.

22.3.2.1 Inorganic Contamination

This is usually associated with the presence of metals, but also includes cyanides and other anions such as chlorides, sulphates and nitrates. Metal contamination is common at former industrial sites, and can be associated with activities such as mining, smelting, steel production, landfarming, scrap metal yards, vehicle maintenance, manufacturing, sewage treatment and the like. In particular, arsenic, cadmium and lead are known to be acutely toxic to both plants and humans, whereas zinc, copper and nickel are primarily phytotoxic (harmful to plants).

Cyanides are commonly associated with mining, manufactured gas plants and waste disposal facilities, whereas sulphates are produced by numerous industrial processes. Both represent a hazard to human and ecological receptors, in addition to acting as corrosive agents on building foundations.

TABLE 22.3. Examples of industries/activities and their contaminants

Inorganic contaminants	Organic contaminants
Examples	Examples
Iron (Fe), manganese (Mn)—usually of concern for aesthetic reasons in water supplies	Polynuclear aromatic hydrocarbons (PAHs)
Heavy metals—silver (Ag), mercury (Hg), cadmium (Cd), zinc (Zn), copper (Cu), lead (Pb), chromium (Cr), arsenic (As)	Phenols
	Pesticides
	Non-chlorinated solvents
Acids and bases	Chlorinated solvents
Cyanide	Polychlorinated biphenyls (PCBs)
	Petroleum hydrocarbons
Typical sources	Typical sources
Mining and smelting operations	Fuel dispensing
Mine tailings and tailings ponds	Degreasing operations
Landfills	Electrical conductors
Sewage	Bulk storage depots
Metal plating operations (Ag, Cr)	Sewage
Paint, painting operations (Pb, Cd)	Metal plating operations
Pulp mills (Hg)	Paint, painting operations
Wood treatment (Cu, Cr, As)	Wood treatment
Auto wrecking yards	Auto industry
Industrial and municipal landfills	Home heating
Thermal power stations	Industrial and municipal landfills
Imported fill soils (when imported to a site from one of the primary source areas)	Power stations

22.3.2.2 Organic Contamination

Organic contaminants include petroleum products (from gasolines to heavy fuel oils to bitumen), coal tar, solvents (including chlorinated compounds), phenols, dioxins and expanded polychlorinated biphenyls (PCBs). These contaminants differ widely in their chemical composition and inherent toxicity. In considering the risks associated with these contaminants, it is important to recognize that organic contaminants are usually mixtures of many individual chemical compounds (gasoline, for example, comprises several thousands of compounds), each exhibiting different chemical, physical, biological and toxicological properties, and that this can result in multiphase environmental partitioning following a spill event. Many organic contaminants, most notably benzene and polynuclear aromatic hydrocarbons (PAHs), are considered to be carcinogenic to humans.

Common vapor phase contaminants include methane, carbon dioxide, carbon monoxide, hydrogen sulphide and sulphur dioxide, as well as low molecular weight volatile organic compounds (VOCs), which exist either in vapor or liquid phases. These substances present a variety of risks, including combustion, asphyxiation, corrosion and health impacts following inhalation.

22.3.3 THE LINK BETWEEN CONTAMINANT BEHAVIOR AND RISK

In considering the various types of contaminant that may be faced by the risk assessor, and how these contaminants should be characterized in order to begin the problem identification process, it is important to understand what happens to a contaminant following its release into the natural environment. In particular, because the toxicity and exposure characteristics (and, therefore, the risks) associated with a particular contaminant are influenced by the chemical form of the contaminant and its distribution in the environment, attention should be given to elucidating how contaminants partition in the environment, and how their composition or chemical form could change following a release (Sims 1990). This is also important in establishing the routes by which exposure could occur. In broad terms, the characteristics that govern contaminant environmental fate and partitioning can be considered as inherent properties of the released material and the properties acquired as a result of contaminant weathering and subsurface attenuation. These are physical, chemical and biological processes that alter the chemical form and, therefore, the subsurface behavior of chemical contaminants. Both inherent and acquired characteristics must be considered in order to obtain a complete picture of the exposure scenarios presented by a particular contaminant.

22.3.3.1 Contaminant Partitioning

The partitioning of contaminants between the available air, water, soil, natural organic matter (NOM), mineral and free-product phases during their passage through the subsurface is governed by the fundamental physical, chemical and biological properties of the contaminants and the geology and hydrogeology of the subsurface.

Different contaminants will partition throughout the subsurface in a variety of ways. For metals, it is important to consider the conditions of the subsurface, especially parameters such as pH and redox potential. These parameters affect the chemical form of metals, which in turn influences the extent to which the contaminants are bound in the soil phase or mobilized in groundwater. Arsenic, for example, can exist in over 30 different forms, each exhibiting its own particular toxicity (Hrudey & Pollard 1993).

Organic contaminants can be complex mixtures of individual compounds and often partition into several phases. For organic non-aqueous-phase liquids (NAPL), for example,

it is generally possible to identify six phases into which the mixture may partition (Hrudey and Pollard 1993):

- immobilized pockets of residual contamination in the unsaturated soil;
- individual contaminant constituents adsorbed to soil natural organic matter (NOM) or mineral matter;
- the vapor phase, through evaporation;
- the aqueous phase, dissolved in groundwater in the saturated or interstitial zone;
- mobile component, termed non-aqueous-phase liquid (NAPL), which is either less dense (LNAPL) or more dense (DNAPL) than water; and
- a contaminated water emulsion.

Figure 22.4 illustrates simplified partitioning behavior of four common environmental contaminants.

Knowledge of the subsurface partitioning allows appropriate risk management technologies to be selected and targeted accordingly. The application of groundwater pump and treatment technologies that focus only on removing contaminants from groundwater, for example, fail to target the residual contamination in the vadose zone and can result in an unnecessarily prolonged, expensive and ultimately ineffective site clean-up operation. Sources of fate and partitioning information for organic contaminants are provided by MacKay *et al.* (1992–1997); Howard (1989, 1991).

22.3.3.2 Contaminant Weathering

A variety of interactions take place between the contaminant and the soil–groundwater matrix that affect the chemical composition of the contaminants and their partitioning throughout the subsurface. These processes are collectively referred to as weathering. Specifically, weathering can be divided into four main processes: (i) volatilization, (ii) dissolution, (iii) chemical alteration, and (iv) biotransformation (Brady *et al.* 1998). Since these processes affect the chemical form of the con-

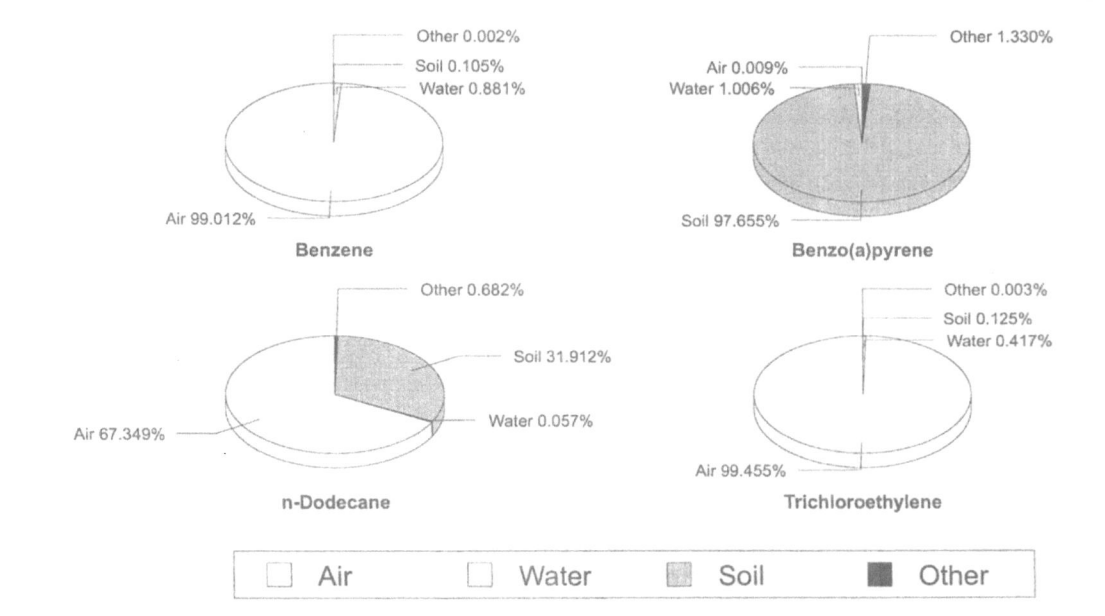

FIGURE 22.4. Fate and partitioning of common organic contaminants based on contaminant "fugacity" or escape potential (see MacKay *et al.* 1991, 1992, 1993).

taminant, they have significant bearing upon how a particular contaminant may be sampled and analyzed, the hazard that it represents, the routes via which exposure can occur and the manner in which it should be treated. For example, trichloroethylene, though itself a fairly mobile, noncarcinogenic hazard, can be degraded into vinyl chloride, a more mobile and reportedly more potent carcinogen.

22.3.4 SELECTION OF ANALYTICAL PARAMETERS

The selection of analytical parameters directed by the nature of the contaminants are identified as being of greatest concern during the phase I site assessment/contaminant identification process. Often, however, risks at contaminated sites can be estimated by considering the presence of one or more individual compounds (usually those of greatest toxicological significance), known as surrogate compounds. The analysis of surrogates may require interrogation of the contaminant matrix beyond conventional "prescriptive" analytical parameters, i.e. standard sets of analytical parameters. Thus, it is advisable to arrange for analyses to be completed by one of the many professionally certified analytical laboratories to ensure that the analytical procedures and techniques used during analyses will be in compliance with standard procedure.

The range of analytical techniques available varies considerably depending upon the type of contamination being examined. Analysis is most effectively accomplished by a phased analytical approach, in which non-specific screening techniques, such as are used to obtain information on the total concentrations of organics, and more sophisticated techniques are then applied to target individual contaminants of greatest toxicological concern.

Screening techniques usually involve measurement of total oil and grease, total petroleum hydrocarbons, total metal loading and total extractable hydrocarbons by gas chromatography (GC) and solvent extraction methods. These techniques allow the amounts of broad collective "groups" of contaminants to be estimated. While this is useful for elucidating the extent of contamination, for the purposes of the risk-based site management it is essential that an assessment of the contaminant extends beyond the measurement of bulk containment concentrations to provide information on specific contaminants and weathering. Individual chemicals are usually detected by gas chromatography–mass spectrometry (GC–MS) or inductively coupled plasma–mass spectrometry (ICP–MS) methods. Target compounds should be those of toxicological significance: benzene, toluene, ethylbenzene and xylene (BTEX); polynuclear aromatic hydrocarbons (PAHs); surrogate compounds; and, special categories such as phenols, volatile aromatic compounds, ketones and so on. Depending on how these latter techniques are employed, it is also possible to shed light on the movement, age and source of a particular contaminant, although this requires the input of a specialist analytical chemist.

It is crucial that appropriate analytical parameters be selected if the results of the ESA are to match the expectations of the risk characterization. In particular, because the chemical form of a contaminant often has considerable bearing upon the severity of the hazard it represents and its mobility in the environment, analytical techniques should be selected according to their ability to impart information on the specific chemical form of greatest concern. The use of generic analytical criteria, such as total extractable hydrocarbons or total metal analyses, does not discriminate between the chemical form or the toxic potential of the chemical under analysis, and thus provide only marginal insight into the true nature of the risk to human and ecological health (Whittaker *et al.* 1995). Similarly, the reporting of chemicals as "non-detects" should be treated with caution by the risk assessor, as analytical methodologies are sometimes unable to resolve highly complex

mixtures of contaminants (absence of proof is not proof of absence). Importantly, method detection limits for analytical techniques employed should be below the maximum acceptable contaminant concentration. This point is elaborated upon in Section 22.3.4.

22.4 Problem Formulation

Problem formulation, as the initial stage of the risk assessment process, provides the foundation for the subsequent stages of the risk assessment. This is sometimes referred to as a screening level risk assessment. The aim is to devise a conceptual model that correctly simulates the anticipated usage of the site. This model can be used to understand and discriminate between significant and insignificant environmental risks by considering the three risk components: the contaminants present at the site of interest, the humans and ecological receptors that use the site, and the exposure pathways of contact that are possible between the contaminants and the receptors. These contaminants, receptors and exposure pathways are examined in some detail to identify the combinations that are possible and contribute the most potential risk. Where no contaminant-exposure route-receptor link can be reasonably established, the risk to the receptor in question is deemed insignificant. These insignificant risks can then be eliminated from further consideration or "screened" out. The significant risks that remain are, therefore, derived from any potential link between the three risk components: contaminant, exposure, receptor. These risks can then be examined and quantified in more detail through an in-depth exposure, toxicology and risk characterization (the subsequent stages of the risk assessment), after which they too may prove to be insignificant, or at least "tolerable." At the qualitative screening level, distinguishing between significant and insignificant risks relies heavily on the experience of the risk assessor: in some cases, comparing site soil and groundwater quality to generic criteria provides a useful indication of where risks may be significant or not.

22.4.1 CONTAMINANTS

The presence, type, amount and distribution of contaminants across the site must be estimated as described in Section 22.3, and this is usually achieved through a site investigation as detailed in Section 22.2. Correct identification of the type and amount of contaminants present at a site is clearly pivotal to establishing the hazard (and, therefore, the risk) associated with a contaminated site. Also important are the details of the fate and partitioning of contaminants. The linkages between contaminant type, fate and partitioning, and the associated human health and environmental risks are described in Section 22.3.

22.4.2 RECEPTORS

Receptor screening involves determining the identity, sensitivity and behavioral characteristics of all site users that could potentially be

exposed to the contaminants of concern. This may include humans, plants, animals, fish, soil invertebrates and other valued ecosystem components. For the purposes of risk assessment, the term "site users" should include all areas potentially affected by the contamination (both on and off site) and taking into account the current or intended usage of the site.

The goal of this task is to develop a list of potential receptors, their toxicological sensitivities and site use characteristics, which can be used as a basis for more detailed receptor analysis in subsequent stages of the risk assessment. Sensitive receptors, such as endangered species, would also have to be identified and included within the assessment. Figure 22.5 illustrates a conceptual framework for receptor screening.

22.4.3 PATHWAYS

Screening of pathways requires consideration of the potential exposure routes by which receptors identified in the previous task could be exposed to risk. There are three funda-

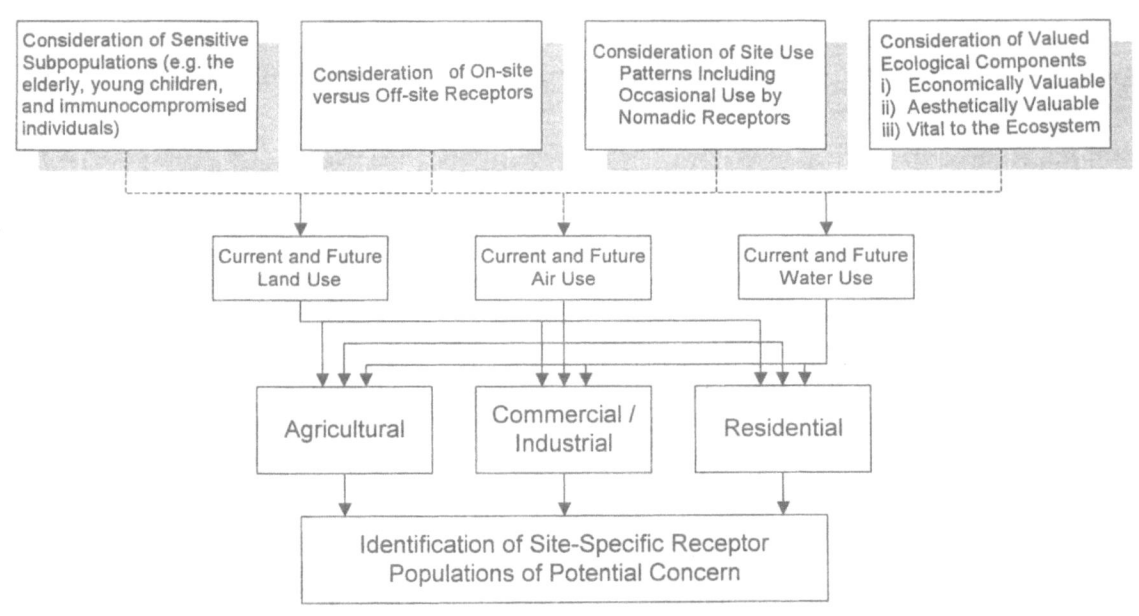

FIGURE 22.5. Receptor screening process.

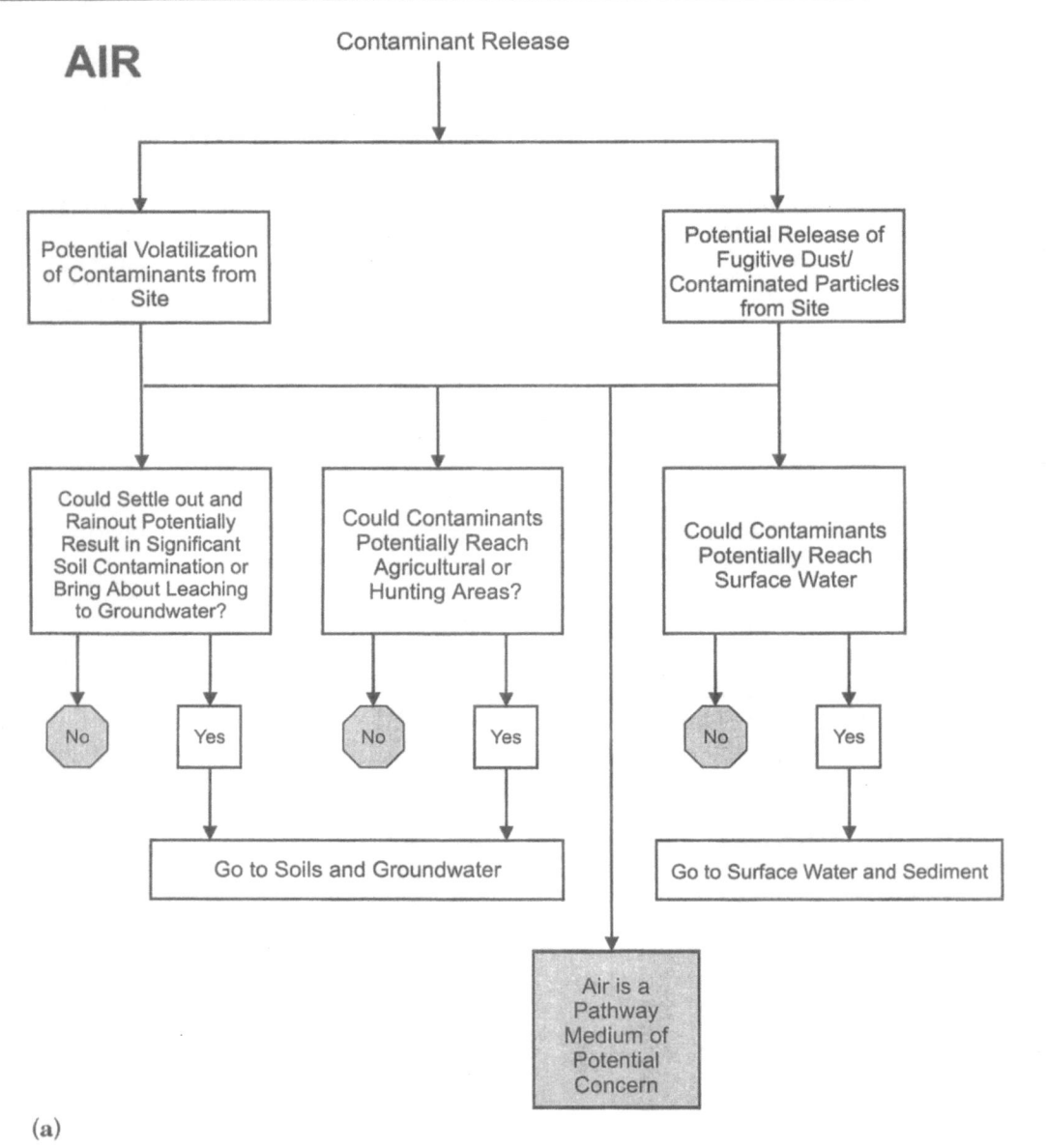

FIGURE 22.6. (a) Pathway process.

mental exposure routes to be examined: ingestion of contaminants (either directly, or through consumption of contaminated materials, including garden produce); inhalation of volatile contaminants, or contaminants carried on dust particles; and dermal contact, where the receptor is exposed by direct contact with contaminants. Receptor exposure to contaminants is usually based on some combination of these pathways. At this stage, the detailed features of the exposure need not be considered or quantified: the goal is to compile a descriptive list of all potential exposure routes (see Fig. 22.6a, b).

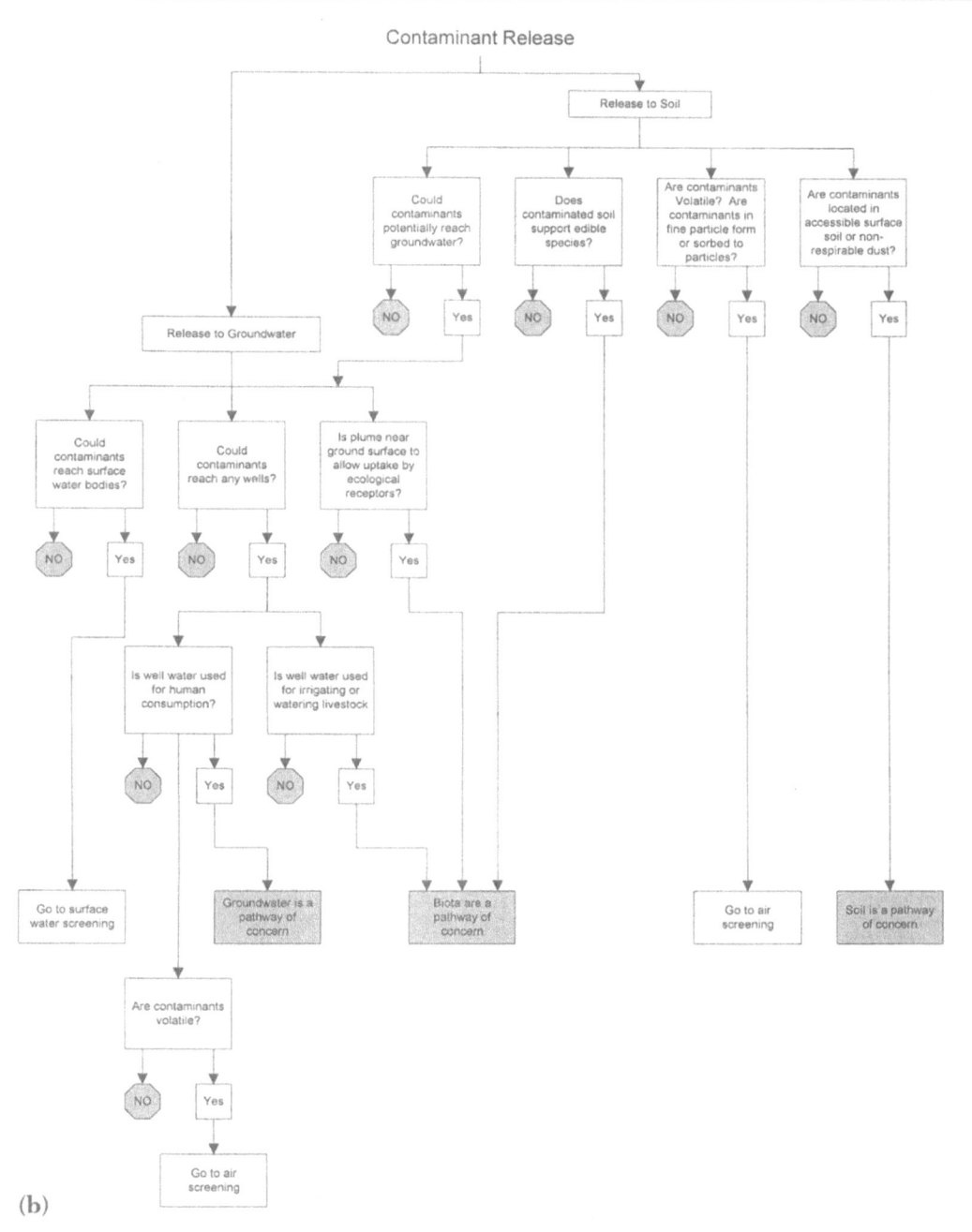

FIGURE 22.6 (Continued). (b) Soil and groundwater screening process (modified from USEPA 1989).

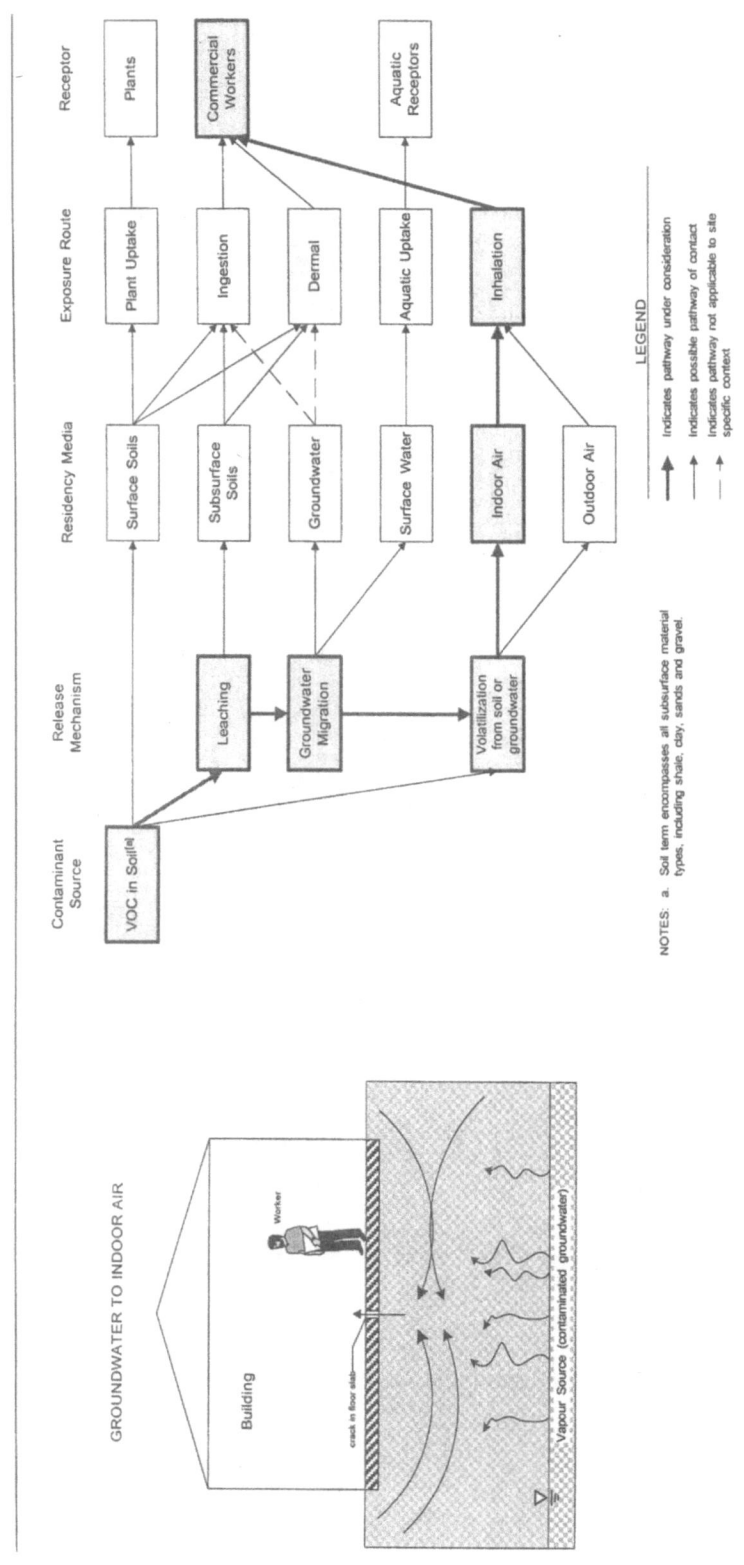

FIGURE 22.7. Conceptual exposure pathways model.

22.4.4 CONCEPTUAL PATHWAY MODEL

The combination of contaminant–pathway–receptor connections (termed environmental components) that emerge from the screening process form the basis of the conceptual model, which simulates the site condition and potential areas of environmental concern. The conceptual model not only provides a qualitative examination of the potential risk scenarios but guides the subsequent quantitative risk assessment, if required. It should also be used to guide additional refinements of the site assessment. Figure 22.7 provides an example of a conceptual pathway model.

In developing the conceptual model, the risk assessor is always seeking to summarize the relationship between contaminants, exposure pathways and receptors taking into account the technical requirements of the project.

Questions answered by the conceptual model include:

- Is it possible for the contaminants to come into contact with the site users?
- What substances are on the site? Where are they in the environment? How much is there?
- What/who are the site users (current or future)? How do/will they use the site?

If by examination of the potential risks there is no route by which the contaminants can come in contact with the receptors, then the contamination does not present a risk (no exposure, no risk) and the risk is deemed insignificant. In such cases, an acceptable environmental risk scenario is considered to prevail, i.e. the risks identified were not considered significant, and the risk assessment should terminate at this stage with recommendations for environmental management controls that maintain the conditions under which the "acceptable risk" situation is prevalent. However, if risks to users are considered to be potentially significant, i.e. there is a possibility

that exposure could occur now or sometime in the future, then the remaining quantitative stages of the risk assessment should be completed. These are described in the following sections.

22.5 Exposure Assessment

22.5.1 OBJECTIVES AND LINK TO RISK ASSESSMENT PROCESS

Exposure assessment is the first stage in the assessment following problem formulation, and is carried out for all contaminating substances, receptors and pathways of exposure considered to be of potential significance during the initial problem formulation/risk screening stage.

In the exposure assessment stage, the magnitude of the potential exposures for the combinations of substances, site users and pathways identified in the conceptual model are evaluated using specified technical procedures, equations and models. Contact rates between contaminants and receptors are quantified by estimating the amount of substances at the site, the physical and chemical properties of the substances of concern and the biological characteristics of the site users, and the estimated duration of exposure for given land use scenarios; this is usually expressed as an amount of chemical intake by the receptor over a specified period of time. Estimated exposures may then be compared to documented exposure limits (or potency factors) to reveal whether acceptable limits have been exceeded and action is required.

Questions answered by the exposure analysis include:

- How much contact between the substance and the site users can occur?
- Which pathway is contributing the most contact?
- What is the acceptable level of exposure of the contaminant?

The exposure assessment is usually carried out in tandem with the toxicity assessment stage of risk assessment described in Section 22.6, since information from these components form the basis of the risk evaluation of exposure and tolerance. For example, the toxicity data of a chemical may be specific to a certain route of exposure intake; likewise, toxicity data may indicate that a certain receptor is particularly sensitive to a contaminant, requiring specific exposure assessment for this receptor.

22.5.2 BASIC ELEMENTS OF EXPOSURE ASSESSMENT

Exposure assessment is achieved by the application of mathematical exposure models and equations that have been developed to simulate exposure route. Thus, models exist for quantifying migration of contaminants through soil to indoor air and subsequent inhalation by humans; or ingestion of contaminated soil by children or small animals, etc. These models are continually being upgraded by researchers and risk assessors to improve their accuracy in predicting the exposure conditions. Selection of appropriate exposure models and choice of input parameters and assumptions require judgment on behalf of the risk assessor as well as communication with other stakeholders, e.g. regulators, the public. Simulation of exposure routes should only be carried out by suitably qualified and experienced practitioners.

The input for the exposure assessment may be obtained from: studies of the concentrations of various contaminants through affected environmental media (air, soil, groundwater and surface water), which may be accomplished through modeling of contaminant fate and distribution (see Chapter 25); an assessment of contaminant bioavailability (ability to participate in biological interactions); an analysis of receptor characteristics (to determine when and how often exposure can occur); and an assessment of

the potential media through which exposure may occur, e.g. soil characteristics, groundwater flow, air flux patterns. The last analysis for human receptors requires gaining an understanding of receptor physiological characteristics, lifestyle and habits in context of the contaminated site.

In conducting the exposure assessment, exposures for each potential receptor are estimated for each individual pathway (see Fig. 22.7), as well as for the background (noncontaminated) scenario. Thus, for a human receptor, contaminant exposure may be estimated by any or all of the three principal routes of ingestion, inhalation or direct contact, and in some cases may be summed to produce a total exposure value. It should be noted that a distinction is sometimes made between the contact exposure to a particular chemical, termed the potential dose, and the internal dose, which is the amount of chemical absorbed into the body. Most exposure models use the potential dose to describe exposure, i.e. a bioavailability of 100%, as the most conservative scenario. A distinction is also sometimes made between chemicals that act systematically, i.e. they must be adsorbed into the body to cause adverse effects, and those that act locally, i.e. do not require adsorption. For the former, which encompasses the majority of common contaminants, the exposure can be expressed as potential or critical dose rate, e.g. milligrams of chemical per kilogram of body weight per day. For locally acting contaminants, exposure rate is expressed as the duration or frequency of contact. This approach is most appropriate for gases, acid/ alkalis and so on. In most cases, the exposure scenarios of greatest significance will become evident, and can then be appropriately simulated.

Many jurisdictions across Canada and the USA have specified exposure models for conducting risk assessments. These approaches may require emphasis to be placed on certain exposure scenarios, or on particular transport media of concern, e.g. sensitive potable water supplies, depending on their unique circumstances. Most are based upon USEPA guidelines (e.g. EPA 1997) and the reader is directed toward these documents as a source of additional information.

Equation 22.1 describes a typical calculation for ingestion exposure.

LADD =

$$\frac{C_{soil} \times IR_{soil} \times RAF_{ing} \times EF \times ED}{BW \times AT} \quad (22.1)$$

where LADD = lifetime average daily dose (mg kg^{-1} per day^{-1}), C_{soil} = concentration of chemical in soil (mg kg^{-1}), IR_{soil} = ingestion rate of soil (mg day^{-1}), RAF_{ing} = relative absorption factor (fraction, usually 1.0), EF = exposure frequency (days y^{-1}), ED = exposure duration (y), BW = receptor body weight (kg), and AT = averaging time (days). Equation 22.1 describes a typical calculation for ingestion exposure.

22.6 Toxicity Assessment

22.6.1 OBJECTIVES AND LINK TO RISK ASSESSMENT PROCESS

Toxicity assessment is conducted for all chemicals of concern identified during the problem formulation, and involves identifying the toxicity associated with the different exposure scenarios. The aims of the toxicity assessment are to identify potential toxicological effects associated with contaminants of concern, and provide an estimate of the amount of chemical exposure that could be considered acceptable without any adverse health effects (for noncarcinogenic chemicals) or within acceptable degrees of risk (for carcinogenic chemicals). In assessing the toxicity of a contaminant, it is important to classify the substances of concern as being either carcinogenic or noncarcinogenic, depending on their mode of toxic action.

22.6.2 BASIC ELEMENTS OF TOXICITY ASSESSMENT

To achieve the objectives of the toxicity assessment as described previously, information is needed on such factors as:

- The mode of toxicological action of the contaminant, i.e. is the contaminant a threshold noncarcinogenic chemical, where adverse effects are noted only above a certain "threshold" dose, or is it a nonthreshold (carcinogenic) chemical, where adverse effects are assumed across the entire dose range.
- The exposure limit for contaminants of concern. This is expressed as either a reference dose (RFD), an allowable daily intake (ADI) or tolerable daily intake (TDI) for threshold chemicals, and as a potency factor, or risk-specific dose (RSD), which expresses the dose rate associated with a specific lifetime cancer risk, for non-threshold chemicals. The dose associated with an increase in lifetime cancer risk of one in one million is typically used as the limit of acceptable risk in the USA and Canada, though one in 100 000 and one in 10 000 risks may be employed by certain other jurisdictions.
- Consideration of other factors relating to toxicity, such as the sensitivity of various receptors, the chemical form of the contaminant (e.g. metal speciation or organic isomer presence), the presence of multiple exposure/toxicity components associated with chemical mixtures and simultaneous exposures.

Toxicology information is available from a variety of sources including literature sources, databases, e.g. Integrated Risk Information System (IRIS 1997), USEPA Health Effects Assessment Summary Table (HEAST 1997) and through suitably designed toxicity studies. In all cases, it is recommended that the acquisition/selection of toxicity information be only carried out by an experienced and qualified toxicologist.

Further considerations on contaminant toxicity include the chemical form of the contaminant, the importance of which is described in Section 22.3, the contaminant bioavailability, i.e. its ability to participate in biological activities, and the synergistic effects on toxicity caused by mixtures of contaminants. It is again recommended that interpretations in regard to the above only be carried out by experienced and appropriately trained personnel.

Questions answered by the toxicity analysis include:

- What types of adverse health effects may substances cause after exposure?
- What is the relationship between the magnitude of exposure and the probability or frequency of occurrence of the adverse health effects?
- What level of exposure can be tolerated without adverse effects?

22.6.3 DERIVATION OF EXPOSURE LIMITS

For chemicals with threshold-type, i.e. highly non-linear, dose–response relationships, the safety factor approach is used to estimate the threshold dose that would not produce measurable adverse effects in humans. The procedure used involves the division of the no-observed-adverse-effect-level (NOAEL) of exposure, usually determined from laboratory studies on animals, by appropriate uncertainty factors to derive an exposure limit (USEPA 1989f; Klaassen & Eaton 1991; Health Canada 1993). Typically, uncertainty factors of 100-fold are applied to the NOAEL:

- ten-fold extrapolation between laboratory species and humans; and
- ten-fold for interindividual differences in the human population, i.e. to account for sensitive subpopulations.

If the toxic effect upon which the exposure limit is based is judged to be severe, then additional uncertainty factors of five- or ten-fold, or larger, may be applied to the NOAEL (Health Canada 1993). If the available data are less than ideal, e.g. no information is available following chronic or long-term exposure, further uncertainty factors of two to ten-fold may be applied, depending on a scientific judgment of the weight-of-evidence available (Klaassen & Eaton 1991). The exposure limits developed by such procedures have been called "acceptable daily intakes" (ADIs) by agencies such as Health Canada and the World Health Organization, and "reference doses" by the USEPA (Barnes & Dourson 1988; Klaassen & Eaton 1991). The adoption of exposure limits based on the above procedures assumes that there is a rate of exposure to the chemical, below which no adverse effects would be observed, i.e. where the risk or probability of adverse effects is zero, for practical purposes.

For non-threshold chemicals, i.e. genotoxic carcinogens, the exposure limit is based on cancer potency factors (CPF), which are derived from extrapolating high exposure rates (from experimental studies) where effects can be measured to low exposure rates considered to pose acceptable risks to human health. The mathematical models used to extrapolate the exposure rates are conservative. A commonly used model is the linearized multistage model (Armitage & Doll 1961). A value for negligible risk is selected to calculate a risk-specific dose as the exposure limit. A common value for acceptable negligible risk is one in one million (1×10^{-6}) and corresponds to a risk that is insignificant. The exposure limit can be calculated by:

$$EL = \frac{ILCR}{CPF} \qquad (22.2)$$

where ILCR = incremental lifetime cancer risk, i.e. 1×10^{-6}; CPF = cancer potency factory; EL = exposure limit. Detailed discussion of applying uncertainty factors for risk assessment can be found in Faustman and Omenn (1996).

Table 22.4 summarizes a list of sources

TABLE 22.4. Sources of toxicity information

Source	Agency/government	Comment
Existing Data and Information		
Integrated Risk Information System (IRIS)	US Environmental Protection Agency	Website: www.epa.gov/iris
Health Effects Assessment Summary Tables (HEAST)	US Environmental Protection Agency	Published annually
Canadian Environmental Protection Act (CEPA)	Environmental Canada, Health Canada	Publications available through Canadian Federal Government
Canadian Council of Ministers of the Environment (CCME)	Environment Canada, Health Canada	Publications available through Canadian Federal Government
National Institute of Health (NIH) National Toxicology Program (NTP)	NIH/NTP	Website: ntp-db.niehs.nih.gov
Canadian Centre of Occupational Health and Safety (CCOHS)	CCOHS	Website: www.ccohs.ca
Agency for Toxic Substances and Disease Registry (ATSDR)	ATSDR	Hardcopy and CD-ROM published periodically. Website: atsdr1.atsdr.cdc.gov:8080/
American Conference of Governmental Industrial Hygienists (ACGIH)	ACGIH	Published guidelines and rationale to the development of guidelines
National Institute for Occupational Safety and Health (NIOSH)	NIOSH (Centers for Disease Control, US Department of Health and Human Services)	Published guidelines available. Website: www.cdc.gov/niosh
Occupational Safety and Health Association (OSHA)	US Department of Labor	Website: www.osha-slc.gov
Toxicology Excellence for Risk Assessment (TERA)	TERA	Website: www.tera.org
Ontario Ministry of Environment (OMOE)	Ontario Provincial Government	Publications available through Provincial Government
Ontario Ministry of Labour (OMOL)	Ontario Provincial Government	Publications available through Provincial Government
British Columbia Ministry of the Environment, Lands and Parks (BCE)	BC Provincial Government	Website: www.env.gov.bc.ca
Massachusetts Department of Environmental Protection (MDEP)	Massachusetts State Government	Website: www.magnet.state.ma.us/dep/
California Environmental Protection Agency (Cal EPA)	California State Government	Website: www.calepa.cahwnet.gov
World Health Organization (WHO)	WHO	Published monographs on wide-ranging issues, health information may be hard to retrieve. Website: www.who.org
US Air Force Total Petroleum Hydrocarbon Criteria Working Group (TPHCWG)	US Government	Website: voyager.wpafb.af.mil
Region 3 and Region 9 USEPA	Regional USEPA	EPA website (above) has links to Regional Information
***De novo* synthesis of toxicity information**		
Specialist toxicology and risk assessment consultants		
Specialist toxicity testing laboratories		

and/or agencies that provide peer-reviewed toxicity data through published literature or on-line databases (including World Wide Web addresses).

22.7 Risk Characterization

22.7.1 OBJECTIVES AND LINK TO RISK ASSESSMENT PROCESS

The final stage of the risk assessment process is risk characterization. This stage integrates the information from the toxicity analysis and exposure analysis stages to estimate the magnitude and nature of the risk (a probability), if any, to which the site users may be subjected.

Risk characterization involves two steps: (i) estimating risks by comparing estimated exposure (as described in Section 22.5) with the acceptable exposure limit (as described in Section 22.6), followed by (ii) a description of the estimated risks in the context of the site-specific factors, background conditions, future land use, uncertainties, and other economic and societal factors. The results of the risk characterization form the basis for subsequent risk management decisions.

22.7.2 BASIC ELEMENTS OF RISK CHARACTERIZATION

Risk estimation may be achieved by either correlating exposure and toxicity assessments to establish whether the site specific exposure is acceptable, or by "back-calculating" maximum allowable soil and groundwater concentrations starting from acceptable exposure limits for each exposure scenario. In this latter methodology, the lowest soil and groundwater concentration across all exposure scenarios is considered to be most protective of human and ecological health, and may therefore be used as acceptable criteria for the site.

Correlation of exposure and toxicity assessments may be expressed as either a hazard quotient (HQ, for threshold chemicals), a numerical cancer risk (NCR) estimate (for

non-threshold chemicals) or an exposure ratio (ER) value (which can be applied to both chemical types). The choice of methodology is usually determined by the regulatory agency, in consideration of risk communication requirements. For the HQ approach, the risk is estimated by comparing predicted exposure and exposure limit according to:

$$ HQ = \frac{\text{Dose rate}}{\text{Exposure limit (threshold)}}. \quad (22.3) $$

An HQ value exceeding one, or in some cases 0.1 (reference value depends on regulatory requirements), implies an unacceptable exposure condition.

An NCR estimate is calculated according to:

$$ NCR \text{ estimate} = \text{dose rate} \\ \times \text{ cancer potency factor.} \quad (22.4) $$

The NCR estimate produces a value for the incremental increased risk of acquiring cancer expressed as a probability, e.g. one chance per one million. The USEPA maximum acceptable incremental risk has been set at one in 1 000 000 and this is the most widely adopted value.

The ER value is determined by:

$$ ER = \frac{\text{Dose rate}}{\text{Exposure limit (threshold or non-threshold)}}. \quad (22.5) $$

It is generally considered that the ER approach is preferable for communicating risks to the public because threshold and non-threshold are placed on a similar basis, and because the negative connotations associated with the terms "hazard" or "cancer risk" are reduced, although this latter consideration has no technical substance.

22.7.3 RISK DESCRIPTION

Evaluated risks are most effectively described by making a comparison between risks associated with different scenarios, e.g. site-specific versus background conditions, one exposure route versus another, or exposure-specific risks versus estimated "safe" exposure-risk scenarios, which invariably include one or more conservative uncertainty factors.

Concerns over the description of risks range from the overestimation of risks due to the incorporation of overly conservative factors into exposure limits, to the underestimation of risks due to insufficient accounting for information uncertainties, synergistic effects of multiple exposures, and the toxicity of chemical mixtures.

22.7.4 REPORTING OF UNCERTAINTY

It should be noted that the results of risk assessments should not be provided in absolute terms, without a description of the inherent assumptions, judgments and potential variations that unavoidably arise during the risk assessment process. For example, data obtained from a single sample taken from a single borehole at a site suspected of being contaminated provides information on the identity and concentration of contamination at the point of sampling only. The variation of contaminant identity and concentration with time, areal and vertical extent, and subsurface media can only be assumed, based upon experience and with knowledge of site geology, hydrogeology and so on.

The quantitative results of a risk assessment can usually be presented in one of three ways: point estimates, as a range, or as a probability distribution (see Fig. 22.8).

Where an analysis has been carried out with input parameters or a range of input values, specifically chosen to simulate the exposure conditions, the results of analysis are respectively provided as a single point or range estimated. Discussions relating to uncertainty

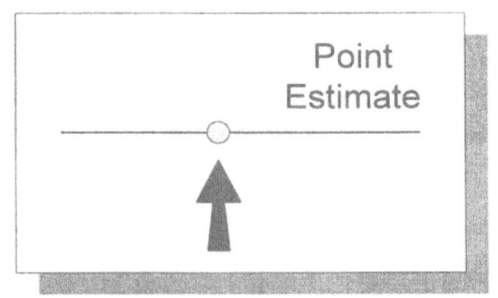

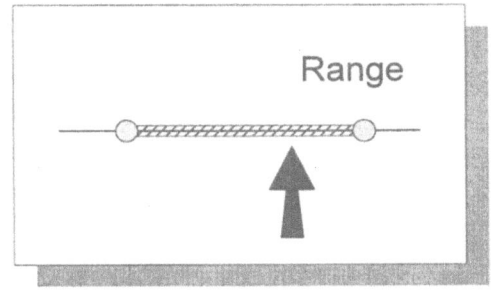

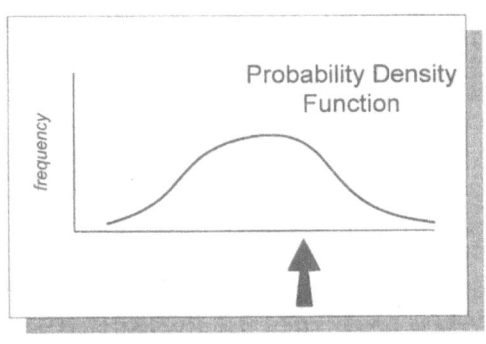

FIGURE 22.8. Methods of presenting results of risk calculations.

focus on the reasons why specific values were chosen.

When the analysis is required to more realistically reflect actual conditions, distributions that better describe the variable, e.g. the distribution of body weight for the study population, can be used in the analysis. The analysis is completed with random selection of input parameter combination and the results can be presented as a distribution.

Assessments that do not acknowledge such uncertainties risk being laid open to misinterpretation. Presentation of this type of information, or the description of uncertainty, should not be seen as admissions of ineptitude on behalf of the proponent; rather, full reporting of uncertainty provides a mechanism by which the realities of the site's environmental condition can be more fully understood and communicated to all project stakeholders.

22.8 Risk Communication

22.8.1 THE NEED FOR EFFECTIVE COMMUNICATION

Risk assessment/risk management projects differ from more conventional environmental management projects in that they invariably involve a broad number of stakeholders, including regulators (from a variety of jurisdictional levels), site owners, developers, lawyers, financial institutions, environmental consultants and the public. These stakeholders have varied interests in the project, and are likely to have differing views, perspectives and understandings of what risk assessment is, what is acceptable and what it can achieve. It is important that there is a clear and mutual understanding between the project team, the stakeholders and the regulatory body(ies) with regards to the scope of the risk assessment and the required level of protection for human and ecological site users. This can only be achieved through communication. Identifying and agreeing upon risk-based restoration, scope and ultimate restoration goals early in the process not only speeds up the remediation process (to everybody's benefit), but also builds and sharpens confidence that the risk assessment has taken into account a broader range of issues to ensure adequate protection for human and ecological site users. In summary, effective communication helps to ensure that all project stakeholders move in unison toward the optimal goal of site restoration. Effective communication resulting in stakeholder commitment is, therefore, critical to the successful application of a risk-based approach to contaminated land management. Two "forms" of communication should be formally accounted for: external communication and internal communication.

22.8.2 EXTERNAL COMMUNICATION

External communication encompasses the necessary public communications program associated with most projects where environmental issues are required to be addressed. This form of communication is often characterized by an atmosphere of scepticism and potential confrontation. Communication regarding environmental issues can become emotionally charged and should be co-ordinated/facilitated by a communication specialist. The primary initial goal of external communication should be to develop a working platform that is founded on a credible, confident and common understanding with the public. The underlying role of risk assessment as a framework for identifying, prioritizing and managing human and ecological health risks. When communicating risks to the general public, it is particularly crucial that the risk assessment option is explained as a legitimate means of providing protection to human health and the environment, and not as a "high tech" means of "risking away" the real issues. It is often forgotten that a risk-based approach was used to develop many generic soil and groundwater criteria, which the remediation industry and the community have accepted.

22.8.3 INTERNAL COMMUNICATION

Internal communication requirements stem from a growing recognition that the management of contaminated land within a risk-based context requires effective communication not just between project stakeholders, but also between the various scientific and technical disciplines that may be represented in the project team (Whittaker *et al.* 1998). Key disciplines are engineering (for risk management and remediation activities), chemistry (for understanding the fate and partitioning of contaminants), geology and hydrogeology (for site assessment, including soil types, and groundwater flow patterns), environmental microbiology (for understanding the effects on spilled oily wastes on soil microbial consortia) and toxicology (for defining the ecological and human health risks presented by contaminants). In problem identification for risk assessment, it is imperative that dialogue is initiated between the toxicologist, chemist and the field engineer during the planning and design stages of the field investigation program, rather than after the field work has been completed. This communication will ensure that the field sampling is directed not just by contaminant abundance and/or availability, but also by toxicology, pathways and exposure concerns.

22.9 Risk Management

As discussed above, completion of the problem formulation and risk assessment will quantify the environmental risk associated with a site and provide the proponent with an understanding of how and what level of significance environmental risks may be presented at a site. Clearly, if the risks are evaluated to be at unacceptable levels for any valued receptor, then the proponent must carry out work to mitigate these exposures. On this basis, the objective of an environmental risk management strategy is to provide suitable protection to the human and ecological site uses from the potential environmental exposure.

The implemented risk-based measures should specifically address the dominant components of the environmental risk and reduce the exposure to acceptable levels.

22.9.1 SELECTION OF RISK MANAGEMENT OPTIONS

Typically, a range of mitigative risk management measures are required to address issues identified in the risk assessment. At a conceptual level, risk management measures are directed to either removing or reducing contaminant levels, blocking exposure pathways or changing the habits or presence of the receptor (Fig. 22.9). Such measures include, but are not limited to, extraction (contaminant removal), encapsulation (barrier installation), avoidance (by modifying land use to restrict receptor exposure), treatment (by physical, chemical or biological remedial methods) or a combination of these alternatives. It is generally recognized that a single remedial method is often insufficient to satisfactorily remediate more complex sites that have required the use of risk-based approach. Effective risk management should, therefore, endeavor to apply a combination of mitigative measures or in sequence (as sometimes termed the "treatment train") (Sims 1990).

Evaluation of the preferred restoration alternatives (e.g. see Chapters 30 and 31) should involve all stakeholders and should take into account evaluation/selection criteria that encompass all aspects of the restoration. The relative significance of the evaluating criteria usually becomes apparent following interactive discussion between all project stakeholders, and should be weighted to reflect the relative site-specific importance of the criteria. Key factors for selection of the preferred risk management options can include:

1. Technical effectiveness: considers the suitability of the proposed risk management strategy within the context of the site management goals.

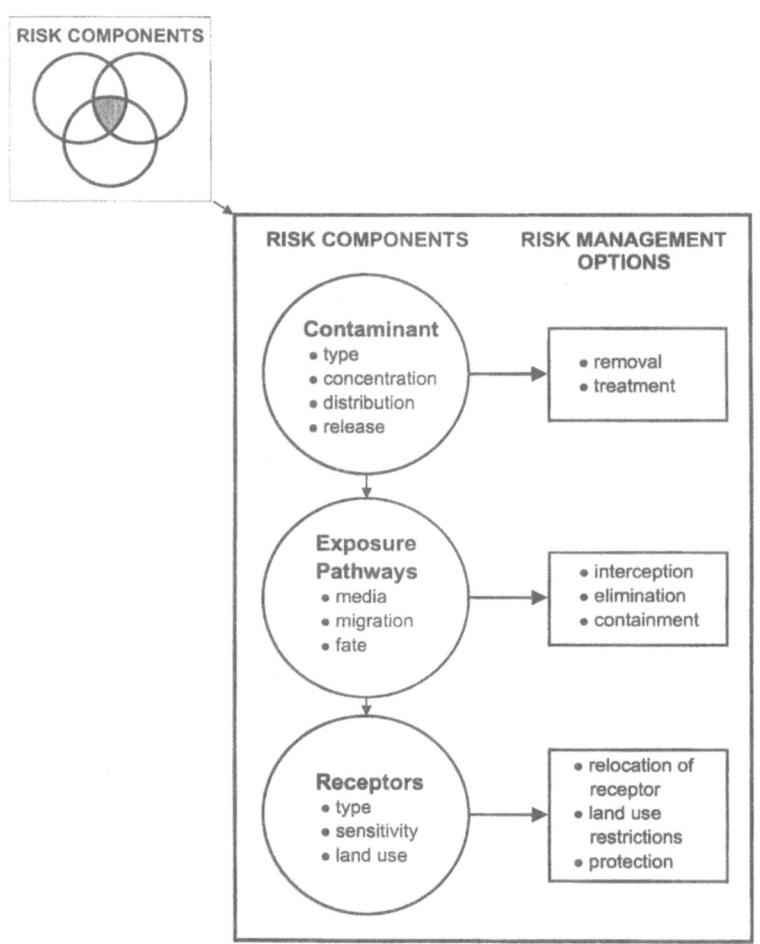

FIGURE 22.9 Risk management options.

2. Long-term performance: considers the longevity of the proposed risk management approach.

3. Degree of protection: examines the degree of protection that the strategy provides, assuming that the strategy has been effectively implemented.

4. Regulatory approvals: examines the approval requirements for the strategy to be implemented.

5. Environmental assessment requirements: considers the need for public environmental assessment of the proposed strategy.

6. Construction constraints: examines the difficulties associated with installation and maintenance of the proposed risk management measures.

7. Community acceptance: considers factors relating to community acceptance of risk management options, including both the installation and construction period, and the long-term operation.

8. Implementation time: may be important if rapid risk management controls are required.

9. Operational maintenance: considers the

need for operational maintenance of proposed risk management controls.

10. Capital costs: examines the total cost of installing or setting up of risk control measures.
11. Long-term operation and maintenance costs: considers the on-going cost of the strategy to continuously provide the required protection.
12. Potential costs associated with residual liabilities (e.g. costs associated with incomplete clean-up, legal fees).

22.9.2 MONITORING RISK MANAGEMENT SUCCESS

When the remedial strategy has been chosen it is important to evaluate the proposed measures in the context of the initial risk assessment to ensure that suitable protection has been or will be achieved. When contaminant removal or extraction has been selected, for example, this evaluation is a straightforward verification of post-extraction conditions. In other situations where the contamination has been encapsulated by a soil or synthetic barrier (e.g. see Chapters 26–28), for example, the verifying risk assessment would have to confirm the integrity of the environmental control barrier as well as confirm its effectiveness in containing the contaminants of concern (e.g. see Chapters 25 and 29). Where contaminant remediation has been carried out, evaluation could entail monitoring compliance with the risk based clean-up criteria through additional confirmatory sampling and analysis.

Finally, in situations where risks have been managed through engineered environmental control measures that allow the contaminants to remain in place, e.g. encapsulation, it is necessary to have in place some mechanism for ensuring that current and future land owners are informed of the site-specific conditions that environmental measures impose on future site operation. Specifically, these limitations relate to any on-going monitoring

that may be required to ensure compliance, the constraints placed on land use and, as a consequence of these constraints, the depreciation in value of the land due to the continued presence of contaminants thereon. In most cases this notification can be achieved by a description of the limitations on some component of the title of the property. Though this type of notice of restriction can create a stigma for the site in question, it should be remembered that good science has been used to establish the specific conditions under which the environmental protection is provided and that the total cost of implementing the risk-based strategy can be significantly lower than the cost of full extraction of contamination. The whole simulation process is based on real, probable land use scenarios and even though various levels of conservatism may be included in the assessment, the results and consequential measures are specific only to the site and the land use scenarios under assessment. In the event of the land use changes, the revised land uses are to be examined in the context of the initial risk assessment to ensure that adequate protection continues to be provided.

The "parking lot" example is best used to illustrate land use dependency and the need for notification to future land owners/users. Consider a contaminated site where the on-site impact is such that exposure is caused by inhalation of frequent dust and/or direct contact with the soil. Combined use of the site can be provided by the installation of a barrier in the form of a parking lot pavement, paved walkways, swimming pool, tennis courts, layer of soil for ornamental gardens, etc. These remedial measures, though sufficient for the intended land use are not sufficient for any land use such as residential, where the exposure would occur. Consequently, notification of constraints must remain accessible for future land owners for all situations where risk management measures have been installed to control/manage the risk due to environmental impact.

22.9.3 SUMMARY

The risk assessment/risk management process should be viewed as a versatile, powerful analytical process that can be used to examine simulated aspects of an environmental issue for specific-site conditions. The range of complexity of a risk assessment can vary from a simple screening comparison with existing criteria to a full quantitative risk assessment. In all cases, the complexity of the assessment should be appropriate for the problem and "common sense" should prevail in order to develop an acceptable solution for all stakeholders.

23. PHYSICOCHEMISTRY OF SOILS FOR GEOENVIRONMENTAL ENGINEERING

J. K. Mitchell

23.1 Introduction

The goals of geoenvironmental engineering include: (1) safe containment of wastes, (2) clean-up of contaminated ground and groundwater, (3) prevention of contamination and pollution, and (4) enhancement of the natural and man-made environment. Attainment of these goals inevitably involves activities on, in, or with the earth. Therefore, soils and their properties and behavior over a range of conditions are of major importance in geoenvironmental engineering. The purpose of this chapter is to identify and summarize aspects of soils that are important in dealing with geoenvironmental problems, such as those considered in subsequent chapters of this handbook. They include contaminant hydrogeology and contaminant migration (Chapter 24), clay liners (Chapter 25), geosynthetics in waste containment systems (Chapter 26), landfill covers (Chapter 27), *in situ* containment and treatment of contaminated soil and groundwater (Chapter 29), and contaminated soil management (Chapter 30). Much of the information in this chapter is covered in more detail in Mitchell (1993).

23.2 Soil Mineralogy and Composition

A soil is typically composed of several types of solid particles, water, dissolved materials and gases. The solid particles include tremendous range of sizes (Fig. 23.1). Although attention is focused on the solid phase in most geotechnical studies, properties of water and the influences of system chemistry must also be taken into account in geoenvironmental engineering.

23.2.1 MINERAL COMPOSITIONS AND STRUCTURES

Gravel, sand and silt-size particles in a soil are usually silicate minerals of the same type as the rock from which they are derived. The most abundant elements in these minerals are oxygen (91.8 vol % and almost 50 wt % in igneous rocks), silicon (about 25 wt % and 0.8 vol %), and aluminum. Magnesium, iron, calcium, potassium and many other elements may be present in small amounts. Clay-size particles are usually hydrous aluminum silicate minerals of various types.

23.2.2 SILICATE MINERALS

Silicate minerals, the dominant minerals in rocks, are composed of silica tetrahedra $(SiO_4)^{4-}$ held together and arranged in various ways. A silicon tetrahedron and silicon tetrahedra arranged in a hexagonal sheet pattern are shown in Fig. 23.2. Many arrangements are possible owing to the small size and high charge of the silicon atom. This accounts

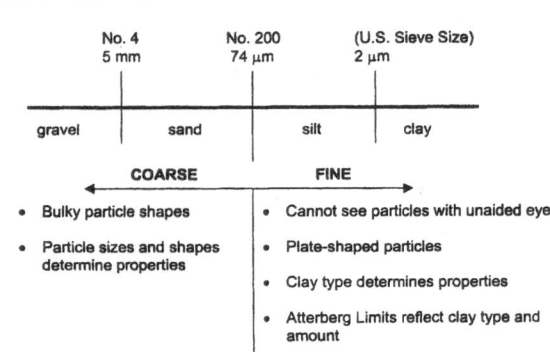

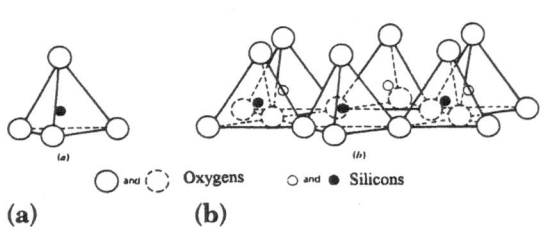

FIGURE 23.1. Soil particle sizes and characteristics.

FIGURE 23.2. Silicon tetrahedron (a) and silica tetrahedra arranged in a hexagonal network (b).

for the many different silicate mineral types in nature. Tetrahedra may be independent or associate in pairs, in rings, in chains, in bands, in sheets and in three-dimensional frameworks. The greater and more complex the association, the more stable and resistant to breakdown is the structure.

23.2.3 NON-CLAY MINERALS IN SOILS

Because soil particles are products of rock weathering and because different silicate minerals have different resistances to weathering, the relative proportions of specific minerals in soils differ greatly from those in the source rocks. Quartz, which accounts for only 12% of igneous rocks, is the most abundant mineral in soils because of its resistant framework structure. On the other hand, feld-spar, which accounts for 59% of igneous rocks, is usually present only in small amounts in soils. Micas, although present in igneous rock in only relatively small amounts, are quite resistant to weathering; thus they are often found in both the coarse and fine fractions of a soil.

Carbonates (such as calcite), sulfates (such as gypsum) and oxides, (especially iron oxides), may also be present in a soil, and these materials require special attention in geoenvironmental engineering. Carbonates and sulfates are sufficiently soluble that they may react with wastes, thus influencing such things as waste containment barrier effectiveness and *in situ* treatment technologies. Iron oxides, usually identifiable by the yellow–red colors imparted to a soil, are left behind from weathering processes. They are found most commonly in wet, warm–tropical climates, and they can act as cementing agents within the soil.

23.2.4 CLAY MINERAL STRUCTURES

Clay minerals are characterized by: (1) small particle size, generally <2 μm; (2) platy morphology (except for halloysite, which has tubular particles, and attapulgite, which has lath-like particles); (3) plasticity when mixed with water; (4) high resistance to weathering, which reflects high structural stability; and (5) net negative particle charge. The net negative charge results from isomorphous substitution during mineral formation, wherein a replacing cation has a valence lower than the replaced cation. The magnitude of the electro-negativity is measured in terms of the cation exchange capacity (equivalents kg^{-1}).

Different clay mineral structures are formed from different associations of silica tetrahedra and aluminum or magnesium octahedra. An octahedron unit and a sheet structure formed by octahedra are shown in Fig. 23.3. A synthesis pattern for the major clay mineral groups is shown in Fig. 23.4. Silicon tetrahedral sheets and octahedral sheets are stacked in regular ways to form the unit

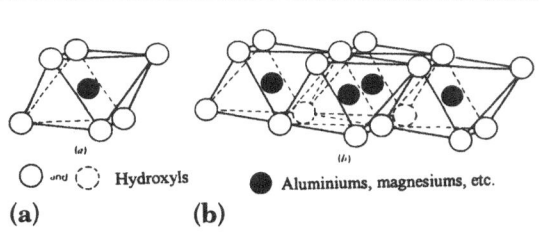

FIGURE 23.3. Octahedral unit (a) and sheet structure of octahedra (b).

cell layers shown in the figure. The actual mineral particles, with the exception of halloysite, consist of stacks of many hundreds or thousands of layers. Pyrophyllite is a 2:1 mineral structure without isomorphous substitution. It is not a clay mineral because it is not found in the very small particle sizes typical of clay, it is electrically neutral, and it is not plastic when mixed with water. Some salient features of the clay minerals likely to

be encountered on geoenvironmental projects are:

- *Kaolinite* consists of 1:1 unit layers held together by hydrogen bonds. It occurs in relatively large particles, has low plasticity and does not swell.
- *Halloysite* in its hydrated state consists of two 1:1 unit cell layers separated by a single layer of water molecules. Particles are hollow tubes. The water layer is irreversibly removed by drying, and possibly also as a result of pressure, with collapse to a structure similar to kaolinite.
- *Smectite minerals* consist of stacked 2:1 unit layers and differ in the isomorphous substitution in both the tetrahedral and octahedral sheets. In montmorillonite, the most common mineral in the group, one of every six aluminum atoms in the octahedral layer is replaced by magnesium. Because the unit layers are not strongly bonded, water and cations can enter between them,

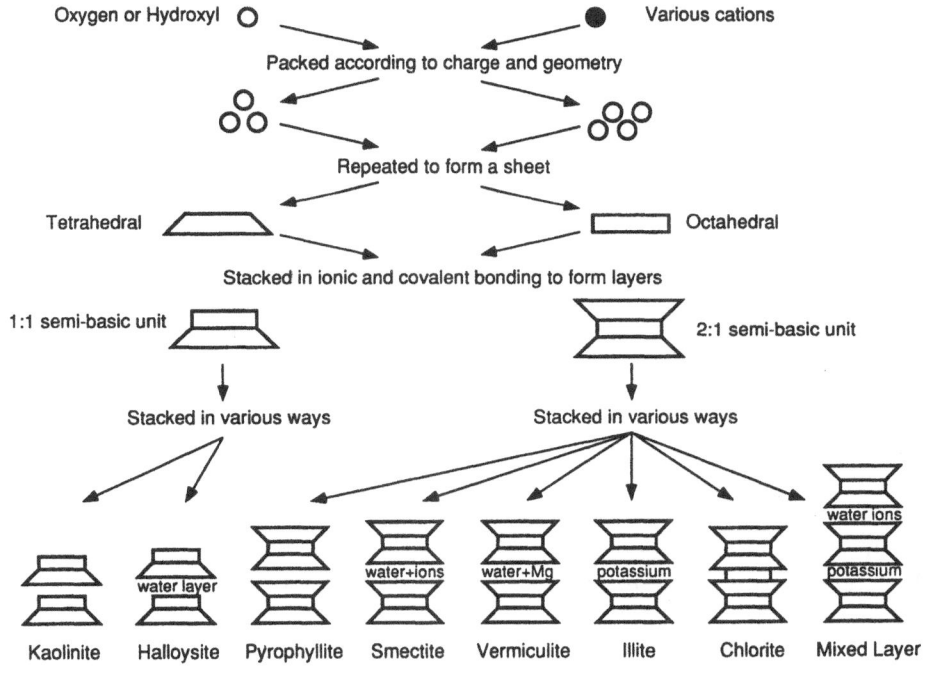

FIGURE 23.4. Synthesis pattern for the clay minerals (reproduced from *Fundamentals of Soil Behavior*, J.K. Mitchell © 1993 John Wiley & Sons, Inc. Reprinted by permission of John Wiley & Sons, Inc.).

accounting for the highly expansive character of the smectite minerals. Bentonite, which is widely used in drilling mud, contaminant containment barriers, landfill liners (see Chapter 25), and covers (see Chapter 27), is a montmorillonite of volcanic origin.

- *Vermiculite*, which is less common than smectite, has a more organized interlayer structure, with Mg^{2+} or Ca^{2+} and water between the layers. It is not as expansive as smectite.
- *Illite* or *hydrous mica* contains interlayer K^+, which fits well into the holes in the silica sheets. The potassium ions bond the unit layers together strongly enough that illite is non-expansive. Illite is the most frequently encountered clay mineral in sedimentary soils in temperate climates.
- *Chlorite* has an octahedral-type layer holding the 2 : 1 layers together. It is neither expansive nor very abundant in soils.
- *Mixed-layer clays* are composed of regularly or randomly stacked layers of different unit cell types, e.g. montmorillonite–illite. Mixed-layer clays are common in soils.
- *Attapulgite* has a double silica chain structure, and occurs as lath or thread-like particles. It is useful for drilling-mud in chemical and saline environments because of its high stability in suspensions in comparison to bentonite.
- *Allophane* is clay-like material without definable crystal structure, i.e. a material that is amorphous to X-rays. It is often found in areas of intense weathering, such as wet, tropical climates.

23.2.5 CLAY MINERAL CHARACTERISTICS

The special, and often unusual, properties of clays are a direct consequence of: (1) their very small particle size and, therefore, very large specific surface; (2) unbalanced surface forces; (3) net negative electrical charge; and (4) platy, tubular and lath-like particle shapes. The possible ranges in each physical, chemical and mechanical property of a given soil are determined by the type and amount of each mineral, the particle sizes and shapes, the types of cations adsorbed to balance the electro-negativity, and the composition of the porewater. Data on these characteristics can be obtained from representative, but disturbed samples. However, the actual values of a property in any case depend on the water content and unit weight, the state of stress, the temperature and the availability of water. Undisturbed samples or appropriate *in situ* tests are needed to determine these properties.

Table 23.1 provides information on some characteristics and properties of common clay mineral types. The properties of smectite vary over the widest range. Because of its very high specific surface, corresponding to its very small particle size, its properties are very susceptible to changes in the composition of the pore fluid, e.g. salt concentration, cation type, dielectric constant and pH.

Real soils are seldom composed solely of one mineral type. The specific values of different properties of a soil cannot be assumed to be in proportion to the amounts of the different minerals that are present because of physical and chemical interactions. Nonetheless, a little clay goes a long way in influencing soil properties, because the specific surface of the clay phase contributes the greatest proportion of the total surface area of the soil, and water is adsorbed approximately in proportion to the specific surface. As a consequence, a relatively small amount of clay in a wet soil is sufficient to coat larger particles, fill the voids between the larger particles, and produce physical properties that are more characteristic of the clay than of the silt and sand that may be present. Thus, identification of the clay minerals present is of great value in understanding and predicting probable behavior.

23.2.6 ORGANIC MATTER IN SOILS

Organic matter occurs in soils in a variety of forms, ranging from large, visible items such as roots, twigs and leaves, to finely divided

TABLE 23.1. Characteristics and properties of common clay minerals

	Kaolinite	Halloysite	Smectite	Illite	Mixed layer
Structure	1:1	1:1	2:1	2:1	2:1
Basal spacing, nm	0.72	1.0	1.0↑	1.0	1.0–1.7
Cation exchange capacity, eq. kg^{-1}	0.03–0.15	0.05–0.40	0.8–1.5	0.1–0.4	Variable
Specific surface, m^2 g^{-1}	10–20	35–70	Up to 800	65–100	Variable
Particle shape	Plates	Tubes	Plates/films	Plates	Plates
Usual location	Warm, humid	Wet, volcanic	Warm, dry	Temperate	Anywhere
Relative particle size	Large	Small	Very small	Medium	Variable
Liquid limit, %	30–75	35–70	100–900	60–120	50–150(?)
Plastic limit, %	25–40	30–60	50–100	35–60	30–60(?)
Activity	< 0.5	0.1–0.5	1–7	0.5–1	0.5–1.5(?)
Shrink/swell	Negligible	No swell, collapse on drying	Very high	Low	Moderate
Hydraulic conductivity, m s^{-1}	10^{-7}–10^{-9}	10^{-7}–10^{-9}	10^{-9}–10^{-11}	10^{-8}–10^{-9}	10^{-8}–10^{-11}
Compression index	0.2–0.3	—	1.0–2.6	0.5–1.1	—
Shear strength, ϕ', deg	24–30	—	5–17	17–26	—

material and colloidal-size particles of carbohydrates, proteins, fats, resins, waxes and hydrocarbons, which are products of biological decay. Decomposing organic material usually imparts a dark color to a soil, often an odor, and results in increased soil plasticity and compressibility. Soils with significant organic matter content will usually exhibit irreversible decreases in plasticity and compressibility if dried.

Organic compounds that are disposed or spilled into the ground react with the organic matter in the soil. Hydrocarbon wastes that are adsorbed onto soil organic matter are difficult to remove. The organic phase plays an important role in *in situ* biodegradation processes, and is the source of carbon dioxide and methane from aerobic and anaerobic decomposition processes.

23.3 Soil Water

The water phase of a soil may comprise a substantial proportion of the total soil volume. For example, in a saturated soil at a water content of 50%, the volume of water will be about 1.35 times the volume of solid material. Accordingly, the properties of soil water and the interactions of the water and solid phases of a soil are important to the overall physical and chemical behavior of the material.

23.3.1 ICE AND WATER STRUCTURE

The eight outer shell electrons in a water molecule are associated in pairs located in a tetrahedral arrangement. This results in separations between the centers of positive and negative atomic charges, making each molecule a strong dipole that can be oriented in an electrical field. Individual water molecules are attracted to each other by hydrogen bonds. In ice a sufficient number of hydrogen bonds is formed to produce a solid structure. Ice breaks down to water when about 15% of the possible hydrogen bonds are ruptured. At room temperature about half the possible hydrogen bonds are broken. Without hydrogen bonding and the open structure that results, the density of water would be about 1.84 g cm^{-3}, rather than 1.0 g cm^{-3}.

23.3.2 THE INFLUENCE OF DISSOLVED IONS

As a result of their uneven charge distribution and dipolar character, water molecules are attracted to ions in solution, resulting in ion hydration. Water molecules move from their normal structure into positions in the hydration shell of an ion, provided their energy state is less as water of hydration than as normal water. Ions disrupt normal water structure. A schematic model for ion–water interaction is shown in Fig. 23.5.

23.3.3 WATER ADSORPTION BY SOILS

Given that water consists of oxygen ions held together by hydrogen, and silicate mineral surfaces are composed of oxygen ions held together by silicon, water is attracted to and strongly adsorbed on soil particle surfaces. All details of the mechanisms of attraction are not known. Some proposed mechanisms are shown in Fig. 23.6. Of these, it seems most likely that hydrogen bonding is important in wet soils, and surface attraction of hydrated ions is significant for drier soils.

Water is strongly attracted to a distance of about 1.0 nm, i.e. three molecular layers, from a silicate layer surface (Sposito 1984)

and has a structure different from that of normal water. Surface forces influence water structure out to 10 nm from particle surfaces. To put this in perspective, consider a low plasticity clay soil having a specific surface area of 20 m^2 g^{-1}, and a high plasticity clay with a specific surface area of 200 m^2 g^{-1}. Each has a water content corresponding to an average water layer thickness on the particle surfaces of 5 nm. The water content of the low plasticity clay, e.g. a kaolinite, would be 10%; whereas, the water content of the high plasticity clay would be 100%. These water contents are in the ranges characteristic of compacted clays (10–20%) and saturated, natural clays (50–100%). Thus, most compacted fine-grained soils contain water that is strongly influenced by the surface properties of the particles, and water properties in more plastic materials can be influenced by surface adsorption forces to relatively high water contents.

23.3.4 PROPERTIES OF ADSORBED WATER

Some characteristics and properties of adsorbed and free water in soil pores that are important in geoenvironmental engineering are:

- Available data suggest that adsorbed water has a hydrogen bonded structure that includes random networks and five-coordinated molecular rings.
- Energy is released during the adsorption of water on soil surfaces.
- The structure of the water adsorbed on soil particle surfaces differs from that of ice and normal water.
- Adsorbed water exhibits both freezing point depression and supercooling.
- There is no evidence for abnormal water viscosity or failure of Darcy's law in soils of the type usually encountered. For practical purposes, the viscosity and diffusion properties of soil water are the same as for pure water.

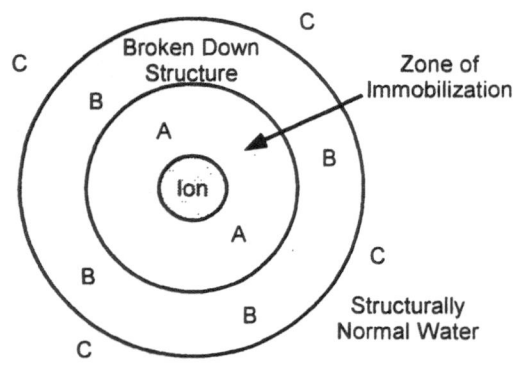

FIGURE 23.5. Ion–water interaction (as postulated in Frank and Wen 1957).

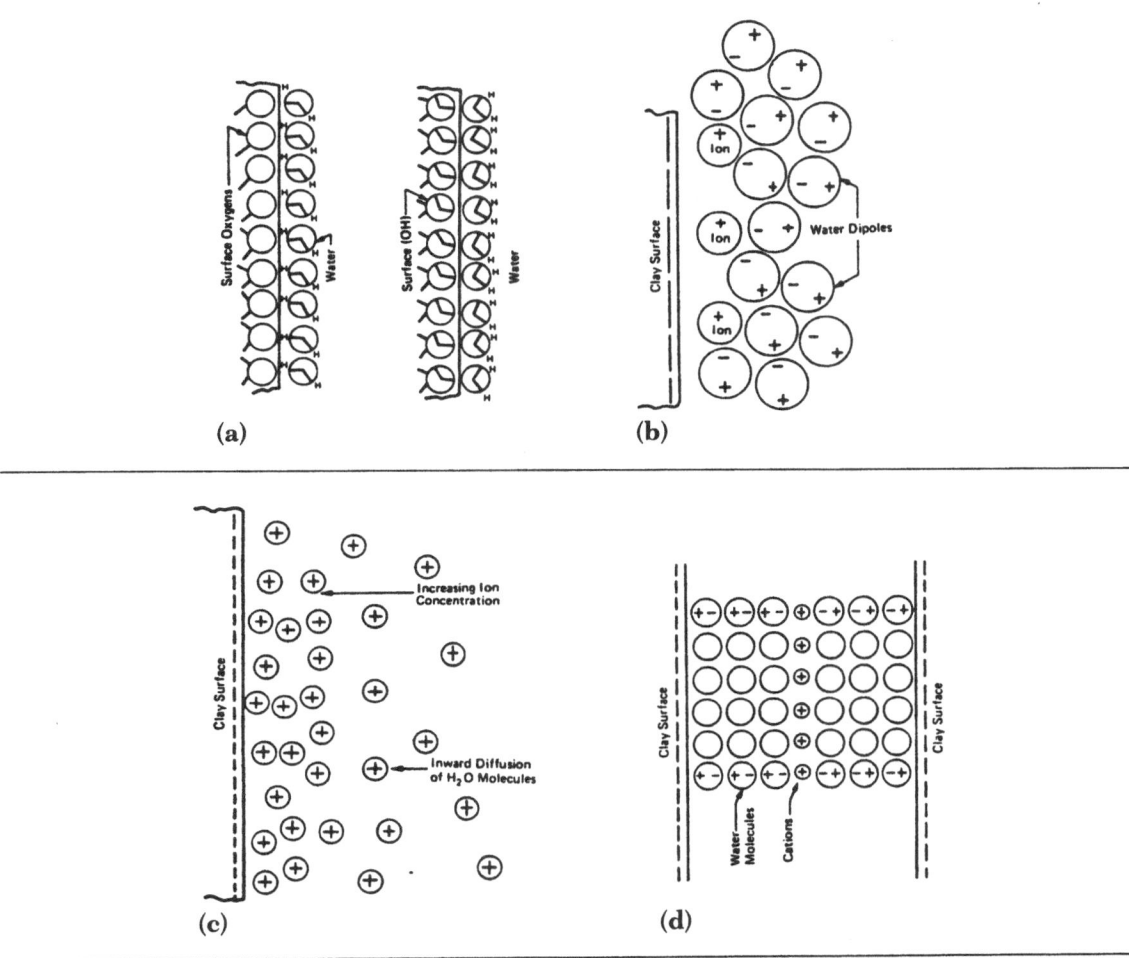

FIGURE 23.6. Possible mechanisms of water adsorption by silicate mineral surfaces: (a) hydrogen bonding, (b) ion hydration, (c) attraction by osmosis, and (d) dipole attraction (modified from Mitchell 1993).

- The dielectric constant of strongly adsorbed water may be less than that of free water.

23.4 Clay–Water–Electrolyte System

The clay particles in a soil behave as lyophobic colloids. In liquid suspensions they are small, solid particles that: (1) are two-phase systems with a large interfacial surface area, (2) exhibit behavior that is dominated by surface forces, and (3) can flocculate in the presence of small amounts of salt. Electrical double layer theory provides a basis for understanding many aspects of the behavior of such colloidal systems. This understanding can be used for explanation of many aspects of soil properties, soil–waste interactions and their effects on engineering properties, ion exchange phenomena, and adsorption processes. It is also useful for development and evaluation of different methods for remediation of contaminated ground.

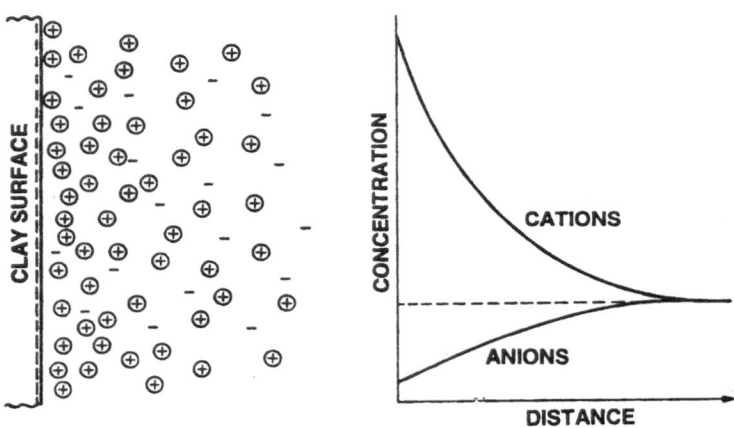

FIGURE 23.7. Distribution of ions adjacent to a clay surface according to the concept of the diffuse double layer.

23.4.1 ION DISTRIBUTIONS: THE DOUBLE LAYER

A representative distribution of cations and ions in solution near the surface of a clay particle is shown in Fig. 23.7. Cation and anion distributions depend on the combined effects of two forces: (1) the electrostatic attraction (cations) or repulsion (anions) of the negatively charged particle surface, and (2) the escaping tendency of the cations and the inward diffusion tendency of the anions owing to the spatial variations in concentration. The charged surface and the distributed charge in the adjacent phase are together termed the diffuse double layer.

The differential equation used to compute ion concentration and electrical potential versus distance in the double layer is obtained by combining: (1) the Boltzmann equation for ion concentration as a function of potential; and (2) the Poisson equation for relating potential, charge and distance.

The Boltzmann equation is:

$$n_i = n_{i0} \exp(-v_i e \psi / kT) \qquad (23.1)$$

where n_i = the concentration of an ion species, i, at a point within the double layer (ions m^{-3}); n_{i0} = reference ion concentration at a point where ions are no longer influenced by the particle surface forces; v_i = valence of ion species; e = electronic charge (1.602×10^{-19} coulomb); ψ = electrical potential (V) at the point (defined as the work to bring a positive unit charge from an infinite distance to the specified point in the electric field); k = Boltzmann constant (1.38×10^{-23} JK^{-1} per molecule), where k = the gas constant (R) per molecule, ($k = R/N$, where N is Avogadro's number); and T = temperature (K).

The Poisson equation relating potential, charge, and distance is:

$$\frac{d^2 \psi}{dx^2} = -\frac{\rho}{\varepsilon} \qquad (23.2)$$

where x = distance from the surface in meters, ρ = charge density at point (ρ = charge per unit volume = $e \Sigma v_i n_i$ in C m^{-3}), ε = electrical permittivity of the medium (C^2 J^{-1} m^{-1}).

The electrical permittivity measures the ease with which molecules can be polarized and oriented in an electric field. Permittivity can be expressed in terms of dielectric constant, D, and permittivity of vacuum, ε_0, as $\varepsilon = \varepsilon_0 \times D$. The dielectric constant is equal to the ratio of the electrostatic capacity of condenser plates separated by a given material to that of the same condenser with vacuum between the plates. The dielectric constant of water at 20 °C is about 80 and the dielectric constant of vacuum is 1.0. Dielectric constants for non-polar hydrocarbons are generally less than ten. The permittivity of vacuum, ε_0, is 8.8542×10^{-12} C^2 J^{-1} m^{-1}.

Substitution for n_i from Eq. 23.1 into $\rho = e\Sigma v_i n_i$ gives:

$$\rho = e\Sigma v_i n_{i0} \exp(-v_i e\psi/kT) \qquad (23.3)$$

which, when substituted into Eq. 23.2 gives:

$$\frac{d^2\psi}{dx^2} = -\frac{e}{\varepsilon}\Sigma v_i n_{i0} \exp(-v_i e\psi/kT).$$
$$(23.4)$$

This is the differential equation that is solved to determine the electrical potential and ionic distributions in the diffuse part of the double layer. ψ is negative because of the negative particle surface charge. ψ_0, the value of ψ at

a particle surface, can be up to a few tens of millivolts. Van Olphen (1977) and Mitchell (1993) give examples of solutions.

23.4.2 DOUBLE LAYER INTERACTIONS

The practical importance of the ionic distributions in the diffuse layer is that when the elevated cation concentration extends a large distance from particle surfaces, i.e. a thick diffuse layer condition, interparticle repulsive forces are high. This condition is associated with deflocculation of suspensions and swelling of expansive clays. If the cations balancing the negatively charged clay particles are compressed against the particle surfaces, i.e. a thin diffuse layer condition, suspensions flocculate and expanded clays may shrink.

Figure 23.8 shows curves of repulsion and attraction versus distance from a particle surface. Repulsion is caused by double layer interactions, i.e. cations from overlapping double layers repel each other. Attraction is caused by van der Waals' forces, which are universal forces of attraction caused by fluc-

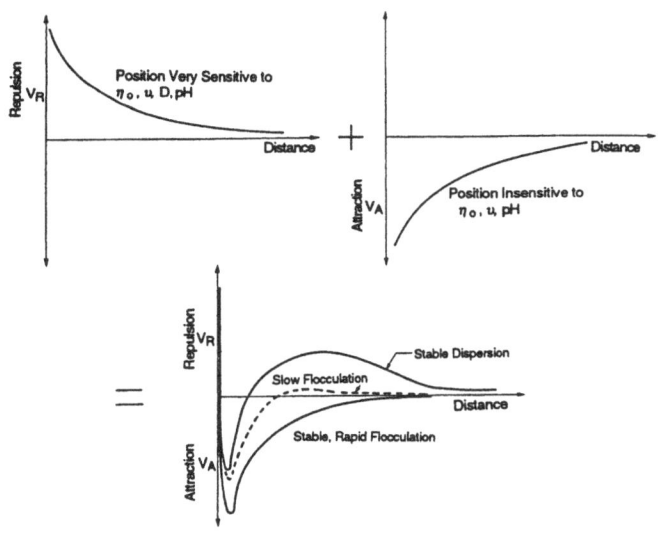

FIGURE 23.8. Energies of repulsion, attraction and net interaction for interacting clay particle minerals (reproduced from *Fundamentals of Soil Behavior*, J.K. Mitchell © 1993 John Wiley & Sons, Inc. Reprinted by permission of John Wiley & Sons, Inc.).

tuating dipoles. Not shown on the repulsion figure is the short range repulsion (Born repulsion), which resists interpenetration of matter.

Combining the two curves gives a net curve of interaction. Net curves of interaction change as the double layer changes. The van der Waals' attraction is rather insensitive to system variables. The character of the interaction curve has a major influence on flocculation–deflocculation and particle arrangements and their stability in sedimented, consolidated or compacted deposits of clay. High repulsion barriers in suspensions prevent particles from getting close to each other and thus cause or maintain dispersion. When the repulsion barrier is small or does not exist, the attractive force causes or maintains flocculation.

23.4.3 FACTORS INFLUENCING DOUBLE LAYER REPULSION

A useful measure of the "thickness" of the double layer is the distance to the center of gravity of the diffuse layer charge distribution, i.e. the distance where there is equal balancing charge both toward and away from the particle surface. Designating this distance by $1/K$:

$$1/K = \left(\frac{\varepsilon_0 D k T}{2 n_0 e^2 v^2} \right)^{1/2}. \qquad (23.5)$$

Equation 23.5 provides a basis to examine the effects of changing some of the system variables, because an increase in $1/K$ means an increase in interparticle repulsion. Flocculation–deflocculation behavior is considered in non-active (low PI) clays such as kaolinite, while shrink–swell behavior is considered for active (high PI) clays such as bentonite. Recall that water and ions are mostly free in non-active clays such as kaolinite (low plasticity soils), whereas they are mostly under the influence of particle surface forces in active clays such as smectite (high plasticity soils).

In inactive clays, increasing $1/K$ causes an increased tendency for deflocculation and dispersion. In active clays, increasing $1/K$ causes swell. The influences of changes in pore solution chemistry, as given by Eq. 23.5 are:

- As the reference ion concentration, n_0, increases, $1/K$ decreases, interparticle repulsion decreases and the tendency for flocculation increases.
- As cation valence, v, increases, $1/K$ decreases and the flocculation tendency increases.
- As dielectric constant, D increases, $1/K$ increases, causing deflocculation or swelling.
- According to Eq. 23.5, an increase in temperature increases $1/K$. However, an increase in temperature also causes a decrease in dielectric constant, and the net effect of temperature change on $1/K$ is small.

Although not directly accounted for in Eq. 23.5, changes in pH may have a large influence on the colloidal behavior of clays. This is because changing the pH will cause changes in the particle surface charge density.

Clay particles have either O^{2-} or $(OH)^-$ exposed on their surfaces and alumina, $Al(OH)_3$, exposed on the edges of the octahedral layers. Two characteristics of these surfaces influence flocculation–deflocculation and swelling behavior. First, the $(OH)^-$ on the surfaces and edges can dissociate, with some H^+ ions moving into solution. More dissociation will occur at high pH because of the smaller concentration of H^+ already in solution. The net negative charge of the particle surface or edge increases as the H^+ leaves the particle. The increased negative surface charge attracts more cations, increasing the thickness of the double layer. Second, alumina is amphoteric; it ionizes positively at low pH and negatively at high pH. As a consequence, positive diffuse layers can develop at the edges of some clay particles, especially the 1:1 minerals, at low pH.

As a result, low pH promotes a positive edge–negative surface interaction, often leading to flocculation. High pH promotes stable suspensions or dispersions. Minerals like kaolinite and halloysite (1:1 minerals) are more sensitive to pH changes than the 2:1 minerals, because more surface (OH) are exposed. In addition, the thickness–diameter ratio of the 1:1 minerals is larger, thus exposing more alumina and (OH) on their edges. The behavior of the 2:1 minerals is not nearly as influenced by pH changes.

23.5 Ion Exchange

Double layer cations and anions are exchangeable. Changes in the type of adsorbed cation, in particular, may produce changes in physical properties, e.g. a smectite with calcium as the adsorbed cation balancing the particle electro-negativity, is much less plastic and expansive than the same clay with sodium as the exchangeable cation. Bentonite with sodium may have a liquid limit around 600%, whereas with calcium it may only be around 100%.

Ion exchange reactions are stoichiometric. Particle structure is not affected by ion exchange. The most commonly found cations in surficial soils are usually in the following decreasing order of abundance: Ca^{2+}, Mg^{2+}, Na^+ and K^+. The most common anions, in order of decreasing abundance are SO_4^{2+}, Cl^-, PO_4^{3+} and NO_3^-. In marine clays and many sediments, however, Na^+ and Cl^- are the most common cations and anions.

Isomorphous substitution is the most important source of ion exchange capacity, with broken bonds at particle edges and replacement of the hydrogen of an exposed hydroxyl by another type of cation contributing to the total exchange capacity in some cases.

Ion exchange usually occurs in an aqueous environment, and it is almost instantaneous for low plasticity clays like kaolinite, but may take hours for clays like illite, and a day or more for high plasticity clays like smectite. In general, as the valence increases and the size of the ion decreases, the replacing power of an ion increases.

A general cation replacement series, in order of increasing replacement power is:

$$Na^+ < Li^+ < K^+ < Rb^+ < Cs^+ < Mg^{2+}$$
$$< Ca^{2+} < Ba^{2+} < Cu^{2+} < Al^{3+}$$
$$< Fe^{3+} < Th^{4+}.$$

Replacement can occur in the other direction, against the general trend shown above, by the law of mass action if the replacing ion is at high concentration.

The development of generally applicable quantitative theories for ion exchange has been difficult because of the great complexity of the system and the large number of variables involved. Fortunately, for many engineering applications, knowledge of general trends is adequate.

Anions can be adsorbed onto particle edges and can possibly replace some surface $(OH)^-$ ions. Some anion types can be preferentially adsorbed, including phosphate, silicate, arsenate and molybdate. Phosphate, PO^{3-}, and silicate, SiO^{4-}, have tetrahedral structures and are similar in size and geometry. If a phosphate attaches to a clay particle, then the electro-negativity of the clay increases and dispersion increases. This is an important reason why phosphates are effective dispersing agents for soils.

Organophillic clays can adsorb organic contaminants. Organic compounds interact with clays by:

- Adsorption of organic compounds on clay surfaces in aqueous systems. The amount depends on the available surface and the ability of the organic molecules to displace water molecules.
- Cationic organic compounds exchange for inorganic adsorbed cations.

- Intercalation, which involves the entry of organic molecules between silicate layers and may be particularly important in the kaolin minerals.
- Attraction of large organic molecules to clay surfaces by van der Waals' forces.

Organophillic clays are manufactured by attaching an organic compound (such as $(NH_4)^+$–organic) to the clay by an ion exchange reaction. The modified clay can act as an adsorbent for other organic compounds.

23.6 Soil Fabric and Its Relation to Properties

Mineralogy and pore solution chemistry by themselves are insufficient to define the properties of a soil that are needed to properly address projects and problems in geoenvironmental engineering. Volume change properties control compression and swelling; stress-deformation and strength properties determine displacements and stability of geoenvironmental structures such as waste landfills; and conductivity properties control the effectiveness of waste containment barriers and contaminant transport.

Soil–water–electrolyte systems are particulate systems, with the solid particles distributed, arranged and bonded in different ways. The arrangement of particles, particle groups and pore spaces is termed the fabric. The fabric and its resistance and response to changes in ambient conditions define the structure. The structure reflects all facets of soil composition, history, present state and environment. Structure-determining factors and processes are summarized in Fig. 23.9. Extensive description, illustration and discussion of soil fabric and structure, and their relationships to the properties of coarse- and fine-grained soils are given in Chapters 8 and 11 of Mitchell (1993).

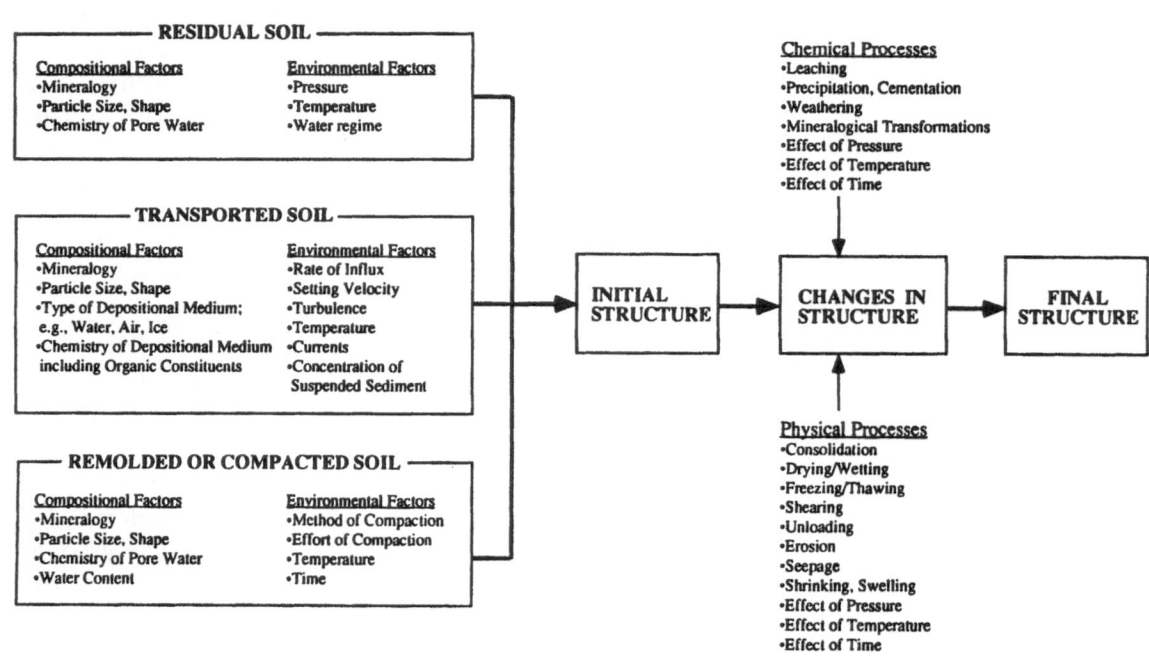

FIGURE 23.9. Structure-determining factors and processes (modified from Mitchell 1993).

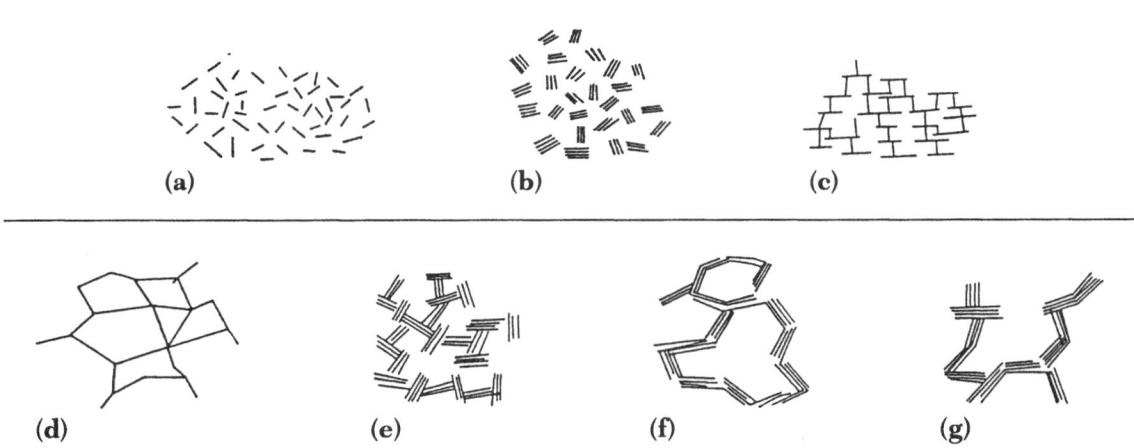

FIGURE 23.10. Particle associations in clay suspensions: (a) dispersed and deflocculated; (b) face-to-face aggregated, but deflocculated; (c) edge-to-face flocculated, but dispersed; (d) edge-to-edge flocculated, but dispersed; (e) edge-to-face flocculated and aggregated; (f) edge-to-edge flocculated and aggregated; (g) edge-to-face and edge-to-edge flocculated and aggregated (modified from van Olphen 1977).

23.6.1 PARTICLE ASSOCIATIONS AND FABRIC SCALE

Soil fabric should be considered relative to three levels of scale:

- *Microfabric*. Regular aggregations of particles and the very small pores between them. Microfabric units are up to a few tens of micrometers across. Types of units in clay suspensions, such as bentonite slurries, used for slurry trenches and cut-off walls are shown schematically in Fig. 23.10.
- *Minifabric*. Aggregations of the microfabric units and the larger pores between them. Minifabric units are up to a few hundred micrometers in size. Particle arrangements in the clay matrix phase of a compacted soil containment barrier can be similar to those shown in Fig. 23.10; however, the individual units will be much closer together, and the inter-unit void spaces will be much smaller.
- *Macrofabric*. A larger volume that may contain cracks, fissures, root holes, laminations, etc. Macrofabric features can have a dominating influence on the gross fluid flow and strength properties of a soil formation. A goal in the engineering of soil-

based cut-off walls, liners and covers for waste containment is to assure that macrofabric features that could act as flow paths do not form during construction or in the future (see Chapter 25).

23.6.2 SINGLE-GRAIN FABRICS

Coarse silt, sand and gravel particles are so large that, in the absence of interparticle cementation, they behave as independent units. Non-clay minerals in soils are in this size range and, with the exception of mica, they are bulky in shape. Clean, cohesionless materials have void ratios typically in the range 0.4–1.0 (porosity in the range 25–50%). As most particles are at least slightly elongated, different fabrics are possible for the same material at the same void ratio. Although the consequences of this are minor as regards hydraulic conductivity, different fabrics are known to be responsible for differences in shear-induced compression–dilation behavior and liquefaction resistance.

23.6.3 MULTIGRAIN FABRICS

Single-grain fabric is rare in soils containing fine silt and clay, because the interparticle attraction and repulsion forces acting on each small particle are of a magnitude comparable to gravitational forces. Accordingly, multi-grain arrangements of the types shown schematically Fig. 23.10 are formed. The stability of aggregates and the ease with which particles can be moved within the material are dependent on both physical and chemical factors. Hence, the variations in properties at a given soil density can be large, with the greatest ranges likely in materials that have a high liquidity index (ratio of the water content less the plastic limit to the plasticity index, i.e. (WC-PL)/PI).

23.6.4 FABRIC–PROPERTY INTERRELATIONSHIPS

Several principles relate the fabric and structure of a soil to its mechanical properties, thus providing a basis for understanding and predicting behavior over a range of existing and future conditions.

1. Under a given effective consolidation pressure, a soil with a flocculated fabric will be less dense than a soil with a deflocculated fabric.
2. At the same void ratio, a flocculated soil with randomly oriented particles and particle groups is more rigid than a deflocculated soil.
3. At stresses greater than the preconsolidation pressure, further stress increase causes a greater change in fabric of a flocculated soil structure than in a deflocculated structure.
4. The average pore diameter and range in pore sizes are smaller in deflocculated soil fabrics than in flocculated fabrics.
5. Shear displacements usually orient platy particles and particle groups with their long axes in the plane of shear.
6. Anisotropic consolidation stresses align platy particles and particle groups with their long axes in the major principal plane.
7. Stresses are unlikely to be distributed

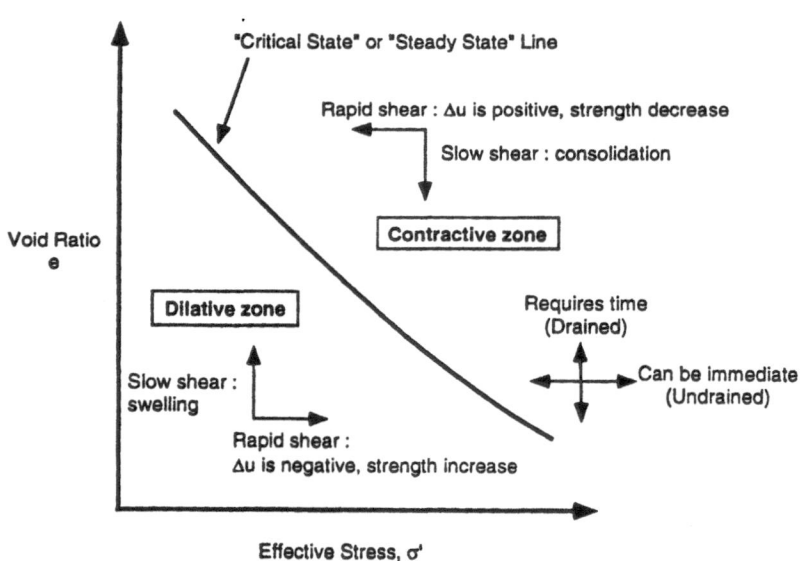

FIGURE 23.11. Soil states in relation to critical or steady state line, and pore pressure and volume changes during deformation where Δu = change in pore water pressure (reproduced from *Fundamentals of Soil Behavior*, J.K. Mitchell © 1993 John Wiley & Sons, Inc. Reprinted by permission of John Wiley & Sons, Inc.).

uniformly among all particles and particle groups.

8. Different stress–deformation histories can produce different structures at the same void ratio.

9. Volume change tendencies determine pore pressure development during undrained deformation.

10. Changes in structure of a saturated soil at constant volume are accompanied by immediate changes in effective stress.

11. Changes in structure of a saturated soil at constant effective stress are accompanied by changes in void ratio. The change in void ratio is time-dependent, owing to the time required for water to enter or drain from the soil.

12. In partially saturated soils, changes in structure can be accompanied by immediate changes in both void ratio and effective stress.

Points 9, 10 and 11 are illustrated and summarized by Fig. 23.11. In general, normally consolidated to slightly overconsolidated clays and saturated loose sands are contractive; whereas, heavily overconsolidated clays and dense sands are dilative.

23.7 Physicochemical Effects on Soil Properties and Their Importance in Environmental Geotechnics

23.7.1 VOLUME CHANGE PROPERTIES

Several compositional and environmental factors control the compression and swelling properties of a soil. Meaningful quantitative predictions are usually only possible if undisturbed samples or in situ tests are used to measure the response to a change in stress, chemical and biological environment, or temperature.

1. *Physical interactions*. Bending, sliding, rolling and crushing of soil particles accompany compression. These processes are most important in soils at low void ratios.

2. *Physicochemical interactions*. These interactions originate in the particle surface forces that are responsible for double layer repulsion, surface and ion hydration, and interparticle attractive forces. They are most important in the formational stages of most fine-grained soil deposits when they are under low pressure and at high water content and void ratio. This concept is embodied in the principle of chemical irreversibility of clay fabric (Bennett & Hurlbut 1986), which states that the chemical environment is critical during the initial stages of fabric formation in sediments in water. After the initial flocculation and deposition of particles, the fabric is much less important, with mechanical energy rather than chemical energy being the dominant factor affecting behavior.

3. *Chemical and organic environment*. Chemical precipitates cement particles together. Organic matter influences surface forces and water and organic chemical adsorption properties.

4. *Mineralogical detail*. Small differences in structural details of the expansive clay minerals, such as lattice charge deficiency (Brindley & MacEwan 1953) and b-dimension of the crystal lattice (Davidtz & Low 1970) have major effects on soil swelling. The formation of hydroxy-cation interlayers of Fe-OH, Al-OH and Mg-OH reduces the cation exchange capacity and restricts swelling.

5. *Fabric and structure*. Compacted expansive soils with flocculent structures swell more than compacted expansive soils with dispersed structures.

6. *Stress history*. An overconsolidated soil is less compressible, but more expansive, than the same material normally consolidated to the same void ratio. Anisotropic consolidation usually results in anisotropic compression and swelling behavior.

7. *Temperature*. An increase in temperature will usually cause a decrease in volume under drained conditions. If drainage is prevented, a temperature increase causes an increase in porewater pressure and a de-

crease in effective stress. Quantitative analysis in terms of the compressibility and thermal expansion properties of soil solids and water is given in Mitchell (1993).

8. *Porewater chemistry*. Any change in the pore solution chemistry that decreases double layer thickness reduces swell and swell pressure in soils containing expansive clay minerals. For soils containing only non-expansive clay minerals, the porewater chemistry has relatively little effect on the compression behavior once the initial fabric has formed and the structure has stabilized under a moderate level of effective stress.

9. *Stress path*. The magnitude of compression and swelling associated with a given change in stress is usually path-dependent. Loading or unloading in stages can give very different volume change behavior than obtained if the stress change is made in one step, as shown, for example, by Seed *et al.* (1962a).

Expansive clays such as bentonite and attapulgite are widely used for geoenvironmental applications including cut-off walls, liners and covers. Thus, the swelling properties of these materials in suspension, in soil–bentonite and cement–bentonite mixes, and in geosynthetic clay liners are important. Both an osmotic pressure theory of swelling, which is based on double layer theory (Bolt 1956; Mitchell 1993), and a water adsorption (hydration) theory, in which swelling is attributed to surface hydration (Low 1987, 1992), have been proposed to explain swelling quantitatively. Each accounts well for some aspects of swelling phenomena; however, neither accounts for all observations. The effective specific surface area, i.e. the relative proportions of fully expandable and partially expandable layers, controls swelling. When the clay content is high and particle dissociation into unit layers is extensive, the effective specific surface area is large, and swelling is large and very sensitive to the magnitude of

double layer repulsive forces. As shown earlier, these forces are very dependent on the pore fluid composition.

23.7.2 DEFORMATION AND STRENGTH PROPERTIES

The stress–deformation and stress–deformation–time behavior of soils must be considered when ground movements and stability are of interest. In geoenvironmental engineering, the deformations and integrity of waste containment barrier components, landfills and tailings dams are important during and following construction.

23.7.2.1 Strength

In the absence of chemical cementation, the strength of sand and normally consolidated clay is given by:

$$\tau_{ff} = \sigma'_{ff} \tan \phi' \qquad (23.6)$$

in which τ_{ff} = the shear stress on the failure plane at failure, σ'_{ff} = the normal effective stress on the failure plane, and ϕ' = the effective stress friction angle. The magnitude of the effective stress friction angle depends on several factors, including: (1) sliding friction between particle surfaces in contact, (2) dilation during shear, (3) particle rearrangements during shear, and (4) grain crushing. Each of these depends on soil mineralogy, gradation, porosity, structure and confining pressure. In reality, ϕ' for most soils is not constant but decreases slightly with increasing effective confining pressures, thus producing curved strength failure envelopes. Mitchell (1993) gives further discussion of the frictional characteristics of soil minerals.

True cohesion is strength in excess of that given by Eq. 23.6. True cohesion is caused by: (1) chemical cementation between particles by carbonates, silica, alumina, iron oxides, and organic compounds; (2) interparticle electro-static and electro-magnetic attractions; and (3) primary valence bonding and adhesion generated as a result of prior compression and unloading (overconsolidation). According to Ingles (1962) chemical cementation in consolidated sediments may account for tensile strengths up to about 500 kPa; whereas, electro-static and electromagnetic attractions (van der Waals' forces) may contribute only a few tens of kilopascals of tensile resistance, even in a very fine-grained soil.

Apparent cohesion exists in partially saturated soils owing to positive interparticle effective stresses generated by capillary pressures (see Chapter 5). The magnitude of this apparent cohesion increases with decreasing particle size, and may range from a few kilopascals in sands to several hundred kilopascals in clays.

For many soils, the maximum or peak strength is greater than the residual strength, which is measured after shear displacements of a few to several tens of millimeters. The difference between the two for a given soil is a function of structural stability, porosity and confining pressure. A composite relationship, derived from Kenney (1967), Lupini *et al.* (1981), Skempton (1985) and others, showing residual friction angle as a function of amount of clay in a soil is given in Fig. 23.12. Knowledge of the residual strength of a soil is important in such geoenvironmental applications as the stability of landfills.

As a change in temperature causes a change in void ratio, a change in effective stress, or a change in both, temperature changes cause changes in strength. For a saturated soil under undrained conditions, temperature increase causes a decrease in effective stress, so strength decreases. Under drained conditions, temperature increase causes consolidation. The effect on strength depends on the relative influences of this consolidation and any subsequent changes in temperature on effective stress (Mitchell 1993). The undrained strength of clays is strain rate dependent, with an increase of

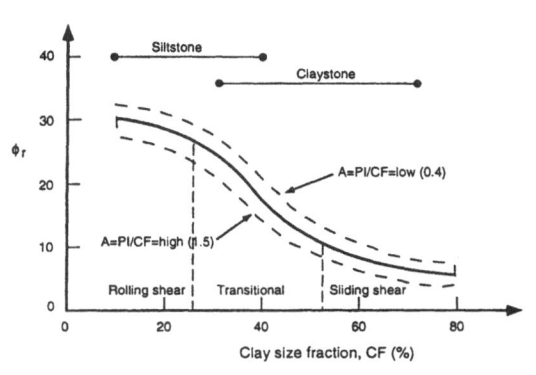

FIGURE 23.12. Composite relationship showing dependence of residual friction angle, ϕ_r, on soil composition: A = activity; PI = plasticity index (from *Fundamentals of Soil Behavior*, J.K. Mitchell © 1993 John Wiley & Sons, Inc. Reprinted by permission of John Wiley & Sons, Inc.).

about 10% for each order of magnitude increase in strain rate.

23.7.2.2 Stress–Strain Behavior

Stress–strain behavior ranges from very brittle for some quick clays, cemented soils, heavily overconsolidated clays and dense sands, to ductile for insensitive and remolded clays and loose sands. The stiffness is both structure and effective stress dependent. Shear deformation of structured saturated soil is accompanied by volume change if drainage is allowed, or change in porewater pressure and effective stress if drainage is prevented, as shown in Fig. 23.13. Under constant total confining stress the magnitude of generated pore pressure is strain dependent. Temperature increase causes a decrease in modulus, i.e. a softening of the soil.

23.7.2.3 Stress–Strain–Time Behavior

Soils undergo both creep (continuing deformation under constant stress) and stress relaxation (continuing reduction in shear stress carried by the soil at constant deformation). The form of behavior is the same for all soil types; however, the magnitude of the effects

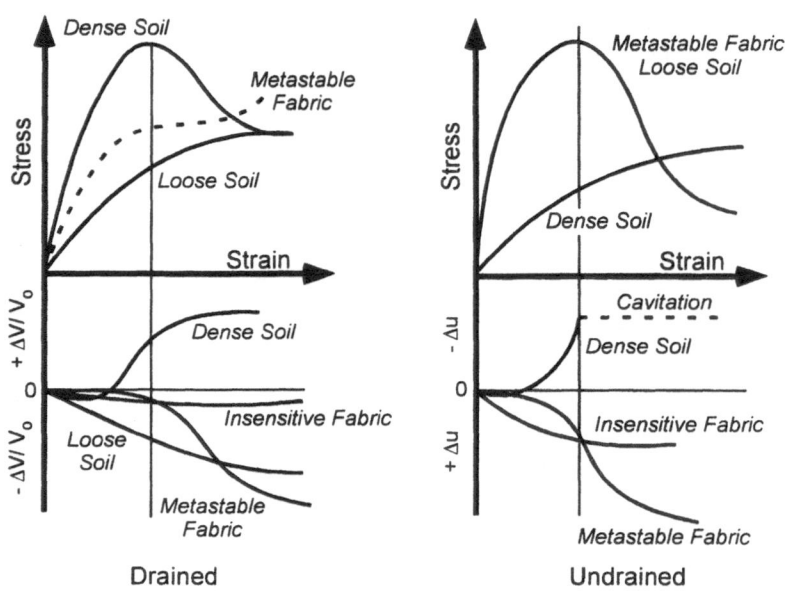

FIGURE 23.13. Volume and pore pressure changes during shear: (a) drained conditions where ΔV = change in volume and V_o = initial volume; and (b) undrained conditions where Δu = change in pore pressure (modified from Mitchell 1993).

increases with plasticity, activity and water content of the soil, and with increases in deviatoric stress. Deformation under sustained stress usually leads to increased stiffness under the action of subsequent increase in the deviatoric stress. Some soils, however, fail under sustained deviatoric stress significantly less (as much as 50%) than the peak stress measured in a normal strength test. This creep rupture is particularly important in overconsolidated soils under drained conditions and sensitive normally consolidated soils under undrained conditions.

23.7.3 CONDUCTIVITY PROPERTIES

Evaluation of fluid and chemical flows is an essential part of geoenvironmental projects involving contaminant transport and waste containment. Heat and electrical flows must be considered in many methods for site remediation. Accordingly, flow processes and conductivity properties of soils must be known. Both direct flows, in which a gradient

of one type causes flow of the same type, and coupled flows, in which a gradient of one type causes a flow of another type, must be considered.

23.7.3.1 Direct Flows

Four types of direct flows through soils are shown in Fig. 23.14. Each is described by a well-known steady state equation, which is valid provided the flow process does not change the state of the soil. The equations shown are for flow in one dimension. The cross-sectional area normal to the flow is designated by A, H is hydraulic head, V is voltage, T is temperature, and C is chemical concentration. The coefficient k_h is the hydraulic conductivity, σ_e is the electrical conductivity, k_t is the thermal conductivity, and D is the diffusion coefficient. Practical ranges of values for these coefficients are given in Table 23.2. Mitchell (1991, 1993) provides detailed analyses of these flow types and coefficients.

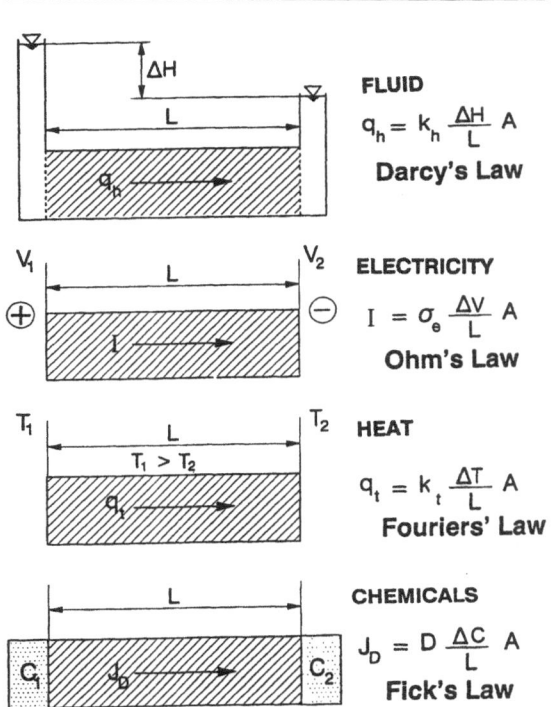

FIGURE 23.14. Four types of direct flow through a soil mass.

FLUID
$$q_h = k_h \frac{\Delta H}{L} A$$
Darcy's Law

ELECTRICITY
$$I = \sigma_e \frac{\Delta V}{L} A$$
Ohm's Law

HEAT
$$q_t = k_t \frac{\Delta T}{L} A$$
Fouriers' Law

CHEMICALS
$$J_D = D \frac{\Delta C}{L} A$$
Fick's Law

23.7.3.2 Coupled Flows

Owing to the internal non-homogeneity of ion distributions, restrictions to ionic movement resulting from electrostatic attraction and repulsion, and the dependence of these interactions on temperature and chemical concentrations, several microscopic and macroscopic effects develop when a wet soil is subjected to flow gradients of different types. Among them are that a flow gradient X, of one type can cause a flow J, of another type, according to:

$$J_i = L_{ij}X_j. \tag{23.7}$$

The L_{ij} are coupling coefficients. The different types of coupled flows involving hydraulic, thermal, electrical and chemical gradients are indicated in Table 23.3.

Advection, the hydraulic transport of dissolved and suspended materials, is the dominant mode of chemical transport in most groundwater flow situations and is a major contributor, along with molecular diffusion, to the contaminant transport across waste containment barriers (see Section 25.5). Of the other coupled flows listed in Table 23.3, those of most importance in fine-grained soils are: (1) thermo-osmosis, the movement of water under a temperature gradient; (2) electro-osmosis, the movement of water under an electrical gradient; (3) chemical osmosis, the flow of water across a clay layer induced by a chemical gradient; (4) isothermal heat transfer, the flow of heat with moving groundwater; and (5) electrophoresis, the movement of charged particles in an electrical field.

Diffusive chemical transport is important in soils having a hydraulic conductivity less than about 1×10^{-9}–1×10^{-10} m s^{-1}. Whereas the hydraulic conductivity, k_h, decreases rapidly with decreasing soil particle size, the effective diffusion coefficient, D, is relatively insensitive to particle size. Detailed analysis of diffusion in soils is given in Sections 24.3 and 25.5 of this handbook, and in Quigley et al. (1987), Shackelford & Daniel (1991a), and Rowe et al. (1995b).

TABLE 23.2. Practical ranges of flow parameters for fine-grained soils[a]

Parameter	Minimum	Maximum
Porosity, n	0.1	0.7
Hydraulic conductivity, k_h (m s^{-1})	1×10^{-11}	1×10^{-6}
Thermal conductivity, k_t (W m^{-1} K^{-1})	0.25	2.5
Electrical conductivity, σ_e (S m^{-1})	0.01	1.0
Electro-osmotic conductivity, k_e (m^2 s^{-1} V^{-1})	1×10^{-9}	1×10^{-8}
Diffusion coefficient, D (m^2 s^{-1})	2×10^{-11}	2×10^{-9}

[a] The above values are for saturated soil. They will be less in partly saturated soil.

TABLE 23.3. Direct and coupled flows through soils

| Flow, J | Gradient, X | | | |
	Hydraulic head	Temperature	Electrical potential	Chemical concentration
Fluid	Hydraulic conduction (Darcy's law)	Thermo-osmosis	Electro-osmosis	Chemical osmosis
Heat	Isothermal heat transfer	Thermal conduction (Fourier's law)	Peltier effect	Dufour effect
Current	Streaming current	Thermoelectricity (Seebeck effect)	Electrical conduction (Ohm's law)	Diffusion and membrane potentials
Ion	Advection	Thermal diffusion of electrolyte (Soret effect)	Electrophoresis	Diffusion (Fick's law)

Quantification of coupled flows requires determination of coupling coefficients applicable to the system, and is formulated using irrevesible thermodynamics (Katchalsky & Curran 1967). Yeung & Mitchell (1992) give a general solution for combined hydraulic, electrical and chemical flows through a clay barrier system.

23.7.3.3 Effects of Flows on Properties

Prolonged application of potential fields and flow of material and energy through a soil may change the state and properties of the material. The changes may be beneficial, as for clay consolidation by electro-osmosis, or detrimental, as for shrinkage cracking of a clay containment barrier as a result of exposure to non-polar organic liquids. Among the effects that may develop are: (1) hydro-consolidation by seepage forces; (2) electro-osmotic consolidation; (3) electrochemical reactions, including gas generation, electrode decomposition, ion exchange and pH changes; (4) thermal drying; (5) swelling and shrinking caused by chemical concentration differences and/or changes in pore fluid composition.

The magnitude of these effects depends on the type and size of potential gradients, soil and pore fluid composition, and initial state and structure of the soil. The finer grained the soil, the more active the clay minerals, and the lower the void ratio, the more important are chemically and electrically induced fluid flows through the soil relative to hydraulic flow. The higher the water content and void ratio, and the greater the compressibility, the larger are the changes in soil state caused by application of the different potential fields.

24. CONTAMINANT HYDROGEOLOGY

R. A. Schincariol

R. K. Rowe

24.1 Hydrogeological Environments

While advances into understanding the complexities of contaminant migration through porous media and fractured rock are continually being made, nothing is more important to understanding contaminant migration, and designing remedial systems, then a complete model of the hydrogeological environment. One must understand the hydrostratigraphy and other factors that control fluid flow at a site before one can hope to understand or control contaminant migration.

24.1.1 SATURATED AND UNSATURATED ZONES

In the subsurface, the region above the water table is referred to as the vadose or unsaturated zone (see Chapter 5), whereas the region below the water table is referred to as the phreatic or saturated zone. However, the terms saturated and unsaturated zones require some clarification because the distinction between the two zones is actually based on fluid pressure and not degree of saturation. Typically, the water table is defined by the water level in a shallow well or piezometer screened across the vadose–phreatic boundary. The water pressure is less than atmospheric in the vadose zone, and greater than atmospheric in the phreatic zone. The

capillary fringe is a zone above the water table where fluid saturations are high, close to 100%, but fluid pressures are less than atmospheric pressure owing to tension of the sediment–surface–water contact (see Chapter 5). The capillary fringe is thicker in fine-grained sediments than coarse-grained sediments because of the greater tension created by the smaller pore openings (Table 24.1). Due to heterogeneity in pore sizes the capillary fringe forms an irregular zone above the water table.

24.1.2 AQUIFERS AND AQUITARDS

An aquifer is a saturated permeable geologic unit that can transmit significant quantities of water to wells under ordinary hydraulic gradients. Unconsolidated sands and gravels, sandstones, limestones and fractured rock are common aquifers. In contrast, an aquitard is a saturated low permeability geologic unit that transmits significantly less water than adjacent aquifers. Clays, shales and unfractured crystalline rock are common aquitards. These definitions are subjective and came into use to suit the needs of water-supply hydrogeology. However, in contaminant hydrogeology, aquifers and aquitards are usually defined within a particular hydrogeologic setting or hydrostratigraphic framework. Gorelick *et al.* (1993) emphasize this by stating that "an aquifer has a clear meaning only when ac-

TABLE 24.1. Height of capillary rise in sediments (adapted from Fetter 1994)

Sediment	Grain diameter (mm)	Pore radius (mm)	Capillary rise (m)
Fine silt	0.008	0.002	7.5
Coarse silt	0.025	0.005	3
Very fine sand	0.075	0.015	1
Fine sand	0.15	0.03	0.5
Medium sand	0.3	0.06	0.25
Coarse sand	0.5	0.1	0.15
Very coarse sand	2	0.4	0.04
Fine gravel	5	1	0.015

companied by a description of the hydrogeo-logic setting, the nature of the contamination problem, and the attendant regulatory and public concerns". Therefore the two terms, aquifer and aquitard, must be viewed as relative; a silt layer might be considered an aquifer in a silt–clay system and an aquitard in a silt–sand system (Gorelick *et al.* 1993). Typically aquitards define the boundaries or act as confining layers of aquifers. If an aquifer is bounded above by an aquitard it is called a confined aquifer. If a water table forms the upper boundary it is an unconfined aquifer.

24.1.3 HYDROGEOLOGIC PARAMETERS: POROSITY, HYDRAULIC CONDUCTIVITY AND PERMEABILITY, TRANSMISSIVITY, STORATIVITY AND SPECIFIC YIELD

Porosity, or more accurately, total porosity, n_t, is defined as the ratio of the volume of voids, V_v, in sediment or rock to the total volume, V_t, of the sample. However, for purposes of quantifying the relative volume of porous media through which groundwater actually flows, the parameter effective porosity, n_e, is more appropriate. Effective porosity, n_e, is defined as the ratio of the volume of interconnected pore space to the total volume of the sample. Thus isolated or dead-end channels that do not contribute to groundwater flow are excluded. Effective porosity implies some connectivity through the

sediment or rock and is more closely related to permeability than total porosity. The total and effective porosity for unconsolidated sediments are usually similar. However, they may not be similar in lithified sediment or rocks (see Table 24.2). It is the effective porosity that determines the advective velocity of groundwater.

Hydraulic conductivity and permeability are the two basic properties that describe the ability of a porous medium to transmit fluid. Hydraulic conductivity was defined in Sections 2.4.2 and 3.3 as a basic soil property. From a hydrogeologic perspective, hydraulic conductivity, k, which has dimensions of length/time (e.g. m s^{-1}), depends on properties of porous medium and properties of fluid. In the geotechnical literature, the term, "coefficient of permeability," is sometimes used to mean the same as hydraulic conductivity. However, this term is being dropped because of potential confusion with the intrinsic permeability, k_i. Intrinsic permeability describes the capacity for flow of any fluid through a porous medium and has units of squared length (e.g. m^2). Unlike hydraulic conductivity, the intrinsic permeability is only a function of porous media properties. The relationship between hydraulic conductivity, k, and permeability, k_i, is:

$$k = \frac{k_i \rho_f g}{\mu_f} \qquad (24.1)$$

TABLE 24.2. Range in values of total porosity and effective porosity

	Total porosity	Effective porosity
Anhydrite[a]	$5 \times 10^{-3} - 5 \times 10^{-2}$	$5 \times 10^{-4} - 5 \times 10^{-3}$
Chalk[a]	$5 \times 10^{-2} - 4 \times 10^{-1}$	$5 \times 10^{-4} - 2 \times 10^{-2}$
Limestone, dolomite[a]	$0 - 4 \times 10^{-1}$	$1.0 \times 10^{-3} - 5 \times 10^{-2}$
Sandstone[a]	$5 \times 10^{-2} - 1.5 \times 10^{-1}$	$5 \times 10^{-3} - 1 \times 10^{-1}$
Shale[a]	$1 \times 10^{-2} - 1 \times 10^{-1}$	$5 \times 10^{-3} - 5 \times 10^{-2}$
Salt[a]	5×10^{-3}	1×10^{-3}
Granite[b]	1×10^{-3}	5×10^{-6}
Fractured crystalline rock[b]	—	$5 \times 10^{-7} - 1 \times 10^{-4}$

[a] Data from Croff *et al.* (1985).
[b] Data from Norton and Knapp (1977).

where ρ_f and μ_f = the density and dynamic viscosity of the fluid, respectively; and g = the gravitational acceleration constant. Given that ρ_f and μ_f vary with the composition and temperature of the fluid, hydraulic conductivity will also vary.

Chapter 3 (Table 3.10) and hydrogeology textbooks (e.g. Domenico & Schwartz 1998; Freeze & Cherry 1979), present tables of hydraulic conductivity for various geologic materials. Most tabulated values assume pure water and a temperature of 15.6°C, i.e. ρ_f = 1000 kg m^{-3}. These tabulated values provide only approximate values and should be used only when site-specific data are not available. When dealing with deep basinal brines, highly contaminated waters, non-aqueous phase liquids, vapors, or any fluid whose viscosity or density varies significantly from pure water (15.6°C) permeability becomes the dominant term of usage.

Hydraulic conductivity and permeability can be measured in the laboratory. These techniques are discussed in Section 3.3.1. While laboratory values are commonly used, care must be exercised in extrapolating laboratory data to the field scale. Large-scale heterogeneities in geologic materials can cause large deviations from local values. If site-wide or field-scale values are required, aquifer pumping tests, which effectively "average" aquifer properties over the area of influence,

give much more representative values. Several hydrogeology texts give an introduction to aquifer tests and hydraulic testing (e.g. Domenico & Schwartz 1998; Fetter 1994). Kruseman & de Ridder (1990) present the most definitive text for the analysis and evaluation of aquifer pumping test data. Permeability values can also be calculated from empirical equations based on porous media characteristics, such as grain and pore size (Bear 1972). These relationships are often imprecise and are usually only applicable to well-sorted, unconsolidated porous media.

Transmissivity and storativity are field-scale properties that define the transmissive and storage properties of geologic units of some specified thickness. Transmissivity, or coefficient of transmissibility, is the product of the average hydraulic conductivity and the saturated thickness of the aquifer, H:

$$T = kH. \tag{24.2}$$

The storativity, or coefficient of storage, is the volume of water released from storage of a saturated confined aquifer of thickness H per unit surface area per unit decline in the hydraulic head. Mathematically, storativity is defined as:

$$S = S'_s H \tag{24.3}$$

where S'_s = the specific storage or the volume of water that a unit volume of aquifer (confined) releases from storage under a unit decline in hydraulic head. The release of water from storage under a decreasing head stems from the reduction in porosity due to increasing effective stress and the expansion of water. The specific storage is defined mathematically as:

$$S'_s = \rho_w g(\alpha_v + n\beta) \qquad (24.4)$$

where α_v = the vertical aquifer compressibility, β = the compressibility of water, and other symbols are as defined earlier. Specific storage has the dimensions of L^{-1} and the storativity is dimensionless, as it is the ratio of the volume of water released per volume of aquifer. Values of storativity for confined aquifers range from 5×10^{-5} to 5×10^{-3} (Kruseman & de Ridder 1990).

Specific yield is the storage property of an unconfined aquifer. It is equal to the volume of water that an unconfined aquifer releases from storage, via gravity drainage, per unit surface area of aquifer per unit decline of the water table. Johnson (1967) has summarized typical values of specific yield for various sediments and rock types. These values range from 0.03 (clay) and 0.08 (silt) to 0.38 (dune sand) and 0.44 (peat).

24.1.4 HYDROGEOLOGIC VARIABLES: HYDRAULIC HEAD, FLUX AND VELOCITY, AQUIFER RECHARGE AND DISCHARGE

Hydraulic or total head, h, pressure head, $h_p = u_w/\gamma_w$, and fluid or porewater pressure, u_w, were defined earlier in Section 2.4. The water table represents the surface where the fluid pressure is atmospheric (gage pressure equals zero) and thus h = water table elevation measured with respect to some datum. Values of hydraulic head, including water table elevation, are required to calculate the directions of groundwater flow. While numerous methods exist to measure hydraulic head in boreholes, wells and piezometers (see Section 4.7 and Nielsen 1991; Fetter 1994; Rowe et al. 1995b), some important considerations regarding the use of hydraulic head data should be recalled:

- The length and position of the screened interval in a well or piezometer can influence the water-level observed in the well. The measured water-level is an integrated average of the hydraulic heads over the screened interval. Thus, large screen lengths for deep monitors could mask important differences in hydraulic head. Furthermore, water table elevations should be taken from wells screened over the vadose–phreatic zone interface. As water-levels rise above the top of the screen, the water-level becomes unrepresentative of true water table elevation.
- Hydraulic heads are often measured across a site to understand spatial and temporal gradients. Measurements of hydraulic head within a single aquifer provide the data to construct a potentiometric map (Fig. 24.1a). One must be careful to ensure that the potentiometric map is related to a single aquifer. It is assumed that flow in the aquifer is horizontal, and vertical gradients are absent. Thus, one cannot make potentiometric maps of low-permeability units as flow typically has a large vertical component. Vertical gradients are typically assessed in hydrogeological cross-sections (Fig. 24.1b). Here a vertical section is aligned parallel to the direction of mean groundwater flow. Domenico & Schwartz (1998) present an excellent discussion of mapping flow in geological systems.
- If significant differences in groundwater density exist hydraulic head measurements should be corrected to a common density before comparing data or calculating gradients (Jorgensen et al. 1982). Methods for correcting data to equivalent freshwater heads are provided in Pickens et al. (1987).

From Darcy's law (Section 2.4.2), the Darcy velocity, v_a, also called the specific discharge, discharge velocity or Darcy flux, may be calculated:

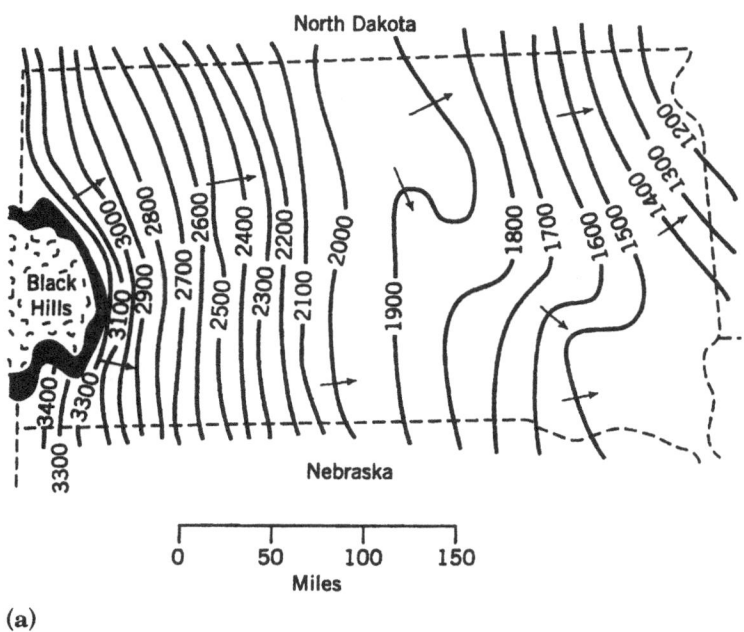

(a)

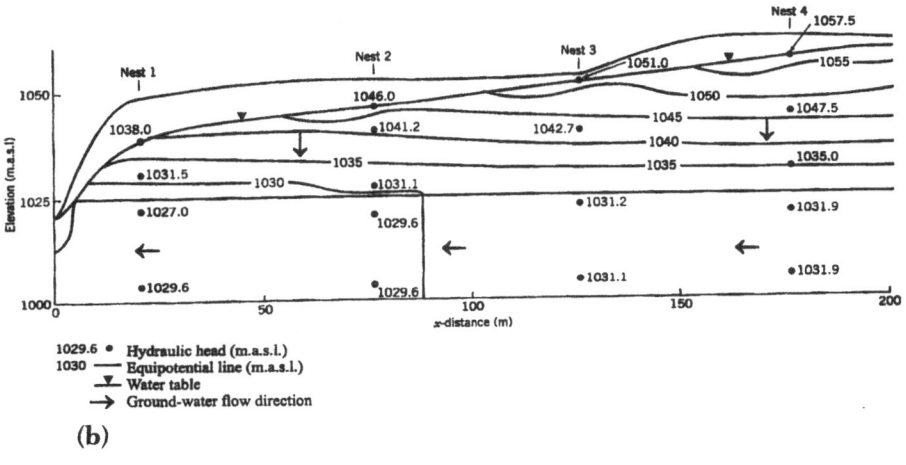

1029.6 ● Hydraulic head (m.a.s.l.)
1030 ——— Equipotential line (m.a.s.l.)
▼ Water table
→ Ground-water flow direction

(b)

FIGURE 24.1. (a) Potentiometric surface of the Dakota Sandstone, contour interval 100 ft (from Darton 1909, U.S. Geol. Surv. Water Supply Paper 227). (b) Example of a hydrogeologic cross-section describing the pattern of groundwater flow (modified from Domenico & Schwartz, 1998, *Physical and Chemical Hydrogeology*, © 1998 John Wiley & Sons, Inc. Reprinted by permission of John Wiley & Sons, Inc.).

$$v_a = -k \frac{\Delta h}{\Delta l} \qquad (24.5)$$

where Δl = the flow distance between points of h measurement. However, as noted in Section 2.4.2, although v_a can have the dimen-

sions of velocity, it corresponds to flow per unit cross-sectional area and it is not the velocity of groundwater. It is best referred to as the Darcy flux.

The average groundwater velocity, which gives an indication of the velocity at which contaminants are advected by the bulk flow of groundwater, is called the average linear groundwater velocity, v, or the seepage velocity. It is defined as the Darcy flux divided by the effective porosity, n_e:

$$v = \frac{v_a}{n_e} = -\frac{k}{n_e}\frac{\Delta h}{\Delta l}. \qquad (24.6)$$

Thus, the average linear groundwater velocity can be calculated indirectly from measurements of hydraulic conductivity, effective porosity, and hydraulic gradient. Alternately, it can be measured directly with tracer tests. Domenico & Schwartz (1998) outline some basic aspects of tracer tests.

Recharge of groundwater depends on infiltration of precipitation through the vadose zone and across the water table, or leakage from surface water bodies. Discharge from unconfined aquifers takes place as evapotranspiration, springs or seeps, and base-flow to effluent streams. Recharge and discharge to deeper confined aquifers occurs via leakage across confining units. Discharge to both unconfined and confined aquifers occurs through the pumping of wells.

Quantifying aquifer recharge and discharge relations is extremely important to many groundwater supply, dewatering, or contaminant transport and remedial action programs. Recharge and discharge rates are controlled by the precipitation, evapotranspiration, and the topography and geology of the basin or watershed. Recharge and discharge rates are very difficult and expensive to measure directly in the field. Furthermore, they are often highly variable both temporally and spatially. Most often, they are obtained by quantifying various components of the water balance for a drainage basin and/or estimating values from the calibration of steady state or transient groundwater models for the basin. Dingman (1994) provides a general review of the water balance and various techniques to quantify components. For a more

detailed and applied approach, the *Handbook of Hydrology* (Maidment 1993) has separate chapters that focus on quantifying each component of the hydrologic cycle. Stephens (1996) provides a detailed review of field techniques for estimating recharge, and Anderson and Woessner (1992) discuss the application and use of numerical models for groundwater problems.

24.1.5 HYDROGEOLOGIC CONTROLS: LITHOLOGY, STRATIGRAPHY AND STRUCTURE

The permeability, porosity and geometry of the aquitard–aquifer systems at a site depend directly or indirectly on the lithology, stratigraphy and structure of the regional and local geologic setting. The geologic setting is usually heterogeneous and complex at all scales and geological training is a prerequisite to understanding the distribution and geometry of aquifers and aquitards.

The lithology of a geologic deposit describes the type of rock or sediment. Rocks and sediments are described in terms of their mineralogy, grain size and orientation, and packing at the local, hand-specimen or outcrop scale. It is the lithology that determines if a geologic deposit is a consolidated, unconsolidated or lithified sediment; a fractured or unfractured rock; and weathered or unweathered. All of these factors and many more control the local porosity and permeability of the geologic deposit.

The stratigraphy of the aquifer–aquitard system is concerned with the arrangement, orientation and chronological order of the lithologic units. Hydrogeologists often group or subdivide traditional stratigraphic units into units that possess similar permeability, called hydrostratigraphic units. Sedimentary systems can range from the dominantly horizontal layered units of lacustrine or marine depositional environments, to the highly heterogeneous and lenticular material found in alluvial or glaciofluvial sediments. Structural disruptions due to folding, faulting, uplift and fracturing can alter the geometry, continuity and hydraulic conductivity of geologic units.

The pattern of groundwater flow depends on the geometry and permeability distribution of the aquifers and aquitards. Typically, if the permeability contrast between aquifer–aquitard systems is two orders of magnitude or greater, flow lines tend to become almost horizontal in the aquifers and almost vertical in the aquitards (Freeze & Cherry 1979). However, actual flow directions depend on the location of recharge and discharge areas in relation to the continuity of the hydrostratigraphic units. Sara (1991) presents a series of figures, shown here as Fig. 24.2a–c, which illustrate this concept. The figures display a landfill over a sedimentary deposit with higher permeability sand lenses within a lower permeability material. Underlying this sedimentary deposit is the regional bedrock aquifer. Figure 24.2a illustrates that even with two orders of magnitude difference between the sand lenses and surrounding sediment, lack of connectivity among the lenses and the influence of the underlying regional aquifer, result in vertical flow in the sand lenses. When the lens continuity is increased, and connection to surface discharge is allowed, flow has a much more pronounced horizontal component in the sand layers (Fig. 24.2b). Only when the permeability of the regional aquifer is reduced does flow become dominantly vertical in the low permeability sediments and horizontal in the sand layers (Fig. 24.2c).

The lithology, stratigraphy and structure of aquitard–aquifer systems are usually established during a site investigation. A site investigation usually involves an examination of air photographs, topographic and geologic maps, and a review of existing geotechnical borings and water well records for the area (see Chapter 4). Following this early phase of the study, the site is usually visited and test borings, detailed mapping of hydrostratigraphic units and fracture traces, and geophysical surveys are carried out. Finally, wells and piezometers may be installed for a complete assessment of on-site aquifer–aquitard relations, permeabilities, gradients and water quality. Chapter 4 discusses the geotechnical and environmental site characterization. Additional references include hydrogeology texts by Domenico and Schwartz (1998) and Fetter (1994), monitoring handbooks by Nielsen (1991), and various regulatory guidelines (e.g. MoEE 1996b; CCME 1994; US EPA 1992a).

24.1.6 FRACTURED AND KARST SYSTEMS: FRACTURE FLOW

In fractured rocks, the interconnected fractures are considered to be the main pathways for fluid flow, while the matrix or solid rock blocks are considered to be much less permeable. Karst systems behave much the same way.

To organize studies of fractured media two approaches are typically followed: equivalent porous medium or discrete fracture. The equivalent porous medium approach assumes that the fractured mass of rock is hydraulically equivalent to a porous medium. In effect, this assumes that a set of porous medium parameters can be defined to characterize the hydraulic and transport behavior. With the equivalent porous medium approach a geologic unit is considered to have a sufficiently high fracture density that the structure of the fracture network need not be described in detail. This simple approach is complicated by the difficulty in justifying that the network of interconnected fractures is of sufficient density to provide an equivalent porous medium. Furthermore, the inherent heterogeneity and evolving scales of fracture networks makes defining a representative elementary volume difficult. Readers interested in learning more about these problems and flow and transport in fractured rock can refer to Domenico & Schwartz (1998), Bear *et al.* (1993) and NRC (1996).

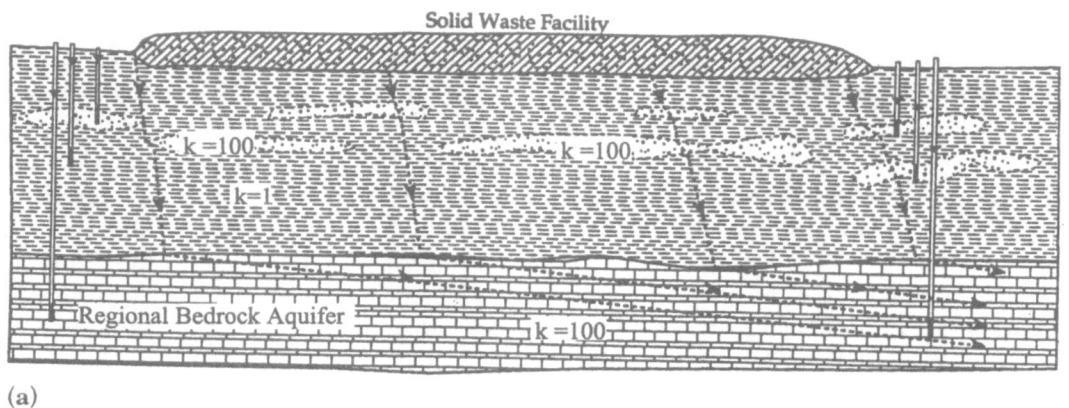

(a)

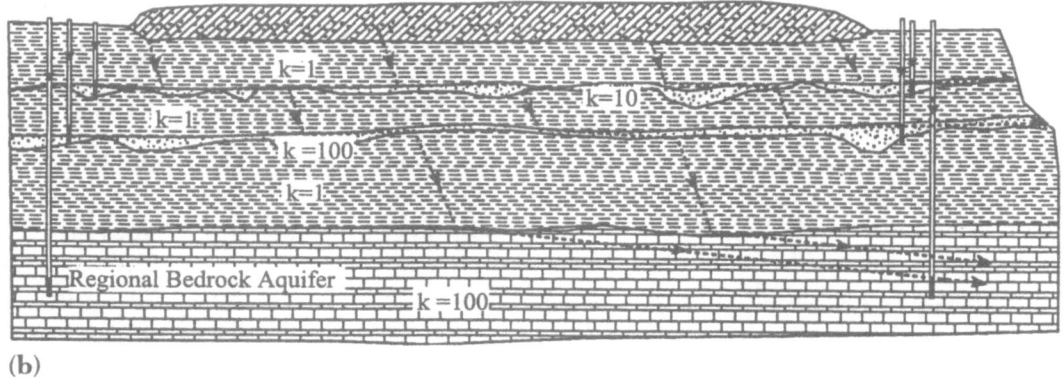

(b)

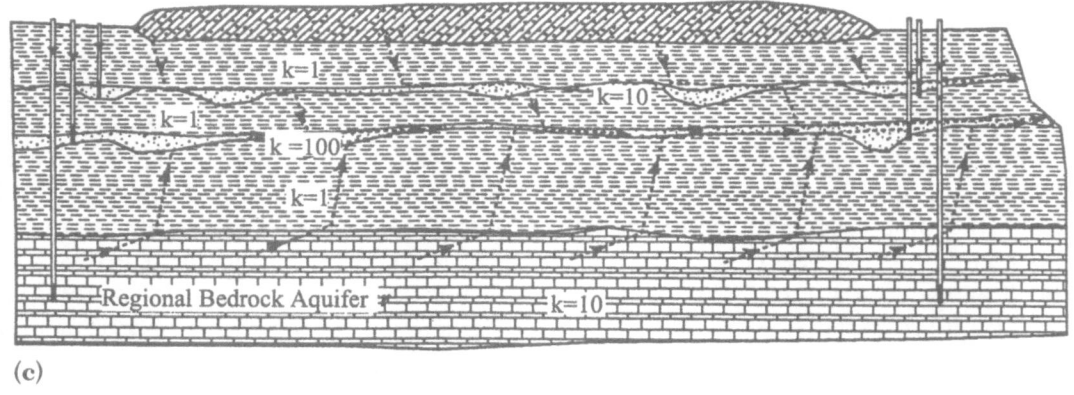

(c)

FIGURE 24.2. Hydrogeologic cross-sections illustrating the importance of hydraulic conductivity contrasts, continuity of units, and recharge and discharge relations on groundwater flow: (a) sand lenses with little horizontal discharge; (b) sand layers with significant horizontal discharge; and (c) sand layers discharging both bedrock and overburden (modified from Sara 1991, *Practical Handbook of Ground-Water Monitoring*, D.M. Nielsen ed., © 1991 CRC Press, Inc. Reprinted by permission of CRC Press, Inc.).

In the discrete-fracture approach, the fracture network is described in terms of individual fractures. This approach also presents several difficulties. First, extensive field testing is required even to begin to characterize the fracture properties, e.g. length, width, aperture, orientation, fracture density, wall smoothness. Second, the fracture apertures, and thus hydraulic conductivity, depend upon the three-dimensional stress field. This coupling has implications both in field testing, where boreholes are pumped or stressed, and in natural changes to the stress field. Finally, in larger apertures, flow could be turbulent, which invalidates Darcy's law.

24.2 Contaminants and Sources

24.2.1 PHYSICAL AND CHEMICAL PROPERTIES OF CONTAMINANTS

A contaminant is generally defined as a miscible (aqueous) or immiscible (non-aqueous) phase liquid, or soluble or insoluble solid phase, which is added to water as a consequence of human activities. The dominant physical properties of a contaminant that would control contaminant transport include density, viscosity and volatility. A key chemical property influencing the abundance of a contaminant in water is the contaminant solubility. Abundance also is controlled by a variety of chemical interactions that may occur among native groundwaters, aquifer materials, and other contaminants once a given contaminant has entered the groundwater flow system. These chemical and geochemical reactions are beyond the scope of this book, but are well-covered in texts such as Stumm & Morgan (1996); Morel (1983); Hounslow (1995). Physical and chemical properties of typical contaminants are found in the chemical literature (Montgomery & Welkon 1990) or from manufacturers' safety data sheets (MSDS).

The solubility defines the maximum quantity of mass that will go into solution per unit volume of solution under specified conditions. However, contaminants are seldom found in groundwater at their theoretical solubilities. This behavior is explained by several factors, such as the heterogeneous nature of flow fields at pore and aquifer scales, the complex kinetics of dissolution, and mixing of waters containing other solutes or contaminants. Fortunately, for many contaminants, their theoretical solubilities are low. For example, non-polar compounds like polychlorinated biphenyls (PCBs), commonly exhibit low solubility due to the polar structure of water.

Another important way contaminants occur are as non-aqueous phase liquids (NAPLs). An NAPL encompasses contaminants present as a distinctly separate liquid phase, like gasoline. NAPLs can migrate as a separate liquid phase, a dissolved phase and a vapor phase.

The two physical properties that largely control how NAPLs migrate are their density and volatility. Dense non-aqueous phase liquids (DNAPLs) are denser than water and, thus, sink through water. Light non-aqueous phase liquids (LNAPLs) are less dense than water and float on water. In unsaturated systems, these chemical compounds can migrate in the vapor phase. The extent of partitioning into the vapor phase, the compound's volatility, is determined by its vapor pressure. The vapor pressure of the compound is a measure of its solubility in a gas phase. The greater the vapor pressure of an organic or NAPL the greater the tendency for volatilization and spreading as a vapor phase in the unsaturated zone. NAPL migration is discussed more fully in Section 24.3.2.

The viscosity of a NAPL is mainly used to assess its relative mobility. As shown by Eq. 24.1, the hydraulic conductivity of a particular fluid–porous media combination is inversely proportional to the viscosity. While viscosity is most important in assessing the relative mobilities of NAPLs it can also be important with miscible contaminants. At high concentrations, as typically found near the source, the contaminant plume viscosity and density may significantly control plume migration (Kimmel & Braids 1980; Van de Molen & Van Ommen 1988; LeBlanc & Celia 1991; Schincariol et al. 1994, 1997; Zhang et al. 1998).

24.2.2 HYDRAULIC CHARACTERISTICS OF SOURCES

The hydraulic characteristics of the source can significantly influence dissolved contaminant transport in the unsaturated and saturated zones. For example, increased recharge at landfills, disposal pits, holding ponds and lagoons, or recharge wells and infiltration galleries often causes a mounding of the water table at these sources. This mounding creates a zone of elevated hydraulic head that perturbs the regional groundwater flow field and results in a local radial spread of contamination. Furthermore, the increased local gradients, and water saturation of some wastes, increases contaminant loadings to the subsurface.

24.2.3 POINT AND NON-POINT SOURCES

The terms point and non-point describe the degree of localization of the source. A point source is characterized by an identifiable, relatively small and fairly well-defined source. Point sources include leaking storage tanks, disposal ponds, landfills, injection wells or leaking pipelines. Usually, these sources produce a reasonably well-defined plume. A non-point source refers to a more spatially distributed source, such as the large-scale loading that results from the atmospheric fallout of airborne contaminants. Alternatively, it could be a large grouping of distributed sources, such as herbicide, pesticide, and nitrate contaminants that result from farming practices, or nutrient loadings from septic beds or leaking sewer systems. In these cases, there are typically no well-defined plumes, but a large enclave of contamination with extremely variable concentrations.

There are also significant differences between point and non-point sources from a remedial perspective. Contaminant plumes from point sources are often candidates for direct remedial action or plume control and remediation (see Chapters 29 and 30). However, non-point plumes are typically candidates for passive remedial action. This includes preventative source control (reduction in fertilizer use, repair of sewer systems) and natural attenuation. The most common approach to groundwater remediation for non-point source contamination has been to treat the groundwater at the well as it is removed for use (Gorelick et al. 1993).

24.2.4 LOADING HISTORY

The loading history describes how the concentration of a contaminant, or its flux, varies as a function of time at the source. A spill or one-time release of contaminants from a storage lagoon or tank is an example of a pulse loading. Here, the source produces contaminants at a fixed concentration for a short period of time. Long-term leakage from a landfill, storage lagoon or tank is termed continuous source loading. In some cases, the continuous source loading could be at a constant concentration. This loading might occur with NAPL contamination, where the NAPL source can dissolve at a slow rate over many years. However, most often the source loadings at industrial sites or operational landfills have a non-constant loading function. In this case, the concentration of chemicals added to a storage lagoon may vary depending on manufacturing needs, techniques, or plant owner. The waste type added to sanitary landfills will also vary with geographic area and clients served. Seasonal factors, such as precipitation and evaporation, will also influence both the concentration and flux of contaminants in waste ponds or landfills. Closed ponds or landfills may exhibit a decaying concentration function, as contaminants are leached from the solid wastes or the waste stream biodegrades (see Rowe *et al.* 1995b).

The source loading function, along with the physical and chemical mass transport processes discussed in Section 24.3, will control the contaminant distributions in plumes. Increasing complexity in the loading function will translate directly into increasingly variable concentration distributions.

24.3 *Principles of Contaminant Transport*

24.3.1 MASS TRANSPORT PROCESSES: ADVECTION, DIFFUSION AND DISPERSION OF SOLUTES

Mass occurring in groundwater as ions, molecules or solid particles is transported by three processes: advection, mechanical dispersion and diffusion. All three processes operate simultaneously in flowing groundwater.

24.3.1.1 Advection

Advection is mass transport, or the movement of dissolved solute, due simply to the flowing groundwater. The direction and rate of transport coincide with the groundwater. The velocity of transport is equal to the average linear groundwater velocity. In aquifers, advection is the dominant process. Therefore, if one understands the flow system or hydrogeological environment, one understands advective transport. Similarly, factors that influence groundwater flow patterns, such as water table configuration, pattern of geologic layering and hydraulic conductivity of geologic units, recharge and discharge relations, control the direction and rate of advective mass transport. If advection is the only process operating, mass added to a streamtube, defined by bounding flowlines, will remain in the streamtube.

24.3.1.2 Diffusion

Whenever contaminants exist in groundwater, there will be zones of low and high concentration. Molecular or ionic diffusion is the process where dissolved mass moves from areas of high concentration to areas of low concentration. Mixing is caused by random molecular motion due to the thermal kinetic energy of the solute. For a simple aqueous non-porous system, at fairly low concentration, Fick's law describes how the chemical mass flux is proportional to the concentration gradient:

$$J = -D_0 \text{grad}(C) \qquad (24.7)$$

where J = the chemical mass flux (negative sign indicating that transport is in the direction of decreasing concentration), D_0 = a proportionality constant called the diffusion coefficient in free solution, and C = the concentration. Diffusion coefficients are temperature dependent. Values for most major ions can be found in Robinson and Stokes (1970).

In a porous medium, the diffusion coefficient must be smaller than that in a fluid environment due to the hindering effect of collisions with solids, and the tortuous fluid pathways. To account for these effects, a porous media diffusion coefficient, D_p, is introduced so Fick's law for diffusion in porous media becomes:

$$J = -D_p \text{grad}(C). \qquad (24.8)$$

There are several relationships for calculating D_p. Greenkorn and Kessler (1972) express D_p as a function of porosity and tortuosity:

$$D_p = \frac{nD_0}{(L_e/L)} \qquad (24.9a)$$

where the tortuosity is defined as the ratio of the length of a flow channel for a fluid particle, L_e, to the length of a porous medium sample, L. Thus the value (L_e/L) is always greater than one in porous media. However, since one is dividing by tortuosity (L_e/L), it has the same effect as more conventional definitions. Bear (1972) also defines D_p as a function of porosity and tortuosity, T°, but where tortuosity has a different definition:

$$D_p = nT^\circ D_0 \qquad (24.9b)$$

and T° is reported to range from 0.56 to 0.8. According to Rowe *et al*. (1995b), the porous media diffusion coefficient is related to D_0 by the relationship:

$$D_p = D_0 W_T \qquad (24.9c)$$

where W_T = a complex tortuosity factor that depends on the porosity of the soil, geometric tortuosity, a fluidity factor related to adsorbed double layer water and electrostatic interaction. These latter factors may be particularly important for diffusion through clayey soils. It is often convenient to replace D_p in Eq. 24.8 by:

$$D_p = nD_e \qquad (24.9d)$$

yielding:

$$J = -nD_e \mathrm{grad}(C) \qquad (24.10)$$

where D_e is defined here as the effective diffusion coefficient.

Unfortunately, the terminology relating to diffusion is somewhat confused with similar words being used to mean different things. For example, Domenico & Schwartz (1998) refer to D_p as the "effective diffusion" coefficient; however, in the geotechnical litera-ture (e.g. Shackelford & Daniel 1991a,b; Rowe *et al*. 1995b) the "effective diffusion" coefficient does not include porosity and is as defined by Eq. 24.10. Thus the reader needs to be very careful in clarifying the definition in a given context prior to using a diffusion coefficient in numerical calculation.

When dealing with contaminant migration through soils, it is important to recognize that diffusive transport is also influenced by electro-osmotic flow and that the effective diffusion coefficient deduced in the laboratory or back-figured from field profiles is really a bulk mass transfer coefficient. It can incorporate the effects of a number of related processes including true diffusion, osmotic flow and, due to the need to maintain an ion balance, electrically induced flow (Rowe *et al*. 1995b). For a given soil, the diffusion of a particular contaminant is temperature dependent. In addition, the diffusion coefficient of even a "conservative" contaminant such as chloride can be influenced as much by the chemical composition of the leachate (due to the effect on both the availability of cations to migrate with chloride and osmotic flow) as by temperature over typical temperature ranges. This is illustrated by the range of observed (deduced) effective diffusion coefficients summarized in Table 24.3.

The nature of the porous media (and its effect on geometric tortuosity) does have

TABLE 24.3. Observed effective diffusion coefficients for chloride through a number of soils (adapted from Rowe *et al*. 1995b and Rowe & Weaver 1997)

Soil type	Method of evaluation	Effective diffusion coefficient, D_e (m² a⁻¹)
Uniform sand or gravel	Laboratory, at 23°C	0.03
Silt and silty sand	Laboratory, at 23°C	0.022–0.028
Low activity clayey soils	Laboratory, at 22°C	0.017–0.021
	Laboratory, at 10°C	0.01–0.025
In situ silty clay below waste	Field	0.006–0.02
Leda clay	Field	0.006
Water-laid clay till	Field	0.01
Freshwater glaciolacustrine clay	Field	0.012–0.018

TABLE 24.4. Effective diffusion coefficients for chloride in porous rock (adapted from Rowe *et al.* 1995b)

Rock	Effective porosity (%)	Effective diffusion coefficient, D_e (m^2 a^{-1})
Sandstone	3.4	0.002–0.003
Mudstone	9.2	0.002–0.003
Shale	10.8	0.005
Mudstone	23.8	0.006

some effect on the effective diffusion coefficient. For example, the diffusion through uniform sand and gravel is greater than through silt, which, in turn, is greater than through clays and clayey tills. Diffusion can also occur through porous rock such as chalk, sandstone and shale. Observed (deduced) effective diffusion coefficients for a number of porous rocks are given in Table 24.4.

Diffusion operates independently of groundwater flow. In aquifers, or primarily high velocity hydrogeological environments, it is secondary in importance to advection and dispersion. However, in aquitards, in which advection or groundwater flow velocities are very small, diffusion can become the dominant transport mechanism (see Rowe *et al.* 1995b).

Matrix diffusion into or out of low permeability clay lenses in aquifers or porous blocks of fractured media plays an important role in contaminant migration. Here, contaminants being transported through the more permeable zone, e.g. aquifer adjacent to a clay lens or a fracture in a clay or porous rock, diffuse into the adjacent low permeability unit, thereby reducing contaminant concentrations in the permeable zone. If not anticipated, this process may lead to misinterpretation of groundwater monitoring data. Matrix diffusion both reduces the magnitude and delays the arrival time of the contaminant peak at a point (Rowe *et al.* 1995b). Field evidence indicates that this effect can be significant (Rowe & Booker

1989). Matrix diffusion can also substantially reduce the effectiveness of conventional "pump and treat" clean-up strategies. This is caused by the slow but persistent release of contaminants back into the permeable zone by diffusion once the concentration in the permeable zone drops below that in the adjacent low permeability material.

Diffusion can also be of importance in unsaturated soils below landfills. The diffusion of dissolved species is reduced with reducing degrees of saturation (volumetric water content), as discussed in Section 25.5.1. Conversely, however, the diffusion of vapor (e.g. volatile organic compounds or oxygen) can be enhanced as the degree of saturation is reduced, see Section 25.5.2.

24.3.1.3 Dispersion

Dispersion, or more formally hydrodynamic dispersion, spreads mass beyond the area it normally would occupy due to advection alone. Stated another way hydrodynamic dispersion will move mass out of flow or stream tubes. Hydrodynamic dispersion occurs as a consequence of two processes, diffusion as discussed above, and mechanical dispersion. These two contributions to hydrodynamic dispersion are represented mathematically as:

$$D = D_{md} + D_e \qquad (24.11)$$

where D = the coefficient of hydrodynamic dispersion and D_{md} = the coefficient of mechanical dispersion. The use of the effective diffusion coefficient, D_e (rather than the porous media diffusion coefficient, D_p) in Eq. 24.11 is important because the coefficient of hydrodynamic dispersion that appears in most formulations of the groundwater transport equation already has the porosity factored out.

Mechanical dispersion is mixing caused by the convoluted pathways that water and contaminants (solutes, molecules, particles) follow while flowing through porous or fractured media. Essentially, mechanical dispersion is caused by local variations in velocity around some mean velocity of flow (average linear groundwater velocity). Thus, mechanical dispersion is purely an advective process related to the statistical variability in the grain-size distribution in the aquifer and is not driven by a chemical potential like diffusion. The more heterogeneous the aquifer is, the greater the irregularity of the flow paths and the deviation about the mean velocity will be, and the greater the mechanical dispersion. This relationship with average linear groundwater velocity is reflected in the equations for the coefficients of mechanical dispersion:

$$D_{\mathrm{md-L}} = \alpha_{\mathrm{L}} v \text{ and } D_{\mathrm{md-T}} = \alpha_{\mathrm{T}} v \quad (24.12)$$

where α_{L} and α_{T} = the longitudinal and transverse dispersivities of the porous medium (units of length), and $D_{\mathrm{md-L}}$ and $D_{\mathrm{md-T}}$ = the longitudinal and transverse dispersion coefficients. $D_{\mathrm{md-L}}$ quantifies mechanical dispersion in the direction of flow, while $D_{\mathrm{md-T}}$ quantifies transverse spreading. In actuality, there are two transverse components of spreading, one in the horizontal plane and one in the vertical plane.

Although Eq. 24.12 is the commonly accepted relation for the coefficient of mechanical dispersion, the order of dependence on velocity is only approximate, i.e. $D_{\mathrm{md-L}} \propto v$. Greenkorn (1983) has explained the reasons for this uncertainty. Furthermore, while it is agreed the dispersivity is a medium property, quantifying dispersivity is difficult. Numerous studies on dispersivity from the column to field scale have been undertaken. Gelhar *et al.* (1992) undertook a critical review of field experiments from 59 sites around the world (Fig. 24.3). Of approximately 106 values for longitudinal dispersivity only 14 values were highly reliable. The reliable values ranged from 10^{-2} to 10^{-1} m for column experiments to 10^{-1} to 2 m for relatively small-scale field studies (over a maximum distance of approxi-

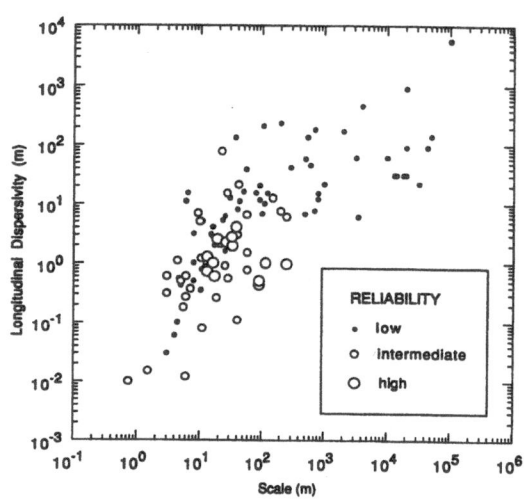

FIGURE 24.3. Longitudinal dispersivity versus scale with data classified by reliability (after Gelhar *et al.* 1992, *Water Res.*, v. 28, p. 1955–1974. © 1992 American Geophysical Union).

mately 100 m). Many published dispersivity values are unrealistically large. They suffer from errors related to the collection and interpretation of field data, which force dispersivity values to be larger than actual values (Domenico & Schwartz 1998). Furthermore, many published estimates of dispersivity are no more than numerical model calibration or curve-fitting values (Anderson 1984), and can be subject to a variety of other errors. While it is known that dispersivity is scale dependent, no larger, properly conducted, field experiments have been carried out to quantify scale dependence.

24.3.2 MULTIPHASE FLOW

Many contamination problems involve more than one fluid phase (e.g. aqueous phase, non-aqueous phase, and vapor phase). Figure 24.4 illustrates a conceptual DNAPL spill, where the non-aqueous phase contaminants occur as residual fluid within the pores and ponded on a low permeability layer. Complexity is added because the DNAPL dissolves into groundwater to form a large aque-

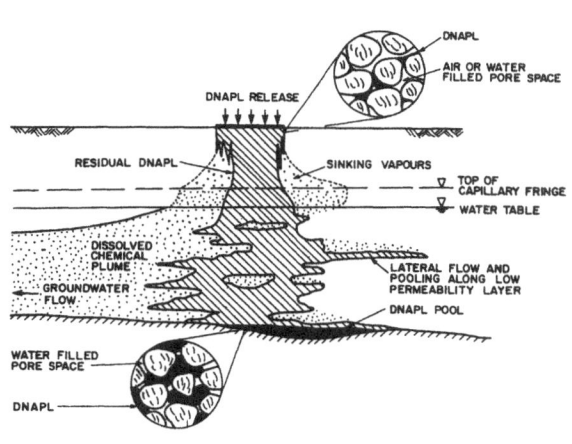

FIGURE 24.4. General groundwater contamination scenario with dense, non-aqueous phase liquids (after Kueper & Frind 1991, *Water Resources Res.*, v. 27, p. 1049–1057. © 1991 American Geophysical Union).

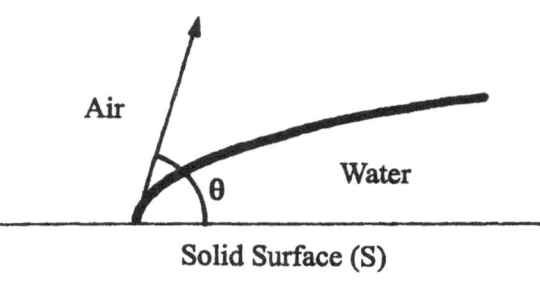

FIGURE 24.5. Interface between air, water and a solid surface (modified from de Marsily, 1986, *Quantitative Hydrogeology.* © 1986 Academic Press, Inc. Reprinted by permission of Academic Press, Inc.).

ous phase plume, and volatilizes into the unsaturated zone to create a vapor plume. Problems involving NAPLs or multiphase flow are much more complex than dissolved plumes.

When several fluids occupy a porous medium, their flow characteristics will be governed by the proportion of each fluid in the medium. Their relative abundance is described by the volumetric saturation, S_i, of the fluids:

$$S_i = V_i/V_v \qquad (24.13)$$

where V_i = the volume of the ith fluid, and V_v = the volume of voids.

When two fluids are in contact with each other, or a fluid is in contact with a solid, there is a free interfacial energy between them created by the difference between the forces that attract the molecules toward the interior of each phase and those that attract the molecules through the contact surface. The interfacial energy is characterized by the interfacial or surface tension, σ_{ij}. Interfacial tension is defined as the work required to separate a unit area of one substance (i) from that of another (j), and is expressed as a force

per unit length. When two or more fluids are present, there is a preference for one fluid to be attracted to a surface over the others. The term wettability describes the preference of fluids to wet a surface. According to Demond and Roberts (1987), the only direct measure of wettability is the contact angle. The contact angle, as in Fig. 24.5, is given by Young's equation (de Marsily 1986):

$$\cos \theta = \frac{\sigma_{sa} - \sigma_{sw}}{\sigma_{aw}} \qquad (24.14)$$

where θ is measured from 0 to 180° in the denser fluid (in Fig. 24.5, the water). If $\theta <$ 90°, the fluid (here water) is said to be wetting with respect to the solid. If $\theta >$ 90°, the fluid is said to be non-wetting. This is the case of the air in Fig. 24.5. Wettability is unique for given types of solids and fluids. However, it is generally found that: water is always the wetting fluid with respect to oil or air on rock-forming minerals; oil is a wetting fluid when combined with air, but a non-wetting fluid when combined with water; and oil is the wetting fluid on organic matter, such as peat or humus, in relation to either water or air (Albertsen *et al.* 1986).

In the fluid domain (Fig. 24.5), on either side of the interface, the pressure is not the same. The pressure discontinuity across any curved interface separating immiscible fluids is called the capillary pressure, P_c, and is given by:

$$P_c = P_{nw} - P_w \qquad (24.15)$$

where P_{nw} = the pressure of the non-wetting fluid, and P_w = the pressure of the wetting fluid. Capillary pressure is a function of the radius of curvature, r, of the interface between the two fluids and of the surface tension, σ_{ij}, existing between them:

$$P_c = \frac{2\sigma_{ij}}{r}. \qquad (24.16)$$

In a fluid sense, capillary pressure is a measure of the tendency of a porous medium to imbibe the wetting phase or to repel the non-wetting phase (Bear 1972). Domenico & Schwartz (1998) explain that capillary pressure can be thought of as the pressure required to move a particle of non-wetting fluid into a pore filled with wetting fluid. Thus, because small pores provide resistance to entry due to capillarity, non-wetting fluids will tend to move through the coarser, more permeable zones in a heterogeneous medium. This behavior traps globules of water. A more detailed understanding of how immiscible fluids displace one another can be obtained by studying capillary pressure curves (Fig. 24.6). Capillary pressure curves illustrate the pressure–saturation relationship as the fluid initially saturating the medium is slowly displaced by the other fluid.

The radius of curvature of the menisci separating the fluids in the porous medium is a function of saturation. Accordingly, capillary pressure depends on the saturations of non-wetting and wetting fluids (Fig. 24.6). Drainage is the displacement of the wetting fluid by the non-wetting fluid and imbibition is the displacement of the non-wetting fluid by the wetting fluid. The curves show the value of capillary pressure that must be exceeded for displacement to take place at each relative

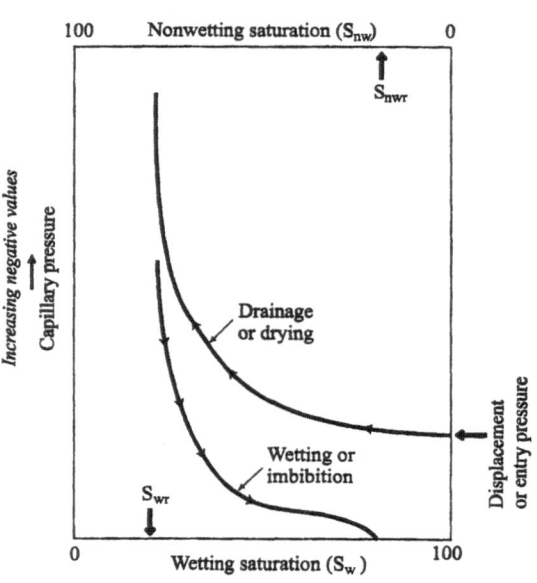

FIGURE 24.6. Capillary pressure curves given as a function of the degree of saturation with respect to the wetting and non-wetting phases (modified from Domenico & Schwartz, 1998, *Physical and Chemical Hydrogeology.* © 1998 John Wiley & Sons, Inc. Reprinted by permission of John Wiley & Sons, Inc.).

saturation. The drainage curve starts off at 100% wetting fluid saturation and some finite value of capillary pressure. For the non-wetting fluid to displace the wetting fluid, this value of capillary pressure, termed the entry pressure, must be exceeded. The curve is essentially a succession of equilibrium states for increasing values of saturation in non-wetting fluids that are very close to each other. As the wetting saturation decreases, the capillary pressure becomes more negative. As the suction increases, the displacement process eventually stops when the capillary pressure tends to infinity. The saturation value at which it stops is known as the residual wetting saturation (designated S_{wr} in Fig. 24.6). At the residual saturation there is no longer a continuum of the wetting fluid and pressures can no longer be transmitted across the pore network. The wetting or imbibition

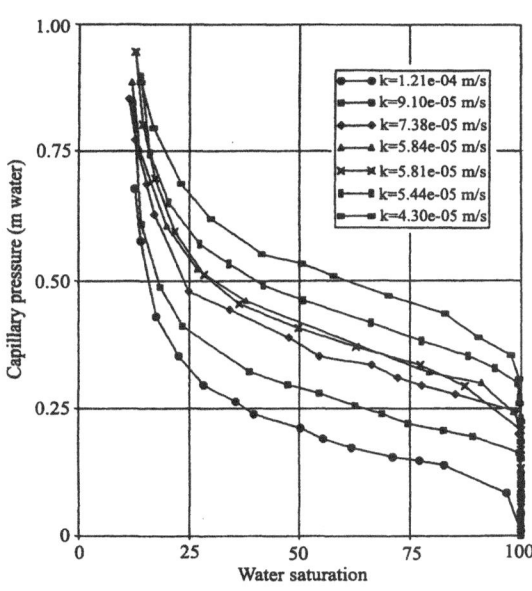

FIGURE 24.7. Capillary pressure–water saturation drainage curves for a tetrachloroethene–
water system for sands of differing hydraulic conductivity (modified from Kueper & Frind, 1991,
Water Resources Res., v. 27, p. 1059–1070. ©
1991 American Geophysical Union.).

curve shows the opposite process. However,
it does not follow the same pathway as the
drying curve since the non-wetting fluid occupies a different pore network during imbibition than during drainage. Note also that
for a zero capillary pressure some saturation
remains in a non-wetting fluid. This is the residual non-wetting saturation (designated S_{nwr}
in Fig. 24.6).

Kueper & Frind (1991) present capillary
pressure curves for a tetrachloroethylene–
water system for sands of differing hydraulic
conductivity (Fig. 24.7). These curves show
that the entry pressure decreases with increasing hydraulic conductivity. Curves like
this can be used at contaminated sites to assess the potential for DNAPL to enter underlying fine-grain material that may serve initially as migration barriers. The reader is
referred to Kueper & McWhorter (1991) for
details.

The capillary pressure curves illustrate
how immiscible fluids tend to interfere with
one another as they flow. In essence, because
some of the pore networks are occupied with
an immiscible fluid these networks are un-
available for the "other" fluid. This concept
is readily illustrated by looking at Darcy's law
for one-dimensional multiphase flow in a homogeneous medium, as given by Demond
& Roberts (1987).

$$q_i = -\frac{k_i k_{ri}}{\mu_i}(\nabla P_i - \rho_i g \nabla h) \qquad (24.17)$$

where q_i = the flow of the ith fluid per unit
area of the medium, k_i = the intrinsic permeability; k_{ri} = the relative permeability to the
ith fluid, h = the elevation, and μ_i, P_i, ρ_i are
the viscosity, pressure and density of the ith
fluid, respectively. The relative permeability
to the ith fluid is defined as:

$$k_{ri} = \frac{k_{eff}}{k_i} \qquad (24.18)$$

where k_{eff} = the effective permeability of the
medium to the ith fluid. Effective permeabilities are generally obtained by measurement
in the laboratory or from published experimental results. A detailed discussion of effective and relative permeabilities can be found
in Corey (1986).

How k_{ri} varies between zero and one depends on many elements including: fluid
densities, viscosities, and interfacial tension;
wetting angle; relative saturation, media
pore-size distribution; and whether the system is undergoing imbibition or drainage.
Figure 24.8 illustrates many features common to most relative permeability curves:

- The simultaneous flow of both immiscible
fluids is only possible if the residual wetting
and non-wetting saturation are exceeded
(in Fig. 24.8 $S_w > 0.2$; $S_{nw} > 0.1$).
- When both fluids are present the relative
permeabilities rarely sum to one.

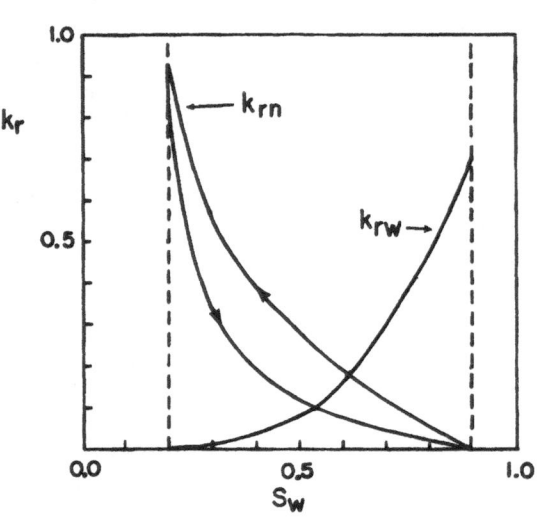

FIGURE 24.8. Typical relative permeability curves (after Demond & Roberts, 1987, *Water Resources Bulletin*, v. 23, p. 617–628. © 1987 American Water Resources Association. Reproduced by permission of American Water Resources Association.).

- When at residual wetting saturation, the relative permeability of the non-wetting fluid is close to one. At residual non-wetting saturation, the relative permeability of the wetting fluid is much less than one. Thus, the presence of a wetting fluid at residual saturation has little influence on the flow of a non-wetting fluid, while the presence of a non-wetting fluid at residual saturation interferes considerably with the flow of a wetting fluid.
- For comparable saturations $k_m > k_{rw}$ (Fig. 24.8). One reason for this is that the wetting fluid occupies the smaller pores that contribute least to flow, while the non-wetting fluid occupies the larger pores that contribute most to flow.

Readers can refer to Demond & Roberts (1987) and texts by Dullien (1979) and Marle (1981) for a more detailed review of capillary pressure or relative permeability curves.

The level of residual saturations has important implications regarding the mobility of

NAPLs and DNAPLs. At residual wetting saturation, the wetting fluid is held by capillary forces in the narrowest parts of the pore space. At residual non-wetting saturation, the non-wetting fluid, in a water-wet aquifer, occurs as an isolated blob in the center of pores. In the unsaturated zone, it is common to find water, air and a NAPL such as oil. Here, water is the wetting fluid, air is the non-wetting fluid, and oil has intermediate wetting properties (non-wetting with respect to water but wetting with respect to air, as discussed by Wilson & Conrad (1984)). Because immiscible fluids behave differently in the saturated zone and unsaturated zone, the residual saturations differ. Residual saturations in the unsaturated zone generally range between 0.1 and 0.2 (Cohen & Mercer 1993). In the saturated zone, where the NAPL is usually the non-wetting fluid, residual saturation values range from approximately 0.1 to 0.5.

Domenico & Schwartz (1998) present many examples of DNAPL and NAPL migration and have an excellent discussion of conceptual models for multiphase migration. In addition, Schwille (1988) presents detailed experimental studies on DNAPL migration in heterogeneous porous and fractured media.

One important key to understanding the migration of NAPLs at a site is understanding the hydrogeological setting or environment (Section 24.1); most importantly a detailed characterization of aquifer and aquitard permeability heterogeneities (hydrostratigraphy). While detailed field characterization is costly, an added problem at a NAPL-contaminated site is that drilling can perforate low permeability zones and enhance downward NAPL migration.

When an LNAPL or DNAPL is released on the ground surface, the free product percolates down toward the water table under gravity driven flow. As the NAPL locally displaces porewater and soil gas, it moves from pore to pore, once saturations exceed residual saturation. Therefore, the rate and area

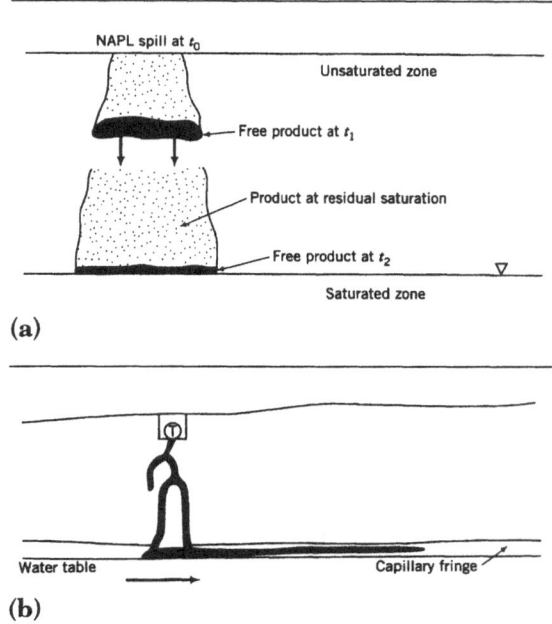

FIGURE 24.9. The downward percolation of a NAPL in the unsaturated zone is determined by the nature of the spill: (a) a sudden loss results in maximum spreading and a large quantity trapped at residual saturation; and (b) a slow leak causes the product to follow a set of channels and a minimal quantity is trapped at residual saturation (after Domenico & Schwartz 1998, *Physical and Chemical Hydrogeology.* © 1998 John Wiley & Sons, Inc. Reprinted by permission of John Wiley & Sons, Inc.).

of NAPL spill influences migration patterns. A large volume release over a short time period will trap a relatively large volume of NAPL at residual saturation (Fig. 24.9a). In comparison, slow leakage over a long time, will greatly increase the volume NAPL reaching the water table (Fig. 24.9b). Thus, if the volume of the spill is relatively small, and includes a sufficiently large area, the NAPL could be trapped in the unsaturated zone at residual saturation. Domenico & Schwartz (1998) discuss quantitative methods to calculate the depth of the residual saturation and other relevant parameters necessary to access NAPL migration in the field.

As the NAPL moves downward, it also tends to spread horizontally due to capillary forces (Fig. 24.9a). However lateral migration is most influenced by media heterogeneities and low permeability zones (Fig. 24.4). As LNAPL encounters the capillary fringe, the water saturations increase and relative permeability decreases. At this point, buoyancy forces become important; LNAPL accumulates and once some critical thickness is achieved, would flow down-gradient. Recent studies (Farr *et al.* 1990; Lenhard & Parker 1990; Huntley *et al.* 1992) show that instead of occurring as a discrete layer at the top of the capillary fringe (the classical model), LNAPL will be distributed throughout and above the capillary fringe. According to Huntley *et al.* (1992), this more realistic model for LNAPL saturation would provide for much less mobility than is suggested for high LNAPL saturation within a single layer. Furthermore, water table fluctuations can redistribute LNAPL across a larger zone. Lehnard *et al.* (1993) show how rapidly rising water tables can trap LNAPL below the water table.

Because DNAPLs are more dense than water, they will continue to move downward, displacing water in the saturated zone. Spreading will continue until the DNAPL is at residual saturation. As with all NAPLs, flow will be very sensitive to permeability variations. DNAPL will accumulate on low permeability units, moving "downhill" with the topography of the boundary (Fig. 24.4). Because most hydrogeological environments are highly heterogeneous on a local scale, DNAPL distributions are often highly complex with numerous isolated pools of DNAPL. DNAPL at residual saturation and in pools also serves as important sources for vapor phase or aqueous (dissolved) phase contamination (Fig. 24.4).

24.3.3 ATTENUATION OF DISSOLVED SOLUTES: BIOGEOCHEMICAL PROCESSES
The concentration of dissolved solutes often is reduced because of chemical reactions.

Such reactions can take place entirely in the aqueous phase, or involve the solid matrix of the porous or fractured medium. Mass transfer reactions that commonly alter contaminant concentrations in groundwater are:

1. acid–base reactions;
2. solution, volatilization and precipitation reactions;
3. surface reactions;
4. oxidation–reduction reactions;
5. hydrolysis; and
6. complexation reactions.

All of these processes probably will not be important at any given site. However, depending on site geology and contaminant type, one or several of these reactions could exhibit significant control on contaminant transport.

24.3.3.1 Acid–base reactions

Acid–base reactions are important in groundwater because they influence the pH and control ion concentrations. One of the most important natural weak acid–base systems is the CO_2–water system. This involves a set of reactions initiated when CO_2 is dissolved in water:

$$\left.\begin{array}{l} CO_2(g) + H_2O = H_2CO_3(aq) \\ H_2CO_3(aq) = H^+ + HCO_3^- \\ HCO_3^- = H^+ + CO_3^{2-} \end{array}\right\} . \quad (24.19)$$

Carbonate species are the most important participants in reactions that control the pH of natural waters (Hounslow 1995). How the relative concentration of the carbonate species changes with pH is shown in Fig. 24.10. In natural waters, the dissolved CO_2 species, bicarbonate (HCO_3^-) and carbonate (CO_3^{2-}), produce most of the alkalinity. Alkalinity is defined as the capacity of a solution to react with, or neutralize, a strong acid. It is determined by titrating a specified volume of water, with a strong acid such as HCl, to a specific end-point (typically pH 4.5). Standard test methods for alkalinity determination can be found in American Society for Testing and Materials (ASTM) designation D 1067. Natural groundwaters are usually alkaline because of the abundance of dissolved CO_2 species, bicarbonate and carbonate. Other non-carbonate contributors to alkalinity, such as hydroxide, silicate, borate and the organic ligands, especially acetate and propo-

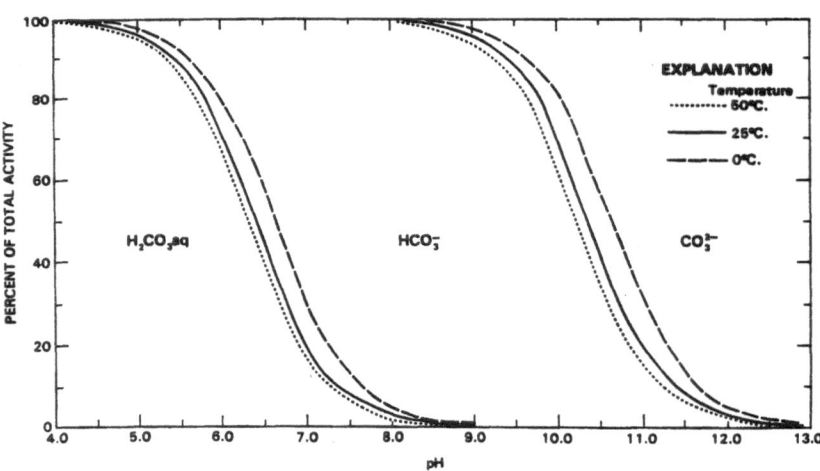

FIGURE 24.10. Percentages of dissolved carbon dioxide species activities at one atmosphere pressure and at (···) 50, (—) 25 and (––)0°C as a function of pH (after Hem 1995, *U.S. Geol. Surv.*, Water Supply Paper 2254).

nate, are only important if present in significant amounts (Hounslow 1995). However, this situation may change as a result of contamination, e.g. oxidation of sulfide minerals from metal mines. Typically, alkalinity is reported in terms of milliequivalents per liter (meq L^{-1}) CaCO$_3$. Example alkalinity calculations, and a much more detailed description of acid–base reactions, carbonate equilibria and alkalinity can be found in Drever (1982).

Carbonate reactions and their control on water chemistry and pH also influence how one samples groundwaters. Exposure of groundwater to atmospheric partial pressures of CO$_2$ often results in the removal of CO$_2$ from the solution. This raises the pH of the solution and can result in various metals being precipitated. In addition, the cation equilibria with solid phases can also change. Thus, the common practice of filtering and adding acid or acidifying groundwater samples for metals analysis.

24.3.3.2 Solution, Volatilization and Precipitation Reactions

Gas solution and exsolution are commonly modeled using equilibrium concepts based on Henry's law. The Henry's law equation for CO$_2$ is:

$$K_H = \frac{P_{CO_2}}{CO_2(aq)} \tag{24.20}$$

where K_H = Henry's law constant, with units like atm m^3 mol^{-1} or bar m^3 mol^{-1}, P_{CO_2} = partial pressure of CO$_2$ (atm) in the gas phase, and CO$_2$(aq) = the molar concentration of CO$_2$ gas in solution (mol m^{-3}). In applications, the concentration of CO$_2$(aq) = H$_2$CO$_3$ (aq).

The Henry's law expression for gases (Eq. 24.21) is the same expression for the volatilization of dissolved organic solutes from water:

$$P_{org} = K_H C_{org} \tag{24.21a}$$

where P_{org} = the partial pressure of the organic compound in the vapor phase (atm), and C_{org}

= the concentration in the aqueous phase (mol m^{-3}). Henry's law can also be expressed as

$$P_{org} = K'_H C_{org} \tag{24.21b}$$

where P_{org} = the solute concentration in air (mol L^{-1}), C_{org} = the aqueous solute concentration (mol L^{-1}) and K'_H = the dimensionless Henry coefficient:

$$K'_H = \frac{K_H}{RK} \tag{24.22}$$

where R = the ideal gas constant (8.205 × 10^{-5} atm m^3 mol^{-1} °K^{-1} = 8.314 × 10^{-5} bar m^3 mol^{-1} °K^{-1}) and K = the temperature of the water (degrees kelvin). The volatilization of pure organic solvents is typically described by Raoult's law:

$$P_{org} = x_{org} P°_{org} \tag{24.23}$$

where x_{org} = the mole fraction of the organic solvent, and $P°_{org}$ = the vapor pressure of the pure organic solvent. The Henry's law constants and vapor pressures of all US EPA priority pollutants can be found in Montgomery & Welkom (1990) along with other physicochemical data such as water solubility, vapor density, octanol-water (K_{ow}) and organic-carbon water (K_{oc}) partition coefficients. A number of these parameters are summarized in Table 24.5. The relative volatility of a substance can be related to the Henry's law coefficient K_H and if $K_H < 10^{-7}$ atm m^3 mol^{-1}, the substance is considered to have low volatility (Lyman et al. 1982). A value of $10^{-7} \leq K_H \leq 10^{-5}$ atm m^3 mol^{-1} is considered to correspond to slow volatilization. Volatilization is usually important for $10^{-5} < K_H < 10^{-3}$ atm m^3 mol^{-1} and the process will occur quickly for $K_H > 10^{-3}$ atm m^3 mol^{-1}. Volatilization is important for problems of NAPL (organic) contamination within the unsaturated and capillary zone (Fig. 24.4). Volatilization is also important for issues related to sample collection and analysis. Volatilization can reduce

TABLE 24.5. Some published Henry's law, log K_{ow}, log K_{oc}, solubility and density values (adapted from Montgomery & Welkom 1990)

Compound	Henry's law constant (atm m^3 mol^{-1})	log$_{10}$ K_{oc} (–)	log$_{10}$ K_{ow} (–)	Solubility in water at 25°C (mg L^{-1})	Density at 20°C (g mL^{-1})
Acetone	3.97×10^{-5}	–0.43	–0.24	∞	0.79
Benzene	5.38×10^{-3}–5.48×10^{-3}	1.69–2.00	1.56–2.15	1780–1800	0.876
bis-(2 ethylhexyl)phthalate	1.1×10^{-5}	5.0	4.2–5.11	0.05–0.4	0.985
2-Butanone (MEK)	4.66×10^{-5}	0.09	0.26–0.29	2.5×10^{5}	0.805
Chlorobenzene	3.6×10^{-3}–4.45×10^{-3}	1.68–2.52	2.71–2.98	295–503	1.1
Chloroethane	8.5×10^{-3}–1.46×10^{-2}	0.51	1.43	4700	0.897
Chloroform	2.9×10^{-3}–5.3×10^{-3}	1.64	1.9–1.97	7200–9600	1.48
p-Cresol	7.9×10^{-7}	1.69	1.67–1.94	18000–23000	1.01–1.03
Di-n-butyl phthalate	6.3×10^{-5}	3.14	4.31–4.79	9–4500	1.046
1,2 Dichlorobenzene	1.2×10^{-3}–2.4×10^{-3}	2.26–3.23	3.38–3.55	93–156	1.3
1,4 Dichlorobenzene	2.7×10^{-3}–4.4×10^{-3}	2.2	3.37–3.62	65–90	1.25
1,1 Dichloroethane	4.3×10^{-3}–5.9×10^{-3}	1.48	1.78	5.060	1.17
1,2 Dichloroethane	9.1×10^{-4}–1.3×10^{-3}	1.15–1.28	1.45–1.48	7890–8650	1.23–1.25
1,1 Dichloroethene	1.5×10^{-2}–2.1×10^{-2}	1.81	1.48–2.13	273–6400	1.21
trans-1,2 Dichloroethene	6.7×10^{-3}–3.8×10^{-1}	1.77	2.09	6300	1.26
Dichloromethane	2.0×10^{-3}–3.2×10^{-3}	0.94	1.25–1.30	13000–19400	1.33
Ethylbenzene	6.4×10^{-3}–8.7×10^{-3}	1.98–2.41	3.05–3.15	77–208	0.867
Naphthalene	3.6×10^{-4}–1.2×10^{-3}	2.74–3.52	3.01–4.70	20–40	1.14–1.16
Pentachloro-phenol	2.8×10^{-7}–3.4×10^{-6}	2.95–2.96	2.75–2.86	20–25	1.98
Phenol	1.7×10^{-7}–4.0×10^{-7}	1.24–1.43	1.46–1.48	67000–93000	1.058
1,1,2,2-Tetrachloroethane	3.8×10^{-4}–4.6×10^{-4}	1.66–2.07	2.39–2.56	2870–2970	1.59
Tetrachloroethene (PCE)	2.9×10^{-3}–1.5×10^{-2}	2.32–2.56	2.10–2.60	150–485	1.62
Toluene	6.7×10^{-3}	2.06–2.18	2.11–2.80	220–627	0.867
1,1,1-Trichloroethane	1.3×10^{-2}–1.8×10^{-2}	2.0–2.18	2.17–2.49	300–1334	1.31–1.34
Trichloroethene (TCE)	9.1×10^{-3}–1.2×10^{-2}	1.81–2.10	2.29–3.30	1100–1470	1.46
Vinyl chloride	5.6×10^{-2}–2.78	0.39	0.60	915–8800	0.91

the concentration of organic contaminants in water samples if exposed to the atmosphere.

The dissolution and precipitation of solids are two of the most important factors controlling the chemistry of groundwater. In natural waters, a recharging groundwater obtains almost all its solute load from the dissolution of minerals along the flow path. If geochemical conditions change along the flow path, mineral precipitation can occur removing some of the dissolved mass. Low pH, metal-rich, contaminant plumes can precipitate minerals as the plume disperses and the pH gradually increases. Because other ions are typically present in groundwaters these mineral dissolution–precipitation reactions are best quantitatively analyzed using geochemical speciation models. For the details of chemical equilibria and rates in natural waters readers are referred to aquatic chemistry texts such as Stumm & Morgan (1996), Morel (1983), and Drever (1982).

24.3.3.3 Sorption

The reaction between solutes and solids plays an extremely important role in controlling the chemistry of groundwater. Sorption reactions can be subdivided into two main types: hydrophobic sorption of organic compounds and ion exchange reactions. For example, non-polar organic molecules partition onto solid organic matter on or within the mineral grains of a soil. Ion exchange typically involves the exchange of cations in solution with other cations on the surface of clay minerals in the soil. In the case of a contaminant sorption, this results in a reduction in concentration in the aqueous phase. This sorption is manifest by a reduction in contaminant velocity relative to the groundwater flow. The term retardation is used to describe this velocity reduction.

The case of ion exchange is somewhat different. The decrease in the concentration of the ion of interest in solution, e.g. K^+, Na^+, Pb^{2+}, Cd^{2+}, Fe^{2+}, Cu^{2+}, is accompanied by an increase in pore fluid concentration of the desorbed ion, e.g. Ca^{2+}, Mg^{2+} as discussed by Rowe et al. (1995b).

Sorption is commonly represented by a simple one-parameter model or K_d approach. The sorption isotherm, describing the equilibrium between solute and sediment, is represented as a simple linear isotherm (isopleth):

$$S = K_d C \qquad (24.24)$$

where S = the mass sorbed per unit mass of dry soil (mg g^{-1}), K_d = the distribution coefficient (cm^3 g^{-1}), and C = the equilibrium solution concentration (mg L^{-1}). In reality, sorption isotherms can take on a complex functional form. However, the linear isotherm is easy to incorporate into mass-transport models and works fairly well for modeling sorption of organic contaminants at relatively low concentrations.

Values of K_d are typically obtained in the laboratory from sorption batch experiments, column tests, diffusion tests (see Rowe et al. (1995b), or estimated from empirical correlations (e.g. Karickhoff et al. 1979; Schwarzenback & Westall 1981):

$$K_d = K_{oc} f_{oc} \qquad (24.25)$$

where K_{oc} = the partition coefficient of a compound between organic carbon and water (cm^3 g^{-1}) and f_{oc} = the weight fraction of organic carbon in the soil (g$_{organic\ carbon}$ g$_{solids}^{-1}$). Values of f_{oc} can be measured in the laboratory. A synthesis of some f_{oc} data at particular sites illustrates how variable the parameter can be (Table 24.6). While K_{oc} values are available for many contaminants (see Table 24.5 and Montgomery & Welkom 1990), a good correlation exists between log K_{oc} and a more often referenced parameter, the octanol-water partition coefficient, K_{ow}. Various regression equations between log K_{oc} and log K_{ow} can be found in Domenico & Schwartz (1998) and Lyman et al. (1982).

The simple linear isotherm (Eq. 24.24) has

TABLE 24.6. Typical f_{oc} values of various sediments

Field site	Type of deposit	Sediment texture	f_{oc}	Reference
Borden, Ontario	Glaciofluvial	Fine–medium sand	0.0002	Jackson et al. (1985)
Gloucester, Ontario	Glaciofluvial	Regressional sands and gravels	0.0001–0.0035	Mackay et al. (1986)
		Stratified clayey silt and silt	0.0029–0.0059	
		Subaqueous outwash gravels, sands and silts	0.0131–0.0015	
Palo Alto, Baylands	Unknown	Silty sand	0.01	Mackay & Vogel (1985)
River Glatt, Switzerland	Glaciofluvial	Sand and gravel	<0.0001–0.01	Schwartzenbach & Giger (1985)
Oconee River, Georgia	River sediment	Sand	0.0057	Karickhoff (1981)
		Coarse silt	0.029	
		Medium silt	0.02	
		Fine silt	0.0226	
Sarnia, Ontario	Glaciolacustrine	Clay	0.0058	Rowe et al. (1995b)
Milton, Ontario	Till	Clayey till	0.0029–0.0045	Rowe et al. (1993)

limitations and can break down under a number of conditions as follows:

- For hydrophilic organic compounds: Karickhoff (1984) recommends that a hydrophobic sorption model (K_d approach) should only be used for organic compounds with solubilities less than 10^{-3} M.
- For low f_{oc} values: sorption of organic compounds on inorganic surfaces, which is normally considered negligible, can be significant. McCarty et al. (1981) have developed an empirical equation to estimate the critical organic fraction, f_{oc}^*, where sorption due to organic and inorganic solids equals:

$$f_{oc}^* = \frac{S_a}{200(K_{ow})^{0.84}} \quad (24.26)$$

where S_a = the surface area of the soil (m^2 g^{-1}) and K_{ow} = the octanol-water partition coefficient. The surface area depends on the clay content and mineralogy (typical values reported (Yong 1985) range from 15 m^2 g^{-1} for kaolinite to 80 m^2 g^{-1} for illites and chlorites, to 800 m^2 g^{-1} for montmorillonites). Values of K_{ow} can be found in Montgomery & Welkom (1990) and some values are given in Table 24.5 (in terms of $\log_{10} K_{ow}$).

- For relatively large organic concentrations: a linear isotherm does not limit the amount of solute that can be sorbed onto the solid. Non-linear isotherms, however, often fit the experimental data better. Two of the most common relationships are the Freundlich and Langmuir isotherm:

$$\text{Freundlich} \quad S = K_f C^m \quad (24.27)$$

$$\text{Langmuir} \quad S = \frac{S_m K_l C}{1 + K_l C} \quad (24.28)$$

where K_f and K_l = empirical coefficients related to the extent of sorption, m = an empirical constant (often ranging from 0.7 to 1.2), and S_m = the maximum sorptive capacity for the soil.

- If the organic compound carries a charge.

Some organic acids and bases ionize to some extent depending on the pH of the groundwater.

Batch tests and empirical correlations are useful in quantitatively describing sorption. However, they do not fully represent in situ conditions and could be subject to error from a number of sources. Particular care is needed in the application of laboratory parameters because they may either underestimate or overestimate the amount of sorption that can occur in the field (e.g. see Rowe et al. 1995b).

The K_d approach has been used to describe metal sorption on mineral surfaces. However, one must be cautious in applying such a simple one-parameter model to what with metals is a much more complex process. Even cation-exchange reactions provide a more sophisticated approach. In these reactions, an exchange of ions takes place between ions in solution and ions sorbed on clay minerals or other oxide surfaces. However, the extent of the exchange depends both on the mineralogy of the clay (or nature of the oxide surface) and the chemical composition of the groundwater. For example, Rowe et al. (1995b) show that the sorption of a cation on a given clay is highly dependent on the presence of competing cations. Thus, either the sorption characteristics must be empirically evaluated under conditions as close as possible to that expected in the field, or rigorously evaluated in terms of multiparameter equilibrium models. These models apply the law of mass action, which accounts for properties of the solution and the solid surfaces.

It is possible to model sorption using more realistic multiparameter equilibrium models. Such models are now being incorporated into state-of-the-art contaminant transport models; however, they are complex and require considerable data. Stollenwerk (1991) uses such an approach to study the sorption of molybdate (MoO_4^{2-}) on ferrihydrite, which coats quartz grains of an outwash aquifer.

Stumm & Morgan (1996) provided a more detailed review of ion exchange and sorption isotherms.

When estimating sorption parameters, e.g. K_d, it is important to recognize that both empirical equations, e.g. Eq. 24.26, and laboratory batch tests or column studies have numerous limitations. For example, in order for a laboratory batch or column test to adequately represent the exchange process, actual field samples of ambient groundwater or leachate (or water with similar major ion concentrations), and porous media are used along with actual contaminant(s) of interest. However, several major difficulties arise with this approach. First, it is very difficult not to alter the geochemical characteristics of the geologic materials as they are sampled and sent to the laboratory. Most basic are the invasion of O_2 and the degassing of CO_2, which can alter the redox condition and pH of the samples. In some cases this can cause the precipitation of iron and manganese hydroxides, which have a strong affinity for the sorption of cationic and anionic trace contaminants. Second, a batch test does not represent the field conditions where the advancing zone of contamination continuously supplies ions to the exchange sites. In the field this can result in some contaminants being less retarded in certain zones due to a change in exchange properties, and other geochemical changes such as a change in the redox state and the precipitation of minerals. Finally, one must be careful not to extrapolate isotherm data. At low concentrations many contaminants exhibit a linear trend, which would appear to support a linear or K_d approach. However, at higher concentrations the type of surface reactions may change substantially shifting the functional form of the isotherm.

24.3.3.4 Oxidation–Reduction Reactions

Oxidation–reduction reactions or redox reactions are very important in contaminant hydrogeology because they can control the mo-

bility of metal ions in solution, and biodegradation or biotransformation reactions. While biotransformation reactions can attenuate organic contaminants, daughter products are not necessarily less toxic. For example, the biotransformation of trichloroethylene can result in the formation of vinyl chloride, which is more toxic than its parent. Micro-organisms act as catalysts speeding up the very sluggish redox reactions. Micro-organisms occur ubiquitously in the subsurface and are most often found as colonies attached as films on porous media or fracture surfaces. Some micro-organisms thrive under aerobic conditions; others prefer anaerobic conditions. The need for biological mediation of most redox processes encountered in natural waters means that approaches to equilibrium depend strongly on biota activities (Stumm & Morgan 1996). While equilibrium calculations and approaches greatly aid attempts at understanding redox patterns in natural waters they only provide boundary conditions and rates of approach toward which the system must be proceeding.

Field electrode measurements of redox potentials present additional difficulties and often these measurements are meaningless. Stumm & Morgan (1996) comment on this:

Values obtained depend on the nature and rates of the reactions at the electrode surface and are seldom meaningfully interpretable. Even when suitable conditions for measurement are obtained, the results are significant only for those components whose behavior is electrochemically reversible at the electrode surface.

The assessment of biodegradation and redox processes in groundwater systems is very difficult because it requires considerable knowledge about both groundwater geochemistry and applied microbiology. Chapelle (1993) nicely bridges this gap and provides a comprehensive text that focuses on the impact of microbial processes on pristine and contaminated groundwater systems. Domenico & Schwartz (1998) present an excellent up-to-date synthesis of how redox processes and microbial systems influence groundwater geochemistry and remediation.

24.3.3.5 Hydrolysis

Hydrolysis is another transformation reaction that can operate on organic compounds; however, unlike biodegradation, it is not catalyzed by micro-organisms. Hydrolysis reactions involve the substitution of a water molecule, or an OH^- ion, for another atom or group of atoms in an organic molecule. Hydrolysis reactions are very prevalent in the aquatic environment, and often the products are less ecologically and toxicologically harmful than the unhydrolyzed reactants (Stumm & Morgan 1996). In addition, the introduction of a hydroxyl group into the parent molecule makes the product more soluble in groundwater, and more susceptible to biodegradation (Neeley 1985).

24.3.3.6 Complexation Reactions

Complexation reactions involve the combination of simpler cations, anions and sometimes organic molecules, to form a "complex". An example is one of the possible hydrolysis reactions of a metal (Cr^{3+}) with a ligand (OH^-) to form a complex $Cr(OH)^{2+}$:

$$Cr^{3+} + OH^- = Cr(OH)^{2+}. \quad (24.29)$$

Sometimes complexes can react themselves with ligands:

$$Cr(OH)^{2+} + OH^- = Cr(OH)_2^+. \quad (24.30)$$

Most inorganic complexation reactions are kinetically fast, with respect to groundwater flow, so they can be examined quantitatively using equilibrium concepts. Calculation of the distribution of a given total metal concentration among various complexes involves the solution of a series of mass law equations. Figure 24.11 illustrates the outcome of such a process for assessing the speciation of Cr as controlled by solution pH.

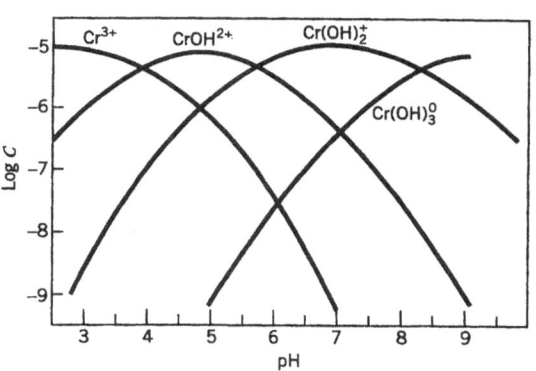

FIGURE 24.11. Log C–pH diagram for chromium hydroxide complexes (after Domenico & Schwartz 1998, Physical and Chemical Hydrogeology. © 1998 John Wiley & Sons, Inc. Reprinted by permission of John Wiley & Sons, Inc.).

Complexation facilitates the transport of toxic metals such as cadmium, chromium, lead and uranium. In general, metals in groundwater are most soluble at low pHs, where most of the metal mass exists as a charged ion. Ignoring the effects of sorption, this mobility declines once the pH increases to a point when equilibrium is reached with a solid phase, such as a metal hydroxide, sulfide or carbonate. Since most natural groundwaters are neutral or alkaline, equilibrium with the solid phase determines that aqueous metal concentrations will be small. It is under these circumstances that complexes can enhance the solubility and mobility of metals.

24.3.4 FACILITATED TRANSPORT OF CONTAMINANTS: COSOLVATION AND COLLOIDAL TRANSPORT

Facilitated transport occurs when the mobility of a contaminant increases as a result of physical, chemical or biological changes in the aquifer or aquitard. Cosolvation and colloidal transport are two major examples of facilitated transport that will be discussed here.

Cosolvation is the process by which the mobility of a contaminant is enhanced by the presence of a solvent (Gorelick *et al.* 1993). Through cosolvation hydrophobic contaminants, such as PCBs, which are practically immobile in the aqueous phase, can become much more soluble when chlorinated solvents are present. Typically, cosolvation is most significant near the source, where solvent concentrations are likely to be high (Gorelick *et al.* 1993).

Colloidal transport occurs when small organic or inorganic particles, generally less than about 1 μm in diameter, move through porous, fractured and karst media. Colloids can be contaminants themselves, such as viruses and bacteria present in waste waters, and organic macromolecules generated in landfill leachates (Gournaris *et al.* 1993). Alternately, harmless colloids, such as clay and humic substances, can become contaminant collectors and facilitators when low mobility contaminants, such as trace metals, radionuclides or organic compounds, sorb onto their large surface areas. As long as colloids do not aggregate, or undergo physico-chemical collection on solid surfaces (McCarthy & Zachara 1989), they can move through the pore networks of gravels, sands and silts. Penrose *et al.* (1990) and Champ *et al.* (1984) have found significant colloid-facilitated radionuclide transport. Additional field studies include the work of Gschwend & Reynolds (1987), who examined colloid formation and transport in relation to a large sewage plume.

If surfactants are present in groundwaters they can also greatly increase the solubility of immiscible organic compounds. The surfactants form a microemulsion in water, trapping immiscible organic compounds, and allow them to stay dissolved and move through an aquifer. These emulsion-trapped organic compounds have similar fate and transport characteristics to that of metals bound up on colloids (USEPA 1992a).

One additional important concept with colloids involves the filtration of groundwater samples. The standard environmental filter has a pore size (0.45 μm) which is the middle of the colloidal size range. The problem arises that this filter does not provide a clean cut-off, between dissolved and colloidal species that are mobile, and colloids and larger particles that are not. Large concentration differences between filtered and unfiltered samples have been observed for metals (Puls & Barcelona 1989; Puls & Powel 1992; Puls *et al.* 1992; Pohlman *et al.* 1994). Therefore, many states, and some programs of the US Environmental Protection Agency, now prohibit filtration for metals analyses. This is due to potential sampling artifacts, created during the purging process, that lead to the production of turbid samples, or artifacts generated during the filtration process itself (Puls & Paul 1995; USEPA 1992a). If field filtration exposes a water sample to the atmosphere, the sample can oxidize dissolved ferrous iron to form a ferric hydroxide precipitate ($Fe(OH)_3$). In turn, other metals in solution may sorb onto the ferric hydroxide colloids, and both the colloid and sorbed metals will be removed from the sample by field filtration (USEPA 1992a). In most instances, proper well construction, and low-flow purging and sampling of monitoring wells, can eliminate the need to field-filter groundwater samples (Puls & Paul 1995).

25. BARRIER SYSTEMS

R. K. Rowe

25.1 Applications, Mechanisms and Scope

Barrier systems are intended to minimize the movement of liquids and/or gases from one location to another. Typical example applications include the use of liners for canals, ponds, dams, waste disposal facilities and spill protection around tanks. The liquids to be "contained" may range from water (e.g. in canals, ponds and dams) to contaminated water (e.g. in landfills; sewage lagoons; brine ponds; or contaminated groundwater) to relatively pure chemicals (e.g. hydrocarbons) stored in tank farms. In these applications the primary objective is to limit the physical escape of the liquid to either surface water or groundwater. In situations where the liner is in contact with contaminated water over long periods of time (e.g. lagoons, landfill liners, etc.), a secondary objective may also be to limit chemical migration by the process of diffusion (Section 24.3.1) whereby contaminants migrate from a point of high concentration (e.g. in the retained fluid) to points of lower concentration (e.g. in groundwater).

Other barriers include covers for solid and mining waste that serve to control the influx of a fluid (e.g. water in landfill covers) or a gas (e.g. oxygen in covers for acidic mine waste). They may also be intended to minimize the outward flux of liquids (e.g. perched leachate in a landfill) or gases (e.g. methane generated in a municipal solid waste landfill).

Barrier systems typically control the flow of liquids by virtue of having a liner component with a low hydraulic conductivity (Section 24.1.3) and, often, by having a transmissive layer adjacent to the liner (e.g. see Fig. 25.1).

This chapter provides basic information concerning obtaining and maintaining (as long as possible):

- high transmissivity of granular layers used to control hydraulic head or gas pressure (Section 25.2),
- low hydraulic conductivity of clay liners and minimizing leakage through composite liner systems (Section 25.3),
- liner compatibility with leachate (Section 25.4),
- minimizing diffusive transport of dissolved contaminants and gases (Section 25.5),
- the effect of temperature on liner performance (Section 25.6),
- assessing equivalency of liner systems (Section 25.7), and
- stability of liner systems (Section 25.8).

Other design considerations related to liners in liquid-containing structures are provided in Chapter 26, while Chapter 27 discusses covers for waste and Chapters 29 and 30 include a discussion of the use of barriers for containment of both contaminated groundwater and soil.

25.2 Transmissive Layers

High transmissivity layers may be used above liners to minimize the hydraulic head act-

740

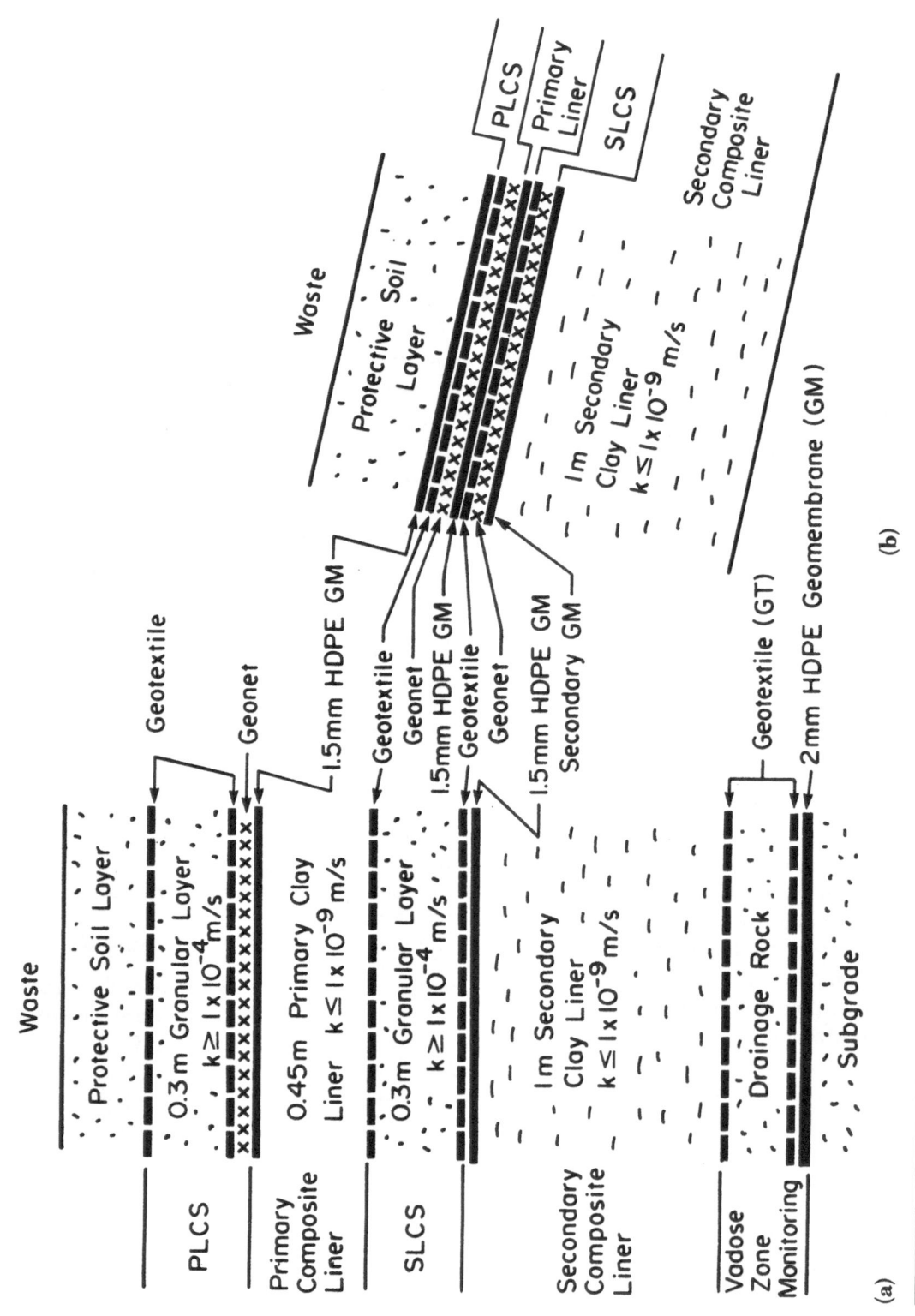

FIGURE 25.1. Schematic showing (a) base and (b) side slope of barrier system for Kettleman Hills landfill: PLCS = Primary leachate collection system; SLCS = Secondary leachate collection system; GM = Geomembrane (modified from Byrne *et al.* 1992).

ing on the liner (and hence minimize flow through the liner) or below liners in covers to allow recovery of gas and control gas pressures that could otherwise cause instability of a low permeability cover. These transmissive layers may be constructed from granular materials or geosynthetics or a combination of both, e.g. see Fig. 25.1. In either case, the primary design objective is to provide a layer with sufficient hydraulic capacity to allow transmission of the fluid to the collection point without allowing a build-up of head (pressure) that exceeds the design value.

25.2.1 HYDRAULIC CAPACITY OF DRAINAGE LAYERS

In the simplest terms, ensuring adequate hydraulic capacity of drainage layers involves consideration of the transmissivity of the layer, T (where $T = kB$ and k is the hydraulic conductivity and B is the thickness of the layer), the hydraulic gradient, i, and the required flow to be accommodated, q_{des}. Thus for a desired factor of safety, F, the required transmissivity of the layer is given by:

$$T_{req} \geq \frac{Fq_{des}}{iw} \qquad (25.1)$$

where w = the width of drainage layer through which the flow q_{des} must pass.

Equation 25.1 is most appropriate when designing a transmissive layer intended to conduct flow q_{des} from one point in the layer to another without significant additional fluid input. However, in many practical situations, the drainage layer is intended to collect percolation that is relatively uniformly distributed (e.g. q_o in Fig. 25.2) and conduct it to drainage pipes (see Fig. 25.2). Typically, a design objective in these cases is to maintain a maximum head, h_{max}, less than the thickness of the transmissive layer, B.

For steady state conditions and the simplest case of a flat base (shown in Fig. 25.2a)

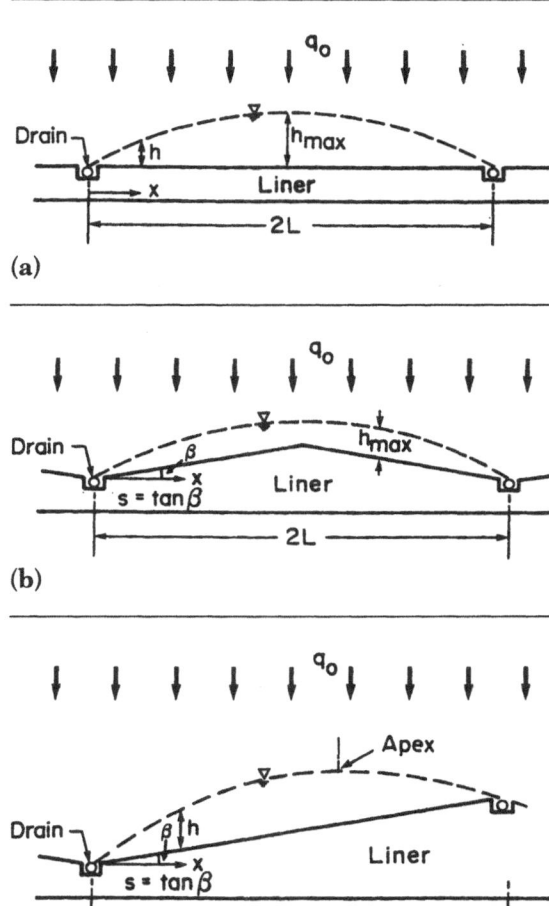

FIGURE 25.2. Schematic showing different collection systems (modified from Rowe 1988).

the maximum head, h_{max}, on the liner is given by:

$$h_{max} = L\left(\frac{q_o}{k}\right)^{1/2} \qquad (25.2)$$

and where L = the half-horizontal distance between drains as shown in Fig. 25.2a and k = the hydraulic conductivity of the material through which the water is flowing laterally to the drains. The average head $h_{avg} = 0.785h_{max}$.

For the situation shown in Fig. 25.2b, the

maximum leachate head is often calculated using Moore's (1983) equation, however, as indicated by Giroud & Houlihan (1995), Moore's equation may result in a significant underestimate of h_{max} (by a factor of two, or more) for slopes of more than a few percent and/or values of $q_0/k < 0.01$ and hence should not be used. They proposed an alternative, more accurate equation for the maximum head, h_{max}:

$$h_{max} = 0.5Lj\,[(\tan^2\beta + 4q_0/k)^{1/2} - \tan\beta] \quad (25.3a)$$

where

$$j = 1 - 0.12\,\exp\{-[\log_{10}(1.6\lambda)^{0.625}]^2\} \quad (25.3b)$$

and

$$\lambda = \frac{q_0}{k\,\tan^2\beta}. \quad (25.4)$$

Equation 25.3 is valid for all the situations shown in Fig. 25.2, reduces to Eq. 25.2 for $\beta = 0$, and is easy to use (it can be further simplified by taking $j = 1$ for an error of 12% or less).

The average head, h_{avg}, acting on the liner varies and is given by:

$$h_{avg} = \Lambda h_{max} \quad (25.5)$$

where typical values of Λ are given in Table 25.1. Other representations and a more detailed discussion are given in Section 26.2.2.

TABLE 25.1. Relationship between average and maximum head on a liner below a drainage layer (adapted from Giroud & Houlihan 1995)

λ	Λ	λ	Λ
0	0.5	1.0	0.80
0.05	0.56	1.2	0.83
0.1	0.6	1.5	0.85
0.15	0.63	3.0	0.87
0.2	0.67	10	0.85
0.35	0.7	30	0.82
0.6	0.75	50	0.81
0.8	0.79	1000	0.79
		∞	0.785

One can readily show that for a value of $k \geq 10^{-5}$ m s^{-1}, $q_0 \leq 0.2\,m$ a^{-1} ($\leq 6 \times 10^{-9}$ m s^{-1}) and drains at a spacing of 20 m, the maximum steady state head is likely less than a typical drainage layer thickness of 0.3 m. For a 1% slope, the spacing between drains as shown in Fig. 25.2b could be 30 m and still maintain a maximum head of less than 0.3 m. As a consequence, these drainage layers are often specified to have a hydraulic conductivity greater than 10^{-5} m s^{-1}. Unfortunately, while this may provide the required drainage immediately after construction, a reduction in hydraulic conductivity due to biological, chemical or particulate clogging may quickly result in an excessive leachate mound as discussed in the next subsection. When there is potential for clogging, special care is required to ensure adequate long-term drainage capacity.

In this chapter the discussion of drainage layers is restricted to granular layers such as the primary (PLCS) and secondary (SLCS) granular collection layers shown in Fig. 25.1a. However, similar considerations also apply to geosynthetic drainage layers as discussed in Section 26.2.

25.2.2 PARTICULATE CLOGGING OF DRAINAGE LAYERS

A number of failures have occurred due to drainage layers having inadequate hydraulic conductivities and the consequent development of excessive seepage forces (e.g. see Rowe 1998a). In a number of cases these failures have occurred due to the accumulation of fines in material that initially had an adequate hydraulic conductivity. In order to avoid particulate clogging it is necessary to: (a) ensure that there is a suitable filter between the drainage layer and adjacent soil or waste, e.g. the geotextile and "protective soil layer" shown in Fig. 25.1a; and (b) develop an appropriate design and construction plan that will minimize the accumulation of fines.

The design of filters is discussed in Section 26.3. However, it is noted here that even a relatively small proportion of fines (0.5–1%) in a sand protective layer separated from an underlying geonet drainage layer (see Section 7.4) by a geotextile filter (see Section 7.2) has the potential to cause significant particulate clogging of the underlying geonet (e.g. see Giroud *et al.* 1998a). The same amount of fines would have a negligible effect on the performance of a 0.3-m thick coarse gravel drainage layer.

25.2.3 BIOLOGICALLY INDUCED CLOGGING OF LEACHATE COLLECTION SYSTEMS

One of the most critical drainage applications is the collection of leachate in landfills. These collection systems typically take the form of toe drains (common in old landfills) around the edge of the waste and/or underdrain systems below the waste. Underdrain systems may take the form of French drains (typically perforated pipes embedded in granular material (e.g. see Figs 25.3a and 25.4) at some spacing, granular drainage blankets with perforated pipes at some specified spacing (see

Fig. 25.3b), and geosynthetic (geonet) drainage layers (e.g. see Fig. 25.1b). Geotextiles are often used as filters between the waste and the drainage layer, especially when either coarse drainage material, e.g. gravel, or geonets are used to provide a drainage blanket. Key issues in the design of these systems are the need to provide adequate drainage, prevent structural failure, e.g. crushing, or other pipe failure, and to minimize clogging.

There is a growing body of evidence that the performance of these systems (especially those shown in Fig. 25.3a) can be greatly impaired by clogging and a number of failures have been reported in the literature (see Rowe 1998a,b for a review of clogging of geotextiles and granular material, respectively).

The clogging problem arises because municipal solid waste leachate contains nutrients that will encourage bacterial growth in geotextile filters, in granular drainage layers, around the perforations in the leachate collection pipes and within the pipes. Clogging of the leachate collection system involves the filling of the void space between solid parti-

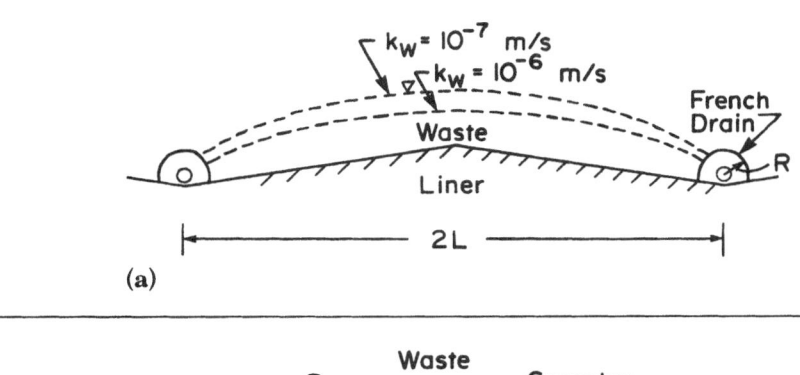

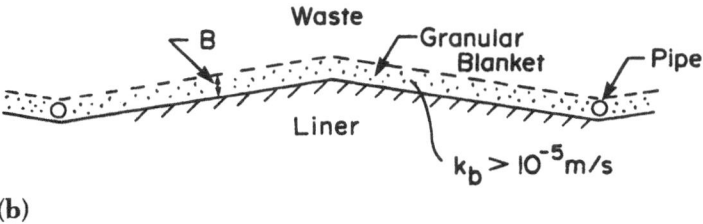

FIGURE 25.3. Two leachate drainage systems: (a) French drains; and (b) blanket drain (modified from Rowe *et al.* 1995b).

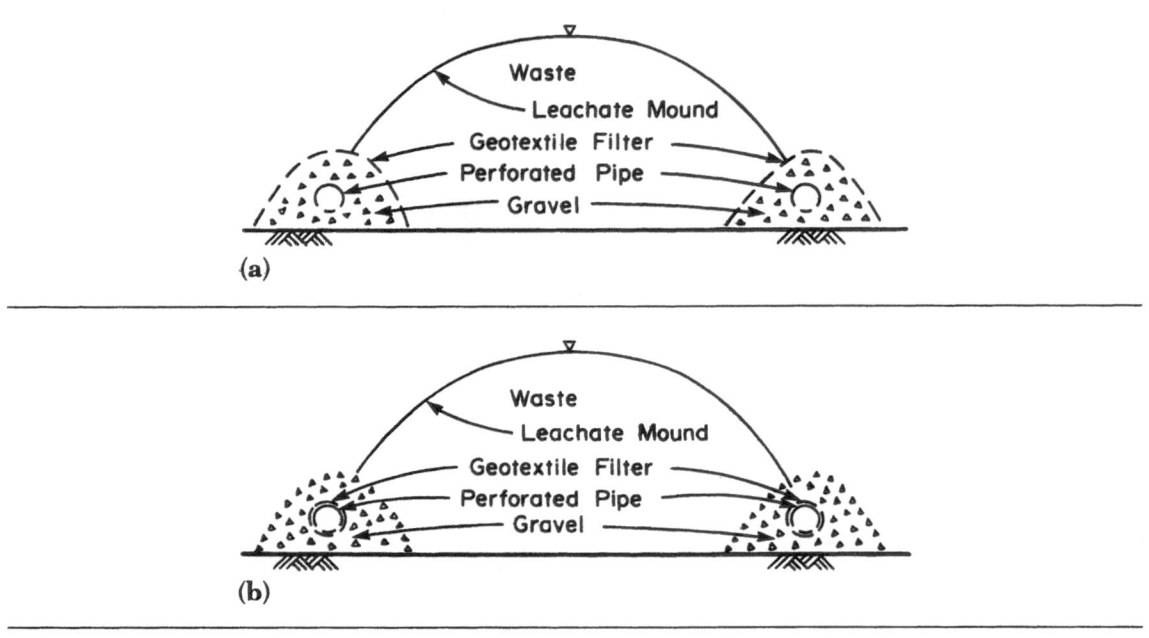

FIGURE 25.4. Schematic showing examples of poor leachate collection system designs: (a) problematic; and (b) even worse. Schematic also shows a leachate mound developed once there is excessive clogging of the geotextile filter and/or the drainage gravel and/or the pipe (modified from Rowe 1992).

cles as a result of a combination of biological, chemical and physical events. Particulate clogging can be a problem in some cases (Section 25.2.2), however, a major component of the clogging is microbiologically related.

Microbiological and chemical studies (Brune *et al.* 1994; Rowe *et al.* 1995c, 1997c,d and Rittmann *et al.* 1996) suggest that the clogging of drainage systems is the result of a mobilization of inorganic constituents of waste, e.g. metals such as calcium, iron, etc., by a process involving fermentative bacteria together with iron- and manganese-reducing bacteria, followed by precipitation processes involving primarily methanogenic- and sulfate-reducing bacteria. The clog typically has a soft (organic) and hard (predominantly $CaCO_3$) component. The rate of clogging is related to the flow through the critical component of the system, the void size, the temperature in the collection system (generally higher temperature implies faster clogging) and the leachate chemistry. The reduction in

void space caused by biofilm growth and chemical precipitation (Brune *et al.* 1994; Vandevivere & Baveye 1992; McBean *et al.* 1993; Rowe *et al.* 1995c, 1997d; Fleming *et al.* 1999) results in a concurrent reduction in the hydraulic conductivity of these drainage systems and hence a reduction in their capacity to laterally transmit leachate. Fleming *et al.* (1999) report a field case where after four years a coarse, uniformly graded gravel (50-mm nominal size) experienced at least a three-order-of-magnitude drop in hydraulic conductivity due to the build-up of clog material. Rowe (1998a) gives experimental data showing a decrease in six–seven orders-of-magnitude in a uniform (6-mm nominal) gravel size material. A decrease in hydraulic conductivity, k, to or around 10^{-7}–10^{-8} m s^{-1} is sufficient to inhibit lateral drainage and cause the build-up of a leachate mound. However, this decrease in k is not sufficient to significantly inhibit vertical movement of contaminants since it is still several orders-

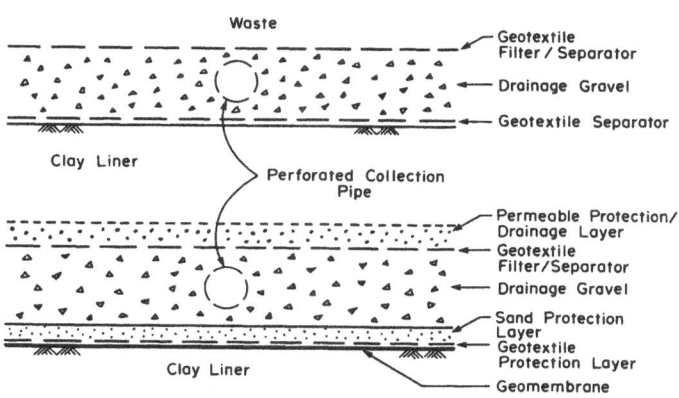

FIGURE 25.5. Schematic showing examples of blanket leachate collection system designs including a geotextile filter layer (modified from Rowe 1992).

of-magnitude higher than the hydraulic conductivity of the underlying liner that will control vertical flow.

Both laboratory experiments and field evidence show that clogging is directly related to leachate mass loading (Koerner & Koerner 1994; Fleming *et al.* 1999; Rowe *et al.* 2000a; Rowe 1998a). The leachate mass loading is a function of: (a) the concentration of volatile fatty acids (represented in terms of chemical oxygen demand, i.e. COD) and metals (especially calcium) in the leachate; (b) the flow rate per unit area; and (c) elapsed time. Thus when leachate that is generated over a large area of the landfill is directed through a filter surrounding isolated drainage gravel (French drains, see Fig. 25.4a) or, even worse, a filter wrapped around the pipe itself (Fig. 25.4b), there is a confluence of flow across the filter and the confluence factor is approximately equal to the area of the landfill over which the leachate going to this pipe is generated, divided by the area of the filter material through which the leachate must flow. Since clogging depends on flow rate per unit area, this confluence of flow increases the mass loading per unit time and hence increases the rate of clogging of the geotextile and the drainage gravel (it is assumed here that clog-

ging of pipes is mitigated by regular cleaning). However, even more important is the effect that clogging has when it does occur. When the resistance to flow into the drainage pipe is increased, a leachate mound will develop in the waste between the drains (see Fig. 25.4). This mound will maintain some flow to the pipe, but will also increase flow out through the underlying liner. Based on this, Rowe (1992) cautioned against designs such as those shown in Figs 25.3a and 25.4. The field investigations discussed by Rowe (1998a,b) confirm the soundness of this recommendation. Geotextiles used to wrap drains or pipes have been found to experience a drop in hydraulic conductivity of between two and five orders-of-magnitude (to as low as 4×10^{-8} m s^{-1}; Rowe 1998a).

25.2.4 GEOTEXTILE "FILTERS" IN LEACHATE COLLECTION SYSTEMS

Many regulatory authorities now recommend or require the use of a blanket drainage layer (see Fig. 25.3b), although details regarding exactly what is required vary substantially. Rowe (1992) recommended the use of a suitably selected geotextile as a blanket filter as part of a blanket underdrain layer (see Fig. 25.5) to reduce clogging of the drainage layer.

TABLE 25.2. Koerner & Koerner's (1995) recommended minimum values
for geotextile filters for use with mild leachate[a] and select waste over the geotextile[b]

Property	Woven monofilament	Nonwoven needle-punched
Mass per unit area, g m^{-2}	200	270
POA, %[c]	10	—
O_{95}, mm[d]	—	0.21
Grab strength, N	1400	900
Trapezoidal tear, N	350	350
Puncture strength, N	350	350
Burst strength, kPa	1300	1700

[a] Total suspended solid (TSS) and biochemical oxygen demand (BOD$_5$) $\leq$ 2500 mg L^{-1}.
[b] No hard or coarse material; for coarse or hard material over geotextile, the strength requirements may need to be increased.
[c] POA, relative open area.
[d] O_{95}, filtration opening size.

The findings of Rowe *et al.* (1995c), Fleming *et al.* (1999) and Koerner & Koerner (1995) provide some evidence that a suitable blanket geotextile will function adequately. These investigations demonstrated that the geotextile provided good protection to the granular drainage material, which experienced relatively little clogging compared to that observed when there was no geotextile filter between the waste and stone drainage medium. Koerner & Koerner (1995) recommended that for mild leachate, a geotextile in contact with the waste should have the properties given in Table 25.2.

Giroud (1996a) has discussed the issue of geotextile clogging as part of a broad review of filter design. He tentatively recommends that sand and nonwoven geotextile filters should not be used even if the waste has been stabilized to produce low strength leachate by pretreatment. Rather, he recommends the use of monofilament woven geotextiles with a minimum filtration opening size (O_{95}) of 0.5 mm and a minimum relative open area (POA) of 15%, with a preference for a POA greater than 30%. The rationale for these recommendations arises from the observations that:

1. the specific surface area for monofilament woven geotextile is much smaller than for nonwoven geotextiles and this decreases the surface area for biofilm growth;

2. the woven filter allows more effective and rapid movement of fine material, i.e. material not intended to be retained, and leachate through the filter; and

3. due to their compressibility, the filtration characteristics of a nonwoven geotextile vary with applied pressure and the critical filtration characteristics should be assessed under design pressures which could be up to 500 kPa.

There is some evidence to suggest that geotextiles selected in accordance with Giroud's (1996a) recommendations are likely to experience less clogging and reduction in hydraulic conductivity with time than geotextiles that simply meet the requirements of Koerner & Koerner (1995 and Table 25.2).

It is important to recognize that Giroud's (1996a) recommendations are based on the premise that one wishes to minimize clogging of the filter. This may indeed be the case for some design situations, e.g. if one insists on using a design such as is shown in Fig. 25.3a even though the use of French drains is generally not recommended. However, while excessive clogging is undesirable, the processes that cause clogging also provide leachate treatment and, in so doing: (a) decrease the potential for clogging at more critical zones, e.g. near collection pipes; and (b) reduce the level of leachate treatment required after removal of leachate from the landfill. Although the subject of ongoing research, it appears desirable to design the leachate collection system to maximize leachate treatment while maintaining its design function (Fleming *et al.* 1999). Under these circumstances, and using a design such as the one shown in Fig. 25.5, a granular or nonwoven geotextile filter may actually be desirable provided that the hydraulic conductivity did not drop to such a point as to cause sufficient perching of leachate to have negative effects such as side seeps.

Based on published data, it appears unlikely that the hydraulic conductivity of the geotextile selected in accordance with either Giroud's (1996a) or Koerner & Koerner's (1995) recommendations (Table 25.2) would be below 4×10^{-8} m s^{-1} (1.3 m a^{-1}) for normal conditions and more likely it would be of the order of 1×10^{-7} m s^{-1} or higher. If the geotextile was used in a blanket drain, e.g. see Fig. 25.5, one can quickly establish that there would be negligible perched leachate on the geotextile for typical rates of leachate generation (less than 3×10^{-8} m s^{-1} or 1 m^3 a^{-1} m^{-2}). Thus, while recognizing that geotextiles will clog, based on the available data it appears that an appropriately designed and selected geotextile used to protect gravel in a blanket drain may improve the performance and the service life of the drainage gravel and, based on available evidence, they do not appear to cause excessive perched leachate mounding.

25.2.5 GRANULAR DRAINAGE MATERIAL IN LEACHATE COLLECTION SYSTEMS

Since the clogging of granular material may be the result of a combination of particulate clogging, clogging due to chemical precipitation and biofilm growth, the rate of clogging can be minimized by:

- maximizing the flow velocity in the drain;
- maximizing the void size; and
- minimizing the surface area available for biofilm growth.

By maximizing the flow velocity, one reduces the residence time for leachate in the collection system, thereby reducing the amount of sedimentation of particulates and chemical precipitation that can occur. The available evidence would suggest that clogging is greater under saturated conditions than for unsaturated conditions (clogging may, in fact, be greatest near the boundary between the saturated and unsaturated zone). Thus, it is desirable to keep drainage systems pumped.

Maximizing the void size within the granular material tends to increase the initial hydraulic conductivity and reduces the likelihood of voids becoming blocked. Since void size is about 20% of the characteristic particle size, it follows that a larger characteristic particle size will result in larger voids (in fact, $O_f \cong 0.2d_{15}$; see Section 26.3.2.2).

Biofilm growth is related to the surface area available on the particles forming the drainage layer. Since the specific surface, i.e. the surface area per unit volume, is inversely proportional to the characteristic particle size of the granular material (Cooke & Rowe 1999), the surface area (and rate of clogging) may be reduced by increasing the characteristic size of the granular material used.

It has been found (in the field and laboratory) that sands with an initial hydraulic conductivity of 10^{-4} to 10^{-5} m s^{-1} clog quickly. While suitably selected sands may serve a purpose as a protection layer or granular filter (see Figs 25.1 and 25.5b) they are not recommended for use in the granular drainage layer.

German investigators (Lechner 1994; Brune *et al.* 1994) recommend that granular drainage material should have a grain-size distribution between 16 and 32 mm. Canadian authorities (MoE 1998) recommend that granular drainage material have $D_{85} \geq 37$ mm; $D_{10} \geq 19$ mm; a uniformity coefficient $C_u = D_{60}/D_{10} < 2$ and no more than 1% finer than 0.075 mm (i.e. passing the US No. 200 sieve) and require that a suitable geotextile or graded granular filter/separator be installed between the drainage layer and both the waste and any underlying liner. Current research would indicate that the 16–32-mm gravel recommended by the German investigators represents a minimum requirement for good, long performance of leachate collection systems. A minimum thickness of 0.3 m of granular drainage material is usually recommended for the drainage blanket (TA Siedlungsabfall 1993; MoE 1998) with a minimum of 0.5 m at the location of perforated leachate collection pipes (MoE 1998). Current research suggests that the service life can be maximized by the use of systems such as that shown in Fig. 25.5b where the granular drainage blanket below the geotextile meets the recommendations in MoE (1998) cited above and the "permeable protection drainage layer" above the geotextile meets the German (Lechner 1994; Brune *et al.* 1994) recommendations of a grain-size distribution between 16 and 32 mm.

The drainage gravel may be either rounded or angular (e.g. crushed stone), however particular care is needed to provide adequate protection to underlying geomembrane liners. Recent research (Tognon *et al.* 1999) indicates that geotextiles alone may not provide adequate long-term protection of geomembranes when there is a combination of a coarse gravel drainage layer and high pressure, i.e. large waste thickness.

25.2.6 SERVICE LIFE OF LEACHATE COLLECTION DRAINAGE MATERIAL

Based on the observations from field and laboratory studies (Sections 25.2.3 and 25.2.5), Rowe & Fleming (1998) developed an approximate method for estimating the service life of drainage gravel used either as part of a French drain system (Fig. 25.3a) or in a blanket drain (Fig. 25.3b). The approach is based on the observation that while clogging is induced by biological processes, the clog material is predominantly composed of inorganic precipitates with calcium representing a proportion, f_{Ca}, of the clog (encrustation) material (see Brune *et al.* 1994; Fleming *et al.* 1999). Thus the rate of clogging can be related to the

change in the concentration of calcium in leachate, $c_L(t)$ between the time the leachate enters the collection system and the time it is collected. The yield of clog material, i.e. the volume of clog material deposited per unit volume of leachate flow, $Y(t)$, at time t is given by:

$$Y(t) = c_L(t)/(\rho_c f_{Ca}) \qquad (25.6)$$

where ρ_c = the density of the clog material. Assuming that the flow, Q, per unit length of pipe into one-half of a collection pipe is given by:

$$Q = q_o L \qquad (25.7)$$

where q_o = the leachate flux through the waste and L = the drainage path length (see Fig. 25.2), the volume of pore space, $V(t)$, that would be occupied by clog material at some time, T, is given by:

$$V(t) = \int_0^T \frac{c_L(t) q_o L}{\rho_c f_{Ca}} \, dt. \qquad (25.8)$$

If the Ca concentration is c_{L1} for $T < T_1$ and then linearly decreases to c_{L2} at $T = T_2$, after which it remains constant at c_{L2}; then the total clog volume for $t > T_2$ is given by:

$$V(t) = 0.5\omega(c_{L1} - c_{L2})(T_1 + T_2) + \omega c_{L2} t \qquad (25.9)$$

where $\omega = q_o L/(\rho_c f_{Ca})$. Thus the time $t = T$ required to obtain a uniform clog through a French drain, such as those shown in Fig. 25.3a, is given by:

$$T = \frac{\pi R^2 v_f}{4\omega c_{L2}} - \frac{(c_{L1} - c_{L2})(T_1 + T_2)}{2c_{L2}} \qquad (25.10)$$

for a French drain of radius R, and where v_f = the reduction in pore space required to cause "clogging." Experimental studies (see Rowe 1998a; Rowe *et al.* 2000a,b) suggest that gravel-size material experiences a significant decrease in hydraulic conductivity (by about six orders-of-magnitude) when $v_f \cong 0.9 n_o$ (n_o is the initial porosity and typically $n_o \cong 0.4$–0.5 for relatively uniformly graded gravel).

Based on field observations at the Keele Valley landfill (Fleming *et al.* 1999), it is considered likely that the clogging of blanket drains (see Fig. 25.6) will not be uniform and that in fact there may be a concentration of clogging toward the collection pipes. Thus it is assumed that the clog filled porosity reduction, v_f', and the portion of the overall drainage blanket thickness subject to clogging, B', both vary linearly from essentially zero at the up-gradient edge of the drainage path ($x = 0$, $Q = 0$) to a maximum value $v_f' = v_f$ and $B' = B$ within a zone defined by $(L - a) \leq x \leq L$ near the collection pipe (see Fig. 25.6). Thus the volume of mineral clog corresponding to a reduction in porosity, v_f', is given by:

$$V_{tot} = \int_0^L v_f'(x)B'(x)dx \qquad (25.11)$$

where

$$v_f'(x) = v_f x/(L - a) \qquad \text{for } x < L - a$$
$$= v_f \qquad L - a < x < L$$
$$B'(x) = Bx/(L - a) \qquad \text{for } x < L - a$$
$$= B \qquad L - a < x < L$$

and hence

$$V_{tot} = \frac{v_f BL(1 + 2a/L)}{3}. \qquad (25.12)$$

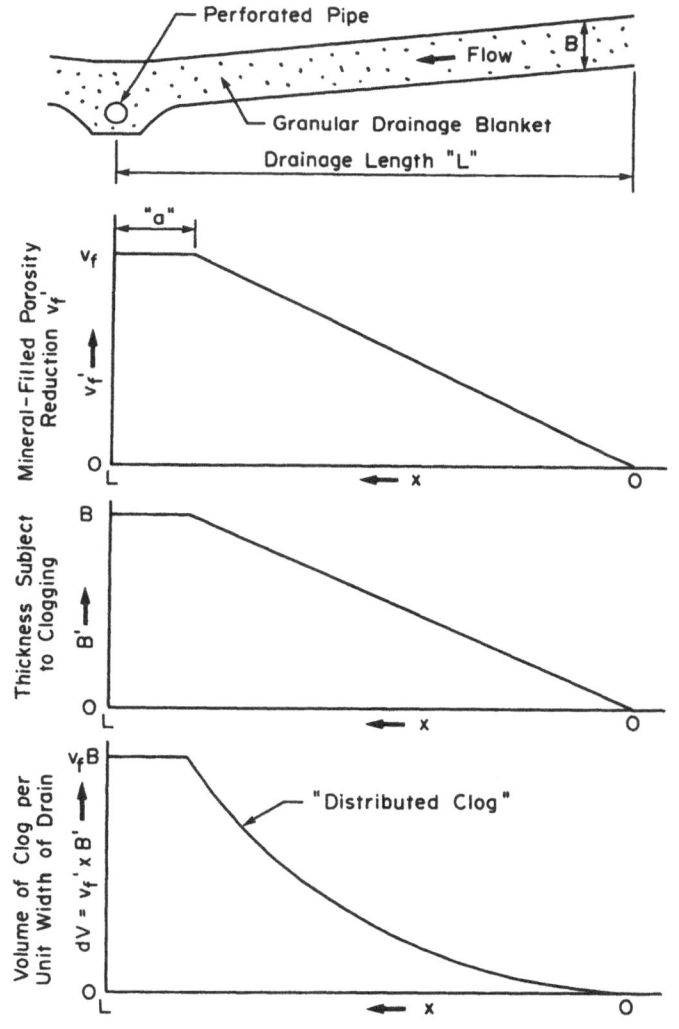

FIGURE 25.6. Distributed clogging model parameters (modified from Rowe & Fleming 1998).

The time to reach the degree of clogging represented by V_{tot} is calculated by solving for the rate of mineral volume growth given by Eq. 25.8 in general terms and Eq. 25.9 for a decreasing concentration of Ca with time. Combining Eqs 25.9 and 25.12 gives the time required for clogging as:

$$T = \frac{(1 + 2a/L)B\rho_c f_{Ca} v_f}{3q_o c_{L2}}$$ (25.13)

$$- \frac{(c_{L1} - c_{L2})(T_1 + T_2)}{2c_{L2}}$$

for a blanket drain of thickness B.

Fleming et al. (1999) found that $f_{Ca} \cong 0.26$, while Brune et al. (1994) found $f_{Ca} \cong 0.21$. The Fleming et al. (1999) study also found $\rho_c \cong 1.5 \text{ g cm}^{-3}$. The decrease in leachate concentration, $c_L (t)$, with time is more difficult to estimate. Data suggest that the calcium concentration in leachate decreases with time as the landfill moves from the acetogenic phase to the methanogenic phase and the average leachate pH increases to greater than seven. Initial concentrations in Canada are typically around 1650 mg L^{-1} (Rowe 1994), while Ehrig & Scheelhaase (1993) report an average of 1200 mg L^{-1} in Europe. Based on European data (Robinson & Gronow 1993; Ehrig & Scheelhaase 1993), the Ca concentration in leachate is expected to decrease to approximately 150 mg L^{-1} or less during the stable methane phase. After about 20 years, two Canadian landfills (each having an average pH > 7) had a Ca concentration of about 360 and 320 mg L^{-1}, respectively (Rowe 1994). It should be recognized that the leachate characteristics and the calcium concentration will vary between the time it leaves the waste, i.e. enters the collection system, and when it is collected. Strictly speaking, the calcium concentrations c_{L1}, c_{L2} should represent the decrease in concentration in the leachate between the time it enters the collection system and when it is collected. However, there is a paucity of data showing this drop and hence generally the estimated calcium concentration in the leachate is used. If this is based on data from within the landfill itself, i.e. before it enters the collection system, then this is a conservative assumption, i.e. will overpredict the rate of clogging. However, if the concentrations are based on data at the "end of pipe," i.e. after the leachate has been through the system, it is not clear whether or not the predicted clogging rate will be conservative and caution is required.

Rowe & Fleming (1998) applied this approach to estimate the time for clogging of a Canadian landfill with a French drain drainage system and obtained predictions that were consistent with the observed period when substantial leachate mounding was observed.

The foregoing represents a simple engineering approach to estimating the time to clogging that can be used in hand calculations. In this context, the service life of the leachate collection system is defined as the period of time that the leachate collection system controls the head to below the design value, e.g. 0.3 m in many cases. Once the head on the liner exceeds the design value, the collection system has "failed" to meet its design objective even though it will still collect some leachate. This is the approach that was used to estimate the service lives of drainage layers given in a Canadian landfill standard (MoE 1998). The approach does not explicitly consider biological processes or factors such as the particle size (it is assumed that the material is gravel). Currently, a more sophisticated computer model is being developed (Rowe et al. 1997b) that may be used to model the biological processes and the progressive development of clogging.

25.2.7 LEACHATE COLLECTION PIPES
Leachate collection pipes need to have: (a) sufficient hydraulic capacity to transmit leachate without the build-up of a leachate

mound exceeding regulatory requirements (often 0.3 m); (b) an internal diameter that will allow TV inspection and cleaning; (c) perforations that will allow leachate to move into the pipe, minimize the potential for clogging, facilitate cleaning and minimize impact on the structural performance of the pipe; and (d) deformations and stresses that will not impair its performance.

The hydraulic design of collection pipes is usually governed by the need to provide adequate drainage during a rainfall event when the cell is in the early stages of development and will be governed by the intensity of rainfall events, the area to be drained by a given pipe, and the slope of the pipe. The pipe diameter can then be selected based on the Manning formula:

$$Q = 1.49AR_H^{0.66}s^{0.5}/N \qquad (25.14a)$$

where Q = the flow in the pipe (in $m^3\ s^{-1}$), A = wetted area (m^2), R_H = the hydraulic radius (m) and is equal to the wetted area divided by the wetted perimeter, s = the pipe slope ($-$) and N is the coefficient of roughness. Koerner (1998) suggests that for plastic pipe with a smooth interior $0.009 \leq N \leq 0.01$ and for plastic pipe with a profiled or corrugated interior $0.018 \leq N \leq 0.025$. For a circular pipe flowing almost full:

$$Q \cong 0.47D^{2.66}s^{0.5}/N \qquad (25.14b)$$

where D = the internal diameter of the pipe (m). To meet the requirement of being able to readily inspect and clean the pipe, a minimum internal diameter of 0.15 m is often specified (e.g. MoE 1998) with clean-out points at regular intervals (usually not exceeding 300 m).

The selection of perforations is a somewhat contentious issue. Some regulatory environments prefer slots, although these can both weaken the pipe and may not be optimal for cleaning. Circular holes (perforations) need

to be large enough to minimize clogging and facilitate cleaning by jetting. A minimum perforation diameter of 12 mm (MoE 1998) may be adequate, but preferably the perforations will be larger and of similar magnitude to the wall thickness of the pipe, subject to the restriction that the holes should not allow movement of the drainage gravel into the pipe. It is essential that leachate collection pipes be regularly inspected and cleaned while the clog material is still "soft." Once the clog material hardens it becomes very difficult to remove.

The deformations and stresses developed in the pipe will depend on the type of pipe, the applied vertical stress and the bedding conditions. This is discussed in more detail in Chapter 18.

25.3 Low Permeability Liners

25.3.1 GEOLOGICAL BARRIERS

When present, a suitable geological barrier represents one of the cheapest and best longterm liners for a wide range of applications (canals, lagoons, landfills, etc.). Typically, the geological barrier consists of a low hydraulic conductivity clayey layer. Near the ground surface, these layers are often weathered and fractured, and it is usually necessary to excavate to some depth below the weathered crust to reach material that will have sufficiently low bulk hydraulic conductivity. As noted by Rowe et al. (1995b), fractures may extend to depths well below the weathered zones and they report cases where fractures have extended to 10–15 m even though the weathered crust only extended to 4–6-m depth.

The suitability of a natural geological unit as a liner can normally only be assessed after a detailed hydrogeologic investigation aimed at identifying the extent and continuity of the unit and its bulk hydraulic conductivity, i.e. large-scale hydraulic conductivity that takes account of local features such as fractures, see Chapter 24. Typical ranges of hydraulic conductivity are given in Table 3.10. Consideration also needs to be given to the thickness of the unit, the potential for future desiccation cracking, and/or chemical interaction with the fluid being retained and geotechnical issues such as the stability of any cut in this unit (see Chapter 14).

The remaining thickness of a natural geologic unit after excavation for construction of the fluid-retaining structure (e.g. lagoon, landfill, etc.) may vary considerably depending on: (a) local geologic conditions; (b) grade and materials balance considerations; (c) the potential for basal heave, e.g. due to the potentiometric head in any underlying aquifer, and other stability considerations; and (d) the thickness required to provide adequate contaminant attenuation (see Section 25.5). The level of attenuation required will depend on the nature of the contaminants, and the nature and reasonable use (either present or future) of the groundwater (e.g. see MoE 1998; Bouchard *et al.* 1992; Rowe *et al.* 1995b). Typically, this thickness is 3 m or more and may be tens of meters. However, with adequate investigation, thicknesses smaller than 3 m may be justified.

25.3.2 COMPACTED CLAY LINERS (CCLs)

Compacted clay liners (CCLs) represent an alternative means of obtaining a low hydraulic conductivity clayey liner when a suitable, natural geologic deposit is not available. This may be necessary because of the absence of suitable clay on site or because the existing clay has a bulk hydraulic conductivity that is too high due to macrostructures (e.g. fractures) but has sufficient plasticity that it can be reworked as a clay liner.

The quality of a CCL will depend on: (a) the characteristics of the clayey soil used; (b) the method of compaction and, in particular, the compaction water content; and (c) the protection against desiccation after construc-

tion. CCLs may range in thickness from 0.3 to 3 m or more, although typical thicknesses range between 0.6 and 1.2 m. The selection of the required thickness may be based on: (a) the need to provide adequate hydraulic resistance, since the hydraulic gradient, and hence the Darcy flux (Section 24.1), across the liner decreases with increasing liner thickness (all other things being equal); (b) the need to provide adequate contaminant attenuation (Section 24.3); (c) the need to minimize the effects of construction or post-construction related defects (e.g. local zones of higher hydraulic conductivity, desiccation, etc.); and (d) the presence of other liner components (e.g. a geomembrane in a composite liner, see Section 25.3.7) or a double liner (Section 25.3.8). Like most engineered liners, CCLs require protection from the elements, e.g. sun and frost, and consideration must be given to the potential long-term effects of differential settlements.

25.3.2.1 Material Selection

CCLs are most commonly constructed from soils that are classified as CL, CH or SC in the unified classification system (see Section 3.2.11 and Table 3.6). The source of the material may include glacial tills, lacustrine, deltaic or aeolian deposits, residual soils or, by crushing, weathered mudstones and shales (e.g. see Lopes *et al.* 1992). Typically, CCLs are expected to have a hydraulic conductivity of 1×10^{-9} m s^{-1} (or less).

With adequate compaction (see Section 13.2.3 for a discussion of compaction tests, compaction curves and definition of terms such as optimum water content), this usually can be achieved for a soil containing a minimum of 15–20% of the particles with a size smaller than 2 μm, a plasticity index, I_p, greater than 7% ($I_p > 7\%$), an activity of 0.3 or greater, and less than 50% gravel, i.e. $D_{50} < 4.75$ mm.

The suitability of a given soil should be assessed by an appropriate laboratory testing program to: (1) establish the variability and potential suitability of the proposed material with respect to grain size and plasticity; and (2) assess the hydraulic conductivity as a function of compaction conditions to establish the acceptable compaction zone with respect to hydraulic conductivity. At a minimum, this may involve:

- standard Proctor compaction tests (ASTM D698) at different water contents to establish standard Proctor optimum water content (and maximum dry density); and
- hydraulic conductivity tests (e.g. ASTM D5084) on these samples to establish the range of water contents and dry densities over which a satisfactory hydraulic conductivity, k, is achieved.

However, caution is required in interpreting the results of these laboratory hydraulic conductivity test results. A laboratory hydraulic conductivity value that exceeds the specified value is usually a good indication of potential problems in the field; however, an acceptable laboratory value does not necessarily imply good field behavior (see discussion in Section 25.3.2.3).

Daniel (1998) recommends a more extensive laboratory program than that described above involving:

- compaction of the soil over a range of compactive energy (this allows one to establish the line of optimums shown in Fig. 25.7);
- performing hydraulic conductivity tests on these compacted samples;
- defining a water content range over which acceptable hydraulic conductivity is achieved;
- limit acceptable dry density–water content relationship to ensure adequate shear strength (and possibly to limit susceptibility to desiccation cracking).

This would result in an "acceptable zone" as indicated schematically in Fig. 25.7. Whether this level of effort is required is still subject to debate.

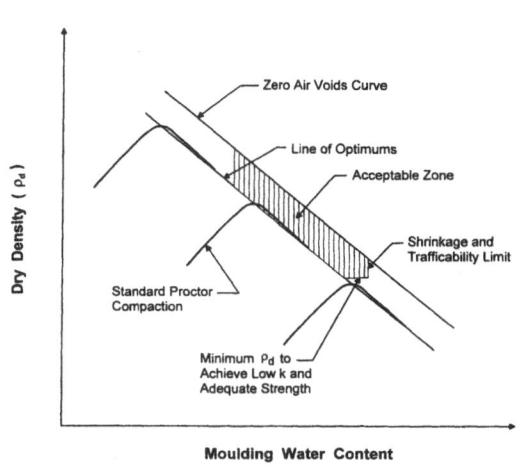

Moulding Water Content

FIGURE 25.7. Method of defining acceptable zone of water content based on compaction, hydraulic conductivity and shear strength considerations (modified from Benson et al. 1999).

Special care is needed with soils of low plasticity ($7 < I_p < 15\%$) to ensure that the desired hydraulic conductivity can be achieved. Care is also required with high plasticity clay ($I_p > 30\%$) since these can form hard clods that may be difficult to break up when constructing the liner and because they are particularly susceptible to shrinkage and cracking if allowed to dry after compaction.

In cases where the *in situ* soil has inadequate clay minerals to achieve the desired hydraulic conductivity, bentonite (a commercially processed soil consisting predominantly of the clay mineral smectite; see Chapter 23) may be mixed with a silty sand or sandy silt soil (as discussed in Section 25.3.3) to reduce the hydraulic conductivity of the amended soil.

25.3.2.2 Compaction Water Content and Density

In order to achieve a low hydraulic conductivity, a CCL should be compacted at a water content that is near or higher than the standard Proctor optimum water content and with a degree of saturation, S_r, greater than that corresponding to the degree of saturation at standard Proctor optimum, $S_{r_{opt}}$, i.e. $S_r \geq S_{r_{opt}}$, which means that the combination of (ω, ρ_d) should plot in the crosshatched zone shown in Fig. 25.7. (See Section 13.2.3 for a discussion of compaction and Mitchell et al. (1965) for a discussion of the effect of water content on hydraulic conductivity.)

Liners with a significant percentage of water contents below the standard Proctor optimum water content will typically have a macrostructure that gives rise to a high hydraulic conductivity (see Fig. 25.8). This can be explained by the observation that in order to provide sufficient plasticity to break up clods and develop a suitable clay fabric in the liner, the soil should be compacted at a water content near, or above, the plastic limit. Since, in North America, the plastic limit is often close to or slightly above the standard Proctor optimum water content, ω_{opt} (Leroueil et al. 1992a,b), standard Proctor optimum is generally a lower limit for the water content in order to obtain a good liner. Increasing water content, ω, above standard Proctor optimum, ω_{opt}, has been found to substantially increase the probability of achieving a hydraulic con-

ductivity of 1×10^{-9} m s^{-1} (Fig. 25.8). An examination of the cases reported by Benson et al. (1994) indicates that the average compacted water content of 11 CCLs with $1 \times 10^{-11} \leq k \leq 1 \times 10^{-10}$ m s^{-1} was PL + 2% while the average for 13 liners with $1 \times 10^{-10} \leq k \leq 1 \times 10^{-9}$ m s^{-1} was PL + 1%, where PL is the plastic limit. In contrast, the average for four liners with $k > 1 \times 10^{-9}$ m s^{-1} was PL − 7.5%.

It is noted that the relationship between optimum water content, ω_{opt}, and PL varies from soil to soil and care is required in the use of any empirical relationship. For example, Morin & Todor (1977) report that for red tropical soils in South America:

$$\omega_{opt} = 0.61 \text{ PL} - 0.84 \text{ (std. dev.} = 3.4\%) \tag{25.15a}$$

and for red tropical soils in Africa:

$$\omega_{opt} = 0.58 \text{ PL} - 4.5 \text{ (std. dev.} = 4.2\%). \tag{25.15b}$$

Well-compacted clay liners will typically have an initial degree of saturation, S_r, exceeding 90%; although some low hydraulic conductivity liners ($k < 1 \times 10^{-9}$ m s^{-1}) have been constructed with initial degrees of saturation as low as 83% (e.g. see Benson et al. 1994; Trast & Benson 1995). Liners with $S_r < 83\%$ typically have $k > 1 \times 10^{-9}$ m s^{-1}. This is consistent with the observation by Benson and Boutwell (1992) that there was a strong correlation between achieving $k < 1 \times 10^{-9}$ m s^{-1} and a combination of dry density and water content that plot on or above the line of optimums, see Figs 25.7 and 25.9. Daniel (1998) has suggested a minimum of 70–80% of (ω, ρ_d) points plotting above the line of optimums, i.e. $P_o > 70$–80%, be specified for CCL, while also cautioning the need to ensure adequate internal shear strength of the CCL and interface strength with adjacent

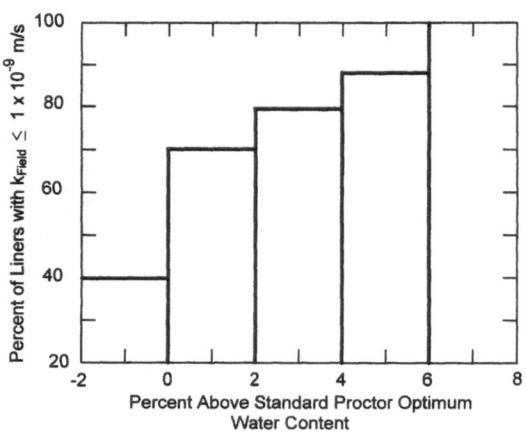

FIGURE 25.8. Percentage of liners with a compaction water content within a given range having $k \leq 1 \times 10^{-9}$ m s^{-1} (modified from Daniel 1998).

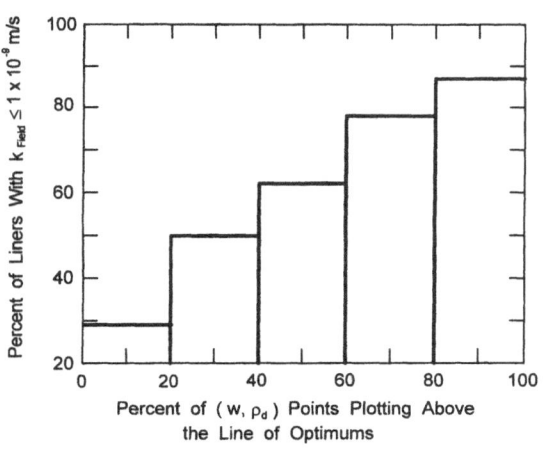

FIGURE 25.9. Percentage of CCLs with $k_{field} \leq 1 \times 10^{-9}$ m s^{-1} versus percentage plotting above the line of optimums (modified from Daniel 1998).

materials, e.g. geomembrane liners—see Sections 25.8 and 27.6.

Although the compacted dry density, ρ_d, may be important in terms of where the liner lies relative to the line of optimums (the degree of saturation) as discussed above, there is no significant correlation between the field hydraulic conductivity and the percent compaction. Thus, specifying ρ_d to be 95% of that obtained by standard Proctor compaction (or other arbitrary percentage) does not provide any assurance of a low hydraulic conductivity liner.

Although it is generally agreed that clay should be compacted such that the water content and dry density points (ω, ρ_d) generally plot above the line of optimums, there is not complete freedom in the choice of water content. If the water content is too high there may be problems with trafficability and obtaining adequate compaction with heavy equipment (Daniel 1998). Furthermore, difficulty may be experienced in obtaining both a sufficiently smooth surface on the liner or the shear strength may not be adequate to ensure stability. Leroueil et al. (1992a) proposed an empirical relationship for estimating the undrained shear strength, s_u, of clays that plot above the A-line (see Fig. 3.4) and have a value of $(\omega - \omega_{opt})/I_p \geq -0.06$, namely:

$$s_u(kPa) = 140 \exp[-5.8(\omega - \omega_{opt})/I_p]. \quad (25.16)$$

This equation can be used to make a preliminary estimate of shear strength for both stability and trafficability considerations. Like all empirical relations, it should be used with caution and appropriate laboratory tests should be conducted to ascertain the actual shear strength on important projects or where there are significant cost or safety implications if the shear strength is not correctly evaluated. This may be especially critical when constructing composite liners (see Sections 25.3.7, 25.3.8 and 25.8). Ensuring adequate shear strength should be considered when specifying the allowable water content–density relationship for a CCL.

25.3.2.3 Estimating the Hydraulic Conductivity of CCLs

Benson et al. (1994) have examined the hydraulic conductivity achieved for soils from 67 landfills in North America where natural clay liners, i.e. excluding soil bentonite mixtures, etc., were used and developed an empirical relationship between the hydraulic conductivity and a number of soil and compaction variables. This equation can be rewritten as:

$$k_F = \exp\left[-22.96 + \left(\frac{894}{w}\right) - 0.08I_p \right.$$

$$\left. - 0.029S + 0.32(G)^{1/2} \right. \quad (25.17a)$$

$$\left. + 0.02C \right]$$

where k_F = the estimated field hydraulic conductivity in m s^{-1}, w = the weight of the compactor in kN, I_p = the plasticity index in percent, S = the initial degree of saturation in percent, G = the percent gravel, and C =

the percent clay. Like all empirical equations, this equation should not be used without independent verification for a particular project (see Section 25.3.2.4).

Equation 25.17a may be useful for estimating hydraulic conductivity under low stress conditions, e.g. as might be obtained using a sealed double-ring infiltrometer (SDRI). In many practical (e.g. landfill) applications the liners will be consolidated under the weight of the overlayer material (e.g. waste) and this has been shown (e.g. see King et al. 1993) to cause a reduction in hydraulic conductivity. By comparing hydraulic conductivity values, k_σ, obtained from tests on samples consolidated to different effective stress levels, σ', with the field value, k_F, obtained under low stress conditions using a SDRI, Trast and Benson (1995) obtained the following empirical relationships.

For soils with $k_F > 1 \times 10^{-9}$ m s^{-1} (σ' in kPa):

$$k_\sigma/k_F = 1 - 0.018(\sigma' - 10)$$

$$\text{for } 10 \leq \sigma' < 44 \text{ kPa} \quad (25.17b)$$

$$k_\sigma/k_F = 0.4 - 0.0027(\sigma' - 44)$$

$$\text{for } 44 \leq \sigma' \leq 147 \text{ kPa} \quad (25.17c)$$

For soils with $k_F < 1 \times 10^{-9}$ m s^{-1}:

$$k_\sigma/k_F = 1 - 0.0059(\sigma' - 10)$$

$$\text{for } 10 \leq \sigma' < 44 \text{ kPa} \quad (25.17d)$$

$$k_\sigma/k_F = 0.8 - 0.0017(\sigma' - 44)$$

$$\text{for } 44 \leq \sigma' \leq 147 \text{ kPa.} \quad (25.17e)$$

Equation 25.17 cannot always be applicable since hydraulic conductivity will also depend on compaction conditions not considered in Eq. 25.17, the compressibility of the clay, the void ratio–hydraulic conductivity relationship, etc. Thus, the equation should only be used as an initial guide to estimate the hydraulic conductivities that might be achieved

for a given soil, weight of the compactor and effective stress level. The hydraulic conductivity that can be achieved in the field can only be properly evaluated by constructing a test pad where the soil is compacted in the field using the proposed construction specifications and then performing tests to confirm that the desired hydraulic conductivity was achieved (see Sections 25.3.2.4–25.3.2.6).

Table 25.3 provides a comparison between measured hydraulic conductivity under a number of stress conditions for two Canadian landfill liners with values predicted using Eq. 25.17. It can be seen that in these cases the predicted values exceed the observed values. In most cases, but not all (e.g. see Section 25.3.5), overpredicting the hydraulic conductivity would be conservative.

25.3.2.4 Construction of CCLs

Successful construction of a low hydraulic conductivity CCL using soil that meets the requirements outlined in Section 25.3.2.1 is highly dependent on: (a) water content control; (b) breakup of clods of soil and homogenization of non-uniform soils; (c) lift thickness; and (d) method of compaction and equipment used.

The soil used to construct a CCL needs to be broken up to minimize clod size and especially for dry, hard, clay-rich soil it is essential that these clods be wetted to the point that they can be readily remolded by the compaction equipment as discussed in Section 25.3.2.2. Daniel & Koerner (1995) recommend a maximum particle size of between 25 and 50 mm stones generally cannot be broken up. Low hydraulic conductivity liners have been sucessfully constructed with stones up to 100 mm in size provided that the liner is of a thickness exceeding 0.9 m. Large stones could be a problem with thin liners (<0.6 m) due to the fact that these stones can have voids adjacent to them due to movement during compaction (see Rowe et al.

TABLE 25.3. Comparison of calculated values of hydraulic conductivity based on Eq. 25.17 and measured values for two Ontario landfills

Soil[a]	Compactor weight (kN)	Predicted k_F	Hydraulic conductivity (m s^{-1})				Reference
			Predicted at $\sigma' = 25$ kPa	Observed at $\sigma' = 25$ kPa[b]	Predicted at $\sigma' \cong 100$ kPa	Observed at $\sigma' = 100$ kPa[b]	
Halton landfill							Rowe *et al.* (1993)
WHT	180	2×10^{-9}	1.5×10^{-9}	1.4×10^{-10}	5×10^{-10}	0.8×10^{-10}	
WHT	300	3×10^{-10}	2.7×10^{-10}	1.1×10^{-10}	2×10^{-10}	0.8×10^{-10}	
UHT	300	3×10^{-10}	2.7×10^{-10}	1.4×10^{-10}	2×10^{-10}	0.8×10^{-10}	
Keele Valley landfill				Field[c]	$\sigma' \cong 145$	$\sigma' = 145$	King *et al.* (1993)
WHT	300	3×10^{-10}		0.7×10^{-10}	1.9×10^{-10}	0.8×10^{-10}	

[a] WHT, weathered Halton till; UHT, unweathered Halton till.
[b] Based on flexible wall permeameter test consolidated to σ'.
[c] Based on field lysimeter below liner.

1993). Oversize particles need to be removed using a mechanized stone picker or by hand.

Prior to placing a lift, the surface of the compacted lift should be rough and may need to be scarified using an industrial disk before constructing the new lift. The loose thickness of lifts used to construct CCLs should generally not exceed 230 mm and the compacted thickness should not exceed about 150 mm. The sheepsfoot/padfoot compactor used should have long feet (>150 mm), which can fully penetrate the lift being compacted into the underlying lift to break up interfaces between the lifts (see Rowe *et al.* 1993). The compactor should have a weight of 180 kN or greater, and typically five (or more) passes are required to achieve the desired compaction.

When placing the first lift over an underlying engineered system it may be necessary to compact a sacrificial layer using a lightweight smooth drum roller to avoid causing damage to the underlying material (e.g. secondary leachate collection/leak detection layer). The sacrificial layer compacted with the smooth drum roller is unlikely to meet the hydraulic conductivity requirement and should not be considered as contributing to the effective thickness of the CCL (see Rowe *et al.* 1993). Also, a smooth drum roller will usually be required to smooth the surface of the final lift after it has been compacted with the sheepsfoot/padfoot compactor.

CCLs may experience an increase in hydraulic conductivity with time of several orders-of-magnitude if not adequately protected against desiccation cracking or frost damage (e.g. Othman & Benson 1993; Bouchard *et al.* 1992). Potentially, this is particularly problematic for covers where it is difficult to provide adequate long-term protection. However, care is also required with the protection of any liner that may be exposed to drying or freezing conditions. For base liners in landfills, protection against desiccation may involve frequent watering using a water truck or covering the liner with a damp geotextile or, most effectively, 150–300 mm of soil and keeping the geotextile–soil damp. The CCL can also be covered with a geomembrane; however, particular care is needed in this case to avoid desiccation of the CCL below the geomembrane due to thermal gradients established in the liner. The most effective means of protecting landfill liners against frost is to schedule liner construction such that an adequate thickness of waste can be placed over the liner before the onset of freezing conditions (see Chapter 20 for a discussion of frost penetration).

25.3.2.5 Required Thickness of CCLs

The required thickness of a CCL varies from one regulatory environment to another and the rational basis for the choice is often hard to find. Thickness may be important in terms of: (a) providing adequate contaminant attenuation potential; or (b) ensuring that the liner has adequate hydraulic conductivity. An assessment of attenuation capacity requires consideration of advective–diffusive contaminant transport as well as appropriate retardation mechanisms (see Sections 24.3 and 25.7) and the thickness required may include part CCL and part lower quality compacted soil or an existing (natural) attenuation layer. The issue of thickness versus ensuring adequate hydraulic conductivity has been the subject of some debate as discussed below.

Benson & Daniel (1994) have performed statistical studies considering the potential for advective transport through CCLs constructed using different numbers of lifts. Based on this they recommend a minimum thickness of 0.6–0.9 m for a CCL constructed in four–six lifts of 0.15-m compacted thickness. This conclusion has been questioned (see Fuleihan & Wissa 1995 and closure by Benson & Daniel 1995) from the perspective that if a liner is well-compacted then there is no trend of increasing hydraulic conductivity with decreasing liner thickness.

As noted in Section 25.3.2.2, one of the most important factors related to construction of a good quality CCL from material meeting the requirements of Section 25.3.2.1 is the water content control and use of appropriate compaction equipment that will bring (ρ_d, ω) above the line of optimums. Liners compacted at water content dry of optimum, i.e. at a water content below standard Proctor optimum, generally give a hydraulic conductivity greater than 1×10^{-9} m s^{-1} (e.g. see Fuleihan & Wissa 1995 or Benson & Daniel 1995). Liners reported to have been constructed wet of optimum using heavy (weight $\geq$ 180 kN) sheepsfoot/padfoot rollers with long penetration feet ($>$150 mm) generally, but not necessarily, had hydraulic conductivities less than 1×10^{-9} m s^{-1} (loc. cit.). The reasons for values exceeding 1×10^{-9} m s^{-1} are unclear, but are likely related to quality control that was not as good as might have been expected; in the writer's experience this is usually associated with lack of attention to water content control.

With appropriate material compacted at the correct water content using the right equipment and with appropriate quality control and assurance it should be possible to construct a low hydraulic conductivity liner that is relatively thin. However, recognizing that field conditions and people involved with the project are not always perfect, there is no question that increasing the number of lifts reduces the probability of a local defect significantly impacting on the liner performance. Two lifts are much better than one. The extent to which more than two lifts are required to achieve the desired hydraulic conductivity will, as shown by Benson & Daniel (1994) depend on variability that is likely to occur in a given layer. Thus there is an economic trade-off between the level of attention that must be paid to construction, the number of lifts required and the accept-

able risk of the liner not meeting the design hydraulic conductivity specifications, which can only properly be evaluated on a site-specific basis. Increasing thickness provides some level of increased assurance of meeting the specified hydraulic conductivity, within limits, but is no substitute for ensuring the soil is properly homogenized, brought to the correct water content and the clay clods are kneaded into a low hydraulic conductivity liner using appropriate compaction equipment. As noted by Fuleihan & Wissa (1995):

Surprisingly, many practitioners were still not aware that construction quality and molding conditions have a much greater effect than thickness on the quality of a clay liner, even though Lambe (1958) showed as early as 1958 the several order-of-magnitude influence that molding conditions can have on the hydraulic conductivity of a compacted clay liner.

Once the specified hydraulic conductivity of the liner has been achieved, the required thickness will depend on factors like the potential for contaminant transport (see Section 25.7). In all cases, attention must be paid to preventing desiccation cracks and frost damage.

25.3.2.6 Construction Quality Control and Assurance

Construction specifications typically indicate the water content and compacted clay density to be achieved; here, the critical issue is that the combination of water content, ω, and dry density, ρ_d, measured for a liner should plot above the line of optimums for a well-constructed liner, see Section 25.3.2.2 and Fig. 25.7.

A well-trained engineer/technician can usually identify whether the water content is correct by eye and feel (as noted earlier, the desired water content is often close to the plastic limit). A soil that appears and feels dry will usually give poor performance if compacted into the liner. Tests to establish the

TABLE 25.4. Recommended minimum testing frequencies: one test for each per volume or area noted below (adapted from Daniel & Koerner 1995)

Parameter	Method ASTM	Soil source[a] (m³)	In loose lift (m³)	After compaction (m² per lift)
Water content (field)	D-3017[b]	—	—	750
	D-2216	2000	800	750
Density (field)	D-2922[b]	—	—	750
	D-1556/2167	—	—	15 000
Atterberg limits	D-4318	5000	800	750
Per cent clay	D-422	5000	800	750
Per cent gravel	D-422	5000	800	750
Compaction curve	D-698	5000	4000	—
Hydraulic conductivity	D-5084	10 000	—	2000[c]
Construction oversight			Continuous	Continuous

[a] Or each change in material type.
[b] Values are field measurements.
[c] Value is m² of finished liner.

water content and density achieved should be used to document the liner construction; however, there is no substitute for having a well-trained engineer/technician overseeing the liner construction on a full-time basis and ensuring that adjustments are made to water content as needed.

Quality control procedure involves checking the soil meets the specifications in terms of Atterberg limits, percentage of clay and gravel and that the CCL meets the specifications in terms of water content, dry density and hydraulic conductivity. Table 25.4 gives recommended testing methods and minimum testing frequency (partially based on Daniel & Koerner (1995) and partially based on the writer's experience).

In the USA, SDRIs are often used to establish the field hydraulic conductivity. However, provided that adequate water content control is achieved and the liner is compacted to a degree of saturation of 87% or higher (e.g. see Trast & Benson 1995), and provided there is no visible evidence of macrostructure, e.g. desiccation cracks, flexible wall hydraulic conductivity tests performed on

Shelby tube samples provide a good indication of the hydraulic conductivity of the liner.

If Shelby tube samples are taken, then this area of the liner must be stripped to below the depth of the Shelby tube penetration over a width exceeding the width of the compactor and the liner must be reconstructed according to the original procedure.

Liner construction procedures are best verified by construction of a test pad prior to construction of the actual liner. This provides a means of acquainting the contractor with potential problems and calibrating the quality control and assurance procedures. Test pads will typically be 10–15-m wide and 15–30-m long.

Due to the inherent variability of soil it may be anticipated that at times parameters evaluated in accordance with Table 25.4 may not meet the specifications. Daniel & Koerner (1995) recommend allowing up to 3% of water content and dry density results falling outside the specified range in any one lift or area provided that no water content is less than 2% below or more than 3% above the specified values and no dry density is less than 80 kg m^{-3} below the required value. Other specifications may also require that no water content may be below standard Proctor optimum. If the liner is found to be out of specifications, additional tests will likely be required to establish the extent of the problem area requiring repair.

If the water content is within range, but the dry density is too low, the problem may be rectified by more passes of the compactor.

25.3.3 SOIL–BENTONITE LINERS

In situations where suitable natural soils are not available for use in a compacted clay liner, bentonite may be added to a noncohesive soil, e.g. silty sand, to achieve a liner with the required hydraulic conductivity. Key considerations in the selection of bentonite and design of these mixed-soil liners is the grain-size distribution of the base soil, the amount of bentonite and the mineralogy of the bentonite. Typically, the bentonite used is a sodium bentonite, calcium bentonite, or sodium activated calcium bentonite (see Chapter 23 for a discussion of clay mineralogy), although a wide range of modified bentonites, e.g. organobentonites, are also available. Both powdered and granular bentonites have been used, although Goldman *et al.* (1990) recommend that only powdered bentonite be used since it is said to mix more uniformly throughout the soil.

The proportion of bentonite in the soil mixture typically ranges between 4 and 10%, and this may give hydraulic conductivities of $10^{-9}–10^{-11}$ m s^{-1}. However, it is important to establish the optimum proportions of bentonite and water content on a site-specific basis. This is usually achieved by compacting different soil mixes with different amounts of bentonite at different water contents and then performing laboratory hydraulic conductivity tests on the compacted samples. Chapuis (1990) describes a simple approximation technique for estimating hydraulic conduc-

tivity and provides suggestions regarding the associated laboratory study. Care is needed to examine the hydraulic conductivities under the maximum gradient that could occur (e.g. with the development of a leachate mound) to ensure that there will not be an increase in hydraulic conductivity due to internal erosion of the bentonite, i.e. transport of the bentonite out of the liner by water flow.

Successful construction of low hydraulic conductivity soil–bentonite liners is highly dependent on: (a) obtaining and maintaining a homogeneous mixture of the base soil and bentonite, and avoiding segregation prior to and during placement; (b) compaction and water content control; and (c) lift thickness. If the soil–bentonite liner is to be used as a landfill liner, it is recommended that the liner be at least 200-mm thick and either special effort be devoted to ensuring uniform mixing of bentonite and negligible segregation prior to placement or, like other CCLs, the soil–bentonite liner be compacted in a number of layers. Good success has been achieved constructing soil–bentonite liners using padfoot compactors with a weight of 180 kN (or greater). Smooth drum rollers have also been used with success. In all cases, it is desirable to establish the required construction procedure based on the construction and evaluation of a test pad prior to liner construction.

A key consideration in the selection and design of soil–bentonite liners is the potential for chemical interaction between the clay and the fluid to be retained (see Section 25.4). Although sand–bentonite specimens have been shown to be more resistant to freeze-thaw cycles than clay or till specimens in laboratory tests (Wong & Haug 1991), as with normal CCLs, it is still prudent to protect the sand–bentonite liners against desiccation and frost. Testing for construction quality control may follow that given in Table 25.4.

25.3.4 GEOSYNTHETIC CLAY LINERS (GCLs)

GCLs (see Section 7.6) provide a convenient and potentially economical low permeability alternative to CCLs both in covers and base liners in many situations. Due to the fact that it is a manufactured product, typically produced using either powdered or granular sodium bentonite, a high level of quality control can be achieved. Since they are relatively thin (≈ 10 mm), there is also the potential for either minimizing the excavation required prior to placing the GCL or alternatively the use of a thin GCL may allow more air space, e.g. for waste disposal. However, the fact that they come in thin sheets that are seamed by overlapping does mean that considerable care is required during construction to avoid tearing the GCL sheets or opening the seams, especially when cover soil is being placed over the GCL. GCLs will usually need to be placed on a prepared foundation layer. Careful consideration must be given to the stability of liner systems involving GCLs and other geosynthetics (see Sections 25.8, 26.5 and 27.6).

25.3.4.1 Hydration of GCLs and Hydraulic Conductivity

As will be shown in subsequent sections, both the hydraulic conductivity, k, and diffusion coefficient, D, of a GCL are highly dependent on the bulk GCL void ratio, e_B (defined by Petrov & Rowe (1997) to be the ratio volume of void to the volume of solids in the GCL including both the geotextiles and bentonite). This may depend on: (a) the method of GCL manufacture; and (b) the confining stress at which hydration of the GCL occurs. Petrov et al. (1997b) showed that the final bulk GCL void ratio of a needle-punched GCL and an otherwise similar GCL without needle punching, i.e. fiber-free, varied substantially with the values for the needle-punched geotextile being much lower (better) under otherwise similar conditions (compare columns 2 and 3 of Table 25.5). Thermal locking of the needle-punched fi-

TABLE 25.5. Final bulk GCL void ratios obtained in confined swell tests (adapted from Rowe & Lake 1999)

Confining stress (kPa)	Final bulk void ratio, e_B		
		With needle punching[b]	
	No needle punching[a]	No thermal locking[a]	Thermal locking[b]
6	7.58	5.12	3.98
25	4.04	3.23	2.97
100	2.58	2.26	2.25
200	1.96	1.68	1.69

[a] Based on data from Petrov et al. (1997b).
[b] Data from Lake & Rowe (2000a).

bers can also have a beneficial influence on the bulk GCL void ratio as noted below.

Thermal locked GCLs are needle-punched GCLs where the bottom geotextile has been heated after needle punching. This bonds the needle-punched fibers to the carrier geotextile (i.e. the bottom geotextile during the manufacturing process), minimizing pull-out of the fibers from the geotextile. Columns 3 and 4 of Table 25.5 compare the final bulk GCL void ratio for two otherwise similar needle-punched GCLs, except that in one case there is no thermal locking (column 3) and in the other there is thermal locking of the fibers (column 4). As can be seen, at lower confining stresses needle punching and thermal locking give a lower value of e_B and hence may be expected to improve the performance of a GCL (other things being equal) in terms of both hydraulic conductivity and diffusion.

The results presented in Table 25.5 are for confined, constant stress swell (CSS) tests, where the stress is applied prior to hydration. The effect of thermal locking is even more evident in post-hydration confinement where the sample is allowed to hydrate at a low confining stress (6 kPa) and then, after hydration, the stress is applied. This might correspond to the case where a GCL used as part of a composite liner is left to hydrate (e.g. from the underlying soil) with only the leachate collection system in place and with little or no waste being placed until after hydration had occurred. As shown by Lake & Rowe (2000a), in this case the beneficial effect of thermal locking is manifest throughout the stress history.

The relationship between the effective confining stress, σ'_v, and the bulk GCL void ratio, e_B, under confined swell conditions depends on the type of GCL manufacturing process. Recognizing that for a given method of manufacturing there will be variability between similar samples of the same GCL at the same confining stress, it is still useful to have some idea of how the bulk void ratio varies with confining stress. These relationships need to be developed for each type of GCL, and for a given GCL will vary with hydrating conditions (e.g. the fluid used to hydrate the GCL and whether the stress was applied before or after hydration). To illustrate the effect of the type of manufacturing process, Rowe & Lake (1999) fitted regression curves to the data obtained for the three types of GCLs indicated in Table 25.6, yielding the following relationships for confined swell condition and hydration with distilled water:

$$\log_{10} \sigma'_v \cong 3.25 - 0.57 e_B$$
$$\text{for the NWNWT GCL} \quad (25.18a)$$

$$\log_{10} \sigma'_v \cong 3.5 - 0.69 e_B$$
$$\text{for the WNWT GCL} \quad (25.18b)$$

$$\log_{10} \sigma'_v \cong 2.9 - 0.32 e_B$$
$$\text{for the WNWBT GCL.} \quad (25.18c)$$

Variability in a given manufactured product may result in some deviation from these relationships and hence they only serve to illustrate general trends and should not be used for detailed design without independent verification.

Extensive research has been conducted into the hydraulic conductivity of GCLs as a function of confining stress and permeant type and concentration as summarized by Rowe (1998a). In particular, it has been shown by Petrov & Rowe (1997) that for a given permeant, there is a log-linear relationship between the hydraulic conductivity, k (in m s^{-1}), and the bulk GCL void ratio, e_B. For example, for the needle-punched NWNWT GCL and conditions they examined, the hydraulic conductivity with respect to distilled water was given by:

$$\log_{10} k(\text{m s}^{-1}) \cong -11.8 + 0.29 e_B. \quad (25.19)$$

The coefficients in this relationship can be expected to vary from GCL to GCL; however, the important point here is that the hydraulic conductivity decreases with decreasing bulk GCL void ratio (which is, in turn, related to stress as implied by Eq. 25.18).

More generally, it can be concluded that when hydrated and permeated with water, GCLs have a low hydraulic conductivity, e.g. 7×10^{-12}–7×10^{-11} m s^{-1} depending on hydrating and applied stress conditions during permeation. However, since this arises from the presence of bentonite, careful consideration must be given to the potential for an increase in hydraulic conductivity due to clay

TABLE 25.6. Basic characteristics of three thermal locked needle-punched GCLs tested

Generic symbol used in paper	Bentonite	Carrier geotextile	Cover geotextile	GCL mass (g m^{-2})[a]
NWNWT	Granular	PP nonwoven	PP nonwoven	5270
WNWT	Granular	PP woven	PP nonwoven	5380
WNWBT[b]	Powdered	PP woven	PP nonwoven	4990

[a] Minimum values.
[b] Cover geotextile filled with a minimum of 600 g m^{-2} of powder bentonite.

leachate interaction as discussed in more detail in Section 25.4. However, it is noted here that the relationship implied by Eq. 25.19 may change when the GCL is permeated with leachate. For example, for the same NWNWT GCL used to obtain the results summarized in Eq. 25.19, the hydraulic conductivity was bounded by:

$$-11.4 + 0.42e_B < \log_{10} k \text{ (m s}^{-1})$$
$$< -11.2 + 0.42e_B \quad (25.20)$$

when hydrated with water and subsequently permeated using a synthetic leachate based on the typical leachate chemistry at the Keele Valley landfill (see Petrov & Rowe 1997).

For typical values of e_B, the difference between the hydraulic conductivity with respect to water (Eq. 25.19) and synthetic leachate (Eq. 25.20) is about one order-of-magnitude. This means that when considering the use of GCLs in landfill liner systems, one should use a hydraulic conductivity for the anticipated leachate when performing contaminant impact calculations to assess equivalency (see Section 25.7).

The results used to obtain Eq. 25.20 were for a GCL hydrated with water and then permeated with synthetic leachate. Petrov & Rowe (1997) showed that a sample hydrated with leachate and then permeated with leachate had twice the hydraulic conductivity of an otherwise similar sample initially hydrated with water. Thus, careful consideration needs to be given to the hydrating conditions when evaluating the hydraulic conductivity.

Since the hydration of GCLs with water prior to permeation with leachate generally results in a lower hydraulic conductivity as discussed above and in Section 25.4, the level of hydration that can be achieved prior to contact with leachate is an important question. Daniel & Shan (1992) indicated that in contact with soil at the wilting point, sodium bentonite can be expected to have a water

TABLE 25.7. Uptake of water by bentonite (interpreted from Daniel & Shan 1992)

Sand initial water content (%)	Bentonite water content at 40–45 days (%)
1	50
2	75
3	88
5	128
10	156
17	193

content rise to about 50%. They performed a number of tests where the bentonite from a GCL was in contact with sand at different water contents. Table 25.7 summarizes the bentonite-water content after 40–45 days for bentonite placed in contact with (0.1–0.2 mm) sand at different water contents (at 14 kPa pressure). It can be seen that under these circumstances, the bentonite is very efficient in its uptake of water from the underlying soil.

Bonaparte et al. (1996) performed hydration tests on three different GCL products in contact with a compacted clay (PL = 22%, I_p = 19%, ω_{opt} = 20%) placed at three different water contents ($\omega_{opt} - 4\%$, ω_{opt}, $\omega_{opt} + 4\%$). Even for the driest soil ($\omega \sim 16\%$) the GCL moisture content increased from initial values of 15–20% up to about 40% within 20 days. At the higher water content there was greater hydration. For compacted soil at ω = 24%, after 75 days the GCL water content had increased to 70–90% depending on the product. Their test suggested that the uptake of moisture over a period of 25 days was not significantly affected by the applied pressure over the stress range of 5–390 kPa. Eberle & von Maubeuge (1997) also report an example where a GCL installed "dry" ($\omega \sim 9\%$) over a sand with a moisture content of 8–10% was hydrated to a moisture content over 100% in less than 24 h and increased to 140% after 60 days.

25.3.4.2 Effect of Holes in GCLs

Shan & Daniel (1991) examined the effect of puncturing a GCL by cutting holes in 152-mm diameter GCL specimens. The samples were allowed to hydrate under a confining stress of 14 kPa. Under these test conditions the bentonite swelled to completely fill the 12-mm and 25-mm diameter holes and the hydraulic conductivity only increased from 2×10^{-11} m s^{-1} with no punching to 5×10^{-11} m s^{-1} with the 25-mm puncture. However, two out of the three 75-mm holes were reported to have not sealed and left an opening of 12-mm diameter giving a very high hydraulic conductivity (the upper limit measurable with the equipment used). These tests suggest that in this test the GCL had the capacity to effectively self-heal small holes but not large holes or tears. However, it should be recognized that this type of test does not simulate many practical situations. If a stone punched a hole in the GCL, the stone will likely stay in the hole. Also, the potential for puncturing will depend on the robustness of the cover geotextile. The GCL should be carefully installed in a manner that will avoid holes in the GCL.

25.3.4.3 Desiccation and Frost Action

GLCs are subject to potential changes in hydraulic conductivity due to desiccation; however, preliminary work by Boardman & Daniel (1996) indicates that while the bentonite in GCLs did form open cracks upon drying, these cracks closed due to swelling upon re-wetting. Thus although the initial hydraulic conductivity was high ($\sim 10^{-7}$ m s^{-1}), after a few days it had reduced to a value similar to the original value ($\sim 7 \times 10^{-11}$ m s^{-1}). Eith et al. (1991) reported laboratory tests where the hydraulic conductivity was found to decrease from an initial value of 4×10^{-12} to 1.5×10^{-12} m s^{-1} after ten freeze–thaw cycles. While these results are encouraging, they were performed under a limited range of conditions and more field data as required to verify these findings.

GCLs subjected to limited freeze–thaw conditions (Hewitt & Daniel 1997; Kraus et al. 1997) were found to perform well with no evidence of cracking, and no significant increase in hydraulic conductivity was observed related to the freeze–thaw of the GCL sheets. Again, there is a need for more field data to verify these encouraging findings. These findings are in contrast to those of Othman & Benson (1993) who observed a significant (one–two or more orders-of-magnitude) increase in the hydraulic conductivity of a conventional CCL subjected to freeze–thaw cycles. Thus the, albeit limited, available evidence would suggest that GCLs may be preferable to CCLs where the liner cannot be protected from desiccation and freeze–thaw, e.g. in covers.

25.3.5 HYDRAULIC CONTAINMENT

In most liner applications the hydraulic head acting on top of the liner exceeds that below the liner and hence there is a gradient tending to cause fluid flow from inside to outside the liner. The magnitude of the outward flow is given by Darcy's law (Sections 2.4 and 24.1) and is controlled in part by the low hydraulic conductivity of the liner. However, in some cases one can take advantage of the presence of a high water table or artesian conditions to induce an inward flow. Thus, if the head in the leachate collection layer surrounding the waste is less than that in the surrounding hydrogeologic environment then the consequent inward flow can be used to resist outward diffusion of contaminants (see Fig. 25.10). This is called hydraulic containment or, sometimes, a natural "hydraulic trap." Joseph & Mather (1993) have advocated a form of hydraulic containment in the UK. The Gallatin National Balefill in Illinois, USA, was approved in 1987 on the basis of its operation as a hydraulic trap (Burke & Haubert 1991). In Ontario, Canada, there was considerable debate in 1987–1988 concerning the proposal to operate the Halton

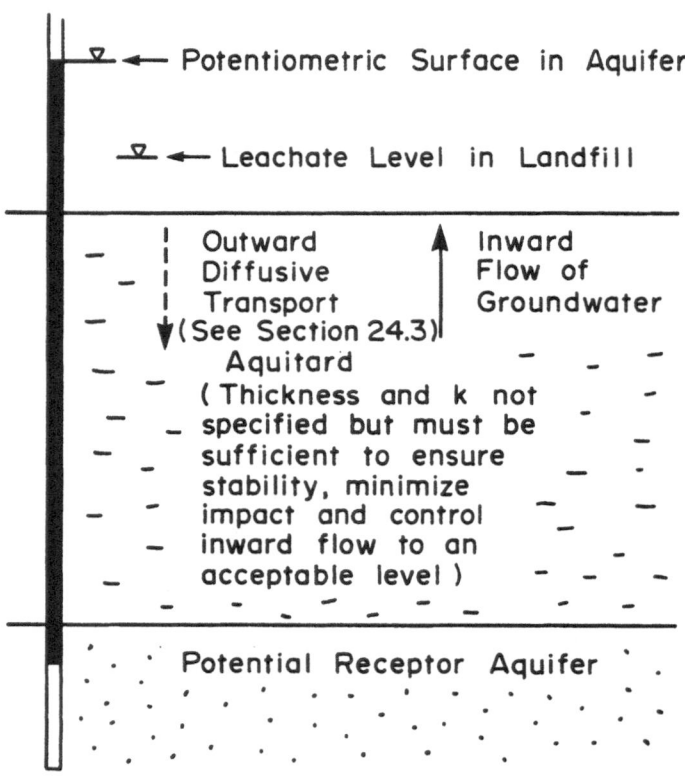

FIGURE 25.10. Hydraulic containment ("hydraulic trap") inward flow but outward diffusion (modified from Rowe 1997b).

landfill as a hydraulic trap. Approval was granted in 1989 and the landfill has been operating as a hydraulic containment site since 1992 (Rowe *et al.* 1993, 1996a, 1998b, 2000c). This approval was a landmark decision, which was followed by the approval, construction and operation (since 1995) of a number of hydraulic containment landfills. The Halton landfill has a CCL over a granular "subliner contingency layer" (see Fig. 25.11) overlying a relatively low hydraulic conductivity till layer. Under normal operations the subliner contingency layer is saturated and operated in a passive mode, i.e. no human intervention. In the event of an unexpected drop in water levels in the underlying aquifer or an unexpected increase in leachate head acting on the compacted liner, this layer can

be operated to control impact on the very sensitive underlying aquifer. The Grimsby landfill, which has been in operation since 1995, also has a hydraulic containment design. It has a reworked till liner (CCL) directly on a fractured till layer, which is underlain by a fractured limestone aquifer.

Subsequent to the approval of the Halton and Grimsby landfills, the Essex–Windsor and North Simcoe landfills have also been approved for operation as hydraulic containment sites using an intact natural clayey barrier (Section 25.3.1); and the "Adams Mine," a worked-out open-pit quarry, has been conditionally approved as a hydraulic containment facility, located in fractured hard rock, without any liner as such.

Like all approaches, hydraulic contain-

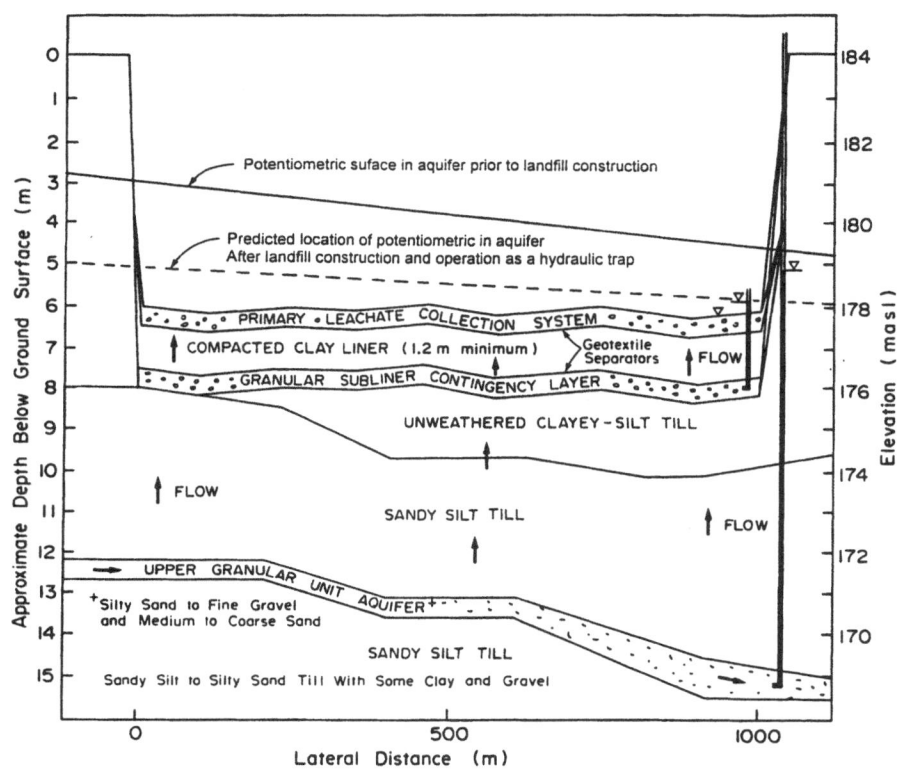

FIGURE 25.11. Hydraulic trap landfill (Halton) showing potential shadow effect due to landfill construction. Also shown is the subliner contingency layer (modified from Rowe *et al.* 1995b).

ment has both advantages and disadvantages as summarized by Rowe (1992, 1995a, 1997b; Rowe *et al.* 1995b). The primary advantages are: (a) the relative ease of construction, development and operation; (b) relatively low construction costs; and (c) inward flow provides some resistance to outward diffusion. For hydraulic containment to work it is essential to control the leachate head on the base of the landfill to a level that will induce inward flow. This is not as easy as it may first appear. Thus the primary disadvantages of this approach are: (a) the design is sensitive to the effects of clogging of leachate collection systems, see Section 25.2.3; (b) careful consideration must be given to the potential for the landfill construction causing a change in the natural water levels (i.e. the "shadow ef-

fect"; see Rowe 1992; Rowe *et al.* 1995b, 1998b); and (c) the success of the design may be sensitive to future natural changes in water levels. There are ways of addressing all of these issues (e.g. see Rowe 1995a; Rowe *et al.* 1995b, 1998b); however, they do increase the complexity and cost of the design.

Irrespective of whether one is using physical containment involving a geomembrane over clay or hydraulic containment, it is necessary to control the leachate mound in the landfill and collect and treat the leachate that is generated. For landfills with a soil cover, the increased amount of leachate requiring collection and treatment due to groundwater coming into the landfill (typically of the order of 0.001 m^3 a^{-1} m^{-2} for landfills in southern Ontario) is small and hence there is little impact on the operating cost relative to a land-fill with a composite liner system and no contribution of groundwater to the volume of leachate generated and collected. For land-fills with a low infiltration cover, the groundwater flow may represent in the order of 25% of the leachate generated; however, this is for relatively small volumes of leachate being generated.

Although hydraulic containment is not always an option, in the right hydrogeologic environment it represents a viable alternative to more conventional barrier designs and should not be eliminated as an option by regulations.

25.3.6 GEOMEMBRANE LINERS

Geomembranes are planar, relatively impermeable polymeric sheets used in contact with soil or rock (Section 7.5). Due to their low permeability, they make excellent liners for fluid-retaining structures. This application is discussed in detail in Chapter 26. When used over a permeable layer, there is considerable potential for leakage through holes in the geomembrane (Section 26.4.3). Geomembranes are often combined with an underlying low permeability soil layer to form a composite liner as discussed in Sections 25.3.7 and 26.4.2. When the geomembrane is used to contain fluids other than water, careful consideration should be given to potential interaction of the geomembrane with the fluid (Section 25.4.3) and the potential for contaminant escape due to diffusion (Section 25.5.3). The service life of a geomembrane liner will be highly dependent on the exposure condition and temperature of the liner (Section 25.6).

25.3.7 SINGLE COMPOSITE LINERS

Single composite liners consist of a geomembrane (Section 25.3.6) over a low hydraulic conductivity soil liner/aquitard (Section 24.1.1 and 25.3.1–25.3.4). In landfill applications this typically involves an HDPE geomembrane over a CCL or GCL. Usually, but not

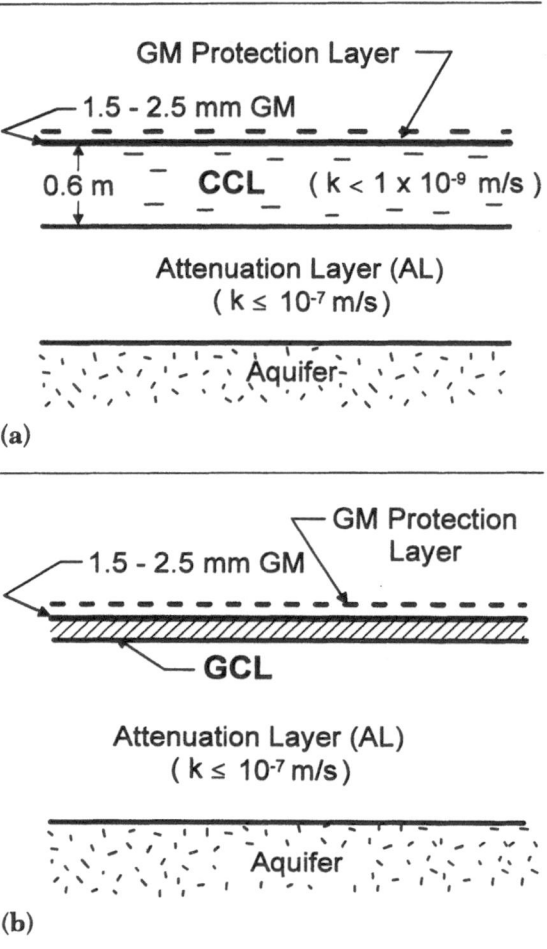

(a)

(b)

FIGURE 25.12. Schematic showing two single composite liner systems: (a) geomembrane (GM) over compacted clay liner (CCL) over an attenuation layer (AL) over an aquifer; and (b) geomembrane (GM) over a geosynthetic clay liner (GCL) over an attenuation layer (AL) over an aquifer.

always, there will be an attenuation layer of natural soil below the CCL or GCL before reaching a saturated aquifer (see Section 24.1.1 for definition of an aquifer) as shown in Fig. 25.12.

The geomembrane provides the primary barrier to the movement of fluid; however, the presence of the low permeability layer in contact with the geomembrane serves to substantially reduce the leakage through defects/holes in the geomembrane. Leakage through composite liners has been reviewed by Rowe

(1998a), and simple equations for calculating leakage through holes in geomembranes are given in Section 26.4. The low permeability liner and underlying attenuation layer also play an important role as a diffusion barrier and zone for attenuation of certain organic contaminants, which can readily diffuse through geomembranes (Section 25.5). Other key considerations, e.g. stability and deformations, are discussed in Chapter 26 and 27.

25.3.8 DOUBLE COMPOSITE LINERS

Double composite liners, e.g. see Fig. 25.1, consist of two single composite liners separated by a drainage layer. Both geonets (Section 7.4) and granular layers (Section 25.2) may be used provided that the geomembranes are adequately protected. The issues related to each composite liner are similar to those discussed in Section 25.3.7. The primary objective of the drainage layer between the liners is to control the head acting on the second liner and hydraulically isolate the flows in the second composite liner from any leachate mounding above the primary geomembrane and any flow through the primary composite liner. Thus, it is essential that the secondary liquid collection layer (sometimes called the "leak detection system," since under some circumstances it can be used to detect flow or leakage through the primary liner) be designed to have sufficient long-term transmissivity to control the head on the secondary geomembrane for as long as required (Section 26.2.3).

25.3.9 VERTICAL CUT-OFF WALLS

There is a growing movement toward "containment" as a primary "remediation" strategy for many contaminated sites where alternative approaches have proven ineffective or are too expensive (see Chapters 22, 28, 29 and 30 for a discussion of issues related to contaminated groundwater and soil). This "containment" will frequently involve the construction of a vertical cut-off wall around

part or all of the contaminated site and possibly (but rarely) a retrofitted "base liner" below the site (see O'Donnell *et al.* 1995). Vertical cut-off walls may consist of slurry trench walls, geomembrane walls or sheet pile walls. A detailed discussion of the state-of-practice with respect to these walls is provided in O'Donnell *et al.* (1995). Daniel & Koerner (1995) provide a discussion of construction quality assurance and construction quality control (CQA/CQC) for these walls. The key concerns with all of these walls is: (a) ensuring sufficient uniformity and quality of construction that leakage is controlled to an acceptably low level; (b) the difficulty of detecting small leaks; (c) long-term durability, e.g. the potential effects of interaction with contaminants. The significance of these concerns can be reduced, but not eliminated, by combining the use of a vertical cut-off wall with hydraulic containment where the head in the contained zone is maintained below that outside the walled area to induce an inward flow (Section 25.3.5). The disadvantage of this is that it requires active control (pumping and treatment of contaminated groundwater). When combined with hydraulic containment, the primary functions of the cut-off wall are: (a) to reduce the volume of groundwater that needs to be pumped to maintain the inward gradient; (b) to limit density driven contaminant migration, e.g. movement of dense nonaqueous phase liquids, (DNAPLs) that might not be otherwise controlled by an inward gradient; (c) to provide a diffusion barrier (see Rowe 1996).

25.3.9.1 Slurry Trench Walls
Slurry trench walls are usually constructed by excavating existing soil using a backhoe or clamshell. To provide stability during construction, the trench is filled with a slurry, which is maintained with a slurry level above the groundwater level.

The slurry is typically a clay–water mixture (usually bentonite and water) or a cement–

bentonite–water mixture. Slurry trenches are then backfilled with either: (a) a mixture of bentonite, soil and water with a consistency of high slump concrete; or (b) a plastic concrete. Cement–bentonite–water slurry walls are left in place and the mix hardens to form the cut-off wall.

These walls may be used to contain, capture and redirect flow of contaminated water, gases or free-phase liquids. There are practical limits to the depths that can be achieved and the type of ground conditions in which they can be constructed. The short- and long-term hydraulic conductivity that can be achieved requires careful consideration. In particular, construction quality control and its effect on initial hydraulic conductivity, the effect of changes in groundwater condition and freeze–thaw, i.e. desiccation and cracking, the effect of interaction with the contaminants to be retained and the long-term hydraulic conductivity all need to be considered. In addition, it must be recognized that even if there is no outward flow there is still potential for contaminants to diffuse across the barrier (see Rowe 1996).

25.3.9.2 Geomembranes

Due to the very low permeability of geomembranes (Sections 7.5 and 25.3.6) they provide an excellent barrier to water and some contaminated waters. The geomembrane may either be used alone or with soil–bentonite and cement–bentonite backfills to create a composite liner. Typically, HDPE geomembranes are used and, as in the case of landfill liners, the greatest challenge is placing the geomembrane with a minimum of holes and providing good interlock between the sheets of geomembrane such that leakage is minimized (see O'Donnell *et al.* 1995). As in the case of landfill liners, consideration must be given to potential interaction between the "leachate" to be contained and the geomembrane (Section 25.4) as well as the potential for diffusion through the geomembrane (Section 25.5).

25.3.9.3 Sheet Pile Walls

Although there is extensive experience in the use of steel sheet pile walls in conventional geotechnical engineering applications, the use of this technique in geoenvironmental applications has been limited. This is primarily due to concerns regarding: (a) providing an interlock that will adequately control movement of contaminated fluids through the wall; and (b) the service life of the wall, i.e. the potential for corrosion in chemically aggressive environmental applications. As discussed in O'Donnell *et al.* (1995), sealable interlocks have been developed that use a sealant that is chemically compatible with the applicable environmental conditions. Resistance to corrosion can be provided by using: (a) thicker steel sections; or (b) sections with an organic coating, e.g. pitch that has been made more damage resistant than normal pitch by combining with vinyl or epoxy resins.

Field data on leakage across sheet pile walls are limited. Based on limited data, O'Donnell *et al.* (1995) suggest a permittivity, Ψ, similar to that of a 0.6-m thick soil–bentonite wall (for $k \sim 10^{-9}$ m s^{-1}; $\Psi \sim 1.7 \times 10^{-9}$ s^{-1}) with values in the range 6×10^{-9}–1×10^{-9} s^{-1} being reported for walls sealed with an organic polymer sealant. Due to the paucity of field experience, there is a need for long-term, large-scale field monitoring to provide better data regarding the long-term effectiveness of sheet piles as a contaminant barrier. Ideal applications for sheet pile walls may include: (a) reducing leakage from reservoirs; (b) reducing volumes of fluid that need to be pumped in hydraulic containment applications, especially when the sheet piles can be driven to a low hydraulic conductivity unit below the more permeable unit to be isolated; (c) allowing construction of reactive walls (see Section 29.5); and (d) temporary support during the excavation of contaminated soils.

25.4 Liner Compatibility with Leachate

There is potential for interaction between contaminated water or other liquids, e.g. hydrocarbons, when they come into contact with and migrate through all types of liners. The effects, if any, will depend on factors such as the chemical characteristics of the fluid, the characteristics of the liner, period of exposure, applied stress and temperature.

25.4.1 CLAY–PERMEANT INTERACTION

Clay minerals are the primary contributor to the low hydraulic conductivity of natural geologic barriers (Section 25.3.1), CCLs (Section 25.3.2), soil–bentonite liners (Section 25.3.3), GCLs (Section 25.3.4), soil–bentonite vertical cut-off walls (Section 25.3.9) and they contribute significantly to the performance of composite liner systems (Sections 25.3.7 and 25.3.8). Clay mineralogy is discussed in Chapter 23. The potential for interaction needs to be understood in terms of the potential for: (a) a change in crystal lattice volume, e.g. K^+ or NH_4^+ fixation by vermiculates can cause a significant decrease in crystal lattice volume and cation exchange capac-

ity (see Rowe et al. 1995b); and (b) double layer expansion or contraction. As indicated in Fig. 25.13, typically interaction with a clay double layer is due to the effect of ions in the permeant fluid (see Fig. 25.13a; Section 23.4 and 23.5) and due to the effect of organic contaminant on the dielectric constant, ε, of the pore fluid (see Fig. 25.13b, Section 23.4 and Rowe et al. 1995b).

Rowe et al. (1995b) indicate that most inactive soils whose minerals consist of illites and chlorites are relatively insensitive to typical municipal solid waste leachate. They may in fact experience a decrease in hydraulic conductivity (Griffin et al. 1976) due to Na^+ adsorption (accompanied by desorption of Ca^{2+} and Mg^{2+}) and the consequent double layer expansion (see Fig. 25.13a) coupled with possible bacterial clogging. In contrast, soils containing significant amounts of swelling minerals (e.g. vermiculite, montmorillonite, interlayered illite–smectite; see Rowe et al. 1995b) may experience c-axis contraction or expansion with consequent increases or decreases in hydraulic conductivity, respectively. For example, K^+ fixation by vermiculite may cause a 28% decrease in crystal volume as well as a contraction in double layer

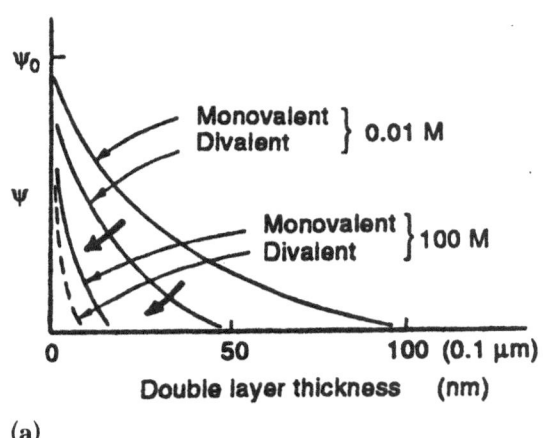

(a)

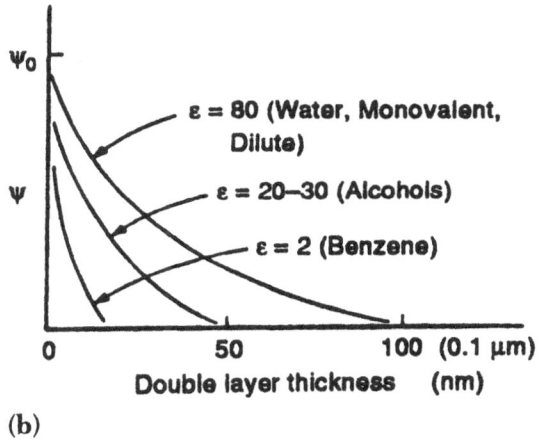

(b)

FIGURE 25.13. Double layer thickness variations: (a) concentration and valency effects; and (b) dielectric effects (modified from Rowe et al. 1995b).

thickness (Rowe *et al.* 1995b). Sodium bentonite (composed primarily of sodium montmorillonite) may experience a considerable increase in hydraulic conductivity when permeated with a solution containing Ca^{2+} or Mg^{2+} or even a solution with a high concentration of Na^+ due to the effect of valence and concentration on double layer thickness.

Organic compounds with a low dielectric constant have been shown (e.g. Fernandez & Quigley 1985; Madsen & Mitchell 1988) to cause double layer contraction and higher hydraulic conductivity through macropores in the flocculated clay. When dealing with organic contaminants, an additional complicating factor is the effect of the organic compound on the viscosity of the permeating fluid, which may cause a decrease in hydraulic conductivity at some concentrations even though the intrinsic permeability remains the same (see Section 24.1.3 for definitions of these terms and see Rowe *et al.* 1995b for an extensive discussion of this effect). At higher concentrations of the organic compound, both the hydraulic conductivity and intrinsic permeability may increase.

The effects of clay–permeant interaction can be reduced by the presence of an applied effective stress, especially if the stress was applied prior to permeation (e.g. see Foreman & Daniel 1986; Fernandez & Quigley 1988; Rowe *et al.* 1995b; Petrov *et al.* 1997a,b; Petrov & Rowe 1997). This reduced effect is due to consolidation, which decreases the void ratio and the size of the macropores.

All clays have some potential for a change in hydraulic conductivity when permeated. While some clays, e.g. those rich in smectite, vermiculite or interlayered illite–smectite, have greater potential for interaction than others, e.g. illite, the only way to assess whether there is a potential problem is to examine the effect of the particular contaminant of interest on the proposed liner material.

Techniques for assessing potential clay–permeant interaction may include both direct and inferential (index) tests as described by Rowe *et al.* (1995b). Direct methods involve permeating a sample of the proposed liner material with the permeant of interest and monitoring the change in hydraulic conductivity. Both flexible wall (fixed gradient) and rigid wall (fixed flow) permeameters may be used and each has advantages and disadvantages (e.g. see Daniel *et al.* 1985). To properly assess the potential for interaction, the permeant should be passed through the clay until chemical equilibrium is reached (or nearly reached). However, this may require many pore volumes of permeation and for a low hydraulic conductivity soil the tests may take an excessively long time using flexible wall permeameters where limitations of the applied gradient result in a relatively small Darcy flux. Small-scale fixed-ring permeameters operated with a constant flow (e.g. Fernandez & Quigley 1985) allow higher gradients for low permeability soils and hence shorter testing times. However, concern might be raised regarding the potential for side wall leakage and the effect of seepage induced stresses when testing samples at low applied stresses, which are applicable to some liner applications (e.g. ponds, lagoons, etc.). Petrov *et al.* (1997a) compared both techniques when examining clay–leachate interaction for GCLs and found that similar results could be obtained by both approaches.

Indirect approaches of assessing potential clay–leachate compatibility problems may involve an examination of the effects of the liquid of interest or the Atterberg limits of the clay (e.g. Acar & Seals 1984; Daniel *et al.* 1985). Another approach is to examine the effect of mixing the fluid and clay on the mineralogy of the clay (see Rowe *et al.* 1995b). However, while these techniques may be a useful screening tool, they should still be coupled with an appropriate hydraulic conductivity testing program (with appropriate controls) if there is any concern about possible interaction.

25.4.2 GCL–PERMEANT INTERACTION

Rowe (1998a) provided a detailed review of GCL hydraulic conductivity testing. It has been found that the hydraulic conductivity of a GCL is highly dependent on the hydrating conditions and the applied effective stress during permeation. These factors combined with the type of GCL manufacturing process used, water content prior to hydration and mass of bentonite all significantly influence the GCL thickness and while there is some correlation between hydraulic conductivity, k, and thickness, H, there is a great deal of scatter. Petrov et al. (1997b) showed that by plotting the bulk void ratio, e_B, rather than thickness, H, against hydraulic conductivity, k, one obtains a much better correlation and much less scatter of the data for a given permeant. This is consistent with the conventional geotechnical engineering concept of a strong correlation between the void ratio and hydraulic conductivity.

The hydraulic conductivity to water of a number of commercially available GCLs ranges between about 5×10^{-11} m s^{-1} at "low" (3–4 kPa) confining stress to 1×10^{-11} m s^{-1} at "intermediate" (34–38 kPa) confining stress, and 7×10^{-12} m s^{-1} at "high" (109–117 kPa) confining stress. It is of interest to note that at a confining stress of about 35 kPa, the six products tested (see Rowe 1998a) had a very similar hydraulic conductivity to water (7×10^{-12}–1×10^{-11} m s^{-1}).

TABLE 25.9. Effect of permeating fluid on hydraulic conductivity of a particular GCL product at 35 kPa (adapted from Ruhl & Daniel 1997)

Permeating fluid	Hydraulic conductivity (m s^{-1})
Tap water	7×10^{-12}
Real leachate	$<1 \times 10^{-12}$
Synthetic leachate	1×10^{-11}
High strength synthetic leachate	2×10^{-8}

Table 25.8 shows the change in hydraulic conductivity of a GCL with the concentration of NaCl and ethanol (note that the decrease in hydraulic conductivity for a 25 and 50% ethanol–water mix is due to decreased viscosity; see Petrov et al. 1997b). Table 25.9 summarizes changes in hydraulic conductivity of a GCL permeated with different real and simulated leachates. The hydraulic conductivity with respect to real leachate obtained by Ruhl & Daniel (1997) (Table 25.9) was lower than that obtained with tap water. Grant et al. (1997) also reported low GCL hydraulic conductivity values for tests with a real municipal solid waste leachate. It can be seen that in general the changes in hydraulic conductivity for modest strength leachates (Table 25.9), salt solutions (<0.6 N; Table 25.8) or ethanol–water mix ($<25\%$ ethanol; Table 25.8) are all modest

TABLE 25.8. Effect of concentration of permeating NaCl solution and ethanol–water mix on hydraulic conductivity of a particular GCL product at 33–36 kPa

Concentration NaCl (N)	Hydraulic conductivity (m s^{-1})[a]	Proportion ethanol % (by mass)[b]	Hydraulic conductivity (m s^{-1})[b]
0	1.3×10^{-11}	0	1.6×10^{-11}
0.1	2.0×10^{-11}	25	7.3×10^{-12}
$\cong 0.13$[c]	7.3×10^{-11}	50	6.0×10^{-12}
0.6	9.3×10^{-11}	75	4.1×10^{-11}
2	4.7×10^{-10}	100	2.0×10^{-9}

[a] Data from Petrov & Rowe (1997).
[b] Data from Petrov et al. (1997b).
[c] Synthetic (simulated Keele Valley landfill) leachate without bacteria.

and the hydraulic conductivity is still low ($<10^{-10}\,\text{m s}^{-1}$). However, a high concentration permeant can result in very significant changes in hydraulic conductivity.

The confining stress at the time of hydration and the hydrating fluid can also have an effect on the final hydraulic conductivity. For example, Fig. 25.14 shows an increase in the hydraulic conductivity of a GCL when permeated with synthetic leachate. At a low confining stress (3–4 kPa) the increase is by almost an order-of-magnitude. At an intermediate confining stress (30–35 kPa) the increase is by a factor of six. The increases in hydraulic conductivity evident in Fig. 25.14 need not be a problem if the liner system has been designed based on the higher values, which reflect interaction with leachate.

Daniel *et al.* (1993) examined the effect of partial hydration on the hydraulic conductivity of a GCL. They found that when the partially saturated GCL was permeated with concentrated hydrocarbons, the hydraulic conductivity was high (see Table 25.10) for a low initial water content (17 and 50%). It was close to that of water (except for trichloroethylene) at a water content of 100%, and less than that due to water at higher water contents. Thus it was not necessary for the GCL to be fully saturated to have a low hydraulic conductivity.

25.4.3 GEOMEMBRANE–PERMEANT INTERACTION

Geomembranes are used to "contain" fluids because of their very high resistance to fluid

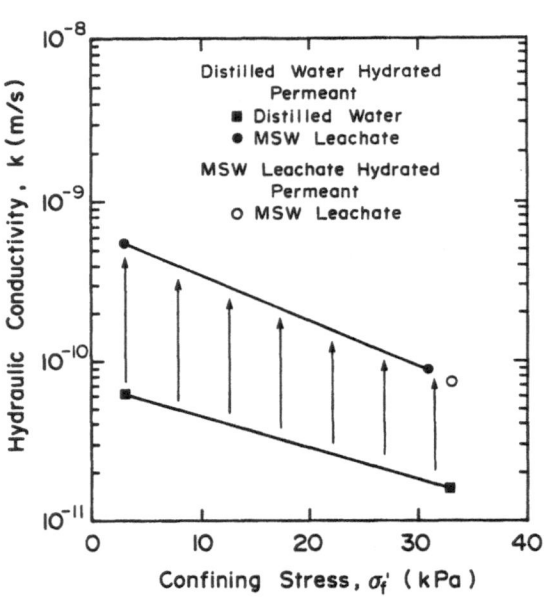

FIGURE 25.14. Hydraulic conductivity versus static confining stress for permeation of synthetic municipal solid waste (MSW) (simulated Keele Valley landfill) leachate (modified from Petrov & Rowe 1997). Distilled water-hydrated, permeant: (■) distilled water; (●) MSW leachate. MSW leachate-hydrated, permeant: ○ MSW leachate.

flow. However, it is essential that the geomembrane have sufficient compatibility with the fluid to maintain this resistance for the service life of the facility.

Based on extensive testing, general chemical resistance guidelines have been developed for a number of geomembranes, e.g.

TABLE 25.10. Summary of results of hydraulic conductivity tests on partially saturated bentonite (adapted from Daniel *et al.* 1993)

Permeating liquid, w_o	Hydraulic conductivity (m s^{-1})			
	17%	50%	100%	125%
Benzene	3×10^{-7}	2×10^{-7}	5×10^{-11}	No flow
Gasoline	4×10^{-7}	4×10^{-7}	4×10^{-11}	No flow
Methanol	3×10^{-7}	3×10^{-7}	3×10^{-11}	No flow
Trichloroethylene	4×10^{-5}	4×10^{-7}	3×10^{-10}	No flow
Water	$1\text{–}2 \times 10^{-11}$	—	—	—

TABLE 25.11. Chemical resistance guidelines for some commonly used geomembranes at 38 °C (adapted from Koerner 1998)

Chemical	Geomembrane type[a,b]		
	HDPE	PVC	CSPE
General			
Aliphatic hydrocarbons	✓		
Aromatic hydrocarbons	✓		
Chlorinated solvents	✓		
Oxygenated solvents	✓		
Crude petroleum solvents	✓		
Alcohols	✓	✓	
Acids			
Organic	✓	✓	✓
Inorganic	✓	✓	✓
Heavy metals	✓	✓	✓
Salts	✓	✓	✓

[a] HDPE = high density polyethylene, PVC = polyvinyl chloride, CSPE = ethylene interpolymer alloy.
[b] ✓ = generally good resistance.

see Table 25.11. However, for materials not listed (e.g. very flexible polyethylene, polypropylene, etc.) or for chemical solutions and mixtures not listed or for all critical projects where geomembrane–chemical interaction could have significant consequences, it is necessary to perform chemical resistance tests on the geomembrane of interest using the fluid of interest. Typically (e.g. see Koerner 1998), this involves immersing samples ("coupons") of the geomembrane in the fluid (using an appropriate sealed container and temperature) and removing coupons for testing at 30, 60, 90 and 120 days according to ASTM D5322 and D5496. The coupons are then tested to identify changes in physical, mechanical and possibly diffusion characteristics. Procedures such as in ASTM D5747 specify the test methods to be used.

At present, there are no well-established rules for assessing what is an acceptable variation from other original geomembrane properties after immersion in the chemical of interest. Koerner (1998) has suggested that a geomembrane is "resistant" to chemical interaction if the properties listed in the first column of Table 25.12 vary from the original

TABLE 25.12. Suggested limits of different test values for incubated geomembranes

Thermoset and thermoplastic polymers, e.g. PVC, CSPE (after Little 1985)[a]	
Permeation rate, g m^{-2} h^{-1}	<0.9
Change in	
Weight, %	<10
Volume, %	<10
Tensile strength, %	<20
Elongation at break, %	<30
100 or 200% modulus, %	<30
Hardness, points	<10
Semicrystalline polymers, e.g. HDPE (after Koerner 1998)	
Permeation rate, g m^{-2} h^{-1}	<0.9
Change in	
Weight, %	<2
Volume, %	<1
Yield strength, %	<20
Yield elongation, %	<30
Modulus, %	<30
Tear strength, %	<20
Puncture strength, %	<30

[a] For low crystallinity polymers (e.g. very flexible polyethylene, flexible polypropylene) Koerner (1998) recommends that the properties should be the same as those for thermoset and thermoplastic polymers and the values for resistance should be slightly more restrictive. PVC = polyvinyl chloride, CSPE = ethylene interpolymer alloy, HDPE = high density polyethylene.

values by less than the values given in the second column of Table 25.12.

25.5 Diffusion through Barriers

As discussed in Section 24.3.1.2, diffusion is a process whereby mass moves from areas of high concentration (potential) to areas of low concentration. This is often modeled using Fick's first law:

$$J = -D_p \frac{dC}{dz} \qquad (25.21)$$

where J = the mass flux at a point (mass per unit area per unit time), D_p = a bulk diffusion coefficient (also called the porous media diffusion coefficient for porous media as discussed in Section 24.3.1.2), and dC/dz = the concentration gradient at that point. Diffusion may occur in gas, liquids and solid phases, and all three are of potential interest to the geoenvironmental engineer.

In general terms, the diffusion coefficient in a gaseous phase is many orders-of-magnitude higher than in water; which, in turn, is orders-of-magnitude higher than in a solid. For example, the bulk diffusion coefficient, D_p, for the volatile organic compound dichloromethane is of the order of 10^{-5} m^2 s^{-1} in air, 10^{-10} m^2 s^{-1} in the porewater of a CCL, and 10^{-12} m^2 s^{-1} in an HDPE geomembrane. Thus in the liner system shown in Fig. 25.1 contaminant migration through an intact composite liner will involve:

- diffusion through the geomembrane (controlled by a low diffusion coefficient but also a potentially high concentration gradient due to the small 1.5–2.5-mm thickness of typical geomembrane liners);
- diffusion through the pore structure of an essentially saturated CCL (controlled by a higher diffusion coefficient but also by a smaller gradient than for the geomembrane such that the resistance of the two components of the system to the diffusion of dichloromethane (DCM) is of a similar order-of-magnitude); and

- diffusion predominantly through the gas-filled pores of the granular secondary leachate collection (SLCS) drainage layer (controlled by diffusion coefficient in air and the volumetric gas content, θ_a, for a drainage layer that is kept pumped and at a low degree of saturation).

The process is then repeated through the secondary liner.

25.5.1 DISSOLVED-PHASE DIFFUSION THROUGH POROUS MEDIA

Diffusion of dissolved contaminants through a porous medium will occur through the water-filled pores of the medium and so it is convenient to separate out the effect of porosity (or volumetric water content) from the effective diffusion coefficient such that for a saturated soil Eq. 25.21 becomes:

$$J = -n_e D_e \frac{dC}{dz} \qquad (25.22a)$$

or more generally for an unsaturated soil (see Section 25.5.1.3 for more discussion):

$$J = -\theta_e D_{e\theta} \frac{dC}{dz} \qquad (25.22b)$$

where n_e = the effective porosity of a saturated soil; D_e = the effective diffusion coefficient; θ_e = the effective volumetric water content ($\theta_e = n_e$ for $S = 100\%$); $D_{e\theta}$ = the effective diffusion coefficient for an unsaturated soil at a volumetric water content, θ ($D_{e\theta} = D_e$ for $S = 100\%$).

Several mechanisms influence the "diffusive" movement of ions or molecules through a soil. These include diffusion itself, electrical flow (e.g. where the rate of movement of a cation is influenced by rate of movement of an anion in order to maintain electrical neutrality and vice versa), and osmotic flow (where there is water flow from the area of low contaminant concentration to the area of high contaminant concentration). Thus when an effective diffusion coefficient is measured by typical laboratory means (e.g. see Rowe *et al.* 1988; 1995b), it is in fact a bulk mass transfer coefficient that includes all of these effects. Fortunately, experience has shown that parameters evaluated in this way do provide a good prediction of observed field behavior (Rowe *et al.* 1995b; 1998b).

Various texts (e.g. Bear 1979) attempt to relate the diffusion coefficient in soil to the diffusion coefficient in free solution, D_o (Section 24.3.1.2) by a geometric "tortuosity" factor to account for the more tortuous path the contaminant must take in soil. This may have some justification in sands, but fails to account for a number of other important factors such as the decreased fluidity of adsorbed double layer water, electrostatic interaction, etc. which influence the effective diffusion coefficient through clayey soils. Since these effects may vary from soil to soil and contaminant to contaminant, the ratio (D_e/D_o) of the effective diffusion coefficient to free solution coefficient will vary from soil to soil and even from one contaminant to another in the same soil depending on the contaminant being considered, the soil mineralogy and porewater chemistry, and the leachate composition (see Rowe *et al.* 1995b).

In addition to being influenced by soil mineralogy, porewater chemistry, soil fabric and leachate chemistry, the effective diffusion coefficient will also depend on temperature. Although corrections can be made to try to adjust for these factors they fail to account for interaction effects. Hence, if a diffusion coefficient is important for a particular project, it is best to obtain an experimental value using the proposed soil, a leachate composition as close as practical to that expected and at a temperature as close as practical to the expected field temperature. Fortunately, despite all the complicating factors involved, the diffusion coefficient varies over a relatively narrow range (relative to other parameters such as hydraulic conductivity) and provided a reasonable attempt is made to pick appro-

priate parameters, quite good predictions of field behavior can be achieved (see Rowe *et al.* 1995b).

For many soils and contaminants, the effective porosity, n_e, will be very close to the total porosity, n, (e.g. Rowe *et al.* 1988; 1995b). However, there are circumstances where the effective porosity may be less than the total porosity (e.g. due to anion exclusion in some soils). The effective porosity may vary from one contaminant to another (e.g. the effective porosity for chloride may not be the same as for dichloromethane) and may not be the same as the effective porosity used to calculate average linearized groundwater velocity when considering advective contaminant transport (Section 24.1.4 and Eq. 24.6). While this is an unfortunate complication, in practical terms generally either advection dominates (in which case n_e can be taken to be the effective porosity with respect to advection) or diffusion dominates (in which case n_e can be taken to be the effective porosity with respect to diffusion).

When dealing with systems where diffusive transport is significant (typical of liners that are performing their design function of controlling advective transport to an advective flux of 0.03 m a^{-1} or less; see Rowe *et al.* 1995b), both the effective porosity and diffusion coefficient can be established from appropriate laboratory tests. In situations where one is considering diffusive transport through liner systems it will also, generally, be necessary to consider the effect of biodegradation and geochemical processes, which will also influence the dissolved concentrations, e.g. see Section 24.3.3.

25.5.1.1 Diffusion through Intact and Compacted Clays
There is now a body of field evidence showing the importance of diffusion in low permeability materials that range over a geological time-scale, i.e. over 10 000–15 000 years since the last Ice Age, and over a period of

TABLE 25.13. Range of effective diffusion coefficients for selected contaminants in low activity natural geologic barrier or compacted clay (porosity = 0.31–0.39)

	Effective diffusion coefficient, D_e[a]	
	$(m^2\ s^{-1})$	$(m^2\ a^{-1})$
Chloride[b,c]	$2–8 \times 10^{-10}$	0.006–0.25
Potassium[b]	$5–7.5 \times 10^{-10}$	0.016–0.024
Sodium[b,c]	$2–4.8 \times 10^{-10}$	0.006–0.015
Zinc[c]	$3.7–7 \times 10^{-10}$	0.012–0.022
Acetone[c]	5.6×10^{-10}	0.018
Aniline[c]	6.8×10^{-10}	0.021
Benzene[c]	$2.5–3.6 \times 10^{-10}$	0.008–0.011
Chloroform[c]	7×10^{-10}	0.022
Dichloromethane[c]	$2.5–8.5 \times 10^{-10}$	0.008–0.027
1,4 Dioxane[c]	4×10^{-10}	0.013
Toluene[c]	$0.6–3 \times 10^{-10}$	0.002–0.01
Trichloroethene[c]	$2.5–3.5 \times 10^{-10}$	0.008–0.011

[a] Data from Barone *et al.* 1989; Barone 1990; Rowe *et al.* 1988; Myrand *et al.* 1987; Karickhoff *et al.* 1979; Rowe & Barone 1991; McKay & Trudell 1989) and unpublished reports.
[b] Some values obtained over the temperature range 7–10 °C.
[c] Some values obtained over the temperature range 20–23 °C.

about 25 years since contaminant migration has been monitored beneath landfills with a low permeability liner (see Rowe *et al.* 1995b). Typical field and laboratory diffusion coefficients for chloride are given in Table 24.3 and vary from about 2×10^{-10} to $8 \times 10^{-10}\ m^2\ s^{-1}$ depending on soil, leachate and temperature. Table 25.13 gives the reported effective diffusion coefficients for a number of selected contaminant diffusions through low activity clays. Values were obtained over the two temperature ranges as noted. While the diffusion coefficient increases with increasing temperature (all other things being equal), other factors such as leachate chemistry often have had a greater effect than temperature. For example, the values reported for sodium include two of $2 \times 10^{-10}\ m^2\ s^{-1}$ at 10 °C. Based on the influence of temperature on the free solution diffusion coefficient (see Rowe *et al.* 1995b), a value of 2×10^{-10} $m^2\ s^{-1}$ at 10 °C should increase to 2.7×10^{-10} $m^2\ s^{-1}$ at 22 °C. However, other experimental data include one diffusion coefficient of $3.7 \times 10^{-10}\ m^2\ s^{-1}$ at 7 °C for industrial waste

leachate, one at $4.6 \times 10^{-10}\ m^2\ s^{-1}$ at 10 °C for municipal solid waste leachate and one at $4.8 \times 10^{-10}\ m^2\ s^{-1}$ at 22 °C for a simple NaCl solution (see Rowe *et al.* 1995b for details). Thus, it is best to establish parameters for the appropriate leachate and temperature if reliable parameters are needed.

Table 25.14 gives a number of published diffusion coefficients for soil–bentonite liners.

25.5.1.2 Diffusion through GCLs
Similar to the hydraulic conductivity of GCLs, the diffusion coefficient of GCLs depends in large part on the bulk void ratio of the GCL; which, in turn, depends on the applied stress conditions. The diffusion coefficient can be estimated from relatively simple tests (see Rowe *et al.* 1997a, 1998a) by monitoring the mass transfer across the GCL. This mass transfer is primarily controlled by the porous media diffusion coefficient, D_p, i.e. $D_p = n_e D_e$.

Lake & Rowe (2000b) examined the diffusion of inorganic contaminants through a GCL and found a linear relationship between the porous media diffusion coefficient, D_p,

TABLE 25.14. Some effective diffusion coefficients in soil–bentonite liners (22 °C)

Bentonite (%)	Porosity, $n(-)$	Contaminant	Diffusion coefficient	
			$(m^2\ s^{-1})$	$(m^2\ a^{-1})$
0	0.3–0.33	Bromide[a]	$10\text{–}11 \times 10^{-10}$	0.033–0.035
		Trichloroethene[a]	$6.6\text{–}7.3 \times 10^{-10}$	0.021–0.023
4	0.37	Bromide[a]	6×10^{-10}	0.02
		Trichloroethene[a]	4.4×10^{-10}	0.014
5	0.4	Chloride	7.9×10^{-10}	0.025

[a] Data from Gullick (1998).

and the bulk GCL void ratio, e_B. For a simple 3–5 g L^{-1} NaCl solution, the relationship between D_p and e_B for chloride ($1 \le e_B \le 3.5$) was given by:

$$D_p \cong (1.02e_B - 0.89) \times 10^{-10}(m^2\ s^{-1}). \quad (25.23)$$

It should be noted that the diffusion coefficient, e.g. as given by Eq. 25.23, is dependent on the chemical species, the void ratio and the other chemical constituents in the leachate, and the background chemistry of the soil below the GCL. Thus, Lake & Rowe (2000b) showed that the chloride porous media diffusion coefficient obtained at a given void ratio increased from $D_p \cong 1 \times 10^{-10}$ m^2 s^{-1} for a 4.6 g L^{-1} NaCl solution to $D_p \cong 3 \times 10^{-10}$ m^2 s^{-1} for a 114 g L^{-1} brine solution. For a given permeant, the relationship between bulk GCL void ratio and diffusion coefficient was relatively independent of the test type (constant volume or constant stress) or hydrating conditions (i.e. pre- or post-confinement), or method of GCL manufacture (at least for the three GCLs examined, see Table 25.6). However, the relationship between applied stress and bulk GCL void ratio did depend on the type of manufacturing process and hydrating conditions (see Section 25.3.4).

The relationship between applied stress and diffusion coefficient will vary depending on the GCL manufacturing process, fluid used to hydrate the GCL, and the leachate characteristics. That said, a preliminary estimate of the diffusion coefficient for chloride through the three GCLs listed in Table 25.6 can be obtained using Eq. 25.18 (depending on the GCL) and Eq. 25.23 (chloride), recognizing that these results were obtained for GCLs hydrated with distilled water under confined conditions and using a 3–5 gl^{-1} NaCl solution as the source "leachate." Equations such as these may be useful in preliminary design calculations; however, if these calculations indicate that the diffusion coefficient is a key parameter, then the diffusion coefficient should be measured directly from a constant stress diffusion test using the proposed GCL, hydrating conditions and leachate (see Rowe et al. 1998a).

Tests have also been conducted to examine sorption and diffusion of organic contaminants through GCLs and it has been found (Rowe & Lake 1999) that:

1. The sorption of organic compounds, e.g. volatile organic compounds (VOCs), in conventional GCLs is low. However, this sorption can be substantially increased by combining activated carbon with the bentonite.
2. The diffusion coefficients for organic contaminants through GCLs are of a similar order-of-magnitude to that for inorganic contaminants at a given bulk GCL void ratio.
3. Diffusion tests for volatile organic compounds are much more difficult than those for inorganic contaminants.

Diffusion can be an important consideration with respect to contaminant impact of well-

TABLE 25.15. Diffusion coefficient in two GCLs as inferred by Lo (1992)

Liner material	Effective diffusion coefficient, D_e (m^2 s^{-1})		
	Chloride	Lead	1,2 dichlorobenzene
Normal GCL	2.4×10^{-10}	5.9×10^{-10}	9.8×10^{-11}
Organo-clay	4.9×10^{-10}	9.0×10^{-10}	1.5×10^{-10}

designed and operated landfill facilities (see Rowe *et al.* 1995b). It can also impact on laboratory hydraulic conductivity tests where diffusion of chloride ahead of the advective contaminant front may be misinterpreted as indicating a macrostructure and preferential flow paths.

The observation that there will be an "early" diffusive movement of contaminant ahead of the advective front in hydraulic conductivity tests can be used to obtain an estimate of diffusion characteristics. For example, Lo (1992) used an early version of the program POLLUTE (Rowe & Booker 1997a) to fit the concentration profile in the effluent for tests performed using the GCL "CL" and obtained diffusion coefficients as summarized in Table 25.15. It was inferred that the diffusion coefficient for lead was greater than that for chloride, while the diffusion coefficient for 1,2 dichlorobenzene was reported to be less than for chloride. Diffusion coefficients for the modified organo-clay were reported to be up to 50% greater than for conventional bentonite.

Compared to conventional bentonite, the organo-clay tested by Lo (1992) provided much less "sorption capacity" for lead but much more sorption for 1,2 dichlorobenzene.

25.5.1.3 Diffusion through Unsaturated Soil

Suitable natural clay liners will usually be saturated. Good CCLs may have a degree of saturation between 87 and 100% and often between 95 and 100% (see Section 25.3.2). If the liner is loaded, e.g. due to placement of waste over the liner, then it may be expected that due to consolidation the degree of saturation will increase. For these soils, the effective diffusion coefficient is essentially the same as the saturated diffusion coefficient discussed in Section 25.5.1.1. However, if necessary, the diffusion through these liners may be modeled using Eq. 25.22b and a value of $D_{e\theta}$ as discussed below.

Compacted clay and composite liners are often underlain by an unsaturated zone, which may include natural unsaturated silts, sands or gravels or an engineered secondary leachate collection system, e.g. see Fig. 25.1. For these soils the volumetric water content, θ, is less than the porosity. In addition, the effective diffusion may be substantially smaller than for saturated soil (Klute & Letey 1958; Porter *et al.* 1960). Rowe and Badv (1996a,b) examined dissolved phase diffusion through unsaturated silt, sand and fine gravel and found that the diffusion coefficient, $D_{e\theta}$, for the unsaturated soils they examined was given by:

$$D_{e\theta} = \left(\frac{\theta - \theta_{min}}{n - \theta_{min}}\right) D_e \qquad (25.24a)$$

where $D_{e\theta}$ = the effective diffusion coefficient in the unsaturated soil at a volumetric water content, θ; n, D_e = the volumetric water content (total porosity) and effective diffusion coefficient in the same soil under saturated conditions; and θ_{min} = the water content at which there is no interconnected water through which diffusion can occur. For their tests, Rowe & Badv (1996a,b) found that $\theta_{min} \cong 0$ and Eq. 25.24a reduces to:

$$D_{e\theta} = \left(\frac{\theta}{n}\right) D_n \qquad (25.24b)$$

(where $D_e \cong 0.028$–0.033 m^2 a^{-1} in their test on silt, sand and gravel with $n = 0.34$–0.38).

Rowe & Badv (1996a,b; Badv & Rowe 1996) demonstrated that advective–diffusive transport through a CCL underlain by unsaturated silt, sand or gravel could be well-predicted using finite layer theory (Rowe & Booker 1997a). It was also shown that mechanical dispersion (Section 24.3.1.3) was very small for clay, silts, sands and fine gravels at volumetric water contents greater than 0.04 and a Darcy flux less than 6×10^{-10} m s^{-1}, i.e. 0.02 m a^{-1}. However, for a coarse gravel drainage layer (20–45-mm particle size) with a volumetric water content of 0.027–0.035, advective transport and mechanical dispersion was dominant over diffusion through the unsaturated coarse gravel at Darcy fluxes of between 6×10^{-10} and 8×10^{-8} m s^{-1} (0.02–2.5 m a^{-1}). For this case, they found that the coefficient of hydrodynamic dispersion (Section 24.3.1.3), D (m^2 s^{-1}), could be given by:

$$D = D_{e\theta} + \alpha v \text{ (m}^2 \text{ s}^{-1}) \quad (25.25a)$$

where $D_{e\theta}$ (m^2 s^{-1}) is given by Eq. 25.24, α (m) = the dispersivity given by:

$$\alpha = 35200v + 0.013 \text{ (m)} \quad (25.25b)$$

the average linearized groundwater velocity, v (m s^{-1}), is given by:

$$v = v_a/\theta \text{ (m s}^{-1}) \quad (25.25c)$$

where v_a (m s^{-1}) = the Darcy flux through the liner and θ (−) is the volumetric water content of the coarse gravel. Equation 25.25b must be used with the units defined above.

The unsaturated diffusion coefficient as defined above will be important with respect to the transport of non-volatile contaminants. It may also control transport of volatile contaminants when the degree of saturation is sufficiently high that there is no significant interconnected gas-filled void. However, for low degrees of saturation, the migration of volatile organic compounds will be dominated by diffusion in a gaseous phase as discussed in Section 25.5.2.

25.5.2 GASEOUS-PHASE DIFFUSION THROUGH POROUS MEDIA

Unsaturated soils may have continuous gas-filled pore space and under these circumstances the movement of volatile contaminants and other gases of potential environmental concern (e.g. the diffusion of oxygen through a cover and into sulphidic mine tailings as discussed in Chapter 27, and the migration of methane generated by municipal solid waste) will be predominantly through the continuous gas-fill pores. In these cases, the diffusive flux is given (again by Fick's law) as:

$$J = -\theta_a D_\theta \frac{dC}{dz} \quad (25.26)$$

but where θ_a = the effective air porosity; D_θ = the diffusion coefficient of the gas of interest through a porous medium with a volumetric water content, θ (or degree of saturation, S_r); and C = the concentration in the gaseous phase. There will be a relationship between the gaseous phase concentration and the dissolved phase concentration, which is given by Henry's law (see Section 24.3.3.2, Eq. 24.22 and Table 24.5).

The gaseous diffusion coefficient through porous media, D_θ, is often related to the diffusion of gas through air, D_a. van Bavel (1952) found that for dry soil and sand mixtures with porosities between 0.1 and 0.6, $D_\theta/D_a = 0.58$. Many studies have been conducted to examine the variation in the ratio D_θ/D_a with the degree of saturation, and these results are summarized in Fig. 25.15. The results cover a range of materials (sands, silts, sand–bentonite, loams, till and clay) with no particular pattern being noticed due to the material itself. Figure 25.15 shows that at low degrees of saturation the data are fairly consistent and approach van Bavel's ratio of

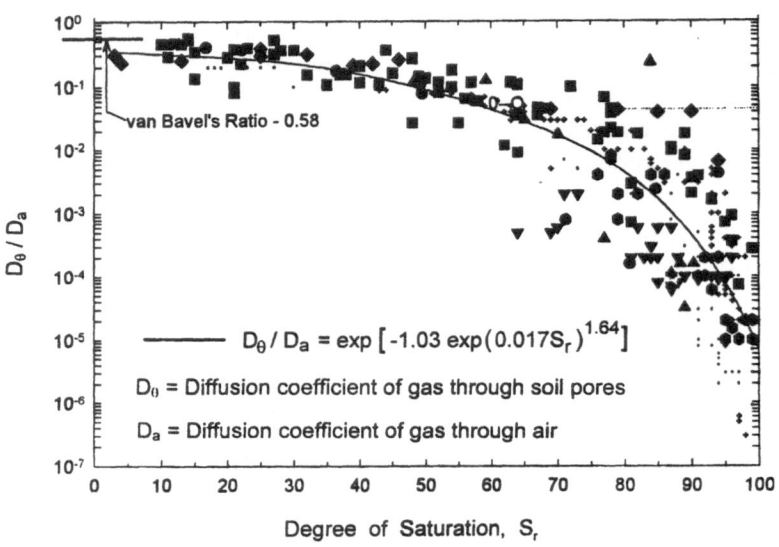

FIGURE 25.15. Results of regression analysis on diffusion–saturation data (modified from MacKay 1997): (—) D_θ/D_a = exp [−1.03 exp (0.017 S_r)$^{1.64}$]; where D_θ and D_a = diffusion coefficient of gas through soil pores and air, respectively.

0.58. As the degree of saturation increases, the value of D_θ/D_a decreases significantly and the scatter increases. The high scatter at high degrees of saturation is likely due to variability in the effective porosity where in two samples with the same degree of saturation and total porosity but different grain distribution may have different amounts of connected gaseous pore space. If one sample has many continuous gaseous pores the value of D_θ will be much larger than for the second sample with the same number of gaseous pores but where the paths between pockets of gaseous pores are blocked by water-filled pores.

Based on the data presented in Fig. 25.15, the "best-fit" relationship between D_θ/D_a and the degree of saturation, S_r, is given by:

$$D_\theta/D_a = \exp[-1.03 \exp(0.017 S_r)^{1.64}]. \quad (25.27)$$

S_r is expressed in percent. It can be seen from Table 25.16 that the diffusion coefficient in air, D_a, and water, D_o, typically differs by about four orders-of-magnitude. As

was the case for diffusion in the dissolved phase, the diffusion coefficient in air increases with increasing temperature. In both cases, the variation is approximately linear with temperature.

The foregoing has focused on the migration of gases through the primary pore structure of the soil. This is likely to be valid where there is no significant secondary porosity, e.g. fractures. However, soil covers may desiccate and crack due to both heating and frost action (Section 25.3). If cracks occur there may be much greater gas migration through the secondary pores (cracks) than through the primary gas-filled pores (intact soil).

25.5.3 DIFFUSION THROUGH GEOMEMBRANES

Water and contaminants dissolved in water or in other solvent can diffuse through a geomembrane. Considering diffusion within the geomembrane itself, Fick's law can be written as shown in Eq. 25.21 where $C = C_g$ is the mass of the penetrant per unit volume of geomem-

TABLE 25.16. Some published diffusion coefficients in air and water (adapted from Thibodeau 1996; Barone *et al.* 1992)

	In air @ $D_a \times 10^5$ (m^2 s^{-1})	T Temperature, (°C)	In water $D_o \times 10^9$ (m^2 s^{-1})	T Temperature, (°C)
Acetic acid	1.33	25	0.88	20
Acetone	1.09	0	1.15	22
Ammonia	2.8	25	1.76	20
Aniline	0.72	25	0.98	22
Benzene	0.88	25	1.15	25[a]
Butyric acid	0.76	25[a]	0.92	25[a]
Carbon dioxide	1.64	25	1.77	20
Chlorine	0.93	0	1.45	25
Chloroform	0.91	0	1.06	22
p-cresol	0.77	25	0.87	25
1,2 dichloroethane	0.90	25[a]	1.07	25[a]
1,4 dichlorobenzene	0.68	25[a]	0.81	25[a]
Dichloromethane	1.04	25[a]	1.26	25[a]
Ethylbenzene	0.75	25[a]	0.91	25[a]
Hydrogen	4.1	25	5.85	25
Methyl ethyl ketone	0.85	25[a]	0.99	25[a]
Nitrogen	1.3	0	2.0	20
Propionic acid	0.95	25[a]	1.01	25[a]
Oxygen	2.06	25	2.35	25
Toluene	0.85	25[a]	0.96	25[a]
Water	2.56	25		
m-xylene	1.23	25[a]	0.79	25[a]

[a] Calculated as per Yaws (1995).

brane. This may be related to the concentration in fluid by a Henry's law relationship:

$$C_g = S_{gf}C_f \qquad (25.28)$$

where S_{gf} = the solubility or partitioning coefficient, or Henry's coefficient; C_f = the equilibrium concentration in the fluid. Thus with the same fluid on either side of the geomembrane (e.g. air or water) Fick's law for mass flux across the geomembrane can be written as:

$$J = -D_g \frac{dC_g}{dz} = -S_{gf}D_g \frac{dC_f}{dz} = -P_g \frac{dC_f}{dz} \qquad (25.29)$$

where P_g = the permeability of the geomembrane (not to be confused with hydraulic conductivity or intrinsic permeability discussed

in Section 24.1.3). Diffusive transport across the geomembrane is described by:

$$\frac{\partial C_f}{\partial t} = D_g \frac{\partial^2 C_f}{\partial z^2}. \qquad (25.30)$$

The diffusion coefficient, D_g, and solubility coefficient, S_{gf}, will depend on the nature of the source fluid and the geomembrane and may be obtained using techniques described in Rowe (1998a). Table 25.17 summarizes published values of S_{gf}, D_g and P_g for a range of contaminants through polyethylene geomembranes.

25.5.3.1 Factors Affecting Diffusion through a Geomembrane

Diffusion coefficient D_g and solubility S_{gf} of a contaminant for a given geomembrane in-

TABLE 25.17. Typical ranges of partitioning coefficient, S_{gf}, and diffusion coefficient, D_g, and permeability, P_g, for polyethylene geomembrane (see Rowe 1998a for full details and sources of data)

Chemical	Solution	S_{gf} (−)	$D_g \times 10^{12}$ (m² s⁻¹)	$P_g \times 10^{15}$ (m² s⁻¹)
Acetic Acid	Aqueous	0.015	0.11–0.29	1.65–4.35
	Pure	0.0086	0.52–0.58	4.47–5.0
Acetone	Aqueous	0.013–0.2	<0.0001	<0.02–0.27
	Pure	0.01–0.012	0.51–0.91	6.1–10.9
Benzene	Aqueous	54.3–57.2	0.037	2010–2120
	Pure	0.08	2.0–4.2	160–340
Carbon tetrachloride	Aqueous			48–57
	Pure	0.18–0.2	0.66–2.4	432
Chlorobenzene	Pure	0.083–0.11	2.2–3.6	200–350
Chloroform	Aqueous			17–25
	Pure	0.134–0.153	1.8–5.9	790
Chloride	Aqueous	0.0008	0.1–0.3	0.08–5
Cyclohexane	Aqueous	2378	0.012	2.8×10^4
	Pure		0.61	
Dichloromethane	Aqueous	1.8–5.6	0.58–2.28	1000–6600
	Pure	0.06	2–10	294
1,1 Dichloroethane	Aqueous			1000–2500
1,2 Dichloroethane	Aqueous	7.2	3.6–6.8	3000–6000
Ethylbenzene	Pure	0.1	2.8	280
Heptane	Pure	0.065–0.066	1.52–1.74	100–115
n-Hexane	Pure	?	2.08–3.6	
Methyl ethyl ketone	Aqueous	>0.025		300–800
	Pure	0.018–0.019	0.75–0.86	13–16
n-Octane	Pure	0.08	1.9	150
Propanoic acid	Pure	0.021	0.3–0.32	6.3–6.7
Tetrachloroethane	Aqueous			7800–8700
Tetrachloroethene	Pure	0.19–0.22	3.8	720
Toluene	Aqueous	63.5–192	0.2–0.56	22 000–98 000
	Pure	0.08–0.09	0.18–4.4	230–490
Trichloroethene	Aqueous	94–189	0.2–0.76	38 000–70 000
	Pure	0.11–0.2	7.7–85	1320–1810
m-xylene	Aqueous	193–370	0.31–0.36	60 000–112 000
	Pure	0.093	1.5–3.7	340
o-xylene	Aqueous	422		
	Pure		0.94	
p-xylene	Aqueous	387		
	Pure		1.6	
Xylene	Aqueous	499–556	0.2–1.8	110 000–500 000
	Pure	0.083–0.095	4.7	390
Water		<0.0003–0.001	0.29–9.0	0.1–0.9

crease with temperature. The product of these two parameters, the permeability ($P_g = S_{gf}D_g$) is highly dependent on the similarity of the penetrant and polymer. Based on research by August & Tatzky (1984) and Rowe *et al.* (1995a, 1996b), the permeability through HDPE geomembranes has the following order:

ions (Cl$^-$, Na$^+$, Zn^{2+}, Ni^{2+},

$$Mn^{2+}, \ Cu^{2+}, \ Cd^{2+}, \ Pb^{2+})$$

and alcohols < acids < nitroderivatives

< aldehydes < ketones

< esters < ethers

< hydrocarbons.

Chlorinated hydrocarbons (e.g. dichloromethane; 1,1 dichloroethane; 1,2 dichloroethane) can diffuse through an HDPE geomembrane and be detected in the porewater on the other side in a matter of days (Rowe *et al.* 1995a, 1996b), whereas the diffusion of salts (Na$^+$, Cl$^-$, and heavy metal salts) was negligible in tests conducted over five years at room temperature.

The "permeability" P_g also depends on the type of geomembrane and decreases with increasing crystalinity. For example, Park *et al.* (1995) reported that the permeability ($P_g = S_{gf}D_g$) of xylene in very low density polyethylene (VLDPE) was almost twice that through HDPE.

25.5.3.2 Diffusion of Water through HDPE Geomembrane

Eloy-Giorni *et al.* (1996) performed a detailed study of water movement through in-

tact geomembranes and concluded that the concept of hydraulic conductivity (as defined by Darcy's law) is not appropriate for describing water transport (and by inference the advective transport of contaminants) through hydrophobic geomembranes. In particular, they demonstrated that a difference in pressure head across a geomembrane of up to 200 m had no significant effect on the movement of water molecules across a 1.7-mm thick HDPE geomembrane ($\rho_g = 940 \ kg \ m^{-3}$). Likewise, a head difference of up to 400 m had no significant effect on water movement across a 1- and 1.6-mm thick, t_{GM}, PVC geomembrane ($\rho_g = 1260 \ kg \ m^{-3}$). As discussed by Rowe (1998a), one can expect negligible diffusion of water through intact geomembranes used as fluid-retaining structures. This is in part due to the very low "permeability" $P_g = S_{gf}D_g$ as evident from the parameters given in Table 25.18 and in part due to the fact that in most cases the relative humidity is very high on both sides of the geomembrane.

25.6 Liner Temperature

The temperature of the liner can have a profound effect on: (a) the service life of the liner; (b) hydraulic conductivity; and (c) diffusion through the liner. In landfill applications, the temperature of the liner may be related to the level of leachate mounding above the liner (see Fig. 25.16) as discussed by Rowe (1998a).

The service life of geomembrane liners will depend, in part, on the rate of antioxidant depletion (see Hsuan & Koerner 1995b;

TABLE 25.18. Solubility and diffusion coefficient for water in three types of geomembrane (adapted from Eloy-Giorni *et al.* 1996)

	S_{gf} (−)	D_g (m^2 s^{-1})	ρ_g (kg m^{-3})	t_{GM} (mm)
HDPE	8×10^{-4}	2.9×10^{-13}	940	1.7
PVC	7×10^{-2}	4.4×10^{-13}	1260	1.0, 1.6
Bituminous	9×10^{-3}	8×10^{-13}	1150	5

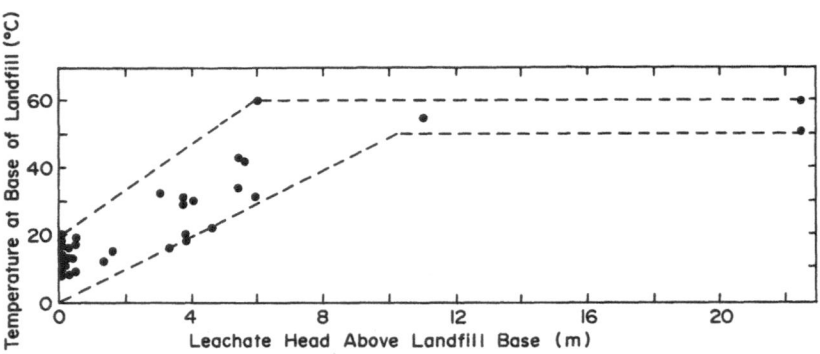

FIGURE 25.16. Variation in temperature at landfill base with leachate head for a number of landfills (modified from Barone *et al.* 1997).

TABLE 25.19. Calculated time for antioxidant depletion based on test data by Hsuan and Koerner (1995b) (see Rowe 1998a for details)

Temperature (°C)	Series I[a] (years)	Series III[b] (years)
10	111	453
15	98	299
20	59	200
25	44	135
30	33	93
35	25	65
40	19	45
45	14	32
50	11	23
55	9	17
60	7	12

[a] Series I: Moving water on both surfaces of geomembrane.
[b] Series III: Stationary water on only one surface.

TABLE 25.20. Effect of temperature on diffusion coefficient, D_T, hydraulic conductivity, k_T, in a liner at temperature, T, relative to values at 10 °C (adapted from Rowe 1998a)

Temperature (°C)	D_T/D_{10}	k_T/k_{10}
10	1.0	1.0
20	1.4	1.3
25	1.6	1.5
35	2.0	1.8
50	2.7	2.4
65	3.5	2.9

Rowe 1998a), and Table 25.19 illustrates how the projected time for antioxidant depletion changes with temperature and the potential for water moving past the surface to increase the leaching of antioxidants. Temperature also has the potential to decrease the service life of clay liners if it is sufficient to cause desiccation of the liner (see Rowe 1998a).

Temperature can influence the hydraulic conductivity and diffusion coefficients, as shown in Table 25.20, which gives the ratio of the values at temperature, T, to that at 10

°C. Domenico and Schwartz (1998) indicate that the groundwater temperature at a depth of 10–20 m is typically 1–2 °C higher than local mean temperature and hence a temperature of 5–10 °C would be typical groundwater temperature in the northern USA and southern Canada. From Table 25.20, it can be seen that diffusive and advective transport are 40 and 30%, respectively, higher at 20 °C than at 10 °C, and 100 and 80%, respectively, higher at 35 °C. If the liner temperature is to remain relatively constant, then constant diffusion and hydraulic conductivity parameters relevant to that temperature may be used. However, if the temperature of the liner changes with time, e.g. due to leachate mounding (see Fig. 25.16), then some consideration may need to be given to the effect of the change in these parameters with temperature on the impact calculations. The approach proposed by Rowe & Booker (1997b) readily models change in D and the effect of a change in k with time as appropriate.

25.7 Contaminant Impact Assessment and Equivalence of Liner Systems

25.7.1 IMPACT ASSESSMENT

The assessment of the suitability of a liner system should, in principle, be related to the potential impact on water quality (especially groundwater) due to the proposed landfill and liner system. This assessment typically involves the solution of the advection–dispersion equation subject to appropriate boundary conditions (e.g. see Rowe *et al.* 1995b for a detailed discussion). Factors that influence the potential impact include: (a) the landfill source concentration and its decay characteristics; (b) the advective flux (leakage) across the barrier system; (c) the thickness of any "attenuation" layer between the base of the low permeability liner(s) and the receptor aquifer (or distance to the water table if the liner is constructed in the unsaturated portion of the aquifer); (d) diffusion (especially for VOCs) across the barrier system; (e) sorption/retardation in the liner and underlying soil; (f) biodegradation; (g) dilution in the aquifer; (h) the service life of the engineered components of the landfill (e.g. covers/caps, leachate collection systems, liners); (i) the mode of operation of the landfill; (j) landfill size and geometry; and (k) consolidation of the liner and any underlying compressible soils.

Consolidation (item k) is usually not considered on the grounds that it is transient and provides water for "dilution" of contaminants; however, it should be noted that con-

solidation may be significant if one is calculating travel times and, in particular, first arrival times. It may be hypothesized that the "premature" arrival of contaminant in leakage detection systems is, in part, attributable to failure to consider consolidation effects in assessing the expected travel times. More research is required into this aspect.

Items (a)–(j) can be readily modeled using existing contaminant transport theory and codes (e.g. Rowe & Booker 1997a,b). However, the factors considered in modeling impact tend to be regulatory driven and vary from one country to another, and hence the assessment of what represents an "adequate" barrier system will vary from regulatory environment to regulatory environment. Thus one cannot generalize about "equivalency" of liner systems since what is "equivalent" depends on what is being compared and how it is being compared.

25.7.2 EQUIVALENCE OF GCL AND CCL SYSTEMS

Rowe (1998a) illustrates the evaluation of the equivalence of a liner system involving a GCL with one involving a CCL. When comparing liner systems, consideration should be given to factors such as:

1. The hydraulic resistance over the contaminating lifespan of the landfill, including the effects of leachate mounding and clay–leachate interaction for both compacted clays and GCLs. Consideration should also be given to the potential for internal erosion when there may be high gradients.
2. The contact conditions between the geomembrane and "clay." The geotextile component of the GCL may control transmissivity at the interface between the GCL and geomembrane and any assessment of leakage should be based on experimental data for the proposed geomembrane and GCL under anticipated stress conditions.
3. The presence of wrinkles may be a problem with respect to the long-term perfor-

mance of composite liner systems in land-fills.

4. The quality of a CCL is highly dependent on the construction control, e.g. compaction equipment, water content, etc. These CCLs are also prone to desiccation cracking. GCLs are easier to place and their long-term performance is less sensitive to desiccation effects; however, care is required to ensure that overlapped seams do not move and open up when overlying layers are placed.

5. The stability of the liner system and the potential for shear failure either at the interface between materials or within a layer itself both during construction and during placement of the waste.

6. The potential for diffusion, sorption and biodegradation. Since CCLs are often not specified to have a minimum organic carbon content, sorption cannot be relied on without independent tests. If sorption is required to control impact to a regulatory driven level, it is possible to specify a minimum organic carbon content and/or use modified clays in the GCL to achieve the desired level of sorption. For both CCL and GCL systems, some consideration should be given to the migration of metals, organic compounds and organo-metallic complexes that may be encountered in leachate.

25.8 Liner Stability

A detailed discussion of liner stability is beyond the scope of this chapter. Stability issues are dealt with in Chapter 14 and Chapters 26 (Section 26.5) and 27 (Sections 27.6 and 27.7) deal with specific issues related to stability of liner systems and cover systems, respectively. However, it is worth noting here that there have been a number of failures associated with liner construction (see Rowe 1998a). These cases illustrate the need to give careful consideration to the potential for sliding: (a) during barrier construction; (b) during expansion of existing landfills; and (c) during landfilling over the barrier system. The likelihood of failure occurring can be minimized by:

1. appropriate design and materials selection (including appropriate laboratory tests, selection of reasonable parameters, and appropriate stability analyses);

2. good CQC/CQA to ensure that the barrier system is installed as designed;

3. development plans for expanded landfills that limit toe excavation and overfilling and define allowable conditions for construction of the expansion area and a means of monitoring adherence to the development plans;

4. operations plans that include consideration of stability as the waste is placed and means of monitoring adherence to the operations plans;

5. contingency plans in the event of changed conditions occurring during construction (e.g. excessive rain, unexpected foundations conditions, etc.); and

6. disposal alternatives so that waste can be diverted if expansion schedules are not met.

26. GEOSYNTHETICS IN LIQUID-CONTAINING STRUCTURES

J. P. Giroud

R. Bonaparte

26.1 Introduction

26.1.1 TYPES OF LIQUID-CONTAINING STRUCTURES

The liquid-containing structures considered herein are those constructed with soil; thus, concrete dams, concrete reservoirs, cavities excavated in rock and steel tanks are not considered. The liquid-containing structures considered include: structures used for liquid storage (e.g. embankment dams, liquid impoundments excavated in soil or surrounded with dikes), structures used for liquid conveyance (canals) and structures used to prevent liquid from migrating into the ground (lined landfills). This chapter addresses analysis and design methods for liquid-containing structures. Required geosynthetic material properties and engineering parameters are addressed. The geosynthetic materials themselves are not discussed, however, and the reader is referred to Chapter 7 for information on this subject.

26.1.2 USES OF GEOSYNTHETICS IN LIQUID-CONTAINING STRUCTURES

Geosynthetics are extensively used in liquid-containing structures. These include: geomembrane liners, geosynthetic clay liners, geonet drainage layers, geotextile filters and geogrid reinforcement in liner systems; geotextile filters in embankment dams and bank protection systems; geonets and geocomposites in the internal drainage system of embankment dams; geotextile and geogrid reinforcement used to construct embankments with steep slopes for dams, levees, landfills, etc.; and geomats, geocells and other specialized geosynthetics used for erosion control.

26.1.3 GEOSYNTHETIC LINER SYSTEMS

Many liquid-containing structures include one or several low permeability layers, referred to as liners (see Section 25.3). These low permeability layers are generally incorporated into liner systems where low permeability materials are associated with high permeability materials. These two categories of materials are associated to perform complementary functions: the function of the liners, i.e. the low permeability materials, is to prevent or minimize the migration of liquid; and the function of liquid collection layers, i.e. the high permeability materials, is to collect and remove liquid. The removal of liquid is important because it reduces the head of liquid on top of the underlying liner, which reduces the rate of liquid migration through that liner (see Section 26.2.1.4 for more details on head).

In Chapter 26, the terminology "liner system" is used to designate liner systems where at least one component is a geosynthetic.

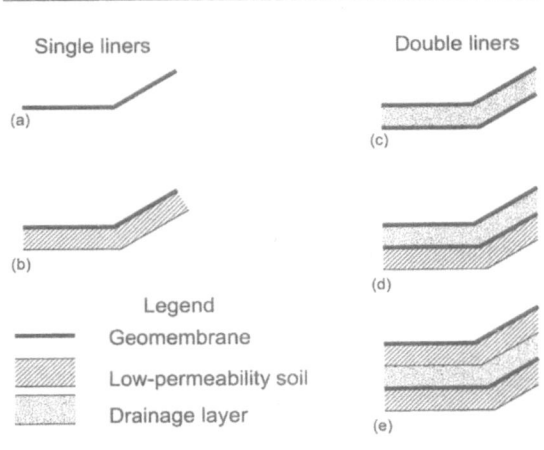

FIGURE 26.1. Five examples of liner systems: (a) single geomembrane liner, (b) single composite liner, (c) double geomembrane liner, (d) double liner with geomembrane primary liner and composite secondary liner, (e) double composite liner.

Chapter 26 complements Chapter 25, which provides information regarding the natural soil components of liner systems, by providing additional information regarding the geosynthetic components of liner systems.

Examples of liner systems are given in Fig. 26.1. A composite liner is a liner composed of two or more components, typically a geomembrane on a layer of compacted clay (or clayey soil), or a geomembrane on a geosynthetic clay liner. A double liner is a liner system that includes two liners (the primary liner on top, and the secondary liner underneath) separated by a drainage layer called the "leakage detection and collection layer" or "secondary liquid collection layer". The function of the secondary liquid collection layer is to detect liquid migration and, more importantly, to remove liquid that migrates through the primary liner, thereby reducing the head of liquid on top of the secondary liner, which reduces the rate of liquid migration through the secondary liner, as indicated above. In lined landfills, there is a primary liquid collection layer (called primary leachate collection layer) on top of the primary liner (in the case of a double liner) or on top of the liner (in case of a single liner) to collect leachate and remove it.

26.1.4 DESIGN CONSIDERATIONS

Many of the design issues addressed in Chapter 26 are related to liner systems. Essentially, a liner system must minimize liquid migration, and must perform this function in spite of the *in situ* stresses and environmental conditions to which it is exposed. As indicated in Section 26.1.3, liquid migration is minimized by decreasing the hydraulic head on top of the liner, which requires effective liquid collection. Effective liquid collection requires effective liquid collection layers and effective filters. There are also many instances where liquid collection layers and filters are not used in conjunction with liner systems. This is the case, for example, of the internal drainage system of embankment dams.

Based on the foregoing discussion, four categories of design considerations will be addressed: liquid collection layer design, filter design, liquid migration evaluation, and resistance of liner systems to instability and deformations.

26.2 Design of Liquid Collection Layers

26.2.1 INTRODUCTION TO LIQUID COLLECTION

26.2.1.1 Liquid Collection

The liquid collection layers discussed herein include the liquid collection layers associated with liner systems, as well as liquid collection layers located inside soil masses, such as the liquid collection layers used for the internal drainage of embankment dams, e.g. toe blankets, chimney drains.

Liquid collection plays an important role in the performance of liquid-containing structures because it is essential to promptly remove liquids from areas where their presence may have a detrimental impact on the performance of the structure. In the case of liner systems, as mentioned in Section 26.1.3, the removal of liquid decreases the liquid head on the liners, thereby decreasing the rate of liquid migration. In the case of liquid collection layers located inside a soil mass, the removal of liquid decreases porewater pressure in the soil mass, thereby increasing the stability of the soil mass; also, the controlled removal of liquids decreases the risk of internal erosion of the soil.

26.2.1.2 Characteristics of Liquid Collection Layers

Liquid collection layers are characterized by their geometry and their hydraulic characteristics. The geometry of a liquid collection layer is characterized by its thickness, its slope and its length (length measured along the slope or, more generally, horizontal projection of the length). The relevant hydraulic characteristic of the material used in the liquid collection layer is its hydraulic conductivity. The relevant hydraulic characteristic of a liquid collection layer is its hydraulic transmissivity. The following relationship exists:

$$\theta_{LCL} = k_{LCL} t_{LCL} \qquad (26.1)$$

where θ_{LCL} = hydraulic transmissivity of the liquid collection layer material, k_{LCL} = hydraulic conductivity (also called coefficient of permeability) of the liquid collection layer material, and t_{LCL} = thickness of the liquid collection layer measured perpendicular to the slope. Equation 26.1 can be used with any set of coherent units. The relevant basis SI units are: θ_{LCL} (m^2 s^{-1}), k_{LCL} (m s^{-1}), and t_{LCL} (m). Equation 26.1 is used with the subscript 1 for the primary liquid collection layer and the subscript 2 for the secondary liquid collection layer. If the liquid collection layer is a geosynthetic, θ_{LCL} is the hydraulic transmissivity of the geosynthetic.

It should be noted that Eq. 26.1 is based on the assumptions that the flow of liquid in the liquid collection layer is laminar, is parallel to the slope, and is not impeded by capillarity. All three assumptions are generally accepted in current design practice. It should also be noted that Eq. 26.1, as well as all the equations given in Sections 26.2.2 and 26.2.3, is valid for granular as well as geosynthetic liquid collection layers.

26.2.1.3 Materials Used in Liquid Collection Layers

Materials used in liquid collection layers are materials with a high hydraulic conductivity, k_{LCL}. These include granular materials such as gravel and sand, and geosynthetic materials such as geonets and drainage geocomposites. More information on granular materials can be found in Section 25.2 and on geosynthetics in Section 7.4. In the case of geosynthetics, it is easier to characterize the material using the hydraulic transmissivity than the hydraulic conductivity, because the test (see Section 7.4 and 7.7) directly gives the hydraulic transmissivity, and deriving the hydraulic conductivity from the hydraulic transmissivity may not be accurate due to the lack of precision on the measurement of geosynthetic thickness. As pointed out by Williams et al. (1984), it is important that testing to measure hydraulic transmissivity be performed under conditions that properly simulate field conditions, in particular the overburden stress, the hydraulic gradient and boundary conditions.

26.2.1.4 Hydraulic Head, Depth and Thickness of Liquid

The rate of liquid migration through a liner is governed by the hydraulic head above the liner. If the liquid does not move (e.g. in the case of a dam or a liquid impoundment), the hydraulic head on top of the liner is equal to the depth of liquid on top of the liner, i.e. $h = D$ in Fig. 26.2a. In the case where the liquid is unconfined and flowing along a

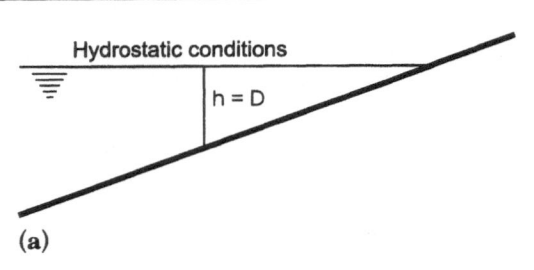

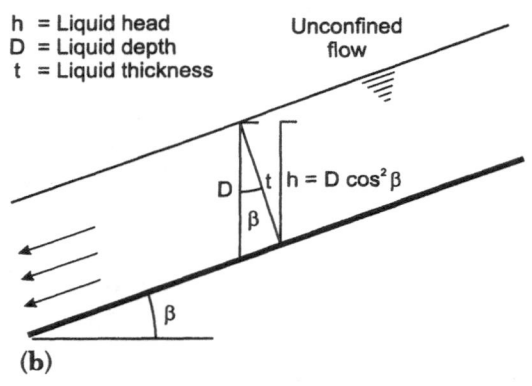

FIGURE 26.2. Head of liquid on top of the liner in the case of a liner on a slope: (a) hydrostatic conditions; and (b) unconfined flow along the slope (after Giroud 1997, © Geosynthetics International, reproduced with permission).

slope, i.e. parallel to a liner resting on a slope, the relationship between the liquid head on top of the liner, h, and the liquid thickness, t, is (Fig. 26.2b):

$$h = t \cos \beta = D \cos^2 \beta \qquad (26.2)$$

where h = hydraulic head on top of the liner; t = thickness of liquid measured perpendicular to the slope; D = depth of liquid, measured vertically; and β = slope angle. The relationship between thickness and depth is:

$$t = D \cos \beta. \qquad (26.3)$$

26.2.1.5 Design of Liquid Collection Layers
The basic principle of the design of liquid collection layers is that the liquid collection layer must be capable of conveying the flow of the collected liquid. This is expressed by the fol-

lowing equation derived from Darcy's equation:

$$Q^{\circ} = k_{\text{LCL}} t_{\text{LCL}} i_{\text{LCL}} = \theta_{\text{LCL}} i_{\text{LCL}} = \theta_{\text{LCL}} \sin \beta \qquad (26.4)$$

where Q° = flow rate per unit width of the liquid collection layer, i_{LCL} = hydraulic gradient in the liquid collection layer (equal to sin β), and β = slope angle of the liquid collection layer. Equation 26.4 can be used with any set of coherent units. The relevant basic SI units are Q° (m² s⁻¹), θ_{LCL} (m² s⁻¹), k_{LCL} (m s⁻¹), t_{LCL} (m) and β (°); i_{LCL} is dimensionless.

For the case of liquid collection layers located inside a soil mass, as for the internal drainage of an embankment dam, the design engineer should evaluate the rate of flow of the collected liquid using geotechnical engineering methods. Methods specific to the liquid collection layers associated with liner systems have been developed and are presented in Sections 26.2.2 and 26.2.3. However, it is important to recognize that the methods presented in Sections 26.2.2 and 26.2.3 are consistent with Eq. 26.4.

26.2.2 DESIGN OF PRIMARY LIQUID COLLECTION LAYERS

26.2.2.1 Analysis of Liquid Flow in a Primary Liquid Collection Layer
At a given time, there is a certain amount of liquid in a primary liquid collection layer. This amount of liquid can be expressed by the depth, thickness or head of liquid on top of the liner (see Section 26.2.1.4). The thickness of liquid is used herein because it can be readily compared to the thickness of the liquid collection layer. The thickness of liquid in the primary liquid collection layer varies along the length of the liquid collection layer. The liquid thickness has a maximum value, $t_{1\text{max}}$, that occurs at a horizontal distance x_{m} from the top of the liquid collection layer slope (Fig. 26.3). The value of x_{m} depends on a parameter λ, defined by:

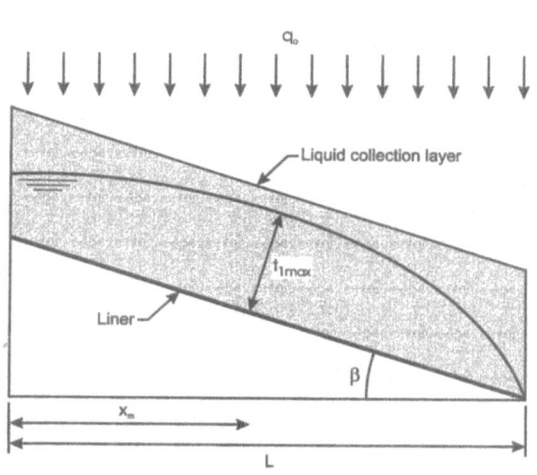

FIGURE 26.3. Liquid flow in a primary liquid collection layer (modified from Giroud & Houlihan 1995).

$$\lambda = \frac{q_o}{k_{LCL1} \tan^2 \beta} \qquad (26.5)$$

where q_o = liquid impingement rate, i.e. rate at which liquid impinges the liquid collection layer; k_{LCL1} = hydraulic conductivity of the primary liquid collection layer material; and β = slope angle. Equation 26.5 can be used with any set of coherent units. The relevant basic SI units are: q_o (m s^{-1}) k_{LCL1} (m s^{-1}) and β (°); λ is dimensionless. Herein, q_o is assumed to be uniformly distributed over the entire area of the primary liquid collection layer. Techniques for calculating q_o in landfills are described in Chapter 27.

As shown by Giroud & Houlihan (1995), x_m/L (where L = the horizontal projection of the length of the primary liquid collection layer) is equal to or greater than 0.8, i.e. the maximum of the liquid thickness occurs close to the toe of the slope, if λ is less than 0.15. Such small values of λ are frequent in geosynthetic liner system applications.

Referring to Fig. 26.3, an excellent approximation of the maximum liquid thickness in a primary liquid collection layer under steady-state flow conditions, t_{1max}, is given by the following equation (Giroud & Houlihan 1995):

$$t_{1max} = Lj \frac{(\tan^2 \beta + 4q_o/k_{LCL1})^{1/2} - \tan \beta}{2 \cos \beta}$$

$$= Lj \frac{[(1 + 4\lambda)^{1/2} - 1] \tan \beta}{2 \cos \beta} \qquad (26.6)$$

where L = horizontal projection of the length of the primary liquid collection layer, and j = dimensionless coefficient given by:

$$j = 1 - 0.12 \exp\left\{-\left[\log\left(\frac{8q_o}{5k_{LCL1} \tan^2 \beta}\right)^{5/8}\right]^2\right\}$$

$$= 1 - 0.12 \exp\left\{-\left[\log\left(\frac{8\lambda}{5}\right)^{5/8}\right]^2\right\}. \qquad (26.7)$$

Values of j are between 0.88 and 1.00. Therefore, the maximum error made by neglecting the dimensionless coefficient j is 12%. As a result, in many practical applications, the following equation, known as Giroud's equation, can be used:

$$t_{1max} = L \frac{(\tan^2 \beta + 4q_o/k_{LCL1})^{1/2} - \tan \beta}{2 \cos \beta}$$

$$= L \frac{[(1 + 4\lambda)^{1/2} - 1] \tan \beta}{2 \cos \beta}. \qquad (26.8)$$

For small values of λ (e.g. $\lambda < 0.15$), Eqs 26.6 and 26.8 tend toward:

$$t_{1max} = \frac{q_o L}{k_{LCL1} \sin \beta}. \qquad (26.9)$$

Equations 26.6–26.9 can be used with any set of coherent units. The relevant basic SI units are: t_{1max} (m), q_o (m s^{-1}), k_{LCL1} (m s^{-1}), L (m) and β (°); λ is dimensionless.

It is important to note that Eq. 26.9 could have been obtained directly by assuming that the maximum liquid thickness occurs at the toe of the slope (which is a good approximation for $\lambda < 0.15$, as noted above). Then, the flow rate in the liquid collection layer under steady state flow conditions is $q_o L$, i.e. the impingement rate multiplied by the horizontal projection of the length of the liquid collection layer, and Eq. 26.4, where the maximum liquid thickness is used instead of the thickness of the liquid collection layer, gives the following equation, which is identical to Eq. 26.9:

$$q_oL = k_{LCL}it_{1max} = k_{LCL}t_{1max} \sin \beta. \quad (26.10)$$

26.2.2.2 Design of a Primary Liquid Collection Layer

The design of a primary liquid collection layer consists of checking that the maximum thickness of liquid, t_{1max}, calculated using Eqs. 26.6, 26.7 or 26.8 is less than the thickness of the primary liquid collection layer:

$$t_{1max} \leq t_{LCL1}. \quad (26.11)$$

If λ is small (e.g. $\lambda < 0.15$), $t_{1\,max}$ can be calculated using Eq. 26.9. Hence:

$$\frac{q_oL}{\sin \beta} \leq k_{LCL1}t_{LCL1} = \theta_{LCL1}. \quad (26.12)$$

Therefore, if λ is small (e.g. $\lambda < 0.15$), the design of a primary liquid collection layer consists of checking that $q_oL/\sin \beta$ is less than the hydraulic transmissivity, θ_{LCL1}, of the primary liquid collection layer. For the calculation of t_{LCL1} at the design stage, a factor of safety may be applied to q_o.

It is important to note that Eqs 26.6–26.9 are only valid for steady state flow conditions. Calculations for the case of transient flow conditions are more complex. Practical guidance in the case of transient flow is provided by Giroud & Houlihan (1995).

26.2.2.3 Average Head of Liquid on Top of the Primary Liner for Liquid Migration Evaluation

The rate of liquid migration through the primary liner depends on the head of liquid on top of the primary liner. Since the head of liquid (which is related to the thickness of liquid, as indicated by Eq. 26.2) varies along the slope, it is convenient to use an average liquid

head to perform the liquid migration rate calculations. The calculation of the exact value of the average liquid head is complex (Giroud & Houlihan 1995). However, based on numerical values provided by Giroud & Houlihan (1995), it is possible to recommend the following values for the average head of liquid on top of the primary liner:

$$h_{1avg} = 0.8t_{1max} \quad \text{if } \lambda > 0.5 \quad (26.13a)$$

$$h_{1avg} = 0.7t_{1max} \quad \text{if } 0.15 < \lambda \leq 0.5 \quad (26.13b)$$

$$h_{1avg} = 0.6t_{1max} \quad \text{if } 0.01 < \lambda \leq 0.15 \quad (26.13c)$$

$$h_{1avg} = 0.5t_{1max} \quad \text{if } \lambda \leq 0.01. \quad (26.13d)$$

In the last case (i.e. for $\lambda \leq 0.01$), combining Eqs 26.2, 26.9 and 26.13d gives:

$$h_{1avg} = \frac{q_oL}{2k_{LCL1} \tan \beta}. \quad (26.14)$$

26.2.3 DESIGN OF SECONDARY LIQUID COLLECTION LAYERS

26.2.3.1 Analysis of Liquid Flow in a Secondary Liquid Collection Layer

Section 26.2.3 addresses the flow of liquid in the secondary liquid collection layer located between the two liners in a double liner system (Fig. 26.4). Since liquid migration through the primary liner of a double liner system occurs essentially through defects, i.e. small holes, of the primary liner geomembrane, and since the number of defects is generally small, the liquid that has migrated through defects of the primary liner generally flows only in portions of the secondary liquid collection layer. These portions are referred to as wetted zones (see Fig. 26.4b). If the defects in the primary liner geomembrane are sufficiently far apart, the wetted zones produced by the various defects do not overlap, and the boundary of the wetted zone related to one defect is approximately a parabola, as shown by Giroud et al. (1997a).

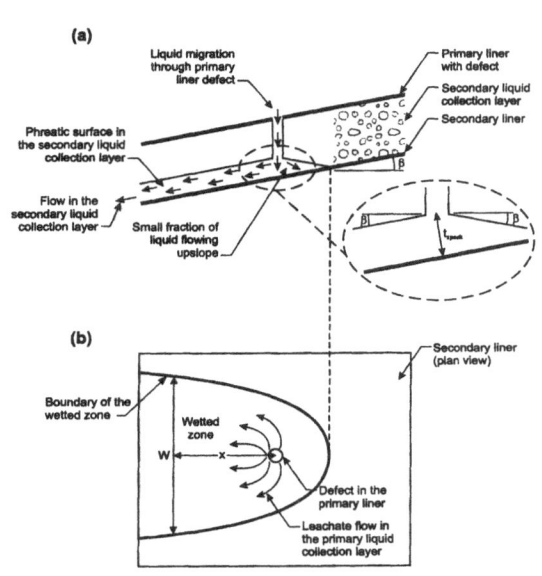

FIGURE 26.4. Liquid flow in the primary liquid collection layer, through a defect in the primary liner, and in the secondary liquid collection layer (leakage detection and collection layer), in the case of a double liner: (a) cross-section; and (b) plan view of the secondary liner (modified from Giroud *et al.* 1997a).

The thickness of liquid in the secondary liquid collection layer has a peak value at the location of each defect in the primary liner. Giroud *et al.* (1997a) showed that the following relationship exists between the rate of liquid migration through a primary liner defect, Q, the hydraulic conductivity of the secondary liquid collection layer, k_{LCL2}, and the thickness of liquid in the secondary liquid collection layer beneath the considered defect, t_{2peak} (Fig. 26.4a):

$$Q = k_{LCL2}\, t_{2peak}^2 \qquad (26.15)$$

where Q = liquid migration rate through the considered defect of the primary liner, k_{LCL2} = hydraulic conductivity of the secondary liquid collection layer material, and t_{2peak} = thickness of liquid in the secondary liquid collection layer at the location of the

considered defect of the primary liner. Equation 26.15 can be used with any set of coherent units. The relevant basic SI units are: t_{2peak} (m), Q_{max} (m^3 s^{-1}) and k_{LCL2} (m s^{-1}).

It should be noted that Eq. 26.15 does not depend on the slope of the secondary liquid collection layer. The maximum thickness of liquid in the secondary liquid collection layer, t_{2max}, occurs at the location of the primary liner defect with the largest rate of liquid migration, Q_{max}. It is given by the following equation derived from Eq. 26.15:

$$t_{2max} = \left(\frac{Q_{max}}{k_{LCL2}}\right)^{1/2} \qquad (26.16)$$

where $t_{2\,max}$ = maximum thickness of liquid in the secondary liquid collection layer, Q_{max} = largest rate of liquid migration through a defect of the primary liner, and k_{LCL2} = hydraulic conductivity of the secondary liquid collection layer material. Equation 26.16 can be used with any set of coherent units. The relevant basic SI units are: $t_{2\,max}$ (m), Q_{max} (m^3 s^{-1}) and k_{LCL2} (m s^{-1}).

26.2.3.2 Design of a Secondary Liquid Collection Layer

The design of a secondary liquid collection layer consists of checking that the maximum thickness of liquid calculated as shown above, t_{2max}, is less than the thickness of the secondary liquid collection layer, t_{LCL2}:

$$t_{2max} \le t_{LCL2}. \qquad (26.17)$$

Combining Eqs 26.16 and 26.17 gives the following condition for the thickness of the secondary liquid collection layer:

$$t_{LCL2} \ge \left(\frac{Q_{max}}{k_{LCL2}}\right)^{1/2}. \qquad (26.18)$$

Equation 26.18 can be used with any set of coherent units. The relevant basic SI units are: t_{LCL2} (m), Q_{max} (m^3 s^{-1}) and k_{LCL2} (m s^{-1}).

It should be noted that Eq. 26.18 is theoretically valid only if the geomembrane defects are sufficiently far apart that the wet zones produced by the different defects do not overlap. From a practical standpoint, however, Eq. 26.18 provides a good approximation of the required secondary liquid collection layer thickness in most typical cases. For the calculation of t_{LCL2} at the design stage, a factor of safety may be applied to Q_{max}. Finally, it should be noted that Eq. 26.18 is only valid for steady state flow conditions.

26.2.3.3 Average Head of Liquid on Top of the Secondary Liner for Liquid Migration Evaluation

To calculate the rate of liquid migration through the secondary liner, i.e. the rate of liquid migration into the ground in the case of a double liner, it is necessary to know the head of liquid on top of the secondary liner. The head of liquid on top of the secondary liner is zero outside the wetted zones, and varies from one point to another inside each wetted zone (see Fig. 26.4). The rigorous calculation of the hydraulic head for each wetted zone and the calculation of the resulting liquid migration rate are very tedious (Giroud et al. 1997a). From a practical standpoint, it is preferable to use an approximate approach that consists of calculating an average value of the liquid head over the entire secondary liner, $h_{2\,avg}$, i.e. as if the entire surface area of the secondary liner was uniformly wetted. This was done by Giroud et al. (1997a), who assumed that all the defects of the primary liner were identical and considered two scenarios of distribution of the defects in the primary liner. They obtained the following values for the average liquid head on the secondary liner:

$$h_{2avgworst} = \frac{FLQ}{k_{LCL2}\tan\beta} \qquad (26.19)$$

$$h_{2avgrand} = \frac{FLQ}{2k_{LCL2}\tan\beta} \qquad (26.20)$$

where $h_{2avgworst}$ = average head of liquid over the entire secondary liner, in the scenario where all the defects are assumed to be at the top of the primary liner slope ("worst" scenario); $h_{2avgrand}$ = average head of liquid over the entire secondary liner, in the scenario where the defects are assumed to be randomly distributed ("random" scenario); F = frequency of defects, i.e. number of defects per unit area; Q = rate of liquid migration through each one of the defects of the primary liner; and k_{LCL2} = hydraulic conductivity of the secondary liquid collection layer material. Equations 26.19 and 26.20 can be used with any set of coherent units. The relevant basic SI units are: h (m), F (m^{-2}), L (m), Q (m^3 s^{-1}), k (m s^{-1}) and β (°).

26.3 Design of Filters

26.3.1 INTRODUCTION TO FILTRATION

26.3.1.1 Use of Filters in Liquid-Containment Structures

Filters are extensively used in liquid-containment structures. They are typically used between soil and a liquid collection layer, i.e. drainage layer, to protect the liquid collection layer against clogging by soil particles. The use of filters with liquid collection layers associated with liner systems has been discussed in Section 25.2.4.

26.3.1.2 Types of Filters

Two types of filters are used: granular filters (generally sand filters) and geotextile filters. Both types of filters perform the same functions as discussed in Section 26.3.1.3. The characteristics of granular filters are presented in Section 26.3.2.2 and the characteristics of geotextile filters are presented in Section 26.3.2.3.

26.3.1.3 Functions of Filters

The function of a filter in a liner system is to retain the soil while allowing the liquid to flow as freely as possible. Therefore, the filter must have openings that are large enough to allow free flow of liquid, but small enough to prevent migration of soil particles.

It is important to note that the function of a filter is not to stop particles that are moving, but to prevent particles from moving. If some fine particles are moving toward the filter, and if the filter stops them, these particles will accumulate on or in the filter and will clog it. Situations where fine particles may be moving toward the filter include: (a) internally unstable soils where liquid flow can carry a significant amount of fine particles; and (b) cases where the filter is not placed in close contact with the soil.

It is important to recognize that there are situations, such as the case of internally unstable soils, where a filter is detrimental. Waste is often internally unstable, and filters placed between waste and leachate collection layers may be expected to experience clogging (see Section 26.3.1.5).

26.3.1.4 Importance of Intimate Contact Between Filter and Soil

For soil retention, it is important to have intimate contact between a filter and the adjacent soil (Giroud *et al.* 1977; Giroud 1989, 1996a), and examples of failures due to lack of intimate contact between a geotextile filter and the adjacent soil have been reported (Giroud 1999). Intimate contact between a filter and soil is usually achieved relatively easily with granular filters, but is sometimes difficult to achieve with geotextile filters. If the soil surface is not smooth, it is recommended to use flexible geotextiles that tend to conform to the shape of the soil; also, it is recommended to use rather small aggregate (e.g. aggregate with particles less than 20 mm) on the other side of the geotextile to exert a uniform pressure on the geotextile, thereby forcing the geotextile to conform to the shape of the soil. When a geotextile filter

is attached to a geosynthetic drainage layer, the geocomposite thus formed is usually rather rigid and may not be able to conform to the shape of the soil, unless the soil surface is smooth. In such cases, it is necessary to place a layer of sand between the geotextile and the irregular soil surface to ensure intimate contact.

26.3.1.5 Clogging of Filters

Clogging is the decrease of permeability of the filter or the soil in the vicinity of the filter. Clogging results from the accumulation of fine soil particles, biological matter, and/or chemicals in or on the filter, or in the soil in the vicinity of the filter. Clogging by formation or precipitation of chemicals (chemical clogging), by growth of biological matter (biological clogging), or both (biochemical clogging) has been discussed in detail (Giroud 1996a; Rowe 1998a). Also, the development of biological clogging in municipal solid waste landfills is discussed in Sections 25.2.3 and 25.2.4.

26.3.2 RELEVANT CHARACTERISTICS OF SOILS AND FILTERS

26.3.2.1 Characteristics of Soils Relevant to Filter Design

The characteristics of the soil in contact with a filter that are relevant to filter design are the following: the soil hydraulic conductivity, k_s; the particle size distribution curve defined by d_m, the particle size such that $m\%$ of the soil particles by mass are smaller than d_m; and the coefficient of uniformity of the particle size distribution curve, C_u, defined by:

$$C_u = \frac{d_{60}}{d_{10}}. \tag{26.21}$$

The linear coefficient of uniformity, C'_u, used in Giroud's retention criterion (see Section 26.3.3.3) is obtained by tracing the tangent to the central portion of the particle size distribution curve of the soil (Fig. 26.5). This tangent intersects the 0% horizontal line for

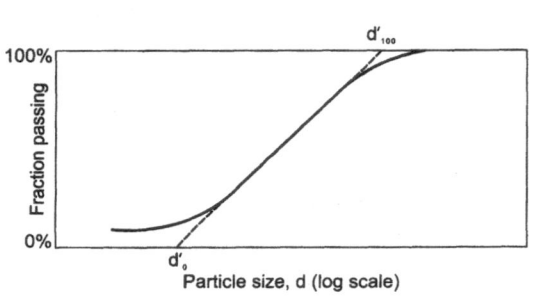

FIGURE 26.5. Determination of the linear coefficient of uniformity.

d'_0 and intersects the 100% horizontal line for d'_{100}. The linear coefficient of uniformity is then given by:

$$C'_u = \left(\frac{d'_{100}}{d'_0}\right)^{0.5}. \qquad (26.22)$$

It should be noted that, if the particle size distribution curve of the soil is approximately linear, the coefficient of uniformity and the linear coefficient of uniformity are approximately equal.

26.3.2.2 Characteristics of Granular Filters

A granular filter is characterized by the following properties: average filtration opening size, O_F; porosity, n; particle size distribution curve defined by d_m, the particle size such that $m\%$ of the filter particles by mass are smaller than d_m; and filter hydraulic conductivity, k_{FGRA}. Measurement of the properties of granular materials is discussed in Sections 3.2 and 3.3. Kenney *et al.* (1985) have shown that the following approximate relationship exists:

$$O_F \approx 0.2 d_{15}. \qquad (26.23)$$

26.3.2.3 Characteristics of Geotextile Filters

A geotextile filter is characterized by the following properties: filtration opening size, O_F, which for practical purposes can be considered approximately equal to the apparent opening size, O_{95}; porosity, n_{FGT} (for nonwoven geotextiles) or relative open area, A_{FGT} (for woven geotextiles); thickness, t_{FGT}; mass per unit area, μ_{FGT}; fiber diameter, d_f; fiber material density, ρ_f; hydraulic conductivity, k_{FGT}; permittivity, ψ_{FGT}. Measurement of geotextile properties is discussed in Section 7.2.

The following basic relationships exist between these properties:

$$k_{FGT} = \psi_{FGT} t_{FGT} \qquad (26.24)$$

$$n_{FGT} = 1 - \frac{\mu_{FGT}}{\rho t_{FGT}}. \qquad (26.25)$$

In addition to the above rigorous relationships, the following approximate relationship has been developed by Giroud (1996a) for nonwoven geotextiles and is in close agreement with experimental data:

$$\frac{O_F}{d_f} \approx \frac{1}{(1 - n_{FGT})^{1/2}} - 1$$
$$\qquad\qquad (26.26)$$
$$+ \frac{10 n_{FGT}}{(1 - n_{FGT}) t_{FGT}/d_f}.$$

Equation 26.26 makes it possible to evaluate the influence on the filtration opening size of a decrease in thickness and porosity due to compressive stresses. Equations 26.24–26.26 can be used with any set of coherent units. The relevant basic SI units are: O_F (m), t_{FGT} (m), μ_{FGT} (kg m^{-2}), d_f (m), ρ_f (kg m^{-3}), k_{FGT} (m s^{-1}), ψ_{FGT} (s^{-1}); n_{FGT} and A_{FGT} are dimensionless.

26.3.3 DESIGN OF FILTERS

26.3.3.1 Filter Criteria

A filter should satisfy three criteria:

- Permeability criterion: the filter must be permeable enough for free liquid flow.
- Retention criterion: the filter opening size must be small enough to retain soil particles.
- No-clogging criterion: the filter should include a large number of openings so the probability for clogging all of them is small.

These three criteria are discussed below for both granular filters and geotextile filters

26.3.3.2 Permeability Criterion

As shown by Giroud (1996a), a complete permeability criterion includes two requirements for free liquid flow. The first requirement is to prevent excessive pore pressure and can be written as follows (Giroud 1996a), regardless of the type of filter, granular or geotextile:

$$k_F > 10 k_s i_s \qquad (26.27)$$

where k_F = filter hydraulic conductivity (k_{FGRA} for a granular filter or k_{FGT} for a geotextile filter), k_s = soil hydraulic conductivity, and i_s = hydraulic gradient in the soil in the vicinity of the filter. Guidance regarding i_s may be found in Table 26.1.

The second requirement is to prevent excessive reduction in flow rate and its expression depends on the type of filter (Giroud 1996a):

$$k_{FGT} > k_s \qquad \text{for a geotextile filter} \qquad (26.28a)$$

$$k_{FGRA} > 25 k_s \qquad \text{for a granular filter} \qquad (26.28b)$$

where k_{FGT} = hydraulic conductivity of a geotextile filter, k_{FGRA} = hydraulic conductivity of a granular filter, and k_s = soil hydraulic conductivity.

Based on the classical Hazen's equation and theoretical considerations (Giroud 1996a), the hydraulic conductivity of a granular soil is proportional to the square of the d_{10} or d_{15} of its particle size distribution curve. Therefore, Eq. 26.28b is equivalent to the classical expression of the permeability criterion for granular filters:

$$d_{15F} > 5 d_{15s} \qquad (26.29)$$

where d_{15F} = d_{15} of the granular filter; and d_{15s} = d_{15} of the soil (see Sections 26.3.2.1 and 26.3.2.2 for the definition of d_{15}).

Based on the foregoing, the classical permeability criterion for granular filters satisfies the requirement to prevent excessive reduction in flow rate. However, as pointed out above, a complete permeability criterion includes a second requirement (expressed by Eq. 26.27): the requirement to prevent the development of excessive pore pressure. Therefore, it is recommended that, for granular filters, the requirement expressed by Eq. 26.27 be used in addition to the classical criterion expressed by Eq. 26.29.

TABLE 26.1. Typical hydraulic gradients (critical applications may require designing for higher gradients than those given; adapted from Giroud (1988, 1996a) and Luettich et al. (1992)

Application	Typical hydraulic gradient
Standard dewatering trench	1.0
Vertical wall drain	1.5
Pavement edge drain	1
Landfill leachate collection/detection removal system	1.5
Landfill leachate collection removal system	1.5
Landfill closure surface water collection removal system	1.5
Dam toe drains	2
Dam clay cores	3 to >10
Inland channel protection	1
Shoreline protection	10
Liquid impoundment with clay liners	>10

TABLE 26.2. Giroud's retention criterion for geotextile filters (adapted from Giroud 1982, 1988, 1994)

Density index of the soil (relative density), I_D		Linear coefficient of uniformity of the soil, C_u'	
		$1 \leq C_u' \leq 3$	$C_u' \geq 3$
Loose soil	$I_D < 35\%$	$O_{95} < (C_u')^{0.3} d_{85}$	$O_{95} < \dfrac{9}{(C_u')^{1.7}} d_{85}$
Medium dense soil	$35\% < I_D < 65\%$	$O_{95} < 1.5 \, (C_u')^{0.3} d_{85}$	$O_{95} < \dfrac{13.5}{(C_u')^{1.7}} d_{85}$
Dense soil	$I_D > 65\%$	$O_{95} < 2 \, (C_u')^{0.3} d_{85}$	$O_{95} < \dfrac{18}{(C_u')^{1.7}} d_{85}$

26.3.3.3 Retention Criterion

The detailed analysis of the mechanism by which geotextile filters retain soils published by Giroud (1982) led to the geotextile retention criterion presented in Table 26.2. An important characteristic of this criterion is the role played by the particle size distribution curve of the soil characterized by its linear coefficient of uniformity, C_u', defined in Section 26.3.2.1. Since 1982, a number of studies have led to a variety of retention criteria for geotextile filters; consistent with the approach proposed by Giroud (1982), most of the proposed retention criteria for geotextile filters include as a parameter the coefficient of uniformity of the particle size distribution curve of the soil.

In the case of granular filters, the retention criterion has been expressed as follows ever since the original work on filters by Bertram and Terzaghi in the 1930s:

$$d_{15F} < 5 d_{85S} \qquad (26.30)$$

where $d_{15F} = d_{15}$ of the granular filter; and $d_{85s} = d_{85}$ of the soil (see Section 26.3.2.1 for the definition of d_{15} and d_{85}).

Combining Eqs 26.23 and 26.30 gives:

$$O_F < d_{85S}. \qquad (26.31)$$

In other words, the meaning of Eq. 26.30, i.e. Terzaghi's classical retention criterion for granular filters, is that the filter opening size should be smaller than the d_{85} of the soil.

The rationale used by Giroud (1982) to develop his retention criterion for geotextile filters is not specific to geosynthetics and can be used for any type of filter. Therefore, Giroud's retention criterion for geotextile filters can be used for granular filters, with the value of the filtration opening size of a granular filter given by Eq. 26.23. It is then possible to compare Giroud's retention criterion applied to soils with Terzaghi's classical retention criterion for granular filters. Comparing Eq. 26.31 with Table 26.2 shows that Terzaghi's classical retention criterion for granular filters (represented by Eq. 26.31, which is equivalent to Eq. 26.30) is equivalent to, or more conservative than, Giroud's retention criterion if the coefficient of uniformity of the soil is less than five. In contrast, Terzaghi's classical retention criterion is less conservative than Giroud's retention criterion for soils having a coefficient of uniformity of the soil greater than five. This results from the fact that Terzaghi's retention criterion for granular filters does not take into account the coefficient of uniformity of the soil. As shown by Giroud (1996a), this potential problem can be prevented by using Terzaghi's classical retention criterion for granular filters only on the fraction of the soil particle size distribution curve related to particles smaller than approximately 5 mm. (This is consistent with the practice often followed by design engineers, which consists of considering only the particles smaller than 4.75 mm when using Terzaghi's filter criteria.)

26.3.3.4 Non-clogging Criterion

To minimize the risk of clogging, the filter should have as many openings as possible. For a given filtration opening size, the number of filter openings is proportional to the porosity of the filter (granular or nonwoven geotextile) or the relative open area (woven geotextile). In the case of granular filters, the porosity is of the order of 0.3; as a result, it is not necessary to include in filter criteria a requirement for the porosity of granular filters. This is why a non-clogging criterion is never mentioned for granular filters. For lack of a better rationale, it has been suggested (Giroud 1985, 1989) that geotextile filters be specified to have a porosity (non-woven geotextiles) or a relative open area (woven geotextiles) greater than 0.3, i.e. the same value as for granular filters. This requirement is easily met by non-woven geotextiles since they typically have a porosity of 0.7–0.9. In contrast, many woven geotextiles do not meet this requirement. Although there are specifications that allow the use of woven geotextile filters with a relative open area as low as 0.04, the authors recommend that woven geotextile filters with a relative open area greater than 0.2 be specified. Woven geotextiles that meet this requirement are available.

26.4 Liquid Migration Through Liners

26.4.1 INTRODUCTION TO LIQUID MIGRATION THROUGH LINERS

26.4.1.1 Types of Liners

All of the liners considered herein are assumed to include a geomembrane. The liners considered are either geomembrane liners or composite liners. In the context of this chapter:

- a composite liner consists of a geomembrane component and a low permeability soil component, i.e. a compacted layer of low permeability soil, or the bentonite layer of a geosynthetic clay liner (GCL); and
- the low permeability soil component of the composite liner is located beneath the geomembrane.

26.4.1.2 The Low Permeability Soil Component of a Composite Liner

When the low permeability soil component of a composite liner is a compacted layer of low permeability soil, this soil generally includes clay and is often referred to as a compacted clay liner (CCL). In the case of a GCL, the low permeability component of the composite liner is a layer of bentonite (which is a variety of clay). There are two types of GCLs: GCLs where the layer of bentonite is encapsulated between two layers of geotextiles, and GCLs where the layer of bentonite is attached to a geomembrane. CCLs are discussed in detail in Sections 25.3.2 and 25.3.3, and GCLs are discussed in detail in Sections 25.3.4, 25.4.2, 25.5.1 and 25.7.2.

The thickness of a CCL is typically between 0.3 and 1.5 m, whereas the thickness of a hydrated GCL depends on the compressive stress applied during hydration and is typically between 5 and 10 mm, i.e. of the order of 100 times less than the thickness of a CCL. The hydraulic conductivity of both CCLs and GCLs depends on the nature of the material, the nature of the liquid, and the applied compressive stress; when the liquid is water or a leachate that does not affect the hydraulic conductivity of clay, including bentonite, the hydraulic conductivity of a CCL is typically between 1×10^{-10} and 1×10^{-9} m s^{-1}, whereas the hydraulic conductivity of a GCL is typically between 5×10^{-12} and 5×10^{-11} m s^{-1}, i.e. of the order of ten–100 times less than the hydraulic conductivity of a CCL.

26.4.1.3 Geomembranes

Geomembrane material can be permeated by liquids as a result of diffusion and solution. The quantities of liquids that can migrate through geomembrane materials are very small; however, the rate of diffusion of certain organic compounds can be significant, as discussed by Rowe (1998a) and in Sections 25.3, 25.5.3 and 25.5.7. Therefore, permeation of certain chemical compounds (e.g. hydrocarbons and solvents) through geomembranes should be carefully evaluated if these compounds are likely to be present in the contained liquid (even in trace amounts).

Since an intact geomembrane allows negligible permeation of water and other polar liquids, most of the migration of these liquids (e.g. water or any aqueous solution such as leachate from municipal or hazardous solid waste landfills) through a geomembrane liner or a composite liner occurs through geomembrane defects. In this chapter, the only mechanism of liquid migration that is considered is advective flow through geomembrane defects. In other words, in this chapter, geomembranes are treated as impermeable layers with defects.

Studies presented by Giroud & Bonaparte (1989a) have shown that geomembrane liners installed with strict construction quality assurance could be considered having a frequency of one–two defects per 4000 m^2, with a diameter of 2 mm, i.e. a defect area of 3.14×10^{-6} m^2. On the other hand, electric leak detection surveys (Laine 1991) have shown that geomembrane liners installed with strict construction quality assurance have five or more defects per 4000 m^2, with a defect diameter less than 0.5 mm. The number of defects may even be greater after placement of a layer of soil on the installed geomembrane, and is certainly greater in the case of geomembranes installed without strict construction quality assurance. Therefore, it is recommended that design engineers consider different scenarios of defect size and frequency when they calculate the rate of liquid migration through geomembrane liners and composite liners. For the sake of conservatism, a defect area of 1 cm^2 is often considered in design calculations evaluating the rate of liquid migration through primary liners as a first step for the design of secondary liquid collection layers.

26.4.1.4 Liquid Migration Evaluation

The evaluation of liquid migration through liners is an essential aspect of the design of a liquid-containment structure. An early discussion of liquid migration through geomembrane liners was presented by Giroud (1984a). Investigations into liquid migration through composite liners have been performed by Faure (1984), Sherard (1985), Fukuoka (1986), Brown et al. (1987), and Jayawickrama et al. (1988). From these investigations, Giroud & Bonaparte (1989a, 1989b) developed a systematic semi-empirical methodology for evaluating the rate of liquid migration through composite liners. Most of the equations presented in this chapter were established using this methodology. A detailed theoretical analysis of liquid migration through composite liners has been presented by Rowe (1998a).

The equations presented in this chapter can be used to compare various types of liners from the viewpoint of advective flow. Systematic comparisons presented by Giroud et al. (1994b, 1998b) show that the rate of liquid migration through composite liners is typically two to four orders-of-magnitude less than liquid migration through geomembrane liners (with the geomembrane having the same number and size of defects in both cases). Clearly, composite liners should be preferred to geomembrane liners, i.e. geomembranes placed directly on a permeable material. However, in liquid impoundments, the risk of geomembrane uplift by liquids accumulating between the geomembrane and clay component of a composite liner should be addressed (see Section 26.5.1.5). Comparisons also show that the rate of liquid migration through a composite liner constructed with a given soil having a hydraulic conductivity less than 1×10^{-6} m s^{-1} is approximately three orders-of-magnitude less than through the soil itself. In other words, placing a geomembrane on a soil decreases the rate of liquid migration by a factor of 1000.

The equations presented in this chapter can also be used to compare various types of liner systems from the viewpoint of advective flow. An example of such comparison is presented by Giroud et al. (1997d). The same comparison was presented by Lake & Rowe (1999) from the viewpoint of diffusion of organic compounds, and the results of the comparison are different from the results of the comparison based on advective flow. The flows calculated using the equations presented in this chapter can be used as input to contaminant transport models (e.g. Rowe & Booker 1997a), as discussed in Section 25.7.

26.4.2 LIQUID MIGRATION THROUGH COMPOSITE LINERS

26.4.2.1 Ideal Case: Perfect Contact

In the ideal situation where the contact between the geomembrane component and the soil component of a composite liner is perfect, the rate of liquid migration through the composite liner, in the case where the geomembrane defect is circular, is given by the equation (Forchheimer 1930):

$$Q = 4rhk_{UM} \qquad (26.32)$$

where Q = rate of liquid migration through the geomembrane defect; r = radius of the geomembrane defect; h = head of liquid on top of the liner; and k_{UM} = hydraulic conductivity of the medium underlying the geomembrane, i.e. the soil component of the composite

liner. Equation 26.32 can be used with any set of coherent units. The relevant basic SI units are: Q (m^3 s^{-1}), r (m), h (m) and k_{UM} (m s^{-1}).

26.4.2.2 Description of Liquid Migration Through a Composite Liner

The contact between a geomembrane and the underlying soil is never perfect. There is a small space between the geomembrane and the underlying soil in which some flow can take place. The mechanism of liquid flow through a composite liner is described as follows by Giroud & Bonaparte (1989b): if there is a defect in the geomembrane component of a composite liner, the liquid passes first through the geomembrane defect, then it flows laterally some distance between the geomembrane and the low permeability soil; finally, it infiltrates into and through the low permeability soil layer, which is the second component of the composite liner. Flow in the space between the geomembrane and the low permeability soil is called interface flow, and the area covered by the interface flow is called the wetted area.

26.4.2.3 Quality of Contact Between the Two Components of a Composite Liner

The quality of the contact between the two components of a composite liner, i.e. the geomembrane and the low permeability soil, is one of the key factors governing the rate of flow through the composite liner, because it governs the radius of the wetted area. Good and poor contact conditions have been defined by Bonaparte et al. (1989) and Giroud (1997) as:

- good contact conditions correspond to a geomembrane installed, with as few wrinkles as possible, on top of a low permeability soil layer that has been adequately compacted and has a smooth surface; and
- poor contact conditions correspond to a geomembrane that has been installed with a certain number of wrinkles, and/or placed on a low permeability soil that has not been well-compacted and does not appear smooth.

Furthermore, as stated by Giroud (1997):

- for good contact conditions, it is assumed that there is sufficient compressive stress to maintain the geomembrane in contact with the low permeability soil layer; and
- in the case of a GCL, good contact conditions may be assumed because GCLs are usually installed flat, and because the bentonite slurry that may exude from a hydrated GCL contributes to establishing a close contact between the geomembrane and the GCL, provided sufficient compressive stress is applied.

Rowe (1998a) has developed a methodology to evaluate the rate of liquid migration through a defect located in a wrinkle of the geomembrane. He has shown that the rate of liquid migration through a defect located in a wrinkle is significantly greater than the rate of liquid migration through a defect located in an area where the geomembrane is in contact with the underlying low permeability material. This finding confirms the requirement for "good contact" given above. This is especially important given that Soong and Koerner (1998) have shown that geomembrane wrinkles do not significantly flatten once buried by placement of overlying soil layers (i.e. at the location of wrinkles, a gap remains between the geomembrane and the underlying surface). Rowe's equations for calculating leakage through a defect in a wrinkle are given in Section 26.4.2.8.

In liquid migration rate equations, contact quality is represented by a dimensionless contact quality factor, C_q. In the case of a circular defect, a square defect, or a defect with similar dimensions in all directions (sometimes referred to as a quasi-circular defect), the contact quality factor is C_{qo} with:

$$C_{qo\,good} \le C_{qo} \le C_{qo\,poor} \qquad (26.33)$$

where $C_{qo\,good}$ = value of C_{qo} in the case of good contact conditions, and $C_{qo\,poor}$ = value

of C_{qo} in the case of poor contact conditions. (The good and poor contact conditions were defined above.) The following values were established semi-empirically (Giroud et al. 1989).

$$C_{qo\,good} = 0.21 \qquad (26.34a)$$

$$C_{qo\,poor} = 1.15. \qquad (26.34b)$$

In the case of a defect of great length with respect to its width, the contact quality factor is $C_{q\infty}$, with:

$$C_{q\infty\,good} \le C_{q\infty} \le C_{q\infty\,poor} \qquad (26.35)$$

where $C_{q\infty\,good}$ = value of $C_{q\infty}$ in the case of good contact conditions, and $C_{q\infty\,poor}$ = value of $C_{q\infty}$ in the case of poor contact conditions. The following values were established semi-empirically (Giroud et al. 1992a):

$$C_{q\infty\,good} = 0.52 \qquad (26.36a)$$

$$C_{q\infty\,poor} = 1.22. \qquad (26.36b)$$

26.4.2.4 Parameters and Units Used in Equations for Liquid Migration Rate Evaluation
The following notations are used in the equations presented in Sections 26.4.2.5 and 26.4.2.6: Q = rate of liquid migration through a geomembrane defect; Q^* = liquid migration rate per unit length of geomembrane defect in the case where the geomembrane defect has an infinite length; h = hydraulic head on top of the liner; t_{UM} = thickness of the medium underlying the geomembrane, i.e. the low permeability soil component of the composite liner; B = length of a rectangular defect; b = side length of a square defect, width of a rectangular defect, or width of an infinitely long defect; d = diameter of a circular defect; a = defect area; k_{UM} = hydraulic conductivity of the medium underlying the geomembrane, i.e. the low permeability soil component of the composite liner; C_{qo} and $C_{q\infty}$ = contact

quality factors, i.e. dimensionless factors that characterizes the quality of contact between the geomembrane and the underlying low permeability soil. Numerical values of C_{qo} and $C_{q\infty}$ are given in Section 26.4.2.3.

It is important to note that the equations for liquid migration rate presented in Sections 26.4.2.5 and 26.4.2.6 are semi-empirical and can only be used with the following SI units: Q (m^3 s^{-1}), Q° (m^2 s^{-1}), h (m), t_s (m), B (m), b (m), d (m), a (m^2) and k_{UM} (m s^{-1}). C_{qo} and $C_{q\infty}$ are dimensionless.

26.4.2.5 General Equations for Liquid Migration Rate

The following equations were developed by Giroud (1997), based on preceding work by Giroud & Bonaparte (1989b) and Giroud et al. (1989, 1992a):

- Circular or quasi-circular defect:

$$Q = C_{qo}[1 + 0.1(h/t_{UM})^{0.95}]a^{0.1}h^{0.9}k_{UM}^{0.74} \quad (26.37)$$

hence, for a circular defect:

$$Q = 0.976C_{qo}[1 + 0.1(h/t_{UM})^{0.95}]d^{0.2}h^{0.9}k_{UM}^{0.74}. \quad (26.38)$$

- Square defect:

$$Q = C_{qo}[1 + 0.1(h/t_{UM})^{0.95}]b^{0.2}h^{0.9}k_{UM}^{0.74}. \quad (26.39)$$

- Rectangular defect:

$$Q = C_{qo}[1 + 0.1(h/t_{UM})^{0.95}]b^{0.2}h^{0.9}k_{UM}^{0.74}$$
$$+ C_{q\infty}[1 + 0.2(h/t_{UM})^{0.95}] \quad (26.40)$$
$$\times (B - b)b^{0.1}h^{0.45}k_{UM}^{0.87}.$$

- Infinitely long defect ($B = \infty$ in Eq. 26.40):

$$Q^\circ = C_{q\infty}[1 + 0.2(h/t_{UM})^{0.95}]b^{0.1}h^{0.45}k_{UM}^{0.87}. \quad (26.41)$$

Equations 26.37–26.41 must be used with the units defined in Section 26.4.2.4. The parameters in Eqs 26.37–26.41 are defined in Section 26.4.2.4. Values of C_{qo} are given by Eq. 26.34, and values of $C_{q\infty}$ by Eq. 26.36.

26.4.2.6 Equations for Liquid Migration Rate for the Case of Small Hydraulic Heads

In the above equations, the analytical expression of the average hydraulic gradient in the low permeability soil component of the composite liner is given by:

$$i_{avgo} = 1 + 0.1(h/t_{UM})^{0.95}$$

for a defect of finite size   (26.42a)

$$i_{avg\infty} = 1 + 0.2(h/t_{UM})^{0.95}$$

for an infinitely long defect.   (26.42b)

Equation 26.42a and b shows that the hydraulic gradient is approximately equal to one if the head of liquid above the liner is less than the thickness of the low permeability soil component of the composite liner (a situation that typically exists in landfills where the leachate head above the liner is small, e.g. 0.3 m or less, and the soil component of the composite liner is a CCL with a thickness of 0.3 m or more). Hence, the following equations can be derived from Eqs 26.37–26.41 (but, in reality, were established first, hence the references given hereafter):

- Circular or quasi-circular defect (Giroud et al. 1989):

$$Q = C_{qo}a^{0.1}h^{0.9}k_{UM}^{0.74}. \quad (26.43)$$

hence, for a circular defect:

$$Q = 0.976C_{qo}d^{0.2}h^{0.9}k_{UM}^{0.74}. \quad (26.44)$$

- Square defect [1875]

$$Q = C_{qo}b^{0.2}h^{0.9}k_{UM}^{0.74}. \quad (26.45)$$

- Rectangular defect [1875]:

$$Q = C_{qo}b^{0.2}h^{0.9}k_{UM}^{0.74}$$
$$+ C_{q\infty}(B - b)b^{0.1}h^{0.45}k_{UM}^{0.87}. \quad (26.46)$$

- Infinitely long defect (B = ∞ in Eq. 26.46) (Giroud et al. 1992):

$$Q^\circ = C_{q\infty}b^{0.1}h^{0.45}k_{UM}^{0.87}. \quad (26.47)$$

Equations 26.43–26.47 must be used with the units defined in Section 26.4.2.4. The parameters used in Eqs 26.43–26.47 are defined in Section 26.4.2.4. Values of C_{qo} are given by Eq. 26.34 and values of $C_{q\infty}$ by Eq. 26.36.

26.4.2.7 Limitations of the Equations for Liquid Migration Through Composite Liners

The limits of validity of the equations presented in Sections 26.4.2.5 and 26.4.2.6 result from considerations such as: the experimental data supporting the work that led to these equations, restrictions to flow imposed by surface tension and the range of applicability of Bernoulli's equation for free flow through an orifice. These limits can be summarized as folows (Giroud *et al.* 1997c):

- If the defect is circular, the defect diameter should be no less than 0.5 mm and not greater than 25 mm. In the case of defects that are not circular, it is proposed to use these limitations for the defect width.
- The liquid head on top of the geomembrane should be equal to or less than 3 m.
- The hydraulic conductivity of the low permeability soil underlying the geomembrane should be equal to or less than a certain value $k_{UM\,max}$. To ensure a smooth transition between liquid migration rates calculated using Eq. 26.37–26.47 and those calculated using Bernoulli's equation (see Section 26.4.3.1.1), Giroud *et al.* (1997c) propose the following value for $k_{UM\,max}$ in the case where the geomembrane defect is circular:

$$k_{UM\,max} = \left[0.3891 d^{1.8} \middle/ \left\{ C_{qo} \left[1 + 0.1 \left(\frac{h}{t_{UM}} \right)^{0.95} \right] h^{0.4} \right\} \right]^{1/0.74} . \quad (26.48)$$

Equation 26.48 must be used with the units defined in Section 26.4.2.4. Values of $k_{UM\,max}$ calculated using Eq. 26.48 with $C_{qo} = 0.21$ (i.e. values of C_{qo} for good contact conditions,

as indicated by Eq. 26.34) and $t_{UM} = 0.6$ m are given in Table 26.3.

26.4.2.8 Leakage Through a Wrinkle

Rowe (1998a) developed the following equation for calculating the flow through a defect in a wrinkle in the geomembrane component of a composite liner:

$$Q = 2L(1 + h/t_{UM})[k_{UM}b + (k_{UM}t_{UM}\theta)^{1/2}] \quad (26.49)$$

where Q = rate of liquid migration through the defect(s) in the wrinkle, L = length of the wrinkle, b = half-width of the wrinkle $(L \gg b)$, h = head on top of the liner, t_{UM} = thickness of the low permeability soil component of the composite liner, k_{UM} = hydraulic conductivity of the low permeability soil component of the composite liner, and θ = hydraulic transmissivity of the interface between the geomembrane and soil components of the composite liner away from the wrinkle.

Equation 26.49 is based on the assumption that the rate of liquid migration through the geomembrane defect(s) is sufficient to fill the volume between the geomembrane wrinkle and the top of the low permeability soil component of the composite liner, and that the pressure head at the bottom of the low permeability soil component of the composite liner is zero. More general cases can also be considered, as discussed by Rowe (1998a). Equation 26.49 can be used with any set of coherent units. The relevant basic SI units are: Q (m³ s⁻¹), L (m), b (m), h (m), t_{UM} (m), k_{UM} (m s⁻¹), θ (m² s⁻¹).

As also indicated by Rowe (1998a), Eq. 26.49 gives an upper limit to the liquid migration rate. A second limit recognizes that the flow may be controlled by: (a) Bernoulli's equation if the hydraulic conductivity of the overlying layer is high enough (see Section 26.4.3.1); and (b) by the capacity of the fluid to drain to the hole. The latter case was examined by Giroud *et al.* (1997b) and can be calculated as discussed in Section 26.4.3.2.

TABLE 26.3. Maximum value, k_{UMmax} (m s⁻¹), of the hydraulic conductivity of the medium underlying the geomembrane for Eqs 26.37–26.47 to be valid with an acceptable approximation in the case where $C_{qo} = 0.21$ (good contact) and $t_{UM} = 0.6$ m (adapted from Giroud et al. 1997c)

Liquid head on top of the geomembrane, h (m)	Geomembrane defect diameter, d (mm)[a]						
	0.5	1	2	3	5	10	11.284
0.01	2.6×10^{-7}	1.4×10^{-6}	7.5×10^{-6}	2.0×10^{-5}	7.0×10^{-5}	3.8×10^{-4}	5.1×10^{-4}
0.03	1.4×10^{-7}	7.7×10^{-7}	4.1×10^{-6}	1.1×10^{-5}	3.8×10^{-5}	2.1×10^{-4}	2.8×10^{-4}
0.1	7.3×10^{-8}	3.9×10^{-7}	2.1×10^{-6}	5.7×10^{-6}	2.0×10^{-5}	1.1×10^{-4}	1.4×10^{-4}
0.3	3.8×10^{-8}	2.1×10^{-7}	1.1×10^{-6}	3.0×10^{-6}	1.0×10^{-5}	5.6×10^{-5}	7.5×10^{-5}
1	1.8×10^{-8}	9.5×10^{-8}	5.1×10^{-7}	1.4×10^{-6}	4.7×10^{-6}	2.6×10^{-5}	3.4×10^{-5}
3	7.1×10^{-9}	3.8×10^{-8}	2.1×10^{-7}	5.6×10^{-7}	1.9×10^{-6}	1.0×10^{-5}	1.4×10^{-5}

[a] For d = 11.284 mm, a = 1 cm².

TABLE 26.4. Minimum value, k_{UMmin} (m s⁻¹), of the hydraulic conductivity of the medium underlying the geomembrane for Eq. 26.51 (i.e. Bernoulli's equation for free flow through an orifice) to be valid with an acceptable approximation (adapted from Giroud et al. 1997c)

Geomembrane defect diameter, d (mm)[a]	0.5	1	2	3	5	10	11.284
k_{UMmin} (m s⁻¹)	2.5×10^{-4}	1.0×10^{-3}	4.0×10^{-3}	9.0×10^{-3}	2.5×10^{-2}	1.0×10^{-1}	1.3×10^{-1}

[a] For d = 11.284 mm, a = 1 cm².

For a geomembrane on a CCL, Rowe (1998a) indicated that the hydraulic transmissivity, θ, can be estimated as follows for good contact away from the wrinkle:

$$\log_{10}\theta = 0.07 + 1.036(\log_{10}k_{UM})$$
$$+ 0.018(\log_{10}k_{UM})^2 \qquad (26.50)$$

where Eq. 26.50 is semi-empirical and must be used with units θ (m^2 s^{-1}), k_{UM} (m s^{-1}). For GCLs, θ may be measured in laboratory tests or estimated from published values, see (Rowe 1998a) for a detailed discussion.

26.4.3 GEOMEMBRANE LINER

26.4.3.1 Geomembrane Overlain and Underlain by Highly Permeable Media

Presentation of the Equation. In the case of a geomembrane liner underlain and overlain by highly permeable media, the rate of liquid migration through a defect in the geomembrane is calculated using the classical Bernoulli's equation for free flow through an orifice, as proposed by Giroud (1984a):

$$Q = 0.6a(2gh)^{1/2} = 0.15\pi d^2(2gh)^{1/2} \quad (26.51)$$

where a = defect area, d = defect diameter in the case of a circular defect, and g = acceleration due to gravity. Equation 26.51 can be used with any set of coherent units. The relevant basic SI units are: a (m^2), d (m), g (m s^{-2}), h (m).

Limit of Applicability of Bernoulli's Equation. As indicated above, Bernoulli's equation is applicable to geomembranes underlain and overlain by highly permeable media. As shown by Giroud *et al.* (1997c), Bernoulli's equation (Eq. 26.51) provides an acceptable value of the liquid migration rate for values of the hydraulic conductivity of the medium underlying the geomembrane that are greater than $k_{UM\,min}$ given by:

$$k_{UM\,min} = 10^3 d^2 \qquad (26.52)$$

with $k_{UM\,min}$ (m s^{-1}) and d(m). Values of $k_{UM\,min}$ calculated using Eq. 26.52 are given in Table 26.4.

Giroud *et al.* (1997b) have shown that Bernoulli's equation gives the rate of liquid migration through a geomembrane defect with an error of less than 5% if the hydraulic conductivity of the medium overlying the geomembrane, k_{OM}, is greater than:

$$k_{OM\,min} = \frac{30d^2}{h^{3/2}} \qquad (26.53)$$

where the following units must be used: d(m), h(m) and $k_{OM\,min}$ (m s^{-1}).

26.4.3.2 Geomembrane Overlain by a Permeable Medium and Underlain by a Highly Permeable Medium

Presentation of the Equation. When a geomembrane is overlain and underlain by highly permeable media, the rate of liquid migration through a geomembrane defect is given by the classical Bernoulli's equation for free flow through an orifice (Eq. 26.51). Bernoulli's equation sometimes grossly overestimates the rate of liquid migration through a geomembrane defect if the geomembrane is overlain by a medium that is not highly permeable (even though the geomembrane is underlain by a highly permeable medium). In this case, the flow of liquid toward the geomembrane defect is hindered and, as a result, the rate of liquid migration is less than in the ideal case of a geomembrane overlain and underlain by infinitely permeable media. Taking into account the fact that liquid does not flow freely toward the geomembrane defect, Giroud *et al.* (1997b) developed the following equation:

$$h = \left\{\frac{aq_i}{2k_{OM}\pi} + \frac{Q}{2k_{OM}\pi}\left[\ln\left(\frac{Q}{aq_i}\right) - 1\right]\right.$$
$$\left. + \frac{1}{4g^2}\left(\frac{Q}{0.6a}\right)^4\right\}^{1/2} \qquad (26.54)$$

where q_i = rate of liquid supply on top of the medium overlying the geomembrane, and k_{OM} = hydraulic conductivity of the medium overlying the geomembrane. Equation 26.54 can be used with any set of coherent units. The relevant basic SI units are: h (m), a (m^2), q_i (m s^{-1}), k_{OM} (m s^{-1}), Q (m^3 s^{-1}) and g (m s^{-2}).

It should be noted that, if k_{OM} is infinite, Eq. 26.54 becomes identical to Equation 26.51, i.e. Bernoulli's equation for free flow through an orifice.

Use of the Equation. Equation 26.54 cannot be solved for Q. Therefore, an iterative procedure is necessary to calculate Q when h, a, k_{OM} and q_i are known. Alternatively, graphical solutions can be used. An example is shown in Fig. 26.6, and a series of similar graphical solutions is provided by Giroud *et al.* (1997b). Figure 26.6 shows that, in general, Bernoulli's equation overestimates the liquid migration rate. However, Fig. 26.6 also shows that, for values of the hydraulic conductivity of the medium overlying the geomembrane that meet the condition expressed by Eq. 26.53, Bernoulli's equation provides an excellent approximation of the liquid migration rate.

26.4.3.3 Geomembrane Overlain by a Permeable Medium and Underlain by a Semi-permeable Medium

Definition of the Case Considered. The case of a geomembrane overlain by a permeable me-

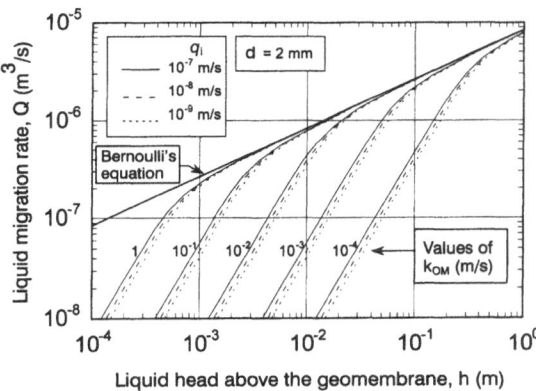

FIGURE 26.6. Graphical solution of Eq. 26.54 for a geomembrane defect having a diameter, d, of 2 mm (after Giroud *et al.* 1997b, © Geosynthetics International, reproduced with permission): $q_i = $ (—) 10^{-7} m s^{-1}, (— —) 10^{-8} m s^{-1}, (····) 10^{-9} m s^{-1}.

dium and underlain by a semi-permeable medium is intermediary between the case of a geomembrane overlain and underlain by infinitely permeable media (Section 26.4.3.1) and the case of a composite liner (Section 26.4.2)

In the context of the liquid migration rate equations discussed herein, a semi-permeable soil is a soil that is not sufficiently permeable to allow the use of Eq. 26.51, i.e. Bernoulli's equation for free flow through an orifice, and too permeable to allow the use of Eqs 26.37–26.47 for composite liners. In other words, the equation presented in this section is valid only for soil hydraulic conductivities of the medium underlying the geomembrane between two limit values, $k_{UM\,max}$ and $k_{UM\,min}$, defined by Eqs 26.48 and 26.52, respectively.

Presentation of the Equation. When a geomembrane is overlain and underlain by infinitely permeable media, the rate of liquid migration through a geomembrane defect is given by Bernoulli's equation for free flow through an orifice (Eq. 26.51 given in Section 26.4.3.1). When a geomembrane is underlain by a low permeability medium, i.e. when the geomembrane and the underlying medium form a composite liner, the rate of liquid migration through a geomembrane defect is given by Giroud's equations (Eqs 26.37–26.47). To evaluate the rate of liquid migration through defects in geomembranes underlain by a semi-permeable medium, i.e. when the hydraulic conductivity, k_{UM}, of the medium underlying the geomembrane is between $k_{UM\,max}$ and $k_{UM\,min}$, Giroud *et al.* (1997c) have developed the following equation:

$$\log Q = 0.3195 + 2 \log d + 0.5 \log h \\ - 0.74 \left(\frac{5 + 2 \log d - \log k_{OM}}{m} \right)^m \tag{26.55}$$

where k_{OM} = hydraulic conductivity of the semi-permeable soil, and m = dimensionless exponent given by:

$$m = 4.6380 - 0.4324 \log d + 0.5405 \log h$$
$$+ 1.3514 \log \left[1 + 0.1 \left(\frac{h}{t_{UM}} \right)^{0.95} \right] \quad (26.56)$$

where t_{UM} = the thickness of the semi-permeable soil. Equations 26.55 and 26.56 were used with $C_{qo} = 0.21$, i.e. assuming good contact between the geomembrane and the underlying medium. Equations 26.53 and 26.54 can only be used with the following units: Q (m^3 s^{-1}), d (m), k_{OM} (m s^{-1}), h (m), t_{UM} (m). Also Equation 26.53 is valid only for hydraulic heads equal to or less than 3 m.

Use of the Equation. Figure 26.7 shows a series of curves that represent the rate of liquid migration through a given geomembrane defect (diameter, d = 2 mm) as a function of the hydraulic conductivity of the medium underlying the geomembrane for various liquid heads. Each curve in Fig. 26.7 comprises three portions: the left-hand portion (straight

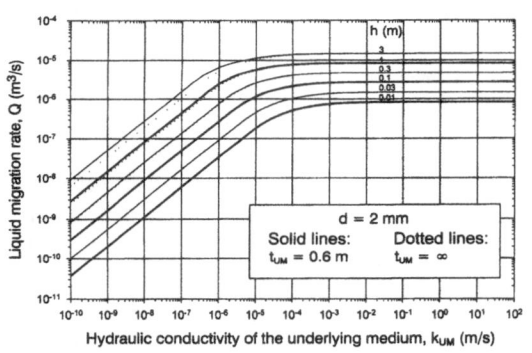

FIGURE 26.7. Rate of liquid migration through a defect having a diameter, d, of 2 mm in a geomembrane underlain by a medium with a hydraulic conductivity, k_{UM}, and overlain by a medium that is significantly more permeable than the underlying medium, for various values of the head of liquid on top of the geomembrane, h (after Giroud *et al.* 1997c, © Geosynthetics International, reproduced with permission): (——) thickness of the medium of hydraulic conductivity, k_{UM}, underlying the geomembrane is t_{UM} = 0.6 m; (····) thickness of the underlying medium is infinite.

line) represents Eq. 26.38; the right-hand portion (plateau) represents Eq. 26.51; and the central portion (curve) was obtained using Eq. 26.55. Both Eqs 26.38 and 26.51 were used with $C_{qo} = 0.21$ (Eq. 26.34a), i.e. assuming good contact between the geomembrane and the underlying medium. The limit value of the hydraulic conductivity between the left-hand portion and the central portion is k_{UMmax}, given by Eq. 26.48 and Table 26.3. The limit value of the hydraulic conductivity between the central portion and the right-hand portion is k_{UMmin}, given by Eq. 26.52 and Table 26.4. Similar graphs for other values of d are given by Giroud *et al.* (1997c).

26.5 Stability and Deformations of Linear Systems

26.5.1 INTRODUCTION

26.5.1.1 Types of Mechanical Actions Applied on Liner Systems

Liner systems are subjected to a variety of mechanical actions that may impact their integrity, such as: (a) instability of soil or waste in contact with the liner system; (b) deformation of soil in contact with the liner system; (c) differential settlement between the soil supporting the liner system and a structure to which the liner system is connected; (d) liner system uplift by fluids; and (e) concentrated stresses applied on the geosynthetic component of a liner system by the materials in contact. These five types of mechanical actions are discussed below.

26.5.1.2 Instability of Soil or Waste in Contact with a Liner System

Instability of soil or waste in contact with a liner system can cause extensive damage to the liner system, in particular to the geosynthetics. Numerous examples of such problems are provided by Rowe (1998a, 1998b) and Giroud (1999). When large masses of soil or waste are involved, general slope stability analysis methods apply (see Chapter 14). Specific slope stability analysis methods were developed for liner systems on slopes. Details can be found in Giroud & Beech (1989), Koerner & Hwu (1991), Giroud et al. (1995a, 1995d), and Koerner & Soong (1998), and in Sections 27.6 (static stability) and 27.7 (seismic stability).

26.5.1.3 Deformation of Soil in Contact with the Liner System

When the medium (e.g. soil, waste) supporting a liner system settles, the liner system may follow the shape of the depressed area or may bridge the depressed area or a portion of it. As a result, the liner system undergoes tensile stresses and strains. This topic is discussed in Section 26.5.2. (Settlement of landfill cover systems is discussed in Section 27.8.4.)

26.5.1.4 Differential Settlement between the Soil Supporting the Liner System and a Structure to which the Liner System is Connected

If differential settlement takes place between the soil supporting the liner system and a structure to which the liner system is connected, the liner system undergoes tensile stresses and strains. This topic is discussed in Section 26.5.3.

26.5.1.5 Uplift of Liner System by Fluids

Liner systems are relatively light and they may be uplifted by pressure applied from underneath by liquids and gas. These situations are addressed by using a drainage system (underdrain) for liquids and/or gas under the liner system. In the case of landfills, a properly designed and operated gas collection system, which is primarily intended to collect gas for energy recovery, prevents the development of high gas pressure in the landfill, thereby minimizing the risk of cover system uplift.

Exposed geomembranes are particularly susceptible to uplift by fluids. The terminology "exposed geomembrane" refers to geomembranes that are not covered with a layer of soil or other material. Exposed geomembranes are used, for example, as liners in many liquid impoundments and as covers in some landfills. The risk of uplifting is high in the case of an exposed geomembrane that is a component of a composite liner for a liquid impoundment. If there is a defect in the geomembrane, which should always be assumed at the design stage, liquid accumulates between the geomembrane and the underlying low permeability soil when the reservoir is full and uplifts the geomembrane in case of drawdown of the reservoir. Therefore, composite liners in liquid impoundments should always be covered with a layer of heavy material, both to minimize the risk of problems due to geomembrane uplift and to protect the geomembrane.

The uplift of exposed geomembrane liners due to suction generated by the wind is an important design consideration. This topic is discussed in Section 26.5.4.

26.5.1.6 Concentrated Stresses Applied on the Geosynthetic Component of a Liner System by the Materials in Contact

Geosynthetics, in particular geomembranes, can be damaged as a result of mechanical actions by materials in contact. For example, the influence of a scratch at the surface of a high density polyethylene geomembrane on its tensile behavior has been quantified in Giroud (1984b, 1994) and Giroud et al. (1994a). Even when geomembranes are not damaged by materials in contact, they may be subjected to high concentrated stresses and strains as a result of the size and shape of the materials in contact and the applied stresses. These high concentrated stresses and strains may be detrimental to the long-term performance of the geomembrane.

Geomembranes are generally protected using geotextiles. Based on the foregoing discussion, the geotextile should not only prevent damage to the geomembrane but also minimize the concentrated stresses and strains induced in the geomembrane. These two aspects of geomembrane protection are discussed by Rowe (1998a). The following references are cited by Rowe (1998a) for discussions of geomembrane damage prevention: Giroud *et al.* (1995b); Narejo (1995a, 1995b, 1996); Bishop (1996); Reddy *et al.* (1996); Wilson-Fahmy *et al.* (1996); Narejo *et al.* (1996); Koerner *et al.* (1996); and Badu-Tweneboah *et al.* (1998). The following references are cited by Rowe (1998a) for discussions of concentrated stresses and strains in geomembranes resulting from materials in contact: Brummermann *et al.* (1994); Saathoff & Sehrbrock (1994); Bishop (1996); Seeger & Müler (1996); and Giroud (1996b). Recent publications on the subject include: Badu-Tweneboah *et al.* (1997); Thiel (1997); Reddy *et al.* (1997); Zanzinger & Gartung (1998); and Tognon *et al.* (1999).

26.5.2 GEOSYNTHETIC STRAIN DUE TO SETTLEMENT OF THE SUPPORTING MEDIUM

26.5.2.1 Relationships between Geosynthetic Deflection and Strain

If a geosynthetic is assumed to have a circular shape when it deflects on an infinitely long depression (trough) or a circular depression of the supporting medium (Fig. 26.8), the fol-

lowing relationship exists between the geosynthetic deflection and the average strain in the geosynthetic (Giroud 1981):

$$\varepsilon = 2\Omega \sin^{-1}\frac{1}{2\Omega} - 1 \qquad (26.57a)$$

where:

$$\Omega = (\tfrac{1}{4})\{(2y/b) + [b/(2y)]\}$$

$$\text{for an infinitely long depression} \qquad (26.57b)$$

$$\Omega = (\tfrac{1}{4})\{(2y/d) + [d/(2y)]\}$$

$$\text{for a circular depression} \qquad (26.57c)$$

where ε = average strain in the geosynthetic, y = geosynthetic deflection, b = width of infinite long depression, d = diameter of circular depression, and Ω = dimensionless factor. It should be noted that Eq. 26.57 is valid only for $y/b < 0.5$, i.e. $\varepsilon < 0.57$ (57%). Numerical values of ε and Ω as a function of y/b or y/d are given in Table 26.5. Equation 26.57b and c can be used with any set of coherent units. The relevant basic SI units are: b (m), d (m) and y (m); Ω is dimensionless.

Giroud (1995) has shown that, for a geosynthetic strain less than 30%, i.e. for a relative deflection (y/b or y/d) less than 0.35, a good approximation of the geosynthetic strain is given by:

$$\varepsilon \approx \frac{8}{3}\left(\frac{y}{b}\right)^2 \quad \text{for an infinitely long depression}$$

$$(26.58a)$$

$$\varepsilon \approx \frac{8}{3}\left(\frac{y}{d}\right)^2 \quad \text{for a circular depression.} \qquad (26.58b)$$

It should be noted that Eqs 26.57 and 26.58 are valid regardless of the cause of the depression, regardless of the magnitude of the pressure applied on the geosynthetic, and whether the geosynthetic is in contact with the supporting medium (Fig. 26.8a) or bridges the depression (Fig. 26.8b).

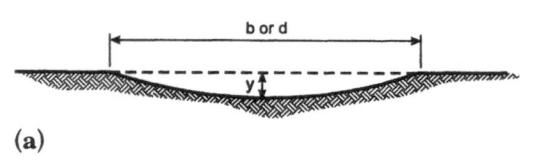

(a)

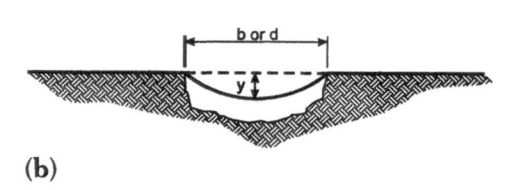

(b)

FIGURE 26.8. Geosynthetic deflection: (a) geosynthetic supported by soil; and (b) geosynthetic overlying a void (materials overlying the geomembrane are not shown).

TABLE 26.5. Corresponding values of relative deflection, y/b or y/d, geosynthetic strain, ε, and dimensionless parameter Ω (adapted from Giroud *et al.* 1990)

y/b or y/d	ε (%)	Ω	y/b or y/d	ε (%)	Ω
0.000	0.000	∞	0.242	15.00	0.64
0.010	0.027	12.51	0.250	15.91	0.62
0.020	0.107	6.26	0.260	17.15	0.61
0.030	0.240	4.18	0.270	18.43	0.60
0.040	0.425	3.15	0.280	19.75	0.59
0.050	0.663	2.53	0.282	20.00	0.58
0.060	0.960	2.11	0.290	21.10	0.58
0.061	1.000	2.07	0.300	22.50	0.57
0.070	1.30	1.82	0.310	23.93	0.56
0.080	1.70	1.60	0.317	25.00	0.55
0.087	2.00	1.47	0.320	25.39	0.55
0.090	2.15	1.43	0.330	26.89	0.54
0.100	2.65	1.30	0.340	28.43	0.54
0.107	3.00	1.23	0.350	30.00	0.53
0.110	3.20	1.19	0.360	31.60	0.53
0.120	3.80	1.10	0.370	33.23	0.52
0.123	4.00	1.08	0.380	34.90	0.52
0.130	4.45	1.03	0.381	35.00	0.52
0.138	5.00	0.97	0.390	36.60	0.52
0.140	5.15	0.96	0.400	38.32	0.51
0.150	5.90	0.91	0.410	40.00	0.51
0.151	6.00	0.90	0.420	41.86	0.51
0.160	6.69	0.86	0.430	43.67	0.51
0.164	7.00	0.84	0.437	45.00	0.50
0.170	7.54	0.82	0.440	45.51	0.50
0.175	8.00	0.80	0.450	47.38	0.50
0.180	8.43	0.78	0.460	49.27	0.50
0.186	9.00	0.76	0.464	50.00	0.50
0.190	9.36	0.75	0.470	51.18	0.50
0.197	10.00	0.73	0.480	53.13	0.50
0.200	10.35	0.72	0.490	55.00	0.50
0.210	11.37	0.70	0.500	57.08	0.50
0.216	12.00	0.69	0.562	70.00	0.50
0.220	12.44	0.68	0.631	85.00	0.51
0.230	13.56	0.66	0.696	100.00	0.53
0.240	14.71	0.64	0.819	130.00	0.56

26.5.2.2 Relationships between Geosynthetic Deflection and Tension

Assuming that the geosynthetic has a circular shape when it deflects over a depression, the following relationship exists between the geosynthetic tension, the width of the depression, and the applied pressure, which is assumed to be uniformly distributed on the geosynthetic (Giroud 1981).

$$T = pb\Omega \quad \text{for an infinitely long depression}$$
(26.59a)

$$T = pr\Omega \quad \text{for a circular depression}$$ (26.59b)

where T = geosynthetic tension, p = pressure normal to the geosynthetic and uniformly distributed over the geosynthetic, b = width of the infinitely long depression, r =

radius of the circular depression, and $\Omega =$ dimensionless factor. Equation 26.59a and b can be used with any set of coherent units. The relevant basic SI units are: T (N m^{-1}), b (m), r (m), and p (Pa); Ω is dimensionless.

As pointed out by Giroud (1981), it should be noted that, if the depression over which the geosynthetic deflects is circular, a uniformly distributed pressure does not correspond to a spherical deflection and, conversely, a spherical deflection does not correspond to a uniformly distributed pressure. Therefore, Eq. 26.59b is only approximate. In contrast, if the depression over which the geosynthetic deflects is infinitely long, and if the geosynthetic has uniform characteristics over its entire area, a uniformly distributed pressure does result in a circular shape of the deformed geosynthetic.

26.5.2.3 Evaluation of the Pressure Applied on the Geosynthetic

If the considered geosynthetic is a geomembrane that is not covered with a layer of soil, the pressure applied on the geomembrane is the pressure applied by the liquid. If the geomembrane is overlain by a layer of soil (Fig. 26.9), only a fraction of the overburden load is applied to the geomembrane due to the development of arching in the soil, as the geosynthetic deflects. Several researchers have presented equations that account for arching including: Bonaparte & Berg (1987), Giroud et al. (1990), and Poorooshasb (1991). The

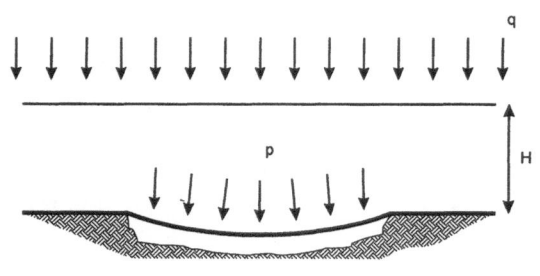

FIGURE 26.9. Geosynthetic deflection under a soil layer.

pressure applied to the geomembrane, as presented by Giroud et al. (1990), is given by:

$$p = 2\gamma b[1 - \exp(-0.5H/b)] + q\exp(-0.5H/b)$$

for an infinitely long depression   (26.60a)

$$p = 2\gamma r[1 - \exp(-0.5H/b)] + q\exp(-0.5H/b)$$

for a circular depression.   (26.60b)

It should be noted that, for $H = 0$, Eq. 26.60a and b gives $p = q$. Combining Eqs 26.59a and 26.60a, and Eqs 26.59b and 26.60b, respectively, gives:

$$T = b\Omega\{2\gamma b[1 - \exp(-0.5H/b)]$$
$$+ q\exp(-0.5H/b)\}$$

for an infinitely long depression   (26.61a)

$$T = r\Omega\{2\gamma r[1 - \exp(-0.5H/r)]$$
$$+ q\exp(-0.5H/r)\}$$

for a circular depression.   (26.61b)

It should be noted that, for $H = 0$, Eq. 26.61a and b is identical to Eq. 26.59a and b, respectively, since, in this case, $p = q$. Equation 26.60a and b can be used with any set of coherent units. The relevant basic SI units are: b (m), r (m), H (m), p (Pa), q (Pa), T (N m^{-1}) and γ (N, m^{-3}); Ω is dimensionless.

26.5.2.4 Use of the Equations

Equation 26.61a and b is used to calculate the required geosynthetic tension, T, for a given value of Ω, when all other parameters (b or r, q, H, and γ) are given. The dimensionless parameter Ω depends either on the allowable geosynthetic strain, ε, or the allowable geosynthetic deflection, y (see Eq. 26.57). The allowable geosynthetic strain is the lesser of the maximum design strains for the considered geosynthetic and the strain beyond which the soil layer would be unacceptably deformed or cracked. The allowable geosynthetic deflection is considered when excessive deflection of the soil surface impairs the serviceability of the system. No method is proposed in this chapter to evaluate the deflection of the soil surface; however, in the case of relatively thin soil layers, the soil surface deflection can be assumed to be of the same order as the geosynthetic deflection. In some instances, both the allowable geosynthetic strain and the allowable geosynthetic deflection may need to be considered. Also, if the geosynthetic deforms to the point that it is in contact with the bottom of the depression (Fig. 26.8a), the value of the deflection is given as being the depth of the depression; in this case, the calculated value of the pressure, p, is the difference between the pressure applied on top of the geosynthetic and the reaction of the bottom of the depression beneath the geosynthetic.

Equations 26.57–26.61 can also be used to calculate any of the parameters when the values of all the other parameters are known. For example, values of H, q or Ω can easily be calculated as a function of the other parameters. The calculation of the depression size, i.e. b or r, requires a trial-and-error approach. Detailed examples solved numerically and graphically are presented by Giroud *et al.* (1990).

Engineers using Eqs 26.59–26.61 should use appropriate factors of safety. The factor of safety can be applied to the geosynthetic tension or the applied loads, or both using a partial factor of safety approach.

26.5.3 DIFFERENTIAL SETTLEMENT AT CONNECTION OF LINER SYSTEM TO RIGID STRUCTURES

26.5.3.1 Tension and Strain in the Geomembrane

Geomembranes are generally supported by a compressible medium, such as soil and/or waste, and are sometimes connected to rigid structures, typically made of concrete, metal, or plastic, such as pipes, intake towers, water gate structures, columns, manhole shafts, and gas vents. The compressible medium that supports the geomembrane settles as a result of its own weight and as a result of the loads due to the weight of the materials overlying the geomembrane (e.g. liquid, soil and/or waste). The differential settlement between the compressible medium and the rigid structure induces a tension (force per unit width) and a strain (elongation per unit length) in the geomembrane. In many instances, geomembranes subjected to such differential settlements have undergone excessive tensions and strains, and have failed next to the rigid structure. Giroud (1973, 1977, 1983) has recommended that concrete structures connected to geomembranes be constructed with battered walls (Fig. 26.10) instead of vertical walls to obtain a settlement transition, thereby minimizing the tension and strain in the geomembrane.

This section presents a method to determine the amount of batter of the walls of the rigid structure required to decrease the tension and strain in the geomembrane to an allowable level. The method can also be used to calculate the tension and strain in a geo-

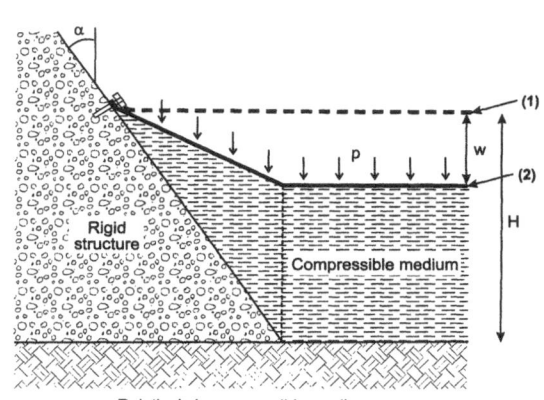

FIGURE 26.10. Geomembrane connected to a battered wall (after Giroud and Soderman 1995, © Geosynthetics International, reproduced with permission): 1 = position of the geomembrane at the time of installation; and 2 = position of the geomembrane after settlement.

membrane for a given wall batter and a given settlement of the medium underlying the geomembrane.

26.5.3.2 Assumptions

The following general assumptions are made:

- The surface of the medium supporting the geomembrane is horizontal and it is uniformly loaded.
- The medium supporting the geomembrane is homogeneous in any horizontal plane, but its properties may vary with depth.
- There is no differential settlement between the structure and the compressible medium at a certain depth, H, below the surface of the compressible medium. This assumption is considered most appropriate if both the structure and the compressible medium rest on a relatively incompressible medium (Fig. 26.10). This assumption is valid with a sufficient approximation if, at the base of the structure, the differential settlement between the structure and the compressible medium is negligible; this situation typically occurs if the compressible medium is a soil whose compression modulus increases as depth increases.
- The geomembrane is mechanically connected to the rigid structure at, or just above, the surface level of the supporting medium at the time of geomembrane installation.
- The problem is two-dimensional, and the geomembrane and the supporting medium are in a plane-strain state.

It should be noted that the last assumption restricts the validity of the equations presented herein to rigid structures of great length perpendicular to the considered cross-section. The equations are not applicable to three-dimensional structures such as circular columns or manholes. Finally, it should be noted that only the settlement of the surface of the compressible medium that occurs after the installation of the geomembrane is considered herein.

26.5.3.3 Presentation of the Method

As shown by Giroud & Soderman (1995) a conservative evaluation of the geomembrane strain at its connection to a wall consists of using the following approximate equation valid for small values of geomembrane strain, e.g. $\varepsilon < 10\%$:

$$\varepsilon_{\text{cons.app}} = \frac{1}{2}\left(\frac{w/H}{\tan \alpha}\right)^2 \qquad (26.62)$$

where $\varepsilon_{\text{cons.app}}$ = conservative and approximate value of the geomembrane strain at its connection to a rigid structure; w settlement of the soil away from the structure; H = thickness of the compressive medium or height of the structure, whichever is less; and α = wall batter, i.e. angle of the wall of the structure with a vertical plane (Fig. 26.10). Equation 26.62 can be used with any set of coherent units. The relevant basic SI units are: w (m), H (m) and α (°); ε is dimensionless.

Similarly, a conservative value of the required wall batter can be obtained from the allowable geomembrane strain, ε_{all}, using the following approximate equation valid for small values of geomembrane strain, e.g. $\varepsilon < 10\%$:

$$\tan \alpha_{\text{cons.app}} = \frac{w/H}{(2\varepsilon_{\text{all}})^{1/2}}. \qquad (26.63)$$

26.5.3.4 Practical Recommendations

Battering the walls of rigid structures is an effective solution to the problem of differential settlement causing excessive tensions and strains in geomembranes next to their connections with rigid structures. Even though the analysis presented herein is limited to two-dimensional problems, the mechanisms involved are such that wall batter is also expected to be an effective solution for three-dimensional problems. Therefore, it is recommended to use battered walls in structures, of all types and shapes, to which geomembranes are connected.

Causes of excessive settlement next to rigid structures should be avoided. The most common cause is inadequate compaction of the soil in the immediate vicinity of the structure. As heavy equipment cannot effectively compact the soil in the immediate vicinity of structures, small equipment should be used. Structures that deform excessively under lateral loads, such as lateral loads due to earth pressure and compaction-induced stresses, may also cause excessive settlements.

As shown by Giroud & Soderman (1995), the interface friction angles between the geomembrane and the materials in contact can have a significant impact on the magnitude of the tension and strain developed in a geomembrane. In general, the lower the interface friction angle, the lower the tension and strain that develops in the geomembrane. In some instances, it may be possible to reduce the effect of differential settlement by reducing the interface friction angle. Reducing the interface friction angle could be accomplished by selecting a smooth geomembrane rather than a textured geomembrane or by placing a piece of smooth geomembrane, i.e. a "slip sheet", between the geomembrane liner and the underlying medium.

26.5.4 ACTION OF WIND ON GEOMEMBRANES

26.5.4.1 Introduction

When the wind blows over an exposed geomembrane, some portions of the geomembrane are subjected to a suction and can be uplifted. The magnitude of the suction exerted by the wind is discussed in Section 26.5.4.2. Whether a geomembrane is uplifted or not depends on its own weight and the wind velocity (see Section 26.5.4.3). Prevention of geomembrane uplift is discussed in Section 26.5.4.4. If the geomembrane is uplifted, tensile stresses and strains are induced in the geomembrane (see Section 26.5.4.5), and the geomembrane must be anchored at its periphery (see Section 26.5.4.6).

26.5.4.2 Suction Due to Wind

The suction that effectively uplifts the geomembrane, S_e, is given by (Giroud et al. 1995d; Zornberg & Giroud 1997):

$$S_e = 0.050\lambda_w V^2 \exp[-(1.252 \times 10^{-4})z] \\ - 9.81\mu_{GM}\cos\beta \tag{26.64a}$$

where S_e = effective suction, λ_w = suction factor, V = wind velocity, z = altitude above sea level, μ_{GM} = mass per unit area of the geomembrane, and β = slope angle. Equation 26.64a can only be used with the following units: S_e (Pa), V (km h^{-1}), z (m), μ_{GM} (kg m^{-2}); λ_w is dimensionless. The right-hand side of Eq. 26.64a comprises two terms: the first term represents the suction applied by the wind on top of the geomembrane, and the second term represents the resistance to uplift resulting from the weight of the geomembrane.

At sea level ($z = 0$), Eq. 26.64a becomes:

$$S_e = 0.050\lambda_w V^2 - 9.81\mu_{GM}\cos\beta \tag{26.64b}$$

with the following units: S_e (Pa), V (km h^{-1}), μ_{GM} (kg m^{-2}).

The suction factor, λ_w, quantifies the fact that the wind suction is not uniformly distributed over the geomembrane. Based on data from wind tunnel tests and the need for extra safety due to gusts of wind, the following values of the suction factor, λ_w, are recommended (Giroud et al. 1995c):

- $\lambda_w = 1.00$ if the crest of a slope or the top part of a landfill cover is considered;
- $\lambda_w = 0.70$ if an entire side slope is considered;
- $\lambda_w = 0.85$ for the top third, $\lambda_w = 0.70$ for the middle third, and $\lambda_w = 0.55$ for the bottom third of a slope decomposed in three-thirds by intermediate benches or anchor trenches as shown in Fig. 26.11c and 26.11d; and
- $\lambda_w = 0.40$ in a flat area such as the bottom of a large empty reservoir.

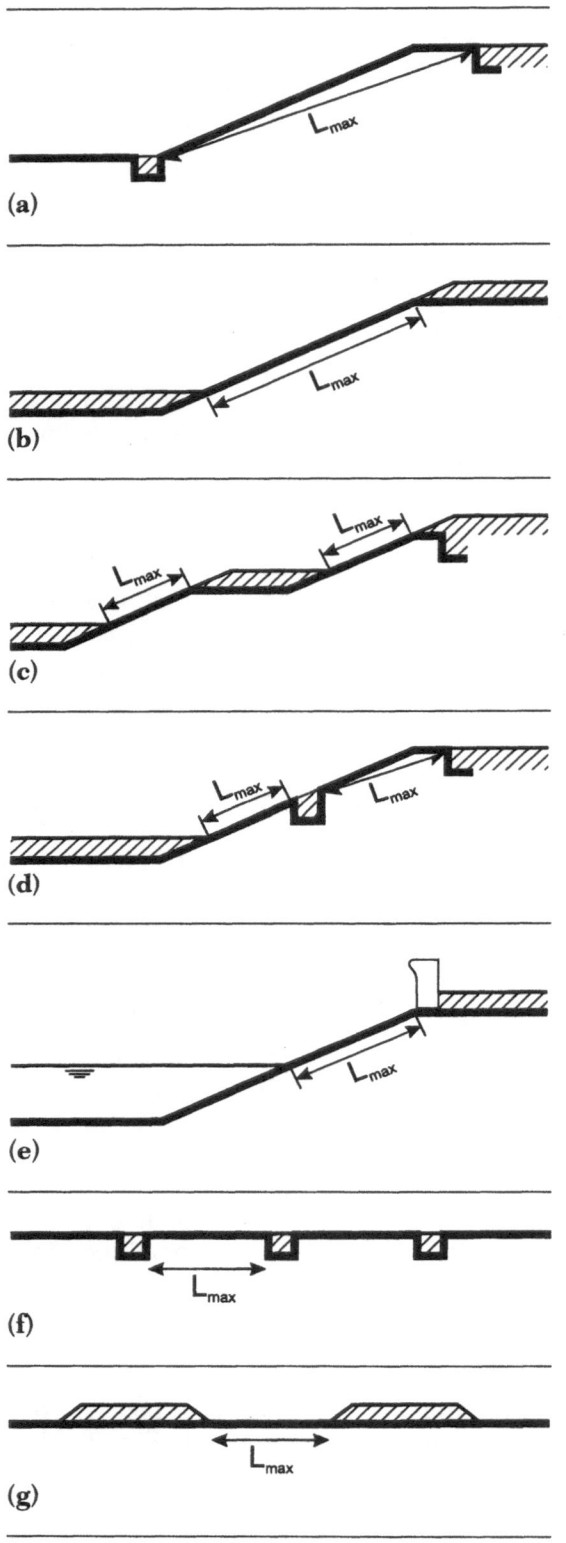

The above values of λ_w can be increased by up to 30% for geomembranes placed on a supporting medium with an unusual geometry likely to increase wind suction. Also, for unusual geometries or large projects for which wind-induced damage of exposed geomembranes may have large financial consequences, wind tunnel tests of reduced-scale models or numerical simulation may be warranted.

26.5.4.3 Resistance to Wind Uplift Due to Geomembrane Self Weight

Based on Eq. 26.64a, it appears that a geomembrane located at altitude z above sea level should not be uplifted by wind if the following condition is met by the geomembrane mass per unit area (Giroud *et al.* 1995c, Zornberg & Giroud 1997):

$$\mu_{GM} \geq \mu_{GM\,req} =$$

$$\frac{0.0051\,\lambda_w V^2 \exp[-(1.252 \times 10^{-4})z]}{\cos\beta} \quad (26.65a)$$

with the following units: $\mu_{GM\,req}$ (kg m^{-2}), V(km h^{-1}) and z (m). At sea level, Eq. 26.65 becomes:

$$\mu_{GM} \geq \mu_{GM\,req} = \frac{0.0051\,\lambda_w V^2}{\cos\beta}. \quad (26.65b)$$

FIGURE 26.11. Typical configurations of a geomembrane exposed to wind: (a) geomembrane anchored in an anchor trench; (b) geomembrane anchored under a pavement or a layer of soil; (c) geomembrane restrained by a soil layer on a bench; (d) geomembrane restrained by an intermediate anchor trench; (e) geomembrane anchored under a structure at the top and restrained by liquid or solids below a certain level; (f) at bottom of reservoir, geomembrane anchored in anchor trenches; (g) at bottom of reservoir, geomembrane anchored by strips of soil or pavement (after Giroud *et al.* 1995c, © Geosynthetics International, reproduced with permission).

TABLE 26.6. Typical density, thickness and mass per unit area
for geomembranes, and relationship between mass per unit area
and threshold wind velocity (adapted from Giroud et al. 1995c)

Type of geomembrane[a]	Geomembrane density, ρ_{GM} (kg m^{-3})	Geomembrane thickness, t_{GM} (mm)	Geomembrane mass per unit area, μ_{GM} (kg m^{-2})	Minimum uplift wind velocity, V_{thresh} (km h^{-1})
PVC	1250[b]	0.5	0.63	11.1
		1.0	1.25	15.7
HDPE	940	1.0	0.94	13.6
		1.5	1.41	16.7
		2.0	1.88	19.2
		2.5	2.35	21.5
CSPE-R	c	0.75	0.9	13.3
		0.90	1.15	15.0
		1.15	1.5	17.2
EIA-R	c	0.75	1.0	14.0
		1.0	1.3	16.0
Bituminous		3	3.5	26.2
		5	6	34.3

[a] PVC = polyvinyl chloride, HDPE = high density polyethylene, CSPE-R = chlorosulfonated polyethylene-reinforced (commercially known as Hypalon), and EIA-R = ethylene interpolymer alloy-reinforced (commercially known as XR5).
[b] PVC geomembranes have densities typically ranging from 1200 to 1300 kg m^{-3}.
c Geomembranes composed of different plies having different densities.

Typical values of geomembrane mass per unit area, density and thickness are given in Table 26.6.

A given geomembrane (defined by its mass per unit area, μ_{GM}) located at altitude z above sea level should not be uplifted if the wind velocity, V, is less than a threshold wind velocity, V_{thresh}, given by the following equation derived from Eq. 26.64a with $S_e = 0$:

$$V_{thresh} = 14.023 \exp[(6.259 \times 10^{-5})z] \times (\mu_{GM} \cos \beta/\lambda_w)^{1/2} \quad (26.66a)$$

with the following units: V_{thresh} (km h^{-1}), z (m) and μ_{GM} (kg m^{-2}). At sea level, Eq. 26.67 becomes:

$$V_{thresh} = 14.023 (\mu_{GM} \cos \beta/\lambda_w)^{1/2} \quad (26.66b)$$

The last column of Table 26.6 gives minimum values of the threshold wind velocity, V_{thresh}, for typical geomembranes, calculated using Eq. 26.66b with $\lambda_w = 1$ and $\beta = 0$, i.e. assum-ing that the geomembrane is located on a horizontal area at sea level, and that the suction factor has a conservative value.

Table 26.6 shows that typical polymeric geomembranes, with masses per unit area ranging between 0.5 and 2 kg m^{-2}, can resist uplift at sea level by winds with velocities ranging between 10 and 20 km h^{-1}; whereas typical bituminous geomembranes, with masses per unit area ranging between 3.5 and 6 kg m^{-2}, can resist uplift at sea level by winds with velocities ranging between 25 and 35 km h^{-1}. It should be noted that the wind velocity values given in Table 26.6 are conservative, i.e. greater wind velocity values would be calculated using a value of λ_w smaller than one and an altitude greater than zero.

26.5.4.4 Prevention of Geomembrane Uplift
Geomembrane uplift by wind can be prevented by covering the geomembrane with a

material that applies a uniform pressure on the geomembrane. The minimum thickness of material required to prevent geomembrane uplift at altitude z above sea level, t_{req}, is given by the following equation derived from Eq. 26.64a with $S_e = 0$ (Giroud *et al.* 1995c; Zornberg & Giroud 1997):

$$t_{req} = \frac{1}{\rho}$$

$$\times \left\{ -\mu_{GM} + \frac{0.0051 \lambda_w V^2 \exp[-(1.252 \times 10^{-4})z]}{\cos \beta} \right\}$$

$$(26.67a)$$

where ρ = the density of the material placed on the geomembrane. Equation 26.67a can only be used with the following units: t_{req} (m), μ_{GM} (kg m^{-2}), V (km h^{-1}) and z (m). At sea level, Eq. 26.67a becomes:

$$t_{req} = \frac{1}{\rho} \left(-\mu_{GM} + \frac{0.0051 \lambda_w V^2}{\cos \beta} \right) \quad (26.67b)$$

Using Eq. 26.67b with $\lambda_w = 1$ and $\cos \beta = 1$, it can be shown that approximately 0.2 m of water or 0.1 m of soil are sufficient to prevent geomembrane uplift for wind velocities up to approximately 200 km h^{-1} (assuming that the water or soil is not removed by the wind). It should be noted that 0.2 m of water is equivalent to a mass per unit area of 200 kg m^{-2}. To mobilize the same mass per unit area using sand bags having a mass of 20 kg would require approximately ten bags per square meter, which is impractical. Sand bags having a mass of 20 kg and installed on a 3-m square grid would be equivalent to a uniform thickness of water of 2.2 mm. Using Eq. 67b with $\lambda_w = 1$ and $\cos \beta = 1$, it can be shown that these bags would prevent uplift of a geomembrane with a mass per unit area of 1 kg m^{-2}, i.e. a typical polymeric geomembrane, for wind velocities up to approximately 15 km h^{-1}. These calculations explain why sand bags are effective in preventing geomembrane uplifting only in the case of low velocity winds.

Geomembrane uplift can also be prevented by applying a suction under the geomembrane equal to the effective suction defined by Eq. 26.64. This can be achieved, in the case of the geomembrane cover of a municipal solid waste landfill, by installing and operating a vacuum-based active gas extraction system. These types of systems are common at landfills. The system can be operated to create an effective suction under the geomembrane equal to or larger than that given by Eq. 26.64. In cases where the geomembrane is installed over a landfill without active gas extraction, a suction can be induced under the geomembrane by "suction vents" located at the top of slopes. As indicated by Giroud *et al.* (1995c):

These vents stabilize the geomembrane by sucking air from beneath the geomembrane when the wind blows, thereby decreasing the air pressure beneath the geomembrane. For the suction vents to work, air located beneath the geomembrane must flow toward the vent when the air vent is exposed to wind-generated suction. If the soil beneath the geomembrane has a low permeability, there is little air beneath the geomembrane. This air will flow toward the vent after the geomembrane has been slightly uplifted. If the soil beneath the geomembrane is permeable there is a significant amount of air entrapped beneath the geomembrane, and while this air is being sucked out by the suction vent, the geomembrane is uplifted. Therefore, in all cases the geomembrane may be uplifted for a short period of time before the suction vents are effective. In cases where the soil beneath the geomembrane has a low permeability, short strips of drainage geocomposites, radiating from the suction vent, have been recommended to help drain the air located beneath the geomembrane. However, the effectiveness of this method has not been evaluated.

Also, according to Giroud *et al.* (1995c):

To the best of the authors' knowledge, no method is available to design suction vents. [. . .] In a number of projects, suction vents have been placed every 15 m along the periphery of ponds. The reason for selecting 15 m as the spacing is not known.

Geomembrane uplift can also be prevented by gluing the geomembrane on the supporting medium. This method is effective only if the supporting medium is able to resist tensile forces. This method has been used in the case of a bituminous geomembrane resting on a layer of porous bituminous concrete on the face of an embankment dam (Saintot *et al.* 1993).

More details on methods for preventing geomembrane uplift by wind are provided in Giroud *et al.* (1995c).

26.5.4.5 Analysis of Geomembrane Uplift

If none of the precautions for preventing geomembrane uplift described in Section 26.5.4.4 is taken, an exposed geomembrane can be uplifted. In the analysis that follows, it is assumed that the geomembrane is sealed around its periphery and, as a result, the wind cannot uplift the geomembrane by reaching beneath it. Therefore, the analysis presented in this section is not applicable to a situation that exists during geomembrane installation where a panel is not seamed at its edge, and is not applicable to geomembranes that are torn open.

It is also assumed that the medium under the geomembrane is permeable enough that uplifting of the geomembrane will not be restricted by a decrease in air pressure beneath the geomembrane due to the sudden increase in volume beneath the geomembrane when uplifting begins. In other words, it is assumed that the geomembrane is free to move away from the supporting medium over a certain length. The geomembrane movements are assumed to be restrained at both ends of the geomembrane by anchor trenches, anchor soil layers, or other means. Typical cases are shown in Fig. 26.11, where a length L_{max} is shown. This is the length of

exposed geomembrane between two locations where its movements are restrained. The actual length, L, of geomembrane subjected to suction due to wind is equal to, or less than, L_{max}.

The geometry of the uplifted geomembrane can be characterized by four parameters: length of geomembrane subjected to suction due to wind measured along the slope, L; geomembrane strain, ε; geomembrane uplift, u; and angle, θ, between any of the extremities of an uplifted geomembrane and the straight line passing through these two extremities (Fig. 26.12). The following relationships exist between these four parameters (Giroud *et al.* 1995c):

$$\varepsilon = \frac{\theta}{\sin \theta} - 1 \qquad (26.68)$$

$$\varepsilon = \frac{1}{2}\left(\frac{2u}{L} + \frac{L}{2u}\right)$$
$$\times \sin^{-1}\left[2 \middle/ \left(\frac{2u}{L} + \frac{L}{2u}\right)\right] - 1 \qquad (26.69)$$

$$\theta = \sin^{-1}\left[2 \middle/ \left(\frac{2u}{L} + \frac{L}{2u}\right)\right]$$
$$= 2 \tan^{-1}\left(\frac{2u}{L}\right) \qquad (26.70)$$

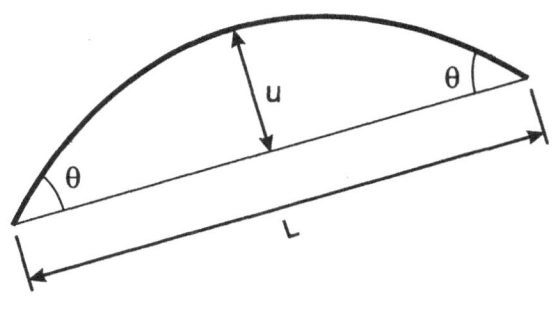

FIGURE 26.12. Geometry of uplifted geomembrane.

$$\frac{u}{L} = \frac{1 - \cos\theta}{2\sin\theta} = \frac{1}{2}\tan\left(\frac{\theta}{2}\right) \quad (26.71)$$

The following relationships exist between effective suction, geomembrane tension and the four parameters defining the geometry of the uplifted geomembrane (Giroud et al. 1995c):

$$\varepsilon = \frac{2T}{S_eL}\sin^{-1}\left(\frac{S_eL}{2T}\right) - 1 \quad (26.72)$$

$$\frac{T}{S_eL} = \frac{1}{2\sin\theta} = \frac{1}{4}\left(\frac{2u}{L} + \frac{L}{2u}\right) \quad (26.73)$$

$$\theta = \sin^{-1}\left(\frac{S_eL}{2T}\right) \quad (26.74)$$

$$\frac{u}{L} = \frac{T}{S_eL} - \left[\left(\frac{T}{S_eL}\right)^2 - \frac{1}{4}\right]^{1/2} \quad (26.75)$$

It should be noted that it is not possible to express $T/(S_eL)$ analytically as a function of ε. Therefore, it is useful to tabulate numerical values. Corresponding numerical values of u/L, ε, $T/(S_eL)$ and θ are given in Table 26.7.

Equations 26.68–26.75 and the values given in Table 26.7 make it possible to solve two important problems:

- If S_e and L are known and a geomembrane is characterized by its allowable tension, T_{all}, and allowable strain, ε_{all}, this geomembrane is acceptable regarding wind uplift resistance if its allowable tension is greater than the tension due to geomembrane uplift, which can be calculated using Eq. 26.72 or using the numerical values of ε and $T/(S_eL)$ given in Table 26.7.
- If S_e and L are known and the tension–strain curve of a geomembrane is known, it is possible to calculate the tension and strain in the uplifted geomembrane by trial-and-error using Eq. 26.72 or the nu-

merical values of ε and $T/(S_eL)$ given in Table 26.7.

Having calculated the tension, T, in the geomembrane, it is then possible to calculate the amount of uplift, u, using Eq. 26.75 and the angle θ using Eq. 26.74. Consistency between the calculated values of u/L, ε and θ can then be checked using Eqs 26.68–26.71. It is useful to calculate the angle θ, because it gives the orientation of the geomembrane tension at both extremities of the geomembrane (Fig. 26.12), which is needed to design the anchor trenches or any other anchor systems.

Equations 26.68–26.75 were established based on the assumption that the geomembrane is laying flat, i.e. with no wrinkles, and without contraction when the wind starts blowing. If this assumption is not met, the geomembrane strain, ε, in Eqs 26.68, 26.69 and 26.72 should be replaced by the apparent strain, ε_{app}, defined as the strain that the geomembrane would undergo if it were uplifted off the same amount from an initial position where it lays flat and with no contraction. In other words, the apparent strain is the strain that is calculated from θ using Eq. 26.68 or from u using Eq. 26.69. At the same time, the tension in Eqs 26.72–26.75 is the total tension, i.e. the tension that results from the initial tension, if any, plus the tension due to wind action. As a result, as pointed out by Giroud et al. (1995c):

If a geomembrane has wrinkles when uplifting begins, it is uplifted more, but with a smaller tension, than if the geomembrane has no wrinkles when uplifting begins [and] if a geomembrane is under tension when uplifting begins, it is uplifted less, but with a greater tension, than if the geomembrane is not under tension when uplifting begins.

More details on this topic are provided in Giroud et al. (1995c) and Zornberg & Giroud (1997).

TABLE 26.7. Values of the geomembrane normalized tension, $T/(S_eL)$, as a function of the geomembrane strain, ε, the relative uplift, u/L, and the uplift angle, θ (adapted from Giroud et al. 1995c)

Relative uplift, u/L (−)	Geomembrane strain, ε (%)	Normalized tension, $T/(S_eL)$ (−)	Uplift angle, θ (°)	Relative uplift, u/L (−)	Geomembrane strain, ε (%)	Normalized tension, $T/(S_eL)$ (−)	Uplift angle, θ (°)
0.000	0.000	∞	0	0.250	15.91	0.63	53.1
0.010	0.027	12.51	2.3	0.260	17.15	0.61	54.9
0.020	0.107	6.26	4.6	0.270	18.43	0.60	56.7
0.030	0.240	4.18	6.9	0.280	19.75	0.59	58.5
0.040	0.426	3.15	9.1	0.2819	20.00	0.58	58.8
0.050	0.665	2.53	11.4	0.2892	21.00	0.58	60.1
0.060	0.957	2.11	13.7	0.290	21.10	0.58	60.2
0.0613	1.000	2.07	14.0	0.2965	22.00	0.57	61.3
0.070	1.30	1.82	15.9	0.300	22.50	0.57	61.9
0.080	1.70	1.60	18.2	0.3035	23.00	0.56	62.5
0.0869	2.00	1.48	19.7	0.310	23.93	0.56	63.6
0.090	2.15	1.43	20.4	0.3105	24.00	0.56	63.7
0.100	2.65	1.30	22.6	0.3174	25.00	0.55	64.8
0.1065	3.00	1.23	24.0	0.320	25.39	0.55	65.2
0.110	3.20	1.19	24.8	0.3241	26.00	0.55	65.9
0.120	3.80	1.10	27.0	0.330	26.89	0.54	66.8
0.1232	4.00	1.08	27.7	0.3307	27.00	0.54	67.0
0.130	4.45	1.03	29.1	0.3373	28.00	0.54	68.0
0.138	5.00	0.97	30.9	0.340	28.43	0.54	68.4
0.140	5.15	0.96	31.3	0.3437	29.00	0.54	69.0
0.150	5.90	0.91	33.4	0.350	30.00	0.53	70.0
0.1513	6.00	0.90	33.7	0.360	31.60	0.53	71.5
0.160	6.69	0.86	35.5	0.370	33.23	0.52	73.0
0.1637	7.00	0.85	36.3	0.380	34.90	0.52	74.5
0.170	7.54	0.82	37.6	0.3806	35.00	0.52	74.6
0.1753	8.00	0.80	38.6	0.390	36.60	0.52	75.9
0.180	8.43	0.78	39.6	0.400	38.32	0.51	77.3
0.1862	9.00	0.76	40.9	0.4096	40.00	0.51	78.6
0.190	9.36	0.75	41.6	0.410	40.08	0.51	78.7
0.1965	10.00	0.73	42.9	0.420	41.86	0.51	80.1
0.200	10.35	0.73	43.6	0.430	43.67	0.51	81.4
0.2064	11.00	0.71	44.9	0.4372	45.00	0.50	82.3
0.210	11.37	0.70	45.6	0.440	45.51	0.50	82.7
0.2159	12.00	0.69	46.7	0.450	47.38	0.50	84.0
0.220	12.44	0.68	47.5	0.460	49.27	0.50	85.2
0.2250	13.00	0.67	48.5	0.4638	50.00	0.50	85.7
0.230	13.56	0.66	49.4	0.470	51.18	0.50	86.5
0.2339	14.00	0.65	50.1	0.480	53.13	0.50	87.7
0.240	14.71	0.64	51.3	0.490	55.09	0.50	88.8
0.2424	15.00	0.64	51.7	0.500	57.08	0.50	90.0

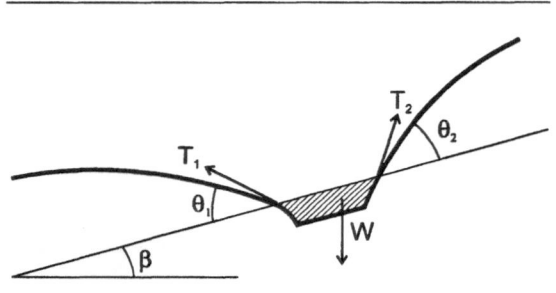

FIGURE 26.13. Geomembrane anchorage to resist uplift by wind.

26.5.4.6 Anchorage of a Geomembrane Subjected to Wind Uplift

The weight of material used to anchor a geomembrane subjected to wind uplift should be greater than the vertical component of the geomembrane tension (Fig. 26.13), hence the following equation (Giroud *et al.* 1999):

$$W = T_1 \sin(\theta_1 - \beta) + T_2 \sin(\theta_2 + \beta) \quad (26.76)$$

where W = weight of material per unit length of anchor trench (kN m^{-1}).

27. COVERS FOR WASTE

R. Bonaparte

E. K. Yanful

27.1 Introduction

Cover systems are used at landfills and other types of waste management units (e.g. waste piles, mine tailings piles, surface impoundments) to contain waste and any waste by-products (e.g. leachate, acid mine drainage, gas), to control moisture and air infiltration into the waste, and to prevent the occurrence of odors, disease vectors, and other nuisances. Cover systems are also used to meet erosion, aesthetic, and other end-use criteria for waste management sites. Cover systems for waste sites may involve only a single soil layer, placed over a mine waste rock pile, for example, or a multicomponent system of soil and geosynthetic layers, placed over a hazardous waste landfill, for example.

This chapter focuses on the hydraulic and geotechnical analysis and design of cover systems and deals with: (a) types of cover systems (Section 27.2); (b) cover system components (Sections 27.3 and 27.4); (c) water balance (Section 27.5); (d) static slope stability (Section 27.6); (e) seismic slope stability (Section 27.7); and (f) settlement (Section 27.8).

27.2 Types of Cover Systems

27.2.1 CLASSIFICATION OF COVER SYSTEMS

At present, cover system designs are based on one or more of three different principles for preventing or minimizing water percola-

tion into waste. Each of these is briefly discussed below.

1. Hydraulic barrier: This type of cover system uses a low permeability physical barrier to impede the downward migration of water into the waste (Fig. 27.1). Hydraulic-barrier materials most commonly include compacted clay layers (CCLs), geosynthetic clay liners (GCLs), geomembranes, and combinations of these materials. These types of barriers are discussed briefly in Section 27.3 (see Chapter 25 for more details). In some countries, e.g. the USA, requirements for cover systems at landfills are developed around the use of hydraulic barriers.

2. Capillary barrier: This type of cover system consists of one or more layers of finer-grained soil overlying one or more layers of coarser-grained soil. Figure 27.2 illustrates the simplest configuration for a capillary barrier: a single clayey soil layer over a sandy soil layer. These systems are discussed in more detail in Section 27.4. At a low degree of soil saturation, i.e. at high matric suction in Fig. 27.2, the hydraulic conductivity of the coarse-grained soil is much less than that of the fine-grained soil. This is the reverse of the condition that occurs when the coarse-grained soil is at a high degree of soil saturation. Capillary barriers either: (i) store water by increased moisture content in the fine-grained soil for subsequent evapotranspiration, or (ii) divert infiltrating water via unsaturated lateral flow

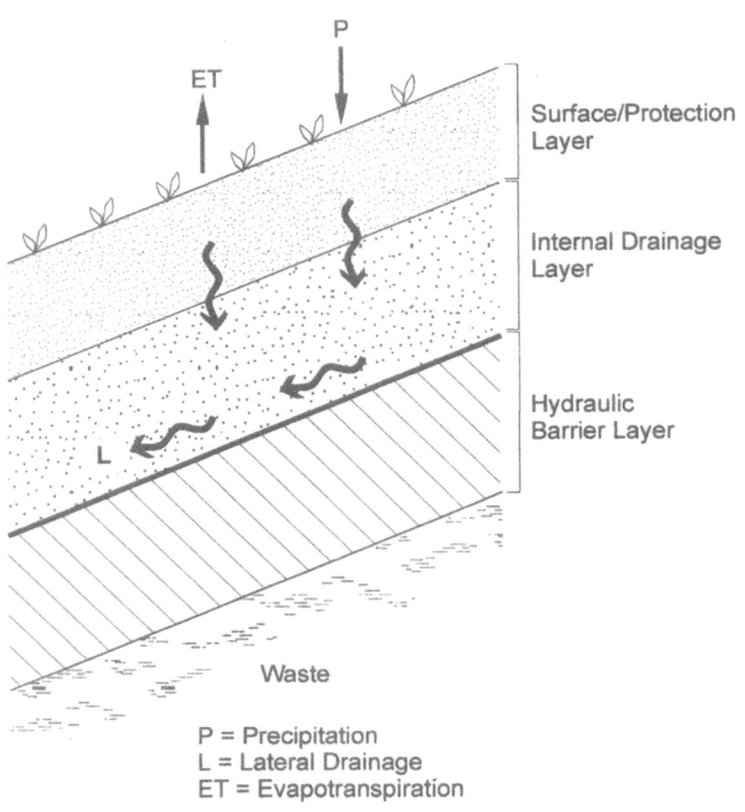

FIGURE 27.1. Hydraulic-barrier cover system.

in the fine-grained soil (above the soil interface). Sometimes a "wicking layer" (with intermediate characteristics to the coarse- and
fine-grained layers) is installed between the
coarse and fine layers to convey lateral flow. At
a high degree of soil saturation in the coarse-
grained soil, the capillary effect breaks down
and percolation through the system can occur.

3. Evapotranspirative barrier: This type of
cover system has also been developed primarily for use at arid and semi-arid sites. Evapotranspirative barriers are covers that consist
of a thick layer of relatively fine-grained soil
capable of supporting vegetation (Fig. 27.3).
Soil thickness can range from about 900
to 1800 mm (Zornberg & Caldwell 1998).
Evapotranspirative barriers exploit two char-

acteristics of fine-grained soils: (a) significant
soil water storage capacity, i.e. they can store
a significant amount of water before gravity
drainage; and (b) low hydraulic conductivity,
even at high degrees of saturation. Low hydraulic conductivity limits advancement of
the soil wetting front during seasonal wet
periods (rainfall or snow melt). High water
storage capacity allows storage of moisture,
which infiltrates until it can later be removed
by evapotranspiration. An evapotranspirative
barrier must be sufficiently thick that changes
in moisture content do not occur near its base,
i.e. all changes in soil water storage should occur in the upper portion of the barrier (Fig.
27.3). Otherwise, percolation will occur. The
required barrier thickness is a function of
the frequency and intensity of precipitation,

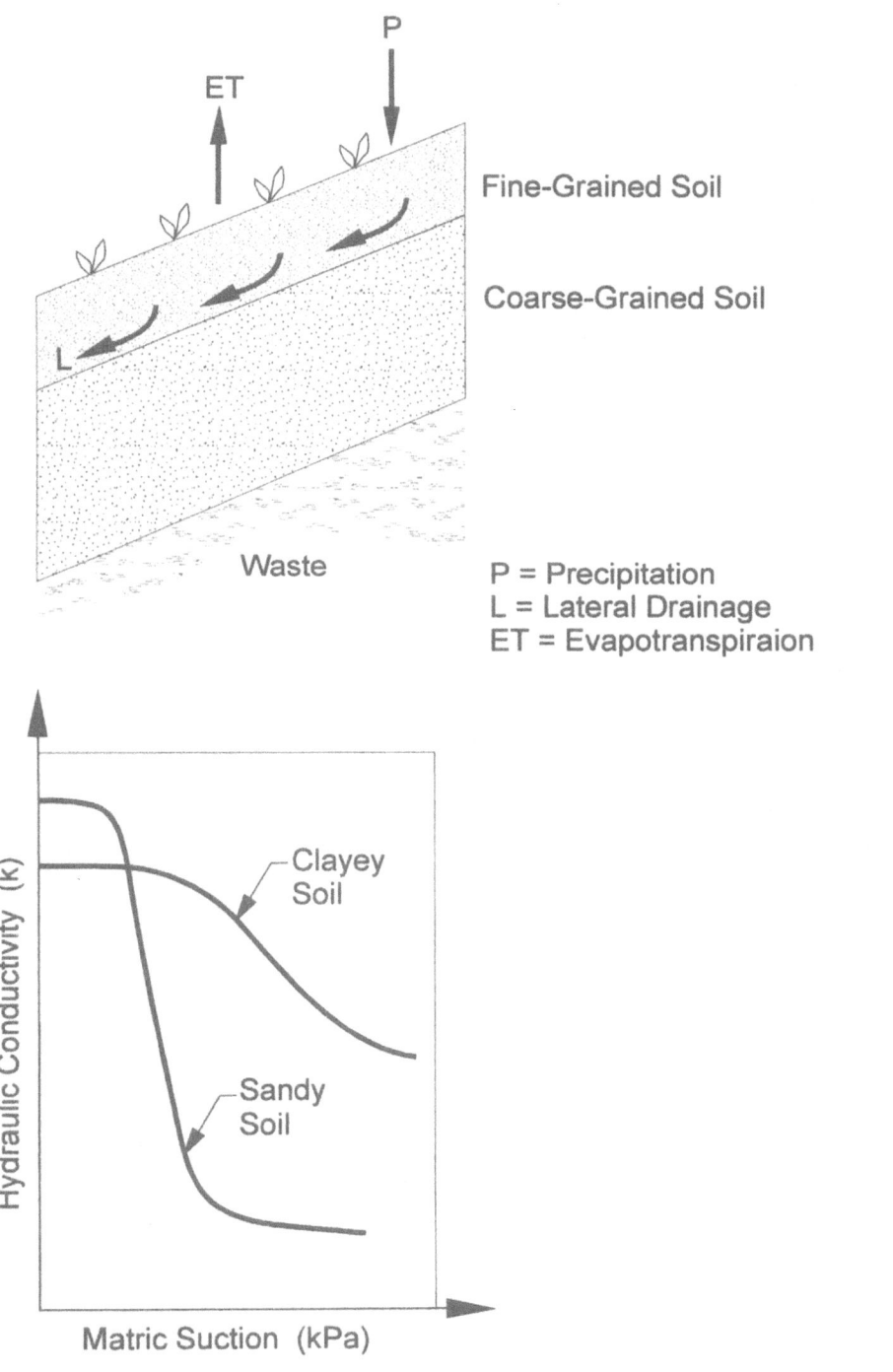

FIGURE 27.2. Capillary-barrier cover system and partially saturated hydraulic conductivity functions.

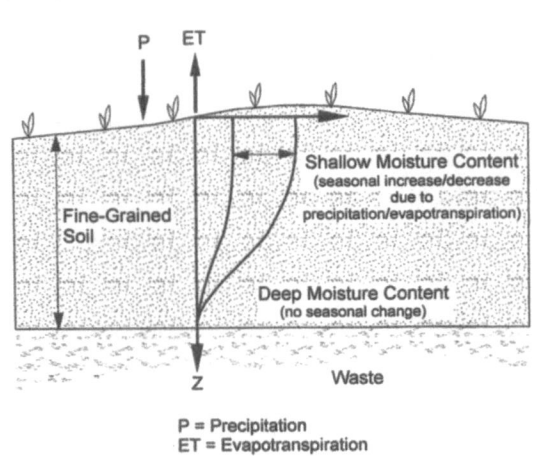

FIGURE 27.3. Evapotranspirative cover system, also called monolayer or monocover (modified from Benson & Khire 1995).

the unsaturated hydraulic properties of the soil, the type and health of cover vegetation, and the rate at which water can be removed by evapotranspiration. Soil types used for construction of evapotranspirative barriers include silty sands, silts and clayey silts.

Different barrier types may be combined in a single cover system. For example, a capillary barrier may be constructed beneath an evapotranspirative barrier.

This chapter focuses primarily on hydraulic-barrier type cover systems, currently the most widely used type of waste cover system, with limited commentary provided on the other two types. It is noted, however, that the use of capillary barriers and evapotranspirative barriers is becoming much more common.

27.2.2 EXAMPLES OF COVER SYSTEMS

Cover systems can be constructed with a wide variety of configurations of soil and geosynthetic layers to satisfy project-specific design criteria. A few examples are presented below. Additional examples can be found in Koerner & Daniel (1997).

Figure 27.4 illustrates two different cover systems for a municipal solid waste (MSW)

landfill, one with a CCL hydraulic barrier, the other with a composite hydraulic barrier. For either example, a GCL can be considered as an alternative to the low permeability soil layer. Soil thicknesses will vary based on project-specific conditions. Figure 27.5 presents the cover system cross-section for a low-level radioactive waste disposal facility in the USA with a minimum design life of 200 years. Figure 27.6 shows the cover system used as part of the remediation of an uncontrolled dump site containing hazardous waste. The site is in a marsh. The low bearing capacity of the foundation soil and waste at the site necessitate the use of this lightweight cover system.

Figure 27.7 illustrates the benefits of cover system installation in preventing or minimizing percolation and thereby reducing the potential for waste liquid generation, e.g. leachate, acid mine drainage. This figure shows leachate generation rates after closure for a number of hazardous waste landfill cells (Gross et al. 1997). On average, leachate generation rates during the first year after closure were approximately 22% of the rates during the year prior to closure. The average rate for a given year after closure ranged from $1150 \, l \, ha^{-1} \, day^{-1}$ at one year after closure to less than $1 \, l \, ha^{-1} \, day^{-1}$ at nine years after closure. Similar data for a MSW landfill cell (in a humid environment precipitation-1000 mm a^{-1}) showed a reduction from $3400 \, l \, ha^{-1} \, day^{-1}$ during filling to $70 \, l \, ha^{-1} \, day^{-1}$ at three years after closure (Bonaparte 1995).

27.2.3 GAS GENERATION AND MANAGEMENT

Gas collection and control is necessary at most MSW landfills, at some other types of landfills and surface impoundments, and at some remediation sites. Anaerobic decomposition of organic material is the principal source of gas in solid waste landfills. Some industrial wastes can generate gas by inorganic chemical reactions.

Gas production rates for MSW vary with

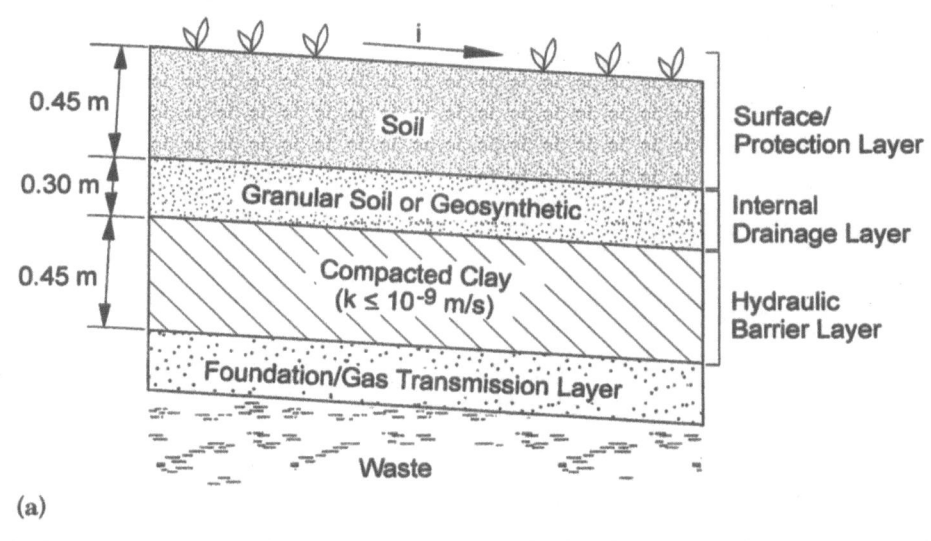

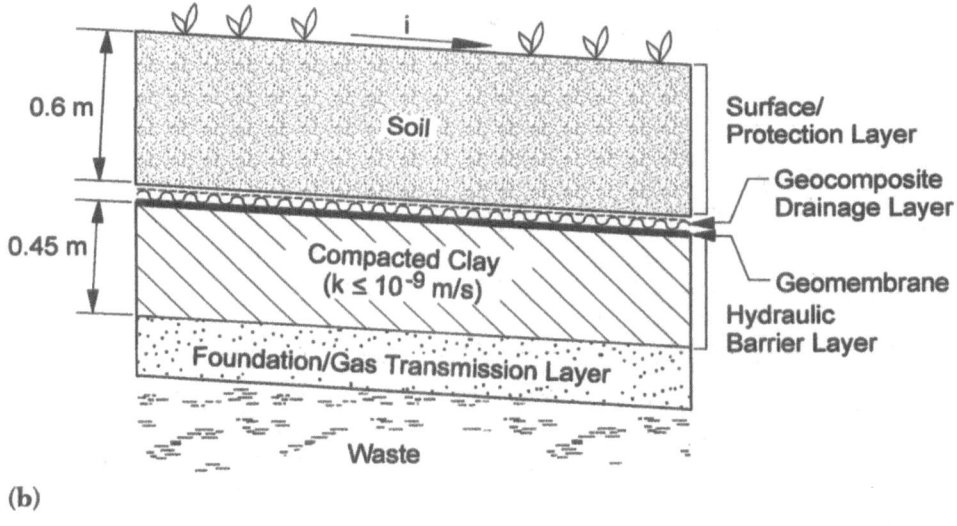

FIGURE 27.4. Example of MSW landfill cover: (a) CCL hydraulic barrier, and (b) composite hydraulic barrier. All thicknesses are typical. Note: i = cover slope (gradient).

the composition and age of waste, waste volume, waste moisture content, and other factors. MSW landfill gas consists mainly of methane and carbon dioxide, with lesser concentrations of nitrogen, oxygen, sulfides, ammonia, and other constituents, and trace concentrations of a variety of volatile organic compounds. Gas generation in a landfill cell can extend over a period of more than 25–50 years, or it can be accelerated through the use of leachate recirculation. Gas generation rates for MSW landfills can be estimated using simplified manual calculations (e.g. Tchobanoglous *et al.* 1993; McBean *et al.* 1995), or computer-based models such as the USEPA landfill air emissions estimation model (USEPA 1996).

Waste-generated gases affect cover sys-

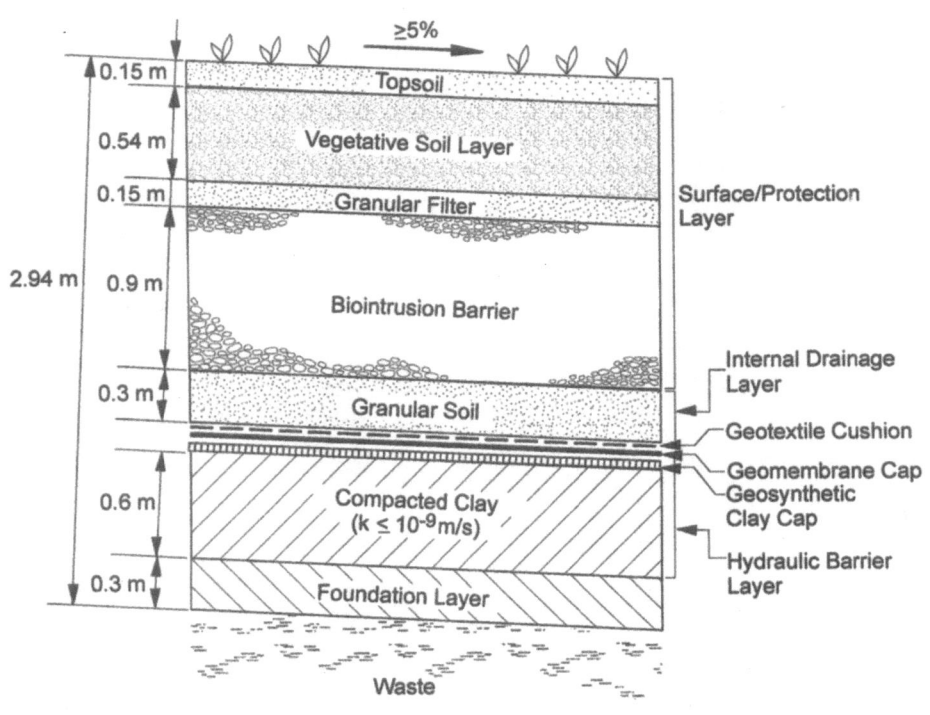

FIGURE 27.5. Example of low-level radioactive waste landfill cover.

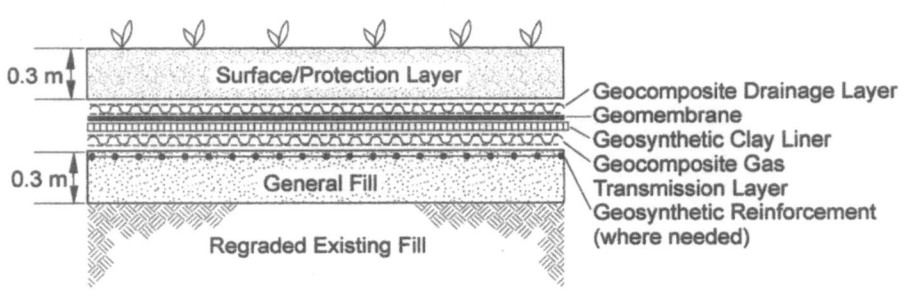

FIGURE 27.6. Example of lightweight site remediation cover over soft waste.

tems in several ways. The presence or absence of gas will influence the selection of the type of hydraulic-barrier layer material. Geomembranes are generally better barriers to gas migration than soils, with the possible exception of intact low permeability soils at or near saturation. Also, it may be necessary to install a gas transmission layer beneath the hydraulic-barrier layer to convey gas to outlets through the cover system. A factor sometimes overlooked in the closure of old landfills and in site remediation is that placement of a cover system will trap any gas being generated by the waste. Gas generation rates at these facilities may be slow enough that gas generation is not even recognized as a design issue. Yet after cover system installation, gas pressure can slowly build up. This process

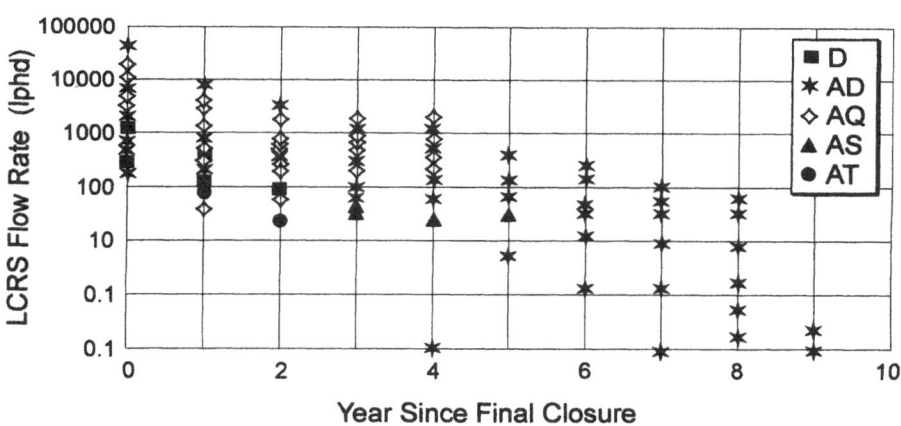

FIGURE 27.7. Leachate generation rates at five hazardous waste landfills in units of liters per hectare per day (lphd = l ha⁻¹ day⁻¹). Covers incorporate geomembrane hydraulic barriers (modified from Gross *et al.* 1997). Note: Symbols and letters identify the five landfills in Gross *et al.* (1997).

may eventually lead to one or more of the following: (a) problems with cover system performance, including a reduction in the factor of safety along interfaces in the final cover system below the hydraulic barrier layer; and (b) for unlined or inadequately lined landfills and contaminated source areas, subsurface gas migration. Subsurface gas migration has caused adverse groundwater quality impacts at many landfills.

27.3 Components of Hydraulic-Barrier Cover Systems

27.3.1 TYPICAL COMPONENTS

Components of a typical hydraulic-barrier type cover system are briefly introduced. The usual sequencing of these typical components is illustrated in Fig. 27.8.

27.3.2 SURFACE LAYER

The primary functions of the surface layer are to resist erosion by water and wind, be maintainable, and, depending on the situation, provide a growing medium for vegetation and satisfy project aesthetic, ecological and post-closure land-use criteria.

Materials that may be used for final cover system surface layers include: (a) topsoil, (b) amended topsoil, (c) lightweight soil, (d) riprap, (e) gravel–soil mixtures, (f) asphaltic concrete, and (g) other materials. Of these materials, topsoil is, by far, the one most commonly used. Suitable topsoil will promote growth of vegetation, thereby maximizing the evapotranspirative component of the cover system water balance. Vegetation also decreases stormwater run-off velocities from cover system slopes and reinforces the topsoil; both of these effects reduce the rate of erosion of topsoil in comparison to a topsoil layer without vegetation. In areas where the amount of precipitation is inadequate to support growth of a vegetative layer, riprap, gravel–soil mixtures, asphaltic concrete, or other materials may be used for the surface layer.

27.3.3 PROTECTION LAYER

A protection layer may serve several functions (Daniel & Koerner 1993b): (a) store water that has infiltrated through the surface layer until the water later returns to the atmosphere through evapotranspiration; (b) serve as a barrier to human, burrowing animal, or plant root intrusion; (c) protect underlying lay-

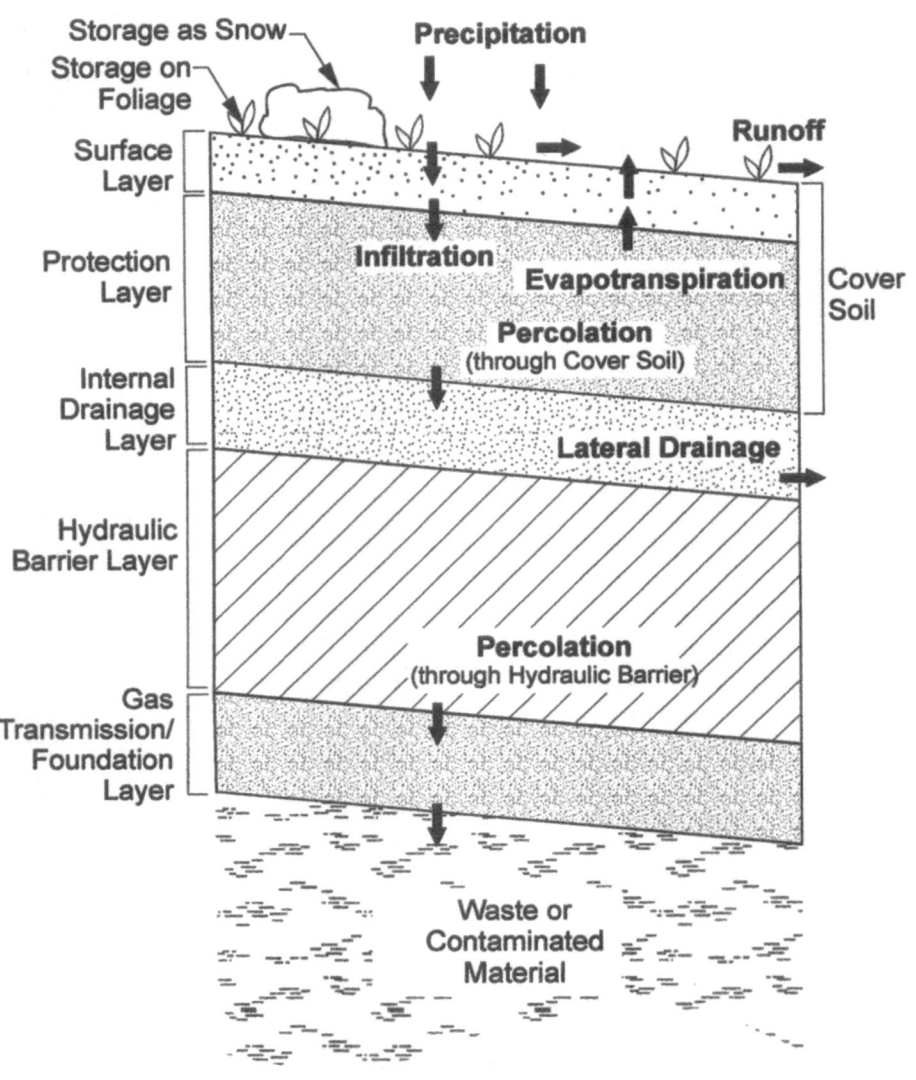

FIGURE 27.8. Identification of typical components of hydraulic-barrier cover system and water movement and storage in cover system.

ers from wet–dry cycles, which could cause cracking of some materials; (d) protect underlying layers from freezing, which could also cause cracking of some materials; and (e) restrict emissions of radon gas for those wastes, such as uranium mill tailings, that emit radon.

On-site or locally available soil is usually suitable for protection layer construction if the primary function of the layer is to serve as a vegetation root zone or for freeze–thaw protection. However, if the primary role of the protection layer is to prevent intrusion by burrowing animals, cobbles, asphaltic concrete or similar materials will typically be required.

27.3.4 INTERNAL DRAINAGE LAYER

In a hydraulic-barrier type cover system, an internal drainage layer may be required above the hydraulic barrier. The functions of this drainage layer are to: (a) limit the buildup of hydraulic head on the underlying barrier layer, which minimizes percolation of water through the barrier; (b) drain the overlying protection and surface layers, which increases the available water-storage capacity of these layers and helps to minimize erosion of these layers by reducing the time during which the surface and protection layer materials remain saturated with water; and (c) prevent excessive seepage forces in surface, protection, and drainage layer materials, which improves cover system slope stability.

Materials used for drainage layers include sand, gravel, geonets and geocomposite drainage materials (see Chapter 7 for details regarding the different geosynthetics). The material used must have adequate hydraulic conductivity to prevent a buildup of liquid head in the slope and adequate hydraulic transmissivity to laterally convey the design flow rate. If gravel or a geonet is used for the drainage layer, a filter layer will typically need to be placed between the drainage layer and the overlying protection layer. Geotextile filter layers are typically used to achieve this function, although soil filter layers can also be used. If the drainage layer consists of gravel and the underlying barrier layer is a geomembrane, a geotextile cushion layer will typically be needed between the geomembrane and gravel. One of the most important aspects of designing an internal drainage layer is providing for free drainage at the layer discharge point.

27.3.5 HYDRAULIC-BARRIER LAYER

The function of a hydraulic-barrier layer is to minimize percolation of water through the cover system. Properly designed barrier layers can virtually eliminate infiltration into waste. Hydraulic barrier layers also restrict migration of gas or volatile constituents from the waste to the atmosphere. Materials used for barrier layer construction include CCLs,

GCLs and geomembranes. These barrier materials may be used alone or in combination.

Historically, CCLs were the most frequently used barrier layer material. Procedures for initial construction of CCL barriers to meet permeability criteria are well-established (see Chapter 25). However, when used alone, CCL barriers in cover systems may not maintain low permeability in the long term. USEPA (1999) describes a number of field case studies where CCL barriers exhibited increasing permeability with time when used alone as waste covers even when overlain by protection and surface layers. This increase is attributed to desiccation cracking, wet–dry and freeze–thaw effects, root penetration, and differential settlement. Field studies by Montgomery & Parsons (1989); Corser & Cranston (1991); Melchior (1997); and the Maine Bureau of Remediation and Waste Management (1997) demonstrate the problem. The USEPA suggests that the best way (USEPA 1999) to maintain low CCL permeability in a cover application is to overlay the CCL with both a geomembrane and an adequate thickness of protection soil.

GCLs are factory fabricated products (see Section 7.6) having attractive features for cover system applications, which include very low saturated hydraulic conductivity (e.g. $k \leq 5 \times 10^{-11}$ m s^{-1}), preservation of low hydraulic conductivity when subjected to differential settlement, and ease of installation. Disadvantages include low hydrated shear strength, potential for increased hydraulic conductivity due to cation exchange reactions under certain conditions, and potential for premature hydration during installation. GCLs are increasingly being used in cover system applications. Additional information on these materials is presented in Chapter 25, Koerner (1998); Daniel & Koerner (1993a); Koerner & Gartung (1995); Koerner & Daniel (1997); and Well (1997). The results of a large-scale test plot program sponsored by the USEPA to evaluate GCL-use in cover systems are presented in Daniel & Scranton (1996) and Daniel *et al.* (1998).

Geomembranes are factory-manufactured polymeric materials (see Section 7.5) that are widely used as hydraulic barriers in cover systems due to their non-porous structure, flexibility, and ease of installation. Spray-on elastomeric and bituminous membranes are also available, but are rarely used in cover system applications.

27.3.6 GAS TRANSMISSION LAYER

Gas transmission layers may be necessary beneath cover system barrier layers for wastes that generate gas. These layers are designed to have adequate in-plane gas transmissivity to convey gas to passive gas vents, active gas wells, or trenches. Gas transmission layers are typically a necessary complement to systems that utilize passive gas vents. Gas transmission layers may not always be needed for landfills with active gas extraction systems, depending on gas generation rates in the landfill, extraction well spacing, presence or absence of horizontal gas trenches, and other factors. Gas transmission layers may be constructed of granular materials or geosynthetics (geotextiles, geonets or geocomposites). When a granular material is used, a separation layer (typically a geotextile) may be needed to separate the granular material from the overlying barrier layer.

27.3.7 FOUNDATION LAYER

The foundation layer forms the bottom-most component of the cover system. The functions of the foundation layer are to provide grade control for cover system construction, adequate bearing capacity for overlying layers, firm subgrade for compaction of overlying layers, smooth surface for installation of overlying geosynthetics, and, in some applications, buffer zone to reduce the potential effects of waste differential settlements on cover system components. Materials most often used for the foundation layer include on-site or locally available soils. Sometimes, intermediate cover soil already in place is used for all or a portion of the foundation layer. In a few situations, waste material can be used to construct the foundation layer. If constructed of granular material, the foundation layer may also serve as a gas transmission layer.

27.4 Capillary Barrier Cover Systems

As noted in Section 27.2.1, a capillary barrier consists of one or more layers of finer-grained soil overlying one or more layers of coarse-grained soil. At a high matric suction (Fig. 27.2), the hydraulic conductivity of coarse-grained soil is much less than that of fine-grained soil. Under unsaturated flow conditions (high matric suction), moisture movement through the fine-grained soil is greatly impeded when the moisture reaches the interface between the fine- and coarse-grained materials. This fact is used to advantage to design capillary soil barriers. Capillary barriers are finding increasing use in two applications: (a) cover systems for arid sites; and (b) mine waste cover systems.

Capillary barriers are being increasingly used at arid sites where the fine-grained surface layer can be maintained at a low degree of saturation. The critical design condition for these barriers is often winter rain or snow melt. The fine-grained soil layer must be designed to prevent saturation of the layer near the soil interface during these periods. This requires that the fine soil have adequate moisture storage capacity above the soil interface. Also, to be effective, fine-grained soil must not undergo desiccation cracking upon drying. The performance characteristics of these types of cover systems in arid site applications have been discussed by Anderson *et al.* (1993), Benson & Khire (1995), Dwyer (1995, 1998), Fayer *et al.* (1992), Gee *et al.* (1997), and Khire *et al.* (1997).

Capillary barriers are also used in soil covers

for controlling acid production in sulfide mine waste. In this application, the cover system is constructed such that the upper portion of the fine-grained soil layer maintains a high degree of saturation to minimize oxygen intrusion, while the lower portion of the layer remains unsaturated to maintain high matric suction at the soil interface. Under this condition, the coarse layer is not able to transmit a significant amount of moisture downward from the overlying fine-grained layer, and essentially prevents the fine-grained layer from draining. Coarser-grained surface layers are often used with the mine waste covers to increase percolation into, and reduce evapotranspiration from, the fine-grained soil layer. The use of capillary barriers for mine waste covers is discussed by Rasmuson & Eriksson (1987), Nicholson *et al.* (1989), and Yanful & Aubé (1993). Investigations of the performance of capillary barriers in field test plots in Quebec, New Brunswick, and Ontario, Canada, are given in Yanful (1993), Yanful *et al.* (1993a,b), Woyshner & Yanful (1995), and Simms & Yanful (1997).

Figure 27.9 shows a cross-section of the capillary-barrier cover system investigated by Yanful (1993) and Woyshner & Yanful (1995) at the Waite Amulet tailings site in Quebec, Canada. The cover was instrumented with time domain reflectometry (TDR) probes and sensors to measure water content and gaseous oxygen in the various layers. Figure 27.10 shows the volumetric water in the tailings, sand base, and the clay during four years of monitoring from 1990 to 1994. The data indicate that the sand base (the capillary barrier) did not attain a volumetric water content of more than 11% (a degree of saturation of 34%) during the four years. This suggests that during the spring and fall months, at least, there was no significant capillary break and that the overall percolation through the cover system was small. Woyshner & Yanful (1995) measured water percolation in the lysimeters and reported an overall annual percolation that was only 4% of the total precipitation. They also showed that the hydraulic conductivity of the sand

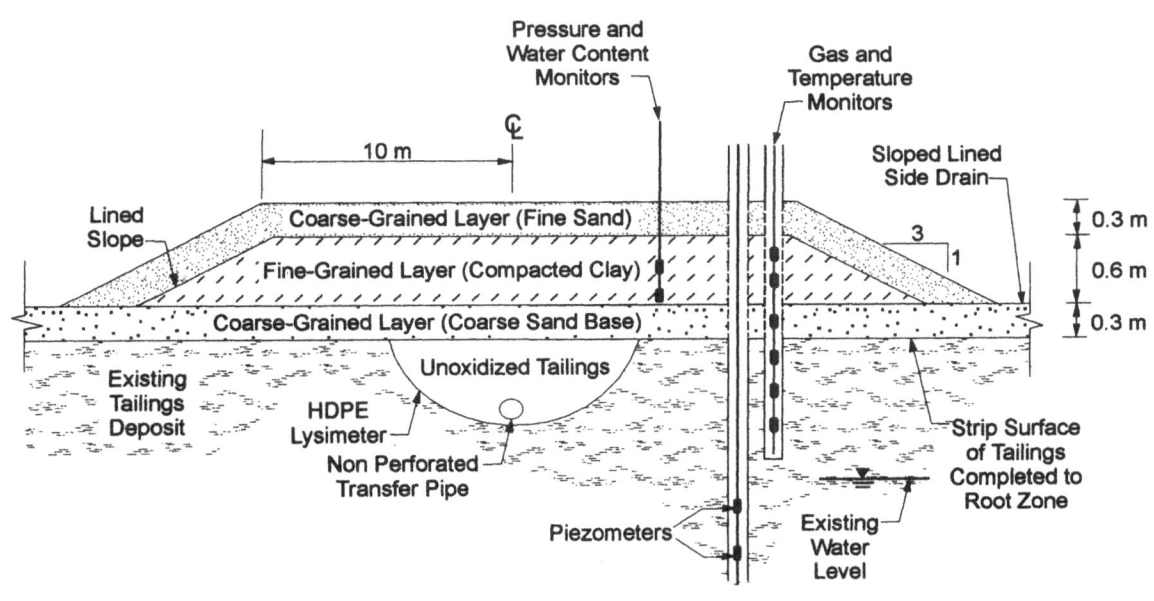

FIGURE 27.9. Capillary-barrier cover system over mine waste at Waite Amulet site, Quebec, Canada (modified from Yanful 1993).

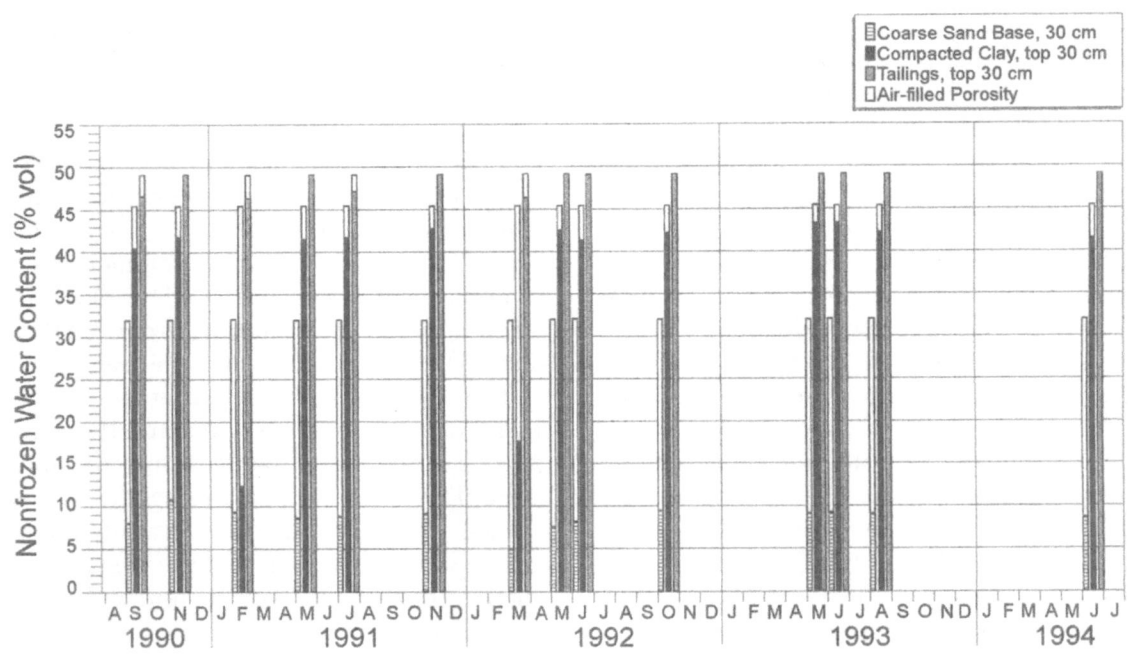

FIGURE 27.10. Water content and air-filled porosity of capillary barrier and mine waste at Waite Amulet site, Quebec, Canada (modified from Woyshner & Yanful 1995).

base capillary barrier would decrease and become lower than that of the clay layer when the suction exceeded 25 kPa.

Figure 27.11 shows oxygen profiles measured in the cover system at different times in 1990 and 1991. The profiles indicate a rapid drop in oxygen concentrations in the clay layer from 12 to about 4%. Yanful (1993) analyzed the data in Fig. 27.11 and calculated an oxygen flux of 0.011 g m^{-2} day^{-1} through the covered tailings. In comparison, the flux through the uncovered tailings was calculated to be 5.83 g m^{-2} day^{-1}, indicating a 99% cover efficiency, based on oxygen diffusion alone.

O'Kane et al. (1995) also presented the results of an investigation into the performance of a soil cover consisting of two till layers placed over an acid-producing mine waste rock pile at the Equity Silver mine site in British Columbia, Canada. The function of the cover was to reduce water infiltration and gaseous oxygen ingress into the rock pile. In

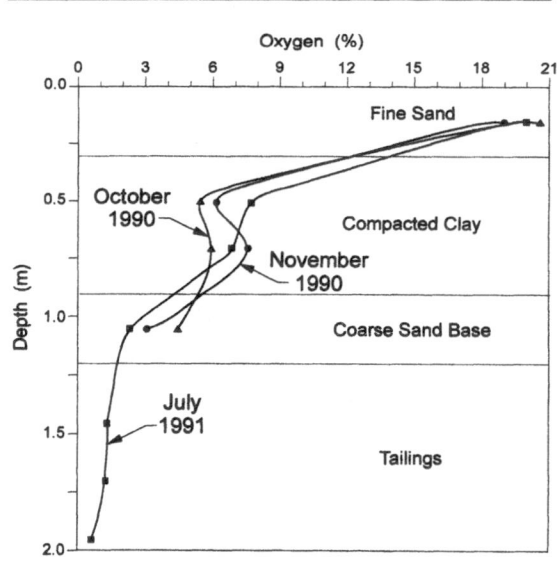

FIGURE 27.11. Field gaseous oxygen profiles in capillary barrier and mine tailings at Waite Amulet site, Quebec, Canada (modified from Yanful 1993).

the design of the cover, the waste rock, being coarse, was relied upon to function as a capillary barrier. The upper 0.3-m thick till layer was loosely placed, while the lower 0.3-m thick till layer was compacted. Suction measurements suggested that the matric suction within the rock pile was similar to the suction at residual saturation. At this suction, the unsaturated hydraulic conductivity of the rock pile would be low and would lead to very little water flow. A water balance evaluation indicated that for these conditions, the overlying compacted till layer would retain a significant portion of its placement water content. The data obtained by O'Kane *et al.* (1995) suggests that the compacted till layer maintained a degree of saturation equal to or higher than the 85% during cyclic wetting and drying.

27.5 Water Balance

27.5.1 OVERVIEW

A water balance analysis is used to quantify the percolation of water into and through the cover system. In addition to estimating percolation, water balance analyses of cover systems are used to: (a) develop an understanding of how the various cover system components will function, and identify which water-routing mechanisms are most important; (b) compare the performance of difference cover system designs; and (c) define the performance criteria for various cover system components (e.g. required storage capacity of surface and protection soil layers, required flow capacity of internal drainage layer) so that these components can be designed.

27.5.2 WATER BALANCE CONCEPT

In a water balance analysis, water is routed into and out of a system using a series of calculations that require conservation of water mass. The potential pathways for water movement into and out of a cover system are illustrated in Fig. 27.8. A cover system water balance is expressed in terms of water inflows

and outflows and storage changes for a unit area of the system over some arbitrary time interval, e.g. day, as:

$$P = R + ET + \Delta W_{surface} + \Delta W_{foliage}$$
$$+ \Delta W_{soil} + L + PERC \qquad (27.1)$$

where P = precipitation (mm day^{-1}), R = run-off (mm day^{-1}), ET = evapotranspiration (mm day^{-1}), $\Delta W_{surface}$ = change in water storage at surface (mm day^{-1}), $\Delta W_{foliage}$ = change in water storage on plant foliage (mm day^{-1}), ΔW_{soil} = change in water storage in cover system soil (mm day^{-1}), L = lateral drainage from internal drainage layer (mm day^{-1}), and $PERC$ = percolation through the cover system (mm day^{-1}). Water is input to the cover system as precipitation in the form of rain or snow, and lost from the cover system by run-off, evapotranspiration, lateral drainage and percolation. Water also is stored on the cover system as ponded water or snow, on plant foliage, and in cover system soils by capillary action. Equation 27.1 is cast above in a time unit of one day; any other time unit could equally well be used.

Infiltration is primarily removed from the cover system by evapotranspiration. Storage of water in soil coupled with removal of water by evapotranspiration are the most important mechanisms for limiting percolation of infiltration. Flow from lateral drainage layers is typically a much smaller component of the water balance than is evapotranspiration. It should be remembered, however, that while the internal drainage layer is typically of secondary importance to the overall cover system water balance, it is of prime importance to cover system slope stability. If even a relatively small amount of potential lateral flow is left undrained in a cover system, hydraulic heads can build up over the hydraulic barrier layer, leading to destabilizing seepage forces on cover system slopes.

Though Eq. 27.1 appears simple, the components of the water balance are dependent on many factors, difficult to quantify and interrelated. It can be especially difficult to quantify percolation in arid and semi-arid environments, where almost all precipitation is consumed by evapotranspiration. Unlike in wetter climates where actual evapotranspiration may approach the magnitude of potential evapotranspiration, i.e. the process is energy limited, in drier climates actual evapotranspiration is generally much smaller than potential evapotranspiration due to the lack of available water. Evapotranspiration is more difficult to accurately estimate under water-limiting conditions. Because the magnitude of percolation in drier climates is so much smaller than the magnitudes of evapotranspiration and precipitation, relatively small errors in estimated evapotranspiration can result in relatively large errors in estimated percolation. Due to the difficulty in performing accurate analytical water balances, field water balances are often performed using weighing or drainage lysimeters or cover system test plots to better assess the water balances components in drier climates.

Water balance calculations are performed for time intervals that may be shorter than one hour or longer than a year. The time interval to use is dependent on the purpose of the water balance analysis.

27.5.3 WATER BALANCE METHODS
A variety of water balance methods are available to analyze and design cover systems. They range in complexity from relatively simple empirical correlations to sophisticated computer-based finite difference and finite element models. This section describes four of the more widely known and used water balance analysis methods:

- simplified manual method (Koerner & Daniel 1997),
- hydrologic evaluation of landfill performance (HELP) model (Schroeder et al. 1994, Schroeder et al. 1994),
- leachate estimation and chemistry model

(LEACHM) (Hutson & Wagenet 1992), and
- UNSAT-H model (Fayer & Jones 1990).

These are all well-documented, public-domain models that consider the significant water balance processes, e.g. precipitation, run-off, and evapotranspiration, and have been used previously for waste cover water balance analyses. The characteristics of these models are compared in Table 27.1.

27.5.4 SIMPLIFIED MANUAL METHOD

27.5.4.1 Description of Method
Koerner & Daniel (1997) present an updated version of a simplified method for performing manual or computer-spreadsheet water balance calculations first presented in Thornthwaite and Mather (1955, 1957); Fenn et al. (1975); and Kmet (1982). In the previous work, only monthly time steps were considered. Koerner & Daniel (1997), however, extended the method to consider a variable time step, e.g. daily, weekly or monthly, to be selected based on the purpose of the analysis.

From the basic concepts of water balance analysis shown in Fig. 27.8 and described by Eq. 27.1, and assuming no water is stored at the surface or intercepted by plants, i.e. $\Delta W_{surface} = \Delta W_{foliage} = 0$, the following relationships are defined for a given time step:

$$P = I + R \qquad (27.2)$$

$$I = ET + \Delta W_{soil} + PERC^{\circ} \qquad (27.3)$$

where I = cover system infiltration (mm day^{-1}); and $PERC^{\circ}$ = percolation through cover soil (mm day^{-1}) in Fig. 27.8.

In the simplified formulation (Eq. 27.2), precipitation is partitioned into run-off or infiltration. Run-off is calculated as a fraction of precipitation using a run-off coefficient appropriate for the cover system soil type and slope, i.e. the "rational formula" is used. According to Fenn et al. (1975), the rational formula will, in most cases, underestimate run-off.

TABLE 27.1. Comparison of some public-domain water balance models

	Simplified manual method	HELP computer code	LEACHM computer code	UNSAT-H computer code
Calculation scheme	Simplified empirical and mechanistic equations	Quasi two-dimensional water-routing model, with multiple uncoupled subroutines Simplified empirical and mechanistic equations Simplified unsaturated flow model with unit hydraulic gradient	Finite difference model with unsaturated flow model based on Richards partial differential equation User-specified boundary conditions	Finite difference model with unsaturated flow model based on Richards partial differential equation User-specified boundary conditions
Advantages	Easy to perform Few data requirements Any time step Considers lateral drainage	Widely accepted Used to design hydraulic barriers Easy to run simulations Default database of climatic, soils and vegetation data Considers lateral drainage	Mechanistic model Solves unsaturated flow equation Used to design evapotranspirative and capillary barriers May give a better estimate of evapotranspiration in arid climates than the other models	Mechanistic model Solves unsaturated flow equation Flexibility in definition of unsaturated hydraulic conductivity–head–moisture content relationships
Disadvantages	Numerous simplifying assumptions must be made Steady state conditions are assumed All calculations are uncoupled Cannot be used for unsaturated flow	Does not solve unsaturated flow equations Demonstrated overprediction of percolation in many cases Limited to daily climatic data	Only models soil profiles that are no more than 2-m thick Does not consider lateral drainage	High computational demands Unsuitable for parametric evaluation Does not consider lateral drainage
Appropriate use	Instructional tool Design of hydraulic barriers Check of computer simulations Parametric evaluations Calculation of peak lateral drainage from cover system	Design of hydraulic barriers Regulatory compliance demonstrations Parametric evaluations Calculation of peak lateral drainage from cover system	Design of evapotranspirative and capillary barriers (no lateral flow) Parametric evaluations Unsaturated flow analysis	Performance assessment of evapotranspirative and capillary barriers (no lateral flow) Calibration with field data prior to making long-term predictions Unsaturated flow analysis

From Eq. 27.3, water infiltrating the cover soil will evapotranspire, become soil water storage, or percolate downward. In the simplified manual method, evapotranspiration is calculated as a function of the potential evapotranspiration, infiltration, and moisture stored in the soil. Potential evapotranspiration is calculated using an empirical method developed by Thornthwaite & Mather (1955). If more water infiltrates the cover system than can potentially evapotranspire, the excess water is first distributed within the root zone until the soil moisture content is at field capacity. The remaining water is then routed as percolation from the cover soil. If evapotranspiration is greater than infiltration, then stored water is lost from the root zone of the cover system soil until the soil moisture content is at the wilting point.

If water does not flow laterally through an internal drainage layer, percolation through the hydraulic barrier is equal to percolation through the cover soil, i.e. $PERC^{\circ} = PERC$. Otherwise:

$$PERC^{\circ} = PERC + L. \qquad (27.4)$$

In the simplified manual method, Eq. 27.4 is solved iteratively since both $PERC$ and L are a function of hydraulic head. Assuming steady state conditions, the maximum flow in the internal drainage layer is calculated as:

$$q_{\mathrm{m}} = lL = l(PERC^{\circ} - PERC) \quad (27.5)$$

where q_{m} = maximum flow in internal drainage layer per unit width perpendicular to the direction of flow (m² s⁻¹), l = slope length (m), and other terms are as previously defined. The hydraulic transmissivity of the drainage layer must be adequate to accommodate this flow. Using Darcy's equation and the DePuit–Forcheimer assumptions, i.e. line of seepage is assumed to be parallel to the cover system slope and the hydraulic gradient is constant and equal to the sine of the slope angle:

$$T = \frac{l(PERC^{\circ} - PERC)}{\sin \beta} FS \qquad (27.6)$$

where T = required hydraulic transmissivity of drainage layer (m² s⁻¹), β = slope angle (degrees), and FS = factor of safety for flow capacity. Koerner & Daniel (1997) recommend use of a FS value of at least five to ten. The maximum hydraulic head in the drainage is given by:

$$h_{\mathrm{m}} = \frac{q_{\mathrm{m}}}{k \tan \beta} \qquad (27.7a)$$

where h_{m} = maximum hydraulic head (m), k = hydraulic conductivity of drainage medium, and q_{m} is as defined by Eq. 27.5. The maximum thickness of flow (measured perpendicular to the slope) is obtained from h_{m} using the equation:

$$t_{\mathrm{m}} = h_{\mathrm{m}}/\cos \beta \qquad (27.7b)$$

where t_{m} = the maximum thickness of flow (m). The design thickness of the internal drainage layer must be larger than t_{m} in order for pressure head not to build up in the layer.

Percolation through CCL or GCL barrier layers is calculated by Koerner & Daniel (1997) using Darcy's equation (see Chapter 25). Percolation through geomembrane and composite liners is calculated by Koerner & Daniel (1997) using the leakage rate equations developed by Giroud & Bonaparte (1989a,b) (see Chapter 26). Hydraulic head is an input parameter to these equations. It is suggested that h_{m} evaluated from monthly average water balance calculations be used to calculate leakage rates through hydraulic barriers.

Input data needs for the simplified manual method are minimal. Only precipitation and mean temperature data are required. Koerner & Daniel (1997) provide guidance for selecting all other parameters, e.g. run-off coefficient, root zone depth, and soil water storage capacity. The advantages of the method are its simplicity, ability to use a variable time step, and ability to calculate lateral flows in cover system internal drainage layers. The main disadvantages of the method are the steady state nature of all calculations and the numerous simplifying assumptions. Nonetheless, when appropriately used, the simplified manual method presents an acceptable approach to the design of hydraulic-barrier type final cover systems. The method is in no way adequate as a simulation or predictive tool, nor is it applicable to the analysis or design of capillary barriers or evapotranspirative barriers.

27.5.4.2 Design of Internal Drainage Layers

Koerner & Daniel (1997) suggest that for sizing the hydraulic requirements of the internal drainage layers, q_m be calculated based on a single storm event. With this approach, the soil above the drainage layer is assumed to be saturated and infiltration into the cover soil is assumed to equal percolation from the cover soil. Accordingly:

$$PERC^\circ = P - R = P(1 - C) \quad (27.8)$$

where: C = run-off coefficient (dimensionless) obtained from Fenn et al. (1975), Table 27.2, or from project-specific information.

Equation 27.5 was developed assuming that:

- the soil is at field capacity before the storm begins,
- there is no evapotranspiration during the storm, and
- the cover soil is sufficiently permeable to accept the calculated infiltration.

TABLE 27.2. Run-off coefficients (adapted from Fenn et al. 1975) suggested by Koerner & Daniel (1997) for simplified manual water balance calculations

Description of soil	Slope		Run-off coefficient
	Description (%)		
Sandy soil	Flat	<2	0.05–0.10
	Average	2–7	0.10–0.15
	Steep	>7	0.15–0.20
Clayey soil	Flat	<7	0.13–0.17
	Average	2–7	0.18–0.22
	Steep	>7	0.25–0.35

Koerner & Daniel (1997) suggest that if the cover soil is not permeable enough to meet this last assumption, $PERC^\circ$ should be calculated using Eq. 27.9:

$$PERC^\circ = P(1 - C)$$
$$\text{when } k_{cs} \geq P(1 - C) \quad (27.9a)$$

$$PERC^\circ = k_{cs} \quad \text{when } k_{cs} \leq P(1 - C) \quad (27.9b)$$

where k_{cs} = the cover soil saturated hydraulic conductivity in the same units as P. Equation 27.9 can be used to develop a conservative estimate of peak flow into a lateral drainage layer during a single storm event, a capability available in none of the other water balance models considered in this chapter.

27.5.4.3 Refinement to Simplified Manual Method

Equations are available to refine those algorithms in the simplified manual method related to calculating hydraulic head in the internal drainage layer and leakage through a hydraulic-barrier layer containing a geomembrane.

An improved estimate (compared to Eq. 27.7) of maximum hydraulic head in the internal drainage layer can be obtained using the equations from Giroud et al. (1992b) and Giroud & Houlihan (1995):

$$h_m = (jl/2)\left[\left(\tan^2 \beta + \frac{4PERC^\circ}{k \cos \beta}\right)^{1/2} - \tan \beta\right] \quad (27.10a)$$

where all terms are as defined previously, and j is given by Eq. 27.10b:

$$j = 1 - 0.12 \exp\{-[\log(8\lambda/5)^{5/8}]^2\} \quad (27.10b)$$

where:

$$\lambda = \frac{PERC^\circ}{k \tan \beta \sin \beta}. \quad (27.10c)$$

Values of average hydraulic head, h_{ave}, for a given value of h_m can be obtained from Fig. 27.12. For the case of $(PERC^\circ/k) < 0.25 \tan^2\beta$:

$$h_{ave} = \frac{PERC^\circ l}{2k \sin \beta \cos \beta}. \qquad (27.11)$$

It is suggested that with the simplified method, h_m be used from single storm event analyses to size the cover system internal drainage layer; h_{ave} can be used to calculate long-term $PERC$ values. Similar to the method in Koerner & Daniel (1997), $PERC^\circ$ to calculate h_m should be derived using hourly water balance calculations for the design storm, and $PERC^\circ$ to calculate h_{ave}

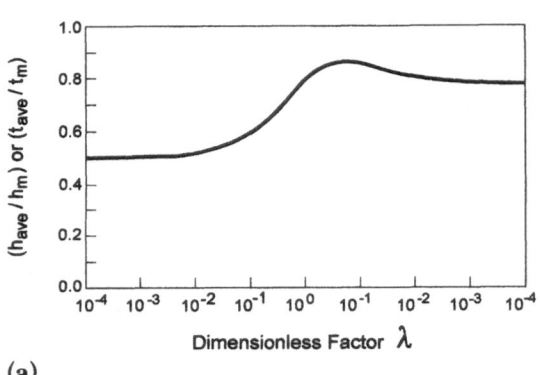

(a)

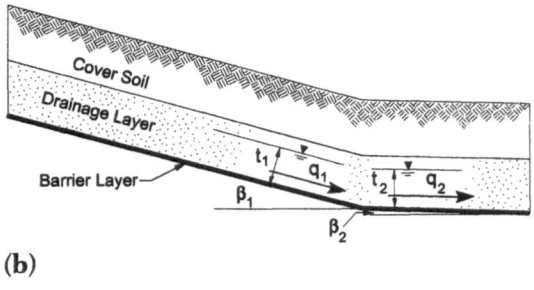

(b)

FIGURE 27.12. Cover system internal drainage layer: (a) values of (h_{ave}/h_m) or (t_{ave}/t_m) modified from Giroud & Houlihan 1995; and (b) flow continuity at slope bench or transition.

should be derived using monthly water balance calculations.

Improved estimates of leakage rates through geomembrane, geomembrane–GCL, and geomembrane–CCL hydraulic barriers can be obtained using the updated equations developed by Giroud and his coworkers, described in Chapter 26.

27.5.4.4 Design of Slope Transitions

Many cover systems are designed with benches and slopes transitions. Cover system failures have occurred due to hydraulic pressure buildup in the internal drainage layer at the transition. To prevent this condition, flow capacity, q, in the internal drainage layer must not decrease across the transition (Fig. 27.12b).

For many cover systems, the hydraulic gradient on the flatter part of the slope transition will be about one order-of-magnitude lower than the hydraulic gradient on the steeper part. For example, the gradient of a 3H : 1V slope is 0.32, whereas the gradient reduces to 0.03 for a 3% slope inclination typical of a cover system bench. For this condition, to prevent backup of flow and buildup of hydraulic head for an internal drainage layer flowing full, the hydraulic transmissivity of the drainage layer on a cover system bench or slope transition will need to be about one order-of-magnitude larger than that of the drainage layer on the sideslope.

More generally, based on Fig. 27.12b, to prevent hydraulic head buildup:

$$T_2 \geq T_1(\sin \beta_1/\sin \beta_2) \qquad (27.12)$$

where T = hydraulic transmissivity ($m^2 \ s^{-1}$), and β = slope angle (degrees). The subscript "1" refers to the portion of the drainage layer on the steeper upslope side of the transition and the subscript "2" refers to the drainage layer on the flatter downslope side of the transition (Fig. 27.12b). Equation 27.12 can be used directly to analyze geosynthetic drainage layers for which hydraulic transmissivities are either known or measured in the laboratory. For granular drainage materials where materials are typically specified in terms of a required hydraulic conductivity and thickness, Eq. 27.12 is recast as:

$$k_2 \geq k_1(t_1/t_2)(\sin \beta_1/\sin \beta_2) \qquad (27.13)$$

where k = hydraulic conductivity of drainage material (m s^{-1}), and other terms are as defined previously. For Eq. 27.13 to be valid, t_1 and t_2 must be less than the total thickness of the drainage layer.

27.5.5 HELP MODEL

The HELP computer code was developed by the US Army Corps of Engineers Waterways Experimental Station (WES) for the USEPA. At the time of writing, the most recent revision (Version 3.07) can be purchased from NTIS or downloaded from the WES website at http://www.wes.army.mil/el/elmodels. The documentation developed by Schroeder et al. (1994a,b) can be downloaded from the USEPA website at http://www.epa.gov/cincl/ or purchased from the US National Technical Information Service, (800) 553-6847. Additional guidance on using the HELP model to evaluate the hydrologic performance of cover systems can be found in USEPA (1991).

The HELP model simulates hydrologic processes for active or closed landfills by performing sequential water balance calculations using a quasi-two-dimensional approach, which considers all flow to be vertical, except at lateral drainage layers, where flow can be vertical or lateral. The simulation is marched forward in time with the water balance processes being considered steady within each time step. The conceptualization of the HELP model is shown in Fig. 27.13. The model can be used to separately evaluate each subsurface profile shown in Fig. 27.13.

The hydrologic processes considered in the model include precipitation, surface water storage, i.e. storage as snow, interception of precipitation by foliage, surface water evaporation, run-off, snow melt, infiltration, plant transpiration, soil water evaporation, soil water storage, vertical flow (saturated and unsaturated) through non-barrier soil layers, vertical percolation (saturated) through soil barrier layers, vertical percolation (saturated) through geomembrane and geomembrane–soil composite barrier layers, and lateral or vertical flow (saturated) through lateral drainage layers.

Run-off in the HELP model is computed using the run-off curve number method of the US Department of Agriculture Soil Conservation Service (USDA-SCS) (USDA-SCS, 1985). Daily run-off is calculated in the model as:

$$R = \frac{(P - 0.2S)^2}{(P + 0.8S)} \qquad (27.14)$$

where S = retention parameter (mm day^{-1}), dependent on SCS curve number; and R and P are daily average run-off and precipitation, respectively (mm day^{-1}). The SCS curve number is a function of soil texture, vegetation quality, and cover system slope length and inclination. McBean et al. (1995) state that use of daily rainfall averages effectively decreases storm intensity (because the duration of most storms is less than 24 h), resulting in a simulation that overpredicts infiltration and underpredicts run-off.

Evapotranspiration is computed in the HELP model using a two-stage modified Penman energy balance method developed by Ritchie (1972). This method uses the potential evapotranspiration concept as the basis for prediction of surface and soil water evaporation and plant transpiration. The potential evapotranspiration demand is first met by evaporation of water or snow on foliage or on the ground, then soil water evaporation, and finally plant transpiration. Due to the assumptions built into the HELP model,

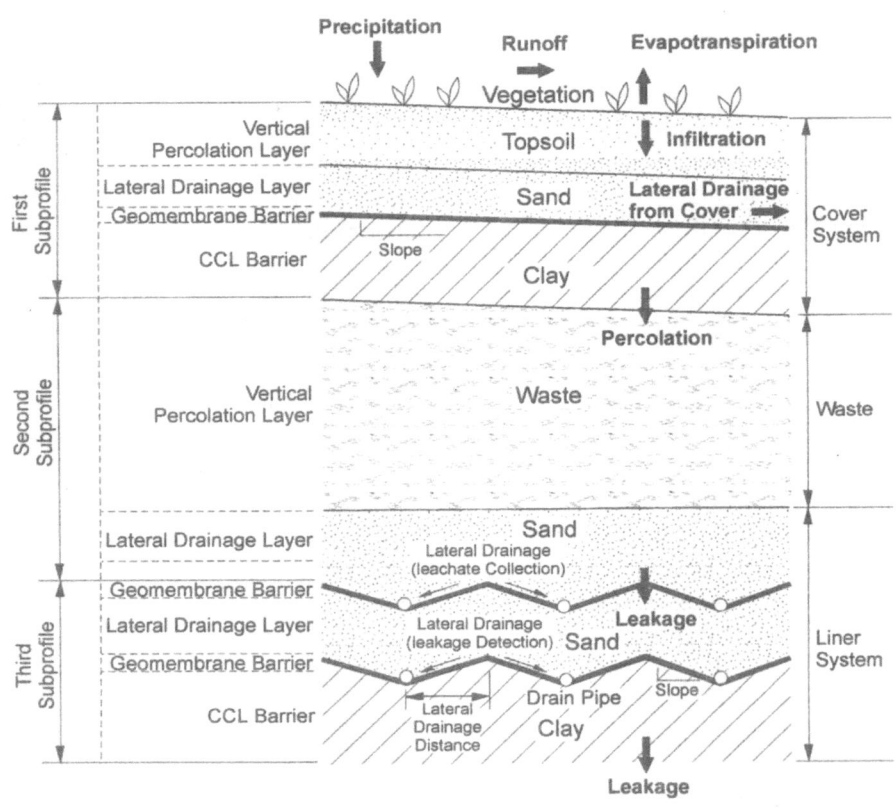

FIGURE 27.13. Conceptualization of HELP water balance model (modified from Schroeder *et al.* 1994a).

evapotranspiration may be underestimated in arid climates. Growth and decay of surface vegetation is modeled using an algorithm taken from the simulator for water resources in rural basins (SWRRB) model (Arnold *et al.* 1989).

Vertical drainage for cover soil, i.e. topsoil and protection, layers for both saturated and unsaturated flow conditions, is computed using Darcy's equation. The HELP model assumes that there is a unit gradient across the vertical percolation layer. The model utilizes an unsaturated hydraulic conductivity and the unit hydraulic gradient to calculate unsaturated flow rates. The unsaturated hydraulic conductivity, k_u (m s^{-1}), is obtained using Campbell's (1974) equation:

$$k_u = k_s[(\theta - \theta_r)/(\theta_s - \theta_r)]^{3+2/\zeta} \quad (27.15)$$

where k_s = saturated hydraulic conductivity of soil layer (m s^{-1}); θ = volumetric moisture content of soil layer (dimensionless); θ_s = volumetric moisture content of soil layer at saturation (dimensionless); θ_r = volumetric moisture content of soil layer at wilting point, typically in the range 0.01–0.10 (dimensionless); and ζ = poresize distribution index (dimensionless), calculated as described in Schroeder *et al.* (1994a,b). As a result of the hybrid formulation given above, the model cannot be used to simulate the physics of water movement through an unsaturated soil layer.

Lateral drainage below a cover soil layer is

modeled by an analytical approximation to the steady state solution of the Boussinesq equation (McEnroe 1993). Vertical percolation through low permeability soil hydraulic barriers is evaluated using Darcy's equation assuming saturated conditions and the calculated average hydraulic gradient across the barrier layer. Vertical percolation through geomembranes and geomembrane–soil composite barriers is evaluated based on the work of Giroud & Bonaparte (1989a,b) and Giroud *et al.* (1992a) (see Chapter 26).

The daily water balance is calculated by a linking process, starting with a surface water balance, then evapotranspiration in the subsurface, and finally subsurface water routing from the surface downward one soil layer at a time. The routing procedure uses a time step that can range from 30 min to 6 h. However, only daily, monthly, annual, and long-term average output data are reported.

Data requirements for Version 3 of the HELP model are summarized in Schroeder *et al.* (1994a,b). The HELP model requires daily and general climatic data, material properties data for the cover system components being modeled, and cover system design data. An important feature of the model is its climatic and material property default data options. Required daily weather data include precipitation, mean temperature, and total global solar radiation.

Due to its method of calculating downward flux and its limiting of upward flux, i.e. no upward flux within or below a barrier layer, the HELP model is not considered a particularly accurate simulation model for cover systems located in arid areas where the subtleties of unsaturated moisture movement can dominate the water balance.

27.5.6 LEACHM MODEL

LEACHM (Hutson & Wagenet 1992) is a one-dimensional finite difference code that was developed to simulate the effects of agricultural management alternatives on the movement of water and chemicals in a shallow soil profile, i.e. to a maximum depth of 2 m. Only the hydrologic component of the model will be discussed further. The code and model documentation may be obtained from the Department of Soil, Crop & Atmospheric Sciences at Cornell University, Ithaca, New York.

LEACHM considers precipitation, runoff, evapotranspiration, soil water storage, and percolation in the water balance. Infiltration of water into the soil profile and vertical drainage are simulated using a finite difference solution to Richards' (Richards 1931) partial differential equation. This equation is obtained by combining the differential form of Darcy's equation for unsteady vertical flow with the one-dimensional differential form of the conservation of mass equation:

$$\frac{\partial \theta}{\partial t} = \frac{\partial}{\partial z}\left[k_{\mathrm{u}}(\theta)\,\frac{\partial \psi(\theta)}{\partial z} - 1\right] - S(z, t) \quad (27.16)$$

where Ψ = matric potential head (negative) due to capillary suction forces (N m^{-2}); θ = soil volumetric moisture content (dimensionless); k_{u} = unsaturated hydraulic conductivity (m s^{-1}); z = vertical co-ordinate, positive downward (m); t = time (s); and $S(z, t)$ = sink term representing uptake by transpiration (s^{-1}).

Unsaturated soil hydraulic conductivity in LEACHM is calculated using the Campbell (1974) relationship. Precipitation in excess of the infiltration capacity of the soil is shed as run-off. Evaporation and transpiration are modeled separately based on the methods of Childs & Hanks (1975). With this method, the potential evaporation and transpiration are first estimated based on the pan evaporation rate, a pan factor, and a crop cover fraction. The actual evaporation is then calculated as the lesser of the potential evaporation and the possible evaporation calculated using Richards' equation and the selected boundary condition. Any remaining potential evapotranspiration demand is applied to transpiration. However, transpiration is not allowed if the matric potential head of the soil is less than −1500 kPa, the potential corresponding approximately to the soil wilting point.

LEACHM requires that climatic data, soil properties, vegetation data, and initial and boundary conditions be input. Unlike the HELP model, there are no default data; the user must specify each input parameter. Required weather data are precipitation magnitude, rate and start time; minimum and maximum air temperatures; and pan evaporation rate. The precipitation option allows rainfall data for short, intense storms to be input. In the absence of pan evaporation rate data, the rate can be calculated by LEACHM using the Linacre equation (Hutson & Wagenet 1992) with site-specific data, i.e. latitude, elevation, temperature and precipitation. Required soil data are bulk dry density, initial moisture content, saturated hydraulic conductivity, and soil water retention curve. If a soil water retention curve is not available, LEACHM contains a routine to compute fitting parameters for Campbell's soil–water characteristic curve from the particle size distribution, bulk density, and organic matter content of the soil. However, there is considerable uncertainty in the use of the regression equations to compute these parameters.

27.5.7 UNSAT-H

UNSAT-H is a one-dimensional finite-difference water balance model developed at Pacific Northwest Laboratory (Fayer & Jones 1990) to assess the water dynamics of waste disposal facilities at the US Department of Energy Hanford site. The model also simulates soil heat flow and non-isothermal vapor flow. Vapor flow may be an important moisture transport mechanism in near-surface soils in arid climates. UNSAT-H was derived from the UNSAT model of Gupta et al. (1978) and has retained many of the same

routines. At the time of this writing, Version 2.0 of UNSAT-H is most current. The code can be obtained from the Energy Science and Technology Software Center, Department of Energy, Oak Ridge, Tennessee.

UNSAT-H considers precipitation, runoff, evapotranspiration, soil water storage and percolation in the water balance. Like the LEACHM model, infiltration of water into, and vertical movement of moisture in, the soil profile is governed by a finite difference solution to Richards' partial differential equation. However, the unsaturated soil hydraulic conductivity term in UNSAT-H is calculated using polynomials, Haverkamp functions, Brooks–Corey functions, or van Genuchten functions, rather than the Campbell equation. Precipitation in excess of the infiltration capacity of the soil is shed as run-off. Evaporation and transpiration are considered separately. Evaporation is calculated using one of two approaches:

1. an integrated form of Fick's law of diffusion, which considers the flow of heat to and from the soil surface, the flow of water from the subsurface to the soil surface, and the transfer of water vapor from the soil surface to the atmosphere; or
2. a Penman-type equation, which is a modification of the diffusion equation and is dependent on net radiation and soil heat flux rather than on soil surface temperature.

Transpiration is calculated using a method based on leaf-area index or cheatgrass data and is limited by potential evapotranspiration.

UNSAT-H requires that climatic data, soil properties, vegetation data, and initial and boundary conditions be input. There are no default data; the user must specify each input parameter. The precipitation option allows rainfall data for short, intense storms to be input. Required soil data are fitting parameters for the soil water characteristic functions and the unsaturated hydraulic conductivity functions.

27.5.8 EVALUATION OF MODELS

A number of researchers have performed studies to evaluate the HELP, LEACHM, and/or UNSAT-H models (Thompson & Tyler 1984; Peters *et al.* 1986; Barnes & Rodgers 1988; Peyton & Schroeder 1988; Nyhan 1989; Nichols 1991; Udoh 1991; Fayer *et al.* 1992; Lane *et al.* 1992; Benson *et al.* 1993; Peyton & Schroeder 1993; Martian 1994; Fleenor & King 1995; Khire 1995; Woyshner & Yanful 1995; Khire *et al.* 1997; Webb *et al.* 1997; Zornberg & Caldwell 1998). These studies either simulated field or laboratory water balance data or investigated trends and magnitudes of the different water balance components. The conclusions of these studies are not always in general agreement. For example, some studies found that certain models overpredicted infiltration and/or percolation, whereas other studies concluded just the opposite.

Of all the available studies, the ones reported in Woyshner & Yanful (1995), Khire (1995), and Khire *et al.* (1997) are perhaps most interesting because of the scope and practical applicability of the study to cover system analysis and design.

The study by Woyshner & Yanful (1995), discussed in Section 27.4, used the HELP model to simulate water percolation through a three-layer soil cover test plot at an inactive tailings site near Rouyn-Noranda, Quebec, Canada. The cover system consisted of a compacted varved clay (0.6-m thick) placed between an upper fine sand layer (0.3-m thick), and a lower coarse sand base (0.30-m thick). A thin gravel surface layer was used for erosion protection and the cover was not vegetated. Both field and HELP model results showed that 4% of the total annual precipitation (rainfall and snow melt) percolated through the cover. The HELP model showed good predictive ability for the field conditions reported by Woyshner & Yanful (1995).

The studies by Khire (1995) and Khire *et al.* (1997) involve field water balance evalua-

tions for three cover system test plots at two landfills, one near Atlanta, Georgia ("Live Oak") and the other near East Wenatchee, Washington ("Wenatchee"). The sites were selected to represent humid and semi-arid climates, respectively. The Live Oak test plot has a cover system with a 0.6-m thick CCL barrier overlain by a 0.15-m thick vegetated soil surface layer. In Wenatchee, one test plot has the same cover system as at Live Oak site and the other test plot has a 0.75-m thick sand capillary barrier in lieu of the CCL. Climate, run-off, percolation, and soil moisture data collected between 1992 and 1995 were reported by Khire (1995) and Khire *et al.* (1997); however, data collection was still ongoing as of 1998. Run-off and percolation are collected in tanks and measured, while soil moisture content is measured by time domain reflectometry.

Khire *et al.* (1997) used their test plot data to assess the predictive capabilities of the HELP and UNSAT-H models and made the following observations from their study:

- Throughout most of the monitoring period, HELP underpredicted run-off for the humid Live Oak site by 740 mm ($\approx$90%) and overpredicted run-off for the semi-arid Wenatchee site with a CCL by 30 mm ($\approx$30%). Cumulative run-off predictions made using UNSAT-H were reasonably accurate for the Live Oak site, i.e. less than 3% error; however, season-to-season differences in run-off amounts were significant. For the Wenatchee site, UNSAT-H underpredicted run-off by 50 mm ($\approx$270%) for the test plot with a CCL, and predicted no run-off for the plot with a capillary barrier. The underpredictions resulted in more water entering the soil in the simulations than in the field. This resulted in higher soil water storage in the simulations than in the field.
- HELP underpredicted evapotranspiration by only 70 mm ($\approx$4%) at the Live Oak site. An accurate prediction of evapotranspiration was not expected, given that more water entered the soil due to the underpre-

diction of run-off. Thus, an overprediction of evapotranspiration was expected unless the potential evapotranspiration demand had already been met. UNSAT-H underpredicted evapotranspiration for the Live Oak site by 300 mm ($\approx$15%). Examination of the water balance equation indicates that underpredicting run-off and predicting evapotranspiration fairly accurately, or vice versa, result in an over-prediction of soil water storage and/or percolation. Both HELP and UNSAT-H overestimated evapotranspiration at the Wenatchee sites by about 20–165 mm ($\approx$20–40%).

- HELP somewhat captured the trends in percolation at the Live Oak site, but overpredicted total percolation by more than 700 mm ($\approx$300%). One reason why percolation was overpredicted was that there was additional water in the soil caused by the underprediction of run-off. Another factor that contributed to the overprediction of percolation by HELP was the assumption that water in the soil flowed vertically downward under a unit hydraulic gradient. Khire (1995) and Khire *et al.* (1997) indicate that under unsaturated conditions the hydraulic gradient in the field rarely equaled "1" and, for most of the time, was oriented vertically upward. UNSAT-H underpredicted percolation for the Live Oak site only slightly, by about 60 mm. Both HELP and UNSAT-H underpredicted percolation for the Wenatchee site with a CCL barrier. However, part of this difference is believed to have been caused by the preferential flow of water and snow melt through cracks and animal burrows during the winter of 1995. Prior to that time, both models had overpredicted percolation. UNSAT-H significantly overpredicted percolation for the Wenatchee site with a capillary barrier. One reason why percolation was overpredicted by over 90 mm ($\approx$2000%) was that there was additional water in the soil caused by the underprediction of run-off.

27.5.9 RECOMMENDATIONS

The specific water balance analysis method and input parameters to use for analysis and design of a cover system should be selected based on the purpose of the analysis and project-specific factors such as climate, type of cover (i.e. hydraulic barrier, capillary barrier or evapotranspirative barrier), and cover system components. Given the inconsistencies in water balance analysis results (e.g. the models sometimes overpredict and sometimes underpredict the various components of the water balance), uncertainties in soil properties and long-term barrier integrity (e.g. CCL hydraulic conductivity may increase over time if the CCL is not adequately protected), and other factors, significant engineering judgment must be applied when performing a water balance analysis for a specific site. The following general recommendations are made regarding the use of water balance methods for the design of cover systems:

- Either the simplified manual method or the HELP model is suitable for analyzing the ability of geomembrane, geomembrane–CCL, or geomembrane–GCL hydraulic barriers to minimize percolation.
- Percolation estimates provided by Gross *et al.* (1997) and summarized in Table 27.3, can be used by design engineers to verify percolation rates calculated on a project-specific basis using either the simplified manual method or the HELP model.
- Either the simplified manual method or the HELP model can be used to design internal drainage layers underlain by hydraulic-barrier layers containing a geomembrane. A discussion of the design storm to use with each method is given at the end of this section.
- Neither the simplified manual method nor HELP is capable of serving as a water balance predictive tool using estimated or default input data. The HELP model has limited capability as a predictive tool when calibrated using site-specific data.
- Any of the water balance analysis methods may be used for evaluating percolation through cover systems with CCL or GCL hydraulic-barrier layers. While methods that incorporate unsaturated flow models are potentially more accurate than methods

TABLE 27.3. Percolation rates through final cover systems with barriers incorporating geomembranes estimated using the HELP model (modified from Gross et al. 1997)[a]

Average annual rainfall (mm)	Average percolation rates			
	Geomembrane barrier mm a^{-1} l ha^{-1} day^{-1}		Geomembrane–CCL or geomembrane–GCL barrier mm a^{-1} l ha^{-1} day^{-1}	
100–300	0–0.05	0–1	0–0.005	0–0.1
300–600	0.002–03	0.06–9	0.0002–0.03	0.005–0.8
600–800	0.1–1	3–30	0.01–0.1	0.3–3
800–1,000	0.3–2	9–60	0.03–0.2	0.8–5
1,000–1,600	1–5	30–100	0.1–0.5	3–10

[a] Assumptions: (i) fair grass vegetation, (ii) sandy loam and silty clay loam topsoil, (iii) 5 and 20% cover system slopes, (iv) coarse sand and geonet internal drainage layers, and (v) 10-year synthetic weather records.

where saturated conditions are assumed for flow through the hydraulic barrier, the latter methods, i.e. simplified manual method and HELP model, are easier to use. These methods are likely to overpredict actual percolation rates for humid sites.

- For capillary-barrier and evapotranspirative-barrier cover systems, a water balance analysis method that can correctly model unsaturated flow is preferred. Thus, either LEACHM or UNSAT-H is preferable to the HELP model for evaluation of these types of systems.
- For cover systems in any climate that rely on enhanced evapotranspiration to minimize percolation, methods that correctly model unsaturated flow and also allow different vegetation scenarios to be input, such as LEACHM or UNSAT-H, are preferred.
- The literature should be consulted for the best available information on the tendencies of the various water balance models to either underpredict or overpredict the various components of the water balance for both wet and arid climatic conditions. This information should be considered in interpreting the results of project-specific water balance analyses.
- Due to the difficulty in performing accurate analytical water balances, field water balances should be performed, whenever possible, to verify the analytical results. This should especially be the case for alternative cover systems.

An important input parameter in the design of internal drainage layers in hydraulic-barrier type cover systems is rainfall intensity and duration. As previously discussed, the HELP model is limited to using daily rainfall data, and thus does not capture short-term intense peaks in storm events. Koerner & Daniel (1997) have suggested that hourly rainfall data be considered along with the simplified manual method to calculate percolation through the cover soil into the internal drainage layer ($PERC^\circ$). They present an example calculation of the sensitivity of $PERC^\circ$ to the use of monthly, daily, or hourly precipitation data. The example assumes a site near Austin, Texas, with a 100-m long 3H:1V slope and a surface run-off coefficient of 0.4. The results of their analysis were as follows:

- $PERC^\circ = 0.011$ mm h^{-1}, using the simplified manual method with the average monthly temperature, duration of sunlight and precipitation data from Austin;
- $PERC^\circ = 1.3$ mm h^{-1}, using the HELP model with default historical daily precipitation data from 1974 to 1977 for San Antonio and all other climatic data generated for Austin; and
- $PERC^\circ = 50$ mm h^{-1}, using Eq. 27.8 with the probable maximum 6-h precipitation event for the project vicinity, i.e. 500 mm.

Koerner & Daniel (1997) noted that the calculated peak flow rate based on hourly storm data is more than one order-of-magnitude larger than the calculated peak flow based on daily precipitation values. Because of this, they recommended that hourly precipitation data be considered to conservatively calculate peak flow rates into the drainage layer and to determine if the drainage layer has adequate capacity to transmit the peak flow during extreme storm events.

The authors of this chapter calculated $PERC^\circ$ for the same example as above using the HELP model with climatic data generated synthetically for Austin for a 20-year simulation period. For the authors' simulation, $PERC^\circ = 3.1\ \mathrm{mm\ h^{-1}}$. This calculated $PERC^\circ$ can be compared with the value of $1.3\ \mathrm{mm\ h^{-1}}$ obtained by Koerner & Daniel (1997) using the historical weather data for 1974–1977 for San Antonio. The higher $PERC^\circ$ obtained by the authors is primarily due to the fact that the HELP precipitation database for the period 1974–1977 reflects unusually dry weather for certain parts of the USA. More generally, short-duration rainfall records may not contain wet weather cycles or intense storm events, which control design. Also, as noted in Koerner & Daniel (1997), the rate of infiltration into a soil cover will be limited by the hydraulic conductivity of the soil materials. If it is assumed in the above example that the cover soil has a saturated hydraulic conductivity of $1 \times 10^{-6}\ \mathrm{m\ s^{-1}}$, then from Eq. 27.9b, the maximum possible rate of infiltration into the cover for a non-ponded surface condition is 3.6 mm $\mathrm{h^{-1}}$, approximately the rate of percolation calculated with the HELP model and daily rainfall data generated synthetically for Austin, Texas, i.e. $3.1\ \mathrm{mm\ h^{-1}}$. Thus, for typical cover systems with low to moderately permeable surface and protection layers, it will often be adequate to use the HELP model and a synthetic rainfall record with a sufficiently long simulation period, e.g. 20 years, to calculate lateral drainage and hydraulic head. Alternatively, Eq. 27.9b can be used directly to obtain a conservative value of $PERC^\circ$ for design.

27.6 Static Slope Stability

27.6.1 OVERVIEW

Slopes on landfills, waste piles, surface impoundments and other waste containment structures are sometimes quite steep. Slope inclinations can range from flatter than 5H : 1V (11.3°) to steeper than 2.0H : 1V (26.6°). This section addresses issues associated with the static stability of cover system components considering internal or interface downslope sliding of one or more of these components.

The frequency of occurrence of cover system stability failures has been high. More than a dozen case studies of past problems of this nature are described in USEPA (1999). One example of a past problem involved sliding of a protection soil layer over a reinforced GCL in a cover system that did not contain an internal drainage layer. Another example involved a sand drainage layer (itself overlain by a 0.3-m thick topsoil/protection layer) sliding over a textured HDPE geomembrane. In this latter case, the internal drainage layer had inadequate flow capacity and the drainage layer outlets were constricted. With geomembranes, GCLs, CCLs, geotextiles, and geocomposite drainage layers commonly used in a variety of configurations, potential low-shear-strength materials or interfaces must be considered for most designs. Significantly, past failures have involved sliding along each of the geosynthetic interfaces listed in Table 27.4.

27.6.2 LIMIT EQUILIBRIUM ANALYSES

27.6.2.1 Overview

The simplest limit equilibrium (LE) formulation for stability of waste cover systems is to assume infinite slope conditions and neglect the stabilizing influences of passive soil resisting forces at the toe of the slope, any true cohesion/adhesion in cover system materials and interfaces, and tension in the geosynthetic layers. More sophisticated LE formulations can account for these factors. Both the infinite slope and more sophisticated LE for-

TABLE 27.4. Interfaces upon which cover system components have undergone sliding

Cover system component interface sliding
Topsoil/protection layer sliding on:
geotextile cushion layer
geomembrane
GCL
Internal drainage layer sand sliding on:
geomembrane
GCL
Internal drainage layer geosynthetic sliding on:
geomembrane
Geotextile sliding on geomembrane
Geomembrane sliding on:
geotextile
GCL
CCL
GCL sliding on a prepared subgrade

mulations are discussed below. In all of the closed-form, two-dimensional LE solutions, force equilibrium is satisfied in the directions normal and parallel to the slope, but moment equilibrium is ignored.

27.6.2.2 Infinite Slope

For cover system geometries where the cover soil thickness is constant, infinite slope equations provide a simple and conservative basis for design. The degree of conservatism decreases as the ratio of the cover soil thickness to slope increases. In keeping with the approach of Giroud *et al.* (1995b), in this chapter the equations are formulated using buoyant unit weights and seepage forces.

Body seepage forces occur in cover systems when water infiltrating the cover system develops a significant flow component in the downslope as opposed to downward direction. This occurs, for example, when infiltration is blocked by a low permeability hydraulic-barrier layer. If the rate of infiltration is sufficient, hydraulic head will build up above the barrier layer and induce downslope flow. Downslope flow of water has a destabilizing

effect on the cover system. The seepage force per unit volume on soil particles in the direction of laminar flow is expressed as:

$$f_w = \gamma_w i \qquad (27.17)$$

where f_w = seepage force per unit volume (kN m^{-3}); γ_w = unit weight of water (kN m^{-3}) and i = hydraulic gradient. The concept of a seepage force, F_w (acting parallel to the slope), and buoyant unit weight, W_b (acting vertically), in an infinite soil slope underlain by a hydraulic barrier layer is illustrated in Fig. 27.14. For a 1-m thick cover system at a 3H:1V slope flowing full, the water induces a downslope body seepage force equivalent to a destabilizing shear stress of 3 kPa.

If there is no water flow in an infinite slope, the slope stability factor of safety is given by (Giroud *et al.* 1995b):

$$FS = \frac{\tan \phi_i}{\tan \beta} + \frac{a_i}{\gamma_t t \sin \beta} \qquad (27.18)$$

where FS = factor of safety, ϕ_i = angle of internal or interface friction for the critical potential slip surface (degrees), a_i = adhesion (for an interface) or cohesion (for internal strength) for the critical potential slip surface (kN m^{-2}), β = slope angle (degrees), γ_t = average total unit weight of material above the

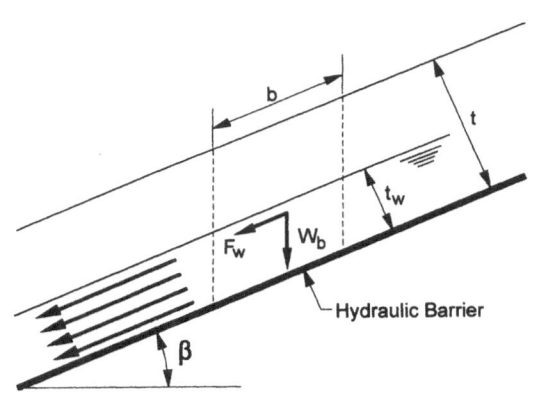

FIGURE 27.14. Seepage force and buoyant unit weight for a soil layer overlying a hydraulic barrier on an infinite slope.

critical potential slip surface (kN m^{-3}), and t = thickness of material above the critical potential slip surface (m). Use of this equation assumes that there is a unique critical potential slip surface in the cover system. For the case of no adhesion or cohesion ($a = 0$), Eq. 27.18 reduces to the classical solution:

$$FS = \tan \phi_i / \tan \beta. \qquad (27.19)$$

For full water flow in an internal drainage layer, two conditions need to be considered, stability above the hydraulic-barrier layer, and stability below the hydraulic-barrier layer. Two conditions must be considered because effective stresses above and below a non-porous hydraulic barrier such as geomembrane are different. The infinite slope factor of safety for "full flow" ($t_w = t$ in Fig. 27.14) parallel to the slope along an interface or internal slip surface above the hydraulic-barrier layer is (Giroud *et al.* 1995b):

$$FS_A = \frac{\gamma_b}{\gamma_{sat}} \frac{\tan \phi_a}{\tan \beta} + \frac{a_a}{\gamma_{sat} t \sin \beta} \qquad (27.20)$$

where FS_A = factor of safety for critical potential slip surface above the hydraulic-barrier layer (dimensionless), ϕ_a = angle of internal or interface friction for the critical potential slip surface above the hydraulic-barrier layer (degrees), a_a = adhesion (for an interface) or cohesion (for internal strength) for the critical potential slip surface above the hydraulic-barrier layer (kN m^{-2}), γ_b = average buoyant unit weight of material above the critical potential slip surface (kN m^{-3}), γ_{sat} = average saturated unit weight of material above the critical potential slip surface (kN m^{-3}), and all other terms are as previously defined. The buoyant unit weight, γ_b, is equal to the average total unit weight, γ_t, minus the unit weight of water, γ_w.

This factor of safety can be compared with the factor of safety expressed by Eq. 27.18 for the case of no water flow. The comparison shows that for typical soils:

- $FS_{A \text{ full flow}}/FS_{no flow} \approx 0.5$ if $a_i = 0$; and
- $FS_{full flow}/FS_{no flow} \approx 0.9$ if $\phi_i = 0$.

Based on these results, for slip surfaces located above the hydraulic-barrier layer, the factor of safety can decrease by a factor of two due to water flow if shearing resistance is generated primarily through friction.

The factor of safety ratios presented above are based on the assumption that the shear strength properties, ϕ_a and a_a, are not influenced by the presence of water. If the presence of water reduces the magnitudes of these parameters, the effects noted in the above comparison would be even more substantial.

The infinite slope factor of safety for "full flow" parallel to the slope along an interface or internal slip surface below a non-porous hydraulic barrier is given by (Giroud *et al.* 1995b):

$$FS_B = \frac{\tan \phi_b}{\tan \beta} + \frac{a_b}{\gamma_{sat} t \sin \beta} \qquad (27.21)$$

where FS_B = factor of safety for critical potential slip surface below the hydraulic-barrier layer, ϕ_b and a_b are the interface or internal shear strength parameters for the critical potential slip surface below the hydraulic-barrier layer, and all other terms are as defined previously. It should be noted that the shear strength parameters ϕ_b and a_b, used in Eq. 27.21, will typically be different from the parameters ϕ_a and a_a, used in Eq. 27.20, as the considered interfaces in the two equations are different. This factor of safety is to be compared with the factor of safety expressed by Eq. 27.18 for the case of no water flow. The comparison shows that for typical soils:

- $FS_{B \text{ full flow}}/FS_{no flow} = 1$ if $a_i = 0$; and
- $FS_{A \text{ full flow}}/FS_{no flow} \approx 0.9$ if $\phi_i = 0$.

Based on these results, the factor of safety along critical potential slip surfaces below the hydraulic-barrier layer is only affected to a relatively minor degree by water flow above the hydraulic-barrier layer.

The final infinite slope case to be considered is for "partial-depth" flow ($t_w < t$ in Fig. 27.14) parallel to the slope. The appropriate equations are (Giroud *et al.* 1995b):

$$FS_a = \frac{\gamma_t(t - t_w) + \gamma_b t_w}{\gamma_t(t - t_w) + \gamma_{sat}t_w}\left(\frac{\tan \phi_a}{\tan \beta}\right) \quad (27.22)$$

$$+ \frac{a_a/\sin \beta}{\gamma_t(t - t_w) + \gamma_{sat}t_w}$$

and

$$FS_B = \frac{\tan \phi_b}{\tan \beta} + \frac{a_b/\sin \beta}{\gamma_t(t - t_w) + \gamma_{sat}t_w} \quad (27.23)$$

where t_w = thickness of water flow parallel to the slope (m), as defined in Fig. 27.14, and all other terms are as defined previously.

27.6.2.3 Slope of Finite Length

Equations for the LE evaluation of sloping geosynthetic–soil layered systems (such as a final cover system) for a slope of finite length have been presented by Giroud & Beech (1989); Koerner & Hwu (1991); McKelvey & Deutsch (1991); Bourdeau *et al.* (1993); Druschel & Underwood (1993); Giroud *et al.* (1995a,b); Soong & Koerner (1997); Koerner & Daniel (1997) and Koerner & Soong (1998), among others. The most detailed treatments of the subject have been presented by Koerner and coworkers and Giroud *et al.* (1995a,b). Giroud *et al.* (1995a) have shown that compared with the method they present, the method utilized by Koerner & Hwu (1991) is more rigorous, but somewhat more complicated to use because it requires solution of quadratic equations. The formulation by Giroud *et al.* (1995a,b) involves an approximation that allows expression of the factor of safety as a closed-form algebraic equation where each term in the equation has a distinct physical meaning and is sufficiently accurate for practical purposes.

The two-part wedge considered by Giroud *et al.* (1995a,b) is illustrated in Fig. 27.15. For this condition, the slope stability factor of safety for a slope with constant soil thickness above the critical potential slip surface and for the case of no water flow, i.e. $t_w = 0$ in Fig. 27.15, is given by:

$$FS = \frac{\tan \phi_i}{\tan \beta} + \frac{a_i}{\gamma_t t \sin \beta}$$

$$+ \frac{t}{h}\left[\frac{\sin \phi_s}{2 \sin \beta \cos \beta \cos(\beta + \phi_s)}\right] \quad (27.24)$$

$$+ \frac{c_s}{\gamma_t h}\left[\frac{\cos \phi_s}{\sin \beta \cos(\beta + \phi_s)}\right] + \frac{T}{\gamma_t h t}$$

where ϕ_s = angle of internal friction for the soil material, i.e. protection layer and/or granular internal drainage layer, above the critical potential slip surface (degrees); c_s = cohesion of soil material above the critical potential slip surface (kN m^{-2}); h = height of slope (m), as defined in Fig. 27.15; T = geosynthetic tension above the critical potential slip surface; and all other terms are as defined previously.

The case of partial-depth and full-depth flow of water for a slope of finite height was addressed by Giroud *et al.* (1995b). For the case of a slope of uniform thickness above the critical potential slip surface, the factor of

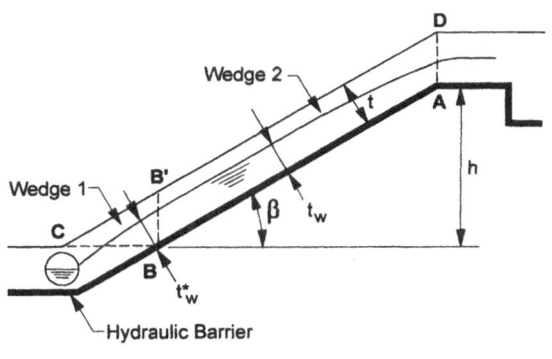

FIGURE 27.15. Definition of two-part wedge and flow thickness for the case of a slope of finite height.

safety above the hydraulic-barrier layer may be calculated using:

$$FS_A = \frac{\gamma_t(t - t_w) + \gamma_b t_w}{\gamma_t(t - t_w) + \gamma_{sat} t_w}\left(\frac{\tan \phi_a}{\tan \beta}\right)$$

$$+ \frac{a_a/\sin \beta}{\gamma_t(t - t_w) + \gamma_{sat} t_w}$$

$$+ \frac{\gamma_t(t - t_w^\circ) + \gamma_b t_w^\circ}{\gamma_t(t - t_w) + \gamma_{sat} t_w}\left(\frac{t}{h}\right)$$

$$\times \left[\frac{\sin \phi_s}{2 \sin \beta \cos(\beta + \phi_s)}\right] \quad (27.25)$$

$$+ \frac{c_s t/h}{\gamma_t(t - t_w) + \gamma_{sat} t_w}$$

$$\times \left[\frac{\cos \phi_s}{\sin \beta \cos(\beta + \phi_s)}\right]$$

$$+ \frac{T/h}{\gamma_t(t - t_w) + \gamma_{sat} t_w}$$

where t_w° = thickness of water in Wedge 1 in Fig. 27.15, and all other terms are as defined previously. For potential slip surfaces below a non-porous hydraulic barrier:

$$FS_B = \frac{\tan \phi_b}{\tan \beta} + \frac{a_b/\sin \beta}{\gamma_t(t - t_w) + \gamma_{sat} t_w}$$

$$+ \frac{\gamma_t(t - t_w^\circ) + \gamma_b t_w^\circ}{\gamma_t(t - t_w) + \gamma_{sat} t_w}\left(\frac{t}{h}\right)$$

$$\times \left[\frac{\sin \phi_s}{2 \sin \beta \cos \beta \cos(\beta + \phi_s)}\right] \quad (27.26)$$

$$+ \frac{c_s t/h}{\gamma_t(t - t_w) + \gamma_{sat} t_w}$$

$$\times \left[\frac{\cos \phi_s}{\sin \beta \cos(\beta + \phi_s)}\right]$$

$$+ \frac{T/h}{\gamma_t(t - t_w) + \gamma_{sat} t_w}$$

When there is full flow of water in Wedge 1 $(t_w = t)$ as well as in Wedge 2 $(t_w = t)$, Eq. 27.25 gives the following expression for the factor of safety for a critical potential slip surface above the hydraulic-barrier layer:

$$FS_A = \frac{\gamma_b}{\gamma_{sat}}\left(\frac{\tan \phi_a}{\tan \beta}\right) + \frac{a_a}{\gamma_{sat} t \sin \beta}$$

$$+ \frac{\gamma_b}{\gamma_{sat}}\left(\frac{t}{h}\right)\left[\frac{\sin \phi_s}{2 \sin \beta \cos \beta \cos(\beta + \phi_s)}\right] \quad (27.27)$$

$$+ \frac{c_s}{\gamma_{sat} h}\left[\frac{\cos \phi_s}{\sin \beta \cos(\beta + \phi_s)}\right] + \frac{T}{\gamma_{sat} t h}$$

and Eq. 27.26 reduces to:

$$FS_B = \frac{\tan \phi_b}{\tan \beta} + \frac{a_b}{\gamma_{sat} t \sin \beta}$$

$$+ \frac{\gamma_b}{\gamma_{sat}}\left(\frac{t}{h}\right)\left[\frac{\sin \phi_s}{2 \sin \beta \cos \beta \cos(\beta + \phi_s)}\right] \quad (27.28)$$

$$+ \frac{c_s}{\gamma_{sat} h}\left[\frac{\cos \phi_s}{\sin \beta \cos(\beta + \phi_s)}\right]$$

$$+ \frac{T}{\gamma_{sat} t h}$$

where all terms have been defined previously.

Another case sometimes encountered is that of a tapered cover soil thickness, as illustrated in Fig. 27.16. For this geometry, the

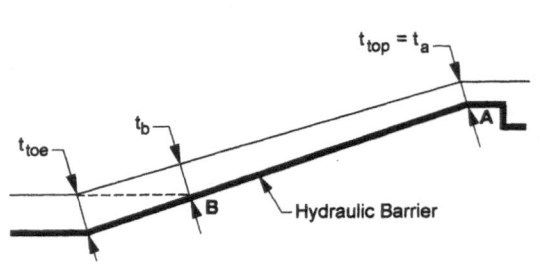

FIGURE 27.16. Definition of slope with a tapered soil layer.

factor of safety for the case of no water pressure is given by the equation:

$$FS = \frac{\tan \phi_i}{\tan \beta}$$

$$+ \frac{t_b}{t_{avg}} \left[\frac{a_i}{\gamma_t t_b \sin \beta} \right.$$

$$+ \frac{t_b}{h} \left[\frac{\sin \phi_s}{\sin(2\beta) \cos(\beta + \phi_s)} \right]$$

$$+ \frac{c_s}{\gamma_t h} \left[\frac{\cos \phi_s}{\sin \beta \cos(\beta + \phi_s)} \right] + \frac{T}{\gamma_t h t_b} \left. \right].$$

$$(27.29)$$

where:

$$t_{ave} = (t_a + t_b)/2 \qquad (27.30)$$

t_a = thickness of soil layer at point A in Fig. 27.16 (m), and t_b = thickness of soil layer at point B in Fig. 27.16 (m).

Equation 27.29 can also be used to calculate the factor of safety for a tapered slope of height h with a break, as illustrated in Fig. 27.17, by calculating an average soil thickness for the entire slope using the equation:

$$t_{ave} = \frac{t_a}{2}\left(1 + \frac{h_u}{h}\right) + \frac{t_b}{2}\left(1 - \frac{h_u}{h}\right) \qquad (27.31)$$

where h_u = height of slope above the break (m), as illustrated in Fig. 27.17.

The equations presented above provide

closed-form solutions to a variety of cover system slope stability situations. Some situations are too complex, however, to address using closed-form solutions and are more easily evaluated using commercially available slope stability computer software, e.g. PCSTABL5 (Achilleous 1988), UTEXAS3 (Wright 1991), XSTABL (Sharma 1994), or SLOPE/W (Geo-Slope Int. Ltd.). Available software has the advantage that it can be applied to non-uniform slope, soil cover, and hydraulic head conditions, and can incorporate a pseudo-static seismic coefficient for use in seismic stability evaluations.

The LE method is useful for evaluating cover system stability under most conditions, but is subject to several limitations. With the LE method, material and interface shearing resistances are assumed to be independent of displacement. For geosynthetic materials and interfaces, however, mobilized shearing resistance is not constant but increases with increasing displacement to a peak value. For many materials and interfaces, the shear resistance decreases with increasing displacement, after reaching the peak, ultimately to a "residual" value (Fig. 27.18). This behavior is sometimes referred to as "strain softening." In using the LE method, judgment must be applied to the selection of shear strength values

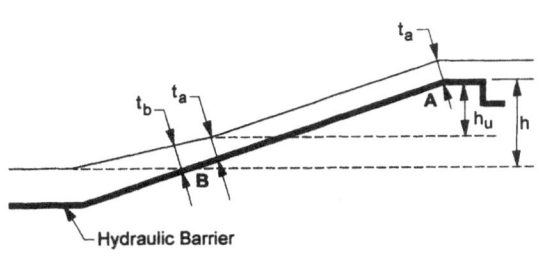

FIGURE 27.17. Definition of slope with a partly tapered soil layer.

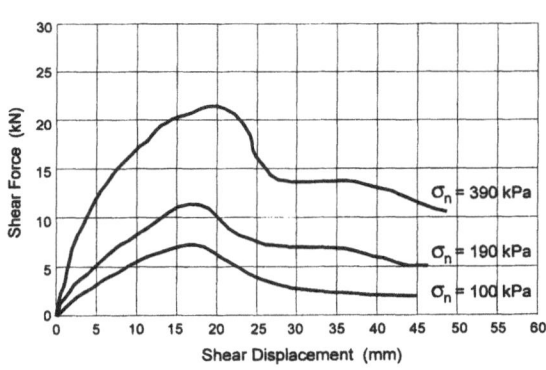

FIGURE 27.18. Result of direct shear test on a GCL illustrating peak and large-displacement shearing resistances.

for strain-softening materials, i.e. peak, residual, or some other value. The LE method is similarly limited with respect to tension forces in cover system geosynthetic components and, therefore, cannot be used to estimate the magnitude and distribution of stresses and deformations in these components.

As a final comment on the LE equations presented in this section, the equations incorporate terms to account for material interface adhesion or internal cohesion, and for geosynthetic tension. Users of these equations should be cautious in selecting adhesion–cohesion and geosynthetic tension values for design. As suggested by Koerner & Daniel (1997), adhesion and cohesion values should be used only when there is clear physical justification. As indicated by the analysis results presented previously, characterization of interface–internal shearing resistance by an adhesion–cohesion term instead of a friction angle will greatly affect the results of slope stability analyses where hydraulic heads are present in the cover system. In general, geosynthetic tension should not be in the equations unless the design includes a geosynthetic reinforcement layer. Other types of geosynthetics, such as geomembranes, geosynthetic drainage layers, etc., are not designed to permanently transmit tensile loads, are potentially subject to significant tensile creep, and typically have a low tensile modulus (which means that the geosynthetic must elongate to generate tension). Even if geosynthetic reinforcement is used, it should only be relied upon for the tensile force, which it can generate at a specified level of deformation. This acceptable level of deformation must be selected considering the overall performance of all system components.

27.6.3 STRESS–DEFORMATION ANALYSES

Stress–deformation analysis methods may be used to account for the stress–strain response of materials and interfaces and, therefore, to predict the distribution of stresses and strains within the cover system components, particularly geosynthetic components. Stress–deformation methods can also account for the effects of construction sequencing. The primary disadvantage of stress–deformation methods is the relatively large effort required to obtain material stress–deformation relationships and perform the calculations compared to the effort required with LE methods.

Several studies have been published on the application of stress–deformation methods to cover systems. For example, Long et al. (1993, 1994) describe a finite difference model (GEOSTRES) that considers stress equilibrium and strain compatibility. GEOSTRES uses inelastic, non-linear springs to model the shear resistance–displacement behavior at each interface and to model the axial load–displacement behavior within each component. Wilson-Fahmy and Koerner (1992, 1993) adopted a two-dimensional finite element model to account for stress equilibrium and deformation compatibility in stability analyses of soil–geosynthetic systems on slopes.

27.6.4 SHEAR STRENGTH PARAMETERS

Laboratory testing, using project-specific materials and testing procedures, and conditions representative of the anticipated field application, is typically performed to establish design shear strength parameters for cover system materials and interfaces. The various methods used for laboratory shear strength testing of soils are well-known and are fully described in a number of geotechnical textbooks and laboratory guides (Lambe 1951; Holtz & Kovacs 1981; and Bardet 1997). The most commonly used methods for laboratory shear strength testing of soils are the triaxial compression test and direct shear test.

Currently there are several types of laboratory devices available for the evaluation of shear strength of geosynthetic materials and interfaces. These laboratory devices include:

- large-scale (300 × 300 mm) direct shear box as specified by ASTM Standard D5321,
- conventional (50–100 mm square of circular) direct shear box with testing generally following ASTM D3080,
- torsional shear device (ASTM standard under development),
- tilt table, and
- large-displacement shear box.

Table 27.5 provides a summary of the advantages and disadvantages of the first four devices. Shallenberger & Filz (1996) describe the capabilities and limitations of the large-displacement shear box.

Project-specific shear strength testing programs are designed to simulate the anticipated field conditions by selecting appropriate values for testing procedures and conditions. These include the soil compaction conditions, i.e. water content and density; soil consolidation stress and time; the wetting conditions for the materials and interfaces; the range of applied normal stresses; the direction of shear for geosynthetic interfaces; and the shear displacement rate and magnitude. The potential effects of many of these testing conditions on measured interface shear strength parameters are reported in the literature, e.g. Martin et al. 1984; Williams & Houlihan 1987; Seed et al. 1988; Giroud et al. 1990; Seed & Boulanger 1991; Swan et al. 1991; Pasqualini et al. 1993; Stark & Poeppel 1994; Bemben & Schulze 1995; Gilbert et al. 1995; Nataraj et al. 1995; Bonaparte et al. 1996; Shallenberger & Filz 1996; Stark & Eid 1996; Stark et al. 1996; Dove et al. 1997; Eid & Stark 1997; Gilbert et al. 1996; Sharma et al. 1997; De & Zimme 1998; Sabatini et al. 1998; and Snow et al. 1998. Particular attention should be given to the following:

- Testing should be performed with materials and boundary conditions representative of the anticipated field conditions.
- Soil materials used in the test should be compacted to representative field conditions. The compaction moisture content for CCLs used in a direct shear testing pro-

TABLE 27.5. Summary of advantages and disadvantages associated with test devices for measuring interface shear strength (adapted from Gilbert et al. 1995)

Test device	Advantages	Disadvantages
Large-scale direct shear box	Industry standard Large scale Large displacement Minimal boundary effects	Machine friction Load eccentricity Limited continuous displacement Limited normal stress Expensive
Conventional direct shear box	Experience with soil Inexpensive Large normal stress	Machine friction Load eccentricity Small scale Limited displacement Boundary effects
Torsional shear device	Unlimited continuous displacement	Machine friction Anisotropic shearing Small scale Expensive
Tilt table	Minimal machine effects Minimal boundary effects Inexpensive	Small experience base Limited continuous displacement Limited normal stress No post-peak behavior Large effort to prepare sample

gram should be near the upper limit of acceptable moisture content and near the lower limit of dry unit weight, allowed by the construction specification.

- For geomembrane–CCL interface shear tests, a variety of opinions exist with regard to the application of additional moisture to the interface just prior to assembly of the test specimen. Options include adding no moisture, lightly or moderately "spritzing" water onto the CCL, or submerging the assembled sample. The rationale for any of these techniques is to simulate suspected installation (e.g. rainfall or moisture conditioning) or post-installation (e.g. condensation at the interface or consolidation-induced water movement to the interface) increases in CCL moisture content at the interface. Bowders et al. (1997) and Shallenberger & Filz (1996) discuss the contributions of other factors such as thermal gradients and post-compaction thixotrophy on interface stability.

- Hydration (soaking) times for GCL samples should be adequate to achieve minimum strength. Gilbert et al. (1996) report hydration times, as determined by cessation of GCL swelling under constant normal stress, for reinforced GCLs of up to 25 days. However, Daniel et al. (1993) have shown that full hydration is not always necessary to achieve minimum shear strength. An acceptable approach to hydration is to monitor vertical deformation of the GCL and continue to hydrate until these deformations have ceased under the applied normal stress (see below). When this procedure cannot be performed, a minimum hydration time of 72 h is recommended for GCLs to be tested in a 300-mm direct shear box; however, the user of the data should also be aware that without full hydration the measured strength may be larger than the fully hydrated strength.

- Testing conditions must adequately reflect the field consolidation conditions of the GCL or CCL components. GCLs hydrated as indicated above will be fully consolidated under the normal stress applied during hydration. This normal stress should be equal to that applied by the cover system. Consolidation requirements for CCL components of the cover may be established using ASTM D3080. Specimen consolidation times of 48 h or more may be required for some materials.

- ASTM D5321 recommends that tests be performed at a minimum of three normal stresses, with each test using new test specimens. The three selected normal stresses should bracket the overburden stress applied by the cover system to the material or interface being tested. This is important because many of the materials used in cover systems exhibit a non-linear relationship between interface or internal shear resistance and normal stress. For cover systems, applicable normal stresses will typically be in the range of about 15–50 kPa. Uniformity of normal stress over the entire test specimen must be maintained during consolidation and shearing so as to avoid stress concentrations.

- Shear displacement rates should be selected considering the type of slope stability analysis to be performed and the types of potentially critical materials or interfaces to be tested. For geosynthetic–geosynthetic interfaces (excluding GCLs), the maximum rate allowed by ASTM D5321 of 0.08 mm s^{-1} will generally be acceptable. For long-term stability conditions where the potentially critical material or interface includes a CCL or GCL component, the shear displacement rate should be as slow as reasonably achievable; the default shear displacement rate given in ASTM of 0.08 mm s^{-1} is too fast to achieve drained shearing conditions for CCL and GCL materials. Procedures for estimating shear rates to obtain fully drained conditions for CCL materials are given in ASTM D3080. Procedures and data for determining shear rates to obtain fully drained conditions for GCLs are given in Shan (1993). It is noted, however, that it may not be necessary to achieve fully drained test conditions to obtain test results suitable for long-term analyses. Available data suggest that for design purposes, a shear displacement rate of not more than 0.0005 mm s^{-1} will produce data appropriate for use in slope stability analyses involving GCL materials and interfaces. In contrast, for the evaluation of seismic stability, shear displacement rates should be as fast as reasonably achievable. For both conditions, testing should be performed using samples fully consolidated under the applied normal stress.

- Tests should be carried out to a shear deformation adequate to evaluate both the peak and large-displacement shear resistance of the interface or material being tested. Many geosynthetic–geosynthetic and soil–geosynthetic interfaces exhibit very significant post-peak reductions in shear strength (Fig. 27.18). ASTM D5321 states that one should "run the test until the applied shear force remains constant with increasing displacement". Torsional ring shear testing (Stark & Poeppel 1994), can be used to evaluate absolute minimum, i.e. residual, shear strengths of soils and geosynthetics for which representative samples can be produced for the small size and torsional shearing mode of this type of test. Alternatively, large-displacement shear box testing (Shallenberger & Filz 1996) can be used to evaluate residual shear strengths for larger-size test specimens in a linear displacement mode. For most practical design applications, true residual strength can be estimated to an acceptable degree of accuracy as 90–95% of the large-displacement direct shear strength obtained from a 300 × 300-mm direct shear test.
- Multicomponent cover systems may have more than one potentially critical slip surface. The shear test program for a project may need to consider several materials or interfaces.
- Some materials exhibit significant manufacturing variability. For example, the degree of texturing on geomembranes and the amount of internal-reinforcing in needle-punched GCLs has been observed to vary significantly from lot to lot. This variability should be considered both in design evaluation of stability and in the selection

of project quality control/quality assurance protocols.
- Test results can be interpreted in terms of a secant friction angle that varies as a function of normal stress, or by a tangent friction angle and apparent adhesion (or cohesion for internal strength) applicable to the range of considered normal stresses. Both of these approaches are illustrated in Fig. 27.19. For cover system applications, interface and internal shear strength parameters should be defined in terms of a secant friction angle for cases where hydraulic heads could develop in the cover soil. Since the apparent adhesion/cohesion may not be a true interface or material property, the use of this parameter with high heads could lead to an overprediction of the true slope stability factor of safety.

All of the foregoing factors should be considered in designing a laboratory shear testing program to evaluate interface and material shear strengths and in using the results of the program in slope stability analyses.

27.6.5 CONSTRUCTION CONSIDERATIONS

The placement of soil over a slope with underlying low shear strength materials or interfaces will induce shear stresses that can reduce slope stability. These shear stresses result from the operation of construction equipment on the slope and, if the soil is pushed down the slope, from the moving soil itself. Construction-induced stresses have been investigated by McKelvey & Deutsch (1991) and Koerner & Daniel (1997). These references present LE equations that can be used to evaluate the effect of construction equipment operation on the stability of the cover system.

The clear recommendation that comes out of the investigations cited in the previous paragraph is that cover soils should be placed over low shear strength materials and interfaces from the bottom of the slope upward

860

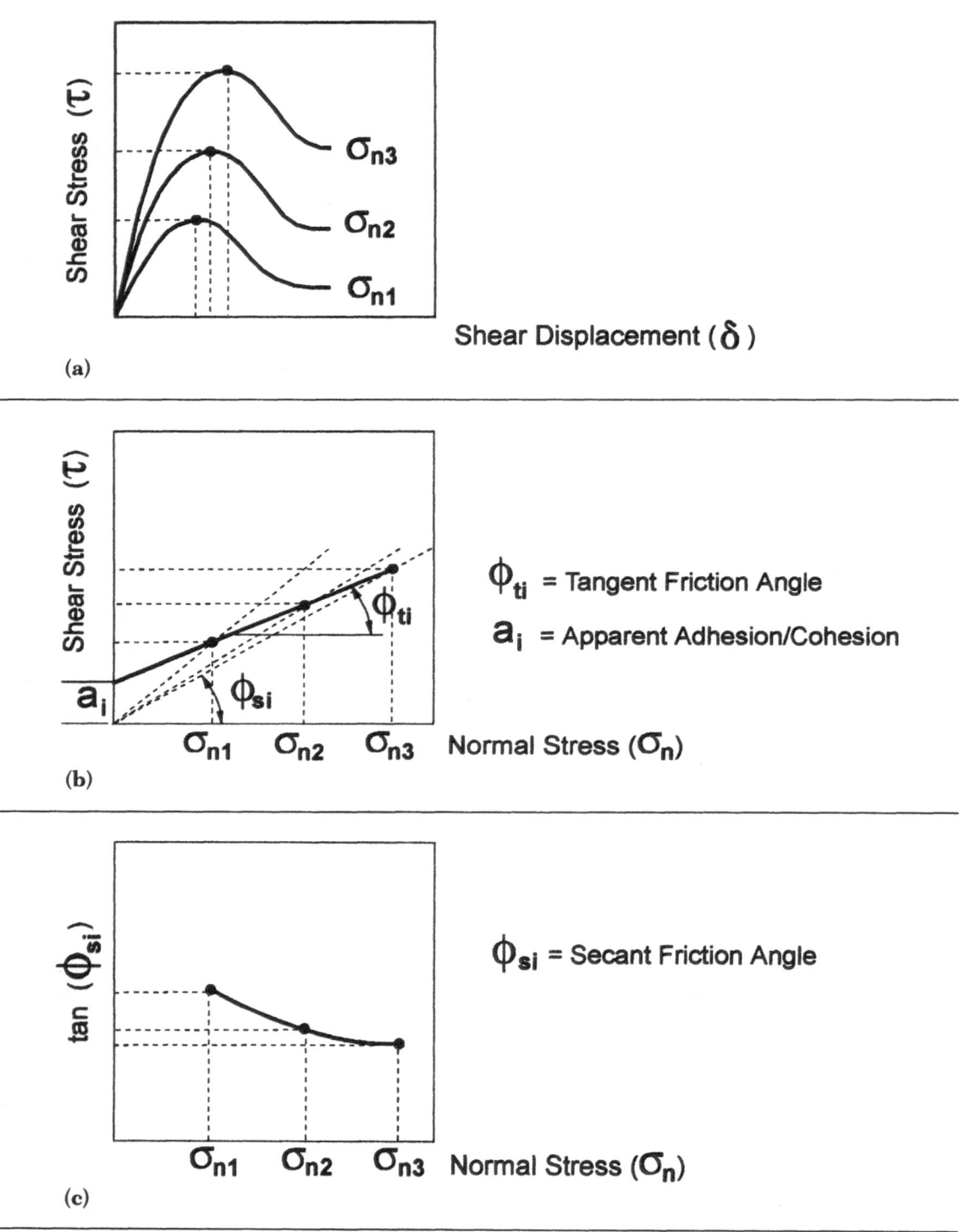

FIGURE 27.19. Interpretation of interface or internal shear test on cover system components: (a) test results for peak strengths; (b) tangent friction angle, ϕ_{ti}, and apparent adhesion/cohesion, a_i; and (c) secant friction angle, ϕ_{si}. Similar interpretations are applied to large displacement and residual conditions.

and not from the top of the slope downward. The following comments are provided with respect to placement of soil materials in final cover systems:

- by placing cover soils from the bottom of the slope upward, a passive, stabilizing soil wedge is established at the toe of slope prior to placement of soil higher on the slope;
- relatively small, wide-track dozers are recommended for placing the soil cover material; and
- downslope dynamic forces can be limited further by limiting the dozer speed on the slope and by instructing the dozer operator to avoid hard breaking, particularly when backing downslope.

By application of the construction procedures described above, construction-induced impacts to the stability of a cover system slope (designed to conventional slope stability factors of safety described in Section 27.6.6) are minor. For other conditions, (e.g. lower factors of safety than recommended in Section 27.6.6, placement of soil from the top of slope, use of large construction equipment) construction-stage stability should be checked using the procedures described in McKelvey & Deutsch (1991) or Koerner & Daniel (1997).

27.6.6 FACTORS OF SAFETY

LE analysis methods provide a calculated slope stability factor of safety (FS). Minimum acceptable FS values for cover systems depend on project-specific conditions and uncertainties. For example, when cover systems include strain-softening components or interfaces, differing minimum factors of safety are often applied for peak strength analyses and analyses based on large-displacement or residual strength. Other criteria may also influence selection of a minimum acceptable FS, including regulatory requirements, reliability of laboratory test methods, similarity between laboratory testing conditions and field conditions, and completeness of laboratory test data.

A number of technical references discuss the selection of FS for slope stability analyses involving geosynthetic interfaces including Bonaparte et al. (1996), Byrne (1994), Koerner & Soong (1998), Liu et al. (1997), and Stark & Poeppel (1994). Based on this available information, the following general guidance is provided:

- A minimum acceptable factor of safety (FS_{min}) for static stability analyses of 1.5 will often be adequate for permanent cover system applications where the design is based on peak interface and internal shear strengths. This FS_{min} is applied to normal operating conditions, e.g. no seismic loading, no live loading, and seepage forces, if any, are accounted for assuming a reasonable "average" operating condition.
- A smaller or larger FS_{min} may be considered based on an evaluation of: (a) consequences of cover system failure, and (b) uncertainty associated with each design parameter.
- If the cover system contains geosynthetic materials that exhibit strain-softening internal or interface shear strengths, FS_{min} for large-displacement conditions should also be checked. A FS_{min} of 1.20 is suggested where large-displacement shear strengths are obtained from testing in a direct shear box in accordance with ASTM D5321. For purposes of this evaluation, 50–75 mm of displacement, coupled with the observation that the shear stress–displacement plot is essentially flat at the end of the test is considered to satisfy the large-displacement condition. If true residual shear strengths are obtained using either a torsional-ring or large-displacement shear apparatus, FS_{min} values as low as 1.15 may be considered.
- Cover system designs should be checked for temporary extreme loading conditions. These conditions need to be identified on a case-by-case basis, but generally include

TABLE 27.6. Engineering measures to increase cover system slope stability factor of safety

---

Use cover system materials that have higher internal or interface shear strengths, as available

Provide for a flatter cover system slope by initially placing waste to a flatter slope (for new facilities) or waste excavation (for existing facilities)

Shorten the slope length through the use of benches or berms

Use perimeter retaining walls or buttresses to achieve a flatter cover system slope angle

Improve cover system internal drainage if hydraulic head buildup is predicted to occur

Utilize geosynthetic reinforcement, but only within the limitations of the approach described in this chapter

---

extreme storm events, live loads and earthquakes. Design for earthquakes is addressed subsequently. It is recommended that design for temporary extreme loading be based on $FS_{min} = 1.25$, 1.1 and 1.05 using peak, large-displacement or residual shear strengths, respectively.

If the factor of safety cannot be achieved for a given set of conditions, the engineer can consider a variety of measures to increase slope stability (Table 27.6).

27.7 Seismic Slope Stability

27.7.1 OVERVIEW

Seismic design of engineering structures is discussed in Chapter 21. In this chapter, specific consideration is given to the evaluation of seismic impacts to cover systems for landfills, waste piles, surface impoundments, and other waste containment structures. Impacts may involve either excessive seismic displacement of one or more cover system components or complete instability of the cover system.

27.7.2 SEISMIC HAZARD EVALUATION

The objective of the seismic hazard evaluation is to characterize the design earthquake with respect to the parameters required for engineering analysis (e.g. magnitude, style of faulting, site-to-source distance, peak ground acceleration, and spectral accelerations). The peak horizontal bedrock acceleration at a project site may be estimated using seismic hazard probability maps or site-specific seismic hazard assessments. In the USA, the most widely used such maps are those developed by the US Geological Survey (USGS) depicting peak and spectral horizontal bedrock accelerations having 10, 5, and 2% probability of exceedance in a 50-year period (corresponding to a 90% probability of not being exceeded for 50-, 100- or 250-year periods). These maps are periodically updated to reflect recent developments in the field of seismology. Maps for the USA can be found at the USGS Web site at http://www.geohazards.cr.usgs.gov/eq./. Background information on the development of these maps is provided by Frankel *et al.* (1996). The 90%, 250-year map corresponds to the USEPA prescriptive regulatory criterion for seismic design of MSW landfills.

Seismic hazard maps like those of USGS discussed above usually present the estimated free-field peak horizontal acceleration for a hypothetical bedrock outcrop on level ground at a particular location. If bedrock is not present at or near the ground surface, the peak acceleration may need to be modified to account for local site conditions. The presence of a waste mass will further modify the earthquake ground motions, as discussed subsequently. The primary difficulty associated with using seismic probability maps is that the maps by themselves do not provide information on the magnitude, site-to-source distance, or duration of the earthquake associated with the map acceleration values. For some types of seismic analyses, information on these variables is necessary. Because they are probabilistically derived, the acceleration values provided on such maps are typically composed of contributions of earthquakes of many different magnitudes from several to many different seismic sources. Each source may be associated with a different site-to-source distance and each magnitude–distance combination with a different duration. The USGS Web site has made information available on the distribution of earthquake magnitudes and site-to-source distances associated with the bedrock accelerations obtained directly from the USGS seismic hazard maps. Using this feature, the peak bedrock acceleration for a given site and recurrence interval is deaggregated by earthquake magnitude and site-to-source distance. Deaggregated spectral accelerations are also provided for spectral periods of 0.2, 0.3, and 1 s. As an example of the information available at the USGS Web site, deaggregated

bedrock acceleration and site-to-source data for Evansville, Indiana, are presented in Table 27.7. This table presents the deaggregated analysis results for a probability of exceedence of 2% in 50 years.

In using deaggregated data, such as given in Table 27.7, the engineer should identify the earthquake magnitude and distance combination that encompasses about two-thirds of the seismic hazard. For example, if more than two-thirds of the seismic hazard for a given site is from a small magnitude, near-source earthquake, then seismic analyses should be performed using input variables, e.g. strong motion records appropriate for this type of earthquake event. In some cases, more than one combination of earthquake magnitude and source distance may need to be considered. The values in Table 27.7 for Evansville illustrates such a case: a significant portion ($\approx 40\%$) of the seismic hazard for Ev-

TABLE 27.7. Deaggregated peak horizontal bedrock accelerations as a percentage of the aggregated peak probabilistic acceleration of 0.328 g for Evansville, Indiana, for a 2% probability of exceedence in 50 years (adapted from USGS Web site)

Hypocentral distance (km)	Earthquake magnitude						
	5.0	5.5	6.0	6.5	7.0	7.5	8.0
25	13.888	13.372	9.591	5.787	2.447	1.584	0.000
50	1.507	3.176	4.624	5.054	2.818	2.440	0.000
75	0.093	0.349	0.927	1.737	1.449	1.757	0.000
100	0.011	0.060	0.247	0.670	0.591	0.938	0.000
125	0.002	0.017	0.095	0.348	0.250	0.488	0.000
150	0.000	0.004	0.030	0.143	0.114	0.269	0.000
175	0.000	0.001	0.008	0.050	0.051	0.146	15.718
200	0.000	0.000	0.002	0.019	0.023	0.083	6.904
225	0.000	0.000	0.001	0.007	0.010	0.042	0.000
250	0.000	0.000	0.000	0.003	0.004	0.020	0.000
275	0.000	0.000	0.000	0.001	0.002	0.012	0.000
300	0.000	0.000	0.000	0.000	0.001	0.006	0.000
325	0.000	0.000	0.000	0.000	0.000	0.003	0.000
350	0.000	0.000	0.000	0.000	0.000	0.002	0.000
375	0.000	0.000	0.000	0.000	0.000	0.002	0.000
400	0.000	0.000	0.000	0.000	0.000	0.001	0.000
425	0.000	0.000	0.000	0.000	0.000	0.001	0.000

ansville is derived from earthquakes less than 25 km from the site with magnitudes between 5 and 6 (though the magnitude of some of the local events contributing to the seismic hazard may be as great as 7.5). However, over 20% of the seismic hazard is from a distant earthquake more than 150 km from the site with a magnitude of 8.0. Therefore, for some projects in Evansville, the impact of both local and distant events may warrant consideration.

Due to the reduced uncertainty and greater accuracy, site-specific seismic hazard analyses are preferred to the use of seismic hazard maps for critical structures in regions of high seismic activity. A site-specific seismic hazard analysis involves:

- identification of the seismic sources capable of strong ground motions at the project site,
- evaluation of the seismic potential for each capable source, and
- evaluation of the intensity of the design ground motions at the project site.

Site-specific seismic hazard analyses may be performed using either a deterministic or probabilistic approach. Detailed discussion of this topic is beyond the scope of this chapter. The reader is referred to Reiter (1990), Krinitsky et al. (1993), Richardson et al. (1995), and Kramer (1996). An example of a site-specific seismic hazard analysis applied to a landfill site (including cover system) in California is given in Kavazanjian et al. (1995a).

27.7.3 SEISMIC RESPONSE ANALYSIS

27.7.3.1 Introduction

The seismic hazard assessment as discussed above provides an estimate of peak horizontal accelerations in bedrock for a given site along with information on the causative earthquake event(s). A response analysis is used to estimate the seismically induced motions (e.g. acceleration, velocities, and/or displacements) at the ground surface or in the waste mass

cover system. Response analyses are needed because soil layers and waste modify the bedrock motions, sometimes in a manner that can significantly increase damage potential.

27.7.3.2 Material Property Selection

The first step in the seismic response analyses is to characterize the soil and waste material properties needed to perform the analysis. For equivalent linear analyses of vertically propagating shear waves (the most common type of seismic response analysis performed for geotechnical and waste management applications), these properties include total unit weight; dynamic shear modulus, G (kPa); and damping ratio for each material through which the waves propagate. Kramer (1996) provides an extensive review of the available technical literature on the selection of soil and rock properties for response analyses. Guidance on selecting MSW waste properties can be found in Sharma et al. (1990), Fassett et al. (1994), Richardson et al. (1995), Kavazanjian et al. (1995b), and Kavazanjian & Matasovic (1995).

Shear modulus and damping ratio curves for the Operating Industries Inc. (OII) site, a large inactive MSW industrial landfill in Monterey Park, California, are presented in Idriss et al. (1995), Augello et al. (1998) and Matasovic & Kavazanjian (1998). The curves independently developed by these three sets of investigators are shown in Fig. 27.20. These three sets of curves represent the best currently available information for use in estimating strain-dependent shear modulus reduction factors, G/G_{max}, and damping ratios for MSW and other solid waste materials for use in equivalent linear response analyses. The strain-dependent damping ratio is obtained directly from Fig. 27.20. The strain-dependent dynamic shear modulus, G, is obtained by multiplying the shear modulus reduction value from Fig. 27.20 by the small strain shear modulus, G_{max} (kPa), which can be calculated using the equation:

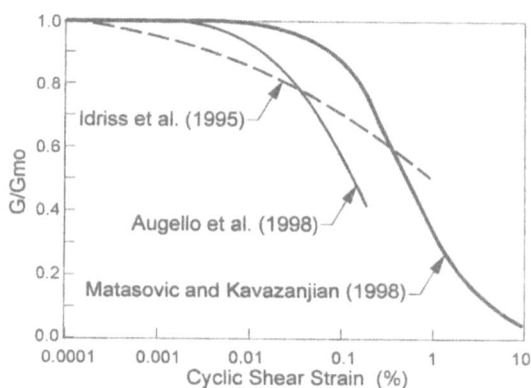

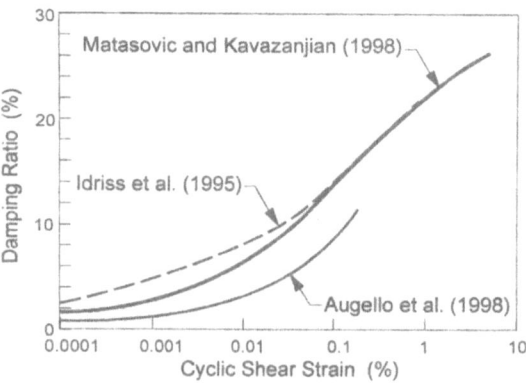

FIGURE 27.20. Estimated OII solid waste landfill shear modulus reduction (G/G_{max}) and damping curves. Data from Idriss *et al.* (1995), Augello *et al.* (1998), and Matasovic & Kavazanjian (1998).

$$G_{max} = \frac{\gamma_t v_s^2}{g} \qquad (27.32)$$

where v_s = shear wave velocity of waste material (m s^{-1}), γ_t = total unit weight of waste (kN m^{-3}) and g = acceleration of gravity (9.8 m s^{-2}). The small-strain shear modulus is ideally obtained from project-specific field testing. For landfills, this type of testing may be conveniently performed with the non-intrusive spectral analysis of surface waves (SASW) technique (Kavazanjian *et al.* 1994, 1996). Intrusive downhole or cross-hole geophysical testing techniques may also be used. In the absence of project-specific testing, the data for

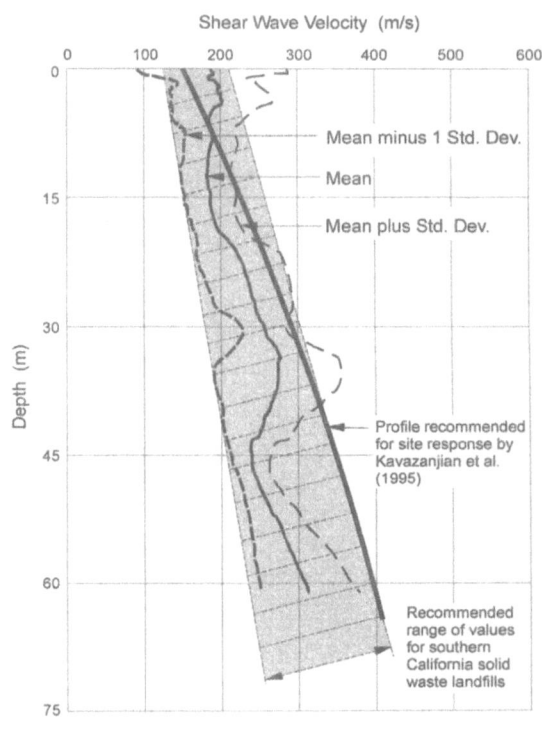

FIGURE 27.21. Shear wave velocities, v_s for southern California solid waste landfills (modified from Kavazanjian *et al.* 1996).

southern California landfills from Kavazanjian *et al.* (1996), presented in Fig. 27.21, can be used. It is noted that results obtained from a limited amount of SASW testing of MSW landfills in the eastern USA (unpublished) suggests that shear wave velocities for waste in these facilities may be lower, on average, than shear wave velocities for waste in relatively dry southern California landfills. In the absence of better information, the lower end of the recommended range of shear wave velocities shown in Fig. 27.21 can be used for MSW landfills located in the eastern USA and other temperate to wet climates.

27.7.3.3 Simplified Response Analysis
Simplified approaches to seismic response analyses involve the empirical correlation of peak horizontal waste mass or cover system acceleration, as applicable, to peak bedrock

TABLE 27.8. Earthquake parameters, corresponding peak horizontal bedrock acceleration estimates and peak horizontal accelerations recorded on the top of the Operating Industries Inc. (OII) landfill, California (adapted from Matasovic *et al.* 1998)

Earthquake	Moment magnitude	Style of faulting	Site-to-source distance (km)	Estimated peak-bedrock acceleration (g)	Peak acceleration at top deck (g)
Pasadena, 3 December 1988	5.0	Strike-slip	13	0.075	0.105
Malibu, 19 January 1989	5.0	Thrust	50	0.018	0.009
Joshua Tree, 23 April 1992	6.1	Strike-slip	163	0.006	0.017
Landers, 28 June 1992	7.3	Strike-slip	140	0.032	0.085
Big Bear, 28 June 1992	6.4	Strike-slip	119	0.015	0.049
Mojave Desert, 11 July 1992	5.5	Strike-slip	131	0.004	0.012
Northridge, 17 January 1994	6.7	Thrust	43	0.104	0.230

acceleration. Correlations of this type were first used in geotechnical engineering to relate peak ground accelerations at a site with soil overlying bedrock to peak bedrock accelerations at the same site (e.g. Seed & Idriss 1982; and Idriss 1990). More recently, Bray *et al.* 1995; Kavazanjian & Matasovic 1995; Singh & Sun 1995; Bray & Rathje 1998; and Matasovic *et al.* 1998 have extended this type of relationship to solid waste landfills.

Matasovic *et al.* (1998) compared estimated horizontal bedrock accelerations to recorded peak horizontal accelerations at the OII site. Table 27.8, taken from Matasovic *et al.* (1998), present peak acceleration values (average of two horizontal components) recorded at the crest of the OII landfill versus the estimated peak horizontal bedrock accelerations for the same site. Based on these results, Matasovic *et al.* (1998) concluded that peak horizontal bedrock accelerations from both near-field and far-field earthquakes with peak horizontal bedrock accelerations up to at least 0.15 *g* can be significantly amplified by solid waste landfills. They observed that the curve developed by Harder (1991) for the upper-bound amplification of seismic accelerations in earth dams, shown in Fig. 27.22, provides a conservative upper bound for amplification of peak accelerations for solid waste landfills. They also suggested that the relationship of Idriss (1990) for soft soil sites provides a reasonable representation of average amplification potential of solid waste landfills.

Independent of Matasovic *et al.* (1998), Bray & Rathje (1998) used the non-linear one-dimensional dynamic response analysis D-MOD (Matasovic 1993; Matasovic & Vucetic 1995) to perform parametric analyses of landfill response for a range of waste properties, waste heights, site conditions, and bedrock ground motions. The results of their arametric evaluation for cover systems are given in Fig. 27.22. This figure presents a plot of peak horizontal acceleration at the landfill crest versus peak horizontal bedrock acceler-

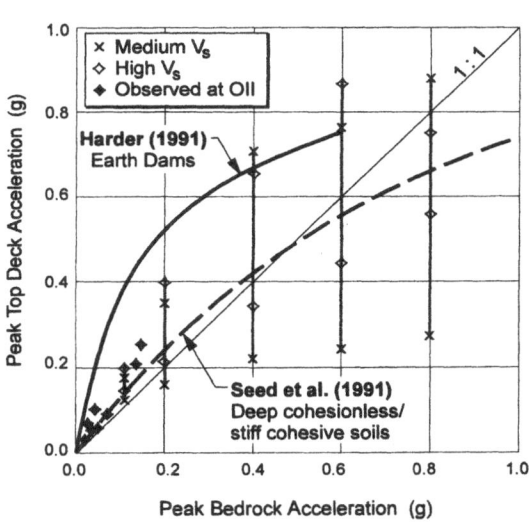

FIGURE 27.22. Results of parametric study comparing calculated peak horizontal acceleration for OII landfill top deck and peak bedrock acceleration (modified from Bray & Rathje 1998). Data from Harder (1991) for earth dams, and Seed *et al.* (1991) for deep cohesionless/stiff cohesive soils. Shear wave velocity v_s: (×) medium, (◇) high, (◆) observed at OII.

ation. They also compared their results to the Harder (1991) and the Seed *et al.* (1991) curves for stiff soil sites. Inspection of Fig. 27.22 shows that the Harder (1991) curve provides a conservative upper bound to the calculated cover system accelerations. Bray & Rathje (1998) note that the large amount of scatter in their parametric analysis results is due in large part to the sensitivity of the results to the input ground motion, i.e. variability among earthquakes. Variability in the assumed foundation conditions and waste profile also influenced the results. These findings are significant and engineers should consider this sensitivity when performing and interpreting the results of seismic response analyses.

Until more data become available, the Harder (1991) curve is recommended as a conservative upper-bound amplification of ground motions for simplified seismic site response of solid waste cover systems. Knowing the peak horizontal bedrock acceleration (from a USGS seismic hazard map, other map or site-specific analysis), the Harder (1991) curve can be used to estimate an upper-bound peak horizontal acceleration at the crest of the landfill. Should the Harder (1991) curve result in excessive cover system accelerations, detailed seismic response analyses can be conducted to assess whether a lower value of peak acceleration can be used on a project-specific basis. Site-specific seismic response analyses should also be used for any site where the average shear wave velocity in the upper 30 m of the foundation is less than 120 m s^{-1}, i.e. soft soil sites.

27.7.3.4 Analytical and Numerical Seismic Response Analyses

A one- or two-dimensional seismic site response analysis may be performed where significant cover system accelerations are anticipated or it is necessary to obtain a better estimate of seismically induced motions in the cover system than can be obtained with the simplified approach. These analyses are also recommended for sites with soft soil foundations and for critical facilities or facilities with special features. Such projects include those in regions where very large earthquakes can occur, where waste thicknesses are relatively large, or where cover system material or interface shear strengths are particularly low. The site response analysis is performed considering both the foundation soils and waste mass.

The computer program SHAKE, originally developed by Seed and coworkers (Schnabel *et al.* 1972) and updated by Idriss & Sun (1992), is perhaps the most commonly used computer program for one-dimensional seismic site response analysis. The SHAKE model idealizes the soil (and waste mass) profile as a system of homogeneous, visco-elastic sublayers of infinite horizontal extent. The response of this system is calculated considering vertically propagating shear waves. An equivalent linear analysis accounts for the strain-dependent non-linearity of soil and waste stiffness and damping using an iterative procedure to obtain modulus and damping values that are compatible with the equivalent uniform strain induced in each sublayer. At the outset, a set of properties (shear modulus, damping and total unit weight) is assigned to each sublayer of the soil or waste deposit. The analysis is conducted using these properties and the shear strain induced in each sublayer is calculated. The shear modulus and the damping ratio for each sublayer are then modified based on the applicable relationship between these two properties and shear strain (see Fig. 27.20).

Basic input to SHAKE includes the soil and waste profile, soil and waste properties, and an earthquake acceleration time history. Soil and waste properties include the maximum shear wave velocity, v_s, or maximum (small-strain) shear modulus, G_{max}, and total unit weight, γ_t, for each soil layer plus shear modulus reduction and damping ratio curves for each soil and waste material.

Computer programs are also available for equivalent-linear and truly non-linear two- and three-dimensional seismic site response analyses. A discussion of these more sophisticated models is provided by Kramer (1996). These models are only occasionally used in cover system design practice. Application of these models to the evaluation of cover system earthquake response often may result in lower-intensity seismically induced cover system motions than obtained using the one-dimensional SHAKE analysis. Use of non-linear analysis methods is recommended (Kavazanjian & Matasovic 1995 and Bray & Rathje 1998) when the peak horizontal bedrock acceleration exceeds 0.4 g. Examples of the use of these models include Idriss *et al.* (1995), Augello *et al.* (1998), and Matasovic & Kavazanjian (1998), which use the equivalent-linear two-dimensional finite element program QUAD4M (Hudson *et al.* 1994) to evaluate the seismic response of the OII landfill, and Kavazanjian & Matasovic (1995), Matasovic *et al.* (1998), and Bray & Rathje (1998), which use the one-dimensional non-linear program D-MOD to evaluate landfill seismic response.

27.7.4. DYNAMIC SHEAR STRENGTH

The dynamic shear strengths of the components and interfaces of a cover system must be estimated to perform seismic slope stability and/or deformation analysis. These estimates are typically based on static or cyclic undrained shear strength tests. Shaking-table laboratory test results and observed earthquake performance of system components are also used to develop information on the performance of these components and interfaces in earthquakes. Information on the cyclic shear strength of soils used in cover systems can be obtained from the geotechnical earthquake engineering literature, e.g. Kramer (1996), Kavazanjian *et al.* (1997). Shear strengths of compacted clays and unsaturated granular soils typically used in cover system construction are not significantly degraded by seismic loading, and cyclic shear strength is assumed to equal static shear strength. Similarly, the limited available data on the cyclic shear strength of interfaces involving geosynthetics (Kavazanjian et al. 1991; Yegian & Lahalf 1992; Augello *et al.* 1995; Yegian *et al.* 1995; and Chaney *et al.* 1997) suggests that cyclic shear strengths can be approximated using the results of static shear strength tests.

27.7.5 SEISMIC STABILITY AND DEFORMATION ANALYSIS

27.7.5.1 Overview

The static LE slope stability analysis methods discussed previously in this document may be adapted for use in the seismic stability evaluation of cover systems. This adaptation can be achieved using a number of different approaches, of which the following three represent the current state of practice:

- the pseudo-static factor of safety method,
- the modified pseudo-static factor of safety method, and
- the permanent seismic deformation method.

27.7.5.2 Pseudo-Static Factor of Safety Method

Due to its simplicity, the pseudo-static factor of safety method remains the most common method of analysis used in practice for seismic design of cover systems. With this approach, the factor of safety for the cover system is calculated using a limit equilibrium analysis that incorporates a specified seismic coefficient, which is applied as a horizontal body force to the potential slide mass. The factor of safety obtained for the calculation is compared to a minimum acceptable factor of safety to determine the adequacy of the design. The seismic coefficient equals the fraction of the weight of the potential failure mass that is applied as a horizontal force to the centroid of the mass in a pseudo-static limit equilibrium stability analysis.

For the case of an infinite slope with no water flow, the pseudo-static factor of safety is given by:

$$FS = \frac{(\cos \beta - k_h \sin \beta)\tan \phi_i}{(\sin \beta + k_h \cos \beta)} \quad (27.33)$$

$$+ \frac{a_i}{\gamma_t t(\sin \beta + k_h \cos \beta)}$$

where k_h = pseudo-static seismic coefficient (dimensionless), and all other terms were defined previously. For the case of a slope of finite length, the pseudo-static factor of safety can be calculated for the case of no water flow using the following approximate solution for sliding on the two-part wedge shown in Fig. 27.15 (Giroud *et al.* 1999):

$$FS = \frac{M}{N} \tan \phi_i + \frac{a_i(1 + \tan^2 \beta)^{1/2}}{N\gamma_t t}$$

$$+ \left(\frac{t}{2h}\right) \frac{(1 + \tan^2\beta)^{3/2} \tan \phi_s \tan \beta}{1 - (N/M)\tan \phi_s} \left(\frac{1 + k_h^2}{MN}\right)^2$$

$$+ \left(\frac{c_s}{\gamma_t h}\right) \frac{\tan \beta(1 + \tan^2\beta)}{1 - (N/M)\tan \phi_s} \left(\frac{1 + k_h^2}{MN^2}\right)$$

$$+ \frac{T \tan \beta}{N\gamma_t ht}$$

$$(27.34)$$

where M and N are dimensionless parameters given by:

$$M = k_h + \tan \beta \qquad (27.35)$$

$$N = 1 - k_h \tan \beta \qquad (27.36)$$

and all other terms were defined previously. Note that if $\phi_s = 0$, $c_s = 0$ and $T = 0$, Eq. 27.34 reduces to Eq. 27.33.

The case of a slope of finite length has also been addressed by Koerner & Daniel (1997), which requires the solving of a quadratic equation. For more complicated geometries and slope conditions, design calculations are more easily performed using one of the LE slope stability computer programs described previously. It is common in performing seismic stability analyses of cover systems to assume no water flow in the slope. The rationale for this assumption is that the probability of occurrence of both a design-level earthquake event and a design-level storm event at the same time is extremely low.

The main drawback of the pseudo-static factor of safety approach lies in the difficulty in relating the value of the seismic coefficient to the characteristics of the design earthquake. Use of the peak acceleration at the top of the waste mass as the seismic coefficient, coupled with a pseudo-static factor of safety of 1.0, results in a very conservative design basis. This result would imply no displacement of the cover system during the design earthquake, not even for the milli-seconds during which the peak accelerations are applied. A seismic coefficient smaller than the peak ground acceleration (expressed on a fraction of gravity) is sometimes used, but the magnitude of cover system displacement in this case is unknown.

27.7.5.3 Modified Pseudo-Static Factor of Safety Method

The problem of selecting an appropriate seismic coefficient for the pseudo-static approach can be addressed by implicitly considering the potential for seismically induced deformations. Based on the work of Hynes & Franklin (1984), Richardson et al. (1995) suggested that seismically induced displacements in a slope will be less than 0.3 m if the yield acceleration, k_y, defined as the k_h producing a factor of safety of 1.0, is no less than 50% of the peak horizontal acceleration of the slope, i.e. cover system. This result represents an upper-bound value for the seismic deformation calculated by Hynes & Franklin using almost 400 earthquake strong-motion records. The value of k_y required to produce 0.3 m of permanent seismic displacement drops to about 15% of the peak horizontal acceleration if the mean plus one standard deviation curve is considered rather than the upper-bound curve. Other values of k_y can be derived from Fig. 27.23.

Kavazanjian (1998) presented a refined procedure for deriving a displacement-based seismic coefficient value specifically for the design of cover systems for solid waste landfills. Seismic coefficient values for specified levels of permanent seismic displacement are calculated by multiplying a ratio, obtained

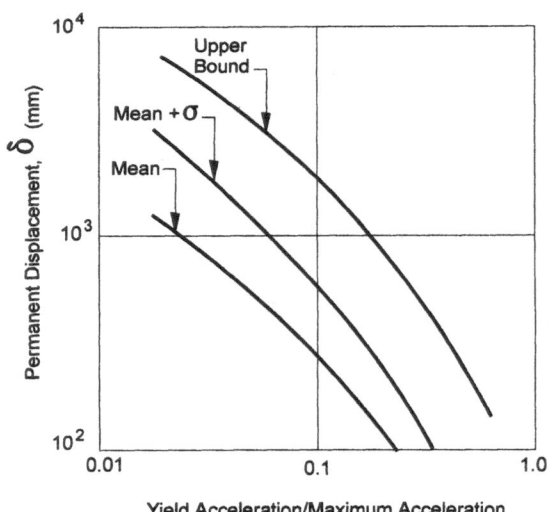

FIGURE 27.23. Hynes & Franklin (1984) permanent seismic displacement chart (modified from Richardson et al. 1995).

TABLE 27.9. Ratio of yield acceleration, k_y, to peak acceleration of cover system as a function of calculated permanent seismic displacement (adapted from Hynes & Franklin (1984) curves, shown in Fig. 27.23)

Calculated displacement (mm)	Mean ratio	Mean + σ ratio[a]
100	0.23	0.35
150	0.17	0.27
300	0.08	0.17
500	0.05	0.11
1000	0.03	0.06

[a] σ = standard deviation.

from Table 27.9, by the peak horizontal acceleration of the cover system obtained using the Harder (1991) curve shown in Fig. 27.22. Kavazanjian suggests that for earthquakes of magnitude less than or equal to 6.5 within 10 km of the project site, and for any earthquake of magnitude less than or equal to 5.5, the ratios in the "mean" column of Table 27.9 be used. Kavazanjian further recommends that for earthquakes of magnitude greater than 6.5, and for earthquakes between magnitude 5.5 and 6.5, i.e. more than 10 km from the project site, the ratios in the "mean + σ" column of Table 27.9 be used. Kavazanjian (1998) recommends that seismic coefficients derived using Table 27.9 be used with a factor of safety of 1.0. It is cautioned that the use of peak shear strength parameters with this approach may be unconservative. Shear strength values should be selected considering the displacement value from Table 27.9 associated with the chosen seismic coefficient. Again, this simplified approach is not recommended for soft soil sites.

27.7.5.4 Permanent Seismic Deformation Method

With the permanent seismic deformation method, cumulative permanent seismic deformations are calculated on the basis of that portion of the acceleration–time history of the cover system that exceeds k_y. For the infinite slope case, k_y is calculated using Eq. 27.33 and FS = 1.0. For the case of a finite length slope with uniform soil thickness above the critical potential slip surface, k_y is calculated using Eq. 27.34 and FS = 1.0. For more complex cases, k_y is calculated using a LE slope stability computer program.

The actual calculation of permanent seismic displacement is usually performed using Newmark's "sliding block on a plane" (1965) method of analysis. In a Newmark analysis (1965), acceleration pulses (in the earthquake acceleration–time history) exceeding k_y are double-integrated to calculate the accumulated "permanent" seismic displacement (Fig. 27.24). Theoretically, this calculated permanent displacement is a rigid body displacement that accumulates everywhere along the critical potential slip surface. Typically, only the horizontal component of the

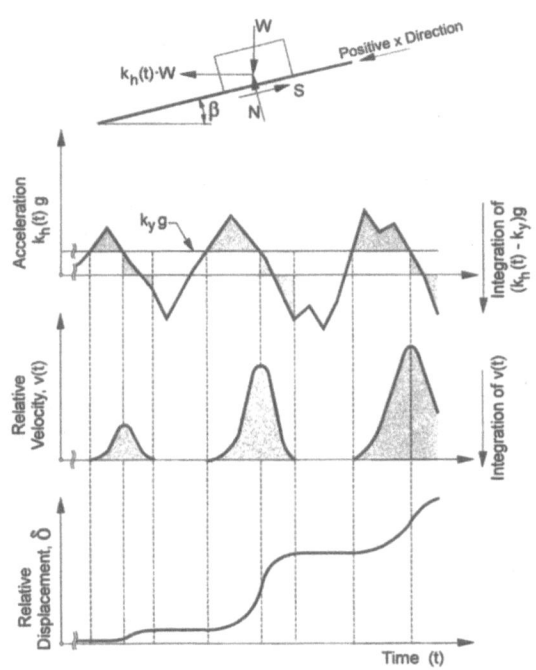

FIGURE 27.24. Basic elements of classical Newmark sliding-block analysis with constant yield acceleration, k_y.

earthquake acceleration–time history is considered in the analysis, and this acceleration is applied parallel to the slip surface. The acceleration–time history of the cover system used in the analysis is obtained from a seismic response analysis. With this approach, the response analysis is "decoupled" from the computation of permanent displacement; i.e. seismic response is calculated assuming no slip displacement between the cover system and landfill, and the cover system displacement is calculated using the results of the seismic response analysis (Bray & Rathje 1998). The decoupled approach is generally conservative (Bray & Rathje 1998) for landfill cover system displacement analyses.

Several PC-based computer programs exist to perform Newmark analyses (Houston *et al.* 1987; Yan *et al.* 1996). These models assume a constant value of k_y. Recognizing that most geosynthetic materials and interfaces exhibit strain-softening shear behavior, a modification to the standard Newmark procedure specifically for cover systems incorporating geosynthetic interfaces was proposed (Matasovic *et al.* 1997). The modified version incorporates a linear k_y degradation model to account for strain-softening materials and interfaces (Fig. 27.25). Matasovic *et al.* (1997) demon-

strated the sensitivity of the calculated permanent seismic deformation for a typical geotextile–compacted soil interface and three differing assumptions regarding k_y: (i) constant, based on peak interface shear strength parameters; (ii) constant, based on residual (or large-displacement) interface shear strength parameters; and (iii) degrading, in accordance with Fig. 27.25. Figure 27.26 presents typical calculation results from Matasovic *et al.* (1997) for the post-peak strain-softening exhibited by a compacted soil–geotextile interface. The sensitivity of the calculation results to the k_y assumption is evident.

27.8 Settlement

27.8.1 MECHANISMS OF SETTLEMENT
Cover systems may be subject to settlements resulting from a variety of mechanisms. For purposes of evaluating cover system performance, settlements can be considered to have one of three sources (see Fig. 27.27):

1. settlement of foundation soil,
2. settlement due to overall waste mass compressibility, and
3. settlement due to localized mechanisms in the waste.

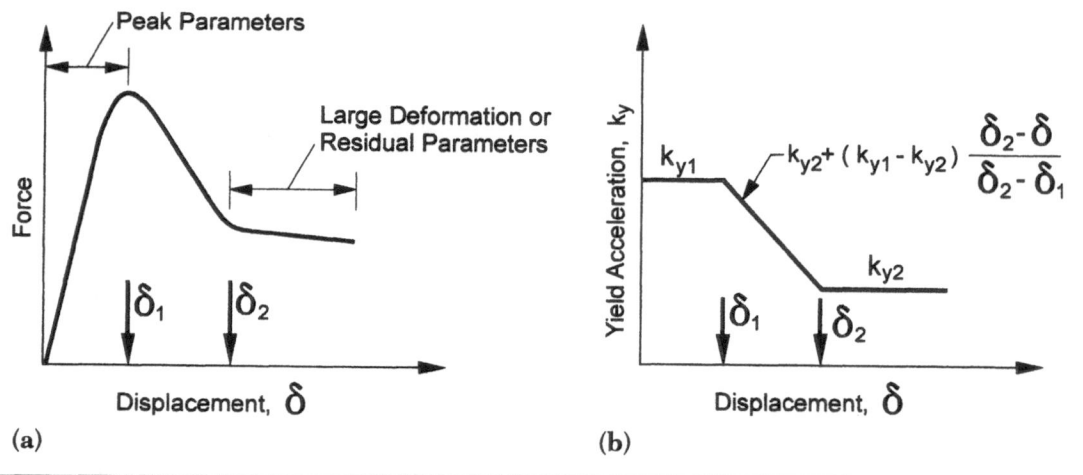

FIGURE 27.25. Yield acceleration degradation model: (a) measured shear force–displacement curve; and (b) yield acceleration degradation (modified from Matasovic *et al.* 1997).

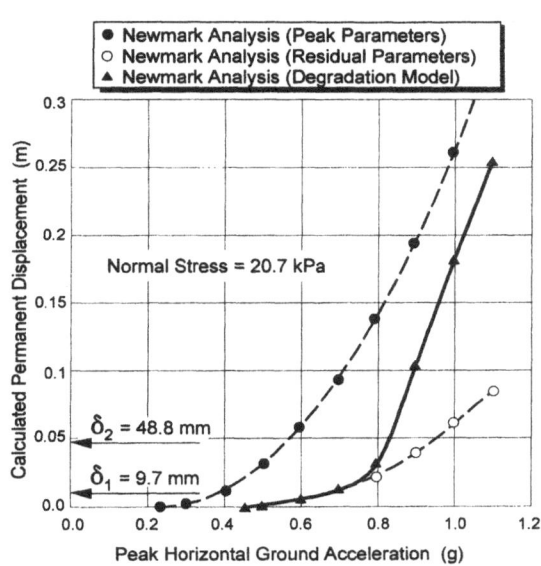

FIGURE 27.26. Results of Newmark seismic deformation analysis for constant and degrading yield acceleration at a normal stress = 20.7 kPa (modified from Matasovic *et al.* 1997): (●) peak parameters; (○) residual parameters; and (▲) degradation model.

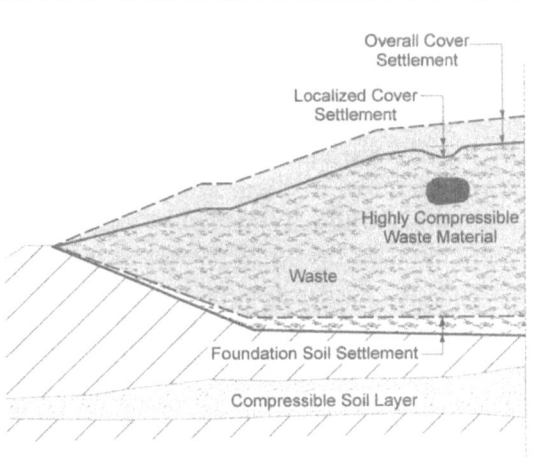

FIGURE 27.27. Sources of cover settlement.

Angular distortion or differential settlement may: (a) induce unacceptable tensile stress and strain in one or more cover system components, which can lead to component tearing or cracking; or (b) cause cover system slopes to

change or reverse; which, in turn, can affect the performance of the cover system drainage layer and/or gas collection layer.

27.8.2 SETTLEMENT OF FOUNDATION SOILS

Impacts of foundation settlement on the performance of a cover system are usually not significant. Occasionally, situations arise where foundation settlements are of sufficient magnitude to affect the cover system design. For example, if the waste mass is underlain by a thick layer of soft clay, consolidation settlements can be large. Both primary settlement and long-term secondary settlement should be considered. Calculations are performed using equations from conventional geotechnical engineering practice (see Chapters 2, 3, 9 and 16) and a timeframe at least equal to the active life and post-closure care period of the facility.

27.8.3 OVERALL WASTE COMPRESSION

Overall waste mass compressibility results in area-wide waste mass settlement. The potential for waste settlement is highly dependent on the type of waste. Relative to most other wastes, MSW is very compressible, due to both its initial compressibility when placed and the additional compressibility induced by the biodegradation of the organic component of the MSW. This latter component creates a significant time dependency to waste settlements. Other types of facilities that can undergo large settlements include impoundments containing high-water-content industrial sludges. Materials such as mine waste, ash and slag, construction and demolition waste, and soil waste have relatively lower settlement potential. The following discussion of overall waste settlement primarily focuses on the settlement potential of MSW waste and other highly compressible waste material. The evaluation of ash and soil waste, or other low compressibility inorganic waste, is typically performed using equations for conventional geotechnical engineering practice (see Chapters 2, 3, 9 and 16).

MSW waste compression results from complex factors (Sowers 1973; Edil *et al.* 1990; Sharma & Lewis 1994), including:

- mechanical compression due to self-weight and surface loads;
- ravelling, i.e. movement of fines into larger voids;
- physiochemical changes, including corrosion, oxidation, and combustion; and
- biochemical decomposition under aerobic and anaerobic processes.

The magnitude and rate of MSW settlement are controlled by many factors, among which are the waste fill height, organic content, age, moisture content, degree of compaction, and temperature. Figure 27.28 presents data from Edgers *et al.* (1992), König *et al.* (1996), and Spikula (1996), indicating that MSW land-fills can settle from about 5 to 20% (and even up to 30%) of the initial landfill thickness (measured from the time the landfill first reaches final grade).

A number of methods have been proposed for evaluating the short-term and long-term compression of waste. Three settlement models that have been adopted from geotechnical engineering and applied to waste are: (a) the one-dimensional compression model, (b) the power creep model, and (c) the Gibson & Lo (1961) model. A discussion of the last two models is presented in Edil *et al.* (1990), and they are not discussed further herein. Presently, there is little experience in applying these last two models, and their applicability to the prediction of long-term settlements is questionable (U.S. EPA 1999).

Conventional one-dimensional compression models, e.g. see Sections 2.7.4, 9.4.3, and 16.4.3, have been used to estimate waste settlements. However, it is often assumed that primary self-weight settlement is complete prior to installation of the cover system. Thus, calculated primary settlements are assumed not to directly influence cover system performance.

Cover system performance will be affected, however, by secondary waste settlement. The secondary waste settlement, $s_{\Delta t}$, that occurs between times t_1 and t_2 is calculated with the one-dimensional model using an equation of the form (see also Section 2.7.5):

$$s_{\Delta t} = C_{\alpha e} H_1 \log \frac{t_2}{t_1} \qquad (27.37)$$

where $C_{\alpha e}$ = modified secondary compression index (dimensionless), H_1 = height of waste at time t_1 (m), t_1 = starting time for the period of secondary compression (s), and $t_2 = t_1$ plus the time duration of secondary compression (s). Use of Eq. 27.37 implies that the magnitude of secondary settlement is independent of the applied stress. A modified form of Eq. 27.37 has been suggested by Bjarngard & Edgers (1990) and Stuglis *et al.* (1995) to account for a variable value of $C_{\alpha e}$ between "intermediate" and "long-term" secondary compression times. Their equation is formulated herein as:

$$s_{\Delta t} = C_{\alpha e 1} H_1 \log \frac{t_2}{t_1} + C_{\alpha e 2} H_2 \log \frac{t_3}{t_2} \qquad (27.38)$$

where $C_{\alpha e 1}$ = modified secondary compression index during the intermediate secondary compression period (dimensionless), $C_{\alpha e 2}$ = modified secondary compression index dur-

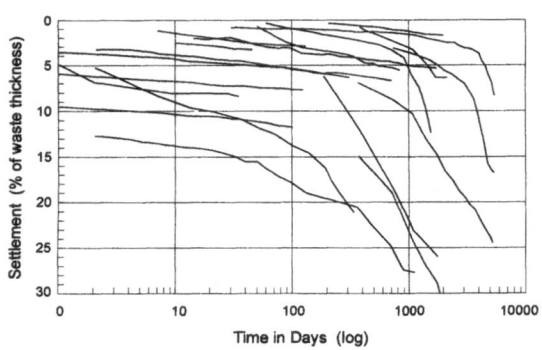

FIGURE 27.28. Total settlement data for MSW landfills, measured from the time the landfill reaches final grade (data from Edgers *et al.* 1992, König *et al.* 1996, and Spikula 1996).

ing the long-term secondary compression period (dimensionless), H_1 = height of waste at time t_1 (m), H_2 = height of waste at time t_2 (m), t_1 starting time for the period of secondary compression(s), $t_2 = t_1$ plus the time duration of intermediate secondary compression (s), and $t_3 = t_2$ plus the time duration of long-term secondary compression. Inspection of Fig. 27.28 suggests that for some MSW materials, $C_{\alpha\epsilon}$ is more or less constant during the period for which data exist, while for other facilities, a variable $C_{\alpha\epsilon}$ could be used to better fit the data.

The reader should be aware that the choice of a value of $C_{\alpha\epsilon}$ cannot be made without consideration of the normalization term t_1. For a given $C_{\alpha\epsilon}$, the calculated value of $s_{\Delta t}$ will be significantly affected by the choice of t_1. Ideally, $C_{\alpha\epsilon}$ and t_1 should be selected by empirically fitting Eqs 27.37 or 27.38 to field settlement data. In the absence of this type of correlation, it is suggested that t_1 be taken as the time period between when waste reaches final grade and when the cover system is installed over the waste.

Since $C_{\alpha\epsilon}$ and t_1 are empirically derived and $s_{\Delta t}$ is assumed to be independent of applied effective stress, and since the primary purpose of calculating $s_{\Delta t}$ herein is to assess potential impacts to the performance of the cover system, it is not necessary to subdivide the waste mass into a series of horizontal layers for purposes of calculating $s_{\Delta t}$. With this approach, calculations are performed for increasing time intervals after closure to obtain a relationship between cover system settlement and elapsed time since closure.

Values of $C_{\alpha\epsilon}$ for MSW reported in the technical literature (Sowers 1973; NAVFAC 1983; Burlingame 1985; Landva & Clark 1990; Fassett et al. 1994; Stulgis et al. 1995) have generally been in the range of 0.01–0.1. Given the empirical nature of $C_{\alpha\epsilon}$ and t_1, it is interesting to compare calculated values of $(s_{\Delta t}/H_1)$ obtained using Eq. 27.37 to the range of observed time-dependent post-clo-

TABLE 27.10. Results of parameter study of calculated post-closure secondary settlements, $S_{\Delta t}$, as a percentage of initial landfill height, H_1

$C_{\alpha\epsilon}$	$(t_2 - t_1)$ (days after closure)		
	100	1000	10 000
0.01	0.30	1.0	2.0
0.03	0.90	3.1	6.0
0.06	1.8	6.2	12.0
0.10	3.0	10.4	20.0

sure settlements for a number of MSW landfills (Fig. 27.28). For the comparison, the waste mass is considered as a single unit with an average value of t_1 of 100 days (approximately three months). Table 27.10 presents calculated values of $s_{\Delta t}/H_1$ (in percent) for values of $C_{\alpha\epsilon}$ ranging from 0.01 to 0.1 and post-closure times of 100, 1000 and 10 000 days (to which 100 days are added to obtain t_2 values).

Based on the calculated values in Table 27.10, $C_{\alpha\epsilon}$ values less than about 0.03, coupled with t_1 values of 100 days, are too small to model MSW time-dependent settlements. Careful review of the source references used to develop Fig. 27.28 suggests that $C_{\alpha\epsilon}$ values in the range 0.04–0.08, coupled with t_1 values of about 100 days provide a reasonable modeling of the settlement trends for modern MSW landfills, which are typically filled fairly quickly and compacted using heavy trash compactors. Larger values of $C_{\alpha\epsilon}$, in the range 0.08–0.12, coupled with $t_1 = 100$ days, are needed to model the settlement trends in some of the older landfills, where possibly more variable waste was placed under conditions less controlled than for modern landfills. Larger values of $C_{\alpha\epsilon}$ would also be expected for landfills undergoing leachate recirculation or otherwise managed to increase biological activity and methane production in the waste mass.

If t_1 is assumed to be 30 days rather than 100 days, calculated $s_{\Delta t}/H_1$ values at 10 000 days would be about 25% larger than given in Table 27.10. Thus, if $t_1 = 30$ days is assumed, $C_{\alpha\varepsilon}$ values should be reduced about 25% from the recommended ranges given above. This calculation exercise clearly points out the sensitivity of calculated $s_{\Delta t}/H_1$ values to the magnitude of t_1.

27.8.4 SETTLEMENT DUE TO LOCALIZED MECHANISMS

Localized settlements, in the form of depressions, can develop within the first several years after cover system installation over MSW. These types of localized occurrences appear to be more common in older waste fills where a number of factors may contribute to the problem, including: (a) little initial waste compaction, (b) variable waste characteristics, (c) placement of sludges in the waste fill, and/or (d) poor surface-water control leading to ponding of water on the waste. Localized differential settlement can lead to excessive stress or strain in cover system components (Gilbert & Murphy 1987). Localized differential settlement of waste is generally attributed to one or more of several mechanisms, namely: (a) deterioration and collapse of objects, e.g. drums, in the waste; (b) settlement associated with a highly compressible zone of waste; and (c) migration or raveling of waste particles into underlying voids.

Several analysis methods are available for use in evaluating the potential effects of localized waste settlements on cover system performance. None of the methods have been field calibrated to any significant degree and selection of input parameters to the analyses is based primarily on engineering judgment.

- Murphy & Gilbert (1985) discuss the application of mine subsidence models to the prediction of waste differential settlements.
- Giroud et al. (1990) presented an approach based on the uncoupled combined use of the tensioned membrane and soil arching theories for analyzing the stresses and strains in geosynthetics (such as geosynthetic layers within a cover system) that lose foundation support after construction due to development of a foundation void or depression. Poorooshasb (1991) used a somewhat different analytical approach to address a similar problem.
- Jang & Montero (1993) used a boundary element formulation to model deformations around a collapsing void within an existing waste mass.
- Carey et al. (1993) performed two-dimensional finite element analyses to evaluate the response of a waste mass containing compressible zones.
- Othman et al. (1995) described application of the displacement method of Sagaseta (1987) to evaluate the response of a cover system over a waste mass containing localized compressible zones.

In the situation where differential settlement is likely to occur and the localized depressions cannot be eliminated, the choices are: (a) continuously grading and maintaining the site, (b) installing a thick buffer soil waste prior to cover system construction, or (c) installing a geosynthetic support beneath the cover system. One or more of the analysis methods described above can be used to design soil buffer or geosynthetic (geogrid or high strength geotextile) support systems. The critical design parameters in any such analysis are the locations and dimensions of the anticipated localized depression or void. Since it is generally not possible to predict where such a feature will occur, any buffer soil or reinforcement layer, if used, will typically need to be installed over the entire waste mass.

27.8.5 IMPACTS OF SETTLEMENT ON COVER SYSTEM

In design, settlement profiles accounting for the various settlement mechanisms are developed to evaluate potential impacts to the final cover system. The evaluation usually considers: (a) post-settlement cover system grades, (b) potential for depressions and ponding in the cover system, and (c) stresses and strains in cover system components. Post-settlement grades should be adequate to shed run-off, prevent ponding, and prevent excessive stress or strain in cover system components, particularly CCL, GCL, and geomembrane hydraulic-barrier layers.

Tensile strains causing cracking in compacted clays have been evaluated by Leonards & Narain 1963; Ajaz & Parry 1975a,b, 1976; Gaind & Char 1983; Chandhari & Char 1985; Jessberger & Stone 1991; and Lozano & Aughenbaugh 1995. Based on these studies, compacted clays tested under unconfined or low confinement conditions exhibit relatively brittle behavior and reach failure at axial extensional strains of 0.02–4%, with most compacted clays exhibiting failure at extensional strains of 0.5% or less. The studies also show that the magnitude of tensile strain causing cracking increases with increasing percentage of fines and water content, and with increasing confining stress.

LaGatta *et al.* (1997) evaluated the impact of differential settlement on the hydraulic conductivity of GCLs overlain by a 600-mm thickness of pea gravel. The results of their evaluation indicate that intact and overlapped samples of needle-punched GCLs can withstand angular distortions of 0.35–0.6, equivalent to tensile strains of up to 5–16%, depending on product, while maintaining a saturated hydraulic conductivity of 1×10^{-9} m s^{-1}, or less. Geotextile-encased, non-reinforced GCL samples (intact and overlapped) were found to be able to maintain a hydraulic conductivity of 1×10^{-9} m s^{-1} up to an angular distortion of 0.1, equivalent to a tensile strain of about 1%. Stitch-bonded geotextile-encased GCL samples (intact) were found to achieve the same hydraulic conductivity criterion up to an angular distortion of 0.35, equivalent to a tensile strain of 5%.

The tensile behavior of geomembranes varies depending on the polymer type, stress-strain characteristics, susceptibility to stress cracking, temperature, and other factors. The present state-of-practice for the design of strain-softening geomembrane barrier layers, e.g. polyethylene geomembranes, is to limit the allowable geomembrane tensile stress (or strain) to the short-term yield value divided by a factor of safety. The allowable tensile stress (or strain) for geomembranes exhibiting strain-hardening behavior is based on the short-term failure (break) value divided by a factor of safety. It should also be remembered that geomembranes are designed to be barrier layers, not tensile inclusions (as is geosynthetic reinforcement, for example). The long-term stress-strain, creep, and brittle fracture behaviors of these materials under stress are not well-understood. To the extent possible, applications should be designed to minimize tensile stresses and elongations in geomembranes.

The authors recommend that when it is necessary to specify allowable geosynthetic tensile stresses and strains, the yield stress and strain of the geomembrane material be determined in a wide-width tension test (for plane deformation) or axisymmetric tension test (for spherical deformation) and that the factor of safety used to calculate the allowable values be at least five. The factor of safety should be applied to the yield values for strain-softening geomembranes and to the failure (break) values for strain-hardening materials. This recommendation should be conservative for virtually all types of commercially available geomembranes used in cover system applications. If a higher value of allowable tensile stress or strain is desired, the engineer must demonstrate that the product to be specified can sustain the allowable values without long-term creep, brittle rupture, or other type of long-term problem. This demonstration must also apply to geomembrane seams.

28. MONITORING OF CONTAMINANTS AND CONSIDERATION OF RISK

E. McBean

K. Schmidtke

W. Dyck

F. Rovers

28.1 Components of a Groundwater Monitoring Program

Groundwater monitoring is a critical component of activities performed at a site where the groundwater may have been adversely impacted by chemicals (i.e. contaminated). The specific purposes of groundwater monitoring will vary depending on the stage of impact identification. The elements of monitoring include:

- initial site characterization using available information;
- refining the site characterization through the installation of boreholes and wells, and sample collection and analysis;
- designing a corrective action; and
- implementing and monitoring a corrective action.

A well-designed sampling and analysis program will:

- provide analytical results of known quality;
- provide a three-dimensional characterization of hydrogeology and chemical presence at the site, including exposure pathways;

- indicate the extent of temporal variability of the chemical concentrations; and
- stay within resource limitations.

The adequacy of a groundwater monitoring program depends upon:

- an adequate characterization of the site hydrogeology,
- the use of appropriate procedures for the collection and analysis of groundwater samples, and
- statistical interpretation of the results.

Inaccurate analytical results cost as much as accurate results initially, but the inaccurate results may later come back to cause more significant problems. The best approach is to perform the program properly from the outset.

This chapter describes the general procedures necessary to ensure that the groundwater monitoring program will achieve the above purposes and some of the statistical procedures frequently used in statistical analyses of monitoring data. Some of the data collection procedures need to be modified based on site-specific factors (e.g. a site with a het-

erogeneous subsurface will require more information than a site with a relatively homogeneous subsurface), and the specifics of the
regulatory standards applicable to a particular site. Chapters 29 and 30 address the issues
of *in situ* containment and treatment of contaminated soil and groundwater, and other
aspects of managing contaminated sites.

28.1.1 DETERMINING MONITORING WELL LOCATIONS AND ANALYTICAL PARAMETERS

28.1.1.1 Initial Site Characterization

The initial step in selecting the location of
monitoring wells and the analytical parameters requires a review of the existing information in the general vicinity of the site, including: site history, geology, hydrogeology,
potential chemical source areas and surface
water drainage. This information is used to
develop a preliminary conceptual model of
the site. The conceptual model should include the following information:

1. Geology underlying the site.
2. Hydrogeology underlying the site (see also
   Section 24.1):
   - the hydrogeologic units present;
   - vertical and horizontal groundwater
     flow paths in each hydrogeologic unit,
     including hydraulic connectivity be-
     tween waterbearing units;
   - hydraulic conductivity of each unit;
   - horizontal and vertical groundwater
     gradient and velocity in each unit;
   - structural discontinuities;
   - outside influences, e.g. tidal fluctua-
     tions, river stage variations, local pump-
     ing influences;
   - local meteorologic conditions, e.g.
     rainfall, evapotranspiration; and
   - seasonal factors, e.g. snow melt.
3. Chemicals produced, used, stored or disposed at the site.
4. Potential chemical source areas, including
   properties adjacent to and upgradient of
   the site.

5. The phase of the chemicals that may be
   present in the subsurface, i.e. dissolved as
   aqueous-phase liquid (APL) or as a separate non-aqueous-phase liquid (NAPL),
   and whether the NAPL is lighter than water
   (LNAPL) or denser than water (DNAPL).

Examples of sources where information regarding the above can be found include: reports from the US Geologic Survey; historical
air photographs; existing water wells or
geotechnical borings at, or in the vicinity of,
the site; plant operational records if the site
was, or is, an operational plant; and local
weather stations. If the existing information
is insufficient to develop a preliminary site
conceptual model, initial activities typically
include the installation of a minimum of
three boreholes set in a triangular pattern
within, and close to, the perimeter of the site.
The boreholes are generally installed until a
confining layer of significant thickness, which
is continuous beneath the site, is encountered.
The information obtained regarding the hydrogeology underlying the site is then used to
develop the site conceptual model and to assist
in selecting monitoring well locations.

Using the conceptual model and assuming
chemically impacted groundwater is identified, a corrective action concept (e.g. groundwater pump and treat, soil excavation/
treatment/disposal, soil vapor extractions, *in
situ* treatment) is then developed.

28.1.1.2 Well Location

During the initial site characterization, a variety of information is obtained and a conceptual model developed. These are then used
to identify gaps in the data and to select
borehole/monitoring well locations for the
subsequent site investigation. These wells
must be located so as to further define the
geology/hydrogeology and to establish potential groundwater flow and chemical migration
pathways underlying the vicinity of the site.
Typically, boreholes/monitoring wells are installed in stages. The benefit of this approach

is that information obtained from previous stages can be used to refine the conceptual site model and to assist in selecting the location of wells for subsequent stages to address the identified data gaps. The first stage is generally limited to:

- vertically, the uppermost water-bearing unit down to the first confining layer; and
- horizontally, the areas of concern (AOCs)/ potential migration pathways, where no or little data exist.

The reason for limiting the first stage to the first confining layer is that the confining layer may have prevented the migration of chemicals from the site into underlying hydrogeologic units. The decision to investigate units underlying the first confining layer would be based on:

- knowledge that the integrity of the confining layer has been breached by either anthropogenic activities (e.g. installation of a deep building foundation, trenches for subsurface utilities) or that a natural discontinuity exists; or
- monitoring, which shows a potential that the next underlying water-bearing unit may have been adversely impacted.

The evaluation of whether to investigate water-bearing units beneath a confining layer also considers the thickness and potential reactivity of the confining layer in contact with site-specific chemicals. Examples of reactivity include: marl, which is chemically reactive with low pH wastes; and illitic clays, which are reactive with some organic chemicals (e.g. xylene (USEPA 1986a)). Other examples of changes in material properties because of contact with site-related chemicals are described in Abdul *et al.* (1990), Acar *et al.* (1985), Gipson (1985) and Lentz *et al.* (1985).

In situations where multiple hydrogeologic units have been adversely impacted, nested wells, described in Section 28.1.5, are often used. Installation details for overburden and bedrock wells are shown on Figs. 28.1 and 28.2, respectively. Figure 28.3 illustrates alternate capping designs for above-grade and flush-mount casings.

Well locations are typically selected to address AOCs, potential migration pathways, and/or exposure locations to human and environmental receptors. Often, possible migration pathways and the exposure locations are examined using groundwater flow and chemical migration models. Additional details relating to the use of models are presented in numerous sources including API (1988), Bear (1972), Charbeneau *et al.* (1989), USEPA (1985, 1989a), Faust *et al.* (1989), Huyakorn *et al.* (1992), Kaluarchchi & Parker (1989), Kaluarchchi *et al.* (1990), Katyal *et al.* (1991), Mendoza & McAlary (1989), and van der Heijde & Elnawawy (1993).

Once the migration pathways and exposure locations are established, well locations can be selected using either a judgmental, geostatistical (kriging) or random design approach. The judgmental approach uses existing knowledge of the site combined with the objectives of the investigation to select the well locations. For example, if it is known that a chemical leak occurred in an area where there are no wells and the objective is to identify areas with elevated concentrations, a well would likely be placed in the area of the spill. The judgmental approach provides for more efficient use of limited resources by focusing the effort on areas which may pose a risk to human health. The judgmental approach may also support a statistical analysis of the analytical results if the specific well locations within the contaminated area are assigned randomly or in a uniform pattern.

The geostatistical approach (kriging) is an interpolation method using spatial correlation structure to obtain either point or spatially averaged estimates of a parameter, e.g. hydraulic conductivity, from irregularly spaced sparse data sets. Furthermore, it provides an estimate of the accuracy of the interpolated values based on the data configura-

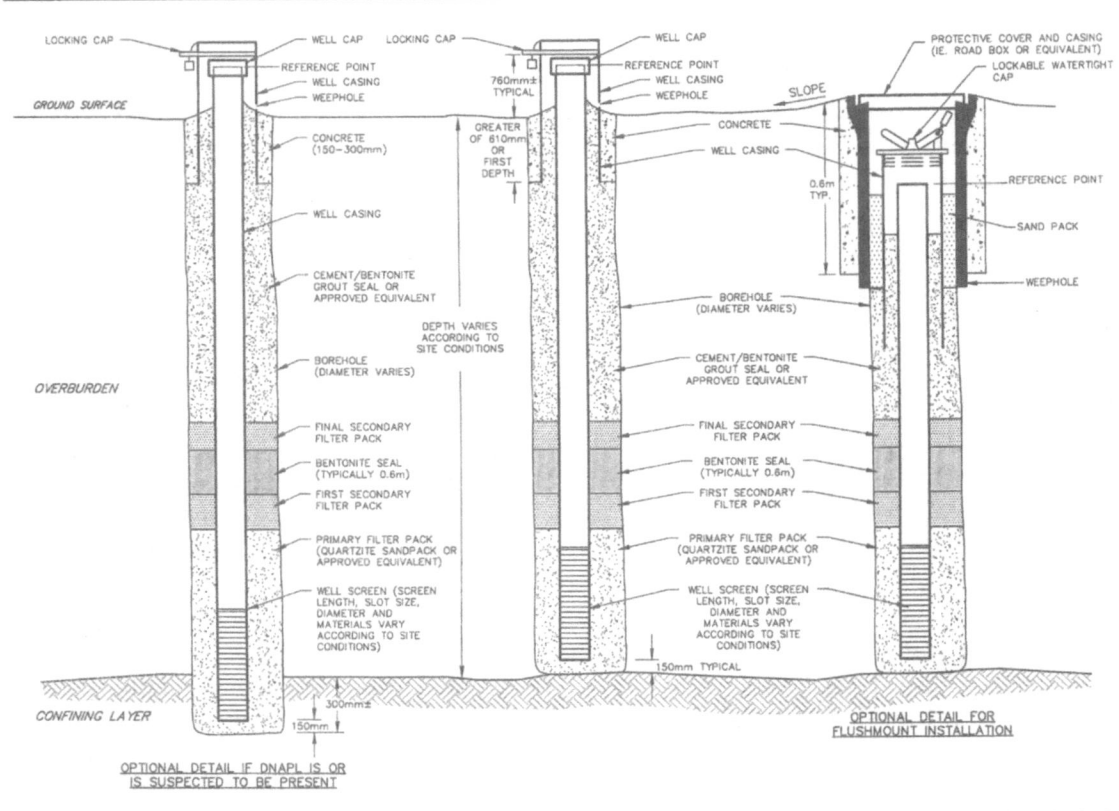

FIGURE 28.1. Typical overburden well installation (center). Left-hand and right-hand figures, respectively show optional detail if DNAPL is, or is suspected, of being present, and flushmount installation.

tion. Areas identified as having a low degree of accuracy are then selected as the next areas from which to obtain information on that parameter. Additional details regarding geostatistical approaches can be found in David (1997), Delhomme (1979), Kafritsas & Bras (1981), and Matheron (1973).

The random approach selects a random pattern of well locations without the use of any previously obtained knowledge.

Systematic grid sampling can be used in both the geostatistical and random design approach. The systematic grid sampling approach uses a defined grid system (e.g. 15 × 15 m) over the site, and can be combined with the random design approach by either randomly selecting which grid block to sample, or by randomly selecting the location within each grid block to sample. If a random design ap-

proach is used, the sample design can often be improved by only sampling a given hydrogeologic unit in order to reduce variability. A detailed discussion is provided in Gilbert (1987), Keith (1987, 1990), and USEPA (1989d).

Wells should be located adjacent to permanent features, i.e. fences, buildings, etc., that offer some degree of protection and a reference point for locating the well. Wells located in high traffic areas, road allowance right-of-ways or low lying areas are undesirable and should be avoided if possible. The well locations should be verified prior to placement to confirm clearance from underground or overhead utilities.

An approach to reduce the number of wells otherwise necessary is to combine multiple AOCs into fewer AOCs. The number of AOCs that can be combined will be controlled by

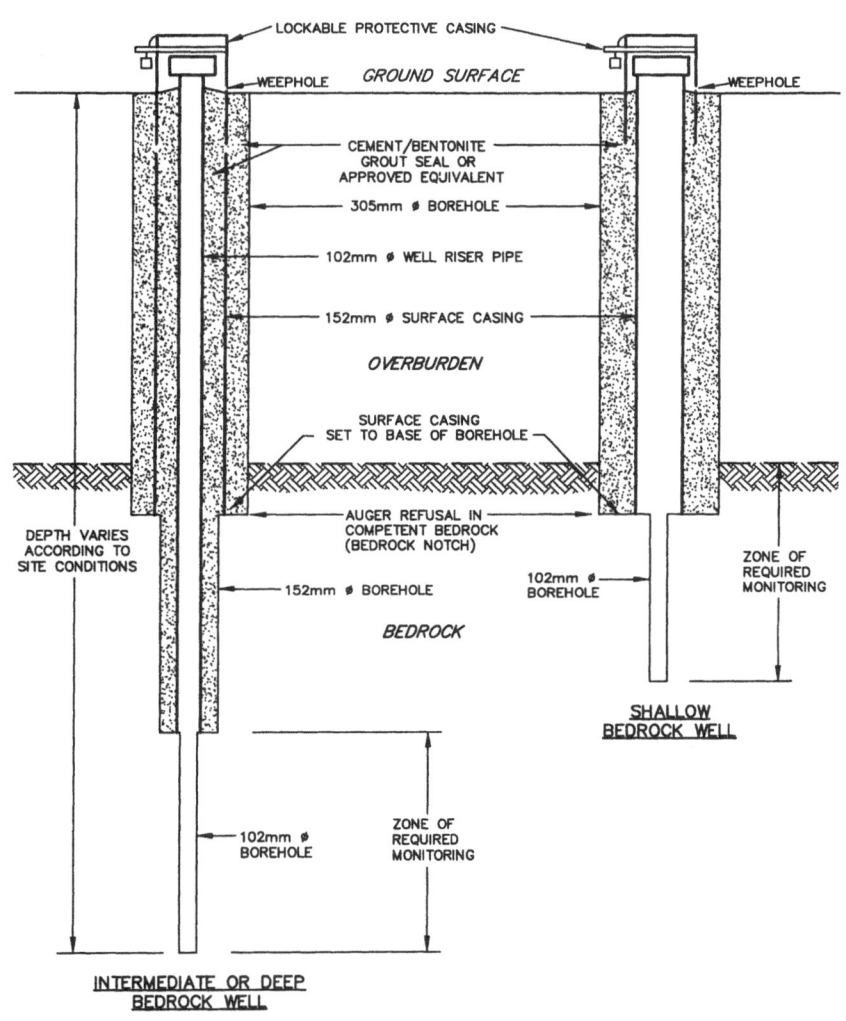

FIGURE 28.2. Typical bedrock well installation: (a) intermediate or deep bedrock well; and (b) shallow bedrock well.

their proximity to each other and the similarity of chemicals which are of potential concern.

The initial number of boreholes/wells for each water-bearing unit underlying an AOC should include one background (upgradient) borehole/well and two boreholes/wells immediately downgradient of the AOC. The background well should be installed upgradient such that it will not be impacted by releases from the AOC. It is expected that information regarding the groundwater flow direction for the site will be available from the

initial site characterization activities described in Section 28.1.1.1.

An important consideration in locating monitoring wells within the vertical thickness of a water-bearing unit is the physical–chemical characteristics of the chemicals of concern. Examples of these considerations are LNAPLs and DNAPLs. LNAPLs spread rapidly in the capillary zone just above the groundwater table. DNAPLs, whose migration is governed largely by gravity (i.e. density) and viscosity, generally have a dominant vertical component until con-

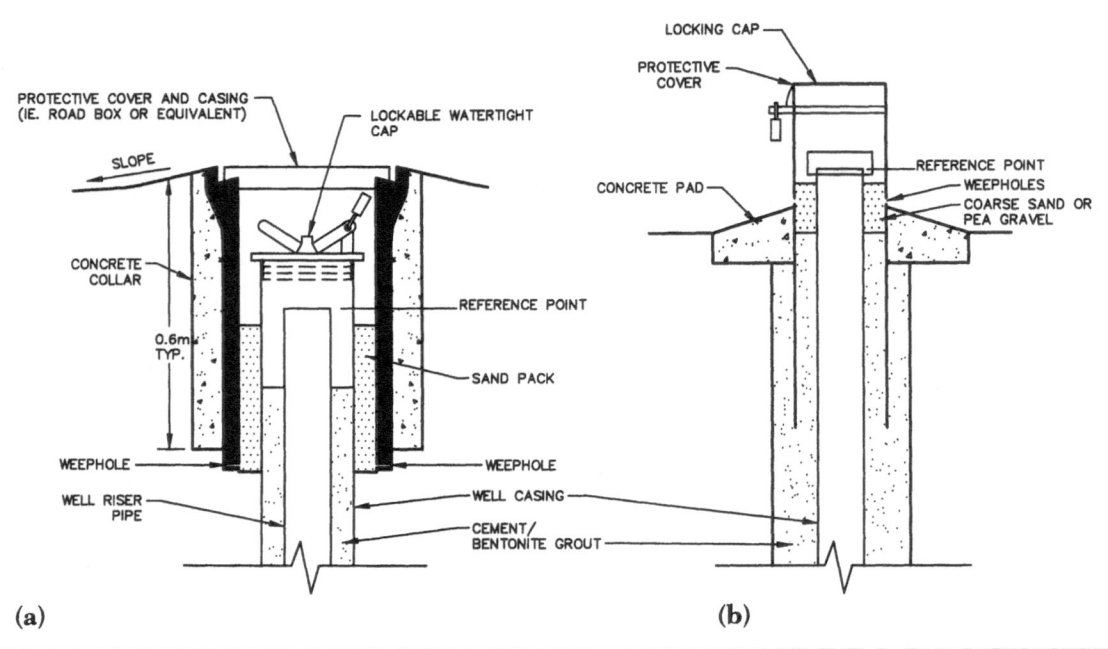

FIGURE 28.3. Protective casing installation: (a) flush mount; and (b) above-grade.

tacting a confining layer. NAPL leaves behind a residual (non-mobile NAPL) as it migrates.

For LNAPLs, the monitoring well-screen should be set so as to straddle the groundwater table. For DNAPLs, the screens should be set at the top of the confining layer and should include a small sump (see Fig. 28.1). The sump will act as a collector for DNAPL. The presence of LNAPL and/or DNAPL may require the installation of more than one well at a sample location (see Section 28.1.5 for nested wells). Additional details regarding NAPL and its effect on well placement can be found in Abdul *et al.* (1989), Blake & Hall (1984), Cohen & Mercer (1993), USEPA (1992d), Gruszczenski (1987), Huling & Weaver (1991), Newell *et al.* (1995), Ostendorf *et al.* (1992), Schwille (1988), and Waterloo Center for Groundwater Research (1989).

For those sites where the background concentrations are close to on-site concentrations, the USEPA (1990) suggests that the number of background samples equals the number of site samples. The reason is to allow the statistical determination of the

"true" difference between the background and site concentrations.

The USEPA method uses confidence levels, power and minimum detectable relative difference to calculate the number of samples required for each AOC/exposure pathway. A concern with the USEPA's approach is that for a site where it is difficult to distinguish site-related chemicals from background conditions (i.e. there has been minimal impact at the site and thus there exists a small detectable relative difference), a large number of samples is required when in fact site-related releases may pose no increased health risk and natural attenuation processes can be relied upon to mitigate these low concentrations.

28.1.1.3 Analytical Parameters

Historical information is used to develop a preliminary list of chemicals of potential concern. Factors evaluated to develop this list include:

- uniqueness of the chemical to the site;
- transport properties of the chemical (e.g. is it mobile in the subsurface or is it highly adsorptive);

- chemical stability;
- human health considerations (e.g. toxicity); and
- availability of reliable and sensitive analytical techniques.

Transport properties are a function of the physical and chemical characteristics of both the chemical compound (e.g. soil partition coefficient, K_{oc}, and vapor pressure) and of the soil and groundwater media (e.g. fraction organic carbon, f_{oc}, and temperature) as discussed in Section 24.3. Information regarding chemical compounds are available from various sources including the USEPA Risk Reduction Engineering Laboratory (RREL) Treatability Database (now called the National Risk Management Research Laboratory, Cincinnati, Ohio), Banerjee (1984), Mercer & Cohen (1990), Miller *et al.* (1990), Piwoni & Keeley (1990), and Rao *et al.* (1991).

Chemical stability refers to the ability of an organic compound/metal to remain in its original state. If there are concerns regarding the possibility of degradation products or metabolites for organic compounds or for oxidation/reduction states of metals, which are toxic compounds, then these must be added to the list of chemicals of potential concern. An example for an organic compound is the degradation of chloroethylene compounds to vinyl chloride (Borden *et al.* 1995). Information regarding oxidation/reduction states of metals can be found in Stumm and Morgan (1996).

Human health information can be obtained from: the Integrated Risk Information System (IRIS) database (USEPA 1989b), USEPA Superfund Health Evaluation Manual (SPHEM) (USEPA 1986b), and the *Health Effects Assessment Summary Tables* (HEAST) (USEPA, issued quarterly).

After a preliminary list of chemicals of potential concern has been established, a decision needs to be made regarding the analytical methods, detection limits and data deliverables. Different analytical methods with varying detection limits are described in 40 CFR Part 136, USEPA (OLM03.2 and

ILM04.0), and Test Methods for Evaluating Solid Waste (USEPA, 1986c).

Confirmation of the preliminary list of chemicals of potential concern would typically require a broad spectrum (e.g. Target Compound List/Target Analyte List) analysis of groundwater samples from a few select wells. The select wells are generally those located in suspected/confirmed source areas, i.e. AOCs, to increase the probability of identifying all compounds attributable to the site.

28.1.1.4 Refinement of Conceptual Site Model/ Selection of Supplemental Well Locations

The information obtained from the initial boreholes/monitoring wells and analytical results is used to refine the site conceptual model and to identify if any significant gaps in the data remain. The need for supplemental wells at an AOC depends on site-specific conditions such as:

- the lateral extent of the AOC perpendicular to the groundwater flow direction,
- the homogeneity/heterogeneity of the water-bearing unit, and
- the isotropy/anisotropy of the water-bearing unit.

The information should also be used to refine the corrective action concept and to determine what, if any, additional sampling and analyses need to be performed to assist in the evaluation of the proposed corrective action technologies.

28.1.1.5 Corrective Action Design

By the time the design of the selected corrective action commences, sufficient analytical results are generally available from the site investigative activities. If sufficient results are not available, it is likely because the technologies being considered to implement the corrective action require either larger amounts or specialized types of results. An example where sufficient data may not be available is where natural attenuation of chlorinated organics is likely to be occurring. The data needs for natural attenuation are more fully described in Borden (1995), Hinchee *et al.* (1995), Klecka *et al.* (1996), and Wiedemeier *et al.* (1995, 1996c).

28.1.1.6 Corrective Action Monitoring

The purpose of corrective action monitoring is to establish whether the corrective action is meeting specified performance criteria. The long-term corrective action monitoring plan should be developed based on site-specific characteristics, the results of chemical migration simulations and the results of exposure pathway/receptor analysis.

1. Chemical concentration performance measures:

- for source control: a decrease in source area concentration to the targeted cleanup level (e.g. maximum concentration limits, MCLs); and no exceedence of MCLs at the downgradient points of compliance.
- for natural attenuation: the rate of concentration decrease between the AOC and exposure location is sufficient to protect the exposure location receptors.

2. Well location: The locations of wells used to monitor corrective action are dependent on the type of corrective action implemented. The locations of these monitoring wells are generally referred to as "points of compliance". Often wells installed for site investigative activities can be used. Examples of the types of corrective action and chemical monitoring wells that are needed are described below.

- Containment: an area of impacted groundwater can be contained physically (e.g. utilize a cap and/or barrier wall), using hydraulics (e.g. extraction wells) or using a combined physical/hydraulic system (e.g. barrier wall with a groundwater collection system within the contained

area). Generally, for a hydraulically controlled system, the point of compliance is typically located at a negotiated distance upgradient from the exposure location, but downgradient from the impacted area. The negotiated distance is often calculated using the advective velocity of groundwater flow for a specified number of years (e.g. two-year travel distance). Generally, two or three wells, spread laterally perpendicular to the direction of groundwater flow, are installed at the point of compliance distance for each water-bearing unit to be monitored. In addition to the point of compliance wells, a background well and a well located within the AOC are typically installed.

- Natural attenuation: can be used either as a component of a corrective action (e.g. use pump and treat for source control and let natural attenuation address the portion of the plume downgradient of the zone of capture) or as the sole remedy. The monitoring requirements for natural attenuation are generally more intensive than for other systems because the groundwater monitoring is used to identify chemical migration, intrinsic abiotic and biotic degradation processes, if any, and the need for future action, if any. As an example, evaluation of the feasibility of complete degradation of chlorinated ethylene by anaerobic reductive dechlorination or anaerobic/aerobic degradation would involve careful mapping of the parent compound(s) (PCE, TCE and/or DCE), the degradation products (DCE, vinyl chloride, ethenes/ethanes) and other parameters (Wiedemeier *et al.* 1996c; Carey *et al.* 1997a, 1998).

3. Analytical parameters: The selection of the parameters to be utilized for monitoring the performance of the corrective action is site specific. Typically, the list of parameters is a subgroup of those used to investigate the

site. In addition to the factors described in Section 28.1.1.3, the frequency of detection and magnitude of the concentrations are also considered. It may also be possible to develop a different parameter list for each AOC and/or for each well which monitors a specific AOC.

28.1.2 MONITORING WELL DESIGN AND CONSTRUCTION

28.1.2.1 General

The design and installation of monitoring wells involves the drilling of boreholes into various types of geologic formations. Designing and installing monitoring wells may require several different drilling methods and installation procedures. It is important that the drilling method minimizes disturbance of subsurface materials, not contaminate the subsurface soils and groundwater, and not provide a hydraulic link between different hydrogeologic units. Samples collected from monitoring wells must not be contaminated by drilling fluids or by the drilling procedures. Numerous publications are available describing current monitoring well design and construction procedures. References include Aller *et al.* (1989), ASTM D5092, and Driscoll (1986). Typical design features are illustrated in Figs. 28.1–28.3.

28.1.2.2 Design Considerations

1. Well diameter: The diameter of the well is primarily dictated by the purpose of the installation. In general, wells installed for groundwater monitoring purposes should be at least 5 cm in diameter. This allows small-diameter bladder pumps to be installed for sampling as well as small-diameter bailers and suction tubes.

2. Screen length and placement: The screen length should be consistent with the desired monitored interval and geologic conditions encountered. For delineation of the vertical profile of a groundwater chemical plume, the number of screened intervals and the screen length will be a function of the thickness of the plume and the desired degree of delineation. A fine degree of delineation will require a relatively large number of small screened intervals. Where cost considerations make this impractical or it is desired to identify where the chemical plume is present, larger well-screens on the order of 1.5–3 m can be used. In cases where the water table is being monitored, e.g. for LNAPL, the screen length should accommodate seasonal fluctuations of the water table to allow any floating layer to be drawn into the well when sampling or to utilize nested wells. If DNAPL is, or is suspected to be, present, the borehole should be installed approximately 0.3 m into the top of the confining layer to provide a sump.

3. Well slot size/filter pack material: The slot size of the well-screen must be sized to match the filter pack material. The filter pack material must be sized to match the geologic strata. Additional details regarding the selection of well slot size and gradation of the filter pack material are presented in Driscoll (1986), Hampton & Heuvelhorst (1989), and Hampton *et al.* (1991).

4. Well materials: The materials selected for the well construction must be compatible with the chemicals present in the groundwater and parameter groups selected for chemical analysis to ensure long-term structural integrity and minimal interference with the quality of the groundwater samples. The following well materials are often used:

- stainless steel well-screen and riser pipe,
- stainless steel well-screen and black steel riser pipe, and
- PVC well-screen and riser pipe.

PVC should not be used where solvents are present at significant concentrations. Additional details regarding well material compatibility are presented in McCaulou *et al.* (1995) and references therein.

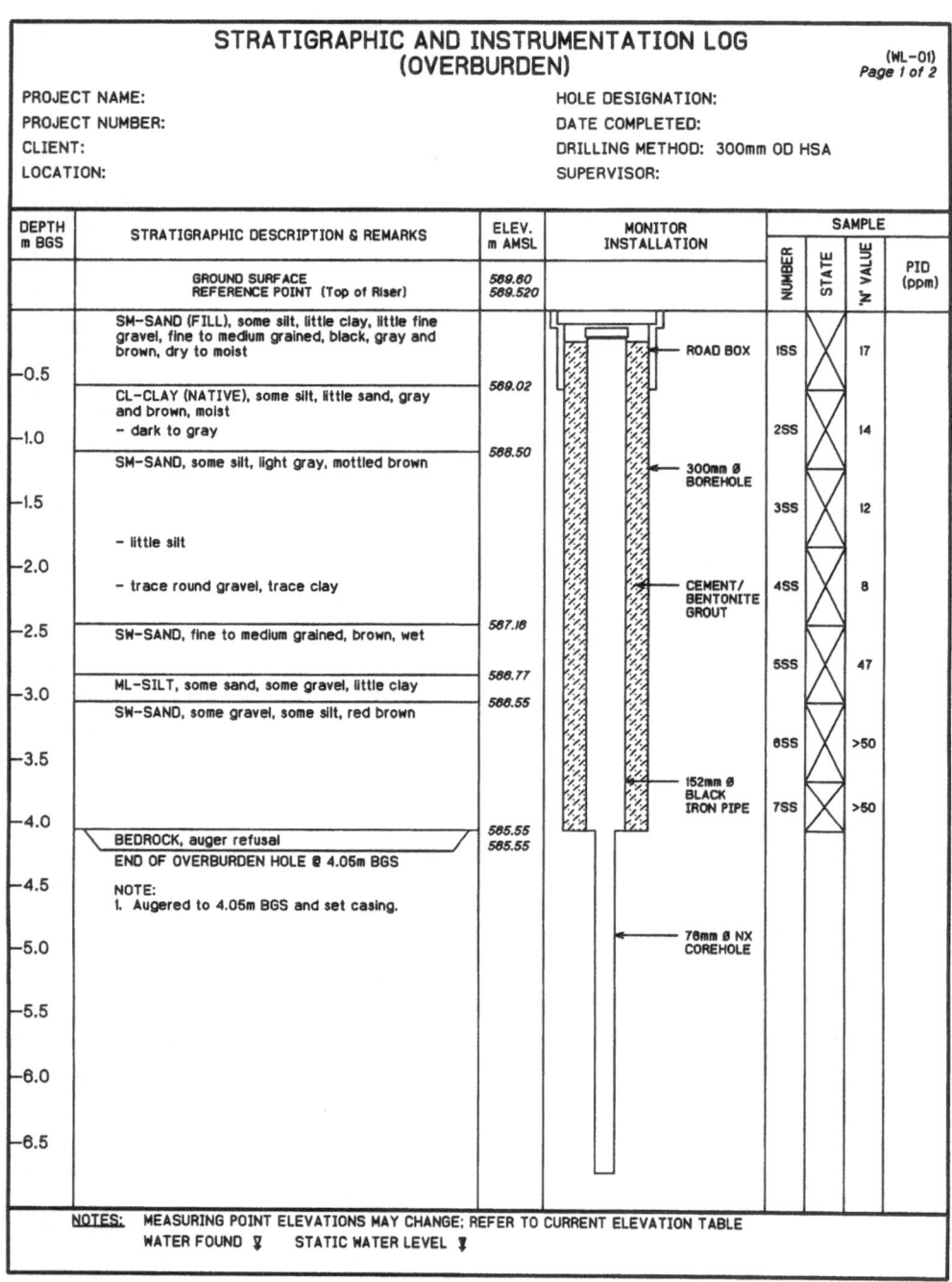

FIGURE 28.4a. Sample stratigraphic and instrumentation logs for overburden well.

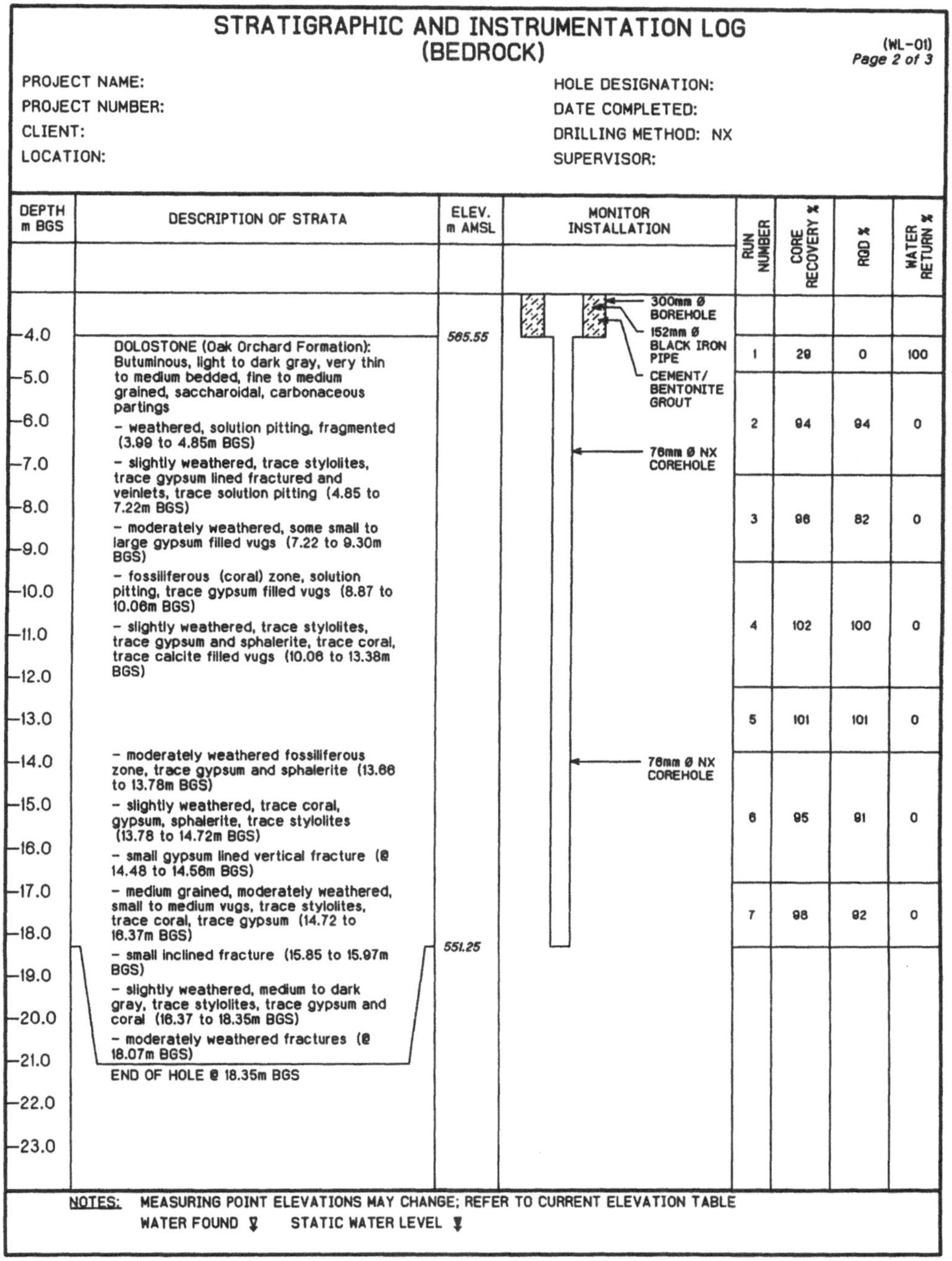

STRATIGRAPHIC AND INSTRUMENTATION LOG
(BEDROCK)

PROJECT NAME: HOLE DESIGNATION:

PROJECT NUMBER: DATE COMPLETED:

CLIENT: DRILLING METHOD: NX

LOCATION: SUPERVISOR:

DEPTH m BGS	DESCRIPTION OF STRATA	ELEV. m AMSL	MONITOR INSTALLATION	RUN NUMBER	CORE RECOVERY %	RQD %	WATER RETURN %
			300mm Ø BOREHOLE				
-4.0		585.55	152mm Ø BLACK IRON PIPE	1	29	0	100
-5.0	DOLOSTONE (Oak Orchard Formation): Butuminous, light to dark gray, very thin to medium bedded, fine to medium grained, saccharoidal, carbonaceous partings		CEMENT/ BENTONITE GROUT				
-6.0	– weathered, solution pitting, fragmented (3.99 to 4.85m BGS)			2	94	94	0
-7.0	– slightly weathered, trace stylolites, trace gypsum lined fractured and veinlets, trace solution pitting (4.85 to 7.22m BGS)		76mm Ø NX COREHOLE				
-8.0	– moderately weathered, some small to large gypsum filled vugs (7.22 to 9.30m BGS)			3	96	82	0
-9.0							
-10.0	– fossiliferous (coral) zone, solution pitting, trace gypsum filled vugs (8.87 to 10.06m BGS)						
-11.0	– slightly weathered, trace stylolites, trace gypsum and sphalerite, trace coral, trace calcite filled vugs (10.06 to 13.38m BGS)			4	102	100	0
-12.0							
-13.0				5	101	101	0
-14.0	– moderately weathered fossiliferous zone, trace gypsum and sphalerite (13.66 to 13.78m BGS)		76mm Ø NX COREHOLE				
-15.0	– slightly weathered, trace coral, gypsum, sphalerite, trace stylolites (13.78 to 14.72m BGS)			6	95	91	0
-16.0	– small gypsum lined vertical fracture (@ 14.48 to 14.56m BGS)						
-17.0	– medium grained, moderately weathered, small to medium vugs, trace stylolites, trace coral, trace gypsum (14.72 to 16.37m BGS)			7	98	92	0
-18.0		551.25					
-19.0	– small inclined fracture (15.85 to 15.97m BGS)						
-20.0	– slightly weathered, medium to dark gray, trace stylolites, trace gypsum and coral (16.37 to 18.35m BGS)						
-21.0	– moderately weathered fractures (@ 18.07m BGS) END OF HOLE @ 18.35m BGS						
-22.0							
-23.0							

NOTES: MEASURING POINT ELEVATIONS MAY CHANGE; REFER TO CURRENT ELEVATION TABLE
             WATER FOUND ⛆   STATIC WATER LEVEL ⛆

FIGURE 28.4b. Sample stratigraphic and instrumentation log for a bedrock well.

28.1.2.3 Well Installation Documentation

Details of each well installation should be recorded on a stratigraphic and well instrumentation log form. Examples of overburden and bedrock well logs are given in Fig. 28.4a and b, respectively.

28.1.3 GROUNDWATER SAMPLING

28.1.3.1 General

The objective of most groundwater quality monitoring programs is to obtain samples that are representative of existing groundwater conditions, i.e. the physical and chemical properties of the groundwater within a water-bearing unit. The program must include consistent sampling and analysis procedures to ensure that the analytical results are legally and technically defensible.

It is common practice to pre-plan the schedule of sampling activities such that sample collection progresses from "clean" to "dirty" areas in an effort to eliminate the potential for cross-contamination.

One of the most important aspects of groundwater sampling is acquiring samples that are free of suspended silt, sediment or other fine-grained particles. These materials often have a variety of chemical components sorbed to the particle or have the ability to sorb chemicals from the aqueous phase to the particle, which will bias the subsequent analytical results (see Section 24.3.4). The additional field expense to extend purging requirements or modify sampling techniques to achieve sediment free groundwater is minor when considering the implications of gathering data that are unfairly biased.

Additional details regarding groundwater sampling procedures are presented in Barcelona *et al.* (1988), USEPA (1986c, 1987, 1989c, 1991c, and 1992a), Keith (1987, 1990), and Puls and Barcelona (1996).

28.1.3.2 Purging/Sampling Equipment

A variety of purging/sampling devices for groundwater sample collection are available. These include peristaltic pumps, suction pumps, submersible pumps, air-displacement pumps, bladder pumps, inertia pumps and bailers. Table 28.1 presents a guide for selecting the method of best fit.

28.1.3.3 Field Quality Control/Quality Assurance (QC/QA)

The following is a brief discussion defining the common types of field-derived quality control samples, which may be required for a groundwater sampling program to ensure that the investigation results can be used with confidence. It should be noted that a specific project may include all or only some of the following quality control samples.

1. Equipment field blanks: Equipment field blanks are used to determine if cleaning procedures are effective and adequate. The blanks are prepared by collecting distilled deionized water, which has been run through or poured over the cleaned sample collection equipment.

2. Field blanks: A field blank is collected to evaluate the influence of field ambient conditions on the sampling process. Field blank collection is typically performed at a sampling site using distilled deionized water poured directly into the sample container.

3. Field duplicates: Duplicates provide an estimate of total measurement error variance, including variance because of sample collection, preparation, analysis and data processing. During groundwater sample collection, the original and duplicate samples are collected simultaneously by partially filling the original and then the duplicate sample containers, and alternating back and forth until both samples have been fully collected.

4. Trip blanks: Trip blanks are submitted for VOCs only, and are intended to determine if the sample shipping or storage procedures influence the analytical results. Trip blanks are prepared before the sampling event by the laboratory and sent to the site in the shipping container(s) designated for the project. The samples should not be opened.

TABLE 28.1. Purging/sampling equipment usability

Purging/sampling equipment	Acceptable for purging	Acceptable for sample collection						Advantages	Limitations
		VOCs	SVOCs	Pest/PCBs	Metals	Conventional parameters	Natural attenuation parameters		
Peristaltic pump	Y	N	N	Y	Y	Y	N	Easily transported and used; Decontamination not required if well-dedicated tubing is used; Minimal maintenance	Limited to a lift of 7.5 m; Limited set of parameters that can be sampled
Suction pump	Y	N	N	Y	Y	Y	N	High flow rate minimizes purging time	Limited to a lift of 7.5 m; Limited set of parameters that can be sampled
Submersible pump	Y	N	N	Y	Y	Y	N	Can pump water from substantial depth; Flow rates vary up to hundreds of $1\ m^{-1}$ minimizing purging time	More difficult to handle because of weight, rigid piping and electrical requirements; Limited set of parameters that can be sampled
Air-displacement pump	Y	N	Y	Y	Y	Y	N	Can pump water from substantial depth; Flow rate range from 8 to $12\ l\ m^{-1}$	Requires pressurized air source; Limited set of parameters that can be sampled
Bladder pump	Y	Y	Y	Y	Y	Y	Y	Can pump water from substantial depth; All parameters can be sampled	Small flow rates limit application to low volume wells
Inertia pump	Y	Y[a]	Y[a]	Y	Y	Y	N	Decontamination not required if well-dedicated tubing is used	Labor-intensive decontamination; Difficult to collect sediment-free sample
Bailer	Y	Y	Y	Y	Y	Y	N		Not practical for deep wells or for removal of large volumes of water; Difficult to collect sediment-free sample

[a] Acceptability is increasing with select state/federal agencies.

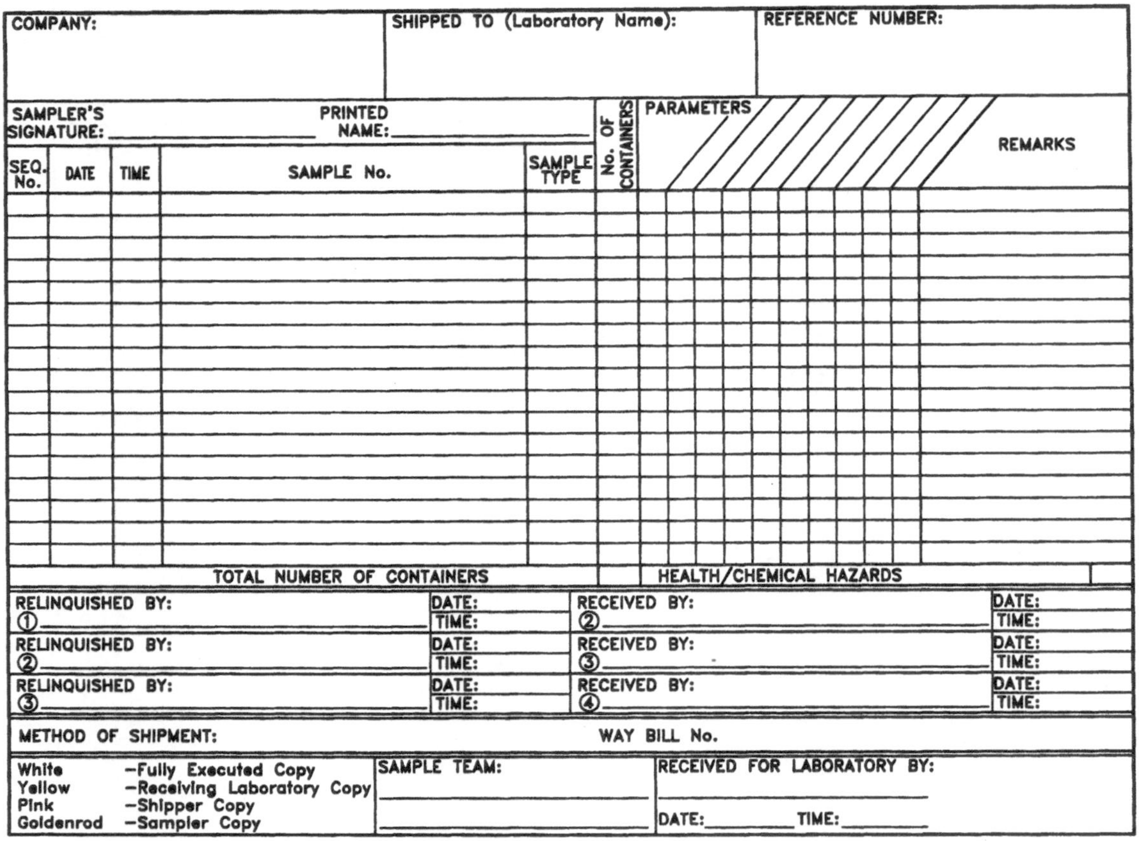

SEQ. No.	DATE	TIME	SAMPLE No.	SAMPLE TYPE	No. OF CONTAINERS	PARAMETERS						REMARKS

COMPANY: **SHIPPED TO (Laboratory Name):** **REFERENCE NUMBER:**

SAMPLER'S SIGNATURE: _____ **PRINTED NAME:** _____

TOTAL NUMBER OF CONTAINERS | **HEALTH/CHEMICAL HAZARDS**

RELINQUISHED BY: ① | DATE: TIME: | RECEIVED BY: ② | DATE: TIME:
RELINQUISHED BY: ② | DATE: TIME: | RECEIVED BY: ③ | DATE: TIME:
RELINQUISHED BY: ③ | DATE: TIME: | RECEIVED BY: ④ | DATE: TIME:

METHOD OF SHIPMENT: **WAY BILL No.**

White — Fully Executed Copy
Yellow — Receiving Laboratory Copy
Pink — Shipper Copy
Goldenrod — Sampler Copy

SAMPLE TEAM: _____

RECEIVED FOR LABORATORY BY: _____
DATE: _____ **TIME:** _____

FIGURE 28.5. Typical chain-of-custody record.

5. Matrix spike/matrix spike duplicate (MS/MSD): MS/MSD sample volumes are additional sample aliquots provided to the laboratory to evaluate the accuracy and precision of the sample preparation and analysis technique. Typically three times the normal sample aliquot is required to conduct MS/MSD procedures.

28.1.3.4 Documentation

Sample collection and analytical procedures must be accurately documented to substantiate the analysis of the sample, conclusions derived from the results, and the reliability of the results. The documentation includes chain-of-custody (see Fig. 28.5) and field notes. The field notes must document all the events, equipment used and measurements collected during the sampling activities. The field notes must be legible and concise, such that the entire sampling event can be reconstructed later for future reference.

28.1.4 SAMPLING FREQUENCY

The frequency of monitoring the performance of a corrective action is dependent on the groundwater flow and chemical migration rates, the resource value of the groundwater, and the potential exposure pathway to humans and the environment.

Groundwater flow and chemical migration rates are controlled by media characteristics (i.e. f_{oc}, hydraulic conductivity, gradient, porosity, etc., see Chapter 24) and the chemical characteristics (i.e. K_{oc}, solubility, etc., see Section 24.3). Generally, chemical migration of dissolved substances occurs relatively slowly. An exception to this could be karst or fractured bedrock where the migration occurs in a few discrete highly conductive channels.

The groundwater chemical concentrations generally change slowly or may in fact be stable under pre-corrective action conditions. Thus, the typical monitoring frequency for the first five years of a corrective action is:

- first year: quarterly,
- second and third year: semi-annually,
- fourth and fifth year: annually.

The frequency may need to be modified based on site-specific factors. The frequency for each subsequent five-year period is determined based on the trends of the analytical results from the previous five years.

Another factor that may influence the frequency of groundwater sampling is change in the groundwater flow direction due to various effects, e.g. seasonal, tidal, etc. This should be taken into account when designing a monitoring program.

28.1.5 MONITORING WELL NESTS

Monitoring well nests may be required for scenarios including the following: (i) the presence of LNAPL and/or DNAPL in a water-bearing unit; (ii) multiple water-bearing units/fracture zones have been or are suspected to be impacted; and (iii) the water-bearing unit is of sufficient thickness that establishing the vertical profile of the plume is necessary to assist in assessing the most effective corrective action.

Generally, the individual wells within a well nest are installed in separate boreholes because of the difficulty in obtaining an effective seal between discrete interval monitors installed within a single borehole. Also, the individual boreholes need to be spaced sufficiently far enough apart so that the effects of well installation at one location do not significantly impact the other locations, e.g. the effect of pH changes due to cement or ce-

ment–bentonite grout. The horizontal spacing between the individual boreholes is a function of the hydrogeologic characteristics of the formation, i.e. the more permeable a formation is, the greater the spacing. Typical spacings in overburden generally range from 3 to 5 m, and in bedrock generally are on the order of 8–10 m. One method to reduce the potential influence of well installation is to place the wells in a line perpendicular to the direction of groundwater flow.

28.2 Statistical Analyses of Monitoring Data

Environmental monitoring data tend to have characteristics that make the application of standard statistical methods difficult and/or inappropriate. Nevertheless, the statistical evaluation of geotechnical and contaminant monitoring data can provide insight into site conditions and discover parameter associations that are not evident in a purely qualitative interpretation. Additionally, there may be governmental or other regulatory mandates requiring the use of statistical methods in the assessment of sites known or suspected to be contaminated with hazardous substances. Therefore, it is important that appropriate statistical analyses be selected and adapted as necessary, to account for constraints of site-specific data sets.

As a prerequisite introduction to the methodology sections following, it is appropriate to note that environmental monitoring data collection is subject to economic, technological and logistical constraints. Since the collection of monitoring data is relatively expensive, monitoring records tend to be brief. As well, the choice of monitoring locations is often intentionally biased toward areas of contamination. For example, if an industrial site is historically known to have had solvent spills in a certain location, then groundwater monitoring wells will likely be installed near the source and downgradient from it in order to characterize the contaminant plume (if present). This choice of sampling location is a sensible allocation of resources, but may not meet an assumption of random sampling for a specific statistical model, and therefore another type of statistical analysis should be chosen. Another problematic characteristic of environmental monitoring data is that the data are often censored. That is, the data have a lower limit or, less frequently, an upper limit, beyond which a monitoring result is reported as "not detected" or "off-scale". Statistically, this causes a change in the shape of the data distribution, which may violate assumptions of specific statistical models and require the use of alternate methods. Finally, natural systems tend to be highly complex and variable, and confounding effects such as seasonality, temperature or matrix stratification may need to be extracted to remove noise from the data.

These features stress the difficulty with statistical interpretation of environmental data. There is a need for the analysis procedure to be sensitive to small changes (e.g. early detection of contamination is desirable) and yet a point of diminishing marginal returns also occurs. The result is that the analyst must be inventive in terms of how statistical analyses proceed. For the variety of reasons previously indicated, there is not a single approach to statistical analyses. Instead, what is frequently needed is a series of approaches each of which possess useful attributes that may be appropriate in addressing a particular question.

28.2.1 SOURCES OF VARIABILITY IN MONITORING DATA

The nature of the variability of environmental quality data will greatly influence how the statistical analyses of the data are undertaken. The specifics of statistical analyses will depend upon the way the phenomenon of interest is defined and sampled. In general, the ability of a sample of environmental quality data to characterize the population from which it is drawn is related to such aspects

TABLE 28.2. Features contributing to the variability in groundwater quality measurements

- Sample collection may involve drilling, sampling and laboratory analysis for many water quality constituents. The expense argues for utilization of brief monitoring records, and thus is representative of only some of the potential conditions.

- Sample collection and laboratory analyses have inherent difficulties, resulting in uncertainties in subsequent interpretation of the findings

- Many important groundwater phenomena take years to evolve, making the available timeframe for sampling programs only statistical "windows" of temporally varying processes

as: (a) the size of the sample, (b) the degree to which it was selected at random, and (c) the degree of independence between the observations comprising the sample.

Samples of environmental quality are just that, namely subsets of the populations that are of interest. To clarify, a limited number of monitoring samples will never fully describe every condition found in the population, i.e. every possible subset of the environment. The challenge then is to acquire a representative sample that can be used to estimate population parameters with a given probability. Some of the features that contribute to the variability and problems of data analysis in groundwater quality phenomena, are briefly described in Table 28.2. The result is that in many cases, estimates of groundwater quality must be developed from only very brief data records, brief being with respect to both temporal and spatial considerations. Even in cases with substantial quantities of data, there may only be a limited amount that is usable for a specific question.

Although much geotechnical data may be acquired directly, certain measurements may be taken from proxy data. For example, remotely sensed satellite data on vegetation class may be used to estimate the soil types present over a wide area. The validity of these "data" depends on such features as one's confidence in a particular measurement method, and the reliability of calibration techniques of the instruments, all of which must then be considered when interpreting the data.

In addition to natural variability, the acquisition of monitoring data is also subject to collection and analysis errors. Errors in sampling procedures, inadequate sample storage and preservation techniques, and laboratory analytical errors are examples of errors in environmental data sets. As a demonstration of the multifaceted initiation points for such errors to exist in a data set, further examples of the sources of error in the collection and analysis of groundwater quality data are listed in Table 28.3.

There is always a degree of uncertainty as-

TABLE 28.3. Examples of sources of error in groundwater monitoring data

- Sampling of a non-homogeneous region in which wells and springs intersect more than one water source
- Piezometers and wells that are inadequately flushed out prior to sampling of groundwater so that a sample of the groundwater is not representative of conditions in the soil environment, i.e. cross-contamination
- Sampling stations that are subject to temporal variations of chemical concentrations can exhibit significant sampling error
- An error in the laboratory protocol of the experiment during the laboratory analysis
- Improper preservation techniques. For example, groundwater samples are often particularly susceptible to changes in the pressure of oxygen and carbon dioxide and improper sample storage. Improper preservation techniques can result in an chemical alteration of the sample as it adjusts to new equilibrium conditions, e.g. the pH levels in groundwater samples have been noted as increasing up to 1.0 pH units due to CO_2 escape to the atmosphere during storage
- Volatilization of biodegradation of VOCs may produce low VOC values for many samples depending on factors such as time and space

sociated with each discrete measurement of environmental quality. In interpreting data, each discrete measurement represents a range of statistically probable values instead of a single value. Uncertainty can be estimated by considering replication and repeatability. Replication is a measure of the variation in results obtained by the same operator in a given laboratory using the same apparatus to make successive determinations on identical test material within a short period of time. Often this is done during quality assurance and quality control testing of a laboratory, to assure that the laboratory results are trustworthy. Replication represents the analytical uncertainty in acquiring monitoring data. Conversely, repeatability is a quantitative expression of the random error associated in the long run with a single operator in a given laboratory obtaining successive results with the same apparatus under constant operating conditions on identical test material. Repeatability is a measure of the natural or sample variability. Note that a systematic error, or bias, in analytical results may be introduced in the field or laboratory due to sampling, handling and/or analysis techniques, and such error may only be detected by making inter-laboratory or inter-technician comparisons using the same samples.

The fact that sampling errors are inherent in random data does not mean, however, that statistical manipulations and sophistication can in any way overcome faulty data. The quality of any statistical analysis is no better than the quality of the data utilized. Furthermore, statistical considerations cannot be used to replace judgment and careful thought in analyzing data. Statistics must be regarded as a tool or an aid to understanding, but never as a replacement for careful thought.

28.2.2 STATISTICAL SIGNIFICANCE TESTING PROCEDURES

28.2.2.1 Types of Hypotheses

The first consideration that should be made in a statistical evaluation of monitoring data is to identify the hypothesis that is to be tested. In fact, if at all possible, this should be identified during the design of the monitoring program, before data collection, and used to modify the investigative design as necessary. There are a number of types of hypotheses that are typically encountered in contaminant monitoring programs, which are generally reflective of governing regulation. The first type of hypothesis is that a contaminant concentration is below a standard. For example, if there is a regulatory requirement for groundwater to be of potable quality at a site, then the monitoring data must be evaluated to determine if all contaminants are below the regulated criteria. Another common type of environmental hypothesis is that a given industrial facility or activity does not cause an impact, i.e. a change from background conditions. In this case, monitoring data from potentially impacted areas are compared with those from background or locations known not to be impacted by the potential source. A third type of hypothesis is that a monitoring parameter is increasing/decreasing/not changing at a location over time. For example, an agricultural facility may monitor the salinity of soils over time to determine if the salt balance is being impacted by farm operations. Finally, a fourth type of hypothesis is the "before" and "after" comparison, where it is postulated that a major event has/has not had an environmental impact, such as the completion of a remedial program at a contaminated site.

These alternative types of hypotheses require the use of different statistical techniques. The selection of an appropriate significance test is considered in Section 28.2.2.3 below. The key requirement for the hypothesis to be investigated is that it be *testable*. That is, the investigator must be able to specify an exact outcome, which the data may be compared against. For example, one could test the hypothesis that the mean concentrations of benzene in the groundwater at

two locations are equal. This is a testable hypothesis, since the condition of equivalency is an exact outcome (i.e. there is only one way in which the two values can be equal), as compared to the alternative condition that the two means are not equal, which is not exact, since there are an infinite number of ways in which one mean could be greater or less than the other. Another type of testable hypothesis is that the monitoring data fit a specific distribution. For instance, a dispersion model of the transport of a contaminant from a point source release may predict changes in concentrations along a gradient, against which actual monitoring results could be compared to test the significance of the model. The statistical term for the hypothesis to be tested is the *null hypothesis*, which is represented by H_0. If H_0 is shown to be a statistically "unacceptable" hypothesis, then the alternate hypothesis (H_a) may be accepted instead.

28.2.2.2 General Steps of Hypothesis Testing

The general procedure to carry out significant difference testing consists of the following steps:

1. Formulate a suitable (in terms of the problem context) null hypothesis, H_0.
2. Formulate a suitable alternative hypothesis, H_a.
3. Calculate an appropriate test statistic.
4. Define an acceptance/rejection region.
5. Locate the test statistic in the acceptance or rejection region.
6. Reject, or do not reject the null hypothesis.
7. Make a final inference (the proverbial "so what?") to draw a conclusion.

Given the statistical nature of the significance tests, there are two possibilities that can cause differences in observed values. These possibilities are: (a) inherent randomness associated with taking a sample from a population; and (b) an actual effect, i.e. a real difference, exists. The task of the significance test is then to establish the probabilities of these two possibilities (by defining acceptance and rejection regions) and thereby draw conclusions with a given probability. When a result is obtained which is very unlikely to have arisen by chance, the result is said to be statistically significant. In other words, it is difficult to ascribe the findings to chance, and the difference must, in all common sense, be accepted as a real difference. There is still the possibility that the result happened by chance; and thus, a significance test result is always reported with a probability statement. For example, "the mean benzene concentrations in groundwater upgradient and downgradient of the source area were found to be statistically different at a significance level of 95%." Further discussion of acceptance and rejection regions is presented in Section 28.2.2.4 below.

If a significance test returns a non-significant result, i.e. the null hypothesis is not rejected, it does not follow that there are grounds for supposing this hypothesis to be true, but merely that there are no grounds for supposing it to be false. If a non-significant result occurs, i.e. the null hypothesis is not rejected, then it is appropriate to consider the *power* of the test, which is the probability that the test would have discovered a significant effect if it had been present. Power is discussed in Section 28.2.6 below. As a final note, even if the result proves non-significant, it is equivalent to a verdict of "not proven", and there is still the opportunity to consider further evidence, i.e. more data, which may become available later on.

In response to the significance testing problem, a number of alternative procedures have been proposed. The selection of the best procedure involves careful scrutiny of the characteristics of the problem-at-hand and the assumptions implicit in the particular significance test being considered. This involves not just the choice of significance level,

but also the choice of a statistical test, and the requirements of the number of samples. The following section describes the selection of a variety of tests applicable to environmental data.

28.2.2.3 Selection of Appropriate Significance Test

Due to the sheer number of statistical procedures available, a comprehensive treatment is not possible within the scope of this chapter. However, of the many statistical procedures available, certain significance tests stand out due to their wide applicability to environmental data and relatively easily understood mathematical bases. A number of these are presented below, and it is intended that these represent a cross-section of frequently used tests, but also recognizing that data from specific monitoring programs may not require the use of further techniques. A more comprehensive listing of statistical procedures can be found in McBean & Rovers (1998) and Helsel & Hirsch (1992).

Student's t-test. The *t*-test is one of the most widely used significance tests. In its various forms, it can be used to compare the observed mean of collected monitoring data against either a single value (e.g. a regulatory criterion), the mean of another data set (e.g. comparing a possible impacted location to a reference location), or the mean of corresponding data from the same sampling locations under different conditions (e.g. before and after a major event). These three forms of this significance test are known, respectively, as the two-sample, one-sample and paired *t*-tests. The *t*-test makes the following assumptions: (a) that the data are independent, i.e. have been randomly sampled; (b) the data are normally distributed; and (c) the variances of the groups of data are equal. The *t*-test is useful in cases where a comparison of a single variable between two groups is to be made. For example, the test could be used to compare the hydraulic conductivities of soil samples from two locations, or to evaluate if a contaminant's groundwater concentrations in a group of monitoring wells exceeds a given standard. The t-test is not useful in cases where more than two groups, more than a single variable, or time series data are to be analyzed. In the first two cases, an ANOVA method is appropriate, while in the third case, a regression or multiple regression approach may be applicable. The t-test is introduced in Section 28.2.3.1 below.

Cochran's Approximation to the Behren's Fisher t-test (CABF). In many cases, field data deviate from the assumptions implicit to the application of the t-test. This may occur for a variety of reasons such as small sample sizes, large natural variability, non-homogeneity of the sample populations or non-random sampling. The Cochran's approximation to the Behren's Fisher t-test (CABF) was developed for cases in which the variances are not equal between the groups to be compared. This occurs relatively frequently in monitoring data, and when the equal variance assumption is shown to be violated, the CABF provides a useful alternative to the standard t-test. The CABF is introduced in Section 28.2.3.2 below.

Analysis of Variance (ANOVA). The analysis of variance (ANOVA) is actually a group of statistical procedures that are used to examine differences in mean values across multiple groups and/or multiple variables. The term ANOVA is used when means for one variable are compared across many groups, whereas in the case of multiple variables compared across many groups, the term "multiple ANOVA" (MANOVA) is used. The ANOVA is actually a generalized form of statistical procedures, of which the t-test is the special case in which a single variable is compared across only two groups. The ANOVA is useful in evaluating monitoring data across a number of locations. For example, if groundwater

from five wells is sampled for a specific contaminant, say benzene, then an ANOVA could be used to test if there are any significant differences in mean benzene concentration between any of the wells. Note that performing a similar comparison by using multiple t-tests to compare the wells one pair at a time is laborious and is much less preferred to the ANOVA method, but is possible through the use of special adjustments such as the Bonferroni technique. Application of ANOVA methods is subject to the same assumptions as for the t-test; i.e. independence, normality and homogeneity of variance. The use of ANOVA is introduced in Section 28.2.4.1 below.

(Linear) Regression. One of the key goals in monitoring is to establish trends in data over time, distance or some other gradient. For instance, one of the lines of evidence of natural breakdown (biodegradation) of a pollutant in the subsurface is the observation of decreasing concentrations over time. The significance of the pattern of this decrease can be statistically assessed using a regression analysis. The most common form of regression analysis uses a linear model, in which a line is fit through the data points and tested for significance. It is also possible to fit non-linear models using regression models, however, many natural processes fit linear or loglinear patterns, and the regression results for higher-order models may be difficult to interpret. The basic assumptions of regression analysis are: (a) for any given value of the independent variable ("x," or in the example above the time), the observed values of the dependent variable ("y", or the contaminant concentration above) are normally distributed; (b) for a given x, the observed y values are independent of one another; and (c) the sample variances are equal. Furthermore, the regression assumes that the true relationship between the variables is linear (or whatever form of model is being used). A linear regression analysis fits a line through

the observed data, and then actually uses an ANOVA to test whether the line is a significant fit to the observed data. As well, the strength of the relationship between the variables is estimated by calculating a regression coefficient. Regression analysis is introduced in Section 28.2.3.3 below.

It should be noted that regression is often confused with correlation analysis. There are many computational similarities between these two methods, but they have fundamentally different bases. In a regression, the independent variable is known or assumed to cause changes in the dependent variable. In comparison, a correlation analysis simply assesses the degree to which two variables are linearly associated (co-vary), with no assumption of causality. Correlation is also useful in examining monitoring program results; e.g. to investigate the association of the concentrations of different contaminants (i.e. the concentrations of two pesticides in groundwater underlying a farmer's field may be very tightly correlated, but one does not cause the other).

Multiple (Linear) Regression. An extension of the regression model is the formulation of multiple regression models. In these models, a dependent (y) variable is assumed to be a function of a number of independent (x) variables. For example, the pH in soil may be influenced by a number of factors such as soil composition, mineral concentrations, temperature, moisture content and contaminant concentrations. It may be possible in this case to analyze the monitoring data using a multiple regression model and determine if a combination of these variables cause the observed pH. The multiple regression shares the same assumptions as the basic regression model. Regression techniques are particularly useful for predictive purposes, in which the effect on the dependent variable caused by changing an independent variable is estimated. Multiple regression is introduced in Section 28.2.4.2 below.

28.2.2.4 Interpretation of Significance Tests: Acceptance and Rejection Regions

In order to interpret the results of a significance test, acceptance and rejection regions for the test statistic must be defined. Typically, a choice is made based on the degree of certainty desired in identifying significant results. That is, if one desires 95% confidence that a result shown by the statistical test to be significant is in fact truly significant, then this significance level (denoted as "$1 - \alpha$") is used to determine the acceptance and rejection regions of the test statistic. Each statistical test defines the acceptance and rejection regions differently in accordance with the method (e.g. a t-test defines a critical t value, whereas an ANOVA defines a critical F value, etc.); however, they are all usually based on the significance level chosen and the number of samples used.

In conducting statistical inference tests, incorrect conclusions sometimes occur. An acceptance rate is assumed, but what that really means is that an error is made a specified percentage of the time. In reality, because population parameters are being estimated from small samples, the comparison of these means involves the risk of making one of two possible errors, denoted type I or a type II. These describe the balance between two conditions:

1. the risk the procedures will falsely show a significant effect, e.g. that a landfill is causing background values or concentration limits to be exceeded, when in fact the landfill is not causing a problem (false positive); and
2. the risk that the procedures will fail to indicate that background values or concentration limits are being impacted when in fact there is an impact, e.g. the landfill is contaminating the groundwater (false negative).

Table 28.4 illustrates the possibilities when the true environmental condition (truth) is

TABLE 28.4. Outcomes and errors in statistical testing

Decision/inference	Truth	
	H_0	H_a
Accept H_0	No error Probability = $1 - \alpha$	Type II error Probability = β
Accept H_a	Type I error Probability = α	No error Probability = $1 - \beta$ = power

interpreted via a statistical significance test (decision/inference):

- The null hypothesis, H_0, is not rejected and is in fact true, thus no error is made.
- The null hypothesis, H_0, is rejected when it is in fact true (and the alternate hypothesis, H_a, erroneously accepted). This is a Type I error, and its probability is denoted by α. This error is also called a false positive (e.g. a "source" is identified as impacting the groundwater when, in fact, it is not impacting the groundwater; or saying something is present when it is not).
- The null hypothesis, H_0, is rejected and H_a is true and thus no error is made.
- The null hypothesis, H_0, is not rejected and H_a is true. This is a Type II error, denoted by β, and represents the failure to detect an effect, i.e. reject H_0, when there is a significant difference. We accept H_0 when in fact H_a is true. This error is also called a false negative (e.g. a "source" is identified as non-polluting when in fact it is polluting). The result of this type of error, for example, is that a remedial action is not implemented on a contaminating source that is, in fact, degrading the groundwater quality.

The α level represents the probability of making a Type I error. By specifying $\alpha = 0.05$, this indicates that in the long term, the Type I error will be made 5% of the time, i.e. one test in 20 will show an effect when there is none. The probability of a Type II error is represented by β. An important fact is that α and β are not independent, in fact they are

related: i.e. if you decrease the possibility of one type of error, you are increasing the possibility of the other. Too often, the desire to make sure that an effect is really happening before flagging an exceedance, i.e. a low α, rate drastically reduces the chances of finding an effect, i.e. β increases. Since the Type II error has environmental consequences, the general approach is to tend to be rather conservative. Further discussion of the power of a statistical test is provided in Section 28.2.6 below.

28.2.3 PROCEDURES FOR SINGLE COMPARISONS

28.2.3.1 Student's t-Test
The steps of the t-test include:

1. The calculation of the t-statistic.
2. Comparison of the calculated value against a critical t value (t^*), which is read off a standard table for a given degree of confidence (α) and number of degrees of freedom (df, a function of the sample size).
3. Making a statistical conclusion, i.e. if $t > t^*$, there is a significant difference, if $t < t^*$, there is no evidence of a statistically significant difference in the means.

The student t-test can be computed in three forms depending on the type of comparison being made. These are the one-sample, two-sample and the paired t-tests.

One-Sample t-Test. The one-sample *t*-test compares a sample mean against a single value, typically either a regulatory standard or population mean. In this case, the *t*-statistic is calculated using:

$$t = \frac{|\bar{x} - \mu|}{S/(n)^{1/2}} \tag{28.1}$$

where $\bar{x}$ = the sample mean, μ = the population mean (or regulatory standard), S = the standard deviation of the observations, and n = the number of observations. For a one-sample *t*-test, the number of degrees of freedom is $n - 1$.

Two-Sample t-Test. The two-sample *t*-test compares the mean of one sample against the mean of another sample. The *t*-statistic for a two-sample comparison is calculated using:

$$t = \frac{|\bar{x}_1 - \bar{x}_2|}{(\text{SE})^{1/2}} \tag{28.2}$$

where $\bar{x}_1$ = the sample mean for the first group, $\bar{x}_2$ = the sample mean for the second group, and SE = the standard error of the difference. If the variances of the two groups are similar, then the standard error is:

$$\text{SE} = \left[\frac{(n_1 - 1)\,S_1^2 + (n_2 - 1)\,S_2^2}{n_1 + n_2 - 2} \right]^{1/2} \tag{28.3}$$

and if the variances are unequal, the standard error is computed as:

$$\text{SE} =$$

$$\left\{ \left[\frac{(n_1 + n_2)}{(n_1 n_2)} \right]\left[\frac{(n_1 - 1)S_1^2 + (n_2 - 1)S_2^2}{n_1 + n_2 - 2} \right] \right\}^{1/2} \tag{28.4}$$

where n_1 and n_2 = the number of observations in the two groups, respectively; and S_1^2 and S_2^2 are the sample variances (squares of the standard deviations). For a two-sample *t*-test, the number of degrees of freedom is $n_1 + n_2 - 2$.

Paired t-Test. The paired *t*-test uses the same computational sequence as the one-sample *t*-test, except that the test is performed on the differences between the paired data. For example, if a certain measurement was taken from a group of monitoring wells before and after a major rainstorm, then for each individual well the measurement before the storm would be subtracted from the measurement after the storm, and these differences would be used to carry out the *t*-test.

Assumption Checking. As noted earlier the assumptions of the *t*-test are: (a) the data are independent of one another, i.e. have been randomly sampled; (b) the data are normally distributed; and (c) the variances of the groups of data are equal.

The assumption of independence of successive monitoring data requires a well-thought-out sampling design. As much as practical, samples should be acquired in a random order, and under similar conditions. In a geotechnical investigation, this is not always possible; e.g. a soil sample from a borehole must be taken during installation, or it may be impractical to randomly traverse a very large site between successive samples and therefore one area at a time is collected. As well, when prior knowledge is available regarding less- and more-contaminated regions at the site, samples are typically collected in order of increasing chemical concentrations to avoid cross-contamination. However, if the assumption of random sampling was of particular concern, the order of sampling the groundwater monitoring wells on a small- to medium-size site could easily be randomized, and then the assumption of independence could be evaluated by plotting the individual data points in order of collection and visually assessed to see if there is any pattern. If the data are independent, the points will have no discernible pattern and fall in a roughly horizontal band across the graph. If instead the data exhibit an increasing or decreasing

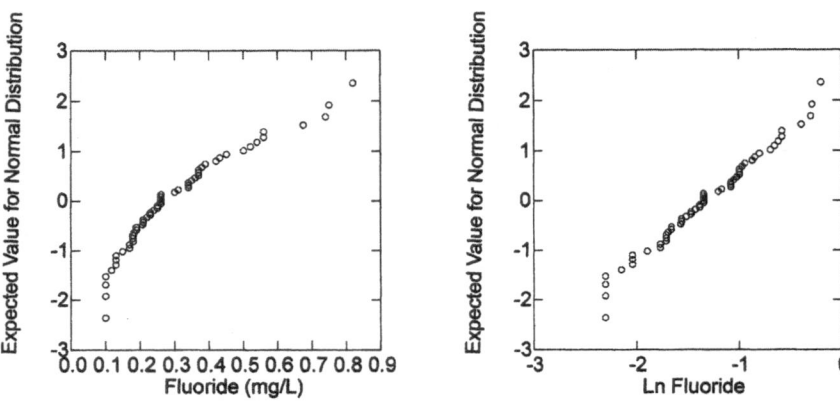

FIGURE 28.6. Cumulative probability plots to assess normality of data.

trend, there is evidence of non-independence.

There are a number of techniques available to assess the assumption of normality for a set of data. The most elegant test of normality is the *cumulative probability plot*, in which the data are plotted using a probability scale for one axis (see Fig. 28.6). Normally distributed data sets are easily identified because they produce a line on such a probability plot. Not only does this plot assess normality, but it can provide useful information that may not be readily apparent from simply looking at the data, such as threshold values, or distinct groups of monitoring results. Another easily performed test of normality is the calculation of the *coefficient of variation* (CV). The CV of a data set is its standard deviation divided by its mean. A CV of greater than one suggests non-normality of the data. A more statistically rigorous alternative is the Shapiro–Wilk test of normality. It is described, along with several other methods, in McBean & Rovers (1998).

The assumption of equal variances, or *homoscedasticity* (from the Greek, "equal scatter"), may be tested by using a number of methods. Like the normality assumption, the simplest test for homoscedasticity is a graphical method: the box plot. A box plot (see Fig.

28.7) is created for each group of data, and the length of the largest box is within about a factor of three of the length of the shortest box, then the variance of the data may be assumed to be equal for the purposes of carrying out a significance test. If not, alternate methods to test for unequal variances should be attempted before concluding that the data are heteroscedastic (unequal variances). These alternates include Levene's F-test, Bartlett's test, Hartley's F_{max} and the log-ANOVA (or Sheffé–Box) test. Detailed descriptions of these tests are given in McBean & Rovers (1998). If a data set is shown to be heteroscedastic, then a data transformation may be needed, using, for example, the log-transformation of the monitoring data.

For further reading about the *t*-test in its various forms, refer to "Student" (1908), Fisher (1925), USEPA (1992c), and McBean & Rovers (1998).

28.2.3.2 Cochran's Approximation to the Behren's Fisher *t*-test

If the variances of the environmental quality data at two locations are not equal, one of the assumptions of the standard *t*-test is violated. In such cases, a useful alternative is the Cochran's approximation to the Behren's Fisher

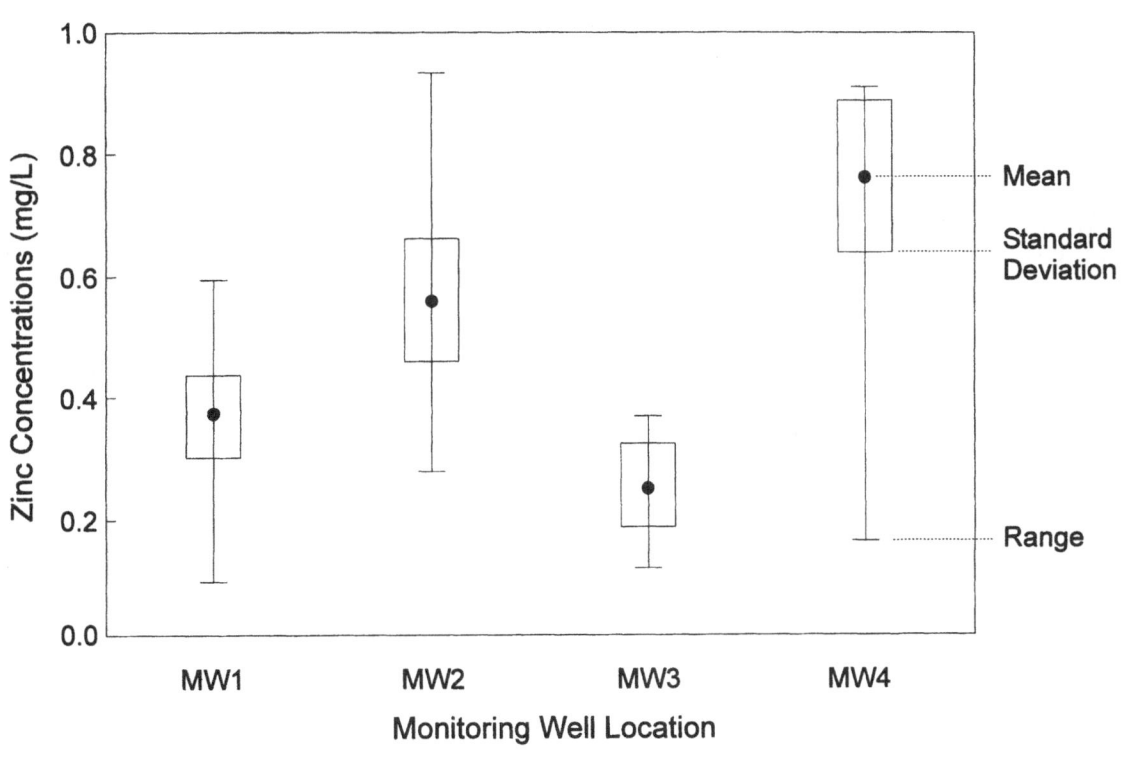

FIGURE 28.7. Example of box plots.

test (CABF). The CABF test functions by re-laxing the requirement that the variance of the data at both sampling locations be of the same magnitude by adjusting the degrees of freedom. Cochran's test adjusts the degrees of freedom (df) by weighting the individual t values for each sample. Cochran's test computes the t-statistic from:

$$t = |\bar{x}_1 - \bar{x}_2| \Big/ \left(\frac{S_1^2}{n_1} + \frac{S_2^2}{n_2}\right)^{1/2} \quad (28.5)$$

The critical t-statistic, t^*, is computed from:

$$t^* = \left[\left(\frac{S_1^2}{n_1}\right)t_1^* + \left(\frac{S_2^2}{n_2}\right)t_2^*\right] \Big/ \left(\frac{S_1^2}{n_1} + \frac{S_2^2}{n_2}\right) \quad (28.6)$$

where:

$$S_1 = \left[\sum \frac{(x_{i1} - \bar{x}_1)^2}{n_1 - 1}\right]^{1/2} \quad (28.7)$$

and

$$S_2 = \left[\sum \frac{(x_{i2} - \bar{x}_2)^2}{n_2 - 1}\right] \quad (28.8)$$

where the t-values t_1^* and t_2^* are determined from t-tables for df of $n_1 - 1$ and $n_2 - 1$, respectively, at a specified significance level, $1 - \alpha$.

As noted above, CABF allows relaxation of the equality of variances requirement. How-ever, if this assumption is *not* violated, utiliza-tion of the approximate test can result in dif-ferent findings, and the standard t-test should be used instead. This concern was described in detail in McBean *et al.* (1988).

28.2.3.3 Linear Regression
In a regression analysis, a line of best fit is computed, which describes the dependent

variable as a function of the independent variable. The form of the linear regression line is:

$$y = ax + b \qquad (28.9)$$

where a = the slope of the line, termed the regression coefficient; and b = the y-intercept of the line (a constant). The equation of the linear regression is usually calculated using the method of least squares, and should be assumed unless specifically stated otherwise. The regression line can be automatically calculated using modern statistical software packages, as well as many spreadsheet programs. If a and b must be manually calculated, the following equations may be used (a mathematical proof for these solutions will not be provided here in the interest of brevity):

$$b = \frac{\Sigma(x_i - \bar{x})(y_i - \bar{y})}{\Sigma(x_i - \bar{x})^2} \qquad (28.10)$$

$$a = \bar{y} - b\bar{x}. \qquad (28.11)$$

Note that these equations apply to the case where there is a single y for each x value, the model to be fit is linear, and the x values are assumed to be known without error (or at least the error in the measurement of x is much lower than the uncertainty in the y measurements). It is possible to fit multiple y values for each x, non-linear, or so-called "random x" models, but techniques for these computations are beyond the scope of this chapter.

The significance of the regression line is tested using an analysis of variance (refer to Section 28.2.4.1) through the calculation of an F statistic.

Interpretation of the linear regression analysis is as follows. First, the significance of the regression model is evaluated using the F-statistic from the ANOVA. If this value is significant with a stated probability of $1 - \alpha$ (typically 95% is used), then the overall re-

gression is said to be a significant fit of the data, i.e. the monitoring data have been shown to exhibit a significant trend. Next, the coefficient of determination, r^2, for the entire model is considered, which yields the proportion of variation in the dependent variable that is explained by the regression, e.g. an r^2 of 0.86 means that the regression explains 86% of the variation in the dependent variable. Following this, a t-test is performed on "b" to test if the regression coefficient of the independent variable is significantly different than zero. If so, this regression coefficient is interpreted as the factor by which the independent variable influences the dependent variable. For example, if a linear regression analysis has yielded a significant fit of depth to groundwater (in mm) over distance (in m), then a value of -10.4 for the regression coefficient means that for every meter of horizontal distance, the water table drops by 10.4 mm.

The assumptions underlying linear regression include those of the t-test, namely: (a) independence, (b) normality, and (c) homoscedasticity. Testing these assumptions should be performed using the methods presented earlier. An additional assumption of linear regression is that the true relationship of the dependent and independent variables is linear. Deviations from linearity will result in a non-significant regression line, and thus this assumption is therefore evaluated as part of the base analysis.

28.2.4 PROCEDURES FOR MULTIPLE COMPARISONS

28.2.4.1 Analysis of Variance (ANOVA)

The various ANOVA procedures are available in modern statistical software packages, as well as some newer spreadsheet programs. It is less common to compute an ANOVA manually; however, the methodology is straightforward as presented below.

The general approach of an ANOVA is to

partition the variation in the data between a number of different terms, and then test these variabilities for significance of the factors (hence the term "analysis of variance"). In order to perform an ANOVA, P mutually exclusive groups of data (e.g. groups of monitoring wells) should have a minimum of three to four samples each, and the total number of samples, n, should exceed the number of groups by at least five (i.e. $n - P \geq 5$).

An ANOVA is performed using the following steps. Note that i denotes the group and j denotes the samples within each group.

1. For each group, the mean value $\bar{x}_i$ is calculated.
2. The grand mean $\bar{x}$ of all data is calculated.
3. For each data point (x_{ij}) in each group, a residual, R_{ij}, is computed as the difference from its group mean, $\bar{x}_i$.
4. Compute the between groups sum of squares:

$$(SS_{between}) = \sum^{P} n_i(\bar{x}_i - \bar{x})^2 \quad (28.12)$$

(note this value has $P - 1$ degrees of freedom).

5. Compute the total sum of squares:

$$(SS_{total}) = \sum^{P} \left(\sum^{n_i} x_{ij} - \bar{x} \right)^2 \quad (28.13)$$

(note this value has $n - 1$ degrees of freedom).

6. Compute the within groups (residual) sum of squares:

$$(SS_{residual}) = SS_{total} - SS_{between} \quad (28.14)$$

(note this value has $n - P$ degrees of freedom).

7. Calculate a mean square for the between groups:

$$(MS_{between}) = \frac{SS_{between}}{df}. \quad (28.15)$$

8. Calculate a mean square for the within groups:

$$(MS_{residual}) = \frac{SS_{residual}}{df}. \quad (28.16)$$

9. Calculate a test statistic:

$$F = \frac{MS_{between}}{MS_{residual}}. \quad (28.17)$$

10. Compare the F statistic to a tabulated critical value with $P - 1$ and $n - P$ degrees of freedom and a given significance level $1 - \alpha$.

If the "between sample variation" is significantly greater than the residual (within) variation then the potential exists that the samples were not, in fact, drawn from the same population, but from populations whose average values differed. If this proves to be the situation, in addition to the "within population variation," there exists also a "between population variation." If the calculated F value exceeds the tabulated value, reject the hypothesis of equal well means. Otherwise conclude there is no significant difference between the concentrations at the P wells (groupings). If a significant difference exists, at least one pair of well means is different. To find out which are different, multiple comparisons must be calculated. These comparisons are called *contrasts* in the ANOVA and multiple comparison framework (USEPA, 1989e). These contrasts are usually performed using *t*-tests with adjusted significance levels to account for error introduced by multiple comparisons. These assure that the total level of significance $(1 - \alpha)$ does not exceed a given limit, e.g. 95%, by adjusting α for the individual comparisons. One such method is the Bonferroni *t*-statistic, which is outlined in USEPA (1989e) and McBean & Rovers (1998).

28.2.4.2 Multiple Regression

Problems involving more than two variables can be treated in a manner analogous to that for two variables. For example, a dependent variable y may be a function of two dependent variables x_1 and x_2:

$$y = a_1 x_1 + a_2 x_2 + b_i \qquad (28.18)$$

where a_1, a_2 = the partial correlation coefficients and b = a constant.

The overall multiple correlation coefficient indicates the strength of the relationship between the dependent and independent variables. The partial correlation coefficients indicate the strength of the relationship between a dependent variable and one or more independent variables, with the effects of other independent variables held constant.

Similarly to the two-variable case described in Section 28.2.3.3, a multiple regression model can be fit to monitoring data and a significance test carried out. The assumptions of the multiple regression are the same as for the bivariate case.

28.2.5 NON-PARAMETRIC PROCEDURES

The previous sections have focused on the available procedures for addressing hypothesis testing questions and the identification of trends. However, as noted above, there are significant difficulties with these procedures when a substantial percentage of censored data exist within the data set and/or when there is a failure to adhere to assumptions implicit within the parametric tests. The result is that many existing environmental quality databases are not amenable to analyses by parametric methods. Techniques may not be suitable because of missing data, censored data and changing laboratory techniques or non-normally distributed samples.

A group of tests have been devised in which no assumptions are made about the distribution of the observations. For this reason, the tests are referred to as distribution-free. An array of these alternative tests exists, collectively referred to as "non-parametric" tests. The fundamental characteristic of these procedures is that the ranks of the data are utilized instead of data values (e.g. assigning a rank of "1" to the smallest value, "2" to the second smallest, and so on up to "n" for the largest, nth, observation). A large body of technical literature exists based on analyses of the ranks of the data. The mathematics involved in non-parametric tests are generally very straightforward. The analyses of ranked data are a direct parallel of the more traditional parametric methods as described in previous sections. For data which do not fulfill the necessary assumptions for the parametric analyses, the non-parametric methods are as powerful or, in fact, more powerful than the equivalent parametric tests. The non-parametric methods usually have the additional advantage that they require less burdensome calculations since they are generally not related specifically to the parameters of a given distribution. The major disadvantage of non-parametric methods is that they may be wasteful of information, and usually they have a smaller efficiency than the corresponding parametric methods provided that the assumptions of the standard (parametric) methods can be met. Hamby (1994) discusses many of the features of the non-parametric tests. In large, normally distributed samples, the non-parametric tests have an efficiency of approximately 95% relative to Student's t-test (Hodges & Lehmann 1956) and in small, normally distributed samples, the signed-ranks test (see Section 28.2.5.3) has been shown to have an efficiency slightly higher than this (Klotz 1963). With non-normally distributed data the efficiency of the non-parametric tests relative to the t-test never falls below 86% in large samples and may be greater than 100% for distributions that have long tails (Hodges & Lehmann, 1956).

Non-parametric analyses are performed on the ranks of the data, which has the effect of equally spacing the datapoints. If two or more observations are tied, each of the observations are assigned the mean of the ranks they would jointly occupy. For example, two observations both having a concentration of 25 mg kg^{-1}, and initially tied for the second and third ranks, will both be assigned a rank of 2.5. Information about relative magnitudes is lost during ranking, but the result is that the assumptions of distributional shape are completely removed. Rank or non-parametric tests are generally tests that do not require any assumptions other than independent samples from continuous populations.

During the last 40 years, the role played by non-parametric tests has undergone considerable change. In the 1950s and 1960s, it was quite common to characterize nonparametric methods as "rough and ready." This attitude has changed now so that practically every introductory statistics text contains a chapter concerning non-parametric, distribution-free or rank methods (Noether, 1981). The non-parametric tests include the Wilcoxon–Mann–Whitney test, the Kruskal–Wallis test, the Wilcoxon signed-ranks test, the Friedman test, Spearman's rank correlation and others. Since the tests are relatively quickly completed, the rank tests are highly useful for the investigator who is doubtful that the data can be regarded as normally distributed.

28.2.5.1 Mann–Whitney (U) Test

The Mann–Whitney (or "U") test is a highly efficient non-parametric test based on rank sums to compare the means of two independent samples. It closely resembles the form of the two-sample t-test, but makes no distributional assumptions.

In order to carry out the Mann–Whitney test, the data in each of the two samples are first ordered from lowest to highest values and then the combined data set is ranked. The ranks are then summed for each group.

For example, the following measurements from two wells could be ranked as:

Well A	Rank	Well B	Rank
15	2	12	1
20	4	17	3
21	5.5	21	5.5
30	8	25	7
Rank sum	19.5		16.5

The null hypothesis is that the two samples are taken from a common population so that there should be no consistent difference between the two sets of rankings. The Mann–Whitney test is typically carried out using a normal approximation by calculating a Z-statistic from the ranks:

$$Z = \frac{U - \bar{U}}{S_u} = \left[\left(nm + \frac{n(n + 1)}{2} - R_1 \right) \right.$$
$$\left. - \frac{nm}{2} \right] \bigg/ \left[\frac{nm(n + m + 1)}{2} \right] \quad (28.19)$$

where U = the calculated U-statistic, $\bar{U}$ = the mean of the U-distribution, S_u = the standard error, n and m = the respective sample sizes of the groups, and R_1 = the sum of the ranks for the first sample. If both n and m are greater than eight, the distribution of the U-statistic is approximated closely by the standard normal distribution and thus the critical Z-score (Z*) can be taken from a Z-table for the desired level of confidence. Comparing the calculated Z to the critical Z*, conclusions can then be made as to whether the null hypothesis of identical population can, or cannot, be rejected. If one sum of the single-sample ranks is sufficiently larger than the other, then the two sample means are considered to be significantly different. For small sample sizes (i.e. n or $m < 8$), the U-statistic is tested directly using a standard reference table of the U-distribution (available in many statistics textbooks).

Use of the Mann–Whitney Test to Test Equality of Variance. If the ranks of the data are assigned in a different way from that outlined above, the U-statistic can be used to test for unequal dispersions or variances. For this type of test, the ranks are assigned "from both ends toward the middle" by assigning Rank 1 to the smallest observation, Ranks 2 and 3 to the largest and second largest observation, Ranks 4 and 5 to the second and third smallest observations and so on. All other aspects of this test for unequal dispersions are identical with those of the Mann–Whitney U test.

28.2.5.2 Spearman's Rank Correlation Coefficient

Spearman's rank correlation coefficient procedure is a very simple ranking technique that can be used to assess inter-laboratory agreement in the order of samples from least to most contaminated. Consider a situation where two laboratories are completing parallel analyses on groundwater quality, and they each analyze the same ten samples for a parameter. To carry out Spearman's rank correlation coefficient test, the samples are ranked from lowest to highest. A difference, d, between the rankings is calculated for each pair of data (e.g., if Laboratory 1 had Sample 5 with a ranking of 3, and Laboratory 2 had Sample 5 ranked as 7, the difference would therefore be $d = 7 - 3 = 4$), and the correlation coefficient is then calculated as:

$$R = 1 - \frac{6 \sum_{i}^{n} d_i^2}{n^3 - n} \qquad (28.20)$$

where n = the number of samples ranked, and d = the rank difference.

To test the significance of the R value, a t-test is conducted (provided $n > 10$), calculating t as:

$$t = R\left(\frac{n - 2}{1 - R^2}\right)^{1/2} \qquad (28.21)$$

with $(n - 2)$ degrees of freedom.

The rank correlation coefficient procedure has been designed so that when the two rankings are identical, R has the value $+1$: when the rankings are as greatly in disagreement as possible, i.e. one ranking is exactly the reverse of the other, R has the value -1.

28.2.5.3 Sign Test for Paired Observations

This test represents a non-parametric alternative to the paired t-test. The sign test does not depend on the size of numerical values of the differences between paired datapoints, but only on their sign of the differences in their rank order. The assumption of normality is not required in order for this test to be valid. If the difference between the values is expected to equal zero, i.e. a null hypothesis of no difference, we would expect approximately one-half of the differences to be negative and one-half positive. If a small proportion of the differences are either negative or positive relative to the other, we suspect that there might be a real difference and a significance test is carried out.

Applications of the sign test for paired observations are the same as for the paired t-test, and its computation is relatively simple. For example, the fit of a groundwater model could be tested using field data as follows. First, for each location the difference, Z_i, between the observed value, x_i, and the predicted model value, y_i, is calculated as $Z_i = y_i - x_i$. Then, the number of differences greater than zero are counted (denoted by S), as are the number of deviations less than zero (denoted s). Note that differences of zero (i.e. the same ranks in both field and model data are ignored. Finally, the probability that $S < s$ is computed using a binomial expansion, as:

$$\Pr(S < s) = \sum_{k=0}^{S} \frac{m!}{k!(m - k)!} \left(\frac{1}{2}\right)^m \qquad (28.22)$$

where m = the number of paired differences Z_i not equal to zero and the counter, k, ranges between zero and S.

With paired observations, the sign test provides a simple test of the null hypothesis that x and y have a common distribution, as opposed to the alternative that x is less than y or x is greater than y.

28.2.5.4 Kruskal–Wallis Test (or Non-parametric ANOVA)

The Kruskal–Wallis H test is the non-parametric analog of the ANOVA. The method is useful when there are three or more groups present. The Kruskal–Wallis H tests the null hypothesis that k groups of samples come from identical populations against the alternative that the populations have unequal means. The Kruskal–Wallis H test is utilized when the data or the residuals have been found to be significantly different from normal and when a log transformation fails to adequately normalize the data. It does not make any assumptions about the underlying distribution. The H test is conducted in a manner similar to the U test. It requires at least three groups with a minimum sample size of three in each group. As with the U test (see Section 28.2.5.1), all observations are ranked from lowest to highest, and if R_1 is the sum of the ranks occupied by the n_i observations of the ith sample, the test is based on the statistic:

$$H = \frac{12}{n(n + 1)} \left(\frac{R_1^2}{n_1} + \frac{R_2^2}{n_2} + \cdots + \frac{R_k^2}{n_k} \right) \quad (28.23)$$

$$- 3(n + 1)$$

where $n_1, n_2, \ldots, n_k$ are the number of observations in each of the k groups.

When $n_i > 5$ for all i, the distribution of the H statistic is well-approximated by the chi-square distribution with $k - 1$ degrees of freedom if the null hypothesis is true. This forms the basis of the significance test. If the monitoring locations are found to differ, *post hoc* contrasts are needed to determine if and where contamination is present. When small

samples are involved ($n_i < 6$), the chi-square approximation can no longer be used as a reliable procedure. In these situations, a table of critical values from the sampling distribution of H must be used. If more than five monitoring locations are involved, the individual comparisons should be performed using $Z_{0.01}$ rather than $Z_{\alpha/(k-1)}$, where $Z_{\alpha/(k-1)}$ is the $[1 - \alpha/(k - 1)]$ percentile of the standard normal distribution.

The Kruskal–Wallis test is performed using the following method. First, defining the number of groups by k and the number of observations in each group by n_i, with n being the total number of well observations, let x_{ij} denote the jth observation in the ith group, where j runs from one to the number of observations in the group, n_i and $i = 1, \ldots, k$ groups. The steps of the procedure are then:

1. Rank all the observations from least to greatest within each group. Let R_{ij} denote the rank of the jth observation in the ith group. As a convention, denote the background monitoring location(s) as Group 1. Ties are given the average rank of the tied values, as described in the introduction to this section (28.2.5).
2. Compute the sum of the rank values (R_i) for each group and the average rank (mean R_i) within each group (k groups).
3. Compute the average rank (mean R) of the overall data set.
4. Calculate the Kruskal–Wallis test statistic, H using Equation 28.23.
5. Compare H with appropriate chi-square critical value χ^2, with degrees of freedom equal to $k - 1$, where $k =$ the number of groups.

Equation 28.23 should be modified if there are any tied ranks in the data set. To allow for tied ranks, the calculated value of H is divided by a correction factor to give $H°$:

$$H° = H \Bigg/ \left[1 - \frac{\Sigma(t^3 - t)}{n^3 - n} \right] \quad (28.24)$$

where t = the number of individual values involved in each set of tied ranks. The summation $\Sigma(t^3 - t)$ must be calculated for each set of tied ranks.

If H° is greater than χ^2, the null hypothesis H_0 is rejected, indicating that there is a significant difference between at least two of the well groups. If the H statistic is significant, then multiple individual comparisons, i.e. contrasts, must be computed between the groups. This is done by calculating the differences between average ranks of each group, and comparing these to a critical difference C_i, defined as:

$$C_i = Z_{\alpha/(k-1)} \left[\frac{n(n+1)}{12} \right]^{0.5} \left(\frac{1}{n_1} + \frac{1}{n_i} \right)^{0.5} \quad (28.25)$$

where $Z_{\alpha/(k-1)}$ = the upper $\alpha/(k-1)$ percentile from the standard normal distribution, with zero mean and unit variance; and $1 - \alpha$ is the selected level of significance (typically, 95%). If a given difference between two groups exceeds the critical difference, than the two groups are shown to be significantly different.

28.2.6 METHODS FOR CENSORED DATA

As mentioned in the introduction to Section 28.2 above, monitoring records are frequently censored due to the technical limitations of modern analytical techniques. Consequently, a number of procedures for examining data records with varying degrees of censorship have been published. The majority of these procedures assume the data are singly censored, e.g. less than five, as opposed to multiply censored data sets, e.g. the monitoring record consists of concentrations reported as less than five as well as less than 15). Procedures for estimating parameters based on singly censored data sets are examined in Gilbert & Kinnison (1981), Gilliom & Helsel (1986), Gleit (1985), Helsel & Gilliom (1986), Kushner (1976), and Owen & deRouen (1980). Often, a censored data set will exhibit a skewed distribution, which violates the assumptions of the typical significance test methods. The following section therefore presents some alternatives for dealing with censored data in monitoring data.

28.2.6.1 Simple Substitution Methods

If a small proportion of the observations are not detected, one approach is to replace the censored data with a small number, for example, the method detection limit (MDL) divided by two. USEPA recommends the MDL/2 approach if less than 15% of all samples are non-detect, but these simple substitution methods tend to perform poorly in statistical tests when the non-detect percentage is higher (Gilliom & Helsel, 1986). Simple substitution procedures provide a degree of quantification, but may seriously affect subsequent utilization of the parameters in the significance testing. Hashimoto & Trussell (1983) compared several estimators of the mean for censored water quality data. Their examples illustrate the bias caused by three commonly used substitution methods: discarding censored observations, setting all censored observations equal to zero, or assigning the detection limit to all censored observations.

28.2.6.2 Test of Proportions

The test of proportions is a non-parametric procedure, i.e. it does not use the actual monitoring data values directly. Instead, each sample is treated as a one or zero depending on whether the measured concentration is above or below the detection limit. The test of proportions relies on a normal probability approximation to the binomial distribution of zeros and ones and assumes that the sample size is reasonably large. Generally, if the proportion of detected values is denoted by P, and the sample size is n, then the normal approximation is adequate, provided that nP and $n(1 - P)$ both are greater than or equal to five.

The test of proportions is appropriate for use when the proportion of quantified values (i.e. detects) is small to moderate (e.g. 10–50%) or, if more than 50% of the data are below detection but at least 10% of the observations are quantified. If very few quantified values are reported (<10%), a method based on the Poisson distribution should be used.

A test of proportions might be used, for example, if none of the background well observations are above the MDL, but if all of the compliance well observations are above the detection limit, one would suspect contamination.

28.2.6.3 Plotting Position Procedure

Use of probability paper provides a simple means of estimating the mean and standard deviation in the presence of censored data (McBean & Rovers, 1984). The limitation of this method arises in that sufficient detected data must exist to convince the analyst that the correct type of probability paper is being utilized and that a reasonable line can be fit to the plotted data. This procedure involves fitting the probability distribution to the data above the detection limit and then extrapolating the line to, in turn, estimate the mean and standard deviation. Potential candidate distributions include the normal and log-normal, since the probability plotting papers are readily available and the distributions are generally appropriate to environmental quality data. Additional useful distributions include the binomial, Poisson, Gumbel and log Pearson Type III, among others.

To be successful, a sufficient portion of the data must be in excess of the detection limit to establish that the data are reasonably described by the assumed probability distribution. As well, there must be sufficient quantities of information from which the fitted line

can be drawn. For example, with a brief data set, one large value can substantially bias the resulting line (McBean & Rovers, 1984). Mage (1982) describes several techniques for plotting ranked observations on probability paper and drawing lines.

A significant advantage of the probability plotting procedure is that it gives a visual assessment of the data. The analyst can downplay the effect of an individual datapoint if there is a valid rationale for presuming the particular point is a data outlier. The disadvantages of probability plots are that there is no commonly accepted rule for plotting the data and drawing lines by eye, and there is no simple objective way to judge how well the datapoints conform to a straight line (Mage 1982).

28.2.6.4 Cohen's Test

Another procedure for estimating the mean and variance of a censored normal distribution was presented by Cohen (1961). Cohen's test may be applied up to the situation of 50% non-detects. This approach assumes the observed data (detects and non-detects) come from the same normal or log-normal population, but that the non-detect values have been censored at their detection limits. The premise of the technique is based on the censored probability model where we estimate new or adjusted mean and standard deviations. Cohen's method adjusts the sample mean and sample standard deviation to account for data below the detection limit, assuming the data are normally distributed and the detection limit remains the same.

Cohen's method provides maximum likelihood estimates of the mean and variance of a censored normal distribution. Cohen's adjustment may not give valid results if the proportion of non-detects exceeds 50%. McNichols & Davis (1988) found the false positive rate associated with the use of t-tests based on Cohen's method rose substantially when the fraction of non-detects was greater than 50%. This occurred because the adjusted estimates of the mean and standard deviation are more highly correlated as the percentage of non-detects increases, leading to less reliable statistical tests.

28.2.6.5 Aitchison's Method

When at least 10% of the groundwater samples have a measurable (detected) value of a particular constituent, the mean and variance of the distribution can be approximated using a method developed by Aitchison (1955). Like Cohen's method, the assumption of a particular probability model for the data leads to adjusted estimates of $\bar{x}$ and S. However, Aitchison's adjustment is constructed on the assumption that the non-detect samples are free of contamination, so that all non-detects may be regarded as zero concentrations. To compute Aitchison's adjustment (Aitchison 1955), it is assumed that the detected samples follow an underlying normal distribution. Alternatively, if the detects are found to be log-normally distributed, the entire computation can be carried out using the logarithms of the data instead.

Aitchison's method results in a lower mean than does Cohen's because of the assumption within Aitchison's method that non-detect samples are free of contamination.

28.2.6.6 Maximum Likelihood Estimator and Alternates

The maximum likelihood estimator (MLE) procedure as a way of incorporating the effect of "less thans" in censored data sets is based on assuming an underlying distribution for the entire data set. MLE estimators of distributional parameters (e.g. mean or standard deviation) can then be derived from the uncensored observations. This procedure is useful for data with censored values, but is considerably more complicated than the preceding alternatives and is best accomplished

when there are large data sets. For small sample sizes, MLEs are not necessarily minimum variance, unbiased estimates.

An evaluation of MLE performance is presented in Owen & deRouen (1981). They used Monte Carlo techniques to evaluate the performance of MLE methods derived for the log-normal distributions. The technical literature also includes Holland and Fitz-Simmons (1982), which describes a computer program for fitting statistical distributions using the maximum likelihood estimation. Sharma et al. (1995a,b) utilize the MLE for parameter estimation in air pathways migration models.

There are several methods for handling regression with censored data. These include the iterative least squares (ILS) method (Schmee & Hahn 1979), the linear unbiased estimate method (Nelson & Hahn 1973), and the MLE method (Dempster et al. 1977; Haas & Jacangelo 1993). The ILS method is simpler to implement than the MLE method, and has good statistical convergence properties (Schmee & Hahn 1979). The ILS method was developed for parameter estimation of a linear model with correct censored data, which arose from life-tests.

28.2.6.7 Poisson Model

If more than 50%, but less than 90% of the monitoring records are non-detects, or the assumptions of Cohen's and Aitchison's methods are not met, parametric statistical intervals should generally be abandoned in favor of non-parametric procedures as described in Section 28.2.5. However, there is one final parametric-based procedure, namely the Poisson model, which may be appropriate for application.

Specifically, when 90% or more of the data values are non-detect, the detected samples may be modeled as rare events by using the Poisson distribution. The Poisson model describes the behavior of a series of independent events over a large number of trials, where the probability of occurrence is low, but remains constant from trial to trial. The Poisson model is similar to the binomial model in that both models represent "counting processes." In the binomial case, non-detects are counted as "misses" or zeros and detects are counted (regardless of contamination level) as "hits" or ones. In the case of the Poisson model, each particle or molecule of contamination (as per the implied assumption implicit in the Poisson distribution model) is counted separately but cumulatively, so that the counts for detected samples with high concentrations are larger than counts for samples with smaller concentrations.

For a detect with concentration of 50 p.p.b., the Poisson count would be 50. Counts for non-detects can be taken as zero or perhaps equal to one-half the detection level (e.g. if the detection level were 10 p.p.b., the Poisson count for a non-detect would be five). Therefore, unlike the binomial model, the Poisson model has the ability to utilize the magnitudes of detected concentrations in statistical tests.

28.2.7 STATISTICAL POWER

The power of a statistical test $(1 - \beta)$ is defined as the probability that the test will successfully detect a statistically significant difference. Alternatively, it is the probability of rejecting a null hypothesis when, in fact, it is false and should be rejected. Recall from Section 28.2.2.4 that the two types of statistical errors (I and II) are dependent on one another. If the critical value α is increased in order to reduce the probability of a Type I error, then the probability of a Type II error will increase. Thus, for a specific alternative hypothesis, this will reduce the power of the significance test, which is of concern in monitoring, since failure to detect significant effects can have environmental consequences.

A major impact with strict environmental regulations is that they tend to increase power by minimizing the false negative rates (i.e. β, the probability that you conclude that there is contamination when there is) at the expense of increasing the false positive rates (i.e. α, the probability that you conclude that there is contamination when there is none). Simply selecting a test based on its false negative rate is a concern, because a test with a high false positive rate will necessarily have a low false negative rate. In the extreme case in which the false positive rate is 100%, the false negative rate is 0%.

It is therefore desirable to identify a reasonable compromise of β and α values. To come to such a compromise, one can develop power curves. A power curve is a graph of $1 - \beta$ versus the true difference in sample means. Power curves are useful because for a given sample they allow the selection of an optimal test, in the sense that it achieves its intended false positive while simultaneously achieving a reasonable false negative rate. It is noteworthy that the power of a test is a function of the magnitude of the difference considered "significant," i.e. it is much more difficult to have high power for a small difference than for a large difference.

The power of a test may be increased in one or more of the following four ways:

- reducing the variance of the measurements to be tested,
- increasing the size of the difference considered significant,
- increasing the number of samples, and
- using a higher α level.

For example, a test that uses an α of 0.10 will have a higher power (i.e. a lower chance of a Type II error), but also a higher probability of a Type I error, than will a test using the 0.05 level.

As an example of the power calculations, the example of the t-test is given for illustrative purposes. For given sample sizes, n_1 and n_2, significance level $(1 - \alpha)$ and effect size (i.e. hypothesized significant difference) between population means, d, the power of the t-test may be estimated (after Zar 1982) by computing:

$$t_{\beta(1)} = \left\{ d \middle/ \left[\frac{2\hat{S}^2}{(2n_1n_2)/(n_1 + n_2)} \right]^{1/2} \right\} - t_\alpha \quad (28.26)$$

where $\hat{S}^2$ = the pooled variance, t_α (one-sided or two-sided depending on the test design) and $t_{\beta(1)}$ (one-sided) = critical values of the t distribution (for $n_1 + n_2 - 2$ degrees of freedom), and n_1 and n_2 = the respective sample sizes.

28.3 Assessment of Risk Associated with Exceedance of Boundary Criteria

Risk assessment and risk management may involve use of monitoring (and potentially fate and transport modeling) to estimate the ecological and human health risks associated with a contaminant plume. As discussed in Chapter 22, risk management has become an important part of geoenvironmental engineering. Risk is generally defined as:

$$\text{Risk} = \text{consequence} \\ \times \text{ probability of occurrence.} \quad (28.27)$$

In the following, consideration will be given to evaluating the probability of occurrence in the context of information gained from monitoring data. Problem identification and risk management involves exposure assessment (Section 22.5), toxicity assessment (dose-response and hazard identification, Section 22.6), risk characterization (Section 22.7), risk communication (Section 22.8) and, finally, risk management (Section 22.9).

Additional detail regarding some of the subcomponents of the risk assessment process that are of particular relevance to envi-

ronmental monitoring are presented in the following sections.

28.3.1 MIGRATION PATHWAYS OF ENVIRONMENTAL CONTAMINANTS

Given environmental contamination, a source of potential exposure risk or hazard exists. However, for an environmental risk to occur, two additional components must exist: (a) one or more pathways by which the contaminants may migrate, and (b) a receptor who will be harmed if the exposure is sufficiently large. If any of these three features are zero, there is no risk. Thus, risk assessment is a systematic process for making estimates of all the significant sources and exposure risk pathways that prevail over an entire range of failure modes and/or exposure scenarios (see Chapter 22). When a contaminant is released into the environment, it moves or partitions into the environmental sectors (gaseous, aqueous, solid) according to its physical properties and the properties of the environment. Subsequently, the determination of mobility of the contaminant to create an exposure risk requires development of quantitative estimates of the extent to which these migration pathways are contributing to exposure risk. The availability of environmental monitoring data for the purposes of calibrating the model mobility of the constituents transfer functions greatly improves the confidence in the modeling results.

For some situations, one migration pathway will clearly be the dominant feature. Conversely, for others, a variety of migration pathways may be relevant, which is suggestive of the degree of difficulty sometimes involved in a comprehensive risk assessment. Examples of migration pathways for different media are provided in Table 28.5. The various pathways may include contaminants being transported via one or more media (including air, soils/sediments, surface water and groundwater) to potential receptors (through, for example, inhalation, dermal

TABLE 28.5. Contaminant exposure pathways by medium

Medium	Exposure pathway
Ground-water	Ingestion from drinking
	Inhalation of volatiles
	Dermal absorption from bathing
Surface water	Ingestion from drinking
	Inhalation of volatiles
	Dermal absorption from bathing
	Ingestion during swimming
	Ingestion of contaminated fish
Soil	Ingestion
	Inhalation of particulates
	Inhalation of volatiles
	Ingestion via plant uptake
	Dermal absorption from gardening

contact and ingestion). The existence of various exposure routes to alternative organs within the body (e.g. inhalation suggests that the target organ is likely the lungs, whereas ingestion may result in the target organ being the stomach) indicates that the pathways assessment and exposure risk calculations may require considerable effort.

28.3.2 EXPOSURE ASSESSMENT PROCEDURES

Once a potential migration pathway that could lead to exposure is identified, the magnitude of the potential exposure of a receptor must be evaluated. The exposure assessment aspect of risk assessment includes characterizing the physical and exposure setting (i.e. contaminant distributions leading from sources of environmental contaminants to the points of exposure), identifying significant migration and exposure pathways, and then calculating chemical intakes for all potential receptors. The total exposure of a receptor to a contaminant is the sum of exposures from all migrational pathways.

Mathematical models to quantify the exposure scenarios are important elements of the assessment. Numerous model classification systems exist, including:

- black box models;
- analytical models, and
- numerical models.

The selection of a model to be used is influenced by at least three conditions (McBean et al. 1990):

1. the judgment of the investigator in the specific application of the various modeling techniques,
2. the extent of the database, and
3. the physical system being modeled.

Examples of common models used to perform risk calculations by summing over the migrational pathways of exposure levels and probabilities of occurrence are AERIS (1990) and API (1994). A detailed discussion of the pros and cons of some of the available models is found in McTernan & Kaplan (1990) and others.

The risk assessment process reduces a problem into logical pieces. However, important limitations may exist in the models used to solve each of these problems (e.g. characterizing the migration of the constituents from the source to the receptor, the data utilized as inputs to the models, and the scientific understanding of the reference doses). There are always *uncertainties* inherent in a risk assessment, and it is critical that these be quantified to assess the *reliability* of the assessment. Since risk assessment is used to inform a process of managing the risks (see Sections 22.9 and 28.3.3), the degree of certainty associated with the results is a key consideration. In fact, uncertainty can be used to determine the level of effort required for the process; if a risk assessment's results have too much uncertainty, then further field work is performed to collect additional monitoring data, and the risk assessment repeated with better estimates of the input parameters. This process may be iteratively repeated until an acceptable level of uncertainty is achieved. In certain cases, the natural variation of a parameter may be such that the desired overall uncertainty level may not be achieved (e.g. the hydraulic conductivity of a soil).

28.3.3 DECISIONS ON RISK-BASED CORRECTIVE ACTION

It is noteworthy that, as Lehr (1990) indicates, "risk" is often interpreted as bad — a thing to be avoided — and yet all economic and technological progress requires that human beings take risks. As Wildavsky (1988) indicates, "there can be no safety without risk." Ultimately, the question of risk reduction becomes one of risk management: what levels of environmental exposure risk are acceptable? The resolution to this question is complicated by the many dimensions to the problems that exist. Resolution of such questions for a particular situation involves both risk assessment and risk management:

• Risk assessment: is mainly an objective process that calculates the risk associated with environmental contamination using factual information to define the health effects of exposure of humans and/or the environment to hazardous materials and situations. For example, this involves the quantification of the extent to which a chemical moves from the zone of contaminated groundwater, to create an exposure scenario for a nearby resident via drinking water, and a determination of the deleterious consequences, if any, of that exposure scenario that is the result.
• Risk management: is of necessity a subjective process of weighing policy and remediation alternatives against the calculated risks. This integrates the results of risk assessment, engineering data, and social, economic and political concerns to reach a decision involving the management of the risk.

The breadth of issues involved with reaching risk management decisions is beyond the scope of this chapter. Readers interested in

questions of how to manage risks are referred to Chapter 22 and other sources for further reading, e.g. Glickman & Gough (1990) and Asante-Duah (1993). Sufficient attention is paid here only to develop an appreciation of the general magnitudes of various tasks that assist in understanding risks associated with environmental concerns. The risk assessment and management processes are intrinsically based on the statistical procedures presented in Section 28.2, and risk assessment is a logical and important extension of statistical methods to assess monitoring data. The field of environmental risk assessment is enormous and evolving, and it is intended that this section has alerted the reader to its key concepts and generic applications.

28.4 The Case for Innovation: A Unique Visualization Method for Demonstrating Intrinsic Remediation

In recent years, acceptance of monitored natural attenuation as a remedial alternative has gradually increased. It is now becoming recognized that for sites where the immediate risk to environmental or human receptors is minimal, and where the subsurface contaminants present are biodegradable by indigenous micro-organisms, an acceptable and cost-effective solution may be to monitor contaminant concentrations as natural processes carry out the remediation. This alternative may be especially attractive with delicate ecosystems where the disturbance caused by an active remedial alternative (e.g. excavation of soil) may cause more real damage than an existing contaminant plume. As knowledge about biodegration processes increases, the role of natural attenuation in contaminant remediation programs will continue to expand.

Demonstrating the occurrence of natural attenuation at a site is a multicomponent analysis based on a "weight-of-evidence" approach, which includes such considerations as trend analysis of contaminant concentrations, groundwater dispersion modeling and the evaluation of biodegradation capability (i.e. reduction–oxidation conditions), etc. Up to this point, such evaluation has often been performed through the use of tabulated data and multiple figures mapping the concentration of each parameter at different locations on the site. Recently, a unique visualization software tool (SEQUENCE) was published in the public domain that allows for the simultaneous mapping of multiple monitoring parameters on a single page, which can greatly assist the evaluation of natural attenuation at a site. This tool is presented here as an example of the type of innovation required in the future of contaminant monitoring.

SEQUENCE produces a *radial diagram* map representing the concentrations of a number of chemical parameters simultaneously. The program takes its name from the sequence of reduction–oxidation (REDOX) zones that are observed downgradient of a biodegradable organic contaminant's source area. Biodegradation typically occurs via biological oxidation processes. These processes sequentially consume oxygen, nitrate, manganese, iron, sulphate and carbon dioxide until the contaminant is completely degraded. Thus, changes in concentrations of these parameters may provide evidence that degradation is taking place in the groundwater. These redox parameters can be plotted by SEQUENCE using a modified radial diagram method. An example radial diagram is shown in Fig. 28.8.

On each radial diagram, the site background concentrations are plotted and form the outer line. One or more inner lines are plotted representing the monitoring results at the location. In this way, the conditions at the monitoring location can very quickly be visually compared. On Fig. 28.8, all of the redox parameters show significant changes from background, indicating that this location

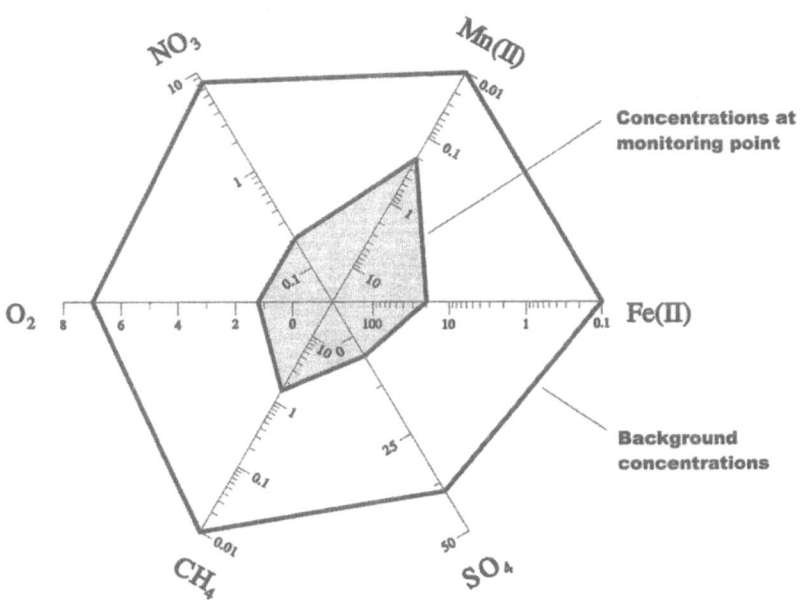

FIGURE 28.8. Example of radial diagram.

918

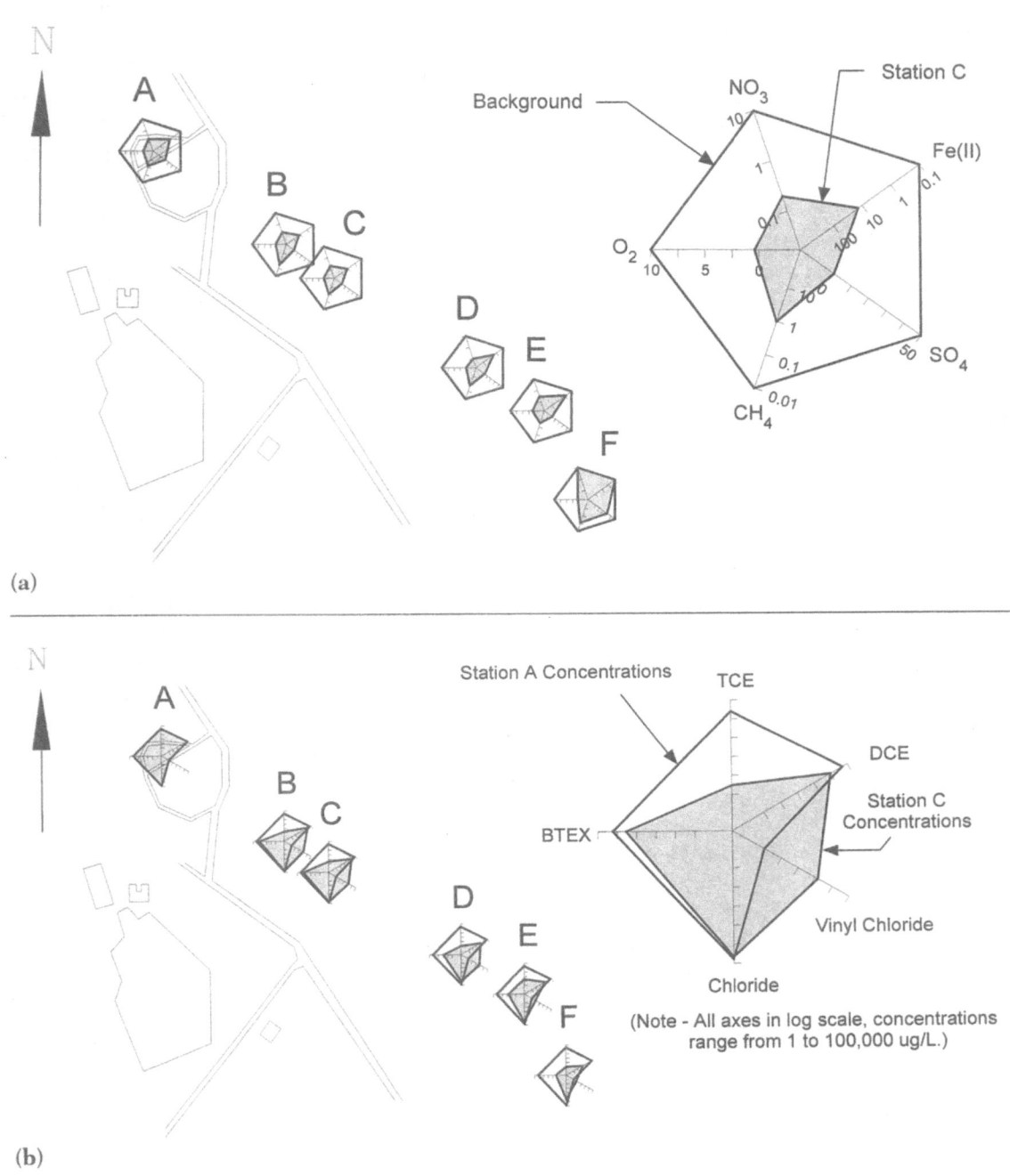

FIGURE 28.9. (a, b) SEQUENCE plots for redox parameters and chemical constituents in groundwater at Plattsburg Air Force Base.

likely is in an advanced stage of biodegradation.

To illustrate the use of this method, consider Fig. 28.9a and b, which presents monitoring data from the Plattsburg Air Force Base (Weidemeir *et al.* 1996). Groundwater flow at the site is in the direction from Station A (upgradient) to Station F (downgradient). Station A is adjacent to a former fire management area that over the years had become contaminated by chlorinated solvents and fuel hydrocarbons in the groundwater. On Fig. 28.9a, notice that at Station A, all of the redox parameters show changes from background concentrations, which is consistent with the occurrence of biodegradation. As the groundwater moves downgradient, the redox parameters gradually return to ambient (although oxygen has not returned to background concentrations by this point) as the contaminant load is decreased by biodegradative removal. This interpretation is supported on Fig. 28.9b, which presents the chemical concentrations at the stations.

Clearly, the concentrations of dichloroethene (DCE) and trichloroethene (TCE) decrease with distance. Vinyl chloride (VC), which is an intermediate product in the biodegradation of these substances, first shows an increase in concentrations (as DCE and TCE start breaking down), and then a decrease to non-detected by the last monitoring Station F (as VC is fully biodegraded). Chloride, which is a byproduct of the degradation of chlorinated organics is highest in the region, with the highest contaminant concentrations suggesting active biodegradation occurring at these locations, and also diminishes as the contaminant mass is used up.

SEQUENCE is an example of an innovative approach in the interpretation of monitoring data. It provides a useful visual tool permitting rapid data interpretation and also presents the data in a manner easily explained to the lay public. References treating its theory and application include Carey *et al.* (1996, 1997a,b, 2000), Murphy (1996) and Harris *et al.* (1998).

29. *IN SITU* CONTAINMENT AND TREATMENT OF CONTAMINATED SOIL AND GROUNDWATER

D. J. A. Smyth

R. W. Gillham

D. W. Blowes

J. A. Cherry

29.1 Introduction to Contaminated Sites

29.1.1 CHARACTERISTICS OF CONTAMINATED SITES

Full restoration of contaminated zones in the subsurface, such that unrestricted use of the land or groundwater resources can occur, has proven to be a challenging, costly and often elusive goal. This has been a consequence of the limitations of available remedial technologies, incomplete investigation and definition of the contaminant problem, and the complex and heterogeneous nature of the subsurface. A remedial program for subsurface contamination arising from widespread non-point sources such as the use of agricultural chemicals and fertilizers generally needs to be approached through the implementation of better management practices, such that the introduction of contamination to the subsurface over extensive areas is either reduced or eliminated. For point-source contamination, however, it may be feasible to implement specific programs to reduce or remove contaminants from the subsurface.

Although the release of contaminated water can create plumes of dissolved contaminants in groundwater, it is not uncommon for plumes to emanate from long-term sources, which were introduced to the subsurface through the inappropriate disposal of wastes or inadvertent chemical spills. The nature of impacts from three conceptual models of point-source subsurface contamination is shown in Fig. 29.1. Each model is characterized by a source zone and mobile plumes of dissolved contaminants in groundwater, or possibly vapour-phase contaminants in the vadose zone from sources of volatile organic LNAPL (light, non-aqueous phase liquids) chemicals, such as petroleum hydrocarbons, or DNAPL (dense, non-aqueous phase liquids) chemicals, such as chlorinated solvents, coal tar and creosote. The complexity of these models, and the difficulty associated with remediation, increases as features such as mixed sources of contaminants, heterogeneity of the subsurface, complicated hydrogeological settings associated with fractured clay or rock, or interactions between the contaminants and the subsurface are incorporated. The mass of contaminants associated with the source zone generally far exceeds that associ-

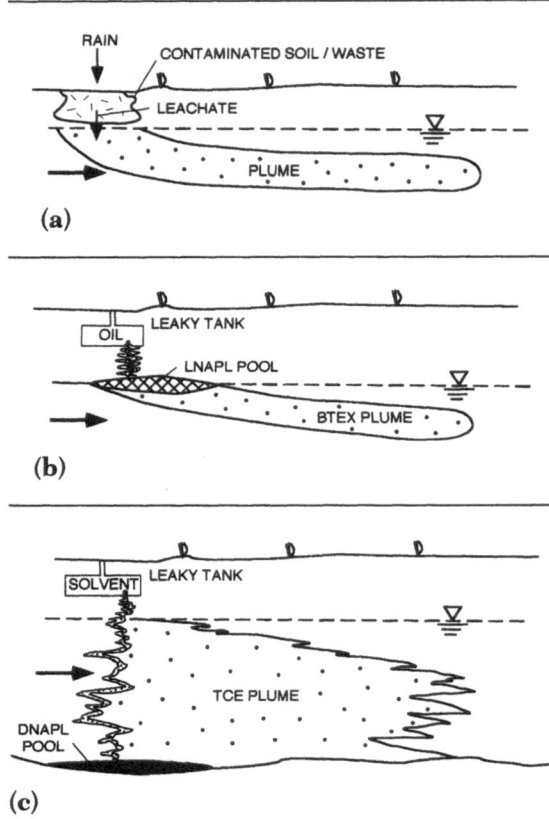

nant hydrogeology and migration/attenuation mechanisms, Chapter 25 for a discussion of liner systems and cut-off walls, Chapter 28 for a discussion of monitoring of contaminated sites and Chapter 30 for a discussion of other techniques for dealing with contaminated soil and sites.

29.1.2 OBJECTIVES OF REMEDIATION

The goals of remediation have a significant influence on the development and implementation of remedial plans for contaminated sites. Full restoration of the subsurface is the most stringent goal. It relies on the complete removal or destruction of the contaminants, such that the subsurface is returned to its original condition and groundwater quality could meet health- and ecologically based criteria such as drinking water standards. Generally full restoration is difficult to achieve as it requires complete remediation of subsurface source zones and the mobile plumes of contaminants in both soil vapor and groundwater. Indeed, for many subsurface contamination problems, in particular those involving DNAPL sources, full restoration is generally not technically feasible. Although full restoration was the desired goal of most of the regulatory programs associated with the pioneering efforts in subsurface remediation in the 1980s, such as the Comprehensive Environmental Response, Compensation and Liability Act (CERCLA), better known as Superfund, and the Resource Conservation and Recovery Act (RCRA) in the USA, there has been growing acknowledgment within recent years that partial restoration may be a more practical, affordable and reasonable goal of subsurface remediation. Partial restoration can imply remediation to risk-based criteria, which accommodate some level of contamination in the subsurface, or long-term management or control options to maintain contaminant concentrations at criteria levels in the absence of complete removal or destruction of contamination.

FIGURE 29.1. Conceptual models for three cases of point-source contamination: (a) the non-NAPL case; (b) the LNAPL case; and (c) the DNAPL case (reproduced from Cherry *et al.* 1996 with permission).

ated with plumes (Mackay and Cherry 1989), but mobile contaminants in plumes generally pose more immediate risks to the environment and receptors. Cherry *et al.* (1996) advocate an approach to remediation that clearly recognizes the distinct characteristics of source zones and plumes. It is a rare circumstance for a single remedial technology to be applicable to both source zones and plumes.

This chapter deals with *in situ* containment and treatment of contaminated sites and hence falls within the broader context of geoenvironmental problem identification and risk management. The reader is also referred to Chapter 22 for a discussion of broader issues, Chapter 24 for a discussion of contami-

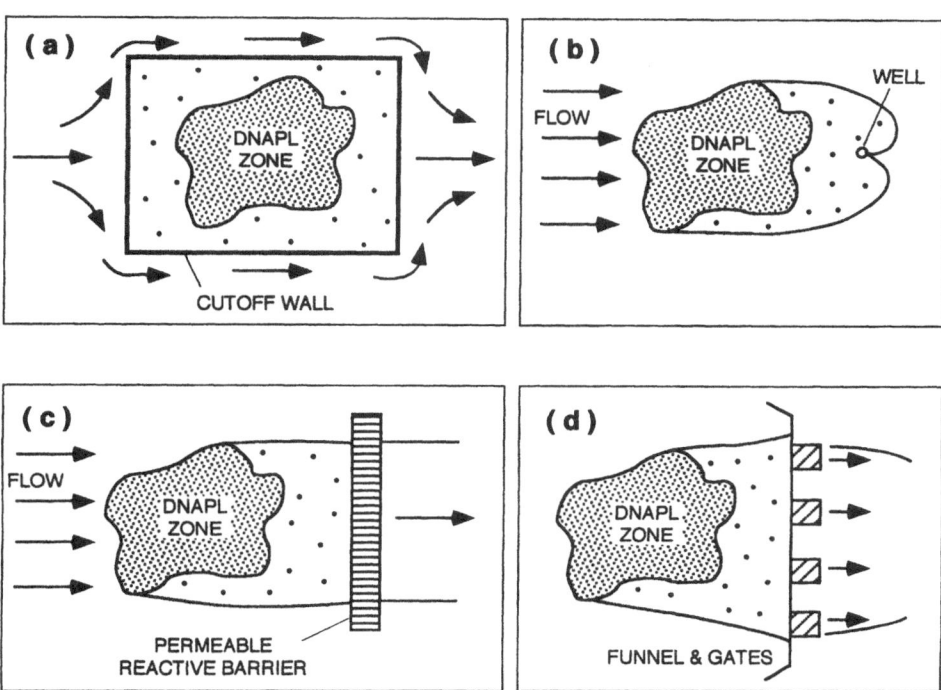

FIGURE 29.2. Engineered remedial actions for subsurface contamination. Actions may include: (a) source-zone isolation, containment or removal; (b) plume control through the use of pump-and-treat systems; (c) passive to semi-passive plume control through the use of permeable reactive walls; or (d) funnel-and-gate systems case (reproduced from Cherry *et al.* 1996 with permission).

A remedial program for contaminated soil and groundwater may involve the implementation of one or a combination of actions to achieve clean-up objectives. At some sites the reliance on natural physical, chemical and biological mechanisms (intrinsic remediation) may reduce contaminant concentrations in the subsurface and achieve clean-up objectives. In other circumstances, as shown in Fig. 29.2, engineered remedial actions, including the following, may also need to be considered:

- Source-zone restoration through the removal or destruction of contaminant mass to eliminate the possibility of further generation of mobile plumes.
- Source-zone isolation or containment through the use of hydraulic or physical barriers in the subsurface.
- Plume control and remediation by active

pumping systems or through the use of *in situ* reactive barriers.

29.2 Intrinsic Remediation

In some cases, natural attenuation of contamination in the subsurface may be sufficient to achieve remedial goals. Referred to as intrinsic remediation, the approach relies on the development of steady state or shrinking distributions of contaminants within and in the vicinity of a source of contamination.

For intrinsic remediation to be applicable at a site, it is important that concentrations of contaminants within a plume decrease to levels compliant with regulatory or other appropriate criteria before the plume reaches critical receptor or discharge points. In groundwater, the processes of natural attenuation that may cause a contaminant plume to attain steady state characteristics are dispersion and mass removal by physical, chemical or biological means (see Section 24.3; Cherry 1996; Ward 1996). The reversible sorption/desorption interaction between dissolved-phase contaminants in groundwater and aquifer solids does not cause permanent removal of contaminant mass from plumes, but it may contribute to the apparent development of steady state conditions as a consequence of delaying the advance of plume fronts. Steady state plume conditions in the subsurface may also arise as a consequence of groundwater discharge to surface water. The ability of surface water to assimilate or dilute contaminated discharge will influence the acceptability of this outcome.

The detailed spatial characterization of the contaminant distribution in a plume is necessary to confirm that natural attenuation mechanisms have the potential to achieve the required level of remediation. Cherry (1996) describes the observed presence of high-concentration core zones and lower-concentration fringe areas extending for several hundreds of meters in plumes emanating from actual spills of solvents. The results of inappropriate or sparse monitoring could be used to infer the apparent decrease of contaminant concentrations in the plume with distance from the source, and provide misleading evidence that intrinsic remediation was occurring. It is thus critical that the specific role of the various natural attenuation mechanisms influencing the fate of contaminants can be identified (see Section 24.3; Ward 1996; Wiedemeir et al. 1996a).

Aromatic hydrocarbons, including benzene, toluene, ethylbenzene and xylenes (BTEX), typically constitute less than 10% of the composition of gasoline, and even less of heavier petroleum hydrocarbon fuels; however, because of their higher volatility, higher solubility, greater mobility in groundwater, and higher toxicity relative to the predominant alkanes in hydrocarbon fuels, the aromatic hydrocarbons have often been considered as the contaminants of most concern at spill or leak sites. The disappearance of BTEX over relatively modest transport distances in groundwater systems has been documented in Barker et al. (1987), Borden (1994), Salanitro (1993), Rifai et al. (1995) and elsewhere. These and other studies have shown that the disappearance can be attributed to biodegradation by native microbial organisms in the subsurface, and that the rates of biodegradation are enhanced under aerobic conditions. Biodegradation of BTEX has also been noted to occur under anaerobic conditions, in the presence of terminal electron acceptors other than oxygen, such as nitrate. Under these conditions, however, benzene is more recalcitrant than under aerobic conditions (Barbaro et al. 1992; Borden et al. 1997a). The persistence of BTEX in plumes at concentrations orders-of-magnitude above drinking water standards more than 200 and 400 m, respectively, from subsurface source zones is documented in Borden et al. (1995) and Davis (1997). Thus, the acceptability of intrinsic remediation for petroleum hydrocarbons will depend on the size of the source zone, the aerobic/anaerobic characteristics within the plume, the rate at which oxygen enters the plume from the adjacent groundwater system, and a travel distance to critical receptor points of sufficient length for biodegradation to achieve the necessary decrease in contaminant concentrations. While the prospects are good for the intrinsic remediation of aromatic and aliphatic petroleum hydrocarbons, it has been shown (Squillace et al. 1997; Borden et al. 1997a) that a gasoline additive, methyl t-butyl ether (MTBE), which has been used extensively in the USA, is resistant to biodegradation in the subsurface, and is a poor candidate for intrinsic remediation.

On the basis of their occurrence in contaminant plumes and drinking water supplies, chlorinated solvents are more recalcitrant than petroleum hydrocarbons (Pankow et al. 1996; Mackay & Cherry 1989). McCarty (1996) and Vogel et al. (1987) provide the conceptual framework for the pathways of degradation for the common chlorinated aliphatic compounds such as trichloroethylene (TCE) and tetrachloroethylene (PCE), through intermediate transformation products, which may also be toxic and persistent, such as cis- and trans- 1,2-dichloroethylene (cis-DCE, t-DCE) and vinyl chloride, to innocuous inorganic products such as chloride, carbon dioxide and water. It is, however, very difficult to predict the initiation, extent and rates of degradation (Wiedemeir *et al.* 1996a). In the reductive dechlorination process, the chlorinated solvents act as electron acceptors, and rely on the metabolism of other natural or anthropogenic sources of carbon such as petroleum hydrocarbons as electron donors.

There are many examples of field evidence of the transformation of chlorinated solvents in groundwater, but cases of natural attenuation to levels compliant with regulatory criteria have not been widely documented. Semprini *et al.* (1995), for example, demonstrated the biotransformation and degradation of TCE to the DCE isomers, vinyl chloride and ethene in a sand aquifer in Michigan. The transformation and significant mass loss of TCE as a consequence of the co-metabolic degradation of petroleum hydrocarbons at a site in Plattsburg, New York, is described in Wiedemeir *et al.* (1996b), but the contaminant plume persisted for more than 1.3 km.

The development of steady state and shrinking distributions for various monaromatic and polyaromatic hydrocarbons as a consequence of biodegradation in groundwater downgradient of an experimental coal tar–creosote source is described in King *et al.* (1995). Raven and Beck (1993) also found contaminant plumes of limited areal extent in the vicinity of coal tar and creosote sources at former coal gas sites in Ontario, providing indirect evidence of intrinsic remediation of plumes. Spain (1996) provides a brief review of the potential biodegradability of chloroaromatic compounds including polychlorinated biphenyls (PCBs), nitro-aromatic compounds and pesticides; but the degree to which intrinsic remediation may occur would need to be confirmed on a site-specific basis.

Natural attenuation of inorganic contaminants may also occur as a consequence of transformation or precipitation reactions in the subsurface. Nitrate, which can originate from septic systems, fertilizers or agricultural wastes, may be removed from groundwater by denitrification (Hem 1995), but the availability of labile carbon to support denitrifying bacteria influences the extent to which denitrification may occur in particular groundwater systems (Starr & Gillham 1993). Heavy metals may be among the most toxic contaminants in landfill leachate, but their migration from landfills is generally restricted by mechanisms such as ion exchange, sorption or precipitation of stable solid phases. Yanful *et al.* (1988), for example, documented the precipitation of copper, zinc and lead carbonates, copper hydroxides and a copper organic-phase in the attenuation of metals in leachate at the base of a landfill underlain by carbonate-rich silty clay. The development of different redox environments in plumes from landfills on granular aquifers are identified in Bjerg *et al.* (1995) and Nicholson *et al.* (1983). They suggest that attenuation processes for many contaminants depend on the redox conditions and for some contaminants, such as sulphate, influence the development of the redox conditions in the plume. The reduction of sulphate to sulphide, and the subsequent precipitation of iron and other metal sulphides, is an example of a redox-controlled attenuation mechanism.

29.3 Source-Zone Restoration

29.3.1 GOALS AND OBJECTIVES

The goal of source-zone remediation, if full subsurface restoration is to be attained, is to remove or treat contamination to the extent that vapor-phase or aqueous-phase plumes do not continue to be generated by the source zone. For reactive, soluble or volatile contaminants, the implication of this goal is that nearly complete removal or destruction of the contamination is required. Partial removal of the contaminant mass in the source zone may do little to decrease the characteristics of plumes emanating from the source, and consequently achieve little in the reductions of risks to potential receptors in the vicinity of the contamination (Freeze & McWhorter 1997; Cherry *et al.* 1997; Cherry *et al.* 1996). Thus, a key issue in the performance of a remedial technology is the mass of contamination remaining in the ground, and not necessarily the rates of mass removal or destruction that the technology can attain.

Many of the technologies being proposed for source-zone remediation are emerging technologies, and data pertaining to experience and performance at the field-scale are sparse. There remains significant uncertainty in the potential performance of these emerging technologies, even under the most ideal conditions. This uncertainty is exacerbated in complex or heterogeneous environments, in cases where source-zone contamination is present in both the vadose zone and below the water table, and in cases involving mixtures of contaminants with differing physical or chemical properties. As indicated in Cherry *et al.* (1996), uncertainty in estimates of the mass of contaminants in typical DNAPL source zones casts further uncertainty on quantitative considerations of the performance of a remedial technology and the duration over which it should be applied.

29.3.2 VADOSE-ZONE CONTAMINATION IN THE SOURCE ZONE

For restoration of typical source zones, some effort in the remediation of contaminants in the vadose zone will often be required. The relative importance of contamination is site specific, and is dependent on the type and volume of contaminants, the site geology, and the hydrogeological setting, including the depth to the water table.

29.3.2.1 Excavation

Excavation of a contaminant source for *ex situ* treatment or off-site disposal has been a common approach to shallow soil contamination for the past two decades. For immobile non-NAPL sources of small to modest size, it can be quite successful, but the approach requires approval for treatment and/or disposal of excavated material in a landfill. In some jurisdictions, the use of landfill capacity for the disposal of contaminated soil is discouraged, and the approach may result in the transfer of a contaminant problem from one location to another.

The removal or replacement of faulty underground storage tanks (USTs) or the active spill source is an appropriate first step in responding to NAPL contamination in the subsurface. Excavation of contaminated soil adjacent to tanks has also been broadly applied in attempts to remediate NAPL sites. While there are some prospects for the remediation of NAPL sources entrapped in the vadose zone by excavation, in cases where the NAPL has penetrated through the vadose zone to the water table, the extent of the immiscible-phase contamination in the source zone may not be easily defined, and hence the limits of excavation required for source removal may be difficult to predict or achieve. In many instances, attempts to excavate NAPL source zones may be only partially successful, and some complementary source-zone or plume control remedial action may be necessary. Patrick and Burgess (1992), for sources of petroleum hydrocarbons and solvents, and Feenstra and Cherry (1996), for a source of solvents at a site in Dayton, New Jersey, describe examples of incomplete remediation of NAPL source zones by excavation. At each site, there was a need for complementary plume control using pump-and-treat systems because remnant contamination source material was left in the subsurface outside and beneath the limits of excavation.

29.3.2.2 Soil Vapor Extraction and Bioventing

Soil vapor extraction (SVE) and bioventing induce the active flow of air through the vadose zone, and have been used to remediate sources of volatile organic contaminants (see Johnson *et al.* 1990; NRC 1994; Fam 1996) and others. Volatile contaminants are transferred to the soil gas from the aqueous, sorbed and immiscible phases of contamination in the soil. The soil gas is drawn under vacuum to the extraction point or points of the SVE system, which induces air flow from the surrounding area through the contaminated zone (see Fig. 29.3). The systems generally incorporate treatment of the effluent vapors, which has important implications to operations, maintenance and costs (Marks *et al.* 1994). The use of SVE to alleviate immediate risks of fires, explosions or the introduction of vapors to buildings from subsurface

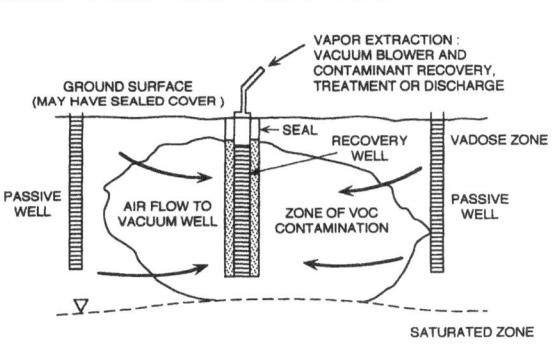

FIGURE 29.3. Elements of a typical soil vapor extraction (SVE)/bioventing system for remediation of VOC contamination in the vadose zone. The key features are the screened well and vacuum pump. Optional features may include passive wells to enhance the flow of air through the zone of contamination, or low permeability ground-surface covers. Upwelling of the water table beneath the vapor extraction wells may inhibit the removal of contaminant mass in these regions.

contamination by volatile organic contaminants has been recognized for more than a decade (Marley & Hoag 1984; O'Connor *et al.* 1984), but SVE is also used extensively for subsurface remediation.

For recalcitrant chemicals such as chlorinated solvents in the vadose zone, the physical transfer of contaminants to the vapor phase is responsible for active mass removal in the SVE process. When SVE is applied to some contaminants such as petroleum hydrocarbons, biodegradation may result in the *in situ* destruction of some contaminant mass, in addition to the physical removal. This combination is referred to as bioventing. In these cases the introduction of oxygen to enhance biodegradation may be of more importance than mass extraction, so air flow requirements may be moderate in comparison to conventional SVE systems (Marks *et al.* 1994). Performance assessment has generally addressed issues associated with the zone of influence of the SVE system, and the rate and mass removed in effluent gases. Intermittent pumping to identify whether there are contaminant residuals remaining in the source zone is a common and recommended practice. Johnson *et al.* (1990) contrast the simple case of SVE in homogeneously contaminated soils with more complex cases in which contaminant mass may be present in pools of immiscible fluids or trapped within low permeability soil lenses with high moisture content. In these latter cases, air flows adjacent to but not through zones of contamination, and the rate of transfer of contaminant mass to the air stream is diffusion-limited (see Section 25.5). Water-table upwelling may occur in the vicinity of vapor extraction wells, and decreases the effectiveness of SVE operation in deeper parts of the vadose zone.

Flynn *et al.* (1994) experienced initial success in the removal of PCE from a fine to medium sand, but longer-term inefficiencies in the performance of SVE were attributed to heterogeneity within the sand, which contributed to variations in PCE distribution, and non-uniform moisture content and distribution of air flow permeability. Buscheck and Peargin (1991), in a review of more than 140 operating SVE systems, found mass removal rates to decrease exponentially with time until they reached asymptotic removal rates near zero for sites underlain by medium to coarse sediments. At other sites underlain by heterogeneous sediments, the exponential decrease in contaminant removal rates was followed by non-zero asymptotic removal rates. Mass removal following the onset of SVE was attributed to air advection and evaporation, but at later times, diffusion-controlled release of mass from the fine-grained sediments contributed to the long-term presence of contaminants in the soil vapor.

Limitations with respect to mass removal from heterogeneous subsurface environments, and from saturated horizons within and below the capillary fringe, have prompted the application of SVE with complementary remedial actions. Johnson et al. (1990) suggest that contaminant mass removal by SVE can be increased if the water table is lowered by pumping, both to counteract the up-welling of the water table induced by lower gas-phase pressures in the immediate vicinity of vacuum extraction wells, and to expose NAPL contamination present within and below the initial capillary fringe.

Air sparging has been used in combination with SVE to remove volatile organic contaminants from water-saturated granular deposits within the capillary fringe and below the water table (Marley et al. 1992; Johnson et al. 1993). Air sparging consists of the continuous or pulsed injection of air under pressure through wells to depths below the water table. This induces air flow upwards through the contamination towards the vadose zone where capture of the soil gases by SVE is achieved. Contaminant mass may be removed by volatilization to the vapor phase or by enhanced biodegradation (Johnson et al. 1993; Hinchee 1993).

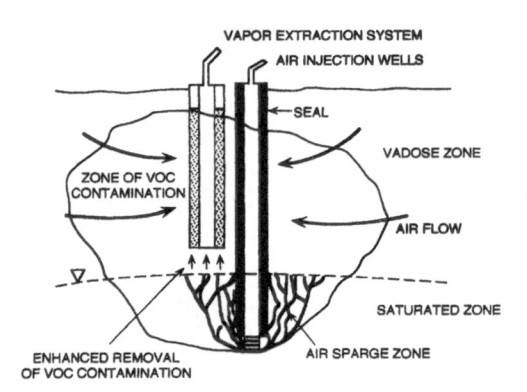

FIGURE 29.4. The elements of a dual soil vapor extraction (SVE)/air sparging system. Air sparging is used to enhance the removal of VOCs within the capillary fringe and beneath the water table. Johnson et al. (1993) suggest that the air-sparge zone is characterized by discrete air flow channels.

As shown in Fig. 29.4, it is generally accepted that air flow occurs through discrete channels and not as bubbles distributed in the porous medium (Johnson et al. 1993). Although the sparge zone may form a parabolic shape centered on the sparge well in coarse and homogeneous sediments, in heterogeneous soils the air flow channels may occupy a zone of irregular radial dimensions (Marley et al. 1992; Ji et al. 1993; Lundegard and LaBrecque 1995).

Evidence of decreasing concentrations of chlorinated solvents and petroleum hydrocarbons in groundwater in proximity to air sparge systems is provided in Grubb and Sitar (1994) and Marley et al. (1992). Tomlinson et al. (1997) provide qualitative evidence for removal of PCE from the subsurface by air sparging, but acknowledge incomplete remediation of a source zone in moderately heterogeneous, fine sands. Within air flow channels, volatilization of contamination to the air phase for transport to the vadose zone occurs rapidly. In areas external to the air channels, which may be large in heterogeneous media, volatilization can only occur after diffusive transport of contaminants from these inaccessible areas to the air flow channels.

29.3.2.3 Soil Flushing

Another approach to the removal of contaminants from the vadose zone is soil flushing. *In situ* soil flushing is undertaken with the intention of accelerating the movement of contaminants through the vadose zone to the water table, to facilitate their ultimate recovery through a groundwater pumping system (Murdoch *et al.* 1990; Lyman and Nomaw 1990). Water, or another treatment fluid, is introduced to the subsurface above the zone of contamination. The mobilization or destruction of the contaminant occurs as the treatment fluid percolates downward through the zone of contamination. Once it reaches the water table, hydraulic containment and collection of the fluid are generally required. The best prospects for successful application of soil flushing are in granular deposits such as sand or gravel. Furthermore, the treatment fluid should be environmentally acceptable and available at reasonable cost. Issues such as hydraulic containment of the treatment fluid, and its treatment once recovered, may add significantly to the cost of soil flushing.

Soil flushing has not been applied widely at the commercial-scale (Marks *et al.* 1994). The main reasons for this are related to the difficulty of predicting and controlling flow in the vadose zone, and the performance of treatment fluids in mobilizing or destroying the contaminants of concern (Murdoch *et al.* 1990). Soil flushing requires direct contact between the treatment fluid and contaminant. The predictability of fluid movement through the vadose zone is poor because of the inherent heterogeneity of unsaturated soils, and the associated variability of hydraulic conductivity characteristics. As demonstrated by Rudolph *et al.* (1996), the density of measurements required to characterize lateral and vertical fluid flow in the vadose zone would be impractical at most contami-

nated sites. It would be difficult, for example, to ensure that contamination did not remain in areas between preferential pathways for vertical flow of the treatment fluid. Furthermore, efforts are often complicated by a limited ability to define the spatial distribution of contaminant sources within the vadose zone and the capillary fringe.

The behavior of the treatment fluid may also be dependent on its composition, and the compositions of both the contaminant and the porous material. Though still complex, the behavior of water being used to flush soluble inorganic salts from the vadose zone is more readily predicted than that of reactive solutions for the flushing of organic contaminants. For reactive treatment fluids such as surfactants, mobility can be decreased through interactions between the fluid and porous media (Lyman and Nomaw 1990). Thus, in formations containing fine-grained silt or clay material, or with organic carbon content approaching 1% or more, soil-flushing performance could be reduced. In their review of the technology, Murdoch *et al.* (1990) indicate that field demonstrations of soil flushing have not achieved the efficiencies of remediation projected on the basis of preliminary laboratory work. This reflects the complexities associated with actual site conditions as a consequence of heterogeneous soil conditions, heterogeneous distribution of contaminants, and mixtures of contaminants with different characteristics.

29.3.3 SOURCE-ZONE RESTORATION IN THE GROUNDWATER ZONE

29.3.3.1 Pumping of Immiscible Fluids

The removal of immiscible, mobile NAPL fluids from the subsurface using pumping to induce gradients in the water and immiscible phases is referred to as waterflooding, and has been used for enhanced oil recovery in the petroleum industry for many years (Gerhard *et al.* 1998). In the remediation of petroleum hydrocarbons and other LNAPL fluids,

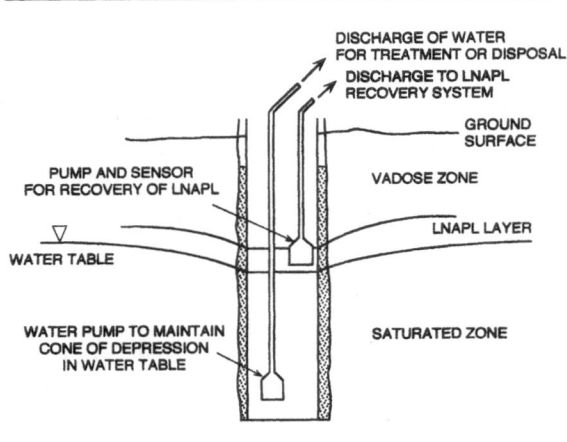

FIGURE 29.5. Typical dual pump system for recovery of LNAPLs such as petroleum hydrocarbons in the subsurface. LNAPL liquid and residual will remain in the subsurface following the removal of the recoverable mobile LNAPL.

the use of free-product recovery pumps in conventional wells or trenches has been adopted for more than 15 years. Dual pumping systems are generally incorporated, one to pump groundwater to create a cone of depression in the water table, and the second to recover the LNAPL as a separate phase (see Fig. 29.5). If removal of mobile LNAPL is successful, residual LNAPL will remain in both the saturated and vadose zones, and may continue to act as a source of dissolved contaminants in groundwater or as a source of contaminant vapors in the vadose zone. In the short term, the concentrations of contaminants in the groundwater and soil plumes may not decrease relative to those present before free-product recovery was undertaken.

Large volumes of DNAPL fluid are not uncommon at coal tar and creosote sites, and free-product recovery by pumping or water-flooding may be feasible (Feenstra *et al.* 1996). At sites with chlorinated solvents, source zones are heterogeneous and direct evidence of DNAPL is rarely encountered in geological cores or monitoring wells. Occasionally, solvent pools have been found in

granular aquifers underlain by aquitards, and recovery of thousands of liters of DNAPL fluid has been possible (Cherry *et al.* 1996). While recovery of free-product by water-flooding using wells or horizontal drains reduces the prospects for future remobilization of DNAPL in the source zone, DNAPL residual will remain. Cherry *et al.* (1996) suggest that as much as two-thirds of the contaminant mass might be removed through free-product recovery under the most favorable circumstances. Sale & Applegate (1997) project recovery of as much as 95% of the free-product associated with an extensive pool of creosote-based wood treatment oil at the base of a sand and gravel aquifer. The waterflood system employs a comprehensive network of parallel injection and recovery wells.

29.3.3.2 *In situ* Mass Removal from Source Zones: Chemical Flushing

Although often successful for plume control and remediation, pump-and-treat technology has been ineffective for source-zone restoration, in particular for sources containing NAPL fluids and residual (Mackay & Cherry 1989; NRC 1994; Cherry *et al.* 1996). The failure of pump-and-treat to achieve remediation of contaminant source zones in practical time frames has prompted research and development of alternate source-zone remedial techniques. These techniques either enhance the mass removal efficiencies of flushing by groundwater, or destroy the contaminant mass *in situ* (Cherry *et al.* 1996). Although these technologies have generally originated in other disciplines such as petroleum engineering or water treatment, their applications and objectives in subsurface environments are new. While significant progress has been made, these are currently emerging rather than proven technologies, and their performance in field applications remains to be confirmed.

In recent years there has been a concerted effort to adapt chemical flushing technologies used to enhance petroleum recovery, to the remediation of subsurface contaminants. The two most common approaches to chemical flushing involve the use of surfactants and co-solvents for enhanced removal of NAPL sources. Although there have been in excess of 20 field demonstrations using surfactants and co-solvents, commercial-scale applications are not yet common (Simpkin *et al.* 1999).

Chemical flushing is achieved by inducing the flow of water containing chemical additives through the contaminant source zone using a system of injection and withdrawal wells (see Fig. 29.6). The water withdrawn from the subsurface requires some form of treatment prior to its recirculation or disposal. Surfactants, or surface active agents, typically act to increase the aqueous solubility of a NAPL, and decrease the NAPL–water interfacial tension, which serves to increase the mobility of the NAPL. Accelerated disso-

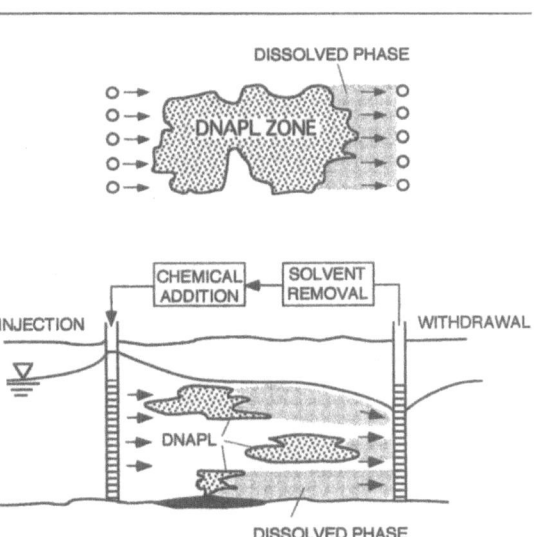

FIGURE 29.6. Chemical flushing for the enhanced removal of NAPL case (reproduced from Cherry *et al.* 1996 with permission). The system includes injection wells for the remedial fluids, and withdrawal wells for the recovery of contaminants and remedial fluids. A forced gradient is maintained between the injection and withdrawal wells.

lution, physical mobilization or both can be enhanced by the surfactant selection process (Simkin *et al.* 1999). Alcohols are mutually miscible in both water and NAPL and like surfactants, they can be used to enhance solubility, mobility, or both, of NAPLs in groundwater. At low concentrations, on the order of 5% alcohol or less, solubility enhancement is achieved and remediation would involve flushing of the source zone with many pore volumes of the alcohol–water mixture. At moderate to high concentrations, the alcohol mixes with the NAPL, resulting in a more mobile NAPL with modified physical and chemical characteristics (Simkin *et al.* 1999). Hydraulic containment and recovery of the flushing and modified NAPL fluid is important to avoid inadvertent spread of the contaminated zone.

The performance of chemical flushing technologies is limited by geological heterogeneity and non-uniform NAPL contaminant distribution. A chemical flush generally removes contaminant mass initially from active zones of flushing, such as coarse granular layers or lenses in porous media, or fractures in fractured porous media. Subsequent to this initial mass removal, contaminants move by diffusion from inactive flow zones within low permeability layers or matrix blocks in porous and fractured porous media, respectively, to the active flow pathways, resulting in the slow but continual release of contaminants to the flushing fluid. The use of polymers or foams in the chemical flushing solutions to improve sweep efficiency in heterogeneous settings has been proposed, but research and applications in this area remain in their preliminary stages (Simkin *et al.* 1999). Processes such as sorption and biodegradation of surfactants in the subsurface may detract from performance of chemical flushing methods.

Fountain *et al.* (1996) conducted a highly controlled flush within a containment cell in a fine to medium sand aquifer using an anionic surfactant mixture to enhance dissolution of a PCE source. Although considerable contaminant mass was removed by flushing more than 20 pore volumes of the surfactant solution through the cell, remnant PCE would quite probably have contributed to the development of dissolved contaminant concentrations in excess of drinking water guidelines in groundwater within the cell in the post-flush period. Costs, and the technical difficulties and uncertainty concerning the degree of success associated with chemical flushing identified by the trials of Fountain *et al.* and others, as documented in Simkin *et al.* (1999), are an impediment to full-scale application of the technology at actual sites.

29.3.3.3 Thermal Flushing Techniques

The petroleum industry has used various thermal flushing techniques to enhance recovery of oil for several decades, but it has only been within the last decade that similar techniques have been applied to subsurface restoration involving organic contaminants (Udell 1997; Sleep 1996). There is a fundamental difference in the objectives of oil-recovery and environmental applications. In subsurface restoration, complete removal of the NAPL source is desired, whereas in the petroleum industry, increased recovery is the objective with minor regard for the amount remaining in the formation. Steam injection is the most common thermal flushing technique, but other approaches such as electrical and radio frequency heating have been explored and may have utility in low permeability media (Udell 1996).

Steam injection is applicable to both LNAPLs and DNAPLs, and is generally operated in combination with water and vapor recovery systems. In steam flushing, steam is injected continuously into the subsurface. Initially the steam condenses, but causes temperatures of soil and fluids to increase. With time a steam front develops, and sweeps away from the injection point. Fluids, including volatile organic contaminants, in the subsurface are volatilized and condense immediately ahead of the steam front. Thus, NAPL that may have been present as immobile residual, may condense as a mobile liquid ahead of the steam front. Ideally this mobile NAPL phase would be recovered through a recovery system, but if capture is incomplete, the contamination could be spread laterally, or to greater depths in the groundwater system if DNAPLs are present. While steam injection enhances potential removal of volatile and semi-volatile NAPL contaminants both below and above the water table, a possible drawback of the technology is the tendency for the steam front to override the contamination as a consequence of the lower density of steam relative to that of water. Other important issues involve the design, operation and maintenance of equipment associated with steam generation and the above-ground handling, treatment or disposal of recovered vapors and fluids (Sleep 1996).

Thermally enhanced subsurface remediation is a recent development, and has no record of commercial application. On the basis of experimental trials, Sleep (1996) reported improved efficiencies for removal of volatile contaminants by steam flushing in sandy soils in comparison to fine-grained and organic-rich soils. Udell (1996) describes two field demonstrations employing thermal enhancements for the removal of solvents and petroleum hydrocarbons from the subsurface. Although the demonstrations achieved considerable mass removal from permeable zones, contaminants remained within, and in proximity to, fine-grained layers. Documentation was not sufficient to determine the degree of cleanup, but clearly full restoration of the sections of the source zones in which the trials were conducted was not achieved.

29.3.3.4 *In situ* Mass Destruction Technologies

In contrast to the use of surfactants or co-solvents for enhanced removal of contaminants from the subsurface, which requires above-ground treatment or disposal of extracted fluids, *in situ* mass destruction technologies employ injected solutions or materials to promote chemical reactions with the contaminants, without the productoin of toxic degradation products. In addition to active flushing through source zones, it may also be possible to use *in situ* destruction techniques in a semi-passive multiple injection approach (Cherry *et al.* 1996). Experience thus far remains restricted to laboratory and small field-scale trials. The technology faces the same obstacles as chemical flushing with respect to the access of the treatment solution to a heterogeneously distributed source in a heterogeneous geological setting.

Schnarr *et al.* (1998) used potassium permanganate in a series of laboratory and field experiments to degrade NAPL sources containing PCE and TCE. Degradation of the chlorinated alkenes occurs by oxidation, resulting in the formation of carbon dioxide, chloride and manganese dioxide. In two field experiments, conducted within a sheet piling enclosure, in excess of 90% of a homogeneous emplaced source of PCE and 60% of a heterogeneous injected source of PCE and TCE were oxidized *in situ*. While the results suggest promise for the technology, the trial remediation of the heterogeneous source indicates that geological heterogeneity may be difficult to overcome, and implies the need to tailor the design of the injection–extraction system to the configuration of the source. This approach to oxidation is applicable to the chlorinated alkene compounds only, and would not be applicable mixed-solvent sources. Furthermore, the consumption of the oxidant by natural organics in the aquifer in addition to the target compounds, and the precipitation of manganese dioxide in the aquifer may also detract from the performance of the technology.

Other approaches to *in situ* destruction of sources are at an early stage of development. On the basis of laboratory experiments (Le-

sage *et al.* 1996), a small-scale field trial of biochemical reduction of a solvent source was undertaken. The treatment solutions were expensive, and the highly reducing conditions required for the reaction could not be easily maintained. Barbaro *et al.* (1997) attempted to promote the biodegradation of petroleum hydrocarbon sources through the addition of solutions amended with oxygen and nitrate, but achieved little mass destruction over a period of months. Cherry *et al.* (1996) indicate that biodegradation has also not yet proven to be a one-step remedial action for chlorinated solvent sources. Extending the use of granular iron for treatment of dissolved plumes, as described in Gillham and O'Hannesin (1994), introduction of granular iron into source zones through soil mixing has been proposed for *in situ* destruction of DNAPLs (Gillham *et al.* 1998). Although in the early stages of development, this approach would avoid the need for flushing, and through the mixing process could overcome difficulties presented by geological heterogeneity.

29.4 Source-Zone Isolation and Containment

29.4.1 SOURCE-ZONE ENCLOSURES

In the absence of the technical capability or financial resources to fully restore contaminant source zones, control of contamination emanating from subsurface sources is often an essential component of environmental management of contaminated sites (Freeze and McWhorter 1997). As shown on Fig. 29.2, source-zone containment can be achieved through interim or long-term isolation within an enclosure constructed of low permeability cut-off walls or barriers, long-term hydraulic containment using an active pump-and-treat system, or *in situ* interception and treatment of contaminated groundwater originating from the source by passive

TABLE 29.1. Description and characteristics of common low permeability barrier technologies[a]

Vertical barrier	Materials	Thickness (m)	Hydraulic conductivity (m s^{-1})	Permittivity[b] (s^{-1})	Construction/installation	Comments
Slurry trench	Soil/bentonite Soil/cement Soil/attapulgite Cement/slag Cement/fly ash	0.5–1.5	1×10^{-9} 1×10^{-8}	$0.7–2 \times 10^{-9}$ $0.7–2 \times 10^{-8}$	Trench excavation and replacement of soil with slurry Backhoe Backhoe/clamshell	Depths of 45 m Removal and disposal of soil may be important issue Difficult in bouldery and bedrock terrain
Soil mixing	Bentonite Cement	1–2	1×10^{-9}	$0.5–1 \times 10^{-9}$	Deep mixing of active soil and grout with augers Overlapping columns	Depths of more than 30 m Minimal removal and disposal of soil Construction possible within confined areas Some native soils may not be suitable
Vibrated beam	Bentonite Cement	0.025–0.1	1×10^{-9}	$1–4 \times 10^{-8}$	Driven H-beam, void filled with grout as beam withdrawn Overlap	Depths of less than 30 m Minimal removal of soil, fluids for disposal Joining of adjacent injections difficult to confirm
Jet grouting	Bentonite Cement Chemical grouts	1–2	1×10^{-8}	$0.5–1 \times 10^{-8}$	Borehole injections of grouts to subsurface Overlapping treated zones or columns	Influence of geological heterogeneity on injection Overlapping of adjacent columns difficult to verify Depth advantages

Barrier type	Materials	Thickness	Hydraulic conductivity	Permittivity[b]	Method of installation	Limitations
Geomembranes	PVC (polyvinyl chloride) HDPE (high-density polyethylene) Others	0.001–0.002	1×10^{-14}	0.5–1×10^{-11}	Thin polymer sheets or panels Installation in slurry trench or as panels within a steel-driving apparatus, which is subsequently removed Adjacent sheets/panels sealed at joints	Removal and disposal of soil Expense Diffusion of organics through polymer may be concern
Conventional steel sheet piling	Steel sheets Some pre-coated epoxy joints	0.005–0.015	1×10^{-7}	0.5–2×10^{-5}	Direct driving using hydraulic-push or vibratory hammers Sheet piles displace soil	Depths of 40 m Strength/confined areas/little waste Difficult in bouldery and bedrock terrain
Sealable-joint steel sheet piling	Steel sheets Grout sealants for joints (bentonite) Cement, modified Cement, chemical polymer	0.75–0.01	1×10^{-12}–5×10^{-11}	1×10^{-14}–1×10^{-8}	Direct driving using hydraulic-push or vibrating hammers Sheet piles displace soil Flushing and sealing of joints following installation	Depths of 40 m Strength/confined areas/little waste Cavity/seal QA/QC Removable Difficult in bouldery and bedrock terrain

[a] Data from Cherry et al. 1996; Mitchell and van Court 1997; Mutch et al. 1997; Starr and Cherry 1990.
[b] Permittivity is the ratio of the hydraulic conductivity and the thickness of a barrier.

reactive barriers (Cherry *et al.* 1996). Barrier installations have generally been restricted to depths of less than 50 m, and may have limitations in difficult geological settings such as bouldery terrain or fractured bedrock.

Barriers have been used to provide hydraulic control for civil engineering projects for decades, but their adaptation for the control and isolation of contamination in the subsurface has only occurred within the past two decades (Starr & Cherry 1992; Mutch *et al.* 1997; Cherry *et al.* 1996). The conventional design of enclosures is to achieve low permeability to minimize water flow and the advective transport of contaminants through the barrier wall. Rowe (1996), Mitchell & van Court (1997) and Devlin & Parker (1996), however, indicate that the potential contaminant storage within and movement through barriers as a consequence of diffusion may be important considerations in the design of containment enclosures. The analysis by Devlin & Parker, which invokes controlled pumping within a containment enclosure, suggests that the outward diffusive flux of contaminants can be countered by the inward advective flux of water through barriers such as slurry walls or sealable-joint sheet piling of relatively low hydraulic conductivity. Rowe (1996), however, indicates that, for typical slurry walls, a significant head difference may be required to achieve this objective. For barriers of exceptionally low hydraulic conductivity, such as plastic geomembranes, it is impractical to develop the necessary hydraulic gradients to induce the sufficient inward advective flux of water to overcome the outward diffusive flux of organic contaminants through the plastic polymer.

The hydraulic performance or degree of containment provided by an enclosure can be optimized in situations where the base of the barrier can be keyed into an underlying aquitard or low permeability layer. Where such geological conditions do not exist, or where the depth to an aquitard is too great,

the enclosure walls could be terminated within the aquifer. Cherry *et al.* (1996) refer to this situation as a hanging enclosure, and suggest that containment of contaminants can be achieved if the barrier walls extend to a depth below the base of the source zone. To maintain hydraulic containment in a hanging enclosure, higher rates of groundwater pumping from within the enclosure would be necessary than for a keyed enclosure.

The major types of barriers are summarized in Table 29.1 and include: slurry trench, geomembrane in slurry trench, vibrated beam walls, jet grout walls, auger mix walls, conventional steel sheet piling and sealable-joint steel sheet piling. Each type of barrier has its advantages and disadvantages, and the performance of a particular type will be dependent on specific site and project conditions. Permittivity incorporates the ratio of hydraulic conductivity to barrier thickness for comparison of the flux of groundwater through different barriers of typical thickness for a constant drop in head across the barrier (Starr & Cherry 1990).

Physiochemical interactions between barrier materials and contaminants may influence the hydraulic performance of source-zone containment enclosures. Middleton & Cherry (1996) suggest that there is unlikely to be significant increases in hydraulic conductivity of natural and compacted clay barriers as a consequence of exposure to chlorinated solvents or petroleum hydrocarbons. Interaction between clays and water-miscible compounds such as co-solvents and surfactants may cause shrinkage of clay minerals. This could induce increases in hydraulic conductivity, but this would be minimized at depth under adequate confining stresses. Both Middleton & Cherry (1996) and Mitchell & van Court (1997) suggest that the soil–bentonite in barriers, with typically high water content and porosity, may be more susceptible to shrinkage and cracking than compacted natural clays in the presence of water-miscible and low solubility organic fluids. No examples of barrier failure at field sites were documented. Mitchell & van Court (1997) note that interactions with low pH or high electrolyte groundwaters may increase the hydraulic conductivity of clay barriers, and may also increase degradation of steel sheet-piling barriers. Polymer materials may also be subject to degradative reactions with organic fluids such as aliphatic and aromatic hydrocarbons and chlorinated and petroleum solvents, and their performance in some barrier applications may be questionable (Mitchell & van Court 1997).

29.4.2 COVERS AND BOTTOM BARRIERS

The degree of isolation provided by an enclosure with vertical barriers can be increased, in some circumstances, through the use of emplaced covers. The covers may serve several functions including reduction of infiltration, control of vapor releases, and enhanced isolation of the contaminated subsurface from the biosphere. The use of covers is described in detail in Chapter 27. The cost, required maintenance and longevity of covers relative to the function they serve in improving the isolation of the contaminated material need to be considered in the decision to incorporate covers with enclosures.

In some situations, covers may serve a geochemical control function in the isolation of source zones. For example, the interaction of oxygen with sulphidic-rich mining waste-rock and tailings wastes contributes to the development of acid-mine drainage problems. Incorporation of oxygen-consuming (Reardon & Moddle 1985) or moisture-retaining materials (Nicholson *et al.* 1989) in covers has been suggested as a remedial control for the management of mine waste areas (see also Section 27.4).

Engineered liners serve as floors and bottom barriers in the control and isolation of contaminants at waste management facilities such as landfills (see Section 25.3). Their con-

struction and use for remedial control at existing contaminated sites, however, has not been common. Although technologies for emplacement of bottom barriers have improved in recent years, they are expensive and verification of their emplacement and performance is difficult (Peterson & Landis 1995).

29.5 *In Situ* Plume Control and Remediation

29.5.1 PUMP-AND-TREAT REMEDIATION

Pump-and-treat technology can rarely achieve full restoration of groundwater systems in the vicinity of contaminated sites within reasonable time frames. NRC (1994), in a review of 77 major pump-and-treat systems in the USA, indicated that clean-up goals had been realized at only eight sites, none of which involved complex hydrogeological settings or DNAPL fluids associated with the contaminant source area. Pump-and-treat systems can be used to remove contaminant mass from a plume, causing the plume to shrink toward its source (Mackay & Cherry 1989; NRC 1994; Cherry *et al.* 1996). Thus, it may be possible to control risks to potential receptors arising from dissolved contaminants in plumes. For control of plumes at depths of 30–50 m or more, pump-and-treat may be the most viable and cost-effective technology. The institutional controls, energy consumption and costs of the treatment of the recovered water prior to its discharge or re-injection are long-term liabilities of the technology. In the absence of specific remediation or isolation of source zones, pump-and-treat operations could be required for decades or more in the vicinity of the source.

Figure 29.7 illustrates the conceptual effluent concentrations with time for an array of pumping wells in a pump-and-treat system associated with a plume emanating from a long-term source of contamination below the

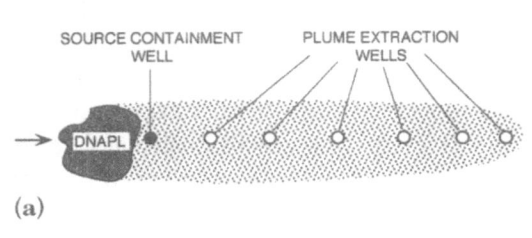

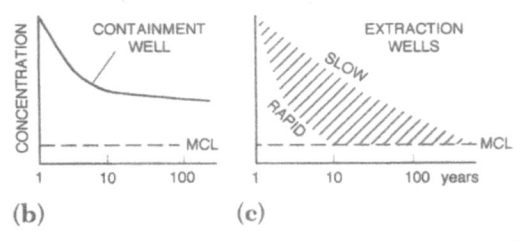

FIGURE 29.7. The conceptual remediation of a source zone and associated plume by pump-and-treat. While remediation of the plume may be achievable over periods of years to decades, the prospects for remediation of the source zone by pump-and-treat are not promising (reproduced from Cherry *et al.* 1996 with permission): (a) plan view; (b) with time at the containment wells; and (c) extraction wells.

water table. For all wells, there is a rapid decline in effluent concentrations immediately following the onset of pumping, followed by a long period of gradually declining concentrations. The quality of the effluent from the source-containment well is influenced by the groundwater quality close to the source, and the effect of dilution by uncontaminated groundwater drawn from surrounding regions as pumping proceeds. The effluent does not, in a practical time frame, attain a quality satisfying environmental guidelines, a direct consequence of continued release of contaminants from the source zone. For the plume extraction wells, contaminant concentrations may decrease to acceptable levels in time frames extending from years to decades or longer. Decreasing contaminant concentrations in effluent from plume extraction wells are a consequence of gradual removal of contaminants from the plume in combina-

tion with the introduction of uncontaminated groundwater from adjacent regions of the aquifer systems.

During the period of plume development, the advective transport of most dissolved contaminants is retarded relative to the rate of groundwater flow because of sorptive interactions between the contaminants and aquifer solids. Furthermore, as a consequence of chemical gradients, contaminants invade low permeability lenses, layers or matrix blocks by diffusion adjacent to active flow zones or fractures. With the onset of pumping, the permeable and active zones of groundwater flow within the plume area may be flushed relatively quickly. The gradual release of contaminants by diffusion and desorption, and retardation of their transport to the pumping well, may continue to contaminate groundwater in the active pathways of flow to unacceptable levels for long periods of time. Wood (1996) suggests that the diffusive release of contaminants from the framework and matrix will continue to provide above-background concentrations to groundwater for a period of at least twice the residence time of the contaminants in the aquifer. Even with optimal placement of pumping wells, the gradual release of contaminants by diffusion and desorption causes the volume of groundwater that must be extracted by pumping to far exceed the volume initially occupied by the plume.

Rivett (1993) used a pump-and-treat system to remediate the leading section of a shallow and narrow plume emanating from a source of TCE, PCE and chloroform in a relatively homogeneous sand aquifer at CFB Borden. With peak concentrations ranging from several tens to more than 100 mg l^{-1} in the core, the plume was remediated to concentrations of several micrograms per liter within a period of approximately one year. The amount of groundwater capture was approximately 60 times the original volume of the plume.

The NRC (1994) review of operating systems indicates that pump-and-treat technology has proven to be successful in arresting the advance of contaminant plumes, even in complex hydrogeological settings, including multiple aquifer–aquitard and fractured bedrock environments. Pump-and-treat technology has not, however, been successful in achieving complete restoration of the subsurface. In the absence of specific source-zone remediation or containment, pump-and-treat systems may need to continue to operate for decades or, in some circumstances, centuries to maintain plume control. Furthermore, the energy consumption and costs for long-term operations and maintenance associated with above-ground treatment or disposal of contaminated groundwater, and the potential for wasting uncontaminated groundwater resources, serve to detract from the utility of pump-and-treat systems.

29.5.2 PERMEABLE REACTIVE BARRIERS

29.5.2.1 General Description of Barriers

An alternative which is increasingly being considered for plume control and management at many contaminated sites involves the interception and *in situ* treatment of contaminated groundwater by permeable reactive barriers (Gillham & O'Hannesin 1992; Cherry *et al.* 1996; Cherry 1996; Shoemaker *et al.* 1995; Gillham & Burris 1997; Powell *et al.* 1998).

Permeable reactive barriers have two essential functions. The barrier must facilitate the interception or capture of a contaminant plume at some distance downgradient of the source, and provide treatment or removal of contaminants to acceptable levels. Treatment is achieved within or downgradient of the barrier by one or a combination of physical, chemical or biological processes. The flux of contaminants in groundwater and the efficiency of the proposed treatment technology will influence design parameters such as the thickness and configuration of the permeable barrier system. For example, a minimum residence time for the contaminated groundwater within the treatment media of the barrier is generally required to achieve the appropriate level of treatment. Potential treatment technologies for application in reactive barriers have been identified for a wide range of organic and inorganic contaminants (Shoemaker *et al.* 1995; Gillham & Burris 1997).

In situ treatment zones, created through the injection of chemicals to modify the reactivity of aquifer materials, have been proposed as a method for controlling contaminant migration in groundwater. Brown & Burris (1996), for example, explored the use of cationic surfactants to increase the sorptive retention of chlorinated solvents on aquifer solids. Fruchter *et al.* (1997) have undertaken a trial involving the injection of a sodium dithionite solution to an aquifer at the Hanford site in Washington: to reduce natural iron minerals in the aquifer, and create an anoxic zone where reduction and precipitation of electroactive metals such as chromium will occur. The distribution of solutions and the overall dimensions of the treatment zone are not easily controlled by injection in heterogeneous environments.

The remaining forms of permeable reactive barriers involve the emplacement of treatment materials in the subsurface (see Fig. 29.8). In a permeable wall, the treatment material is emplaced across the entire width and depth of a plume. The wall may consist of material that replaces the native aquifer materials, or it may be a mixture of treatment and native materials. Possible emplacement techniques include open trenching, polymer slurry trenching, overlapping caissons, vertical hydrofracturing, jetting or soil mixing, but most experience to date has been achieved with open-trench methods (Shoemaker *et al.* 1995). The hydraulic conductivity of the wall should be equal to or greater than that of the surrounding aquifer.

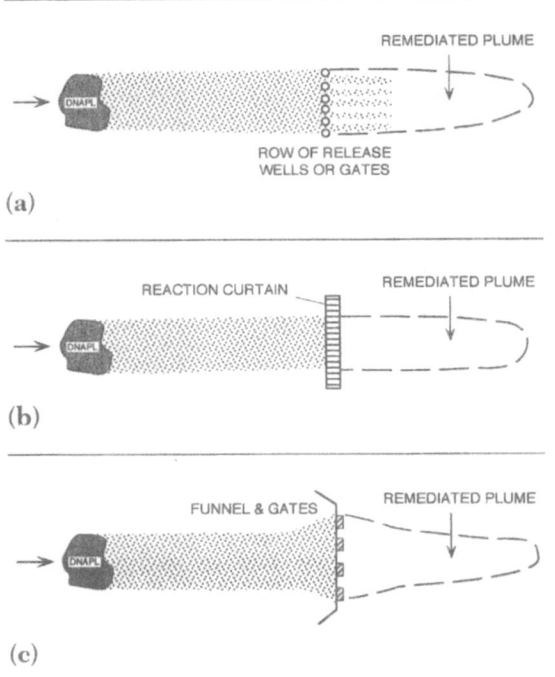

FIGURE 29.8. Possible options for the interception and treatment of contaminant plumes by permeable reactive barriers. Options include: (a) rows of unpumped wells; (b) permeable reactive walls; or (c) funnel-and-gate systems case (reproduced from Cherry *et al.* 1996 with permission).

A second version of the permeable reactive barrier is the funnel-and-gate system (Starr & Cherry 1994). Vertical cut-off walls or barriers of low hydraulic conductivity are installed across a plume to funnel or direct the flow of contaminated groundwater through permeable gates or chambers, in which treatment occurs. The funnels can be installed using any of a number of technologies for the construction of low permeability barriers. The treatment gates can be installed using modifications of the open trenching, polymer slurry trenching or caisson methods applicable to the installation of permeable walls. Gates can be single-chambered, for one-step treatment, or multichambered for situations in which sequential treatment steps may be required to achieve remedial goals (Cherry *et al.* 1996).

A third version of a permeable reactive barrier, which may provide advantages at depth is single or multiple rows of unpumped wells (Wilson & Mackay 1995; Wilson *et al.* 1997). Interception of the plume is achieved through the cumulative effects of convergence of groundwater flow to open wells from the upgradient direction, and subsequent divergent groundwater flow downgradient of the wells. Reactants or nutrients, which induce or enhance degradation reactions, are added to the groundwater in the wells, such that the degradation reactions occur within or downgradient of the wells.

29.5.2.2 Hydraulic Performance of Permeable Barrier Systems

On the basis of numerical simulations for simple hydrogeological systems, Smyth *et al.* (1997) showed that the hydraulic capture for permeable reactive walls, and by inference, passive well systems, appeared to be more favorable than that for funnel-and-gate systems. If the hydraulic conductivity of the treatment material within a permeable reactive wall can be maintained at or in excess of that of the surrounding aquifer, there will be no diversion of groundwater flow around the wall. Thus, the length and depth dimensions of the permeable wall could be established on the basis of width and depth of the plume. The length and depth dimensions for funnel-and-gate systems, however, may need to exceed the width and depth dimensions of the plume (Starr & Cherry 1994; Shikaze *et al.* 1995). Some component of groundwater flow will be diverted around the ends of a funnel-and-gate system, which fully penetrates an aquifer and is keyed into an underlying aquitard, or will be diverted around and beneath a partially penetrating or hanging system in a surficial aquifer (see Fig. 29.9). To ensure the capture of typical plumes, it is likely that funnel-and-gate systems may require several gates across the width of the plume, and need to extend to depths below the base of the

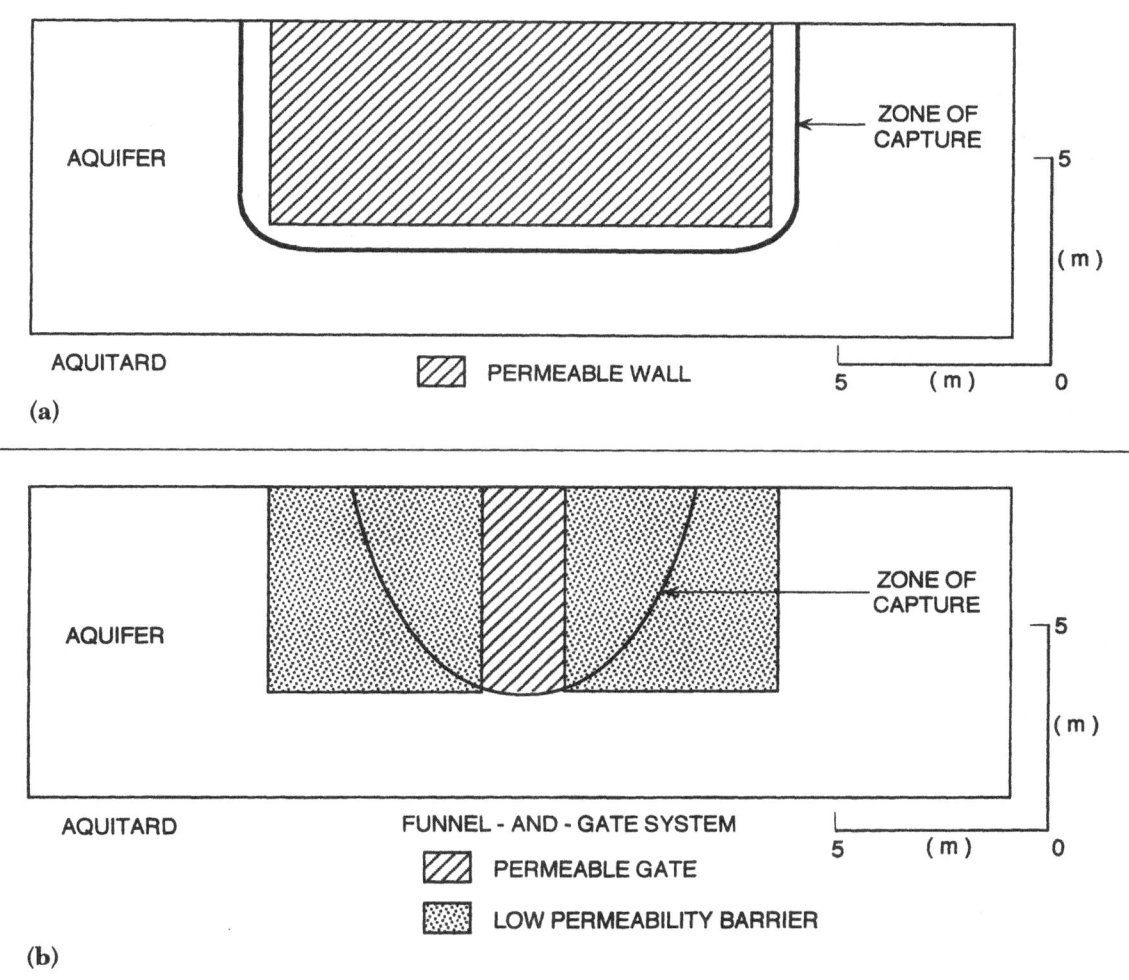

FIGURE 29.9. Sectional views of groundwater flow through permeable reactive barriers. The barrier systems are installed perpendicular to the horizontal groundwater flow direction, and partially penetrate a surficial granular aquifer. For a permeable reactive wall in (a), the zone of capture for groundwater approaching the barrier is slightly greater than the dimensions of the reactive wall if the hydraulic conductivity of the reactive medium exceeds that of the aquifer. In (b), for a funnel-and-gate system with the same reactive medium, the zone of capture is larger than the dimensions of the treatment gate, but some groundwater flow is diverted around the edges and beneath the low permeability funnel walls.

plume at sites where the depth of the plume is not constrained by an underlying aquitard. The flux of groundwater and contaminants through a permeable reactive wall is similar to that through the aquifer. The convergence of flow created by a funnel-and-gate system causes the flux of groundwater and contami-nants through the treatment gates to increase by several-fold over that in the aquifer. Thus, to maintain necessary residence times of con-taminated groundwater within treatment me-dia, it can be anticipated that the thickness of the treatment zones in the direction of groundwater flow will need to be greater for funnel-and-gate systems than for permeable walls.

The performance of permeable reactive barriers in demonstrations and commercial applications has largely been inferred from groundwater quality monitoring; i.e. satisfactory water quality downgradient of a reactive barrier reflects satisfactory performance of the barrier. The hydraulic performance of the barrier with respect to the flow rates through the barrier, subtle variations of hydraulic head in the vicinity of the barrier, or zones of capture has proven to be difficult to measure directly.

29.5.2.3 In situ Treatment Technologies for Contaminated Groundwater

In situ treatment of contaminated groundwater may be achieved by one or a combination of physical, chemical or biological processes within or downgradient of permeable reactive barriers. The greatest advantage of these systems relative to other plume control technologies arises where treatment proceeds passively, or without the continual need for energy inputs, maintenance or replacement of treatment media, and results in the transformation of contaminants to innocuous and stable degradation products. In many circumstances, completely passive treatment may not be feasible, so semi-passive approaches requiring some maintenance, and operational supervision may be implemented.

Pankow et al. (1993) describe the physical treatment of plumes containing VOCs, such as petroleum hydrocarbons and chlorinated solvents, by air sparging in permeable walls or gates. Air is sparged through wells to the base of the treatment wall or gate and, as the resultant bubbles rise toward the water table, the volatile contaminants in the groundwater partition to the vapor phase and their concentrations in the water phase decrease. Collection and treatment of the gas at the water table would normally be required.

A second example of physical treatment of dissolved contamination is a sorptive barrier. This requires the emplacement of a treatment medium in the ground, but designs should incorporate an allowance for periodic replacement. O'Brien & Keyes (1997) implemented groundwater treatment in a modified funnel-and-gate system using activated carbon, which will require replacement at four-year intervals, to remove several organic contaminants including petroleum hydrocarbons, chlorinated solvents and chlorophenols by sorption. Zeolites, which are hydrated aluminosilicates, have also been considered as a sorptive medium. Clinoptilolite has traditionally been used in above-ground operations to remove two radionuclides, cesium and strontium, from water, and has been proposed for in situ treatment of contaminated groundwater (Fuhrmann et al. 1995).

In contrast to physical treatment options, chemical and biological processes offer the potential to achieve in situ treatment through degradation, transformation or precipitation reactions, and avoid the need for aboveground collection or handling of waste materials. Although there are now a range of in situ chemical treatment technologies for a variety of inorganic and organic contaminants, an early initiative in the development of permeable reactive barriers was the use of zero-valent iron for the degradation of many of the common chlorinated solvents (Gillham & O'Hannesin 1992, 1994). It is generally agreed that the mechanism for degradation is abiotic reductive dechlorination of the solvent. The reaction appears to be pseudofirst-order with respect to the concentration of the organic, and results in the generation of dissolved chloride, iron and innocuous organic compounds (Orth & Gillham 1996; Matheson & Tratnyek 1994).

Within the past five years, iron-based reactive barriers have been installed at more than 30 sites primarily in the USA, for in situ treatment of chlorinated solvents in groundwater. O'Hannesin & Gillham (1998) demonstrated that the treatment performance of a reactive wall at CFB Borden, containing a 22:78 iron–sand mixture, did not diminish during five years of continuous operation. The wall continued to achieve degradation of 85–90% of TCE and PCE, which were present in the plume at concentrations of as much as 268 and 58 mg l^{-1}, respectively. Although there was evidence of some inorganic mineral precipitation near the upgradient interface of the barrier, coating of reactive surfaces or plugging of pore spaces did not create observable negative impacts on barrier performance. Focht *et al.* (1996) described three commercial examples in which reactive walls or funnel-and-gate systems were installed. Each achieved remediation of contaminated groundwater to regulatory objectives, which generally were based on drinking-water quality guidelines. Projects to date have generally occurred at depths of less than 15 m, but Hubble *et al.* (1997) explored potential barrier installation techniques at depths in excess of 30 m.

Blowes *et al.* (1995) described potential applications of permeable reactive barriers for the removal of metals from groundwaters. Two mechanisms for removal are suggested. In the first, the contaminant is converted to a lower oxidation state, and is removed from solution by precipitation or co-precipitation reactions. In the second, the oxidation state of another constituent associated with the mobile metal of concern is changed by the reactive barrier, resulting in the co-precipitation of the metal.

Blowes *et al.* (1997) installed a permeable wall consisting of granular iron to remove hexavalent chromium (Cr^{VI}) and TCE from a contaminant plume at a coastal site in North Carolina. The barrier, which was installed using a continuous trencher, was 45 m in length and 7.5 m in depth. Cr^{VI} concentrations decrease from in excess of 6 to less than 0.01 mg l^{-1} as a consequence of chemical reduction and the precipitation of low solutibility

Cr^{III} oxyhydroxides. TCE concentrations decrease from in excess of 5 to less than 0.005 mg l^{-1} by reductive dechlorination as groundwater flows through the barrier.

In an example of indirect treatment of a contaminant, Benner *et al.* (1997) describe the application of a permeable reactive barrier to remove metals from a plume emanating from sulphidic mine tailings in the Sudbury, Ontario, area. Typical of acid-mine drainage, the contaminant plume contains elevated concentrations of sulphate, iron and other metals. Treatment is achieved by the promotion of sulphate reduction and subsequent co-precipitation of iron and other metal-sulphide mineral phases. The sulphate-reduction process is microbially mediated, and is initiated through the addition to groundwater of carbon for the sulphate-reducing bacteria in the permeable wall. Initial indications are that more than 90% of the soluble iron is removed and alkalinity increases ten-fold as the groundwater passes through the reactive wall. The reactive wall has effectively removed the capacity of groundwater to generate acidity following discharge to receiving surface water. It has been estimated that the wall will function passively for at least 15 years. In another example, microbially mediated sulphate reduction and metal-sulphide precipitation has been investigated as a possible mechanism for removing dissolved uranium from mine-waste waters in New Mexico (Thombre *et al.* 1997).

Robertson & Cherry (1995, 1997) have used various configurations of reactive barriers to promote the removal of nitrate from groundwater, landfill leachate, effluent from septic systems and discharge from agricultural drains. The barrier types have included reactive walls for groundwater plumes, horizontal layers beneath septic beds or reactive chambers. The intent of the remediation is to add carbon to the water to act as an energy source for denitrifying bacteria. In trials to date, 50–80% attenuation of nitrate, at influent concentrations of as much as 170 mg l^{-1} (NO_3-N), has been achieved, and evidence suggests that the reactive material could function effectively for several decades. Baker *et al.* (1997) demonstrated the use of a mixture of alkaline metal oxides in a reactive barrier to remove more than 85% of the phosphate associated with a septic-system plume at a rural school property. The mechanisms for removal include adsorption and mineral precipitation.

Permeable barriers may also be used to enhance the bioremediation of organic contaminants by native microbial populations in groundwater systems. Conventional approaches to enhanced bioremediation have involved the injection of limiting chemicals or nutrients to the subsurface through wells. The efforts, however, have been frustrated by the difficulty to achieve mixing between the contaminated groundwater and the injected solutions (Gillham & Burris 1997). Permeable barriers provide a mechanism to overcome this limitation. To date, the approach has been to stimulate native microbial activity by modifying geochemical conditions rather than relying on the introduction of isolated or engineered organisms.

Devlin & Barker (1996) developed a semi-passive scheme for adding nutrients to a contaminated plume in a permeable wall, and promoting the development of conditions favorable for biodegradation of an organic contaminant downgradient of the wall. In their field trial, anaerobic conditions were created through the addition of an acetate solution to promote the biodegradation of selected chlorinated solvents in groundwater by native microbial populations.

Many petroleum and other hydrocarbons biodegrade readily in groundwater under aerobic conditions. Bowles *et al.* (1995) used air sparging in a funnel-and-gate system to increase dissolved oxygen concentrations to several milligrams per liter to promote the biodegradation of petroleum hydrocarbons in a fractured clay–till environment. This approach was successful because groundwater flow rates, contaminant concentrations and oxygen demands were relatively low. Borden *et al.* (1997b) and Chapman *et al.* (1997) used a proprietary magnesium peroxide powder in barriers consisting of arrays of unpumped wells to release oxygen to plumes containing elevated concentrations of gasoline constituents. While the trials showed promise and partial treatment of the plumes was achieved, BTEX concentrations did not decrease to typical regulatory levels such as drinking-water standards. The limited success of the trials could be attributed to insufficient characterizations of the plumes prior to the trials, an insufficient number of wells within the barrier systems, decreasing oxygen-release rates within several weeks of installation of the peroxide powder, some evidence of biofouling of the wells and oxygen sources and oxygen demands from other organic and inorganic constituents in groundwater. In further work, Wilson *et al.* (1997), for example, explored use of diffusive emitter tubes as an approach to introducing oxygen at elevated concentrations to groundwater in permeable barrier systems.

29.6 Summary

Restoration of contaminated soil and groundwater has been a significant environmental issue for only about the last two decades. Progress over this period has been slowed by the poor initial definition of the problem. In particular, the confounding effects of geologic heterogeneity and the occurrence of DNAPLs in the subsurface were not fully appreciated until the late 1980s and early 1990s. With the improved understanding that currently exists, it is reasonable to expect higher rates of development and application of remedial technologies in the future.

At the present time, using developed and emerging technologies, there are few situations in which the goal of unrestricted use of land and groundwater resources in the vicinity of a contaminated site can be achieved. In some cases this can be a consequence of the limitations of a particular technology, but in virtually all cases, heterogeneity in the hydraulic properties of the geologic materials, in the distribution of contaminants and in the physical characteristics of the contaminants are major contributing factors.

Successful application of existing technologies and the development of new technologies require careful consideration of the effects of heterogeneity, as well as a clear distinction between the characteristics and remedial objectives of source zones and dissolved contaminant plumes.

At many contaminated sites, complete removal or destruction of contaminant mass in the source zone is the most difficult challenge. As an alternative to destruction, there have been successful applications of the use of impermeable barriers to facilitate long-term isolation and containment of source zones. In addition, technologies such as pump-and-treat and, more recently, permeable reactive barriers have been implemented to achieve control and treatment of plumes emanating from sources of a broad range of organic and inorganic contaminants. In the absence of source-zone restoration,

source containment and plume remediation, technologies may need to function for prolonged periods. In general more success has been achieved with these technologies for situations where the subsurface source of contaminants occurs within tens of meters of ground surface, and where geological conditions are favorable for the installation of low permeability barriers or permeable reactive barriers. Although improvements in containment and plume control technologies can be anticipated, application to contaminant zones at great depth or in complex geologic settings will continue to be challenging.

Technologies for restoration of source zones, for DNAPL sources in particular, are generally new and not fully developed. With a requirement for complete removal or destruction of contaminant mass, source-zone restoration by flushing technologies will need to overcome limitations arising from heterogeneities in the distribution of contaminant mass and in geological properties. Although there is some promise for *in situ* destruction technologies to overcome these limitations, there have been few full-scale demonstrations to date. Furthermore, the technologies may only be applicable to specific groups of contaminants. Further research, development and demonstration of these technologies are required.

30. MANAGEMENT OF CONTAMINATED SOIL IN ENGINEERING CONSTRUCTION

P. C. Lucia

G. Ford

H. A. Tuchfeld

30.1 Introduction

With the advent of legislation prescribing the handling of contaminated soil and groundwater, the typical scope of a modern site investigation includes consideration of both engineering properties of the soil and groundwater and how to manage any contaminated soil and groundwater in a manner consistent with law, protective of human health and the environment, and in the most cost-effective manner possible. Construction in most urban areas and in some rural areas can lead to the discovery of contaminated soil or groundwater. Linear construction features, such as excavation for pipelines, in an urban area are typically adjacent to numerous sites capable of contaminating soil and groundwater at the construction site, as shown in Fig. 30.1. A business, commercial or industrial building may occupy a site that is the former location of a gas station, dry-cleaning facility, or paint manufacturing or distributing center. The innocent-looking warehouse building in a rural town may be the former site of a pesticide storage or formulating facility. An environmental investigation is required if new construction requires excavation of soil or dewatering where there is reason to believe

contamination may exist. The discovery of contaminated soil initiates what can be a complex legal, regulatory and engineering process if the proposed project is to move forward. The engineer must manage the handling of the contaminated soil in a manner that is consistent with good engineering practice, consideration of the legal and regulatory environment of the project, and the cost efficiency and schedule goals of the client.

This chapter discusses the management of contaminated soil leading to a decision on whether the proposed project can move forward or whether the proposed project may not be feasible. Environmental site characterization was examined in Chapter 4. Chapter 22 discussed problem identification and risk management. Chapter 29 has reviewed techniques for *in situ* containment and remediation. The present chapter reviews approaches to contaminated soil management that range from complete removal of the contaminated soil to containment in place. The options available to meet the goals will be discussed, including legal and regulatory considerations, on-site and off-site treatment options, and their limitations and case histories. Guidance regarding this subject is available

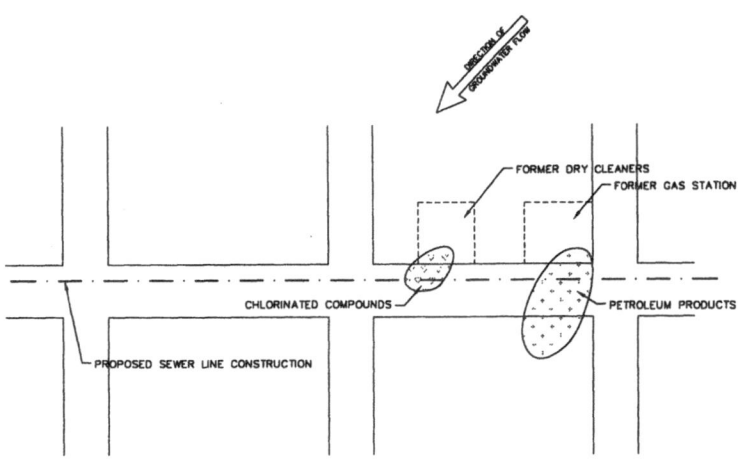

FIGURE 30.1. Example of potential contamination in an urban area.

from a number of USA publications including USEPA (1989f, 1991b, 1992b, 1993a,b, 1994, 1998) and the key governing regulations are given in the US Federal Register (1994).

30.2 Soil Management Goals

When contaminated soil is encountered either during the pre-construction investigation or during construction, a number of important issues arise that include: (a) extent of the soil contamination with respect to the bounds of the proposed project and adjacent properties; (b) legal issues regarding responsibility for the cost of remediation; (c) regulatory issues that must be addressed before initiating remedial actions; and (d) types of options available for managing the contaminated soil and the effects of implementation of such options on the proposed project.

Extending the investigation onto properties not owned by the engineer's client can raise significant legal issues and require site access permission from the adjacent land owner that must be resolved before the engineer can conduct the expanded investigation. The engineer should inform the client that resolution of legal issues beyond the engi-

neer's expertise is needed and then participate with the client in obtaining legal advice from the client's in-house or independent legal counsel. The determination of the source of the contamination can also raise significant issues of legal and financial responsibility and the engineer should work with the client and its counsel to decide if this determination is an objective of the investigation.

If the current owner of the property wishes to recover costs associated with the remediation from prior owners or adjacent property owners, a prescribed investigation and remedial option evaluation methodology and community relations/participation process may be required by law. The engineer should advise the client that determinations and opinions regarding cost recovery, beyond technical matters, are outside of the expertise of the engineer.

An understanding of the regulatory issues and options for managing the contaminated soil are part of developing the soil management goals. The engineer is typically responsible for addressing regulatory issues that may affect remediation and remedial options, often with legal advice regarding interpretation of complex regulatory issues.

One of the most important steps is initially defining the goals of managing the contaminated soil. Many different goals can be considered, including:

- Complete removal: removal of all contaminated soil, possibly including areas beyond those required for the proposed project.
- Non-degradation: partial removal of contaminants or removal of contaminated soil so that the remaining contaminated soil will not contribute to further environmental degradation of present or future groundwater or surface water uses. Non-degradation measures would reduce the potential for significant leaching of contaminants from soil to groundwater.
- Remediation to health-based standards: partial removal of contaminants or contaminated soil so that the remaining contaminated soil does not present a potential existing or future threat to human or ecological health.
- Remediation to the limits of technology-based standards: use of the best available control technology (BACT) to remove as much of the contaminants as technically possible.
- Partial-use restrictions or institutional controls: legal restrictions on areas or resource use, operational measures and/or physical

barriers to prevent access to contaminated soils and other media. Examples of use restrictions include deed restrictions to prevent future residential use of a property or to prevent use of groundwater or surface water at a contaminated soil site from being used as a water supply. Operational measures include 24-h security guards to restrict site access and physical barriers including fences, barriers and caps.

- Containment: use of engineered systems (such as caps or subsurface barriers) for minimizing migration of the soil contaminants to locations where human or ecological receptors could be exposed to the contaminants.

A brief summary of the advantages and disadvantages of each of these goals is presented in Table 30.1. Remediating a site to its planned use or eliminating significant health and environmental risks does not always require restoration to original quality. Accordingly, at a growing number of sites, the soil management goals are either containment or remediation to health-based (i.e. human or ecological health) or resource-based (such as protection of underlying groundwater use) standards. A typical approach for achieving such goals is to evaluate the risk associated

TABLE 30.1. Advantages and disadvantages of various potential soil clean-up goals

Goals	Advantages	Disadvantages
Complete restoration	Eliminates risk and liabilities	Difficult to implement
Non-degradation	Reduction of contaminants to a level that prevents further degradation of environment	May be difficult or not practical for many contaminants
Health-based standards	Mitigates risk to human health or environment	Health-based standards are often controversial
Technology-based standard	Allows treatment to the best capabilities of current technology	May not reduce risk sufficiently
Partial-use restrictions	Prevents contact between contaminants and receptors	Ongoing liability and maintenance
Containment	Relatively predictable and reliable; typically less costly than other remediation	Ongoing liability and requires maintenance and monitoring

with exposure to the contaminated soil and to then design a remediation approach that will provide a sufficient level of protection of human health and the environment, i.e. health-based standards; see Chapter 22.

30.3 Soil Management Options

The management of contaminated soils requires consideration of multiple issues that include, along with typical geotechnical concerns:

- Regulatory requirements that may involve a complex series of local, state and federal requirements governing contaminated soil excavation, remediation and disposal.
- Community concerns regarding degree of site clean-up and potential exposure to contaminants (resulting from on-site treatment, off-site hauling and/or containment in place).
- The cost and time schedule required for soil clean-up (including regulatory approval of soil clean-up plan) relative to client expectations and plans.
- Logistical impacts to site preparation, construction, and final land use.

Most construction projects will require the excavation, transportation and disposal of some volume of soil. Excavation of contaminated soil will require a health and safety plan and training for all workers potentially exposed to the contaminants. The level of protection must be consistent with the level of risk to the workers. The more hazardous and the greater the concentration of contaminants in the soil, the greater the level of protection required and, consequently, the greater the cost (and possible duration) of excavating and handling the soil. If the concentration of contaminants in the soil is of a sufficiently high level that the soil needs to be managed as a hazardous waste, then the contractor transporting the soil must be properly licensed to haul hazardous materials. Landfills are permitted to accept haz-

ardous soil consistent with the level of environmental protection incorporated in the design and construction of the facility. All of the above must be considered by the engineer in the excavation, transportation and disposal of contaminated soil from a construction site.

Use of on-site treatment or containment methods is highly dependent on project-specific conditions, including the planned uses of the site and surrounding neighborhood, soil and groundwater conditions, and whether the contaminated soil was discovered long in advance of the project, or during the initial phases of construction. On-site methods tend to be most feasible and successful at sites that are sufficiently large to allow room to implement the methods, and where the treatment option minimizes or eliminates neighborhood exposure issues, particularly at locations where schools, hospitals, houses, childcare centers and the like are relatively distant, or the surrounding areas are commercial or industrial sites.

Processes such as bioremediation (Section 30.5.2.1) and soil vapor extraction (Section 29.3.2.2) require substantial periods of time to reduce the concentrations of contaminants to appropriate levels. These methods may not be feasible in cases where the contamination is discovered after construction has begun or where there is insufficient time to reduce the concentration prior to the beginning of construction. In most cases where contamination is discovered during construction or found to be more extensive than expected, management of the contaminated soil is generally at an appropriate off-site location. Where space is limited for treatment, it is usually necessary to export material from the site quickly, leaving no time for time-consuming on-site treatment or space-consuming containment.

Over the past 15 years, there has been a trend away from simply digging up contaminated soil and hauling it to a landfill. Presently, soil remediation alternatives are evaluated in the context of risk to human health and the environment (see Chapter 22). In many cases, analysis of both existing and future health risk leads to the conclusion that the contaminated soil presents less risk to the community if it is left relatively undisturbed, or at least not physically hauled from the site. Addressing community concerns regarding contamination left in place is important, and can often be a long and difficult process. Exceptions to this include scenarios in which the contamination has migrated onto adjacent property, where factors such as litigation may control the selection of the remediation alternative, and cases where the contaminants are mobile and may present a threat to groundwater.

In the event that on-site treatment is selected as the preferred alternative for addressing contamination at the construction site, the alternative must also include measures that minimize the risk to the public and local off-site environment during the treatment process. Regulatory and community concerns associated with on-site containment include sufficient demonstration that the mobility of contaminants is adequately reduced such that potential future risk to human health and the environment is considered acceptable.

30.4 Soil Treatment Technologies

The applicability of a soil remedial technology is a function of many factors. Evaluation criteria for remedial technologies are presented in Table 30.2. Specific contaminants and their general groupings are presented in Table 30.3. Remedial technologies generally applicable to the contaminant groups are presented in Table 30.4.

The selection of a remediation technology must be made within the constraints of the project's goals and options. The following sections of this chapter describe the most effective and widely used approaches to incorporating soil remediation within a construction project.

30.5 On-site Treatment Technologies and Containment Options

The use of on-site treatment technologies at a proposed construction site implies that: (a) the extent and nature of contamination can be treated on site to meet the remediation goals; and (b) the time required to achieve the remediation is allowed within the proposed project schedule. The suitability of any on-site treatment method or containment option is highly dependent on the nature of the

TABLE 30.2. Remedial technology evaluation criteria

Contaminants treated	Technology applicability to all contaminants present
Cost	Capital, operation and maintenance cost
Commercial availability	General availability (beyond pilot scale)
Residuals	Evaluation of whether technology produces solid, liquid or vapor residuals that require additional handling and disposal procedures
Effectiveness	Potential for technology to achieve clean-up goals. How technology addresses reduction in soil contaminant toxicity, mobility and/or mass. Potential to provide long-term effectiveness
Acceptability	Acceptability of technology to the regulators and community
Time	Ability of the technology to achieve the desired results in the time available
Site-specific conditions	Technology effectiveness for the soil conditions at the site

TABLE 30.3. Typical contaminant types

Compound category	Typical contaminants
Halogenated volatile organics	Trichloroethylene (TCE)
	Perchloroethylene (PCE)
	1,1-Dichloroethylene (1,1-DCE)
	1,2-Dichloroethylene (1,2-DCE)
	Methylene chloride
	Vinyl chloride
	1,1-Dichloroethane (1,1-DCA)
	1,2-Dichloroethane (1,2-DCA)
	1,1,1-Trichloroethane (1,1,1-TCA)
	Chloroform
	Carbon tetrachloride
Halogenated semivolatile organics	1,2-Dichlorobenzene
	1,3-Dichlorobenzene
	1,4-Dichlorobenzene
	1,2,4-Trichlorobenzene
	2-Chlorophenol
	2,4-Dichlorophenol
	2,4,5-Trichlorophenol
	2,4,6-Trichlorophenol
Non-halogenated volatile organics	Benzene
	Toluene
	Xylene
	Ethylbenzene
	Acetone
	Methyl ethyl ketone (MEK)
	Methyl tertiary butyl ether (MTBE)
Non-halogenated semivolatile organics	Polycyclic aromatic hydrocarbons (PAHs)
	Acenaphthene
	Acenaphthylene
	Anthracene
	Benzo(a)anthracene
	Benzo(b)fluoranthane
	Benzo(k)fluoranthane
	Benzo(g,h,i)perylene
	Benzo(a)pyrene
	Chrysene
	Fluoranthaene
	Fluorene
	Indeno(1,2,3-cd)pyrene
	Naphthalene
	Phenanthrene
	Pyrene
	Phthalate compounds
	Bis(2-ethylhexyl)phthalate
	Di-n-butyl phthalate
	Others
	Dibenzofuran
	Phenol
Fuel hydrocarbons[a] and oxygenates	Fuel hydrocarbons
	Benzene
	Toluene
	Xylene
	Ethylbenzene
	Total petroleum hydrocarbons (indicator of petroleum fuel contamination)

TABLE 30.3. (Continued)

Compound category	Typical contaminants
Fuel hydrocarbons[a] and oxygenates	Fuel oxygenates 　Ethanol 　Acetone 　Tert-butanol 　Methyl tertiary butyl ether (MTBE) 　Butyl ether (ETBE) 　Diisopropyl ether (DIPE) 　Tertiary amyl methyl ether (TAME)
Chlorinated pesticides and PCBs	Polychlorinated biphenyls (PCBs) 　Aroclor 1016, 1221, 1232, 1242, 1248, 1254, and 1260 Chlorinated organic herbicides 　2,4,5-TP (Silvex) 　2,4-Dichloroprop 　Dinoseb Chlorinated organic pesticides 　Aldrin 　Chlordane 　DDT, DDE, DDD 　Dieldrin 　Methoxychlor 　Toxaphene
Inorganics	Arsenic Barium Cadmium Chromium Copper Lead Mercury Nickel Zinc Nitrate Asbestos pH (indicator parameter)

[a] Fuel hydrocarbons are contained in such materials as gasoline, diesel fuel, heating oil, bunker oil and motor oil.

soil and the extent and characteristics of the contamination. An evaluation of the applicability of a treatment method may include a feasibility study, bench-scale testing or even *in situ* pilot scale testing.

30.5.1 REGULATORY CONSIDERATIONS

Commonly encountered soil contaminants during excavation projects (Table 30.3) include petroleum fuel hydrocarbons, metals (including lead), chlorinated volatile organic compounds (including organic solvents), pesticides and herbicides, polycyclic aromatic hydrocarbons (PAHs), polychlorinated biphenyls (PCBs) and asbestos. Different regulatory requirements apply to management of the excavated soil depending on the nature, types and levels of these contaminants, type of site and its history, and regulatory agency (ies) with jurisdiction over the site.

An important first step in assessing the applicable regulatory requirements for the site would be to identify the: (a) locations and history of any present or past underground stor-

TABLE 30.4. Soil treatment technologies generally applicable to compound groups

Compounds	Technology
Halogenated volatile organics	Soil vapor extraction (SVE)
	Soil flushing
	Low temperature thermal desorption
Halogenated semivolatile organics	Thermal enhanced SVE
	Soil washing
	High temperature thermal desorption
	Incineration
Non-halogenated volatile organics	Biodegradation
	Bioventing
	Soil vapor extraction (SVE)
	Soil flushing
	Landfarming
	Low temperature thermal desorption
Non-halogenated semivolatile organics	Biodegradation
	Bioventing
	Thermal enhanced SVE
	Soil washing
	Solvent extraction
	High temperature thermal desorption
	Incineration
Fuel hydrocarbons	Biodegradation
	Bioventing
	Soil vapor extraction (SVE)
	Landfarming
	Soil washing
	Low temperature thermal desorption
Pesticides	Thermally enhanced SVE
	High temperature thermal desorption
	Incineration
Inorganics	Solidification
	Soil washing

age tanks, above-ground storage tanks, waste management units, or disposed waste or spill sites in the subject area; (b) history of past environmental investigations and clean-ups; (c) environmental permitting and regulatory enforcement history; and (d) regulatory agencies that have been involved in the past and those that should be involved in planned excavation of contaminated material and in formulating risk management decisions. This is typically performed through a record search and through discussions with appropriate officials.

The excavation, transport, treatment or containment of contaminated soil typically requires regulatory approval and/or permits. The regulatory process requires definition of the goals of the project and agreement of the regulators on the methodologies and the goals of the project. The client may solicit legal advice to navigate through the complex series of regulations or through problems that may have a high degree of complexity and/ or risk to human health and the environment.

30.5.1.1 Soil Contaminated by Petroleum Fuel Products

Soil at many construction sites is contaminated with petroleum fuel products, such as from existing or past leaking underground gasoline, diesel, and heating or waste oil tanks. When soil is contaminated by petroleum fuel products alone, the regulatory requirements for investigation and management of the contaminated soil are often less stringent than for soil contamination by other potentially hazardous materials, such as chlorinated organic solvents. This is because subsurface contamination by such fuels is believed to typically (depending on the levels of free product and dissolved phase contamination and site setting) pose a lower potential health risk, have limited migration potential in groundwater systems, and have a relatively high biodegradation potential. However, sites where methyl tertiary butyl ether (MTBE) or other anti-knock oxidants associated with gasoline are present may be of greater concern due to the potential higher mobility and toxicity of such compounds.

Sites contaminated only by petroleum fuel products from underground storage tanks typically require a limited site investigation (consisting of installing and sampling soil from several shallow borings in the vicinity of the tank). In the event that soil contaminated by petroleum fuels is identified, such agencies typically require tank removal, removal of free product, remediation of significantly contaminated soil in the immediate vicinity of the site, and possible groundwater contamination assessment. Many state agencies in the USA either directly set minimum clean-up levels and management standards (such as for total petroleum hydrocarbon levels

and/or individual petroleum hydrocarbon compounds, petroleum-related metals, and MTBE) or have a tiered health risk-based process for setting clean-up levels and management requirements for petroleum product contaminated soil. Thus, excavated soil from construction sites with petroleum product contamination below such levels (in the absence of MTBE contamination) may be allowed to be used as fill or replaced on site with no or minimal treatment.

30.5.1.2 Soil Contaminated by Hazardous Wastes

The management of soil that is contaminated by constituents other than, or in addition to, petroleum fuels is often subject to more stringent requirements. The highest level of stringency for management of the excavated soils is associated with soil containing hazardous waste or hazardous waste constituents. The federal statute known as the Resource Conservation and Recovery Act (RCRA) regulates the management of hazardous waste from "cradle to the grave", i.e. generation through disposal, in the USA. The US Environmental Protection Agency (USEPA) or a state agency approved by USEPA implements and enforces RCRA. States are required to be at least as stringent as the USEPA in regulating hazardous wastes, and can be more restrictive.

The requirements for identification and management of hazardous wastes under RCRA Subtitle C are contained in 40 "Code of Federal Regulations", 40 CFR 260–268 (USFR 1994). Under RCRA, a contaminated soil may need to be managed as a hazardous waste if it contains a hazardous waste from a specific waste stream or source listed in 40 CFR 261 or is hazardous based on the following characteristics defined in 40 CFR 261: (a) ignitability; (b) corrosivity; (c) reactivity; or (d) toxicity.

Excavated soil which is to be managed as a hazardous waste under RCRA could have the following implications: (a) on-site burial could require RCRA permitting of a double-lined landfill with a leachate collection system, rigorous final cap requirements, groundwater monitoring system, and post-closure maintenance and monitoring; (b) on-site storage of contaminated soil piles could be regulated by RCRA and on-site re-use as fill may be prohibited; (c) off-site disposal would have to be to a licensed Subtitle C (hazardous waste) landfill, incinerator, or other hazardous waste management facility; and (d) either on-site or off-site disposal of the contaminated soil may require pre-treatment prior to land disposal to meet the requirements of 40 CFR 268 (RCRA Land Disposal Restrictions).

On-site burial of soil regulated as hazardous waste would typically be very expensive because of the possible permitting, bottom liner, capping and long-term monitoring requirements. In addition, an on-site disposal facility could require a long lead time for permitting.

If the contaminated soil is to be managed as a hazardous waste, a careful review of the governing regulations, discussions with the site owner/operator and possibly regulators, and a cost-benefit analysis of management options are recommended. Other regulations that should be evaluated for applicability for management of the contaminated soil include the US Comprehensive Environmental Response, Compensation and Liability Act, as amended (CERCLA, or "Superfund"), US Toxic Substances Control Act (TSCA) (which includes the management of PCBs and asbestos), and state and local requirements. In particular, CERCLA applies to certain sites where hazardous substances have been disposed and environmental contamination has resulted. CERCLA requires a remedial investigation and clean-up process, which can be very time consuming and expensive.

30.5.1.3 Soil Contaminated by Non-Hazardous Solid Waste

Solid non-hazardous waste includes materials such as trash, municipal solid waste, used tires and construction demolition debris. Management and disposal of soil that contains these materials is less stringently regulated than soil containing hazardous waste. Off-site disposal of such soil in the USA can be to a RCRA Subtitle D (non-hazardous) landfill, rather than to a RCRA Subtitle C (hazardous waste) landfill. Depending on the characteristics of the contaminated soil and regulatory requirements, on-site disposal of the soil may require a bottom liner and cap, but one less stringently designed than for soil containing hazardous waste.

30.5.2 TREATMENT TECHNOLOGIES

If on-site treatment is to be a viable option, the proposed technology must be cost effective, able to meet the remediation goals within the project schedule, and meet regulatory approval requirements. Treatment technologies that meet these criteria are those that have significant records of success. A great many technologies are available with varying degrees of development. The technologies discussed below are those most widely considered for typical projects.

30.5.2.1 Bioremediation

Soil bioremediation is the process of using naturally occurring or introduced bacteria to decompose toxic organic contaminants into less toxic compounds. This process occurs naturally in many cases (intrinsic or passive bioremediation), or can be induced or accelerated by human interference (extrinsic or engineered bioremediation). Most soils that have not been chemically or thermally sterilized contain naturally occurring bacteria that are capable of breaking down complex organic molecules. Often this process for aerobic biodegradation is limited by a lack of sufficient oxygen within the soil.

Aerobic bioremediation consists of inducing conditions that foster the growth and reproduction of carbon-consuming aerobic bacteria within the contaminated zone. By introducing moisture, oxygen and nutrients to the contaminated soil, aerobic bacteria can be induced to thrive and degrade the organic contaminants. This process occurs aerobically, and typically works well on petroleum hydrocarbons and benzene, toluene, ethylbenzene and xylene (BTEX) compounds, and works less well on chlorinated solvents, phenols, PCBs, pesticides, dioxins and other organic compounds (some of which may be more susceptible to anaerobic biodegradation). Water is usually added to the soil during the bioremediation process; therefore, it is most effective for soil that does not contain other highly soluble inorganic contaminants that may affect the environment.

It should be noted that anaerobic bacteria are also capable of metabolizing hydrocarbon compounds, but do so at rates that are typically 10–100 times slower than aerobic bacteria. Thus, anaerobic degradation of hydrocarbon contamination is not often economically feasible.

Bioremediation can be performed *in* or *ex situ*. The primary advantage of *in situ* bioremediation is that it can be performed without excavation of the soil. *In situ* bioremediation's main disadvantage is that it is generally a time-consuming process, and it typically does not result in complete detoxification of contaminants. *Ex situ* bioremediation generally offers the advantage of rapid completion, but usually requires excavation of the contaminated soil and is more practical where contamination occurs at relatively shallow depth.

In situ Bioremediation. *In situ* aerobic bioremediation is most commonly performed on contaminated soil deep in the vadose zone,

where the cost to excavate and treat, would be prohibitive. Air and/or moisture is introduced to the contaminated zone by injection through wells. Nutrients may be placed into the contaminated zone through wells, direct injection, or other means. The technique is effective at sites where the soil conditions are favorable for transport of air and fluids, hydraulic conductivities are 10^{-8} m s^{-1} or greater, soil moisture content is in the range 25–85%, and pH values are not extreme.

Aerobic bioremediation in the saturated zone can also be enhanced by increasing the oxygen content in the groundwater through processes such as air sparging or introduction of hydrogen peroxide.

Monitoring of the effectiveness of *in situ* bioremediation is usually accomplished by periodic sampling of the soil to measure contaminant concentrations and bacterial populations. It is typical for the rate of contaminant decomposition to fall off as the process proceeds, such that the decline in concentrations makes an asymptotic approach to detection thresholds. For this reason, *in situ* bioremediation is not the preferred method if the goal is to reduce contaminants to nondetectable levels. The costs of periodic soil sampling may become a significant portion of the total remediation cost.

Ex situ Bioremediation. *Ex situ* bioremediation involves excavating the contaminated soil and treating it above-ground to reduce contaminant concentrations. The treated soil may then be replaced into the ground as compacted fill or disposed off-site, depending upon project requirements. *Ex situ* bioremediation offers the advantage of relatively rapid processing compared to *in situ* bioremediation, ease of monitoring, and it can be used in cases where the site configuration and soil types are not conducive to reasonably efficient *in situ* bioremediation. In particular, where the contamination occurs in highly cohesive clayey soils, it may be impossible to drive oxygen, moisture and nutrients into the *in situ* pore spaces to induce growth of aerobic bacteria. If the soil is removed from the ground, the soil fabric can be broken up and more surfaces exposed, such that bioremediation becomes feasible. The process works most efficiently in cases where the soil is relatively porous or coarse grained, and when the hydrocarbons are light, short-chain molecules such as fuels or solvents. However, the method can be successfully used on heavier, long-chain hydrocarbons by using longer treatment periods or by adding proprietary enzyme admixtures to facilitate the degrading of heavy hydrocarbons. *Ex situ* bioremediation is typically accomplished by one of two methods:

1. Excavating the contaminated soil and placing it into a bioreactor, where temperature, oxygen, nutrient composition and moisture content are held near optimum for the bacteria, whether native or introduced. This form of bioremediation occurs in a closed, temperature-controlled vessel and the process can be closely monitored and performed regardless of season. This process is usually performed by specialty contractors who use proprietary equipment and procedures.

2. Landfarming of excavated soil. Soil is removed from the ground, placed in shallow lifts or windrows, and water, nutrients (typically nitrogen-rich fertilizer and/or enzymes to assist in breaking down long-chain hydrocarbons) and/or bacteria (if necessary) are applied. The soil is then tilled periodically to introduce air to the pore spaces. This process is continued for a period of days, weeks or months until contaminant concentrations have declined to a target level. In some cases, air, water and nutrients may be supplied to the soil by injection pipes. Landfarming bioremediation is temperature sensitive, and is thus best suited to warm-weather sites, or projects where bioremediation can be carried out over several seasons. Sufficient space is

also required for effective landfarming. It should be noted that there are regulatory restrictions and criteria for landfarming of certain wastes in the USA.

The progress of *ex situ* bioremediation is easily monitored by periodic sampling of the soil and testing to determine bacterial density and vigor, and to confirm declining contaminant concentrations. Often, some type of closure sampling and analysis is required to demonstrate that the landfarming operation did not result in contamination of soil beneath the treatment area.

30.5.2.2 Soil Vapor Extraction

As described in greater detail in Section 29.3.2.2, soil vapor extraction is an *in situ* method for removing volatile organic compounds (VOCs) from soil by inducing air flow through the soil using vacuum wells. Contaminant reduction is thought to result from two concurrent processes:

1. The VOCs are volatized and flow through the pore spaces to the wells, where they are carried to the surface for final treatment by combustion, sorption or catalytic decomposition.
2. The increased air flow stimulates the growth of soil bacteria, where present, which consume the VOCs. The basic vapor extraction system is frequently augmented by use of passive soil vents, heating of the soil, surface seals and air sparging (when contaminants are also present in the saturated zone).

Vapor extraction is most applicable when the contamination consists of solvents, gasoline or diesel with a relatively high vapor pressure that is lodged in the vadose zone of a relatively permeable soil. This process may require months to years to substantially reduce contaminant concentrations; therefore, it is most useful in cases where the contaminated soil has been discovered well in advance of construction or where it can be used in conjunction with the finished project. Under the right circumstances, the method can be effective at removing a large fraction of VOCs from the soil, but cannot remove all contaminants because the rate of removal decreases as the residual concentration of VOCs in the soil decreases. The design of vapor extraction systems must be detailed, site-specific and based on a thorough understanding of the site conditions (see Section 29.3.2.2). Vapor extraction systems can induce local mounding of groundwater, causing migration of dissolved contaminants off-site, and they can draw off-site contaminant vapors inward, unless the system design mitigates these tendencies.

Vapor extraction systems are typically monitored by analyzing the system exhaust for VOCs. When VOC concentrations in the vapor stream decline to low levels, closure sampling of the contaminated soil is usually performed to document that project objectives have been reached.

30.5.2.3 Stabilization

In situ and/or *ex situ* stabilization may be performed to reduce the mobility or exposure hazard of contaminants in soil. *In situ* stabilization can be accomplished by various means, including: (a) conventional or chemical grouting to reduce hydraulic conductivity and/or fix contaminants; and (b) thermal stabilization, ranging from steam or air injection at low temperatures to strip volatiles to high temperature *in situ* vitrification, where soil electrodes are used to melt the soil mass into an inert, glassy mass.

Ex situ stabilization is most commonly done using proprietary processes. A processing plant is placed on site, and the contaminated soil is excavated, stockpiled, processed and then re-used as backfill or disposed off-site. Treatments can be designed to mitigate a range of contaminant characteristics, including: reduction of the solubility of metals;

reduction of the mobility and/or toxicity of heavy organics; and removal of petroleum hydrocarbons and/or VOCs. The costs of these *ex situ* processes vary with the type and degree of contamination.

30.5.3 CONTAINMENT OPTIONS

Containment systems (see also Section 29.4) can control (to negligible levels) migration of contaminants to locations where human health or the environment can be impacted. Where contamination is shallow and/or not mobile and migration is not a concern, capping may prevent human or environmental contact and provide acceptable levels of protection. The use of containment can often facilitate the project, but may restrict the project if, for example, excavations were proposed for basements or underground parking. Containment systems may also require long-term monitoring to evaluate their effectiveness and result in additional costs and remaining liability.

30.5.3.1 Capping and Covers

Capping of a contaminated site by the construction of an appropriate cover may be an effective containment option when one or more of the following conditions exists: (a) the contaminants are not mobile in the soil if infiltration of water is mitigated through installation of caps, (b) groundwater is located at a distance from the contaminants such that the reduced infiltration of water due to the cover will mitigate any impacts on groundwater, and (c) the cover provides sufficient protection from contaminant intrusion at the surface and the long-term integrity of the cap can be assured.

The cover must be designed to minimize infiltration, direct surface flows away from the cover surface, and provide long-term durability (see Chapter 27). The decision leading to use of a cover should include an evaluation of the effectiveness of the cap in reducing the mobility of the contaminants, and in reducing the potential risk to human health and the environment.

The cover can be constructed of a number of low permeability materials designed to minimize infiltration. Often a number of different materials are used in conjunction to provide a combination of low permeability, drainage and durability, as described in detail in Chapter 27.

30.5.3.2 Cut-off Walls

Physical barriers are vertical features in the ground that provide containment of waste. Physical barriers can be formed using a variety of materials. Several of the most commonly used physical barriers are listed in Table 30.5, along with their typical dimensions.

The oldest and most commonly used types of physical barriers are cut-off walls (see Section 25.3.9). Cut-off walls limit migration of contaminated groundwater and leaching of contaminants from contaminated soil by forming a physical impediment to groundwater flow. The effectiveness of the wall is a function of its continuity, its resistance to degradation by contaminants and its resistance to physical degradation. A barrier impedes groundwater flow, and therefore the elevation of groundwater may rise (or "mound") on the upgradient side of the barrier. Other controls may thus be needed to prevent overtopping or flanking of the bar-

TABLE 30.5. Types of physical barriers

Barrier type	Width (m)	Maximum depth (m)
Soil–bentonite	0.6–1	25
Cement–bentonite	0.6–1	25
Biopolymer drain	0.6–1	20
Deep mixing	0.75	27
DM structural	0.75	27
Jet grouting	0.5–1	60
Grout curtain	One row	60
Sheet piling	One sheet	45
Geochemical barrier	Varies	Hundreds

rier by contaminated groundwater, such as: (a) groundwater extraction wells or trenches adjacent to the barrier to route mounded groundwater to a discharge point, or (b) permeable segments in the barrier to allow water to pass through controlled locations in the wall.

Soil–bentonite slurry walls are by far the most common type of physical barrier used in geotechnical and environmental remediation projects. Construction of a slurry wall consists of installing a mixture of soil (or other material) and bentonite clay into a vertical trench to form a very low permeability, vertical barrier to groundwater flow (see Section 25.3.9.1).

The effectiveness of a soil–bentonite slurry wall in preventing migration of contaminants is a function of: (a) the environment of the wall, i.e. the nature of the chemicals in groundwater to which the soil–bentonite mix will be exposed; (b) the percentage of bentonite clay used in the soil–bentonite mix; (c) the type and gradation of soil used, which is typically measured as a percentage of the soil that is finer than the No. 200 USA standard sieve; and (4) the quality of wall construction. The performance of a soil–bentonite slurry wall can be adversely affected by exposure of the wall to VOCs, high levels of aqueous salts (such as those contained in sea water), or non-aqueous phase liquids (NAPLs). Sai and Anderson (1992) reported an increase of two to three orders-of-magnitude in hydraulic conductivity for soil–bentonite slurries that were permeated with the VOCs xylene and methanol. This level of increase may not be typical but illustrates the need to consider clay–permeant compatibility as discussed in Section 25.4.1.

30.5.4 CASE STUDY

In the 1960s, a wastewater treatment plant received soil–sludge waste (waste material) from a refinery. Initially the waste material was described as a non-homogeneous mixture of organic sludge, lead sulfate, dirt, sulfuric acid (resulting in soil with a low pH), water, tars and wood debris. The waste material was deposited and incorporated into various areas of the facility as fill or to stabilize soil. This occurred more than a decade prior to the passage of major legislation, i.e. RCRA, pertaining to classification and disposal of waste materials.

Subsequent investigations indicated that substances found in soil at the facility included: lead, cadmium and chromium; petroleum hydrocarbons quantified as diesel (TPH-d); BTEX constituents; PAHs; and oil and grease. Groundwater monitoring indicated the presence of only one chemical constituent, benzene, that exceeded a regulatory criterion.

A preliminary risk assessment concluded that the hazardous substances detected did not pose a significant threat to on-site workers and that off-site risk to human health was minimal. However, since the site owner was considering construction of a potential future expansion project on areas of the facility underlain with contaminated soils, the owner proposed taking actions to make the project area suitable for construction of treatment plant facilities.

The subsequent investigations and evaluations addressed the actions to be taken to prevent, minimize or mitigate damage that might otherwise result from a release of hazardous substances at the site at the time hazardous substances actually are moved during construction activities. The specific project goals include:

- excavate soil in the contaminated soil area as required for future treatment plant expansion,
- handle excavated soils in a manner that does not exceed acceptable risk levels to human health and the environment, and
- to the extent possible, provide on-site containment for the contaminated soils.

A baseline risk assessment indicated that the aggregate risk to human health (as hazard index and total theoretical excess cancer risk)

fell within the range considered by USEPA to be "safe and protective of the public health" (Federal Register 56(20): 3535, 1991). These conclusions were restricted to occupational exposure only and assumed a future industrial land use. The risk estimates and corresponding exposure calculations would not necessarily hold should a different land use, e.g. unrestricted residential, become important. The facility was a wastewater treatment plant and the future land use was considered unlikely to change.

Four alternatives were identified as feasible and potentially appropriate for the removal action and management of contaminated soil. These alternatives were:

- Alternative 1: no action.
- Alternative 2: consolidate (place excavated contaminated soil in one location on-site)/treat low pH soil, construct a cover over consolidated soils.
- Alternative 3: consolidate/treat low pH soil, passive bioremediation, construct a cover over consolidated soils.
- Alternative 4: off-site disposal of excavated soil.

Alternative 1 would not meet the primary goal of excavating soil in the contaminated soil area for future treatment-plant expansion. The other three alternatives would all meet this goal. All four alternatives could be implemented within a schedule that compliments the treatment-plant expansion.

The treatment methods for low pH soils proposed for Alternatives 2, 3 and, potentially, 4 were readily available, and have demonstrated performance. The passive bioremediation technology proposed for Alternative 3 was also readily available. However, bioremediation is not very effective on heavy hydrocarbons or PAHs, and is not effective at all on lead. Further, bioremediation has limited success in the type of clayey soils present at the site.

Alternatives 2, 3, and 4 have health and safety requirements related to dust, heavy equipment and soil contaminants. Risks to workers and the public from dust can be easily mitigated with standard engineering controls, such as keeping soil wet during excavation and consolidation or loading activities. Risks to workers from heavy equipment and soil contaminants can be mitigated by implementing an appropriate health and safety plan, including a hazard analysis, mitigation measures, and appropriate field and medical monitoring in accordance with occupational health and safety requirements.

Alternative 4 posed some additional health and safety risks to the community from the transport of contaminated soil on local streets and highways. However, this risk can be mitigated by using appropriately licensed and experienced haulers.

All alternatives require long-term groundwater monitoring because at least some contaminated soils will remain on site. Alternatives 2 and 3 will also have cover monitoring requirements. However, inspection and maintenance of the cover was anticipated to be easily adapted with minimal impact into the facilities maintenance program. Alternatives 2, 3 and 4 will all require dust and health and safety monitoring during construction. However, this type of monitoring is fairly routine for construction and can be easily implemented.

Alternative 2 was estimated to be the lowest cost alternative. The Alternative 3 cost estimate was higher because of the costs associated with passive bioremediation. Passive bioremediation was not expected to result in a significant benefit over Alternative 2.

Alternatives 1 and 4 were estimated to be the highest cost alternatives. A high estimated cost was associated with the "no action" alternative, due to the anticipated cost of having to locate treatment-plant expansion facilities somewhere other than the proposed expansion area. The high estimated cost of the off-site disposal alternative was associated with transportation and disposal of contaminated materials. Under US law, Alternative 4

would also result in potential long-term liability at the off-site disposal facility.

Alternative 2 (consolidate/treat low pH soil and construct a cover), was the recommended option based on the alternatives evaluation and comparative analysis. Implementation of this alternative would involve the following actions:

- excavate soil in the contaminated soil area,
- segregate and treat low pH soil,
- consolidate excavated material in the on-site disposal area,
- construct a soil cover,
- vegetate the cover,
- perform quality assurance monitoring during construction,
- perform work under an appropriate health and safety plan, and
- perform long-term groundwater monitoring and cover inspection.

The proposed alternative met regulatory requirements, provided a reasonable cost alternative by containing the contaminated soils on site, and provided protection against excessive potential risk to human health and the environment.

30.6 Off-Site Treatment Technologies and Disposal Options

Often, project space and/or schedule constraints dictate that contaminated soil be removed from a site for treatment and final disposal. For instance, in many urban development projects, construction of basements or subsurface parking levels requires that soil be exported from the site to create space underground. In these cases, the decision to perform off-site treatment to reduce contaminants is based on a number of factors, including: (a) cost of treatment versus the cost of disposing of untreated soil, (b) availability of off-site treatment facilities, and (c) regulatory requirements.

Direct hauling and disposal of untreated soil is the simplest alternative for off-site disposal. However, this may also be highest cost choice, owing to the high disposal fees for contaminated waste, and the often great distance (and high transportation costs) to hazardous or regulated waste disposal sites. Treating soil to reduce contaminant concentrations such that it qualifies for non-hazardous or non-regulated disposal will often reduce costs substantially in the following two ways:

- by taking advantage of lower disposal fees, and a generally wider selection of disposal options available for non-hazardous and non-regulated wastes; and
- by reducing transportation costs in most cases, as non-hazardous and non-regulated landfills are more common than hazardous facilities, and may be located closer to the project site.

Information regarding treatment prior to disposal, cost of disposal, and criteria for disposal can be obtained from landfill operators or regulators who permit landfill operations.

30.6.1 REGULATORY CONSIDERATIONS

In the USA, the off-site handling and disposal of contaminated soil are typically governed by three sets of regulations:

- RCRA waste classification and management requirements,
- state and/or local requirements, and
- health and safety requirements to protect both workers handling the soil and the surrounding environment.

Generally, soil classified for management as hazardous waste may only be treated on-site or off-site at a licensed hazardous waste management facility because regulations typically prohibit placing, storing or handling this soil at an interim site. In some cases, it may be cost effective to treat the soil at the project site to reduce the specific contaminants that render it hazardous, then remove the soil to an interim site for additional treatment, followed by final disposal at a permitted non-hazardous waste disposal facility. Frequently, more treatment options are available for soil that is not considered a hazardous waste for management purposes.

Contaminated soil should be hauled to the off-site disposal facility by an experienced and appropriately licensed hazardous waste transporter. In addition, soil that is managed as a hazardous waste must be manifested in accordance with regulatory requirements. Contaminated soil to be disposed at a licensed off-site facility in the USA as a hazardous or non-hazardous waste must not have free liquids. In the event that free liquids are present, the contaminated soil must be stabilized by addition of an appropriate absorbent or solidifying material prior to disposal. There are also regulatory requirements for the manner in which wastes with free liquids are to be transported to disposal sites for stabilization.

30.6.2 DISPOSAL OPTIONS

Disposal options are discussed below for contaminated soil that will be managed as a hazardous waste and for soil that can be managed as non-hazardous waste. Some options, such as incineration or incorporation into asphalt are options for certain oil-contaminated soils, and may, depending upon the constituents and local regulations, be considered regardless of whether the soil needs to be managed as a hazardous waste. Typically, the engineer develops a range of disposal options and facilities from which the client makes a selection based on consideration of long-term liability.

30.6.2.1 Non-Hazardous Waste Landfill

Non-hazardous contaminated soil can be disposed of in a suitable non-hazardous waste landfill. Most jurisdictions have local criteria governing disposal in non-hazardous waste landfills. For example, in California, Class III (non-hazardous waste) landfills frequently accept soil that contains less than 100 p.p.m. petroleum hydrocarbons (gasoline/BTEX, diesel, or oil and grease), as long as the soil meets all other disposal requirements. The soil is frequently used as daily cover for household waste. Class II landfills are authorized to accept soils containing up to 1000 p.p.m. of various petroleum hydrocarbons, but typically charge substantially higher disposal fees. While disposal fees vary with location, level of business activity, and the landfill's need for daily cover soil, Class II and III disposal fees are generally much lower than hazardous waste disposal fees.

30.6.2.2 Hazardous Waste Landfill

Soil that must be managed as a hazardous waste must be disposed in a licensed off-site hazardous waste disposal facility, e.g. RCRA Subtitle C facility. This option is relatively expensive, and in the USA, could have long-term liabilities. Typically, this option is only considered in those cases where the contaminated soil must be managed as a hazardous waste.

Depending upon the nature and level of the soil contaminants, treatment may be required prior to land disposal within the landfill. The RCRA Land Disposal Restrictions of 40 CFR 268 (USFR 1994) describe the types and level of treatment required for the hazardous waste based on the nature of the contaminants. This treatment could be conducted either on- or off-site. If on-site treatment is considered, it is recommended that the governing regulatory agencies be contacted regarding permitting and other regulatory considerations.

30.6.2.3 Incineration

Incineration at an off-site facility is typically the most expensive option for management of a contaminated soil. It is a potentially viable option when the contaminated soil contains certain levels of chemicals of particularly high toxicity and potential persistence in the environment (such as PCBs or dioxins). The RCRA Land Disposal Restrictions of 40 CFR 268 (1991) lists those types and levels of constituents in hazardous waste that require incineration. One potential advantage of incineration over disposal in a landfill is that the contaminants are destroyed. Thus, the incineration option could have lower long-term liabilities.

Licensed mobile incinerators are available for destruction of particular contaminants. The costs of on-site incineration through use of such mobile incinerators are typically less than at off-site incineration facilities due to the cost savings in transportation. Depending on the nature of the contamination, other types of mobile thermal treatment short of incineration may be viable. Public groups often oppose on-site incineration and the process can often be difficult to permit.

30.6.2.4 Other Off-Site Management Options

In recent years, several other management options have become available for contaminated soil. These include: (a) recycling of contaminated soils for use as road base, and (b) use as fill on construction sites.

These management options can potentially be cost effective, but are very dependent on the waste characteristics. Long-term liability can be high if the waste composition is highly variable, resulting in undetected levels of certain contaminants.

Processes for recycling of contaminated soil for use as road base are usually proprietary. Their use is generally limited to mildly contaminated soil. Re-use of contaminated soil as road base is cost effective only when the project site is not too distant from the recycling facility.

In some cases, contaminated soil may be treated to such an extent that it can be used as fill on construction sites. This alternative is normally restricted to soil that is not highly contaminated, and is also subject to some practical and political restrictions and potential future liability considerations, even when present legal restrictions do not govern. For example, it is rarely advisable to use formerly contaminated soil, no matter how extensively treated, as backfill at school, hospital or other public facility sites.

30.6.3 TREATMENT TECHNOLOGIES

30.6.3.1 Aeration

Aeration is the process of reducing the concentrations of organic contaminants in soil by evaporation. The technique is effective on high vapor-pressure organic contaminants, such as gasoline, BTEX and light solvents (VOCs). Under favorable conditions, aeration can reduce VOC concentrations to below detection thresholds, such that the soil may be re-used as on-site backfill or as daily cover in a municipal landfill. Aeration is less effective for diesel, and not effective for heavy hydrocarbons. The efficiency of the aeration process depends on the specific hydrocarbon to be aerated, the characteristics of the soil, ambient temperature and humidity, and space available. Aeration is usually an attractive alternative in cases where the weather is warm and relatively dry, enough space is available to allow the placement of soil in thin lifts, and the soil is relatively porous.

The simplest method of aeration is landfarming, where the contaminated soil is spread in thin lifts and tilled periodically to expose more soil to the air. Aeration can also be accomplished, at higher cost, by injection of air into heaps or windrows of soil. Various methods of covering the soil to retain heat and capture escaping VOCs have also been devised, often involving construction of temporary greenhouse-type enclosures.

In many urban areas, the aeration process is regulated by air quality management authorities. These regulations typically limit the amount of contaminated soil that may be exposed to the air each day on the basis of contaminant concentration. In cases where large volumes of soil are highly contaminated, regulations may render open-air aeration impractical. For example, in the San Francisco Bay Area, soil containing more than 5000 p.p.m. (average) VOCs is effectively precluded from open aeration by rules limiting the quantity of soil to be aerated to 0.1 m^3 per day. In such cases aeration may sometimes be accomplished by constructing temporary enclosures with vapor-trapping equipment or by transporting the contaminated soil to an appropriate interim aeration site.

30.6.3.2 Bioremediation/Landfarming

Off-site bioremediation may be performed at virtually any site that meets the basic physical and regulatory requirements. These typically include:

- sufficient space to spread the soil into workable lifts,
- public access capable of being restricted to reduce human exposure to contaminants, and
- adequate run-off and drainage control is possible so that the nearby environment and subsurface soils are not contaminated by run-off or infiltration from the treatment pile.

Although it is theoretically possible to till soils to a depth of about 1.5 m, bioremediation works most efficiently if the soil is spread in lifts 0.5-cm thick or less.

In some cases, it has been convenient and cost effective to construct a temporary treatment area specifically designed for the bioremediation of soil. Such a treatment area may be constructed at the project site, space permitting, at the landfill or other disposal site, or at an interim site. Public agencies such as redevelopment agencies or large commercial entities with very large or multiple sites to clean up have found it cost effective to construct permanent bioremediation treatment areas that can be used for treating soil from multiple sites.

Stockpiled soil is placed in lifts and tilled periodically to introduce air to the pore spaces. This process is continued until contaminant concentrations have declined sufficiently to permit final disposal of the soil.

Treatment progress is monitored by periodic soil sampling. As with on-site bioremediation, usually closure sampling and analysis is required to demonstrate that the landfarming operation did not contaminate soil beneath the treatment area.

30.6.4 CASE STUDY

A large urban building project provides an example of several on- and off-site treatment techniques that were combined to effect a rapid clean-up of an urban site so that a high-rise office building could be constructed. Contaminated fill containing oil, grease, lead and localized pockets of gasoline was discovered on two city blocks in an urban redevelopment area. The proposed buildings included two basement parking levels, requiring all of the contaminated soil to be removed from the site before construction could begin. The combination of oil, grease and lead in the soil rendered it unsuitable for direct disposal at a Class III (non-hazardous) landfill. A pilot study indicated that the soil was amenable to bioremediation to reduce the concentrations of oil and grease.

Various alternatives for direct disposal of the contaminated fill soil were investigated and rejected because of the high cost and the client's unwillingness to become a generator of hazardous waste. *In situ* bioremediation followed by off-site disposal at a non-hazardous landfill was considered and rejected because the *in situ* process could not be completed quickly enough. Landfarming bioremediation on adjacent vacant lots was also rejected because of lack of space and scheduling constraints. Finally, a program of off-site bioremediation for the fill and on-site aeration of soil containing gasoline was adopted. This program met the client's primary objectives, including:

- removing the contaminated soil from the site quickly so that construction of the building could proceed; and

- avoiding the high transportation and disposal costs and potential long-term liability of placing the soil in a hazardous waste facility, the nearest of which was located over 300 km from the project site.

Beginning in early 1990, approximately 25 000 m³ of contaminated fill were removed from the site and trucked to a Class III landfill in central California. At the landfill, the soil was placed on a specially prepared treatment area constructed of compacted clay, with *in situ* moisture sensors and perimeter drainage channels. Two summer seasons of bioremediation, including weekly tilling of the soil, were required to reduce the oil and grease concentrations from an average of about 500 p.p.m. to less than 100 p.p.m. Bacterial decomposition of the oil and grease essentially stopped during the winter months, owing to excessively high moisture contents and low temperatures. Upon completion of treatment, the soil was used as daily cover in an asbestos disposal unit at the landfill. Closure sampling was performed in the bioremediation area at the landfill to document that the bioremediation process had not resulted in contamination of the treatment area. The total cost to move, treat and dispose of the soil was approximately US\$75 m⁻³, which compares favorably with estimated Class I disposal costs exceeding US\$150 m⁻³.

After the fill containing oil and grease had been removed from the urban site, approximately 1500 m³ of fill and native soil containing gasoline at concentrations up to 4500 p.p.m were excavated, stockpiled, and then spread out and aerated in accordance with local air quality guidelines. When the average gasoline concentrations had declined to non-detectable levels, the soil was placed in a Class III landfill. Finally, closure sampling of the downtown site was performed to document that the contaminated soils had been successfully removed.

REFERENCES

Aas, G., Lacasse, S., Lunne, T. & Hoeg, K. (1986) Use of in-situ tests for foundation design in clay. In *Proceedings of ASCE Geotechnical Engineering Division Specialty Conference, In-Situ '86, on Use of In-Situ Tests in Geotechnical Engineering*, Virginia Tech, Blacksburg, VA. Geotechnical Special Publication No. 6, American Society of Civil Engineer, Reston, VA, pp. 1–30.

AASHTO T-99 (1990a) *The Moisture-Density Relationship of Soils Using a 5.5. lb [2.5 kg] Rammer and a 12 in. [305 mm] Drop*, Standard Specifications for Transportation Materials and Methods of Sampling & Testing, Part II Tests, 15th Ed, American Association of State Highway Transportation Officials, Washington, DC.

AASHTO T-180 (1990b) *The Moisture–Density Relations of Soils Using a 10 lb (4.54 kg) Rammer and a 18 in. (457 mm) Drop*, Standard Specifications for Transportation Materials & Methods of Sampling & Testing, Part II Tests, 15th Ed, American Association of State Highway & Transportation Officials, Washington, DC.

AASHTO (1993) *AASHTO Guide for Design of Pavement Structures*, American Association of State Highway & Transportation Officials, Washington, DC.

AASHTO (1994) *LRFD Bridge Design Specifications, SI Units*, American Association of State of Highway & Transportation Officials, Washington, DC.

AASHTO (1997) *Specification for Highway Bridges*, Interim Report, American Association of State Highway & Transportation Officials, Washington DC.

AASHTO (1998) *DARWin*, American Association of State Highway & Transportation Officials, Washington, DC.

Abdul, A.S., Kia, S.F. & Gibson, T.L. (1989) Limitations of monitoring wells for the detection and quantification of petroleum products in soils and aquifers. *Ground Water Monitoring Review*, 9(2): 90–99.

Abdul, A.S., Gibson, T.L. & Rai, D.N. (1990) Laboratory study of the flow of some organic solvents and their aqueous solutions through bentonite and kaolin clays. *Ground Water*, 28(4): 524–533.

Aboshi, H. & Suematsu, N. (1985) The state of the art on sand compaction pile method. In *Soil Improvement Methods, Proceedings of the Third International Geotechnical Seminar*, Nanyang Technological Institute, Singapore.

Abou-matar, H. & Goble, G.G. (1997) SPT dynamic analysis and measurements. *ASCE Journal of Geotechnical & Geoenvironmental Engineering*, Reston, VA, 123(10): 921–928.

Abrahamson, N.A. & Shedlock, K.M. (1997) Some comparisons between recent ground motion relations. *Seismological Research Letters*, 68(1): 9–23.

Abrahamson, N.A. & Silva, W.J. (1997) Equations for estimating horizontal response spectra and peak acceleration from western North American earthquakes: a summary of recent work. *Seismological Research Letters*, 68(1): 94–127.

Abramson, L.W. (1985) Rock wedge stability analysis on a personal computer. In *Proceedings of the 26th US Symposium on Rock Mechanics*, Rapid City, pp. 675–682.

Abramson, L.W. (1997) Ground treatment. Section 3.0 In Schaefer, V.R. (Ed.) *Ground Improvement, Ground Reinforcement, & Ground Treatment: Developments 1987–1997*, Geotechnical Special Publication No. 69, American Society of Civil Engineers, ASCE Press, Reston, VA, pp. 306–371.

Acar, Y.B. & Seals, R.K. (1984) Clay barrier technology for shallow land waste disposal facilities. *Hazardous Waste* 1: 167–181.

Acar, Y.B., Hamidon, A., Field, S.D. & Scott, L. (1985) The effect of organic fluids on hydraulic conductivity of compacted kaolinite in hydraulic barriers in soil and rock. In Johnson, A.I., Frobel, R.K., Cavalli, N.J. & Petersson, C.B., (Eds) *ASTM STP 874*, American Society for Testing Materials, West Conshohocken, PA, pp. 177–187.

Achilleous, E. (1988) PC STABL5M, User Manual, *Informational Report*, School of Civil Engineering, Purdue University, West Lafayette, 132 p.

ACPA (1992) *Design of Concrete Pavements for City Streets*, American Concrete Pavement Association, Skokie, IL.

Adachi, T., Kimura, M. & Zhang, F. (1994) Analysis of ultimate behaviour of lateral loading of cast-in-place concrete piles by 3-dimensional elasto-plastic FEM. In *Proceedings of the Eighth International Conference on Computer Methods and Advances in Geomechanics*, Morgantown.

Adachi, T., Liu, J., Koike, A. & Zhang, F. (1996) Finite element analysis of Biot's consolidation in slope excavation based on a constitutive model with strain softening. In *Proceedings of the Seventh International Symposium on Landslides*, Trondheim, Vol. 2, pp. 1131–1136.

Adachi, T., Oka, F. & Zhang, F. (1998) An elasto-viscoplastic constitutive model with strain softening. *Soils and Foundations*, **38**(2): 27–35.

Adams, J.I. (1961) Laboratory compression tests on peat. In *Proceedings of the Seventh Muskeg Research Conference*, NRC Technical Memorandum 41, Ottawa, ON, Canada.

Adams, J.I. (1965) The engineering behaviour of a Canadian muskeg. In *Proceedings of the Sixth International Conference on Soil Mechanics & Foundation Engineering*, Montreal, Vol. 1, pp. 3–7.

AERIS Software Inc. (1990) AERIS Model Version 3.0 Technical Manual, Supply & Services Canada, Contract 09SE-DE405-6-6586.

Ahlvin, R. (1991) Origin of developments for structural design of pavements. Technical Report GL9126, US Army Corps of Engineers, Waterways Experiment Station, Vicksburg, MI.

Ahlvin, R.G. (1971). Multiple-wheel heavy gear load pavement tests, Technical Report S-17-17, Vol. 1, U.S. Army Corps Engineers, Waterways Experimental Station, Vicksburg, MS.

Ahmed, S., Lacroix, Y. & Steinbach, J. (1975) Pumping tests in an unconfined aquifer. In *Proceedings of ASCE Geotechnical Engineering Division Specialty Conference on In-Situ Measurement of Soil Properties*, North Carolina State University, Raleigh, NC, June 1–4, 1975, Vol. 1, pp. 1–21.

Ahuja, L.R., Naney, J.W. & Williams, R.D. (1985) Estimating soil–water characteristics from simpler properties and limited data. *Soil Science of America Journal*, **49**: 1100–1105.

Airey, D.W. & Wood, D.M. (1987) An evaluation of direct simple shear tests on clay. *Géotechnique*, **37**(1): 25–35.

AISI (1990) *Modern Sewer Design*, 2nd Edn, American Iron & Steel Institute, Washington, DC.

Aitchison, J. (1955) On the distribution of a positive random variable having a discrete probability mass at the origin. *Journal of American Statistical Association*, **50**: 901–908.

Ajaz, A. & Parry, R.H.G. (1975) Stress–strain behavior of two compacted clays in tension and compression. *Géotechnique*, **25**(3): 495–512.

Ajaz, A. & Parry, R.H.G. (1975) Analysis of bending stresses in soil beams. *Géotechnique*, **25**(3): 586–591.

Ajaz, A. & Parry, R.H.G. (1976) Bending test for compacted clays. *Journal of Geotechnical Engineering Division, ASCE*, Reston, VA, **102**(9): 929–943.

Akagi, T. (1979) Consolidation caused by mandrel-driven sand drains. In *Proceedings of the Sixth Asian Regional Conference on Soil Mechanics & Foundation Engineering*, Singapore, Vol. 1, pp. 125–128.

Aki, K. (1988) Local site effects on strong ground motion, In Van Thun, J.L. (Ed.), *Earthquake Engineering & Soil Dynamic, Recent Advances in Ground-Motion Evaluation*, Geotechnical Special Publication No. 20, American Society of Civil Engineers, Reston, VA, pp. 103–155.

Alberro, J. (1979) Stabilité à long terme des excavations dans la ville de Mexico. *International Symposium of Soil Mechanics*, Oaxaca, Mexico, Vol. 1 pp. 125–143.

Albertsen, M., and an ad hoc Task Force (1986) Beurteilung und Behandlung von Mineralölschadensfällen im Hinblick auf den Grundwasserschutz, Jeil 1, Die wissenschaftlichen Grundlagen zum Verständnis des Verhaltens von Mineralöl im Untergrund. Federal Office of the Environment, LTw S no. 20, 178 p.

Aldrich, H.P. (1956) Frost penetration below highway and airfield pavements. *US Highway Research Board, Bulletin*, **135**: 24–149.

Aller, L., Bennett, T.W., Hackett, G., Petty, R.J., Lehr, J.H., Sedoris, H., Nielsen, D.M. & Denne, J.E. (1989) *Handbook of Practices for the Design and Installation of Ground-Water Monitoring Wells*, EPA/600/4-89/034. National Water Well Association.

Al-Sanad, H. & Aggour, M.S. (1984) Dynamic soil properties from sinusoidal and random vibrations. In *Proceedings of the Eighth World Con-*

ference on Earthquake Engineering, San Francisco, pp. 15–22.

Ambraseys, N.N. (1988) Engineering seismology. Earthquake Engineering & Structural Dynamics, **17**: 1–105.

Amos, E.M., Blakeway, D. & Warren, C.D. (1986) Remote sensing techniques in civil engineering surveys. In Hawkins, A.B. (Ed.) Site Investigation Practice: Assessing BS 5930, Engineering Geology Special Publication No. 2, Geological Society, London, England, pp. 119–124.

Anagnostou, G. & Kovari, K. (1996) Face stability conditions with earth-pressure-balance shields. Tunnelling & Underground Space Technology, **11**(2): 165–173.

Andersland, O.B. & Anderson, D.M. (1978) Geotechnical Engineering for Cold Regions, McGraw-Hill, New York.

Andersland, O.B. & Ladanyi, B. (1994) An Introduction to Frozen Ground Engineering, Chapman & Hall, New York.

Anderson, D.M. & Morgenstern, N.R. (1973) Physics, chemistry and mechanics of frozen ground. In Proceedings of The Second International Conference on Permafrost, Yakutsk, USSR, pp. 257–288.

Anderson, D.M. & Tice, A.R. (1972) Predicting unfrozen water contents in frozen soils from surface area measurements. Highway Research & Record **373**: 12–18.

Anderson, J.E., Nowak, R.S., Ratzlaff, T.D. & Markham, O.D. (1993) Managing soil moisture on waste burial sites in arid regions. Journal of Environmental Quality, **22**: 2–69.

Anderson, J.N. & Lade, P.V. (1981) The expansion index test. Geotechnical Testing Journal, **4**(2): 58–67.

Anderson, L.G. (1968) A modern approach to overburden drilling. Western Miner, November.

Anderson, M.P. (1984) Movement of contaminants in groundwater: groundwater transport-advection and dispersion. In National Research Council (U.S.) Geophysics Study Committee (Eds.) Groundwater Contamination, NRC Studies in Geophysics. National Academy Press, Washington, DC, pp. 37–45.

Anderson, M.P. & Woessner, W.W. (1992) Applied Groundwater Modeling Simulation of Flow and Advective Transport, Academic Press, San Diego, CA.

Andresen, A., Berre, T., Kleven, A. & Lunne, T. (1979) Procedures used to obtain soil parameters for foundation engineering in the North Sea. Marine Geotechnology, **3**(3): 1–18.

API (1984) Recommended Practice for Planning, Designing and Constructing Fixed Offshore Platforms, 14th Edn. APIRP2A, American Petroleum Institute, Dallas, TX.

API (1988) Phase Separated Hydrocarbon Contaminant Modeling for Corrective Action, Publication American Petroleum Institute, 4474, Washington, DC.

API (1993) Recommended Practice for Planning Designing and Constructing Fixed Offshore Platforms, American Petroleum Institute, Dallas, TX.

API (1994) Decision Support System for Exposure and Risk Assessment, Version 1.0, American Petroleum Institute, Nassau, New York.

Arango, I. (1996) Magnitude scaling factors for soil liquefaction evaluations. Journal of Geotechnical Engineering, ASCE, Reston, VA, **122**(11): 929–936.

Arias, A., Sanchez-Sesma, F.J. & Ovando-Shelley, E. (1981) A simplified elastic model for seismic analysis of earth-retaining structures with limited displacements, In Proceedings of The International Conference on Recent Advances in Geotechnical Earthquake Engineering & Soil Dynamics, St Louis, MO, Vol. I, 235–240.

Arman, A. (1978) Current practices in the treatment of soft foundations. In Soil Improvement, History, Capabilities, and Outlook, Report by Committee on Placement and Improvement of Soils, Geotechnical Engineering Division, American Society of Civil Engineers, Reston, VA, pp. 30–51.

Armitage, P. & Doll, R. (1961) Stochastic model for carcinogenisis. In Lecam, W. & Heyman, J. (Eds) Proceedings of the Fourth Berkeley Symposium on Mathematical Statistics & Probability, Vol. 4, University of California Press, Berkeley, CA.

Arnold, J.G., Williams, J.R., Nicks, A.D. & Sammons, N.B. (1989) SWRRB, a Simulator for Water Resources in Rural Basins, Agricultural Research Service, USDA, Texas A&M University Press, College Station, TX.

Arya, L.M. & Paris, J.F. (1981) A physicoempirical model to predict the soil moisture characteristic from particle size distribution and bulk density data. Soil Science of America Journal, **45**: 1023–1030.

Arya, S., O'Neil, M. & Pincus, G. (1979) Design of Structures and Foundations for Vibrating Machines, Gulf Publishing, Houston, TX.

AS 2870.2 (1990) *Residential Slabs and Footings, Part 2: Guide to Design by Engineering Principles*, Australian Standards, Sydney, Australia.

Asante-Duah, D.K. (1993) *Hazardous Waste Risk Assessment*, Lewis, Boca Raton, FL.

Asaoka, A. (1978) Observational procedure of settlement prediction. *Soils & Foundations*, **18**(4): 87–101.

ASCE (1974) *Subsurface Exploration for Underground Excavation & Heavy Construction. Proceedings of the Geotechnical Engineering Division Specialty Conference*, New England College, Henniker, NH, American Society of Civil Engineers, Reston, VA.

ASCE (1975) *In-Situ Measurement of Soil Properties. Proceedings of ASCE Geotechnical Engineering Division Specialty Conference*, North Carolina State University, Raleigh, NC, June 1–4, American Society of Civil Engineers, Reston, VA.

ASCE (1976) *Subsurface Investigation for Design & Construction of Foundations of Buildings*. ASCE Manuals & Reports on Engineering Practice No. 56, American Society of Civil Engineers, Reston, VA.

ASCE (1978) *Site Characterization & Exploration. Proceedings of the Geotechnical Engineering Division Specialty Conference*, Northwestern University, Evanston, IL, American Society of Civil Engineers, Reston, VA.

ASCE (1986) *Use of In-Situ Tests in Geotechnical Engineering. Proceedings of the ASCE Geotechnical Engineering Division Specialty Conference*, Virginia Tech, Blacksburg, Virginia. Geotechnical Special Publication No. 6, American Society of Civil Engineers, Reston, VA.

ASCE (1995) *Rock Foundations*. Technical Engineering and Design Guides as adapted from the US Army Corps of Engineers, No. 16, American Society of Civil Engineers, Reston, VA.

ASCE (1997) Geotechnical baseline reports for underground construction. In Essex, R.J. (ed.) *Guidelines & Practices*, American Society of Civil Engineers, Reston, VA.

Aspe, J.I. (1988) Polyethylene resins for geomembrane applications. In Koerner, R.M. (ed.) *Geosynthetics Research Institute Conference on the Durability & Aging of Geosynthetics*, Elsevier Applied Science, London, England, pp. 159–176.

ASTM A716 Specification for ductile iron culvert pipe, American Society for Testing and Materials, West Conshohocken, PA.

ASTM A746 Specification for ductile iron gravity sewer pipe, American Society for Testing and Materials, West Conshohocken, PA.

ASTM A796 Standard practice for structural design of corrugated steel pipe, pipe-arches, and arches for storm and sanitary sewers and other buried applications, American Society for Testing and Materials, West Conshohocken, PA.

ASTM A798 Standard practice for installing factory-made corrugated steel pipe for sewers and other applications, American Society for Testing and Materials, West Conshohocken, PA.

ASTM C76 Specification for reinforced concrete culvert, storm drain, and sewer pipe, American Society for Testing and Materials, West Conshohocken, PA.

ASTM C78 *Standard Test Method for Flexural Strength of Concrete (Using Simple Beam With Third-Point Loading)*, American Society for Testing & Materials, West Conshohocken, PA, *http://www.astm.org, Annual Book of ASTM Standards* 4.02.

ASTM C14 Specification for concrete sewer, storm drain, and culvert pipe, American Society for Testing and Materials, West Conshohocken, PA.

ASTM C88 *Standard Test Method for Soundness of Aggregates By Use of Sodium Sulfate or Magnesium Sulfate*, American Society for Testing & Materials, West Conshohocken, PA.

ASTM C131 *Standard Test Method for Resistance to Degradation of Small-Size Coarse Aggregate By Abrasion and Impact in the Los Angeles Machine*. American Society for Testing & Materials, West Conshohocken, PA.

ASTM C123 *Standard Test Method for Lightweight Pieces in Aggregate*. American Society for Testing & Materials, West Conshohocken, PA.

ASTM C301 Test method for vitrified clay pipe, American Society for Testing and Materials, West Conshohocken, PA.

ASTM C655 Standard specification for reinforced concrete D-load culvert, storm drain, and sewer pipe, American Society for Testing and Materials, West Conshohocken, PA.

ASTM D413 *Test Methods for Rubber Property-Adhesion to Flexible Substrate*, West Conshohocken, PA.

ASTM D422-63 *Standard Test Method for Particle-Size Analysis of Soils*, American Society for Testing & Materials, West Conshohocken, PA, *http://www.astm.org, Annual Book of ASTM Standards* 04.08, *Soil & Rock* (I).

ASTM D427-93 *Standard Test Method for Shrinkage Factors of Soils by Mercury Method*, American Society for Testing & Materials, West Conshohocken, PA.

ASTM D638 *Test Method for Tensile Properties of Plastics*, West Conshohocken, PA.

ASTM D698 *Standard Test Methods for Moisture–Density Relations of Soils & Soil Aggregate Mixtures Using 5.5-lb (2.49 kg) Rammer and 12 in (305 mm) Drop* 4.08, American Society for Testing & Materials, West Conshohocken, PA.

ASTM D698-91 *Standard Test Method for Laboratory Compaction Characteristics of Soil Using Standard Effort (12,400 ft-lbf/ft³ (600 kN.m/m³))*, American Society for Testing & Materials, West Conshohocken, PA.

ASTM D 751 *Test Methods for Coated Fabrics*, West Conshohocken, PA.

ASTM D854-92 *Standard Test Method for Specific Gravity of Soils*, American Society for Testing & Materials, West Conshohocken, PA.

ASTM D882 *Test Methods for Tensile Properties of Thin Plastic Sheeting*, West Conshohocken, PA.

ASTM D1067-92 *Standard Test Methods for Acidity or Alkalinity of Water*, 04.08, American Society for Testing & Materials, West Conshohocken, PA.

ASTM D1194-72 *Standard Test Method for Bearing Capacity of Soil for Static Load and Spread Footings*, 04.08, American Society for Testing & Materials, West Conshohocken, PA.

ASTM D1238 *Standard Test Method for Flow Rates of Thermoplastics by Extrusion Plastometer*, American Society for Testing & Materials, West Conshohocken, PA.

ASTM D1505 *Standard Test Method for Density of Plastics by the Density-Gradient Technique*, American Society for Testing & Materials, West Conshohocken, PA.

ASTM D1556 *Standard Test Method for Density and Unit Weight of Soil in Place by the Sand-Cone Method*, American Society for Testing & Materials, West Conshohocken, PA.

ASTM D1557 *Standard Test Methods for Moisture-Density Relations of Soils and Soil–Aggregate Mixture Using 10-lb (4.54 kg) Rammer and 18 in (457 mm) Drop*, American Society for Testing & Materials, West Conshohocken, PA.

ASTM D1557-91 *Standard Test Method for Laboratory Compaction Characteristics of Soil Using Modified Effort (56,000 ft-lbf/ft³ (2,700 kN·m/m³))*, American Society for Testing & Materials, West Conshohocken, PA.

ASTM D1883 *Standard Test Method for Bearing Ratio of Laboratory-Compacted Soils*, 4.08, American Society for Testing & Materials, West Conshohocken, PA.

ASTM D2167 *Standard Test Method for Density and Unit Weight of Soil in Place by the Rubber Balloon Method*, American Society for Testing & Materials, West Conshohocken, PA.

ASTM D2216 *Standard Test Method for Laboratory Determination of Water (Moisture) Content of Soil and Rock*, American Society for Testing & Materials, West Conshohocken, PA.

ASTM D2321 Practice for underground installation of thermoplastic pipe for sewers and other gravity-flow applications, American Society for Testing and Materials, West Conshohocken, PA.

ASTM D2412 Standard test method for determination of external loading characteristics of plastic pipe by parallel plastic plate loading, American Society for Testing and Materials, West Conshohocken, PA.

ASTM D2419 *Standard Test Method for Sand Equivalent Value of Soils and Fine Aggregate*, 4.08, American Society for Testing & Materials, West Conshohocken, PA.

ASTM D2434-68 *Standard Test Method for Permeability of Granular Soils (Constant Head)*, American Society for Testing & Materials, West Conshohocken, PA.

ASTM D2487-93 *Standard Test Method for Classification of Soils for Engineering Purposes*, American Society for Testing & Materials, West Conshohocken, PA.

ASTM D2664-86 *Test Method for Triaxial Compressive Strength of Undrained Rock Core Specimens Without Pore Pressure Measurements*, American Society for Testing & Materials, West Conshohocken, PA.

ASTM D2922 *Standard Test Methods for Density of Soil and Soil–Aggregate in Place by Nuclear Methods (Shallow Depth)*, American Society for Testing & Materials, West Conshohocken, PA.

ASTM D2936-84 *Test Method for Direct Tensile Strength of Intact Rock Core Specimens*, American Society for Testing & Materials, West Conshohocken, PA.

ASTM D2938-86 *Test Method for Unconfined Compressive Strength of Intact Rock Core Specimens*, American Society for Testing & Materials, West Conshohocken, PA.

ASTM D2974-87 *Standard Test Methods for Moisture, Ash, and Organic Matter of Peat and Other Organic Soils*, American Society for Testing & Materials, West Conshohocken, PA.

ASTM D3017 *Standard Test Method for Water Content of Soil and Rock in Place by Nuclear Methods (Shallow Depth)*, 04.08, American Society for Testing & Materials, West Conshohocken, PA.

ASTM D3034 Specification for Type PSM poly (vinyl chloride) (PVC) sewer pipe and fittings, American Society for Testing and Materials, West Conshohocken, PA.

ASTM D3080 *Standard Test Method for Direct Shear Test of Soils Under Consolidated Drained Conditions*, 04.08, American Society for Testing & Materials, West Conshohocken, PA.

ASTM D3083 *Specification for Flexible Poly (Vinyl Chloride) Plastic Sheeting for Pond, Canal and Reservoir Lining*, West Conshohocken, PA.

ASTM D3262 Specification for "Fiberglass" (glass-fiber-reinforced thermosetting-resin) sewer pipe, American Society for Testing and Materials, West Conshohocken, PA.

ASTM D3895 *Standard Test Methods for Oxidative-Induction Time of Polyolefins by Differential Scanning Calorimetry*, 04.08, American Society for Testing & Materials, West Conshohocken, PA.

ASTM D4123 *Standard Test Method for Indirect Tension Test for Resilient Modulus of Bituminous Mixtures*, 04.03, American Society for Testing & Materials, West Conshohocken, PA.

ASTM D4253-91 *Standard Test Methods for Maximum Index Density and Unit Weight of Soils Using a Vibratory Table*, American Society for Testing & Materials, West Conshohocken, PA.

ASTM D4254-91 *Standard Test Methods for Minimum Index Density and Unit Weight of Soils and Calculation of Relative Density*, American Society for Testing & Materials, West Conshohocken, PA.

ASTM D4318-95 *Standard Test Method for Liquid Limit, Plastic Limit, and Plasticity Index of Soils*, American Society for Testing & Materials, West Conshohocken, PA.

ASTM D4355 *Standard Test Method for Deterioration of Geotextiles from Exposure to Ultraviolet Light and Water (Xenon-Arc Type Apparatus)*, American Society for Testing & Materials, West Conshohocken, PA.

ASTM D4394 *Test Method for Determining the* in situ *Modulus of Deformation of Rock Mass Using the Rigid Plate Loading Method*, American Society for Testing & Materials. West Conshohocken, PA.

ASTM D4395 *Test Method for Determining the* in situ *Modulus of Deformation of Rock Mass Using the Flexible Plate Loading Method*, American Society for Testing & Materials, West Conshohocken, PA

ASTM D4437 *Practice for Determining the Integrity of Field Seams Used in Joining Flexible Polymeric Sheet Geomembranes*, West Conshohocken, PA.

ASTM D4491 *Standard Test Method for Water Permeability of Geotextiles by Permittivity*, American Society for Testing & Materials, West Conshohocken, PA.

ASTM D4506 *Test Method for Determining the* in situ *Modulus of Deformation of Rock Mass Using a Radical Jacking Test*, American Society for Testing & Materials, West Conshohocken, PA.

ASTM D4595 *Standard Test Method for Tensile Properties of Geotextiles by the Wide-Width Strip Method*, American Society for Testing & Materials, West Conshohocken, PA.

ASTM D4643 *Standard Test Method for Determination of Water (Moisture) Content of Soil by the Microwave Oven Method*, American Society for Testing & Materials, West Conshohocken, PA.

ASTM D4751 *Standard Test Method for Determining Apparent Opening Size of a Geotextile*, American Society for Testing & Materials, West Conshohocken, PA.

ASTM D4791 *Standard Test Method for Flat or Elongated Particles in Coarse Aggregate*, 4.03, American Society for Testing & Materials, West Conshohocken, PA.

ASTM D4829-95 *Standard Test Method for Expansion Index of Soils*, American Society for Testing & Materials, West Conshohocken, PA.

ASTM D4885 *Standard Test Method for Determining Performance Tensile Strength of Geomembranes Using Wide Strip Testing*, American Society for Testing & Materials, West Conshohocken, PA.

ASTM D4971 *Test Method for Determining the* in situ *Modulus of Deformation of Rock Using the Diametrically Loaded 76-mm (3-in) Borehole Jack*, 04.09, American Society for Testing & Materials, West Conshohocken, PA.

ASTM D5084 *Standard Test Method for Mea-*

surement of a Hydraulic Conductivity of Saturated Porous Materials Using a Flexible Wall Permeameter, American Society for Testing & Materials, West Conshohocken, PA.

ASTM D5092 Standard Practice for Design and Installation of Ground Water Monitoring Wells in Aquifer, American Society for Testing & Materials, West Conshohocken, PA.

ASTM D5101 Standard Test Method for Measuring the Soil–Geotextile System Clogging Potential by the Gradient Ratio, American Society for Testing & Materials, West Conshohocken, PA.

ASTM D5199 Standard Test Method for Measuring Nominal Thickness of Geotextiles & Geomembranes, American Society for Testing & Materials, West Conshohocken, PA.

ASTM D5208 Standard Practice for Operating Fluorescent Ultraviolet (UV) Condensation Apparatus for Exposure of Photo-degradable Plastics, American Society for Testing & Materials, West Conshohocken, PA.

ASTM D5261 Standard Test Method for Measuring Mass per Unit Area of Geotextiles, American Society for Testing & Materials, West Conshohocken, PA.

ASTM D5321 Standard Test Method for Determining the Coefficient of Soil and Geosynthetic or Geosynthetic and Geosynthetic Friction by the Direct Shear Method, American Society for Testing & Materials, West Conshohocken, PA.

ASTM D5322 Standard Practice for Immersion Procedures for Evaluating the Chemical Resistance of Geosynthetics to Liquids, American Society for Testing & Materials, West Conshohocken, PA.

ASTM D5397 Standard Test Method for Evaluation of Stress Crack Resistance of Polyolefin Geomembranes using Notched Constant Tension Load Test, American Society for Testing & Materials, West Conshohocken, PA.

ASTM D5496 Standard Practice for Field Immersion Testing of Geosynthetics, American Society for Testing & Materials, West Conshohocken, PA.

ASTM D5617 Standard Test Methods for Multi-Axial Tension Test for Geosynthetics, American Society for Testing & Materials, West Conshohocken, PA.

ASTM D5747 Standard Practice for Tests to Evaluate the Chemical Resistance of Geomembranes to Liquids, American Society for Testing & Materials, West Conshohocken, PA.

ASTM D5885 Standard Test Method for Oxida-tive Induction Time of Polyolefin Geosynthetics by High-Pressure Differential Scanning Calorimetry, American Society for Testing & Materials, West Conshohocken, PA.

ASTM D5887 Standard Test Method for Measurement of Index Flux Through Saturated Geosynthetic Clay Liner Specimens Using a Flexible Wall Permeameter, American Society for Testing & Materials, West Conshohocken, PA.

ASTM D5890 Standard Test Method for Swell Index of Clay Mineral Component of Geosynthetic Clay Liners, American Society for Testing & Materials, West Conshohocken, PA.

ASTM D5993 Standard Test Method for Measuring the Mass Per Unit Area of GCL, American Society for Testing & Materials, West Conshohocken, PA.

ASTM D5994 Standard Test Method for Measuring the Core Thickness of Textured Geomembranes, American Society for Testing & Materials, West Conshohocken, PA.

ASTM E11 Wire–Cloth Sieves for Testing Purposes, American Society for Testing & Materials, West Conshohocken, PA, Website: http://www.astm.org, Annual Book of ASTM Standards Soil & Rock (I).

ASTM E96 Standard Test Method for Water Vapor Transmission of Materials, American Society for Testing & Materials, West Conshohocken, PA.

ASTM E1527-93 Standard Practice for Environmental Site Assessments: Phase I Environmental Site Assessment Process, American Society for Testing & Materials, West Conshohocken, PA.

ASTM F679 Specification for Poly(Vinyl Chloride) (PVC) Large-Diameter Plastic Gravity Sewer Pipe and Fittings, American Society for Testing and Materials, West Conshohocken, PA.

ASTM F714 Standard specification for polyethylene (PE) plastic pipe (SDR-PR) based on outside diameter, American Society for Testing and Materials, West Conshohocken, PA.

ASTM F810 Specification for smoothwall polyethylene (PE) pipe for use in drainage and waste disposal aborption fields, American Society for Testing and Materials, West Conshohocken, PA.

ASTM F892 Specification for polyethylene (PE) corrugated pipe with smooth interior and fittings, American Society for Testing and Materials, West Conshohocken, PA.

ASTM F894 Specification for polyethylene (PE) large diameter profiled wall sewer and drain pipe, American Society for Testing and Materials, West Conshohocken, PA.

ASTM F949 Specification for poly(vinyl chloride) (PVC) corrugated sewer pipe with smooth interior and fittings, American Society for Testing and Materials, West Conshohocken, PA.

ASTM (1988) *Vane Shear Strength Testing in Soils: Field and Laboratory Studies*, Special Technical Publication 1014, Richards, A.F. (Ed.), American Society for Testing & Materials, West Conshohocken, PA.

ASTM (1996) Standard Guide for Risk-Based Corrective Action Applied at Petroleum Release Sites. E1739-95 Annual Book of ASTM Standards, Vol. 11.04, American Society for Testing and Materials, West Conshohocken, PA.

ASTM (1997a) *Environmental Sampling*, 2nd Edn, ASTM Standards, Committee D-18 on Soil & Rock, American Society for Testing & Materials, West Conshohocken, PA.

ASTM (1997b) *Environmental Site Characterization*, ASTM Standards, Committees D-18 on Soil & Rock, D-19 on Water, D-34 on Waste Management, E-47 on Biological Effects and Environmental Fate and E-50 on Environmental Assessment, American Society for Testing & Materials, West Conshohocken, PA.

AS 3725 Loads on buried concrete pipes, Standards Australia International Ltd, Sydney.

AS 46060 Loads on buried vitrified clay pipes, Standards Australia International Ltd, Sydney.

AS/NZS 2566.1 Buried flexible pipeline—structural design, Standards Australia International Ltd, Sydney.

Atkinson, J.H. & Bransby, P.L. (1978) *The Mechanics of Soils–An Introduction to Critical State Soil Mechanics*, McGraw-Hill, Maidenhead, England.

Atkinson, J.H. & Mair, R.J. (1981) Soil mechanics aspects of soft ground tunnelling. *Ground Engineering*, **14**(5): 20–28.

Attewell, P.B. & Farmer, I.W. (1976) *Principles of Engineering Geology*, Chapman & Hall, London, England.

Aubery, D. & Chapel, F. (1981) 3-D dynamic analysis of groups of piles and comparisons with experiments. In *Transactions of the Sixth International Conference on Structural Mechanics in Reactor Technology*, Paris.

Augello, A.J., Matasovic, N., Bray, J.D., Kavazanjian, E., Jr., & Seed, R.B. (1995) Evaluation of solid waste landfill performance during the Northridge earthquake, In *Earthquake Design & Performance of Solid Waste Landfills*, Yegian, M.K. and Finn, W.D.L. (Eds.) Geotechnical Special Publication No. 54, American Society of Civil Engineers, Reston, VA, pp. 17–50.

Augello, A.J., Bray, J.D., Abrahamson, N.A., & Seed, R.B. (1998) Dynamic properties of solid waste based on back-analyses of OII landfill. *Journal of Geotechnical Engineering, ASCE*, Reston, VA, **124**(3): 211–222.

August, H. & Tatzky, R. (1984) Permeability of commercially available polymeric liners for hazardous landfill leachate organic constituents. In *Proceedings of International Conference on Geomembranes*, June 20–24, Denver, CO, Vol. I, Industrial Fabrics Association International, Roseville, MN, pp. 163–168.

Avery, S. (1951) Analysis of groundwater lowering adjacent to open water. *Journal of the Soil Mechanics & Foundation Division, ASCE*, Reston, VA, **77**(106): 1–16.

AWWA C905 Standard for PVC water transmission pipe (nominal diameters 14 in. through 36 in.), American Waterworks Association, Denver.

Badu-Tweneboah, K., Giroud, J.P., Carlson, D.S. & Schmertmann, G.R. (1997) Discussion of "Field evaluation of protective covers for landfill geomembrane liners under construction loading" by Reddy *et al.*, *Geosynthetics International*, **4**(5): 543–544.

Badu-Tweneboah, K., Giroud, J.P., Carlson, D.S. & Schmertmann, G.R. (1998) Evaluation of the effectiveness of HDPE geomembrane liner protection. In Rowe, R.K. (Ed.) *Proceedings of the Sixth International Conference on Geosynthetics*, Vol. 1, March 1998, Atlanta, GA, Industrial Fabrics Association International, Roseville, MN, pp. 279–284.

Badv, K. & Rowe, R.K. (1996) Contaminant transport through a soil liner underlain by an unsaturated stone collection layer. *Canadian Geotechnical Journal*, **33**(2): 416–430.

Baez, J.I. (1995) *PhD thesis*, University of Southern California, Los Angeles.

Baguelin, F. & Frank, R. (1979) Theoretical studies of piles using the finite element method. In *Proceedings of the Conference on Numerical Methods in Offshore Piling*, London, Institution of Civil Engineers, Paper No. 11.

Baguelin, F., Jezequel, J.F., LeMee, E. & LeMelhaute, A. (1972) Expansion of cylindrical probes in cohesive soils. *ASCE Journal of Soil*

Mechanics & Foundations Division, Reston, VA, **98**(11): 1129–1142.

Baguelin, F., Jezequel, J.F. & Shields, D.H. (1978) *The Pressuremeter & Foundation Engineering*, Trans Tech Publications, Clausthal, Germany.

Baker, M.J., Blowes, D.W. & Ptacek, C.J. (1997) Phosphorus adsorption and precipitation in a permeable reactive wall: applications for wastewater disposal systems. In *Proceedings of the 1997 International Containment Technology Conference & Exhibition*, February 10–12, St Petersburg, FL, pp. 697–703.

Baker, T.L. & Marienfeld, M.L. (1995) Correlation of outdoor exposure to xenon-arc weatherometer exposure. In *Geosynthetics '95 Conference, Proceedings*, Nashville, TN, IFAI, Roseville, MN, pp. 829–840.

Baker, W.H. (1982) Planning and performance of structural chemical grouting. In Baker, W.H. (Ed.) *Grouting in Geotechnical Engineering*, Proceedings of the Conference, New Orleans, American Society of Civil Engineers, Reston, VA, pp. 515–540.

Baker, W.H., Cording, E.J. & MacPherson, H.H. (1983) Compaction grouting to control ground movements during tunneling. *Underground Space*, **7**: 205–212.

Balaam, B.E., Poulos, H.G. & Brown, P.T. (1977) Settlement analyses of soft clays reinforced with granular piles. In *Proceedings of the Fifth Southeast Asian Conference on Soils Engineering*, Bangkok, pp. 81–92.

Banerjee, P.K. (1978) Analysis of axially and laterally loaded pile groups. In Scott, C.R. (Ed.) *Developments in Soil Mechanics*, Applied Science, London, England, Ch. 9.

Banerjee, P.K. & Davies, T.G. (1977) Analysis of pile groups embedded in Gibson soil. In *Proceedings of the Ninth International Conference on Soil Mechanics & Foundation Engineering*, Vol. 1, pp. 381–386.

Banerjee, S. (1984) Solubility of organic mixtures in water. *Environmental Science Technology*, **18**(8): 587–591.

Bara, J.P. (1978) Collapsible soils and their stabilization. *Soil Improvement, History, Capabilities, & Outlook*, Report by Committee on Placement & Improvement of Soils, Geotechnical Engineering Division, American Society of Civil Engineers, ASCE Press, Reston, VA, pp. 141–152.

Barbaro, J.R., Barker, J.F., Lemon, L.A. & Mayfield, C.I. (1992) Biotransformation of BTEX under anaerobic, denitrifying conditions: field and laboratory observations. *Journal of Contaminant Hydrology*, **11**: 245–272.

Barbaro, J.R., Barker, J.F. & Butler, B.J. (1997) In situ bioremediation of gasoline residuals under mixed electron acceptor conditions. In Situ & On-Site Bioremediation **4**(5): 22–26.

Barcelona, M.J., Helfrich, J.A. & Garske, E.E. (1988) Verification of sampling methods and selection of materials for groundwater contamination studies. In Collins, A.G. & Johnson, A.I. (Eds) *Ground-Water Contamination: Field Methods*, ASTM STP 963, American Society for Testing & Materials, West Conshohocken, PA, pp. 221–231.

Bardet, J.P. (1997) *Experimental Soil Mechanics*, Prentice Hall, NJ.

Barkan, D.D. (1962) *Dynamics of Bases & Foundations* (translated from Russian), McGraw-Hill, New York.

Barker, J.F., Patrick, G.C. & Major, D. (1987) Natural attenuation of aromatic hydrocarbons in a shallow sand aquifer. *Ground Water Monitoring Review*, **7**: 64–71.

Barksdale, R.D. (1987) *State of the Art for Design and Construction of Sand Compaction Piles*, Technical Report REMR-GT-4, US Army Corps of Engineers, Washington, DC.

Barksdale, R.D. & Bachus, R.C. (1983) *Design and Construction of Stone Columns*, Vol. I, Report No. FHWA/RD-83/026, Federal Highway Administration, Springfield, VA.

Barnes, D.G., & Dourson, M. (1988) Reference dose (RFD): description and use in health risk assessment. *Regulatory Toxicology and Pharmacology*, **8**: 471–478.

Barnes, F.J. & Rodgers, J.E. (1988) *Evaluation of Hydrologic Models in the Design of Stable Landfill Covers*, EPA Project Summary, EPA/600/S2-88/048, US Environmental Protection Agency, Washington, DC.

Barone, F.S. (1990) *PhD thesis*, University of Western Ontario, London, Ontario.

Barone, F.S., Yanful, E.K., Quigley, R.M. & Rowe, R.K. (1989) Effect of multiple contaminant migration on diffusion and adsorption of some domestic waste contaminants in a natural clayey soil. *Canadian Geotechnical Journal*, **26**: 189–198.

Barone, F.S., Rowe, R.K. & Quigley, R.M. (1992) A laboratory estimation of diffusion and adsorption coefficients for several volatile organics in a natural clayey soil. *Journal of Contaminant Hydrogeology*, **10**: 225–250.

Barone, F.S., Costa, J.M.A. & Ciardullo, L. (1997) Temperatures at the base of a municipal solid waste landfill. In *50th Canadian Geotechnical Conference*, October, Ottawa, Vol. 1 pp. 144–152.

Barron, R.A. (1948) Consolidation of fine-grained soils by drain wells, *Transactions, ASCE*, Reston, VA, **113**: 718–747.

Barton, N. (1973) Review of a new shear strength criterion for rock joints. Engineering Geology, **7**: 287–332.

Barton, N. (1976) The shear strength of rock and rock joints. *International Journal of Rock Mechanics, Mineral Science & Geomechanics Abstracts*, **13**: 1–24.

Barton, N. & Bandis, S.C. (1990) Review of predictive capabilities of JRC-JCS model in engineering practice. In Barton, N. & Stephansson, O. *Proceedings of the International Symposium on Rock Joints*, Leon, Norway, pp. 603–610.

Barton, N. & Choubey, B. (1977) The shear strength of rock joints in theory and practice. *Rock Mechanics*, **10**(1): 1–54.

Barton, N., Lien, R. & Lunde, J. (1974) Engineering classification of rock masses for the design of tunnel support. *Rock Mechanics*, **6**: 189–236.

Bathurst, R.J. & Alfaro, M.C. (1996) Review of seismic design, analysis and performance of geosynthetic reinforced walls, slopes and embankments. Invited keynote paper, *IS-Kyushu '96, Third International Symposium on Earth Reinforcement*, Japan, 12–14 November 1996, Fukuoka, Kyushu, Vol. 2, A.A. Balkema, 887–918.

Bathurst, R.J. & Crowe, E.R. (1992) Recent case histories of flexible geocell retaining walls in North America. In Tatsuoka, F. & Leshchinsky, D. (Eds) *Recent Case Histories of Permanent Geosynthetic-Reinforced Soil Retaining Walls*, 6–7 November 1992, Tokyo, Japan, A.A. Balkema, Rotterdam, pp. 3–20.

Bathurst, R.J. & Simac, M.R. (1994) Geosynthetic reinforced segmental retaining wall structures in North America. Invited keynote paper, *Fifth International Conference on Geotextiles, Geomembranes & Related Products*, Vol. 4, 6–9 September 1994, Singapore, pp. 1275–1298.

Bathurst, R.J., Simac, M.R. & Berg, R.R. (1993) Review of NCMA segmental retaining wall design manual for geosynthetic-reinforced structures. *Transportation Research Record*, **1414**: 16–25.

Baziar, M.H. & Dobry, R. (1995) Residual strength and large-deformation potential of loose silty sands. *ASCE Journal of Geotechnical Engineering Division*, Reston, VA, **121**(12): 896–906.

Bear, J. (1972) *Dynamics of Fluids in Porous Media*, American Elsevier, New York.

Bear, J. (1979) *Hydraulics of Groundwater*, McGraw-Hill, New York.

Bear, J., Tsang, C. & de Marsily, G. (1993) *Flow and Contaminant Transport in Fractured Rock*, Academic Press, San Diego, CA.

Becker, D.E. (1996a) Eighteenth Canadian Geotechnical Colloquium: limit states design for foundations. Part I. An overview of the foundation design process. *Canadian Geotechnical Journal*, **33**(6): 956–983.

Becker, D.E. (1996b) Eighteenth Canadian Geotechnical Symposium: limit states design for foundations. Part II. Development for the National Building Code of Canada. *Canadian Geotechnical Journal*, **33**(6): 984–1007.

Becker, D.E., Crooks, J.H.A. & Been, K. (1988) Interpretation of the field vane test in terms of in-situ and yield stresses. In Richards, A.F. (Ed.) Vane shear strength testing in soils: field and laboratory studies. Special Technical Publication 1014, American Society for Testing & Materials, West Conshohocken, PA, pp. 71–87.

Been, K., Crooks, J.H.A., Becker, D.E. & Jefferies, M.G. (1986) The cone penetration test in sands: Part I. State parameter interpretation. *Géotechnique*, **36**(2): 239–249.

Been, K., Jefferies, M.G., Crooks, J.H.A. & Rothenburg, L. (1987) The cone penetration test in sands: Part II. General inference of state. *Géotechnique*, **37**(3): 285–299.

Bell, A.L. (1993) Jet grouting. In Moseley, M.P. (Ed.), *Ground Improvement*, Blackie Academic & Professional, Blackwood, NJ, pp. 149–174.

Bemben, S.M. & Schulze, D.A. (1995) The influence of testing procedures on clay–geomembrane shear strength measurements. In *Proceedings, Geosynthetics '95 Conference*, Nashville, Industrial Fabrics Association International, Roseville, MN, pp. 1043–1056.

Benner, S.G., Blowes, D.W. & Ptacek, C.J. (1997) A full-scale porous reactive wall for prevention of acid mine drainage. *Ground Water Monitoring & Remediation*, **17**(4): 99–107.

Bennett, R.H. & Hurlbut, M.H. (1986) *Clay Microstructure*, International Human Resources Development Corporation, Boston/Houston/London, 161 p.

Benson, C. & Boutwell, G. (1992) Compaction

control and scale-dependent hydraulic conductivity of clay liners. In *Proceedings 15th Annual Madison Waste Conference*, University of Wisconsin, Madison, pp. 62–83.

Benson, C.H. & Daniel, D.E. (1994) Minimum thickness of compacted soil liners: I. Stochastic models and II. Analysis and case histories. *Journal of Geotechnical Engineering, ASCE*, Reston, VA, **120**(1): 129–172.

Benson, C.H. & Daniel, D.E. (1995) Reply to Discussion by N.F. Fuleihan & A.E.Z. Wissa. *American Society of Geotechnical Engineering*, **121**(6): 506–509.

Benson, C.H. & Khire, M.V. (1995) Earthen covers for arid and semi-arid sites. In *Landfill Closures: Environmental Protection & Land Recovery*, Geotechnical Special Publication No. 53, Dunn, R.J. and Singh, R.P. (Eds.) American Society of Civil Engineers, Reston, VA, pp. 201–217.

Benson, C.H., Khire, M.V. & Bosscher, P.J. (1993) *Final Cover Hydrologic Evaluation, Final Cover—Phase II*, Environmental Geotechnics Report No. 93-4, University of Wisconsin, Madison.

Benson, C.H., Zhai, H. & Wang, X. (1994) Estimating hydraulic conductivity of compacted clay liners. *Journal of Geotechnical Engineering, ASCE*, Reston, VA, **120**(2): 366–387.

Benson, C.H., Daniel, D.E. & Boutwell, G.P. (1999) Field performance of compacted clay liners. *Journal of Geotechnical & Geoenvironmental Engineering, ASCE*, Reston, VA, **125**(5): 390–403.

Beredugo, Y.O. & Novak, M. (1972) Coupled horizontal and rocking vibration of embedded footings. *Canadian Geotechnical Journal*, **9**(4): 477–497.

Berezantzev, V.G., Khristoforov, V. & Golubkov, V. (1961) Load bearing capacity and deformation of piled foundations. In *Proceedings of the Fifth International Conference on Soil Mechanics & Foundation Engineering*, Paris, Vol. 2, pp. 11–15.

Bergado, D.T., Anderson, L.R., Miura, N. & Balasubramaniam, A.S. (1996) *Soft Ground Improvement in Lowland and Other Environments*, ASCE Press, Reston, VA.

Bergen, A.T. & Monismith, C.L. (1973) Characterization of subgrade soils in cold regions for pavement design purposes. *Transportation Research Board Record* **431**: 25–27.

Bermingham, P. & Janes, M. (1989) An innovative approach to load testing of high capacity piles. In *Proceedings of the International Conference on Piling & Deep Foundations*, Vol. 1, pp. 409–413.

BGS (1990) British Geological Society *Proceedings of the Third International Symposium on Pressuremeters*, April 1990, Oxford, Thomas Telford Publishers, London, England.

Bickel, J.O., Kuesel, T.R. & King, E.H. (1996) *Tunnel Engineering Handbook*, 2nd Edn, Chapman & Hall, New York.

Bieniawski, Z.T. (1974) Geomechanics classification of rock masses and its application in tunneling. In *Third International Congress on Rock Mechanics*, Vol. 2A, ISRM, Denver, pp. 27–32.

Bieniawski, Z.T. (1976) Rock mass classifications in rock engineering. In Bieniawski, Z.T. (Ed.) *Proceedings of Symposium on Exploration for Rock Engineering*, Vol. 1, Balkema, Rotterdam, pp. 97–106.

Bieniawski, Z.T. (1978) Determining rock mass deformability: experience from case histories. *International Rock Mechanics, Mineral Science & Geomechanics, Abstracts*, **15**: 237–247.

Bieniawski, Z.T. (1984) *Rock Mechanics Design in Mining and Tunneling*, Balkema, Rotterdam.

Bieniawski, Z.T. (1989) *Engineering Rock Mass Classifications*, New York, Wiley.

Biot, M.A. (1941) General theory of three-dimensional consolidation. *Journal of Applied Physics*, **12**: 155–164.

Biot, M.A. (1955) Theory of elasticity and consolidation for a porous anisotropic solid. *Journal of Applied Physics*, **26**(2): 182–185.

Bishop, A.W. (1955) The use of the slip circle in the stability analysis of slopes. *Géotechnique*, **5**(1): 7–17.

Bishop, A.W. & Bjerrum, L. (1960) The relevance of the triaxial test to the solution of stability problems. In *Proceedings, ASCE Research Conference on Shear Strength of Cohesive Soils*, Boulder, Co, pp. 437–501.

Bishop, A.W. & Blight, G.E. (1963) Some aspects of effective stress in saturated and unsaturated soils. *Geotéchnique*, **2**(3): 177–197.

Bishop, A.W. & Henkel, D.J. (1962) *The Measurement of Soil Properties in the Triaxial Test*, 2nd Edn, Edward Arnold, London, England.

Bishop, D.J. (1996) Discussion of "A Comparison of Puncture Behavior of Smooth and Textured HDPE Geomembranes" and "Three Levels of Geomembrane Puncture Protection" by Narejo, D.B. *Geosynthetics International*, **3**(3): 441–443.

Bjarngard, A. & Edgers, L. (1990) Settlements of municipal solid waste landfills. In *Proceedings, 13th Annual Madison Waste Conference, Madison*, WI.

Bjerg, P.L., Rugge, K., Pedersen, J.K. & Christensen, T.H. (1995) Distribution of redox-sensitive groundwater quality parameters downgradient of a landfill (Grinsted, Denmark). *Environmental Science & Technology*, **29**(5): 1387–1394.

Bjerrum, L. (1963a) Discussion on "Proceedings of the European Conference on Soil Mechanics & Foundation Engineering, Vol. III," *Norwegian Geotechnical Institute Publication*, No. 98, Oslo, Norway, pp. 1–3.

Bjerrum, L. (1963b) Discussion on compressibility of soils. In *Proceedings of the European Conference on Soil Mechanics & Foundation Engineering*, Vol. 2, Wiesbaden, pp. 16–17.

Bjerrum, L. (1972) Embankments on soft ground. In *ASCE Geotechnical Engineering Division Specialty Conference on Performance of Earth & Earth-Supported Structures*, Vol. 2, Purdue University, Lafayette, IN, pp. 1–54.

Bjerrum, L. (1973) Problems of soil mechanics and construction on soft and structurally unstable soils (collapsible, expansive and others). In *Proceedings, Eighth International Conference on Soil Mechanics & Foundation Engineering*, Vol. 3, Moscow, pp. 111–159.

Bjerrum, L. & Andersen, K.H (1972) In-situ measurements of lateral pressures in clay. In *Proceedings, Fifth European Regional Conference on Soil Mechanics & Foundation Engineering*, Vol. 1, Madrid, pp. 11–20.

Bjerrum, L. & Rosenqvist, I.T. (1956) Some experiments with artificially sedimented clays. *Géotechnique*, **6**: 124–136.

Bjerrum, L., Moum, J. & Eide, O. (1967) Application of electro-osmosis on a foundation problem in a Norwegian quick clay. *Géotechnique*, **17**(3): 214–235.

Blake, S.B. & Hall, R.A. (1984) Monitoring petroleum spills with wells: some problems and solutions. In *Proceedings, Fourth National Symposium, On Aquifer Restoration & Ground Water Monitoring*, National Ground Water Association, Dublin, OH, pp. 305–310.

Blaney, G.W., Kausel, E. & Roesset, J.M. (1976) Dynamic stiffness of piles. In *Second International Conference on Numerical Methods in Geomechanics*, Vol. 2, ASCE, Reston, VA, pp. 1001–1012.

Blondeau, F. & Queyroi, D. (1976) Rupture de la tranchée expérimentale de la Bosse-Galin (argile molle). *Bulletin de Liaison des Laboratoires des Ponts et Chaussées*. Numéro spécial III: Stabilité des talus—2: Déblais et remblais, pp. 59–69.

Blondeau, F., Christiansen, M., Guilloux, A. & Schlosser, F. (1984) Talren, méthode de calcul des ouvrages en terre renforcée. In *Proceedings, International Colloquium on* in situ *Reinforcement of Soils & Rocks*, ENPC Press, Paris, pp. 219–224.

Bloomquist D. & Townsend, F.C. (1991) *Development of* in situ *Equipment for Capacity Determinations of Deep Foundations in Florida Limestone*. Report to Florida Department of Transportation, U. Florida, Gainesville, FL.

Blowes, D.W., Ptacek, C.J., Cherry, J.A., Gillham, R.W. & Robertson, W.D. (1995) Passive remediation of groundwater using *in situ* treatment curtains. In Acar, Y.B. & Daniel D.E. (Eds) *Geoenvironment 2000, Characterization, Containment, Remediation & Performance in Environmental Geotechnics*. Geotechnical Special Publication No. 46, Vol. 2, American Society of Civil Engineers, Reston, VA, pp. 1588–1607.

Blowes, D.W., Puls, R.W., Bennett, T.W., Gillham, R.W., Hanton-Fong, C.J. & Ptacek, C.J. (1997) *In-situ* porous reactive wall for treatment of Cr(VI) and trichloroethylene in groundwater. In *Proceedings, 1997 International Containment Technology Conference & Exhibition*, February 10–12, St Petersburg, FL, pp. 851–857.

Boardman, B.T. & Daniel, D.E. (1996) Hydraulic conductivity of desiccated geosynthetic clay liners. *Journal of Geotechnical Engineering, ASCE*, Reston, VA, **122**(3): 204–208.

Bohlke, B.M. (1996) Geotech design reports get a litmus test. *ASCE Civil Engineering*, Reston, VA, **66**(12): 47–49.

Bolt, B.A. (1976) *Nuclear Explosions and Earthquakes: the Parted Veil*, W.H. Freeman, San Francisco, CA.

Bolt, G.H. (1956) Physico-chemical analysis of the compressibility of pure clays. *Géotechnique*, **6**(2): 86–93.

Bolton, M.D. (1986) The strength and dilatancy of sands. *Géotechnique*, **36**(1): 65–78.

Bonaparte, R. (1995) Long-term performance of landfills. Acar, Y.B. & Daniel, D.E. (Eds) *Geoenvironment 2000, Geotechnical Special Publication No. 46*, Vol. 1, American Society of Civil Engineers, Reston, VA, pp. 515–553.

Bonaparte, R. & Berg, R.R. (1987) The use of

geosynthetics to support roadways over sink-hole prone areas. In *Proceedings, Second Multidisciplinary Conference on Sink-holes & Environmental Impact of Karst*, Orlando, FL, pp. 437–445.

Bonaparte, R. & Christopher, B.R. (1987) Design and construction of reinforced embankments over weak foundations. *Transportation & Research Record*, **1153**: 26–39.

Bonaparte, R., Holtz, R.D. & Giroud, J.P. (1987) Soil reinforcement design using geotextiles & geomembranes, In *Geotextile Testing and the Design Engineer*, ASTM, STP 952, American Society for Testing & Materials, West Conshohocken, PA, pp. 69–116.

Bonaparte, R., Giroud, J.P. & Gross, B.A. (1989) Rates of leakage through landfill liners. In *Proceedings, Geosynthetics '89*, Vol. 1, February 1989, San Diego, CA, Industrial Fabrics Association International, Roseville, MN, pp. 18–29.

Bonaparte, R., Othman, M.A., Rad, N.S., Swan, R.H. & Vander Linde, D.L. (1996) Evaluation of various aspects of GCL performance. In *Report of 1995 Workshop on Geosynthetic Clay Liners*, Daniel, D.E. & Scranton, H.B. (Eds) EPA/600/R-96/149, US Environmental Protection Agency National Risk Management Research Laboratory, Cincinnati, Appendix F1–F34.

Bonilla, M.G., Mark, R.K. & Lienkaemper, J.J. (1984) Statistical relations among earthquake magnitude, surface rupture length, and surface fault displacement. *Bulletin of the Seismological Society of America*, **74**: 2379–2411.

Booker, J.R. (1991) Analytic methods in geomechanics, In *Proceedings, Seventh International Conference On Computer Methods & Advances in Geomechanics*, Beer, G., Booker, J.R. & Carter, J.P. (Eds), Vol. 1, pp. 3–14.

Booker, J.R. & Small, J.C. (1983) The analysis of liquid storage tanks on deep elastic foundations. *International Journal for Numerical & Analytical Methods in Geomechanics*, **7**: 187–207.

Boone, S.J. (1995) Ground-movement-related building damage. *Journal of Geotechnical Engineering, ASCE*, Reston, VA, **122**(11): 886–896.

Boore, D.M., Joyner, W.B. & Fumal, T.E. (1993) *Estimation of Response Spectra and Peak Accelerations from Western North American Earthquakes: an Interim Report*, part 2, Open File Report, US Geological Survey, pp. 94–127.

Boore, D.M., Joyner, W.B. & Fumal, T.E. (1997) Equations for estimating horizontal response spectra and peak acceleration from western North American earthquakes: a summary of recent work. *Seismological Research Letters*, **68**(1): 128–153.

Boot, J.C. & Husein, N.M. (1991) Vitrified clay pipes subject to jacking forces. In *Proceedings, Conference on Pipejacking & Microtunnelling '91*, Pipe Jacking Association, London. pp. 6.1–6.9.

Borden, R.C. (1994) Natural bioremediation of hydrocarbon-contaminated ground water. *Handbook of Bioremedation*, CRC Press, Boca Raton, FL, pp. 177–199.

Borden, R.C., Gomez, C.A. & Becker, M.J. (1995) Geochemical indicators of intrinsic bioremediation. *Ground Water*, **33**(2): 180–189.

Borden, R.C., Daniel, R.A., Lebrun, L.E. & Davis, C.W. (1997a) Intrinsic biodegradation of MTBE and BTEX in a gasoline-contaminated aquifer. *Water Resources Research*, **33**(5): 1105–1115.

Borden, R.C., Goin, R.J. & Kao, C.M. (1997b). Control of BTEX migration using a biologically enhanced permeable barrier. *Ground Water Monitoring & Remediation*, **17**(1): 70–80.

Borden, R.H., Holtz, R.D., & Juran, I. (Eds) (1992) *Grouting, Soil Improvement, & Geosynthetics*, Proceedings of the ASCE specialty conference, New Orleans, Geotechnical Special Publication No. 30, American Society of Civil Engineers, ASCE Press, Reston, VA.

Bouchard, R. Leroueil, S. & Marchand, G. (1992) *Aspects géotechniques des étangs pour l'épuration des eaux usées municipales*, Report for Techmat (1992) Inc., Quebec, 111 p.

Bouclin, G. (1990) *MS thesis*, Université Laval, Québec, Canada.

Boudali, M., Leroueil, S. & Murthy, B.R.S. (1994) Viscous behaviour of natural soft clays. In *Proceedings, 13th International Conference on Soil Mechanics & Foundation Engineering*, New Delhi, India, Vol. 1, pp. 411–416.

Bourdeau, P.L., Ludlow, S.J. & Simpson, B.E. (1993) Stability of soil-covered geosynthetic-lined slopes: a parametric study. In *Proceedings, Geosynthetics '93*, Vancouver, BC, Industrial Fabrics Association International, Roseville, MN, Vol. 2, pp. 1511–1521.

Bouyoucos, G.J. (1916) The freezing point method as a new means of measuring the concentration of the soil solution directly in the soil. *Michigan Agricultural College Experimental Station, Technical Bulletin*, **24**: 1–44.

Bowders, J.J., Daniel, D.W., Wellington, J. & Houssidas, V. (1997) Managing desiccation cracking in compacted clay liners beneath geomembranes. In *Proceedings, Geosynthetics '97 Long Beach, CA*, Industrial Fabrics Association International, Roseville, MN, Vol. 1, pp. 527–540.

Bowles, J.E. (1996) *Foundation Analysis & Design*, 5th Ed, McGraw Hill, New York.

Bowles, M., Bentley, L., Barker, J. & Rathgeber, J. (1995) *In-situ* remediation of hydrocarbon contaminated groundwater in low hydraulic conductivity media using trench and gate technology. In *Proceedings (Compact Disk), Fifth Annual Symposium on Groundwater and Soil Remediation (GASReP)*, October 2–6, Toronto, Ontario, Environment Canada.

Bozozuk, M. (1974) Minor principal stress measurements in marine clay with hydraulic fracture tests. In *ASCE Geotechnical Engineering Division Specialty Conference on Subsurface Exploration for Underground Excavation & Heavy Construction*, New England College, Henniker, NH, pp. 333–349.

Brabb, E.E. & Harrod, B.L. (Eds) (1989) *Landslides: Extent and Economic Significance*, In *Proceedings, 28th International Geological Congress: Symposium on Landslides*, Washington, DC.

Brachman, R.W.I. (1999) Mechanical performance of landfill leachate collection pipes *PhD thesis*, University of Western Ontario, London, Canada.

Brady, B.H.G. & Brown, E.T. (1993) *Rock Mechanics for Underground Mining*, 2nd Edn, Chapman & Hall, London, England.

Brady, P.V., Brady, M.V. & Borns, D.J. (1998) *Natural Attenuation—CERCCA, RBCAs and the Future of Environmental Remediation*, Lewis, New York.

Brambati, E., Faccioli, E., Carulli, E., Culchi, F., Onofri, R., Stefanini, R. & Uloigrai, F. (1980) Studio de microzonizzacione sismica dell'are do Tarento (Fruili), (in Italina) Edito da Regione Autonoma Fruili-Venezia, Giulia.

Brandl, H. (1983) Improvement of cohesionless soils, state-of-the-art report. In *Proceedings, Eighth European Conference on Soil Mechanics & Foundation Engineering*, Helsinki, Vol. 3, pp. 1009–1026.

Bray, J.D. & Rathje, E.M. (1998) Earthquake induced displacements of solid waste landfills. *Journal of Geotechnical & Geoenvironmental Engineering, ASCE*, Reston, VA, **124**(3): 242–253.

Bray, J.D., Augello, A.J., Leonards, G.A., Repetto, P.C. & Byrne, R.J. (1995) Seismic stability procedures for solid waste landfills. *Journal of Geotechnical Engineering*, **121**(2): 139–151.

Bright, D.G. (1993) Testing for biological deterioration of geosynthetics in soil reinforcement and stabilization. In Cheng, S.C.J. (Ed.) *Geosynthetic Soil Reinforcement Testing Procedures, ASTM STP 1190*, American Society for Testing & Materials, West Conshohocken, PA, USA, pp. 218–227.

Brindley, G.W. & MacEwan, D. (1953) Structural aspects of mineralogy of clays and related silicates. In *Ceramics, A Symposium*, Green, A.T. & Stewart A.H. (Eds.) British Ceramic Society, London, UK, pp. 15–19.

Britto, A.H. & Gunn, M.J. (1987) *Critical State Soil Mechanics via Finite Elements*, Ellis Horwood, Chichester, UK.

Bromhead, E.N. (1992) *The Stability of Slopes*, Blackie Academic & Professional, Glasgow.

Bromhead, E.N., Cooper, M.R. & Petley, D.J. (1998) The Selborne cutting slope stability experiment (CD-ROM "The Selborne data collection cd").

Broms, B.B. (1964a) Lateral resistance of piles in cohesive soils. *Journal of Soil Mechanics & Foundations, ASCE*, Reston, VA, **90**(2): 27–63.

Broms, B.B. (1964b) Lateral resistance of piles in cohesionless soils. *Journal of Soil Mechanics & Foundations, ASCE*, Reston, VA, **90**(3): 123–156.

Broms, B.B. (1966) Methods of calculating the ultimate bearing capacity of piles—a summary. *Sols Soils*, **18–19**: 21–32.

Broms, B.B. (1979) Problems and solutions to construction in soft clay. In *Proceedings, Sixth Asian Regional Conference on Soil Mechanics & Foundation Engineering*, Singapore, Guest Lecture, Vol. II, pp. 3–38.

Broms, B.B. (1991a) Deep compaction of granular soils. In Fang, H.Y. (Ed.) *Foundation Engineering Handbook*, Van Nostrand Rheinhold, New York, NY, pp. 814–832.

Broms, B.B. (1991b) Stabilization of soil with lime columns. In Fang, H.Y. (Ed.) *Foundation Engineering Handbook*, Van Nostrand Rheinhold, New York, NY, pp. 833–855.

Broms, B.B. (1993) Lime stabilization. In Moseley, M.P. (Ed.) *Ground Improvement*, Blackie Academic & Professional, Blackwood, NJ, pp. 65–99.

Broms, B.B. & Anttikoski, U. (1983) Soil stabilization. *Proceedings, Eighth European Conference on Soil Mechanics & Foundation Engineering*, Helsinki, General Report, Specialty Session 9, Vol. 3, pp. 1298–1301.

Broms, B. & Boman, P. (1978) *Stabilization of Soil with Lime Columns, Design Handbook*, 2nd Edn, Royal Institute of Technology, Stockholm.

Broms, B.B. & Wong, I.H. (1985) Embankment piles. In *Proceedings, Third International Geotechnical Seminar on Soil Improvement Methods*, Singapore.

Brooks, R.H. & Corey, A.T. (1964) *Hydraulic Properties of Porous Materials*, Colorado State University, Hydrology Paper No. 3, p. 27.

Brown, K.W., Thomas, J.C., Lytton, R.L., Jayawickarama, P. & Bahrt, S.C. (1987) *Quantification of Leak Rates Through Holes in Landfill Liners*, report for cooperative agreement CR-810940, EPA Risk Reduction Engineering Research Laboratory, Cincinnati.

Brown, M.J. & Burris, D.R. (1996) Enhanced organic contaminant sorption on soil treated with cationic surfactants. *Ground Water*, **34**(4): 734–744.

Brown, P.T. (1969) Numerical analyses of uniformly loaded circular rafts on elastic layers of finite depth. *Géotechnique*, **19**(2): 301–306.

Brown, P.T. (1975) Strip footing with concentrated loads on deep elastic foundations. *Geotechnical Engineering*, **6**: 1–13.

Brown, S.F. (1996) 36th Rankine Lecture: Soil mechanics in pavement engineering. *Géotechnique*, **46**(3): 383–426.

Brown, S.F. & Pappin, J.W. (1981) Analysis of pavements with granular bases. *Transportation Research Board Record*, **810**: 17–22.

Brown, S.F., Brunton, J.M. & Stock, A.F. (1985) The analytical design of bituminous pavements. *Proceedings, Institution Of Civil Engineers*, London, England. **79**(2): 1–31.

Bruce, D.A. (1992) Recent progress in American pin pile technology. *Grouting, Soil Improvement, & Geosynthetics*, Geotechnical Special Publication No. 30, American Society of Civil Engineers, Reston, VA, pp. 765–777.

Bruce, D.A. (1994a) Pin piles. *Geotechnical News*, **12**(4): 33–36.

Bruce, D.A. (1994b) An overview of grouting developments. *Geotechnical News*, **12**(4): 36–40.

Bruce, D.A. & Jewell, R.A. (1987) Soil nailing: applications and practice. *Ground Engineering*, **20**(1): 21–23.

Brummermann, K., Blümel, W. & Stoewahse, C. (1994) Protection layers for geomembranes: effectiveness and testing procedures. In *Proceedings, Fifth International Conference on Geotextiles, Geomembranes & Related Products*, September 1994, Singapore, Vol. 3, pp. 1003–1006.

Brune, M., Ramke, H.G., Collins, H.J. & Hanert, H. (1994) Incrustation problem in landfill drainage systems. In Christensen, T.H. Cossu, R. & Stegmann, R. (Eds) *Landfilling of Waste: Barriers*, E & FN Spon, London, pp. 569–605.

BS 5400 (1990) *Code of Practice for Steel, Concrete and Composite Bridges*, Part 2, *Highway Loading, British Standards Institution*, London, England.

BS 5930 (1981) *Code of Practice for Site Investigations*. British Standards Institution, London, England.

BS 8006 (1995) *Code of Practice for Strengthened/ Reinforced Soils & Other Fills*, British Standards Institution, London, England.

BS 8010 (1987, 1989, 1993) Code of Practice for pipelines. Pipelines on land: design, construction and installation. British Standards Institution, London, England.

Budhu, M. & Davies, T.G. (1987) Nonlinear analysis of laterally loaded piles in cohesionless soils. *Canadian Geotechnical Journal*, **24**: 21–39.

Budhu, M. & Davies, T.G. (1988) Analysis of laterally loaded piles in soft clays. *Journal of Geotechnical Engineering, ASCE*, Reston, VA, **114**(1): 21–39.

Bunce, C.M., Cruden, D.M. & Morgenstern, N.R. (1997) Assessment of the hazard from rock fall on a highway. *Canadian Geotechnical Journal*, **34**(3): 344–356.

Burd, H.J. & Frydman, S. (1997) Bearing capacity of plane-strain footings on layered soils. *Canadian Geotechnical Journal*, **34**: 241–253.

Burdine, N.T. (1953) Relative permeability calculations from pore size distribution data. *Journal of Petroleum Technology*, **5**: 71–78.

Burke, L.M. & Haubert, A.E. (1991) Burying by the bale. *Journal of Civil Engineering, ASCE*, Reston, VA, **61**(8): 58–60.

Burland, J.B. (1973) Shaft friction on piles in clay–a simple fundamental approach. *Ground Engineering*, **6**(3): 30–42.

Burland, J.B. (1990) On the compressibility and shear strength of natural clays. *Géotechnique*, **40**(3): 329–378.

Burland, J.B., Broms, B.B. & de Mello, V.F.B. (1977) Behaviour of foundations and structures, state-of-the-art review. In *IX International Conference on Soil Mechanics & Foundation Engineering*, Tokyo, Vol 3, pp. 495–546.

Burlingame, M.J. (1985) *Construction of a High-*

way on a Sanitary Landfill and Its Long-Term Performance, Transportation Research Record 1031, TRB, Washington, DC: pp. 34–40.

Burns, J.Q. & Richard, R.M. (1964) Attenuation of stresses around buried cylinders. In *Proceedings, Symposium on Soil–Structure Interaction*, Tucson, AZ, American Society for Testing & Materials, West Conshohocken, PA, pp. 379–392.

Burr, J.P., Pender, M.J. & Larkin, T.J. (1997) Dynamic response of laterally excited pile groups. *Journal of Geotechnical & Geoenvironmental Engineering*, **123**1: 1–8.

Buscheck, T.E. & Peargin, T.R. (1991) Summary of nation-wide vapor extraction system performance study. In *Proceedings, Petroleum Hydrocarbons & Organic Chemicals in Ground Water: Prevention, Detection & Restoration*, November 20–22 Houston, TX, National Ground Water Association, pp. 205–220.

Bustamante, M. & Gianeselli, L. (1982) Pile bearing capacity prediction by means of static penetrometer CPT. In *Proceedings*, ESOPT II, Amsterdam, Vol. 2, pp. 492–500.

Butterfield, R. & Banerjee, P.K. (1971) The elastic analysis of compressible piles and pile groups. *Géotechnique*, **21**(1): 43–60.

Butterfield, R. & Douglas, R.A. (1981) *Flexibility Coefficients for the Design of Piles and Pile Groups*. Technical Note 108, CIRIA, London.

Bycroft, G.N. (1956) Forced vibrations of a rigid circular plate on a semi-infinite elastic space and on an elastic stratum. *Philosophical Transactions of the Royal Society*, London, Serial A, **248**: 327–368.

Byrne, P.M., Imrie, A.S. & Morgenstern, N.R. (1994) Results and implications of seismic performance studies for Duncan Dam. *Canadian Geotechnical Journal*, **31**(6): 979–988.

Byrne, P.M., Anderson, D.L. & Jitno, H. (1996) Seismic analysis of large buried culvert structures. *Transportation Research Record*, **1541**: 133–139.

Byrne, R.J. (1994) Design issues with strain-softening interfaces in landfill liners. In *Proceedings, Waste Tech '94 Conference*, National Solid Waste Management Association, Charleston, SC p. 26.

Byrne, R.J., Kendall, J. & Brown, S. (1992) Cause and mechanism of failure of Kettleman Hills Landfill lining material. In *Proceedings of ASCE Conference on Stability & Performance of Slopes & Embankments* II, ASCE Press, Reston, VA, pp. 1–23.

Callanan, J.F. & Kulhawy, F.H. (1985) *Evaluation of Procedures for Predicting Foundation Uplift Movements*. Report, Electric Power Research Institution, No. EPRI EL-4107, Cornell University.

Campanella, R.G. & Weemees, I. (1990) Development and use of an electrical resistivity cone for groundwater contamination studies. *Canadian Geotechnical Journal*, **27**(5): 557–567.

Campanella, R.G. & Robertson, P.K. (1991) Use and interpretation of a research dilatometer. *Canadian Geotechnical Journal*, **28**(1): 113–126.

Campanella, R.G., Davies, M.P., Boyd, T.J. & Everard, J.L. (1994) Geoenvironmental subsurface site characterization using *in-situ* soil testing methods. In *Proceedings, First International Congress on Environmental Geotechnics*, Edmonton, Alberta.

Campbell, G.S. (1974) A simple method for determining unsaturated hydraulic conductivity from moisture retention data. *Soil Science*, **117**(6): 311–314.

Campbell, K.W. (1997) Attenuation relationships for shallow crustal earthquakes based on California strong motion data. *Seismological Research Letters*, **68**(1): 180–189.

CAN/CSA-A5, Portland cement, Canadian Standards Association, Rexdale, Ontario.

CAN/CSA A60 Vitrified clay pipe, Canadian Standards Association, Rexdale, Ontario.

CAN/CSA A257 Concrete culverts, storm drain and sewer pipe, Canadian Standards Association, Rexdale, Ontario.

CAN/CSA B182 Plastic non-pressure pipe, Canadian Standards Association, Rexdale, Ontario.

CAN/CSA C140, Canadian Standards Association, Ottawa. Corrugated steel, Rexdale, Ontario.

Caquot, A. & Kerisel, J. (1966) *Traite de mecanique des sols*, 4th Edn, Gauthier-Villars, Paris.

CARACAS (1998) *Concerted Action on Risk Assessment for Contaminated Sites in the European Union*, European Commission, Brussels, Belgium.

Carey, G.R., Mateyk, M.G., Turchan, G.T., McBean, E.A., Rovers, F.A., Murphy, J.R. & Campbell, J.R. (1996) Application of an innovative visualization method for demonstrating intrinsic remediation at a landfill superfund site. In *Proceedings, API/NGWI Petroleum Hydrocarbons & Organic Chemicals in Groundwater Conference*, Houston, TX.

Carey, G.R., Van Geel, P.J., McBean, E.A. & Rovers, F.A. (1997a) Effect of landfill cap permeability on the natural attenuation of chlorinated solvents below a landfill. In *1997 Canadian Geotechnical Conference Proceedings*, October 20–22, Ottawa, Ontario.

Carey, G.R., Van Geel, P.J., McBean, E.A. & Rovers, F.A. (1997b) An innovative modeling and visualization approach for demonstrating the effectiveness of natural attenuation, presented at the IBC Natural Attenuation '97 Conference, Scottsdale, Arizona, December 8–10.

Carey, G.R., Van Geel, P., Murphy, J., McBean, E., & Rovers, F. (1998) Dimensionality Considerations for a Coupled Biodegradation-Redox Model (BIOREDOX), MODFLOW '98, October 4–8, Denver, Colorado.

Carey, G.R., Van Geel, P.J., Wiedemeier, T.H. & McBean, E.A. (2000) SEQUENCE natural attenuation application for visualizing primary and secondary lines of evidence. *Ground Water Monitoring Review*, in press.

Carey, P.J., Koragappa, N. & Gurda, J.J. (1993) Renewal of old sites, overliner system design, a case study of the Brookhaven Landfill, Long Island, New York. In *Proceedings, Waste Tech '93 Conference*, National Solid Waste Management Association, Marina Del Rey, p. 13

Carrubba, P. (1997) Skin friction of large-diameter piles socketed into rock. *Canadian Geotechnical Journal*, **34**: 230–240.

Carslaw, H.S. & Jaeger, J.C. (1959) *Conduction of Heat in Solids*, 2nd Ed, Oxford University Press, England.

Carter, J.P. & Kulhawy, F.H. (1988) *Analysis and Design of Drilled Shaft Foundations Socketed into Rock*, Report El-5918, Electric Power Research Institute, Palo Alto, CA.

Carter, J.P. & Kulhawy, F.H. (1992) Analysis of laterally loaded shafts in rocks. *Journal of Geotechnical Engineering, ASCE*, Reston, VA, **118**(6):839–855.

Carter, P.G., Pirie, R.M. & Sneddon, M. (1986) Marine site investigations and BS 5930. In Hawkins, A.B. (Ed.) *Site Investigation Practice: Assessing BS 5930*. Geological Society Engineering Geology Special Publication No. 2, Geological Society, London, England, pp. 163–166.

Casagrande, A. (1935) Seepage Through Dams, New England Water Works Association, **2**: 131–172..

Casagrande, L. (1937) Method of hardening soils. U.S. Patent No. 2,099,328.

Casagrande, L. (1952) Electro-osmotic stabilization of soils. *Journal of Boston Society of Civil Engineers*, **39**: 59–83.

Casagrande, L. (1959) *Review of Past and Current Work on Electro-Osmotic Stabilization of Soils*, Harvard Soil Mechanics Series, No. 45, Harvard University Press, Cambridge, MA.

Casagrande, L. (1983) Review of stabilization of soils by means of electro osmosis state of the art. *Geotechnical Section of the BSCE/ASCE*, **69**(2):255–302.

Casagrande, L., Loughney, R.W. & Matich, M.A.J. (1961) Electro-osmotic stabilization of a high slope in loose saturated silt. In *Fifth International Conference of Soil Mechanics & Foundation Engineering*, Paris, Vol. 2, pp. 555–561.

Casagrande, L., Wade, N., Wakely, M. & Loughney, R. (1981) Electro osmosis projects, British Columbia, Canada. In *10th International Conference of Soil Mechanics & Foundation Engineering*, Stockholm, Vol. 2, pp. 607–610.

CCME (1994) *Subsurface Assessment Handbook for Contaminated Sites*, CCME-EPC-NCSRP-48E, Canadian Council of Ministers of the Environment, Winnipeg, Manitoba.

CCME (1996) *A Protocol for the Derivation of Environmental & Human Health Soil Quality Guidelines*. Canadian Council of Ministers of the Environment, Winnipeg, Manitoba.

CDSA (1995) *Dam Safety Guidelines*, Canadian Dam Safety Association, Edmonton, Alberta.

Cedergren, H.R. (1967) *Seepage, Drainage & Flow Nets*, 2nd Edn, Wiley, New York.

Cedergren, H.R. (1989) *Seepage, Drainage, & Flow Nets*, 3rd Edn., Wiley, New York.

Celebi, M. (1987) Topographical and geological amplifications determined from strong motion and aftershock records of the 3 March 1985 Chile earthquake. *Bulletin of the Seismological Society of America*, **77**(4): 1147–1167.

Celebi, M. (1991) Topographical and geological amplification case studies and engineering implications, In *Proceedings, International Workshop on Spatial Variation of Earthquake Ground Motion*, Elsevier Science, B.V. Amsterdam.

CEN (1992) *Geotechnical Design, General Rules*. European Committee for Standardization (CEN), Eurocode 7. Danish Geotechnical Institute, Copenhagen, Denmark.

CEN (1994) *Design Provisions for Earthquake Resistant Structures*, Part 5: Foundations, Retaining Structure and Geotechnical Aspects, Eurocode 8, European Committee for Standardization, Central Secretariat, Brussels.

Cernica, J.N. (1995) *Geotechnical Engineering: Soil Mechanics*, Wiley, New York.

CFEM (1992): see CGS (1992).

CGS (1985) *Canadian Foundation Engineering Manual*, 2nd Edn, Canadian Geotechnical Society, Ottawa.

CGS (1992) *Canadian Foundation Engineering Manual*, 3rd Edn, Canadian Geotechnical Society, BiTech Publishers, Richmond, British Columbia.

Challa, P. & Poulos, H.G. (1991) Behaviour of single pile in expansive clay. Geotechnical Engineering, **22**(2): 189–216.

Champ, D.R., Young, J.L., Robertson, D.E. & Abel, K.H. (1984) Chemical speciation of long-lived radionuclides in a shallow groundwater flow system, *Water Pollution Research Journal of Canada*, **19**: 35–54.

Chamberlain, E.J. (1973) A model for predicting the influence of closed-system freeze–thaw on the strength of thawed soil. In *Proceedings, Symposium on Frost Action Roads*, Oslo, pp. 94–97.

Chamberlain, E.J. (1981) *Frost Susceptibility, Review of Index Tests*, Monograph 81–2. US Cold Regions Research and Engineering Laboratory (CRREL).

Chandhari, A.P. & Char, A.N.R. (1985) Flexural behavior of reinforced beams. *Journal of Geotechnical Engineering, ASCE*, Reston, VA, **111**(11): 1328–1333.

Chandler, R.J. (1984) Recent European experience of landslides in over-consolidated clays and soft rocks. In *Proceedings, Fourth International* Symposium on Landslides, Toronto, Vol. 1, pp. 61–81.

Chaney, R.C., Thiel, R., Richardson, G.N. & Cadwallader, M. (1997) Cyclic response of clay–geomembrane interfaces and their impact on the seismic response of lined landfills. In *Proceedings, Geosynthetics '97 Conference*, Long Beach, Industrial Fabrics Association International, Roseville, MN, Vol. 2, pp. 977–987.

Chapelle, F.H. (1993) *Ground-Water Microbiology & Geochemistry*, Wiley, New York.

Chapman, D.N., Falk, C., Rogers, C.D.F. & Stein, D. (1996) Experimental and analytical modelling of pipebursting ground displacements. *Trenchless Technology Research, Tunnelling and Underground Space Technology*, **11**(1): 53–68.

Chapman, S.W., Byerley, B.T., Smyth, D.J.A. & Mackay, D.M. (1997) A pilot test of passive oxygen release for enhancement of *in-situ* bioremediation of BTEX-contaminated groundwa-

ter. *Ground Water Monitoring & Remediation*, **17**(2): 93–105.

Chapuis, R.P. (1988) Two case histories of major frost heaving in refrigerated buildings: thermal analyses, repairs and prevention. *Canadian Geotechnical Journal*, **25**(3): 535–540.

Chapuis, R.P. (1990) Sand–bentonite liners: predicting permeability from laboratory tests. *Canadian Geotechnical Journal*, **27**(1): 47–57.

Charbeneau, R.J., Wanakule, N., Chiang, C.Y., Nevin, J.P. & Klein, C.L. (1989) A two-layer model to simulate floating free product recovery: formulation and applications. In *Proceedings, Conference On Petroleum Hydrocarbons and Organic Chemicals in Ground Water: Prevention, Detection, and Restoration*, National Ground Water Association, Dublin, OH, pp. 333–345.

Charles, J.A., Tedd, P., Hughes, A.K. & Lovenbury, H.T. (1996) *Investigating Embankment Dams: A Guide to the Identification and Repair of Defects*, Building Research Establishment Report, Construction Research Communications, Watford, England.

CHBDC (1998) *Buried Structures*, Canadian Highway Bridge Design Code, Section 7, Canadian Standards Association, Rexdale, Ontario.

Chen, L. & Poulos, H.G. (1997) Piles subjected to lateral soil movements. *Journal of Geotechnical & Geoenvironmental Engineering, ASCE*, Reston, VA, **123**(9): 802–811.

Cheney, R.S. (1984) *Permanent Ground Anchors*, Report No. FHWA-DP-68-1, Federal Highway Administration, Springfield, VA.

Cherry, J.A. (1996) Conceptual models for chlorinated solvent plumes and their relevance to intrinsic remediation. In *Symposium on Natural Attenuation of Chlorinated Organics in Groundwater*, September 11–13, Dallas, TX, US Environmental Protection Agency 540/R-96-509, pp. 29–30.

Cherry, J.A., Feenstra, S. & Mackay, D.M. (1996) Concepts for remediation of sites contaminated with DNAPLS. In Pankow, J.F. & Cherry, J.A. (Eds) *Dense Chlorinated Solvents and Other DNAPLs in Groundwater*, Waterloo Press, Rockwood, Ontario, pp. 475–506.

Cherry, J.A., Feenstra, S. & Mackay, D.M. (1997) Developing rational goals for *in situ* remedial technologies. In Ward, C.H., Cherry, J.A. & Scalf, M.R. (Eds) *Subsurface Restoration*. Ann Arbor Press, Chelsea, MI, pp. 75–98.

Childs, E.C. & Collis-George, N. (1950) The permeability of porous materials. *Proceedings, Royal Society of London*, **201A**: 392–405.

Childs, S.W. & Hanks, R.J. (1975) Model of soil salinity effects on crop growth. *Soil Science Society of American Journal*, **39**: 617–622.

Choa, V. (1985) Preloading and vertical drains. In *Soil Improvement Methods, Proceedings, Third International Geotechnical Seminar*, Nanyang Technological Institute, Singapore, pp. 87–99.

Chopra, A.K. (1981) *Dynamics of Structures, a Primer*, Earthquake Engineering Research Institute, Oakland, CA.

Choquet, P. & Tanon, D.D.B. (1985) Nomograms for the assessment of toppling failure in rock slopes. In *Proceedings, 26th US Symposium on Rock Mechanics*, South Dakota School of Mines & Technology, Vol. 1, pp. 19–30.

Chow, Y.K. (1986) Analysis of vertically loaded pile groups. *International Journal of Numerical Analytical Methods in Geomechanics*, **10**: 59–72.

Chow, Y.K., Chin, J.T. & Lee, S.L. (1990) Negative skin friction on pile groups. *International Journal of Numerical and Analytical Methods in Geomechanics*, **14**(1): 75–91.

Christian, J.T. & Carrier, W.D. (1978) Janbu, Bjerrum & Kjaernsli's chart reinterpreted. *Canadian Geotechnical Journal*, **15**: 123–128.

Christoulas, S. & Frank, R. (1991) Deformation parameters for pile settlement. In *Proceedings, 10th European Conference on Soil Mechanical and Foundation Engineering*, Florence, Vol. 1, pp. 373–376.

Circeo, L.J. & Mayne, P.W. (1994) *In-situ* thermal stabilization of roads and foundation soils using plasma arc technology. In *Proceedings, Fourth International Conference on Bearing Capacity of Roads and Airfields*. University of Minnesota, Minneapolis.

Clarke, B.G. & Gambin, M.P. (1988) Pressuremeter testing in onshore ground investigations: A report by the ISSMGE Committee TC16. In Robertson, P.K. & Mayne, P.W. (Eds) *Proceedings, First International Conference on Site Characterization, ISC'98*, April 19–22, Atlanta, GA, A.A. Balkema, Rotterdam, pp. 1429–1468.

Clayton, C.R.I. (1990) SPT energy transmission: theory, measurement and significance. *Ground Engineering*, December: 35–43.

Clayton, C.R.I. (1995) *The Standard Penetration Test (SPT): Methods and Use*, Report 143. Construction Industry Research & Information Association (CIRIA), London, England.

Clough, G.W., Smith, E.M. & Sweeney, B.P.

(1989) Movement control of excavation support systems by iterative design. *Foundation Engineering: Current Principles and Practices*, ASCE Press, Reston, VA, **2**: 869–882.

Clough, R.W. & Penzien, J. (1993) *Dynamics of Structures*, 2nd Edn, McGraw-Hill, New York.

Clouterre (1991) Recommendations Clouterre 1991. Presses de l'ENPC, Paris (In French). Also Recommendations Clouterre 1991 (Soil nailing recommendations). Published in English by Federal Hyway Administration, Washington, D.C., 301 p.

CMHC (1994) *Phase I ESA Interpretive Guideline*, The Canadian Mortgage and Housing Corporation, Ottawa.

Cohen, A. (1961) Tables for maximum likelihood estimates from singly truncated and singly-censored samples. *Technometrics*, **3**: 535–541.

Cohen, R.M. & Mercer, J.W. (1993) *DNAPL Site Evaluation*, CRC Press, Boca Raton, FL.

Colas, G. & Locat, J. (1993) Glissement et coulée de La Valette dans les Alpes de Haute-Provence: présentation générale et modélisation de la coulée. *Bulletin de Liaison des Laboratoires des Ponts et Chaussées*, **187**: 19–28.

Colas, G., Simon, A., Payani, M. & Pilot, G. (1976). Les glissements de terrain sur l'autoroute A-7 à Rognac, près de l'étang de Berre. *Bulletin de Liaison des Laboratoires des Ponts et Chaussées*, Numéro spécial III, **2**: 70–80.

Cole, D., Bentley, D., Durell, G. & Johnson, T. (1986) *Resilient Modulus of Freeze–Thaw Affected Granular Soils for Pavement Design and Evaluation*, Report 86–4, CRREL.

Collins, B.D. & Znidarcic, D. (1998) Slope stability issues of rainfall induced landslides. In *Proceedings, 11th Danube–European Conference on Soil Mechanics & Geotechnical Engineering (Geotechnical Hazards)*, Poreč, pp. 791–798.

Colwell, R.N. (1983) *Manual of Remote Sensing*, 2nd Edn, American Society of Photogrammetry, New York, NY.

Comer, A.I., Hsuan Y.G. & Konrath, L. (1988) The performance of flexible polypropylene geomembranes in covered and exposed environments. In *Sixth International Geosynthetics Conference*, R.K. Rowe (Ed), Atlanta, GA, IFAI, Roseville, MN.

Cooke, A.J. & Rowe, R.K. (1999) Extension of porosity and surface area models for uniform porous media for studying clogging of landfill collection systems. *Journal of Environmental Engineering, ASCE*, Reston, VA, **125**(2): 126–145.

Coon, R.F. & Merritt, A.H. (1970) *Predicting* in-situ *Modulus of Deformation Using Rock Quality Indexes*, ASTM Special Technical Publication 477, West Conshohocken, PA, pp. 154–173.

Cooper, M.R., Bromhead, E.N., Petley, D.J. & Grant, D.I. (1998) The Selborne cutting stability experiment. *Géotechnique*, **48**(1): 83–101.

Corey, A.T. (1986) *Mechanics of Immiscible Fluids in Porous Media*, Water Resources Publications, Littleton, CO.

Cornell University (1951) Final report on soil solidification research, Ithaca, N.Y.

Cornforth, D.H. (1964) Some experiments on the influence of strain conditions on the strength of sand. *Géotechnique*, **14**: 143–167.

Corser, P. & Cranston, P. (1991) Observation on long-term performance of composite clay liners and covers. In *Proceedings, Geosynthetic Design & Performance*, Vancouver Geotechnical Society, Vancouver, BC.

Costa, J.E. & Baker, V.R. (1981) *Surficial Geology–Building with the Earth*, Wiley, New York.

Coyle, H. & Reese, L.C. (1966) Load transfer for axially loaded piles in clay. *Journal of Soil Mechanics & Foundations Division, ASCE*, Reston, VA, **92**(2): 1–26.

CPCA (1984)*Thickness Design for Concrete Highway and Street Pavements*, Canadian Portland Cement Association, Toronto, Ontario.

Craig, R.F. (1997) *Soil Mechanics*, 6th Edn, E. & F.N. Spon, London, U.K.

Craig, R.N. (1983) *Pipe Jacking: a State-of-the-Art Review*, Technical Note No. 112, Construction Industry Research and Information Association, London.

Croff, A.G., Lomenick, T.F., Lowrie, R.S. & Stow, S.H. (1985) *Evaluation of Five Sedimentary Rocks other than Salt for High Level Waste Repository Siting Purposes*, Oak Ridge National Lab, ORNL/CF-85/2/V2, Oak Ridge, TN.

Croney, D. & Croney, P. (1997) *Design and Performance of Road Pavements*, 3rd Edn, McGraw-Hill, London.

Croney, D., Coleman, J.D. & Black, W.P.M. (1958) Movement and distribution of water in soil in relation to highway design and performance. In *Water and its Conduction in Soils*, Highway Research Board, Special Report, Washington, DC, No. 40, pp. 226–252.

Crooks, J.H.A., Becker, D.E., Jefferies, M.G. & McKenzie, K. (1984) Yield behavior and con-solidation. 1: pore pressure response. In *Proceedings, ASCE Symposium on Sedimentation Consolidation Models: Predictions and Validation*, pp. 356–381.

Crouse, C.B. & Cheang, L. (1987) *Dynamic Testing and Analysis of Pile–Group Foundation*, Geotechnical Special Publication No. 11, American Society of Civil Engineers, Reston, VA, pp. 79–98.

Cruden, D.M. & Varnes, D.J. (1996) Landslide types and processes. *Landslides–Investigation and Mitigation*, A.K. Turner & R.L. Schuster (Eds.) Special Report 247, Transportation Research Board, Washington, DC, pp. 36–75.

CSA (1994) *Phase I Environmental Site Assessment*, Canadian Standards Association Report Z768–94, Etobicoke, Ontario.

CSA *Phase II Guideline*, Canadian Standards Association, Toronto, ON.

Dafalias, Y.F. & Herrmann, L.R. (1980) A bounding surface soil plasticity model. In *Proceedings, International Symposium on Soils Under Cyclic and Transient Loading*, Balkema, Rotterdam, pp. 335–345.

Dahlberg, R. (1974) Penetration, pressuremeter, and screw plate tests in a preloaded natural sand deposit. In *Proceedings, European Symposium on Penetration Testing*, 2.2, ESOPT-1, Stockholm, Sweden.

Dakoulas, P. & Gazetas. G. (1986) Seismic shear vibration of embankment dams in semi-cylindrical valleys. *Earthquake Engineering and Structural Dynamics*, **14**: 19–40.

Daniel, D.E. (1998) Landfills for solid and liquid wastes. In *Proceedings, Third International Congress on Environmental Geotechnics*, September, Lisbon, A.A. Balkema, Rotterdam.

Daniel, D.E. & Koerner, R.M. (1993a) *Technical Guidance Document: Quality Assurance and Quality Control for Waste Containment Facilities*, EPA/600/R-93/182, Environmental Protection Agency Risk Reduction Research Laboratory, Cincinnati.

Daniel, D.E. & Koerner, R.M. (1993b) Final cover systems. In Daniel, D.E. (Ed.) *Geotechnical Aspects of Waste Disposal* Chapman & Hall, London, pp. 225–272.

Daniel, D.E. & Koerner, R.M. (1995) *Waste Containment Systems: Guidance for Construction, Quality Assurance, and Quality Control of Liner and Cover Systems*, ASCE Press, Reston, VA.

Daniel, D.E. & Scranton, H.B. (1996) *Report of 1995 Workshop on Geosynthetic Clay Liners*,

EPA/600/R-96/149, US Environmental Protection Agency, National Risk Management Research Laboratory, Cincinnati, OH.

Daniel, D.E. & Shan, H.Y. (1992) *Effects of Partial Wetting of Gundseal on Strength and Hydrocarbon Permeability*, Report submitted to Gundle Lining Systems Inc., Austin, TX.

Daniel, D.E., Anderson, D.C. & Boynton, S.S. (1985) Fixed wall versus flexible wall permeameters. In *Hydraulic Barriers in Soil and Rock*, ASTM STP 874, American Society for Testing & Materials, West Conshohocken, PA, pp. 107–123.

Daniel, D.E., Shan, H.Y. & Anderson, J.D. (1993) Effects of partial wetting on the performance of the bentonite component of a geosynthetic clay liner. In *Proceedings, Geosynthetics '93*, Vancouver, BC, Industrial Fabrics Association International, Roseville, MN, Vol. 3, pp. 1483–1496.

Daniel, D.E., Bowders, J.J. & Gilbert, R.B. (1997) Laboratory hydraulic conductivity testing of GCLs in flexible-wall permeameters. In Well, L.W. (Ed.) *Testing & Acceptance Criteria for Geosynthetic Clay Liners*, ASTM STP 1308 American Society for Testing & Materials, West Conshohocken, PA, USA, pp. 208–228.

Daniel, D.E., Koerner, R.M., Bonaparte, R., Landreth, R.E., Carson, D.A. & Scranton, H.B. (1998) Slope stability of geosynthetic clay liner test plots. *Journal of Geotechnical & Geoenvironmental Engineering, ASCE*, Reston, VA, **124**(7): 628–637.

D'Appolonia, D.J., D'Appolonia, E. & Brissette, R.F. (1970) Settlement of spread footings on sand. *Journal of Soil Mechanics & Foundations Division, ASCE*, Reston, VA, **96**(SM2): 754–762.

Darracott, B.W. & McCann, D.M. (1986) Planning engineering geophysical surveys. In Hawkins, A.B. (Ed.) *Site Investigation Practice: Assessing BS 5930*. Geological Society Engineering Geology Special Publication No. 2, Geological Society, London, England, pp. 85–90.

Darton, N.H. (1909) *Geology and Undergroundwaters of South Dakota*, US Geology Survey, Water Supply Paper 227.

Das, B.M. (1990) *Principles of Foundation Engineering*, PWS Kent, Boston, MA.

Das, B.M. (1995) *Principles of Soil Dynamics*. PWS Kent, Boston, MA.

David, M. (1977) *Geostatistical Ore Reserve Estimation*, Elsevier, New York.

Davidtz, J.C. & Low, P.F. (1970) Relation between crystal lattice configuration and swelling of montmorillonites. *Clays and Clay Minerals*, **18**(6): 325.

Davies, T.G., Sen, R. & Banerjee, P.K. (1985) Dynamic behavior of pile groups in inhomogeneous soil. *Journal of Geotechnical Engineering, ASCE*, Reston, VA, **111**(12): 1365–1379.

Davis, E.H. (1980) Some plasticity solutions relevant to the bearing capacity of rock and fissured clay, J.C. Jaeger Memorial Lecture, In *Proceedings, Third Australia–New Zealand Conference on Geomechanics*, Wellington, New Zealand, Vol. 3, pp. 27–36.

Davis, E.H. & Booker, J.R. (1971) The bearing capacity of strip footings from the point of view of plasticity theory. In *Proceedings, First Australia–New Zealand Conference on Geomechanics*, Institution of Engineers, Melbourne, Australia. Vol. 1, pp. 276–282.

Davis, E.H. & Booker, J.R. (1973) The effect of increasing strength with depth on the bearing capacity of clays. *Géotechnique*, **23**(4): 551–563.

Davis, E.H. & Poulos, H.G. (1968) The use of elastic theory for settlement prediction under three-dimensional conditions. *Géotechnique*, **18**(1): 67–91.

Davis, E.H. & Poulos, H.G. (1972) Rate of settlement under three-dimensional conditions. *Géotechnique*, **22**(1): 95–114.

Davis, G.B. (1997) Intrinsic attenuation of BTEX in groundwater: quantification and monitoring issues (abstract). In *International & On-Site Bioremediation Symposium*, New Orleans, LA, Battelle Press, Columbus, OH, Vol. 4(1), p. 19.

De, A. & Zimme, T.F. (1998) Frictional behavior of landfill liner interfaces with geonets. In *Proceedings*, Rowe, R.E. (Ed.) *Sixth International Conference on Geosynthetics*, Atlanta, GA, Industrial Fabrics Association International, Roseville, MN, Vol. 1, pp. 443–446.

Décourt, L. (1982) Prediction of the bearing capacity of piles based exclusively on N values of the SPT. In *Proceedings, ESOPT II*, Amsterdam, Vol. 1, pp. 29–34.

Décourt, L. (1989) The standard penetration test–State-of-the-art report. In *12th International Conference on Soil Mechanics & Foundation Engineering*, August 13–18, Rio de Janeiro, Brazil.

Décourt, L. (1994) Prediction of settlement of spread footings. In Predicted and Measured Behaviour of Five Spread Footings on Sand, Briaud, J.-L. & Gibbons, R.M. Eds. ASCE Geotechnical Special Publication No. 41,

American Society of Civil Engineers, Reston, VA, pp. 210–213.

Décourt, L. (1995) Prediction of load–settlement relationships for foundations on the basis of the SPT-T. *Ciclo de Conferencias Internationale*, Leonardo Zeevaert, UNAM, Mexico, pp. 85–104.

Décourt, L., Belicanta, A. & Quaresma, Filho, A.R. (1989) Brazilian experience on SPT. In *Proceedings 12th International Conference on Soil Mechanics & Foundation Engineering*, Rio de Janeiro, Supplement, Contributions by Brazilian Society for Soil Mechanics, pp. 49–54.

Deere, D.U. (1964) Technical description of cores for engineering purposes. *Rock Mechanics & Engineering Geology*, **1**: 16–22.

Deere, D.U. & Miller, R.P. (1966) *Engineering Classification and Index Properties for Intact Rock*, Air Force Weapons Laboratory, Report AFWL-TR-65-16, Kirtland Air Force Base, NM.

Deere, D.U., Hendron, A.J., Patton, F.D. & Cording, E.J. (1967) Design of surface and near surface construction in rock. In *Eighth US Symposium on Rock Mechanics*, Minneapolis, pp. 237–303.

Delhomme, J.P. (1979) Spatial variability and uncertainty in groundwater flow parameters: a geostatistical approach, *Water Resources Research*, **15**(2): 269–280.

D'Elia, B., Esu, F., Pellegrino, A. & Pescatore, T.S. (1985) Some effects on natural slope stability induced by the 1980 Italian earthquake. In *Proceedings, 11th International Conference on Soil Mechanics & Foundation Engineering*, San Francisco, CA, Vol. 4, pp. 1943–1949.

D'Elia, B., Picarelli, L., Leroueil, S. & Vaunat, J. (1998) Geotechnical characterisation of slope movements in structurally complex clay soils and stiff jointed clays. *Italian Geotechnical Journal*, **XXXII**(3): 5–32.

de Marsily, G. (1986) *Quantitative Hydrogeology Groundwater Hydrology for Engineers*, Academic Press, Orlando, FL.

Demirtas, R., Karakisa, S., Yatman, A., Baran, B., Zunbul, S., Iravut, Y., Altin, N. & Yilmaz, R. (1996) 1 Ekim 1995 Dinar Depremi Mekanizmasi (in Turkish), *Deprem Arastirma Bulteni* **23**(74): 5–45.

Demond, A.H. & Roberts, P.V. (1987) An examination of relative permeability relations for two-phase flow in porous media. *Water Resources Bulletin*, **23**(4): 617–628.

Dempster, A.P., Laird, N.M. & Rubin, D.B. (1977) Maximum likelihood from incomplete data via the EM algorithm. *Journal of the Royal Statistical Society, Series B. Methodological*, **39**: 1–22.

den Hoedt, G. (1986) Creep and relaxation of geotextile fabrics. *Journal of Geotextiles & Geomembranes*, **4**(2): 83–92.

Deniau, A., Cartier, G., Morbois, A. & Virollet, M. (1981) Un nouveau procédé de drainage: la paroi drainante. *Bulletin de Liaison des Laboratoires des Ponts et Chaussées*, Special XI, **F**: 15–18.

Denver, H. (1982) Modulus of elasticity determined by SPT and CPT. *Proceedings, ESOPT II*, **1**: 35–40.

De Nicola, A. & Randolph, M.F. (1993) Tensile and compressive shaft capacity of piles in sand. *Journal of Geotechnical Engineering, ASCE*, Reston, VA, **119**(12): 1952–1973.

De Ruiter, J. & Beringen, F.L. (1979) Pile foundations for large North Sea structures. *Marine Geotechnics*, **3**(3): 267–314.

Desai, C.S. (1974) Numerical design-analysis for piles in sands. *Journal of The Geotechnical Engineering Division, ASCE*, Reston, VA, **100**(6): 613–635.

Desai, C.S. & Faruque, M.O. (1984) Constitutive model for (geological) materials. *Journal of Engineering Mechanics, ASCE*, Reston, VA, **110**(9): 1391–1408.

Devlin, J.F. & Barker, J.F. (1996) Field investigation of nutrient pulse mixing in an *in situ* biostimulation experiment. *Water Resources Research*, **32**(9): 2869–2877.

Devlin, J.F. & Parker, B.L. (1996) Optimum hydraulic conductivity to limit contaminant flux through cutoff walls. *Ground Water*, **34**(4): 719–726.

DI (1965) *Code of Practice for Foundation Engineering*, Dansk Ingeniorforening, Copenhagen, Denmark, DS415.

Diaz-Rodriquez, J.A., Leroueil, S. & Aleman, J.D. (1992) Yielding of Mexico City clay and other natural clays. *Journal of Geotechnical Engineering Division, ASCE*, Reston, VA, **118**(7): 981–995.

DiMaggio, J.A. (1978) *Stone Columns for Highway Construction*, Demonstration Project No. 46, Report No. FHWA-DP-46-1, Federal Highway Administration, Springfield, VA.

DiMaggio, F.L. & Sandler, I.S. (1971) Material model for granular soils. *Journal of Engineering Mechanics Division, ASCE*, Reston, VA, **97**(3): 935–950.

Dingman, S.L. (1994) *Physical Hydrology*, Macmillan, New York.

DM-7 (1971) *Design Manual for Soil Mechanics*, Foundations and Earth Structures, Department of the Navy, Washington, DC.

Dobry, R. & Gazetas, G. (1985) Dynamic stiffness and damping of foundations by simple methods. In Gazetas, G. & Selig, E.T. (Eds) *Vibration Problems in Geotechnical Engineering*, American Society of Civil Engineers, Reston, VA, USA, pp. 77–107.

Dobry, R. & Gazetas, G. (1988) Simple method for dynamic stiffness and damping of floating pile groups. *Géotechnique*, **38**(4): 557–574.

Dobry, R., Vasquez-Herrera, A., Mohammad, R. & Vucetic, M. (1985) Liquefaction flow failure of silty sand by torsional cyclic tests. *Advances in the Art of Testing Soils under Cyclic Loading Conditions, ASCE Convention*, Detroit, Michigan, ed: V. Khosla, 29–50.

Domenico, P.A. & Schwartz, F.W. (1998) *Physical and Chemical Hydrogeology*, Wiley, New York.

Donald, I.B. (1956) Shear strength measurements in unsaturated non-cohesive soils with negative pore pressures. In *Proceedings, Second Australia–New Zealand Conference on Soil Mechanics & Foundation Engineering*, pp. 200–204.

Donaldson, J.J. (1994) Texturing techniques. In Hsuan, Y.G. & Koerner, R.M. (Eds) *GRI 8 Conference on Geosynthetic Resins, Formulations & Manufacturing*, IFAI, Roseville, MN, pp. 113–122.

Dove, J.E., Frost, J.D., Han, J. & Bachus, R.C. (1997) The influence of geomembrane surface roughness on interface strength. In *Proceedings Geosynthetics '97 Conference*, Long Beach, Industrial Fabrics Association International, Roseville, MN, pp. 863–876.

Dreimanis, A. (1962) Quantitative gasometric determination of calcite and dolomite by using Chittick apparatus. *Journal of Sedimentary Petrology*, **32**(3): 520–529.

Drever, J.I. (1982) *The Geochemistry of Natural Waters*, Prentice-Hall, Englewood Cliffs, NJ.

Driscoll, F.G. (1986) *Groundwater & Wells*, 2nd Edn, Johnson Filtration Systems, St Paul, MN.

Drnevich, V.P. (1976) *Drnevich Resonant Column Apparatus Operating Manual*, Soil Dynamics Instruments, Lexington, KY.

Drucker, D.C., Gibson, R.E. & Henkel, D.J. (1957) Soil mechanics and work-hardening theories of plasticity. *Transactions, ASCE*, Reston, VA, **122**: 338–346.

Drumright, E.E. (1989) *PhD thesis*, Colorado State University, CO.

Druschel, S.J. & Underwood, E.R. (1993) Design of lining and cover system sideslopes. In *Proceedings Geosynthetics '93 Conference*, Vancouver, BC, Industrial Fabrics Association International, Roseville, MN, Vol. 3, pp. 1341–1355.

Dullien, F.A.L. (1979) *Porous Media: Fluid Transport and Pore Structure*, Academic Press, New York.

Duncan, J.M. (1979) Behaviour and design of long-span metal culverts. *Journal of Geotechnical Engineering, ASCE*, Reston, VA, **105**(3): 399–418.

Duncan, J.M. (1992) State-of-the-art: Static stability and deformation analysis. In *Proceedings ASCE Specialty Conference on Stability and Performance of slopes and Embankments – II*, Berkeley, Geotechnical Special Publication No. 31, Vol. 1, pp. 222–266.

Duncan, J.M. (1996) Soil slope stability analysis. In *Landslides—Investigation and Mitigation*, A.K. Turner & R.L. Schuster (Eds.) Special Report 247, Transportation Research Board, Washington, DC, pp. 337–371.

Duncan, J.M. & Buchignani, A.L. (1976) *An Engineering Manual for Slope Stability Studies*, Department of Civil Engineering, University of California, Berkeley.

Duncan, J.M. & Seed, R.B. (1986) Compaction-induced earth pressures under K_o-conditions. *ASCE Journal of Geotechnical Engineering*, Reston, VA, **112**(1): 23–43.

Duncan, J.M. & Wright, S.G. (1980). The accuracy of equilibrium methods of slope stability analysis. *Engineering Geology*, **16**: 5–17.

Duncan, J.M., Williams, G.W., Sehn, A.L. & Seed, R.B. (1991a) Estimation of earth pressures due to compaction. *ASCE Journal of Geotechnical Engineering*, Reston, VA, **117**(12): 1833–1847.

Duncan, J.M., Williams, G.W., Sehn, A.L. & Seed, R.B. (1993b) Discussion. *ASCE Journal of Geotechnical Engineering*, Reston, VA, **119**(7): 1162–1177.

Dunnicliff, J. (1988) *Geotechnical Instrumentation for Monitoring Field Performance*, Wiley, New York.

Dwyer, S.F. (1995) Alternative landfill cover demonstration. In *Landfill Closures: Environmental Protection and Land Recovery*, Geotechnical Special Publication No. 53, Dunn, R.J. and Singh, U.P. (Eds.) American Society of Civil Engineers, Reston, VA, pp. 19–34.

Dwyer, S.F. (1998) Alternative landfill covers pass the test. *Civil Engineering, ASCE*, Reston, VA, **68**(9): 50–52.

Dysli, M. (1990) *Le gel et son action sur les sols et les fondations*, Presses Polytechniques et Universitaires Romandes.

Ebeling, R.M. & Morrison, E.E. (1992) *The Seismic Design of Waterfront Retaining Structures*, Technical Report ITL-92-11, Waterways Experiment Station, Vicksburg, MS.

Eberle, M.A. & von Maubeuge, K. (1997) Measuring the *in-situ* moisture content of geosynthetic clay liners (GCLs) using time domain reflectometry (TDR). In Rowe, R.K. (Ed.) *Proceedings, Sixth International Conference on Geosynthetics*, Atlanta, GA, Industrial Fabrics Association International, Roseville, MN, Vol. 1, pp. 205–210.

Edgers, L., Noble, J.J. & Williams, E. (1992) A Biologic Model for Long-Term Settlement in Landfills, Proc., Conf. on Environmental Technology, Balkan: 177–184.

Edil, T.B. & Dhowian, A.W. (1981) At-rest lateral pressure of peat soils. *Journal of Geotechnical Engineering Division, ASCE*, Reston, VA, **107**(2): 201–217.

Edil, T.B., Ranguette, V.J. & Wuellner, W.W. (1990) Settlement of municipal refuse. In *Geotechnics of Waste Fills – Theory & Practice*, STP 1070, Landua, A. and Knowles E.D. (Eds.) American Society for Testing & Materials, West Conshohocken, PA, pp. 225–239.

Editorial reports on seismic risk. *International Journal on Hydropower and Dams*, **5** (1, 2).

Edlefsen, W.H. & Anderson, A.B.C. (1943) Thermodynamics of soil moisture. *Hilgardia*, **15**(2): 31–298.

Eggestad, A. (1983) Improvement of cohesive soils–state-of-the-art report. In *Proceedings, Eighth European Conference on Soil Mechanics & Foundation Engineering*, Helsinki, Vol. 3, pp. 991–1007.

Ehrig, H.-J. & Scheelhaase, T. (1993) Pollution potential and long term behaviour of sanitary landfills. In *Proceedings, Fourth International Landfill Symposium*, Sardinia, pp. 1203–1255.

Eid, H.T. & Stark, T.D. (1997) Shear behavior of an unreinforced geosynthetic clay liner. *Geosynthetics International*, **4**(6): 645–659.

Eigenbrod, K.D. (1975) Analysis of the pore pressure changes following the excavation of a slope. *Canadian Geotechnical Journal*, **12**(3): 429–440.

Einsele, G. (1992) *Sedimentary Basins*, Springer-Verlag, Berlin.

Einstein, H.H. & Schwartz, C.W. (1979) Simplified analysis for tunnel supports. Journal of *Geotechnical Engineering, ASCE*, Reston, VA, **105**(4): 499–518.

Einstein, H.H., Bischoff, N. & Gofmann, E. (1972) Behavior of invert slabs in swelling shale. In *International Symposium on Underground Openings*, Lucerne, Switzerland, pp. 296–329.

Eith, A.W. & Koerner, R.M. (1992) Field evaluation of geonet flow rate (transmissivity) under increasing load. *Journal of Geotextile & Geomembranes*, **11**(5–6): 153–166.

Eith, A.W., Boschuk, J. & Koerner, R.M. (1991) Prefabricated bentonite clay liners. *Geotextiles & Geomembranes*, **10**: 575–599.

Elias, V. & Christopher, B.R. (1996) *Mechanically Stabilized Earth Walls and Reinforced Soil Slopes; Design and Construction Guidelines*, Report No. FHWA-SA-96-071, US Department of Transportation, Federal Highway Administration, Washington, DC.

Elias, V. & Juran, I. (1991) *Soil Nailing for Stabilization of Highway Slopes and Excavations*, US Department of Transportation, Federal Highway Administration Publication No. FHWA-RD-89-193, Washington, DC.

El-Marsafawi, H., Han, Y.C. & Novak, M. (1992) Dynamic experiments on two pile groups. *Journal of Geotechnical Engineering, ASCE*, Reston, VA, **118**(4): 576–592.

El Naggar, M.H. & Novak, M. (1995a) Nonlinear lateral interaction in pile dynamics. *Journal of Soil Dynamics & Earthquake Engineering*, **14**(2): 141–157.

El Naggar, M.H. & Novak, M. (1995b) Effect of foundation nonlinearity on modal properties of offshore towers. *Journal of Geotechnical Engineering, ASCE*, Reston, VA, **121**(9): 660–668.

El Naggar, M.H. & Novak, M. (1996) Influence of foundation nonlinearity on offshore tower response. *Journal of Geotechnical Engineering, ASCE*, Reston, VA, **122**(9): 717–724.

Eloy-Giorni, C., Pelte, T., Pierson, P. & Margarita, R. (1996) Water diffusion through geomembranes under hydraulic pressure. *Geosynthetics International*, **3**(6): 741–769.

Elsabee, F. & Morray, J.P. (1977) *Dynamic Behavior of Embedded Foundations*, Research Report R77-33, Civil Engineering Department, Massachusetts Institute of Technology, Cambridge, MA.

Emrich, W.J. (1971) Performance study of soil sampler for deep-penetration marine borings.

In *Sampling of Soil & Rock*, ASTM Special Technical Publication 483, American Society for Testing & Materials, West Conshohocken, PA, pp. 30–50.

EN 1401 Plastic piping systems for non-pressure underground drainage and sewerage—unplasticized polyvinyl chloride (PVC-U). European Standard, European Committee for Standardization, Brussels.

EN 12666 Plastics piping systems for non-pressure underground drainage and sewerage-polyethylene (PE), European Standard, European Committee for Standarization, Brussels.

ERES (1998) *Catalogue of Current State Pavement Design Features*, Draft Final Report on NCHRP Project 1-32, ERES Consultants, Champaign, IL, pp. 61821–63915.

Escario, V. & Juca, J. (1989) Strength and deformation of partly saturated soils. *Proceedings, 12th ICSMFE*, Rio de Janeiro, Vol. 3, pp. 43–46.

Esrig, M.I. (1968) Pore pressures, consolidation and electrokinetics. *Journal of Soil Mechanics & Foundation Engineering Division, ASCE*, Reston, VA, **94**(4): 899–921.

Esrig, M.I. & Bachus, R.D. (Eds) (1991) *Deep Foundation Improvements: Design, Construction, and Testing*, Presented at a symposium, Las Vegas, ASTM STP 1089, ASTM, West Conshohocken, PA.

Evans, S.G. (1997) Fatal landslides and landslide risk in Canada. In *Proceedings, International Workshop on Landslide Risk Assessment*, Honolulu, HI; pp. 185–196.

FAA (1987) Federal Aviation Administration Advisory Circular.

FAA (1995a) *LEDFAA User's Manual*, Federal Aviation Administration Advisory Circular ISO/5320-16 Airport Pavement Design for the Boeing 777 Airplane, US Department of Transportation, FAA, Washington, DC.

FAA (1995b) *Airport Pavement Design & Evaluation*, Federal Aviation Administration Advisory Circular ISO/5320-60, US Department of Transportation, FAA, Washington, DC.

Faccioli, E. (1991) Seismic amplification in the presence of geological and topographic irregularities. In *Proceedings, Second International Conference on Recent Advances in Geotechnical Earthquake Engineering & Soil Dynamics*, St Louis, MO, Vol. II, pp. 1779–1797.

Fam, S. (1996) Vapor extraction and bioventing. In Nyer, E.K. (Ed.) In Situ *Treatment Technology*, CRC Lewis, Boca Raton, FL, pp. 101–148.

Fang, H.-Y. (1991) *Foundation Engineering Handbook*, 2nd Edn, Chapman & Hall, New York.

Farmer, I.W. (1968) *Engineering Properties of Rocks*, E. & F.N. Spon Limited, London, England.

Farr, A.M., Houghtalen, R.J. & McWhorter, D.B. (1990) Volume estimation of light nonaqueous phase liquids in porous media. *Groundwater*, **28**: 48–56.

Fassett, J.B., Leonards, G.A. & Repetto, P.C. (1994) Geotechnical properties of municipal solid wastes and their use in landfill design. In *Proceedings, Waste Tech '94 Conference*, National Solid Waste Management Association, Charleston, SC.

Faure, R.M. (1985) Analyse des contraintes dans un talus par la méthode des perturbations. *Revue Française de Géotechnique*, **33**: 49–59.

Faure, R.M., Pham, M., Robinson, J.C. & Jolly, P. (1996) Three dimensional slope stability by the perturbation method. In *Proceedings, Seventh International Symposium on Landslides*, Trondheim, Vol. 2, pp. 1207–1212.

Faure, Y.H. (1984) Design of drain beneath geomembranes: discharge estimation and flow patterns. In *Proceedings, International Conference on Geomembranes*, Denver, CO, Industrial Fabrics Association International, Roseville, MN, Vol. 2, pp. 463–468.

Faust, C.R., Guswa, J.H. & Mercer, J.W. (1989) Simulation of three-dimensional flow of immiscible fluids within and below the unsaturated zone, *Water Resource Research*, **25**(12): 2449–2464.

Faustman, E.M. & Omenn, G.S. (1996) Risk assessment. In Klaassen, C.D. (Ed.) *Casarett & Doull's Toxicology: The Basic Science of Poisons*, 5th Edn, McGraw-Hill, New York, pp. 80–83.

Fay, J.J. & King, R.E. III (1994) Antioxidants for geosynthetic resins and applications, Hsuan, Y.G. & Koerner, R.M. (Eds) *GRI 8 Conference on Geosynthetic Resins, Formulations & Manufacturing*, IFAI, Roseville, MN, pp. 74–91.

Fayer, M. & Jones, T. (1990) *Unsaturated Soil Water and Heat Flow Model, Version 2*, Pacific Northwest Laboratory, Richland, Washington, DC.

Fayer, M.J., Rockhold, M.L. & Campbell, M.D. (1992) Hydrologic modeling of protective barriers: comparison of field data and simulation results. *Soil Science Society of America Journal*, **56**(3): 690–700.

Feenstra, S. & Cherry, J.A. (1996) Diagnosis and assessment of DNAPL sites. In Pankow, J.F. &

J.A. Cherry (Eds) *Dense Chlorinated Solvents and Other DNAPLs in Groundwater*, Waterloo Press, Rockwood, Ontario, pp. 395–474.

Feenstra, S., Cherry, J.A. & Parker, B.L. (1996) Conceptual models for the behavior of DNAPLs in the subsurface. In Pankow, J.F. & J.A. Cherry (Eds) *Dense Chlorinated Solvents and Other DNAPLs in Groundwater*, Waterloo Press, Rockwood, Ontario, pp. 53–88.

Federal Register (1991), **56**(20): 3535.

Fell, R. (1994) Landslide risk assessment and acceptable risk. *Canadian Geotechnical Journal*, **31**(2): 261–272.

Fellenius, B.H. (1989) *Unified Design of Piles and Pile Groups*, Transportation Research Board, Washington, TRB Record 1169, pp. 75–82.

Fellenius, W. (1927) *Erdstatische Berechnungen (Calculation of Stability of Slopes)*, Ernst, Berlin (revised edn 1939).

FEMA (1997) *NEHRP Recommended Provisions for Seismic Regulations for New Buildings, Part 2—Commentary, Federal Emergency Management Agency*, 1994 Edition, Building Seismic Safety Council, Washington, DC.

Fenelli, G.B. & Picarelli, L. (1990) The pore pressure field built up in a rapidly eroded soil mass. *Canadian Geotechnical Journal*, **27**(3): 387–392.

Fenelli, G.B., Silvestri, F. & Picarelli, L. (1992) Deformation process of a hill shaken by the Irpinia earthquake in 1980. In *Proceedings, French–Italian Conference on Slope Stability in Seismic Areas*, Bordighera, pp. 47–62.

Fenn, D.G., Hanley, K.J. & DeGeare, T.V. (1975) *Use of the Water Balance Method for Predicting Leachate Generation from Solid Waste Disposal Sites*, US Environmental Protection Agency, EPA/530/SW-168, Washington, DC.

Fernandez, F. & Quigley, R.M. (1985) Hydraulic conductivity of natural clays permeated with simple liquid hydrocarbons. *Canadian Geotechnical Journal*, **22**: 205–214.

Fernandez, F. & Quigley, R.M. (1988) Viscosity and dielectric constant controls on the hydraulic conductivity of clayey soils permeated with water-soluble organics. *Canadian Geotechnical Journal*, **25**(3): 582–589.

Fernando, N.S.M & Carter, J.P. (1998) Elastic analysis of buried pipes under surface patch loadings. *Journal of Geotechnical & Geoenvironmental Engineering, ASCE*, Reston, VA, **124**(8): 720–728.

Fetzer, C. (1967) Electro-osmotic stabilization of West Branch Dam. *Journal of Soil Mechanics & Foundation Division, ASCE*, Reston, VA, **93**(4): 85–104.

Fetter, C.W. (1994) *Applied Hydrogeology*, Macmillan College Publishing, New York.

FHWA (1987) *Pavement Design, Principles and Practices, FHWA Short Course*, 3rd Revision, Federal Highway Administration & ERES Consultants, Champaign, IL, pp. 61821–63915.

FHWA (1993a) *Recommendations Clouterre 1991* (English translation), *Soil Nailing Recommendations 1991*, Report FHWA-SU-026, Federal Highway Administration, Washington, DC.

FHWA (1993b) *FHWA International Scanning Tour on Soil Nailing*, Federal Highway Administration, US Department of Transportation, Washington, DC.

FHWA (1996) *Mechanically Stabilized Earth Walls and Reinforced Soil Slopes Design and Construction Guidelines*, Federal Highway Administration Demonstration Project 82, Washington, DC.

Finlay, P.J. & Fell, R. (1997) Landslides: risk perception and acceptance. *Canadian Geotechnical Journal*, **34**(2): 169–188.

Finn, W.D.L. (1988) Dynamic analysis in geotechnical engineering. In *Proceedings, Earthquake Engineering & Soil Dynamics II – Recent Advances in Ground Motion Evaluation*, American Society of Civil Engineers, Geotechnical Engineering Division, Park City, UT, pp. 523–591.

Finn, W.D.L. (1990) Analysis of post-liquefaction deformations in soil structures. Invited paper, In *Proceedings, H. Bolton Seed Memorial Symposium*, University of California, Berkeley, Duncan, J.M. (Ed.) Bi-Tech Publishers, Vancouver, Vol. 2, pp. 291–311.

Finn, W.D.L. (1993) Seismic safety evaluation of embankment dams. In *Proceedings, International Workshop on Dam Safety Evaluation*, 26–28 April, Grindewald, Switzerland, Vol. 4, pp. 91–135.

Finn, W.D.L. (1995) Foundation factors for seismic design. In *Proceedings, Seventh Canadian Conference on Earthquake Engineering*, Montreal, pp. 197–204.

Finn, W.D.L. (1999) State of the art of geotechnical earthquake engineering practice. Invited Keynote Lecture, *Seventh International Conference on Soil Dynamics and Earthquake Engineering*, August 10–15, Bergen, Norway.

Finn, W.D.L. & Yogendrakumar, M. (1989) *TARA-3FL: A Program for Analysis of Flow*

Deformations in Soil Structures with Liquefied Zones, Soil Dynamics Group, Department of Civil Engineering, University of British Columbia, Vancouver, BC.

Finn, W.D.L., Yogendrakumar, M., Yoshida, N. & Yoshida, H. (1986) *TARA-3: A Program for Nonlinear Static and Dynamic Effective Stress Analysis*, Soil Dynamics Group, University of BC, Vancouver, BC.

Finn, W.D.L., Ledbetter, R.H., Fleming, R.L. Jr., Templeton, A.E., Forrest, T.W. & Stacy, S.T. (1991) Dam on liquefiable foundation: safety assessment and remediation. In *Proceedings, International Workshop on Remedial Treatment of Potentially Liquefiable Soils*, January, Tsukuba Science City, Japan.

Finn, W.D.L., Wu, G. & Ledbetter, R.H. (1994) Problems in seismic soil–structure interaction. In *Proceedings, Eighth International Conference on Computer Methods and Advances in Geomechanics*, Morgantown, WV, Vol. 1, pp. 139–151.

Finn, W.D.L., Byrne, P.M., Evans, S. & Law, T. (1996) Some geotechnical aspects of the Hyogo-ken-Nanbu (Kobe) earthquake of January 17, 1995. *Canadian Journal of Civil Engineering*, **23**(3): 778–796.

Fisher, R.A. (1925) Applications of Student's distribution. *Metron* **5**: 90–104.

Flaate, K. & Rygg, N. (1964) Sawdust as embankment fill on peat bogs. In *Proceedings, Ninth Muskeg Research Conference*, NRC, ACSSM Technical Memorandum 81, pp. 136–149.

Fleenor, W.E. & King, I.P. (1995) Identifying limitations on use of the HELP model. In *Landfill Closures: Environmental Protection and Land Recovery*, Geotechnical Special Publication No. 53, Dunn, R.J. Sing, U.P. (Eds) American Society of Civil Engineers, Reston, VA, pp. 121–138.

Fleming, I.R., Rowe, R.K. & Cullimore, D.R. (1999) Field observations of clogging in a landfill leachate collection system. *Canadian Geotechnical Journal* **36**(4): 685–707.

Fleming, W.G.K., Weltman, A.J., Randolph, M.F. & Elson, W.K. (1985) *Piling Engineering*, Surrey University Press, Halsted Press, New York.

Fleming, W.G.K., Weltman, A.J., Randolph, M.F. & Elson, W.K. (1992) *Piling Engineering*, 2nd Edn, Halsted Press, New York.

Flynn, D.J., Thomson, N.R. & Farquhar, G.J. (1994) The influence of heterogeneity on the mass removal efficiency on conventional soil vacuum extraction technology: a field study. In *Proceedings, Eighth National Outdoor Action Conference*, May 23–25, Minneapolis, MN, pp. 41–56.

Focht, J.A. & Koch, (1973) Rational analysis of the lateral performance of offshore pile groups. In *Proceedings, Fifth Offshore Technical Conference*, Houston, TX, Vol. 2, pp. 701–708.

Focht, R., Vogan, J. & O'Hannesin, S. (1996) Field application of reactive iron walls for *in situ* degradation of volatile organic compounds in groundwater. *Remediation*, **6**(3): 81–94.

Folkes, D.J. & Crooks, J.H.A. (1985) Effective stress paths and yielding in soft clays below embankments. *Canadian Geotechnical Journal*, **22**: 357–374.

Forchheimer, P. (1930), *Hydraulik*, 3rd Edn, B.G. Teubner, Leipzig und Berlin.

Foreman, D.E. & Daniel, D.E. (1986) Permeation of compacted clay with organic chemicals. *Journal of Geotechnical Engineering, ASCE*, Reston, VA, **112**: 669–681.

Forsyth, R.A. & Egan, J.P. (1976) *Use of Waste Materials in Embankment Construction*, Transportation Research Record 593, Washington, DC, pp. 3–8.

Fountain, J.C., Starr, R.C., Middleton, T., Beikirch, M., Taylor, C. & Hodge, D. (1996) A controlled field test of surfactant-enhanced aquifer remediation. *Ground Water*, **34**(5): 910–916.

Fox, P.J., Edil, T.B. & Lau, L.T. (1992) C_α/C_c concept applied to compression of peat. *ASCE Journal of Geotechnical Engineering*, Reston, VA, **118**(8): 1256–1263. See also Discussion, **120**(4): 764–770.

Fox, P.J., Roy-Chowdhury, N. & Edil, T.B. (1999) Secondary compression of peat with or without surcharging, Discussion of paper by Mesri *et al. ASCE Journal of Geotechnical & Geoenvironmental Engineering*, Reston, VA, **125**(2): 160–162.

Frank, R. & Magnan, J.-P. (1995) Cone penetration testing in France: national report. In *Proceedings, CPT '95*, Linkoping, Swedish Geotechnical Society, Vol. 3, pp. 147–156.

Frank, H.S. & Wen, W.Y. (1957) Structure aspects of ion-solvent interaction in aqueous solutions. A suggested picture of water structure. *Faraday Society Discussions*, **24**: 133–140.

Frankel, F., Mueller, C., Bernhard, T., Perkins, D., Leyendecker, E.V., Dickman, N., Hanson, S. & Hopper, M. (1996) *Interim National Hazard Maps: Documentation*, Draft Report, US Geological Survey, Denver, CO.

Franklin, G. & Chang, F. (1977) *Earthquake Resistance of Earth and Rockfill Dams. 5: Permanent Displacement of Earth Embankments by Newmark Sliding Block Analysis*. Miscellaneous Paper S-71-17, Soils & Pavements Laboratory, US Army Engineer Waterways Experiment Station, Vicksburg, MS.

Fraser, R.A. & Wardle, L.J. (1976) Numerical analysis of rectangular rafts on layered foundations. *Géotechnique*, **26**(4): 613–630.

Freden, S. (1965) Some aspects on the physics of frost heave in mineral soils. *Surface Chemistry*, 79–90.

Fredlund, D.G. (1964) *MSc thesis*, Department of Civil Engineering, University of Alberta, Edmonton.

Fredlund, D.G. (1979) Second Canadian Geotechnical Colloquium: appropriate concepts and technology for unsaturated soils. *Canadian Geotechnical Journal*, **16**(1): 121–139.

Fredlund, D.G. (1992) Background, theory and research related to the use of thermal conductivity matric suction measurements. In *Advances in Measurement of Soil Physical Properties: Bringing Theory into Practice*, San Antonio, Texas, SSSA Special Publication No. 30, Soil Science of America, Eds. G.C. Top, W.D. Reynolds and R.E. Green, Madison, WI, pp. 249–262.

Fredlund, D.G. (1995) Prediction of unsaturated soil functions using the soil–water characteristic curve. In *Proceedings, Bength B. Broms Symposium on Geotechnical Engineering*, December 13–15, Nanyang Technological University, Singapore, River Edge, NJ, World Scientific, pp. 113–133.

Fredlund, D.G. & Krahn, J. (1977) Comparison of slope stability methods of analysis. *Canadian Geotechnical Journal*, **14**(3): 429–439.

Fredlund, D.G. & Morgenstern, N.R. (1977) Stress state variables for unsaturated soils. *Journal of Geotechnical Engineering*, ASCE, Reston, VA, **103**: 447–466.

Fredlund, D.G. & Rahardjo, H. (1993) *Soil Mechanics for Unsaturated Soils*, Wiley, New York.

Fredlund, D.G. & Xing, A. (1994) Equations for the soil–water characteristic curve. *Canadian Geotechnical Journal*, **31**: 521–532.

Fredlund, D.G., Bergan, A.T. & Sauer, E.K. (1975) Deformation characterization of subgrade soils for highways and runways in northern environments. *Canadian Geotechnical Journal*, **12**(2): 213–223.

Fredlund, D.G., Morgenstern, N.R. & Widger, A. (1978) The shear strength of unsaturated soils. *Canadian Geotechnical Journal*, **15**(3): 313–321.

Fredlund, D.G., Xing, A. & Huang, S. (1994) Predicting the permeability function for unsaturated soils using the soil–water characteristic curve. *Canadian Geotechnical Journal*, **31**(4): 533–546.

Fredlund, M.D., Sillers, W.S., Fredlund, D.G. & Wilson, G.W. (1996a) Design of a knowledge-based system for unsaturated soil properties. In *Proceedings, Third Canadian Conference on Computing in Civil Engineering*, August 26–28, Montreal, Quebec, pp. 659–677.

Fredlund, M.D., Xing, A., Fredlund, M.D. & Barbour, S.L. (1996b) The relationship of the unsaturated soil shear strength to the soil–water characteristic curve. *Canadian Geotechnical Journal*, **33**(3): 440–448.

Fredlund, M.D., Fredlund, D.G. & Wilson, G.W. (1997a) Prediction of the soil–water characteristic curve from grain-size distribution and volume–mass properties. In *Proceedings, Third Brazilian Symposium on Unsaturated Soils, NONSAT '97*, April 22–25, Rio de Janeiro, **1**: 13–23.

Fredlund, M.D., Fredlund, D.G. & Wilson, G.W. (1997b) Estimation of unsaturated soil properties using a knowledge-based system. In *Proceedings, Fourth Congress on Computing in Civil Engineering*, June 16–18, American Society of Civil Engineers, Reston, VA, pp. 501–510.

Freeze, R.A. & Cherry, J.A. (1979) *Groundwater*, Prentice-Hall, NJ.

Freeze, R.A. & McWhorter, D.B. (1997) A framework for assessing risk reduction due to DNAPL mass removal from low permeability soils. *Ground Water*, **35**(1): 111–123.

French Standard (1992) *Soil Reinforcement-Backfilled Structures with Inextensible and Flexible Reinforcing Straps or Sheets*, NF, Paris, P 94-220 (English translation).

Fruchter, J.S., Cole, C.R., Williams, M.D., Vermeul, R., Teel, S.S., Amonette, J.E., Szecsody, J.E. & Yabusaki, S.B. (1997) Creation of a subsurface permeable treatment barrier using *in situ* redox manipulation. In *Proceedings, 1997 International Containment Technology Conference & Exhibition*, February 10–12, St Petersburg, FL, pp. 704–710.

Fuhrmann, M., Aloysius, D. & Zhou, I. (1995) Permeable, subsurface sorbent barrier for

^{90}Sr: laboratory studies of natural and synthetic materials. *Waste Management*, **15**(7): 485–493.

Fukuoka, M. (1986) Large Scale Permeability Tests for Geomembrane-Subgrade System. In *Proceedings, Third International Conference on Geotextiles, Geomembranes, & Related Products*, The Hague, The Netherlands, Vol. 3, pp. 917–922.

Fuleihan, N.F. & Wisse, A.E.Z. (1995) Minimum thickness of compacted soil liners: II. Analysis and case histories–Discussion. *Journal of Geotechnical Engineering, ASCE*, Reston, VA, **121**(6): 504–506.

Fung, Y.C. (1977) *A First Course in Continuum Mechanics*, 2nd edn, Prentice-Hall, Englewood Cliffs, NJ.

Gaind, K.J. & Char, A.N.R. (1983) Reinforced soil beams. *Journal of Geotechnical Engineering, ASCE*, Reston, VA, **109**(7): 977–982.

Gallavresi, F. (1992) Grouting improvement of foundation soils (keynote lecture). In *Grouting, Soil Improvement, & Geosynthetics*, Geotechnical Special Publication No. 30, American Society of Civil Engineers, ASCE Press, Reston, VA, pp. 1–38.

Gan, J.K.-M., Fredlund, D.G. & Rahardjo, H. (1988) Determination of the shear strength parameters of an unsaturated soil using the direct shear. *Canadian Geotechnical Journal*, **25**: 500–510.

Gardner, W.R. (1958) Some steady state solutions of the moisture flow equation with application to evaporation from a water table. *Soil Science*, **85**: 228–232.

Gässler, G. & Gudehus, G. (1983) Soil nailing-statistical design. In *Proceedings, VIII ECS-MFE*, Helsinki.

Gazetas, G. (1991) Foundation vibrations. Fang, H.Y. (Ed.) In *Foundation Engineering Handbook*, 2nd Edn, Van Nostrand Reinhold, New York.

Gazetas, G. & Dobry, R. (1984) Horizontal response of piles in layered soils. *Journal of Geotechnical Engineering, ASCE*, Reston, VA, **110**(1): 20–40.

Gazetas, G. & Makris, N. (1991) Dynamic pile-soil–pile interaction, part I: Analysis of axial vibration. *Earthquake Engineering & Structural Dynamics*, **20**: 115–132.

Gazetas, G., Fan, K. & Kaynia, A.M. (1993) Dynamic response of pile groups with different configurations. *Journal of Soil Dynamics & Earthquake Engineering*, **12**(4): 239–257.

Gazioglu, S.M. & O'Neill, M.W. (1984) Evaluation of P–Y relationships in cohesive soils. In Meyer, J.R. (Ed.) *Analysis & Design of Pile Foundations*, American Society of Civil Engineers, Reston, VA, pp. 192–213.

Gee, G.W., Anderson, L.W. & Fayer, M.J. (1997) Surface barrier research at the Hanford site. In *Proceedings, International Containment Technology Conference*, February, St Petersburg, FL, pp. 305–311.

Gelhar, L.W., Welty, C. & Rehfeldt, K.R. (1992) A critical review of data on field-scale dispersion in aquifers. *Water Resources Research*, **28**(7): 1955–1974.

Geller, M. (1982) *Compaction Equipment for Asphalt Mixtures*, ASTM STP829, American Society for Testing & Materials, West Conshohocken, PA.

Gemperline, M.C. (1988) Centrifuge modelling of shallow foundations, *Proceedings, ASCE Spring Convention*, ASCE Press, Reston, VA.

Gens, A., Hutchinson, J.N. & Cavounidis, S. (1988) Three-dimensional analysis of slides in cohesive soils. *Géotechnique*, **38**(1): 1–23.

Geological Society (1986) *Site Investigation Practice: Assessing B5 5930*. Geological Society Engineering Geology Special Publication No. 2., Hawkins, A.B. (Ed.) The Geological Society, London, England.

Geo-Slope International Ltd (undated) Slope/W, Calgary, Alberta, Canada (E-mail: *info@geo-slope.com*).

Geosynthetics Research Institute GC-1 (1996) *Soil–Filter Core Combined Flow Test*, GSI, Folsom, PA.

Geosynthetics Research Institute GG-1 (1996) *Geogrid Rib Tensile Strength*, GSI, Folsom, PA.

Geosynthetics Research Institute GG-2 (1996) *Geogrid Junction Strength*, GSI, Folsom, PA.

Geosynthetics Research Institute GM-12 (1996) *Asperity Measurement of Textured Geomembranes using a Depth Gage*, GSI, Folsom, PA.

Gerhard, J.I., Kueper, B.H. & Hecox, G. (1998). Waterflooding for the removal of pooled DNAPL. *Ground Water*, **36**(2): 283–292.

Ghaboussi, J. & Momen, H. (1982) Modelling and analysis of cyclic behavior of sands, In Pande, G.N. & Zienkiewicz O.C. (Eds) *Soil Mechanics–Transient & Cyclic Loads*, Wiley, London, England, pp. 313–342.

Ghiassian, H., Hryciw, R.D. & Gray, D.H. (1997) Stabilization of coastal slopes by anchored geosynthetic systems. *Journal of Geotechnical &*

Geoenvironmental Engineering, ASCE, Reston, VA, **123**(8): 736–743.

Ghosh, R.K. (1980) Estimation of soil–moisture characteristics from mechanics properties of soils. *Soil Science Journal,* **130**(2): 60–63.

Gibson, R.E. (1963) An analysis of system flexibility and its effect on time-lag in porewater pressure measurements. *Géotechnique,* **13**(1): 1–11.

Gibson, R.E. & Lo, K.Y. (1961) A theory of soils exhibiting secondary compression. *Acta Polytechnica Scandinavica, C,* **296**(10): 1–15.

Gilbert, P.A. & Murphy, W.L. (1987) *Prediction/ Mitigation of Subsidence Damage to Hazardous Waste Landfill Covers,* US Environmental Protection Agency, Hazardous Waste Engineering Research Laboratory, Cincinnati, OH, EPA/ 600/12/87-025.

Gilbert, R.B., Liu, C.N., Wright, S.G. & Trautwein, S.J. (1995) A double shear test method for measuring interface strength. In *Proceedings, Geosynthetics '95 Conference,* Nashville, Industrial Fabrics Association International, Roseville, MN, Vol. 3, pp. 1017–1029.

Gilbert, R.B., Fernandez, F. & Horsfield, D.W. (1996) Shear strength of reinforced geosynthetic clay liner. *Journal of Geotechnical Engineering, ASCE,* Reston, VA, **122**(4): 259– 266.

Gilbert, R.O. (1987) *Statistical Methods for Environmental Pollution Monitoring,* Van Norstrand Reinhold, New York.

Gilbert, R.O. & R.R. Kinneson. (1981) Statistical methods for estimating the mean and variance from radionuclide data sets containing negative, unreported or less-than values. *Health Physics,* **40**: 377–390.

Gillham, R.W. & Burris, D.R. (1997) Recent development in permeable *in situ* treatment walls for remediation of contaminated groundwater. In Ward, C.H., Cherry, J.A. & Scalf, M.R. (Eds) *Subsurface Restoration,* Ann Arbor Press, Chelsea, MI, pp. 343–356.

Gillham, R.W. & O'Hannesin, S.F. (1992) Metal-catalysed abiotic degradation of halogenated organic compounds. In *Proceedings, IAH Conference on Modern Trends in Hydrogeology,* May 10–13, Hamilton, ON, pp. 94–103.

Gillham, R.W. & O'Hannesin, S.F. (1994) Enhanced degradation of halogenated aliphatics by zero-valent iron. *Ground Water,* **32**(6): 958– 967.

Gillham, R.W., Major, L., Wadley, S.L.S. & Warren, J. (1998) Advances in the application of zero-valent iron for treatment of groundwater containing VOCs. In *Groundwater Quality: Remediation & Protection Proceedings, GQ'98 Conference,* IAHS Publication No. 250, September 21–25, Tubingen, Germany, pp. 475– 481.

Gilliom, R.J. & Helsel, D.R. (1986) Estimation of distributional parameters for censored trace level quality data. 1: Estimation techniques. *Water Resources Research,* **22**: 135–146.

Gipson, A.H., Jr, (1985) Permeability testing on clayey soil and silty sand–bentonite mixture using acid liquor. In *hydraulic barriers in soil and rock,* Johnson, A.I., Frobel, R.K., Cavalli, N.J. & Petersson, C.B. (Eds) ASTM STP 874, American Society for Testing & Materials, West Conshohocken, PA, pp. 140–154.

Girault, D.P. (1987) Analysis of foundations failures in the Mexico earthquakes, 1985: factors involved and lessons learned. In Cassaro, M.A. & Romero, E.M. (Eds.) American Society of Civil Engineers, ASCE Press, Reston, VA, pp. 178–192.

Giroud, J.P. (1973) L'étanchéité des retenues d'eau par feuilles déroulées. *Annales de l'ITBTP,* **312**(TP161): 94–112 (in French).

Giroud, J.P. (1977) Conception de l'étanchéité des ouvrages hydrauliques par géomembranes. In *Proceedings, First International Symposium on Plastic & Rubber Waterproofing in Civil Engineering,* Session 3, Paper 13, June, Liège, Belgium, Vol. 1, pp. 1–17 (in French).

Giroud, J.P. (1981) Designing with geotextiles. *Matériaux et Constructions,* **14**(82): 257– 272.

Giroud, J.P. (1982) Filter criteria for geotextiles. In *Proceedings, Second International Conference on Geotextiles,* August, Las Vegas, NV, Vol. 1, pp. 103–108.

Giroud, J.P. (1983) Rapport général sur les bassins revêtus de géomembranes. In *Comptesrendus du colloque sur l'étanchéité superficielle des bassins, barrages et canaux,* February, Paris, France, Vol. II, pp. 75–90d (in French).

Giroud, J.P. (1984a) Impermeability: the myth and a rational approach. In *Proceedings, International Conference on Geomembranes,* June Denver, CO, Vol. 1, pp. 157–162.

Giroud, J.P. (1984b) Analysis of stresses and elongations in geomembranes. In *Proceedings International Conference on Geomembranes,* June, Denver, CO, Industrial Fabrics Association International, Roseville, MN, Vol. 2, pp. 481–486.

Giroud, J.P. (1985) Tomorrow's designs for geotextile applications. In Fluet, J.E. (Ed.) *Geotextile Testing & the Design Engineer*, ASTM STP 952 American Society for Testing & Materials, West Conshohocken, PA, pp. 145–158.

Giroud, J.P. (1988) Review of geotextile filter criteria. In *Proceedings, First Indian Geotextiles Conference*, December, Bombay, India, pp. 1–6.

Giroud, J.P. (1989) Panelist contribution on geotextile filters. In *Proceedings, 12th International Conference on Soil Mechanics & Foundation Engineering*, August Rio de Janeiro, Brazil, Vol. 5, pp. 3105–3106.

Giroud, J.P. (1994) Quantification of geosynthetics behavior (special lecture). In *Proceedings, Fifth International Conference on Geotextiles, Geomembranes & Related Products*, September Singapore, Vol. 4, pp. 1249–1273.

Giroud, J.P. (1995) Determination of geosynthetic strain due to deflection. *Geosynthetics International*, **2**(3): 635–641.

Giroud, J.P. (1996a) Granular filters and geotextile filters. In Lafleur, J. & Rollin, A.L. (Eds) *Proceedings, GeoFilters '96*, May, Montréal, pp. 565–680.

Giroud, J.P. (1996b) Workshop on testing of geosynthetic materials for landfill liners and covers. *IGS News*, **12**(2): 3–4.

Giroud, J.P. (1997) Equations for calculating the rate of liquid migration through composite liners due to geomembrane defects. Special Issue on Liquid Migration Control Using Geosynthetic Liner Systems, *Geosynthetics International* **4**(3–4): 335–348.

Giroud, J.P. (1999) Lessons learned from failures associated with geosynthetics, keynote lecture. In *Proceedings, Geosynthetics '99*, April, Boston, MA, IFAI, Industrial Fabrics Association International, Roseville, MN, Vol. 1, pp. 1–66.

Giroud, J.P. & Beech, J.F. (1989) Stability of soil layers on geosynthetic lining systems. In *Proceedings, Geosynthetics '89*, February San Diego, CA, Industrial Fabrics Association International, Roseville, MN, Vol. 1, pp. 35–46.

Giroud, J.P. & Bonaparte, R. (1989a) Leakage through liners constructed with geomembranes, Part I: geomembrane liners. *Geotextiles & Geomembranes*, **8**(1): 27–67.

Giroud, J.P. & Bonaparte, R. (1989b) Leakage through liners constructed with geomembranes, Part II: composite liners. *Geotextiles & Geomembranes*, **8**(2): 71–111.

Giroud, J.P., Gleason, M.H. & Zornberg, J.G. (1999) Design of geomembrane anchorage against wind action. *Geosynthetics International* **6**(6): 481–507.

Giroud, J.P. & Houlihan, M.F. (1995) Design of leachate collection layers. In *Proceedings, Fifth International Landfill Symposium*, S. Margherita di Pula, Cagliari, Italy, Vol. 2, pp. 613–640.

Giroud, J.P. & Soderman, K.L. (1995) Design of structures connected to geomembranes. *Geosynthetics International*, **2**(2): 379–428.

Giroud, J.P., Gourc, J.P., Bally, P. & Delmas, P. (1977) Comportement d'un textile non tissé dans un barrage en terre (Behaviour of a nonwoven fabric in an earth dam). In *Proceedings, International Conference on the Use of Fabrics in Geotechnics*, April, Paris, France, Vol. 2, pp. 213–218, (in French).

Giroud, J.P. Bonaparte, R., Beech, J.F. & Gross, B.A. (1988) Load carrying capacity of a soil layer supported by a geosynthetic overlying a void. In *Theory and Practice of Earth Reinforcement*, Yamanouchi, T., Miura, N. and Ochiai, H. (Eds.), A.A. Balkema, Rotterdam, pp. 185–190.

Giroud, J.P. Khatami, A. & Badu-Tweneboah, K. (1989) Evaluation of the rate of leakage through composite liners. *Geotextiles & Geomembranes*, **8**(4): 337–340.

Giroud, J.P., Bonaparte, R., Beech, J.F. & Gross, B.A. (1990) Design of soil layer–geosynthetic systems overlying voids. *Geotextiles & Geomembranes*, **9**(1): 11–50.

Giroud, J.P., Badu-Tweneboah, K. & Bonaparte, R. (1992a) Rate of leakage through a composite liner due to geomembrane defects. *Geotextiles & Geomembranes*, **11**(1): 1–28.

Giroud, J.P., Gross, B.A. & Darasse, J. (1992b) *Flow in Leachate Collection Layers–Steady-State Conditions*, GeoSyntec Consultants Internal Report, Boca Raton, FL.

Giroud, J.P., Beech, J.F. & Soderman, K.L. (1994a) Yield of scratched geomembranes, *Geotextiles and Geomembranes*, Vol. 13, No. 4, pp. 231–246.

Giroud, J.P., Badu-Tweneboah, K. & Soderman, K.L. (1994b) Evaluation of landfill liners. In *Proceedings, Fifth International Conference on Geotextiles, Geomembranes & Related Products*, Singapore, Vol. 3, pp. 981–986.

Giroud, J.P., Bachus, R.C. & Bonaparte, R. (1995a) Influence of water flow on the stability of geosynthetic–soil layered systems on slopes. Special Issue on Design of Geomembrane

Applications, *Geosynthetics International*, **2**(6): 1149–1180.

Giroud, J.P., Badu-Tweneboah, K. & Soderman, K.L. (1995b) Theoretical analysis of geomembrane puncture. Special Issue on Design of Geomembrane Applications, *Geosynthetics International*, **2**(6): 1019–1048.

Giroud, J.P., Pelte, T. & Bathurst, R.J. (1995c) Uplift of geomembranes by wind. Special Issue on Design of Geomembrane Applications, *Geosynthetics International*, **2**(6): 897–952 (Erratum, **4**(2): 187–207).

Giroud, J.P., Williams, N.D., Pelte, T. & Beech, J.F. (1995d) Stability of geosynthetic–soil layered systems on slopes. Special Issue on Design of Geomembrane Applications, *Geosynthetics International*, **2**(6): 1115–1148.

Giroud, J.P., Gross, B.A., Bonaparte, R. & McKelvey, J.A. (1997a) Leachate flow in leakage collection layers due to defects in geomembrane liners. Special Issue on Liquid Migration Control Using Geosynthetic Liner Systems, *Geosynthetics International*, **4**(3–4): 215–292.

Giroud, J.P., Khire, M.V. & Soderman, K.L. (1997b) Liquid migration through defects in a geomembrane overlain and underlain by permeable media. Special Issue on Liquid Migration Control Using Geosynthetic Liner Systems, *Geosynthetics International*, **4**(3–4): 293–321.

Giroud, J.P., King, T.D., Sanglerat, T.R., Hadj-Hamou, T. & Khire, M.V. (1997c) Rate of liquid migration through defects in a geomembrane placed on a semi-permeable medium. Special Issue on Liquid Migration Control Using Geosynthetic Liner Systems, *Geosynthetics International*, **4**(3–4): 349–372.

Giroud, J.P., Soderman, K.L. & Badu-Tweneboah, K. (1997d) Optimal configuration of a double liner system including a geomembrane liner and a composite liner. Special Issue on Liquid Migration Control Using Geosynthetic Liner Systems, *Geosynthetics International*, **4**(3–4): 373–389.

Giroud, J.P., Houlihan, M.F., Bachus, R.C. & Qureshi, S. (1998a) Clogging potential of geosynthetic leachate collection layers by fine particles from sand protective layers. In Rowe, R.K. (Ed.) *Proceedings, Sixth International Conference on Geosynthetics*, March, Atlanta, GA, Industrial Fabrics Association International, Roseville, MN, Vol. 1, pp. 185–190.

Giroud, J.P., Soderman, K.L., Khire, M.V. & Badu-Tweneboah, K. (1998b) New developments in landfill liner leakage evaluation. In Rowe, R.K. (Ed.) *Proceedings, Sixth International Conference on Geosynthetics*, March, Atlanta, GA, Industrial Fabrics Association International, Roseville, MN, Vol. 1, pp. 261–268.

Giroud, J.P., Bonaparte, R., Matasovic, N. & Kavazanjian, E., Jr (2000) Approximate solution for pseudo static seismic analysis of landfill cover systems, *Geosynthetics International*, (submitted).

Gleit, A. (1985) Estimation of small normal data sets with detection limits. *Environmental Science & Technology*, **19**: 1201–1206.

Glickman, T.S. & Gough, M. (Eds) (1990) Readings in Risk, *Resources for the Future*, Washington, DC.

Goh, A.T.C., The, C.I. & Wong, K.S. (1997) Analysis of piles subjected to embankment induced lateral soil movements. *Journal of Geotechnical & Geoenvironmental Engineering, ASCE*, Reston, VA, **123**(9): 792–801.

Goldman, L.J., Greenfield, L.I., Damle, A.S., Kingsbury, G.L., Northeim, C.M., and Truesdale, R.S. (1990) *Clay Liners for Waste Management Facilities, Design, Construction & Evaluation*. Noyes Data, NJ.

Goodman, R.E. & Bray, J.W. (1976) Toppling of rock slopes. In *Proceedings, Specialty Conference on Rock Engineering for Foundation & Slopes*, Boulder, CO, American Society of Civil Engineers, ASCE Press, Reston, VA, Vol. 2, pp. 201–234.

Gorelick, S.M., Freeze, R.A., Donohue, D. & Keely, J.F. (1993) *Groundwater Contamination Optimal Capture and Containment*, Lewis, Boca Raton, FL.

Gounaris, V., Anderson, P.R. & Holsen, T.M. (1993) Characteristics and environmental significance of colloids in a landfill leachate. *Environmental Science & Technology*, **27**: 1381–1387.

Graf, E.D. (1992a) Compaction grout. In *Grouting, Soil Improvement, and Geosynthetics*, Geotechnical Special Publication No. 30, American Society of Civil Engineers, ASCE Press, Reston, VA, Vol. 1, pp. 275–287.

Graf, E.D. (1992b) Earthquake support grouting in sands. In *Grouting, Soil Improvement, & Geosynthetics*, Geotechnical Special Publication No. 30, American Society of Civil Engineers, ASCE Press, Reston, VA, Vol. 2, pp. 879–888.

Grant, B.L., McKenna, R.W. & von Maubeuge, K.P. (1997) Investigations of liner leakage rates and EPA guidelines. In *Environmental Geotechnics*, Bouzza, A., Kodikara, J. and Parker, R. (Eds.) A.A. Balkema, Rotterdam, pp. 315–324.

Grant, R., Christian, J.T. & Vanmarcke, E.H. (1974) Differential settlement of buildings. *Journal of Geotechnical Engineering, ASCE*, Reston, VA, **100**(9): 973–991.

Grassie, N. & Scott, G. (1985) *Polymer Degradation and Stabilization*, Cambridge University Press, England.

Graves, R.W., Pitarka, A. & Somerville, P.G. (1998) Ground motion amplification in the Santa Monica area: effects of shallow basin edge structure. *Bulletin of Seismological Society of America*, **88**(5): 1224–1242.

Gray, D.H. & Leiser, A. (1982) *Biotechnical Slope Protection and Erosion Protection*, Van Nostrand Reinhold, New York, NY.

Gray, D.H. & Sotir, R.B. (1992) Biotechnical stabilization of cut and fill slopes. In *Specialty Conference on Stability & Performance of Slopes and Embankments-II*, Berkeley, Geotechnical Special Publication No. 31, American Society of Civil Engineers, Reston, VA, Vol. 2, pp. 1395–1410.

Gray, D.H. & Sotir, R.B. (1995) *Biotechnical Stabilization of Steepened Slopes*, Transportation Research Record No. 1474, Washington, DC, pp. 23–29.

Gray, D.M. (1970) *Handbook on the Principles of Hydrology*, Canadian National Committee for the International Hydrological Decade, National Research Council of Canada, Ottawa, Canada.

Greenkorn, R.A. (1983) *Flow Phenomena in Porous Media: Fundamentals and Applications in Petroleum, Water and Food Production*, Marcel Dekker, New York.

Greenkorn, R.A. & Kessler, D.P. (1972) *Transfer Operations*, McGraw-Hill, New York.

Gress, J.-C. (1992) Siphon drain: a technic for slope stabilization. In *Proceedings, Sixth International Symposium on Landslides*, Christchurch, Vol. 1, pp. 729–734.

Gress, J.-C. & Faure, R. (1996) Induced streaming electrical potential and water flows through landslides. In *Proceedings, Seventh International Symposium on Landslides*, Trondheim, Vol. 2, pp. 739–742.

GRI GM-10 (1996) *Specification for the Stress Crack Resistance of HDPE Geomembrane Sheet*, GSI, Folsom, PA.

GRI GM-13 (1997) *Specification for HDPE*

Smooth and Textured Geomembranes, GSI, Folsom, PA.

Griffin, R.A., Cartwright, K., Shimp, N.F., Steel, J.D., Ruch, R.R., White, W.A., Hughes, G.M. & Gilkeson, R.H. (1976) Attenuation of pollutants in municipal landfill leachate by clay minerals: Part I—column leaching and field verification. Illinois State Geological Survey, Environmental Geology Notes No. 79.

Griffith, A.A. (1921) The phenomena of flow and rupture in solids. *Philosophical Transactions of the Royal Society, London*, **A221**: 163–198.

Griffith, A.A. (1924) Theory of rupture. In *Proceedings, First International Congress Applied Mechanics*, Delft, pp. 55–63.

Gross, B.A., Bonaparte, R. & Othman, M.A. (1997) Inferred performance of surface hydraulic barriers from landfill operational data. In *Proceedings, International Containment Technology Conference*, St Petersburg, FL, pp. 374–380.

Grubb, D.G. & Sitar, N. (1994) *Evaluation of Technologies for In Situ Cleanup of DNAPL Contaminated Sites*, EPA/600/R-95/120, US Environmental Protection Agency, Ada, OK, 173 p.

Gruszczenski, T.S. (1987) Determination of a realistic estimate of the actual formation product thickness using monitor wells: a field bailout test. In *Proceedings, Conference On Petroleum-Hydrocarbons and Organic Chemicals in Ground Water: Prevention, Detection, and Restoration*, National Ground Water Association, Dublin, OH, pp. 235–253.

Gschwend, P.M. & Reynolds, M.D. (1987) Monodisperse ferrous phosphate colloids in an anoxic groundwater plume. *Journal of Contaminant Hydrology*, **1**: 309–327.

Gudehus, G. (1972) Lower and upper bounds for stability of earth-retaining structures. In *Proceedings, Fifth European Conference of International Society for Soil Mechanics and Foundation Engineering*, Madrid, **1**: 21–28.

(1990) *Guideline for a Professional Engineer's Duty to Report*, Professional Engineers Ontario, Toronto, ON.

Gullà, G. & Sorbino, G. (1996) Soil suction measurements in a landslide involving weathered gneiss. In *Proceedings, Seventh International Symposium on Landslides*, Trondheim, Vol. 2, pp. 749–754.

Gullick, R.W. (1998) *PhD thesis*, University of Michigan.

Guan, Y. & Fredlund, D.G. (1997) Direct measurement of high soil suction. In *Proceedings,*

Third Brazilian Symposium on Unsaturated Soils, NONSAT '97, April 22–25, Rio de Janeiro, Brazil, **2**: 543–550.

Gumbel, J.E. (1981) Discussion of Pender (1980) *Géotechnique*, **31**(3): 434–436.

Gupta, S.C. & Larson, W.E. (1979) Estimating soil water retention characteristics from particle size distribution, organic matter percent, and bulk density. *Water Resources Research Journal*, **15**(6): 1633–1635.

Gupta, S.K., Tanji, K.K., Nielsen, D.R., Biggar, J.W., Simmon, C.S. & MacIntyre, J.L. (1978) *Field Simulation of Soil-Water Movement with Crop Water Extraction*, Water Science & Engineering Paper No. 4013, Department of Land, Air, & Water Resources, University of California, Davis.

Gutenberg, B. & Richter, C.F. (1944) Frequency of earthquakes in California. *Bulletin of the Seismological Society of America*, **46**: 105–145.

Gysi, H.J., Linder, A. & Leoni, E. (1975) Prestressed diaphragm walls. In *Proceedings, European Conference of International Society for Soil Mechanics and Foundation Engineering*, Vienna, Austria.

Haas, C.N. & Jacangelo, J.G. (1993) Development of regression models with below-detection data. *Journal of Environmental Engineering*, **119**(2): 214–230.

Haas, R., Hudson, W.R. & Zaniewski, J (1994) *Modern Pavement Management*, Krieger, Krieg Malabar, FL.

Hachey, J.E., Plum, R.L., Byrne, J., Killian, A.P. & Jenkens, D.V. (1994) Blast densification of a thick, loose debris flow at Mt St Helens, Washington. *Proceedings, Settlement '94*, Geotechnical Special Publication No. 40, American Society of Civil Engineers, ASCE Press, Reston, VA, Vol. 1, pp. 502–512.

Haimson, B.C. (1978a) The hydrofracturing stress measuring method and recent field results. *International Journal of Rock Mechanics, Mineral Science & Geomechanics, Abstracts*, **15**: 167–178.

Haimson, B.C. (1978b) Underground nuclear power station study. Hydrofracturing stress measurements. Hole UN-1 Darlington G.S. Report No. 78250. Geotechnical Engineering Department, Ontario Hydro.

Haimson, B.C. (1985) SAB-NIAGARA GS NO. 3. Hydrofracturing stress measurements and permeability tests in Holes NF3 and NF4. Report No. 85357. Geotechnical & Hydraulic Engineering Department, Ontario Hydro.

Haimson, B.C. & Fairhurst, C. (1969) *In-situ* stress determination at great depth by means hydraulic fracturing. In *11th US Symposium on Rock Mechanics*, pp. 559–584.

Haimson, B.C. & Lee, C.F. (1980) Hydrofracturing stress determinations at Darlington, Ontario. In *13th Canadian Rock Mechanics Symposium* (The H.R. Rice Memorial Symposium), pp. 42–50.

Haimson, B.C., Tunbridge, L.W., Lee, M.Y. & Cooling, C.M. (1989) Measurement of rock stress using the hydraulic fracturing method in Cornwall, UK – Part II. Data reduction and stress calculation. *International Journal of Rock Mechanics, Mineral Science & Geomechanics, Abstracts*, **26**(5): 361–372.

Hakulinen, M. (1991) Measured full-scale dynamic lateral pile responses in clay and in sand. In *Proceedings, Second International Conference on Recent Advances in Geotechnical Earthquake Engineering & Soil Dynamics*, University of Missouri at Rolla, Rolla, MO, pp. 201–206.

Halse, Y., Koerner, R.M. & Lord, A.E. (1987a) Effect of high levels of alkalinity on geotextiles – Part 1: $Ca(OH)_2$ solution. *Journal of Geotextiles & Geomembranes*, **5**(4): 261–282.

Halse, Y., Koerner, R.M. & Lord, A.E. (1987b) Effect of high levels of alkalinity on geotextiles – Part 2: Na(OH) solution. *Journal of Geotextiles & Geomembranes*, **6**(4): 295–306.

Halton, G., Loughney, R & Winters, E. (1965) Vacuum stabilization of subsoil beneath runway extension at Philadelphia International Airport. In *Sixth International Conference of Soil Mechanics & Foundation Engineering*, Montreal, Vol. 2, pp. 61–65.

Hamby, D.M. (1994) A review of techniques for parameter sensitivity analysis of environmental models. *Environmental Monitoring & Assessment*, **32**: 135–154.

Hampton, D.R. & Heuvelhorst, H.G. (1990) Designing gravel packs to improve separate-phase hydrocarbon recovery: laboratory experiments. In *Proceedings, Conference On Petroleum Hydrocarbons and Organic Chemicals in Ground Water: Prevention, Detection and Restoration*, National Ground Water Association, Dublin, OH, pp. 195–209.

Hampton, D.R., Smith, M.M. & Shank, S.J. (1991) Further laboratory studies of gravel pack design for hydrocarbon recovery wells. In *Proceedings, Conference On Petroleum Hydrocar-*

bons and Organic Chemicals in Ground Water: Prevention, Detection and Restoration, National Ground Water Association, Dublin, OH, pp. 615–629.

Hampton, M.A., Lee, H.J. & Locat, J. (1996) Submarine landslides. Review in Geophysics, 34(1): 33–59.

Handa, S.C. (1984) Foundation performance of very old structures. In Proceedings, International Conference on Case Histories in Geotechnical Engineering, Rolla, MO, Vol. I, pp. 201–208.

Handy, R.L., Remens, B., Moldt, S., Lutenegger, A.J. & Trott, G. (1982) In-situ stress determination by Iowa stepped blade. Journal of Geotechnical Engineering, ASCE, Reston, VA, 108(11): 1405–1422.

Hanna, T.H. (1982) Foundations in Tension, Trans Tech Publications, McGraw-Hill, New York, NY.

Hanna, T.H. (1985) Field Instrumentation in Geotechnical Engineering, Trans Tech Publications, Clausthal-Zellerfeld, Germany.

Hanna, A.M. & Meyerhof, G.G. (1980) Design charts for ultimate bearing capacity of foundations on sand overlying soft clay. Canadian Geotechnical Journal, 17: 300–303.

Hanrahan, E.T. (1964) A road failure on peat. Géotechnique, 14(3): 185–202.

Hansbo, S. (1979) Consolidation of clay by band-shaped prefabricated drains. Ground Engineering, 12(5): 16–25.

Hansbo, S. (1981) Consolidation of fine-grained soils by prefabricated drains. In Proceedings, Tenth International Conference on Soil Mechanics & Foundation Engineering, Stockholm, Vol. 3, pp. 12–22.

Hansbo, S. (1983) Techno-economic trend on subsoil improvement methods in foundation engineering (special lecture). In Proceedings Eighth European Conference on Soil Mechanics & Foundation Engineering, Helsinki, Vol. 3, pp. 1333–1343.

Hansbo, S. (1987) Design aspects of vertical drains and lime column installations. In Proceedings, Ninth Southeast Asian Geotechnical onference, Bangkok, Thailand, Vol. 2, pp. 8–12.

Hansbo, S. (1993) Band drains. In Moseley, M.P. (Ed.) Ground Improvement, Blackie Academic & Professional, Blackwood, NJ, pp. 40–64.

Hansen J.B. (1970) A Revised and Extended Formula for Bearing Capacity, Danish Geotechnical Institute Bulletin No. 28.

Hanzawa, H. (1989) Evaluation of design parameters for soft clays as related to geological stress history. Soils & Foundations, 29(2): 99–111.

Hanzawa, H., Kishida, T. & Matsuda, E. (1982) Stability analysis with the effective stress method for embankments constructed on an alluvial marine clay. Soils & Foundations, 22(3): 32–46.

Harder, L.F. & Boulanger, R.W. (1997) Application of K_σ and K_α correction factors. In Youd, T.L. & Idriss, I. (Eds) Proceedings, NCEER Workshop on Evaluation of Liquefaction Resistance of Soils, December 31, Technical Report, NCEER-7-0022.

Harder, L.F. & Seed, H.B. (1986) Determination of Penetration Resistance for Coarse-grained Soils Using the Becker Hammer Drill. Report No. UCB/EERC-86/06, Earthquake Engineering Research Center, University of California, Berkeley.

Harder, L.S., Jr (1991) Performance of earth dams during the Loma Prieta earthquake. In Proceedings, Second International Conference on Recent Advances in Geotechnical Earthquake Engineering & Soil Dynamics, St Louis, pp. 11–15.

Hardin, B.O. (1978) The nature of stress–strain behavior of soils. In Proceedings, Earthquake Engineering & Soil Dynamics, American Society of Civil Engineers, Reston, VA, Vol. 1, pp. 3–90.

Hardin, B.O. & Drnevich, V.P. (1972) Shear modulus and damping in soils: design equations and curves. Journal of Soil Mechanics & Foundations Division, ASCE, Reston, VA, 98(7): 667–692.

Harris, S.M., Day, S., Roberts, E., Rovers, F.A. & Carey, G.R. (1998) Applications of an innovative method for visualizing natural attenuation. In Proceedings, First International Conference on Remediation of Chlorinated and Recalcitrant Compounds, CA, May 18–21 Monterey, Battelle Press, Columbus, OH.

Hartford, D.N. (1997) Dam management in Canada—a Canadian approach to dam safety. In Proceedings, International Workshop on Risk Based Dam Safety Evaluations, Trondheim, Norway, Norwegian National Committee of the International Committee on Large Dams.

Hartlén, J. & Wolski, W. (1996) Embankments on Organic Soils, Elsevier, Amsterdam.

Hashimoto, L.K. & Trussel R.R. (1983) Evaluating water quality data near the detection limit, In Proceedings, American Water Works Association *Advanced Technology Conference*, June 5–9, Las Vegas, NV.

Haslem, R.F. (1986) Pipe jacking forces; from theory to practice. In *Proceedings, Infrastructure, Renovation and Waste Control Centenary Conference*, Manchester, England, North West Association, Institution of Civil Engineers, pp. 173–180.

Hassan, M.H., O'Neill, M.W., Sheikh, S.A. & Ealy, C.D. (1997) Design method for drilled shafts in soft argillaceous rock. *Journal of Geotechnical & Geoenvironmental Engineering*, **123**(3): 272–280.

Hausmann, M.R. (1990) *Engineering Principles of Ground Modification*, McGraw Hill, New York, NY.

Haverkamp, R. & Parlange, J.Y. (1986) Predicting the water-retention curve from particle-size distribution: 1. Sandy soils without organic matter. *Soil Science Journal*, **142**(6): 325–339.

Hayhoe, H.N. & Balchin, D. (1990) Field frost heave measurement and prediction during periods of seasonal frost. *Canadian Geotechnical Journal*, **27**: 393–397.

He, Z., Kennepohl, G., Haas, R. & Cai, Y. (1996) OPAC 2000: a new pavement design system. In *Proceedings Transportation Association of Canada Annual Conference*, Charlottetown, PEI. Proc. CD-ROM Format (no hard copy) ISBN 1-55187-106-8.

Head, K.H. (1986) *Manual of Soil Laboratory Testing*, Volume 3, *Effective Stress Testing*, Pentech Press, London, England.

Health Canada (1993) *Health Risk Determination*. Health Canada, Ottowa, Ontario.

Hearmon, R.F.S. (1961) *An Introduction to Applied Anisotropic Elasticity*, Oxford University Press, England.

HEAST (1997) *Health Effects Assessment Summary Tables FY-1997 Update*, Office of Research & Development, US Environmental Protection Agency, Washington, DC, EPA 540/R-97-036.

Heerten, G., von Maubeuge, K., Simpson, M. & Mills, C. (1993) Manufacturing quality control of geosynthetic clay liners–A manufacturer's perspective. In *Proceedings, Sixth GRI Seminar, MQC/MQA and CQC/CQA of Geosynthetics*, St Paul, MN, IFAI, Roseville, MN, pp. 86–95.

Hefny, A.M. & Lo, K.Y. (1992) The interpretation of horizontal and mixed-mode fractures in hydraulic fracturing tests in rock. *Canadian Geotechnical Journal*, **29**: 902–917.

Hefny, A.M. & Lo, K.Y. (1995) Interpretation of initial stresses from hydraulic fracturing tests at AECL's underground research laboratory, Manitoba. *Canadian Tunnelling Journal*, 123–134.

Hefny, A., Lo, K.Y. & Huang, J.A. (1996) Modeling of long-term time-dependent deformation and stress-dependency of Queenston Shale. *Canadian Tunnelling Journal*, 115–146.

Helenelund, K.B. (1977) *Methods for Reducing Undrained Shear Strength of Soft Clay*, Report No. 2, Swedish Geotechnical Institute, Linköping.

Helsel, D.R. & Gilliom, R.J. (1986) Estimation of distributional parameters for censored trace level water quality data. 2. Verification and applications. *Water Resources Research*, **22**(2): 147–155.

Helsel, D.R. & Hirsch, R.M. (1992) *Statistical Methods in Water Resources* (Studies in Environmental Science 49), Elsevier, Amsterdam.

Hem, J.D. (1995) *Study and Interpretation of the Chemical Characteristics of Natural Water*, United States Geological Survey Water-Supply Paper 2254, 3rd Edn, US Government Printing Office, Washington, DC.

Henkel, D.J. (1956) The effect of overconsolidation on the behaviour of clays during shear. *Géotechnique*, **6**(4): 139–150.

Hewitt, R.D. & Daniel, D.E. (1997) Hydraulic conductivity of geosynthetic clay liners after freeze–thaw. *Journal of Geotechnical & Geoenvironmental Engineering, ASCE*, Reston, VA, **123**(4): 305–313.

Heyman, L. (1965) Measurement of the influence of lateral earth pressure on piles. In *Proceedings, Sixth International Conference on Soil Mechanics & Foundation Engineering*, Montreal, Vol. 2 pp. 257–260.

Herzog, M. (1985) Die Pressenkrafte bei Schildvrieb und Rohrvorpressung im Lockergestein. *Baumaschine & Bautechnik*, **32**: 236–238.

Hilf, J.W. (1991) Compacted fill. In Fang, H. (Ed.) *Foundation Engineering Handbook* 2nd Edn, Van Nostrand Reinhold, New York.

Hill, R. (1983) *The Mathematical Theory of Plasticity*, Oxford Science Publications, Oxford.

Hinchberger, S.D. & Rowe, R.K. (1998) Modelling the rate sensitive characteristics of the Gloucester foundation soil. *Canadian Geotechnical Journal*, **35**(5): 769–789.

Hinchee, R.E. (1993) Bioventing of petroleum hydrocarbons. In *Handbook of Bioremediation*, Robert S. Kerr Environmental Research Laboratory, CRC Press, Boca Raton, FL, pp. 39–60.

Hinchee, R.E., Wilson, J.T. & Downey, D. (1995) *Intrinsic Bioremediation*, Battelle Press, Columbus, OH.

Hirayama, H. (1991) Pile group settlement interaction considering soil nonlinearity. *Computer Methods & Advances in Geomechanic*, Beer, G., Booker, J.R., and Carter, J.P. (Eds.), A.A. Balkema, Rotterdam **1**: 139–144.

Hirt, R.C. & Searle, N.Z. (1964) Wavelength sensitivity or activation spectra of polymers. In *Regional Technical Conference*, Society of Plastics Engineers, Washington, DC, pp. 286–302.

Hittinger, M. (1978) *PhD thesis*, University of California, Berkeley.

Ho, D.Y.F., Fredlund, D.G. & Rahardjo, H. (1992) Volume change indices during loading and unloading of an unsaturated soil. *Canadian Geotechnical Journal*, **29**(2): 195–207.

Hodge, R.A.L. & Freeze, R.A. (1977) Groundwater flow systems and slope stability. *Canadian Geotechnical Journal*, **14**(4): 466–476.

Hodges, J.J., Jr & Lehmann, E.L. (1956) The efficiency of some nonparametric competitors to the *t*-test. *Annals of Mathematical Statistics*, **27**: 324–335.

Hoeg, K. (1968) Stresses against underground structural cylinders. *Journal of Soil Mechanics & Foundation Engineering, ASCE*, Reston, VA, **94**(4): 833–858.

Hoek, E. & Bray, J. (1981) *Rock Slope Engineering*, 3rd Edn, Institute of Mining & Metallurgy, London.

Hoek, E. & Brown, E.T. (1980) Empirical strength criterion for rock masses. *Journal of Geotechnical Engineering Division, ASCE*, Reston, VA, **106**(GT9): 1013–1035.

Hoek, E. & Brown, E.T. (1988) The Hoek–Brown failure criterion–a 1988 update. In Curran, J.C. (Ed.) *Rock Engineering for Underground Excavations, 15th Canadian Rock Mechanics Symposium*, Toronto, pp. 31–38.

Hoek, E., Wood, D. & Shah, S. (1992) A modified Hoek–Brown criterion for jointed rock masses. In Hudson, J.A. (Ed.) *Proceedings, Rock Characterization, Symposium of the International Society of Rock Mechanics, Eurock '92*, British Geological Society, London, pp. 209–214.

Hoek, E., Kaiser, P.K. & Bawden, W.F. (1995) *Support of Underground Excavations in Hard Rock*, Balkema, Rotterdam.

Hoekstra, P. (1969) Water movement and freezing pressures. *Soil Science Society of America, Proceedings*, **33**: 512–518.

Holland, D. & Fitz-Simmons, T. (1982) Fitting statistical distributions to air quality data by the maximum likelihood method. *Atmospheric Environment*, **16**(5): 1971–1076.

Holloway, P.M., Clough, G.W. & Vesic, A.S. (1975) *The Mechanics of Pile–Soil Interaction in* Cohesionless Soils. Soil Mechanic Series No. 39, Duke University School of Engineering, Durham, NC, USA.

Holtz, R.D. (1989) *Treatment of Problem Foundations for Highway Embankments*, NCHRP Synthesis of Highway Practice 147, Transportation Research Board, Washington, DC.

Holtz, R.D. (1990) Design and construction of geosynthetically reinforced embankments on very soft soils. State-of-the-Art Paper, Session 5, Performance of Reinforced Soil Structures. In *Proceedings, International Reinforced Soil Conference*, Glasgow, British Geotechnical Society, pp. 391–402.

Holtz, R.D. (1991) Stress distribution and settlet of shallow foundations. In Fang, H.Y. (Ed.) *Foundation Engineering Handbook*, Van Nostrand Reinhold, New York, NY, pp. 166–222.

Holtz, R.D. & Christopher, B.R. (1987) Characteristics of prefabricated drains for accelerating consolidation. In *Proceedings, Ninth European Conference on Soil Mechanics & Foundation Engineering*, Dublin, Vol. 2, pp. 903–906.

Holtz, R.D. & Holm, G. (1973) Excavation and sampling around some sand drains at Ska-Edeby, Sweden. In *Proceedings, Sixth Scandinavian Geotechnical Meeting*, Trondheim, NGI.

Holtz, R.D. & Holm, G. (1979) Test embankment on an organic silty clay. In *Seventh ECSMFE*, Brighton, Vol. 3, pp. 79–86.

Holtz, R.D. & Kovacs, W.D. (1981) *An Introduction to Geotechnical Engineering*, Prentice Hall, Englewood Cliffs, NJ.

Holtz, R.D. & Schuster, R.L. (1996) Stabilization of soil slopes. *Landslides: Investigation and Mitigation*, Turner, A.K. and Schuster, R.L. (Eds) Transportation Research Board, Washington, DC, Special Report 247, pp. 439–473.

Holtz, R.D. & Wager, O. (1975) Preloading by vacuum-current prospects. *Transportation Research Record*, **548**: 26–29.

Holtz, R.D., Jamiolkowski, M.B., Lancellotta, R. & Pedroni, S. (1991) Prefabricated vertical drains; design and performance. In *CIRIA Ground Engineering Report: Ground Improvement*, Butterworths-Heinemann, London, England.

Holtz, R.D., Christopher, B.R. & Berg, R.B.

(1997) *Geosynthetic Engineering*, BiTech, Vancouver.

Hong Kong Government (1996) *Geoguide 2; Guide to Site Investigation*, Government Publications Centre, Hong Kong Government, Hong Kong.

Hooker, V.E. & Bickel, D.L. (1974) *Overcoring Equipment and Techniques Used in Rock Stress Determinations*, Information Circular 8618, US Bureau of Mines.

Hooper, J.A. (1973) Observations on the behaviour of a piled raft foundation on London clay. *Proceedings, Institute of Civil Engineers*, **55**(2): 855.

Horvath, J.S. (1995) *Geofoam Geosynthetic*, Horvath Engineering, PC, Scarsdale, NY.

Horvath, R.G. (1989) Load–displacement behaviour of socketed piers–Hamilton General Hospital. *Canadian Geotechnical Journal*, **26**: 260–268.

Horvath, R.G., Kenney, T.C. & Kosicki, P. (1983) Method of improving the performance of drilled piers in weak rock. *Canadian Geotechnical Journal*, **20**(4): 758–772.

Houlsby, G.T. (1998) Advanced interpretation of field tests. In Robertson, P.K. & Mayne, P.W. (Eds) *Proceedings First International Conference on Site Characterization – ISC' 98*, April 19–22, Atlanta, GA, Vols. 1 & 2, A.A. Balkema, Rotterdam, pp. 99–112.

Houlsby, G.T. & Jewell, R.A. (1988) Analysis of unreinforced and reinforced embankments on soft clays by plasticity analysis. In *Proceedings, Sixth International Conference on Numerical Methods in Geomechanics*, Innsbruck, Vol. 2, pp. 1443–1448.

Hounslow, A.W. (1995) *Water Quality Data Analysis and Interpretation*, CRC Press, Boca Raton, FL.

Houston, S.L., Houston, W.N. & Padilla, J.M. (1987) Microcomputer-aided evaluation of earthquake-induced permanent slope displacements. *Microcomputers in Civil Engineering*, **2**: 202–222.

Houston, W.N. & Mitchell, J.K. (1969) Property interrelationships in sensitive clays. *Journal of Soil Mechanics & Foundations Division, ASCE*, Reston, VA, **95**(4): 1037–1062.

Howard, A.K. (1977) Modulus of soil reaction values for buried flexible pipe. *Journal of Geotechnical Engineering, ASCE*, Reston, VA, **103**(1): 33–43.

Howard, P.H. (Ed.) (1989–1997) *Handbook of Environmental Fate and Exposure Data for Organic Chemicals*, Lewis, Chelsea, MI.

Howard, P.H. (Ed.) (1991) *Handbook of Environmental Degradation Rates*, Lewis, Chelsea, MI.

Hrudey, S.E. & Pollard S.J.T. (1993) The Challenges of Contaminated Sites; Remediation Approaches in North America, *Environmental Review*, **1**: 55–72.

Hryciw, R.D. (Ed.) (1995) Soil improvement for earthquake hazard mitigation. In *Proceedings, Sessions at the ASCE Convention*, San Diego, Geotechnical Special Publication No. 49, ASCE Press, Reston, VA, p. 141.

Hsuan, Y.G. & Guan, Z. (1997) Evaluation of the oxidation behavior of polyethylene geomembranes using oxidative induction time tests. In Riga, A.T. & Patterson, G.H. (Eds) *Oxidative Behavior of Materials by Thermal Analytical Techniques*, ASTM STP 1326, American Society for Testing & Materials, West Conshohocken, PA, pp. 76–90.

Hsuan, Y.G. & Koerner, R.M. (1995a) The single point–notched constant tension load test: a quality control test for assessing stress crack resistance. *Geosynthetics International*, **2**(5): 831–843.

Hsuan, Y.G. & Koerner, R.M. (1995b) *Long Term Durability of HDPE Geomembranes: Part I–Depletion of Antioxidants*. GRI Report 16, Geosynthetic Research Institute, Philadelphia, PA.

Hsuan, Y.G. & Koerner, R.M. (1998) Antioxidant depletion lifetime in HDPE geomembranes. *Journal of Geotechnical & Geoenvironmental, ASCE*, Reston, VA, **124**(6): 532–541.

Hsuan, Y.G., Koerner, R.M. & Lord, A.E. Jr (1993) A review of the degradation of geosynthetic reinforcing materials and various polymer stabilization methods. In Cheng, S.C.J. (Ed.,) *Geosynthetic Soil Reinforcement Testing Procedures*, ASTM STP 1190, American Society for Testing & Materials, West Conshohocken, PA, pp. 228–243.

Huang, S. (1994) *PhD thesis*, University of Saskatchewan, Saskatoon, Saskatchewan, Canada.

Huang, S., Barbour, S.L. & Fredlund, D.G. (1994) A history of the coefficient of permeability function. In *Sino-Canadian Symposium on Unsaturated/Expansive Soils*, June 7–8, Wuhan, China, pp. 57–80.

Huang, Y.H. (1993) *Pavement Analysis and Design*, Prentice Hall, Englewood Cliffs, NJ.

Hubble, D.W., Gillham, R.W. & Cherry, J.A. (1997) Emplacement of zero-valent metal for remediation of deep contaminant plumes. In

Proceedings, 1997 International Containment Technology Conference & Exhibition, February 10–12, St Petersburg, FL, pp. 872–878.

Hudson, M., Idriss, I.M. & Beikae, M. (1994) *QUAD4M: A Computer Program to Evaluate the Seismic Response of Soil Structures Using Finite Element Procedures and Incorporating a Compliant Base*, University of California, Davis.

Hudson, W.R., Haas, R. & Uddin, W. (1997) *Infrastructure Management*, McGraw-Hill, New York.

Hughes, C.S. (1989) *Compaction of Asphalt Pavement*, National Cooperative Highway Research Program (NCHRP) Synthesis of Highway Practice No. 152, Transportation Research Board, Washington, DC.

Huling, S.G. & Weaver, J.W. (1991) *Ground Water Issue, Dense Nonaqueous Phase Liquids*, EPA/540/4-91/002.

Humphrey, D.N. & Holtz, R.D. (1986) Reinforced embankments–a review of case histories. *Geotextiles & Geomembranes*, 4(2): 129.

Humphrey, D.N. & Rowe, R.K. (1991) Design of reinforced embankments: recent developments in the state-of-the-art. *Proceedings, ASCE Geotechnical Engineering Congress*, ASCE Special Publication No. 27, American Society of Civil Engineers, ASCE Press, Reston, VA, pp. 1006–1020.

Hungr, O. (1997) Some methods of landslide hazard intensity mapping. In *International Workshop on Landslide Risk Assessment*, Honolulu, HI, pp. 215–226.

Hungr, O., Salgado, F.M. & Byrne, P.M. (1989) Evaluation of a three-dimensional method of slope stability analysis. *Canadian Geotechnical Journal*, 26(4): 679–686.

Hungr, O., Sobkowicz, J. & Morgan, G.C. (1993) How to economize on natural hazards. *Geotechnical News*, 11(1): 54–57.

Huntley, D., Hawk, R.N. & Corley, H.P. (1992) Non-aqueous phase hydrocarbon saturations and mobility in a fine-grained, poorly consolidated sandstone. In *Proceedings, 1992 Petroleum Hydrocarbons and Organic Chemicals in Groundwater: Prevention, Detection, and Restoration*, National Groundwater Association, Columbus, OH, pp. 223–237.

Hutchinson, J.N. (1977) Assessment of effectiveness of corrective measures in relation to geological conditions and types of slope movement. *Bulletin of the International Association of Engineering Geologists*, 16: 131–155.

Hutchinson, J.N. (1987) Mechanisms producing large displacements in landslides on pre-existing shears. *Memoir of the Geological Society of China*, 9: 175–200.

Hutchinson, J.N. (1988) General report: morphological and geotechnical parameters of landslides in relation to geology and hydrogeology. In *Proceedings, Fifth International Symposium on Landslides*, Lausanne, Vol. 1, pp. 3–35.

Hutson, J.L. & Wagenet, R.J. (1992) *LEACHM Leaching Estimation and Chemistry Model*, Version 3, Department of Soil Crop and Atmospheric Sciences Research Series No. 92–3, New York State College of Agricultural and Life Sciences, Cornell University, Ithaca, NY.

Huyakorn, P.S., Wu, Y.S. & Panday, S. (1992) A comprehensive three-dimensional numerical model for predicting the fate of petroleum hydrocarbons in the subsurface. In *Proceedings, Conference On Petroleum Hydrocarbons and Organic Chemicals in Ground Water: Prevention, Detection, and Restoration*, National Ground Water Association, Dublin, OH, pp. 239–253.

Hvorslev, M.J. (1937) *Über die Festigkeitseigenschaften gestörter bindinger Böden* (København: Danmarks Naturvidenskabelige Samfund) Ingeniørvidenskabelige Skrifter, A45. English translation (1969) *Physical Properties of Remolded Cohesive Soils*, Waterways Experiment Station, Vicksburg, MS, No. 69–5.

Hvorslev, M.J. (1948) *Subsurface Exploration and Sampling of Soils for Civil Engineering Purposes*, US Army Waterways Experiment Station, Vicksburg, MS.

Hvorslev, M.J. (1951) *Time Lag and Permeability in Groundwater Observations*. Bulletin No. 36, US Army, Waterways Experiment Station, Vicksburg, VA.

Hwu, B.-L., Sprague, C.J. & Koerner, R.M. (1990) Geotextile intrusion into geonets. In *Proceedings, Fourth International Conference On Geotextiles, Geomembranes, and Related Products*, G. den Hoedt (Ed). A.A. Balkema, Rotterdam, pp. 351–356.

Hynes, M.E. & Franklin, A.G. (1984) *Rationalizing the Seismic Coefficient Method*, Miscellaneous Paper GL-84-13, US Army Engineer Waterways Experiment Station, Vicksburg, MS.

ICAO (1983) *Aerodrome Design Manual*, Document 9157, *Part 3—Pavements*, International Civil Aviation Organization, Montreal, Canada.

ICBO (1997) *1997 Uniform Building Code, 2, Structural Engineering Design Provisions*, International Conference of Building Officials, Whittier, CA.

Idriss, I.M. (1990) Response of soft soil sites ring earthquakes. In *Proceedings, Symposium to Honor Professor H.B. Seed*, Bitech, Vancouver, BC, J.M. Duncan (Ed.) Vol. 2, pp. 273–290.

Idriss, I.M. & Sun, J.I. (1992) *SHAKE91: A Computer Program for Conducting Equivalent Linear Seismic Response Analyses of Horizontally Layered Soil Deposits. User's Manual*, University of California, Davis.

Idriss, I.M., Lysmer, J., Hwang, R. & Seed, H.B. (1973) *QUAD-4: A Computer Program for Evaluating the Seismic Response of Soil Structures by Variable Damping Finite Element Procedures*, Report No. EERC 73-16, University of California at Berkeley.

Idriss, I.M., Fiegel, G., Hudson, M.B., Mundy, P.K. & Herzig, R. (1995) Seismic response of the operating industries landfill. *Earthquake Design and Performance of Solid Waste Landfills*, Geotechnical Special Publication No. 4, Yegian, M.K. and Finn, W.D.L. (Eds.), American Society of Civil Engineers, Reston, VA, pp. 83–118.

Imai, G. & Tang, Y.X. (1992) A constitutive equation of one-dimensional consolidation derived from inter-connected tests. *Soils & Foundations*, **32**(2): 83–96.

Imai, T. & Tonouchi, K. (1982) Correlation of N value with S-wave velocity an shear modulus. In *Proceedings, European Symposium on Penetration Testing, ESOPT-II*, A.A. Balkema, Amsterdam, Vol. 1, pp. 67–72.

Ingles, O.G. (1962) Bonding forces in soils, Part 3: A theory of tensile strength for stabilised and naturally coherent soils. In *Proceedings, First Conference of the Australian Road Research Board*, Vol. 1, pp. 1025–1047.

Inglis, C.E. (1913) Stresses in a plate due to the presence of cracks and sharp corners. *Transactions of the Institution of Naval Architecture*, London, **55**: 219–241.

Ingold, T.S. (1979) The effects of compaction in retaining walls. *Géotechnique*, **29**(3): 265–83.

Institution of Civil Engineers (1996) *Advances in Site Investigation Practice*. Craig, C. (Ed.) Thomas Telford Services, London, England.

International Journal on Hydropower and Dams, Editorial Reports on Seismic Risk, Vol. 5, Issues 1 and 2, Aqua-Media, Sutton, Surrey, UK.

International Standards Organization/DIS 11058 *Determination of Water Permeability Charac-*teristics to the Plane, Without Load, Brussels, Belgium.

International Standards Organization/DIS 12956 *Determination of the Characteristic Opening Size of Geotextile, Brussels, Belgium.*

International Standards Organization/DIS 12958 *Determination of Water Flow Capacity in their Plane, Brussels, Belgium.*

International Standards Organization/DIS 13431 *Determination of the Tensile Creep and Creep Rupture Behavior, Brussels, Belgium.*

International Standards Organization 10319 *Wide Width Tensile Test, Brussels, Belgium.*

IRIS (1997) *Integrated Risk Information System*, US Environmental Protection Agency. Online search (Internet: http://www.epa.gov/docs/ngispgm3/iris/index.html).

ISC (1998) Geotechnical site characterization. In Robertson, P.K. & Mayne, P.W. *Proceedings, First International Conference on Site Characterization–ISC'98*, April 19–22, Atlanta, Ga, Vols. 1 & 2, A.A. Balkema, Rotterdam.

Ishibashi, I. (1992) Discussion to 'Effects of soil plasticity on cyclic response,' by M. Vucetic & R. Dobry. *Journal of Geotechnical Engineering, ASCE*, Reston, VA, **118**(5): 830–832.

Ishihara, K. (1993) Liquefaction and flow failure during earthquakes, *Géotechnique*, **43**(3): 351–415.

Ishihara, K. (1996) *Soil Behaviour in Earthquake Geotechnics*, Oxford University Press, New York.

Ishihara, K. & Li, S. (1972) Liquefaction of saturated sand in triaxial torsion shear tests. *Soils & Foundations*, **12**(2): 19–39.

ISO 4435 Unplasticized poly(vinyl chloride) (PVC-U) pipes and fittings for buried drainage and sewerage systems; specifications, International Standards Organization, Geneva.

ISO 7186 Ductile iron products for sewage applications, International Standards Organization, Geneva.

ISO 7370 Glass fiber reinforced thermosetting plastics (GRP) pipes and fittings—nominal diameters, specified diameters and standard lengths, International Standards Organization, Geneva.

ISO 8772 High density polyethylene (PE-HD) pipes and fittings for buried drainage and sewerage systems; specifications, International Standards Organization, Geneva.

ISOPT (1988) De Ruiter, J. (Ed.) *Penetration Testing 1988, Proceedings, First International Symposium on Penetration Testing, ISOPT-1,*

March 20–24, Orlando, FL, Vols. 1 & 2, A.A. Balkema, Rotterdam.

ISO/TR 7074 Performance requirements for plastic pipes and fittings for use in underground drainage and sewage, International Standards Organization, Geneva.

ISO/TR 7073 Recommended techniques for the installation of unplasticized poly (vinyl chloride) (PVC-U) buried drains and sewers, International Standards Organization, Geneva.

ISO/TR 10465 Underground installation of flexible glass-reinforced thermosetting resin (GRP) pipes; Part I: Installation procedures, International Standards Organization, Geneva.

ISRM (1978a) Suggested methods for the quantitative description of discontinuities in rock masses. Commission on Standardization of Laboratory and Field Tests. Coordinator: N. Barton. *International Journal of Rock Mechanics, Mineral Science & Geomechanics, Abstracts*, **15**: 319–368.

ISRM (1978b) Suggested methods for determining tensile strength of rock materials. Suggested method for determining indirect tensile strength by the Brazil test. Commission on Standardization of Laboratory and Field Tests. Z.T. Bieniawski and I. Hawkes. *International Journal of Rock Mechanics, Mineral Science & Geomechanics, Abstracts*, **15**: 102–103.

ISRM (1978c) Suggested methods for determining tensile strength of rock materials. Suggested method for determining direct tensile strength. Commission on Standardization of Laboratory and Field Tests. Z.T. Bieniawski and I. Hawkes. *International Journal of Rock Mechanics, Mineral Science & Geomechanics, Abstracts*, **15**: 101–102.

ISRM (1978d) Suggested methods for determining the strength of rock materials in triaxial compression. Commission on Standardization of Laboratory and Field Tests. U.W. Volger and K. Kovari. *International Journal of Rock Mechanics, Mineral Science & Geomechanics, Abstracts*, **15**: 47–51.

ISRM (1979a) Suggested methods for determining in situ deformability of rock. Part 1. Suggested method for deformability determination using a plate test (superficial loading). Commission on Standardization of Laboratory and Field Tests. J.H. Coulson (Co-ordinator) *International Journal of Rock Mechanics, Mineral Science & Geomechanics, Abstracts*, **16**: 197–202.

ISRM (1979b) Suggested methods for determining in situ deformability of rock. Part 2. Suggested method for field deformability determination using a plate test down a borehole. Commission on Standardization of Laboratory and Field Tests. J.H. Coulson (Co-ordinator) *International Journal of Rock Mechanics, Mineral Science & Geomechanics, Abstracts*, **16**: 202–208.

ISRM (1979c) Suggested methods for determining in situ deformability of rock. Part 3. Suggested method for measuring rock mass deformability using a radial jacking test. Commission on Standardization of Laboratory and Field Tests. J.H. Coulson (Co-ordinator) *International Journal of Rock Mechanics, Mineral Science & Geomechanics, Abstracts*, **16**: 208–214.

ISRM (1979d) Suggested methods for determining the uniaxial compressive strength and deformability of rock materials. Commission on Standardization of Laboratory and Field Tests. Co-ordinator: Z.T. Bieniawski. *International Journal of Rock Mechanics, Mineral Science & Geomechanics, Abstracts*, **15**: 135–140.

ISRM (1986) Suggested method for deformability determination using a large flat jack technique. Commission on Testing Methods. J. Loureiro-Pinto (Co-ordinator) *International Journal of Rock Mechanics, Mineral Science & Geomechanics, Abstracts*, **23**: 131–140.

ISRM (1987a) Suggested methods for rock stress determination. Commission on testing Methods. Co-ordinators: K. Kim and J.A. Franklin. *International Journal of Rock Mechanics, Mineral Science & Geomechanics, Abstracts*, **24**(1): 53–73.

ISRM (1987b) Suggested method for deformability determination using a flexible dilatometer. Commission on Testing Methods. B. Ladanyi (Co-ordinator) *International Journal of Rock Mechanics, Mineral Science & Geomechanics, Abstracts*, **24**: 123–134.

ISRM (1989) Suggested methods for laboratory testing of argillaceous swelling rocks. Commission on Standardization of Laboratory and Field Tests. Co-ordinator: H. Einstein. *International Journal of Rock Mechanics, Mineral Science & Geomechanics, Abstracts*, **26**: 415–426.

ISRM (1996) Suggested method for deformability determination using a stiff dilatometer. Commission on Testing Methods. J.L. Yow, Jr. (Co-ordinator) *International Journal of Rock Mechanics, Mineral Science & Geomechanics, Abstracts*, **33**: 733–741.

ISTT (undated) *Trenchless Technology Guidelines*, International Society for Trenchless Technology, London, England.

Itasca (1996) *FLAC* (Version 3.3), *FLAC—Fast Lagrangian Analysis of Continua*, Itasca, Minneapolis, MN.

Ivan, B. (1993) Thermal stability, degradation, and stabilization mechanisms of poly(vinyl chloride). In Clough, R.L., Billingham, N.C. & Gillen, K.T. (Eds) *Polymer Durability*, Advances in Chemistry Series 249, American Chemical Society, New York, NY, USA, pp. 19–32.

Jackson, R.E., Patterson, R.J., Graham, B.W., Bahr, J., Belanger, D., Lockwood, J. and Priddle, M. (1985) *Contaminant Hydrogeology of Toxic Organic Chemicals at a Disposal Site, Gloucester, Ontario: 1. Chemical Concepts and Site Assessment*, Environment Canada, *National Hydrological Research Institute*, Paper No. 23, Ottawa.

Jaegar, J.C. & Cook, N.G.W. (1969) Fundamentals of Rock Mechanics, p. 135, Methuen Press.

Jaky, J. (1948) Pressure in silos. In *Proceedings, Second International Conference on Soil Mechanics & Foundations Engineering*, Rotterdam, Holland, Vol. I, pp. 103–107.

Jamiolkowski, M. & Garassino, A. (1977) Soil modulus for laterally loaded piles. In *Proceedings, Special Session No. 10, Ninth International Conferences on Soil Mechanics & Foundation Engineering*, Tokyo, pp. 43–58.

Jamiolkowski, M., Lancellotta, R. & Wolski, W. (1983) Precompression and speeding up consolidation. General Report, Specialty Session 6, *Proceedings, Eighth European Conference on Soil Mechanics & Foundation Engineering*, Helsinki, Vol. 3, pp. 1201–1226. Summary of discussion, pp. 1242–1245.

Jamiolkowski, M., Ladd, C.C., Germaine, J.T. & Lancellotta, R. (1985) New developments in field and laboratory testing of soils. State-of-the-Art Report, *11th International Conference on Soil Mechanics & Foundation Engineering*. San Francisco, CA, Vol. 1, pp. 57–153.

Jamiolkowski, M., Ghionna, V., Lancellotta, R. & Pasqualini, E. (1988) New correlations of penetration tests for design practice. *Proceedings ISOPT-I*, **1**: 263–296.

Jamiolkowski, M., Lancelotta, R. & Lo Presti, D.C.F. (1994) Remarks on the stiffness at small strains of six Italian clays. In Shibuya, S., Mitachi, T. & Miura, S. (Eds) *Proceedings, First International Conference on Pre-Failure Deformation Characteristics of Geomaterials* Sapporo, Japan, Balkema, Rotterdam, Vol. 2, pp. 817–836.

Janbu, N. (1954a) *Stability Analysis of Slopes with Dimensionless Parameters*, Harvard Soil Mechanics Series, Harvard University, Cambridge, MA, No. 46.

Janbu, N. (1954b) Application of composite slip surfaces for stability analysis. In *Proceedings, European Conference on Stability of Earth Slopes*, Stockholm, Vol. 3, pp. 45–49.

Janbu, N. (1963) Soil compressibility as determined by oedometer and triaxial tests. In *Proceedings European Conference on Soil Mechanics & Foundation Engineering*, Wiesbaden, Vol. 1, pp. 19–25.

Janbu, N. (1968) *Slope Stability Computations*, Soil Mechanics & Foundation Engineering Report. Technical University of Norway, Trondheim.

Janbu, N. (1977) Slopes and excavations in normally and slightly overconsolidated clays. In *Proceedings, Ninth International Conference on Soil Mechanics & Foundation Engineering*, Tokyo, Vol. 2, pp. 549–566.

Janbu, N. (1996) Slope stability evaluation in engineering practice. In *Proceedings, Seventh International Symposium on Landslides*, Trondheim, Vol. 1, pp. 17–34.

Janbu, N. & Sennesest, K. (1973) Field compressometer: principles and applications. In *Proceedings, Eighth International Conference on Soil Mechanics & Foundation Engineering*, Moscow, Russia, Vol. 1, pp. 191–198.

Jang, D.J. & Montero, C. (1993) Design of liner systems under vertical expansions: an alternative to geogrids. In *Proceedings, Geosynthetics '93 Conference*, Vancouver, BC, Industrial Fabrics Association International, Roseville, MN, Vol. 3, pp. 1497–1510.

Janssen, H.A. (1895) Versuche uber Getreidedruck in Silozellen. *Zeitschrift des Vereines Deutscher Ingenieure*, **39**(35): 1045–1049.

Japan Highway Public Corporation (1987) *Guide for Design and Construction on Reinforced Slope with Steel Bars*, Japan Highway Public Corporation, Tokyo, Japan.

Jardine, R.J. (1992) Some observations on the kinematic nature of soil stiffness. *Soils & Foundations*, **32**(2): 111–124.

Jardine, R.J. & Hight, D.W. (1987) The behaviour and analysis of embankments on soft clay. In *Embankments on Soft Clays*, Special Publication Bulletin of the Public Works Research Centre, Athens, pp. 159–244.

Jardine, R.J., Potts, D.M., Fourie, A.B. & Burland, J.B. (1986) Studies of the influence of non-linear stress–strain characteristics in soil–structure interaction. *Géotechnique*, **36**(3): 377–396.

Jayawickrama, P., Brown, K.W., Thomas, J.C. & Lytton, R.L. (1988) Leakage rates through flaws in geomembrane liners. *Journal of Environmental Engineering, ASCE*, Reston, VA, **114**(6): 1401–1420.

Jefferies, M.G. (1988) Determination of horizontal geostatic stress in clay with self-bored pressuremeter. *Canadian Geotechnical Journal*, **25**(3): 559–573.

Jefferies, M.G. & Davies, M.P. (1991) Soil classification by the cone penetration test: discussion. *Canadian Geotechnical Journal*, **28**(1): 173–176.

Jefferies, M.G. & Davies, M.P. (1993) Use of CPTU to estimate equivalent SPT N_{60}. *ASTM Geotechnical Testing Journal*, West Conshohocken, PA, **16**(4): 458–468.

Jefferies, M.G., Crooks, J.H.A., Becker, D.E. & Hill, P.R. (1987) Independence of geostatic stress from overconsolidation in some Beaufort Sea clays. *Canadian Geotechnical Journal*, **24**(3): 342–356.

Jessberger, H.L. (Ed.) (1979) *Ground Freezing, Developments in Geotechnical Engineering*, Vol. 26, Elsevier, New York.

Jessberger, H.L. & Stone, K. (1991) Subsidence effects on clay barriers. *Géotechnique*, **41**(2): 185–194.

Jewell, R.A. (1982) A limit equilibrium design method for reinforced embankments on soft foundations. In *Proceedings, Second International Conference on Geotextiles*, Las Vegas, NV, Industrial Fabrics Association International, Roseville, MN, Vol. 2, pp. 671–676.

Jewell, R.A. (1996) *Soil Reinforcement with Geotextiles*, Special Publication No. 123, CIRIA London, England.

Jezequel, F.J. & Mieussens, C. (1975) *In-situ* measurement of coefficients of permeability and consolidation in fine soils. In *Proceedings, ASCE Geotechnical Engineering Division Specialty Conference on* In-Situ *Measurement of Soil Properties*, June 1–4, North Carolina State University, Raleigh, NC, Vol. 1, pp. 208–224.

Ji, W., Dahmani, A., Ahlfeld, D.P., Lin, J.D. & Hill III, E. (1993) Laboratory study of air sparging: air flow visualization. *Ground Water Monitoring & Remediation*, **13**(4): 115–125.

Joffe, Z. & Ranby, B. (1976) Some weathering and UV aging problems of PVC contaminated with Fe-complexes and organic additives: long-term properties of polymers and polymeric materials. In Editors: Ranby, B. & Rabek, J.F. *Journal of Applied Polymer Science: Applied Polymer Symposium 35*, Interscience, New York, NY, USA, pp. 307–320.

Johnson, A.I. (1967) *Specific-Yield-Compilation of Specific Yields for Various Materials*, Water Supply Paper 1662-D, US Geological Survey.

Johnson, K.A. & Sitar, N. (1990) Hydrologic conditions leading to debris-flow initiation. *Canadian Geotechnical Journal*, **27**(6): 789–801.

Johnson, P.C., Stanley, C.C., Kemblowski, M.W., Byers, D.L. & Colthart, J.D. (1990) A practical approach to the design, operation and monitoring of *in-situ* soil venting systems. *Ground Water Monitoring Review*, **10**(2): 159–178.

Johnson, R.L., Johnson, P.C., McWhorter, D.B., Heinchee, R.E. & Goodman, I. (1993) An overview of *in situ* air sparging. *Ground Water Monitoring & Remediation*, **13**(4): 127–135.

Johnson, S.J. (1970a) Precompression for improving foundation soils. *Journal of Soil Mechanics & Foundations Division, ASCE*, Reston, VA, **96**(1): 111–144.

Johnson, S.J. (1970b) Foundation precompression with vertical sand drains. *Journal of Soil Mechanics & Foundations Division, ASCE*, Reston, VA, **96**(1): 145–175.

Johnson, T.C., Cole, D. & Chamberlain, E.J. (1978) *Influence of Freezing and Thawing on the Resilient Properties of a Silt Soil Beneath an Asphalt Concrete Pavement*, Report 78–23, Cold Regions Research and Engineering Laboratory (CRREL).

Johnston, G.H. (1981) *Permafrost. Engineering design and construction*, Associate Committee for Geotechnical Research National Research Council of Canada, Wiley, Toronto.

Jones, C.J.F.P. (1996) *Earth Reinforcement and Soil Structures*, Thomas Telford, London, England.

Jones, C.J.F.P., Fakher, A., Hamir, R. & Nettleton, I.M. (1996) Geosynthetic materials with improved reinforcement capabilities. In Ochiai, M., Yasafuku, N. & Omine, K. (Eds) *Earth Reinforcement*, A.A. Balkema, Rotterdam, Vol. 2, pp. 865–883.

Jorstad, F. (1968) *Snesmelting som arsak til jordskred pa ostlandet ved manedsskiftet april–mai 1966*, Norwegian Geotechnical Institute, Publication No. 75, Oslo, pp. 33–38.

Jorgensen, D.G., Gogel, T. & Signor, D.C. (1982)

Determination of flow in aquifers containing variable-density water. *Groundwater Monitoring Review*, **2**: 29–38.

Joseph, J.B. & Mather, J.D. (1993) Landfill–Does current containment practice represent the best option. In *Proceedings, Fourth International Landfill Symposium*, Cagliari Sardinia, October, pp. 99–107.

Journal of Geotextiles and Geomembranes, Special Issue, **15**(1–3), 1997.

Jumikis, A. (1971) *Foundation Engineering*, Intext Educational, Scranton, PA.

Jung, F. & Phang, W.A. (1974) *Elastic Layer Analysis Related to Performance in Flexible Pavement Design*, Research Report 191, Ministry of Transportation, Ontario, Canada, M#M 1J8.

Juran, I. & Elias, V. (1991) Ground anchors and soil nails in retaining structures. In Fang, H.Y. (Ed.) *Foundation Engineering Handbook*, Van Nostrand Reinhold, New York, NY, pp. 868–905.

Jurgenson, L. (1934) The application of theories of elasticity and plasticity to foundation problems. Boston Society of Civil Engineers, *Contributions to Soil Mechanics 1925–1940*, 184–226.

Kabbaj, M. (1985) *PhD thesis*, Université Laval, Québec.

Kabbaj, M., Tavenas, F. & Leroueil, S. (1988) *In situ* and laboratory stress–strain relationships. *Géotechnique*, **38**(1): 83–100.

Kafritsas, J. & Bras, R.L. (1981) *The Practice of Kriging*, Ralph M. Parsons Laboratory, Report No. 263, Massachusetts Institute of Technology, MA.

Kalteziotis, N., Zervogiannis, F.R., Seve, G. & Berche, J.-C. (1993) Experimental study of landslide stabilization by large diameter piles. In A. Anagnostopoulos, *et al.* (Eds) *Geotechnical Engineering of Hard Soils–Soft Rocks*, Anagnostopoulos, A., Schlosser, F., Kalteziotis, N., and Frank, R. (Eds.) Balkema, Rotterdam, pp. 1115–1124.

Kaluarachchi, J.J. & Parker, J.C. (1989) An efficient finite element model for modeling multiphase flow in porous media. *Water Resources Research*, **25**(1): 43–54.

Kaluarachchi, J.J., Parker, J.C. & Lenhard, R.J. (1990) A numerical model for areal migration of water and light hydrocarbon in unconfined aquifers. *Advanced in Water Resources*, **13**: 29–40.

Kana, D.D., Boyce, L. & Blayney, G.W. (1986)

Development of a scale model for the dynamic interaction of a pile in clay. *Journal of Energy Resources Technology, ASME*, Vol. **108**: 254–261.

Karabalis, D.L. & Beskos, D.E. (1985) Dynamic response of 3-D embedded foundations by the boundary element method. In *Second Joint ASCE/ASME Conference*, Albuquerque, p. 34.

Karickhoff, S.W. (1981) Semi-empirical estimation of sorption of hydrophobic pollutants on natural sediments and soils. *Chemosphere*, **10**(8): 833–846.

Karickhoff, S.W. (1984) Organic pollutant sorption in aquatic systems. *Journal of Hydraulic Engineering, ASCE*, Reston, VA, **110**(6): 707–735.

Karickhoff, S.W., Brown, D.S. & Scott, T.A. (1979) Sorption of hydrophobic pollutants on natural sediments. *Water Resources Research*, **13**: 241–248.

Karol, R.H. (1982) Chemical grouts and their properties. In Baker, W.H. (Ed.) *Proceedings, Grouting in Geotechnical Engineering*, New Orleans, American Society of Civil Engineers, ASCE Press, Reston, VA, pp. 359–377.

Karpurapu, R.G. & Bathurst, R.J. (1992) Numerical investigation of controlled yielding of soil retaining wall structures. *Geotextiles & Geomembranes*, **11**(2): 115–131.

Karube, D. (1988) New concept of effective stress in unsaturated soil and its proving test. In *Advanced Triaxial Testing of Soil and Rock*, ASTM STP 977, Eds. R.T. Douglas, R.C. Chaney and M.L. Silver, American Society for Testing & Materials, West Conshohocken, PA, pp. 539–552.

Katchalsky, A. & Curran, P.R. (1967) *Nonequilibrium Thermodynamics in Biophysics*, Harvard University Press, Cambridge, MA.

Katona, M.G. (1978) Analysis of long-span culverts by the finite element method. *Transportation Research Record*, **678**: 59–66.

Katyal, A.K., Kaluarachchi, J.J. & Parker, J.C. (1991) *MOFAT: A Two-Dimensional Finite Element Program for Multiphase Flow and Multicomponent Transport*, Program Documentation and User's Guide, EPA 600/2-91-020.

Kauschinger, J.L., Perry, E.B. & Hankour, R. (1992) Jet grouting: state-of-the-practice, grouting. In *Soil Improvement, & Geosynthetics*, Geotechnical Special Publication No. 30, American Society of Civil Engineers, ASCE Press, Reston, VA, Vol. 1, pp. 169–181.

Kausel, E. & Ushijima, R. (1979) *Vertical and*

Torsional Stiffness of Cylindrical footing, Civil Engineering Department Report R79-6, MIT, Cambridge, MA.

Kausel, E., Roesset, J.M. & Waas, G. (1975) Dynamic analysis of footings on layered media. *Journal of the Engineering Mechanics Division, ASCE*, Reston, VA, **101**(5): 679–693.

Kavazanjian, E., Jr (1998) Current issues in seismic design of geosynthetic cover systems. In R.K. Rowe (Ed.) *Proceedings, Sixth International Conference on Geosynthetics*, Atlanta, GA, Industrial Fabrics Association International, Roseville, MN, Vol. 1, pp. 219–226.

Kavazanjian, E., Jr & Matasovic, N. (1995) Seismic analysis of solid waste landfills. In *Geoenvironment 2000*, Geotechnical Special Publication No. 46, Acar, Y.B. and Daniel, D.E. (Eds.) American Society of Civil Engineers, Reston, VA, Vol. 2, pp. 1066–1080.

Kavazanjian, E., Jr, Hushmand, B. & Martin, G.R. (1991) Frictional base isolation using a layered soil–synthetic liner system. In *Proceedings, Third US Conference on Lifeline Earthquake Engineering*, Los Angeles, Monograph No. 4, pp. 1140–1151.

Kavazanjian, E., Jr, Snow, M.S., Matasovic, N., Poran, C. & Satoh, T. (1994) Non-intrusive Rayleigh wave investigations at solid waste landfills. In *Proceedings, First International Conference on Environmental Geotechnics*, Edmonton, Alberta, pp. 707–712.

Kavazanjian, E., Jr, Bonaparte, R., Johnson, G.W., Martin, G.R. & Matasovic, N. (1995a) Hazard analysis for a large regional landfill. In *Earthquake Design & Performance of Solid Waste Landfill*, Geotechnical Special Publication No. 54, Yegian, M.K. and Finn, W.D.L. (Eds.) American Society of Civil Engineers, Reston, VA, pp. 119–141.

Kavazanjian, E., Jr, Matasovic, N., Bonaparte, R. & Schmertmann, G.R. (1995b) Evaluation of MSW properties for seismic analysis. In *Geoenvironment 2000*, Geotechnical Special Publication No. 46, Acar, Y.B. & Daniel, D.E. (Eds.) American Society of Civil Engineers, Reston, VA, Vol. 2, pp. 1126–1141.

Kavazanjian, E., Jr, Matasovic, N., Stokoe, K.H. & Bray, J.D. (1996) *In situ* shear wave velocity of solid waste from surface wave measurements. In *Proceedings, Second International Conference on Environmental Geotechnics*, Osaka, Japan, Vol. 1, pp. 97–102.

Kavazanjian, E., Jr Matasovic, N., Hadj-Hamou, T. & Sabatini, P.J. (1997) *Geotechnical Earth-quake Engineering for Highways, 1. Design Principles*, Geotechnical Engineering Circular No. 3, US Federal Highway Administration, Publication No. FHWA-SA-97-076.

Kawai, T. (1985) *Summary Report on the Development of the Computer Program DIANA–Dynamic Interaction Approach and Non-linear Analysis*, Science University of Tokyo.

Kaynia, A.M. & Kausel, E. (1982) Dynamic behavior of pile groups. In *Second International Conference on Numerical Methods in Offshore Piling*, Austin, TX, pp. 509–532.

Keith, L.H. (1987) *Principles of Environmental Sampling*, American Chemical Society, Washington, DC.

Keith, L.H. (1990) *Environmental Sampling and Analysis*, American Chemical Society, Washington, DC.

Kempton, G., Russell, D., Pierpont, N.D. & Jones, C.J.F.P. (1998) Two- and three-dimensional numerical analysis of the performance of piled embankments. In *Sixth International Conference on Geosynthetics*, Atlanta, GA, R.K. Rowe (Ed.) International Fabrics Association International, Roseville, MN, Vol. 2, pp. 767–772.

Kenney, TC. (1967) The influence of mineralogical composition on the residual strength of natural soils. In *Proceedings, Oslo Geotechnical Conference on the Shear Strength Properties of Natural Soils & Rocks*, Vol. I pp. 123–129.

Kenney, T.C. & Lau, K.C. (1984) Temporal changes of groundwater pressure in a natural slope of nonfissured clay. *Canadian Geotechnical Journal*, **21**(1): 138–146.

Kenney, T.C., Chahal, R, Chiu, E., Ofoegbu, G.I., Omange, G.N. & Ume, C.A. (1985) Controlling constriction sizes of granular filters. *Canadian Geotechnical Journal*, **22**(1): 32–43.

Kersten, M.S. (1949) *Thermal Properties of Soils*, University of Minnesota, Engineering Experiment Station, MN, Bulletin 28.

Kezdi, A. (1975) Lateral earth pressure. In Winterkorn, H.F. & Fang, H-Y. (Eds) *Foundation Engineering Handbook* Van Nostrand Reinhold, New York, pp. 57–220.

Khire, M.V. (1995) *Field Hydrology and Water Balance Modeling of Earthen Final Covers for Waste Containment*, Environmental Geotechnics Report No. 95-5, University of Wisconsin, Madison, WI.

Khire, M.V., Benson, C.H. & Bosscher, P.J. (1997) Water balance modeling of earthen final covers. *Journal of Geotechnical & Geoenviron-*

mental Engineering, ASCE, Reston, VA, **123**(8): 744–754.

Kim, S.K., Hong, W.P. & Kim, Y.M. (1992) Prediction of rainfall-triggered landslides in Korea. In *Proceedings, Sixth International Symposium on Landslides*, Christchurch, Vol. 2, pp. 989–994.

Kimmel, G.E. & Braids, O.C. (1980) Leachate plumes in groundwater from Babylon and Islip landfills, Long Island, New York, Water-Supply Paper 2218, US Geological Survey.

King, K.S., Quigley, R.M., Fernandez, F., Reades, D.W. & Bacopoulos, A. (1993) Hydraulic conductivity and diffusion monitoring of the Keele Valley Landfill liner, Maple, Ontario. *Canadian Geotechnical Journal*, **30**(1): 124–134.

King, M.W.G., Barker, J.F. & Hamilton, K.A. (1995) Natural attenuation of coal tar organics in groundwater. *Intrinsic Bioremedation*, Hinchee, R.E., Wilson, J.T., and Downey, D.C. (Eds.) Battelle Press Inc. Columbus, OH, **3**(1): 171–180.

Kishida, H. & Nakai, S. (1977) Large deflection of single pile inder horizontal load. In *Proceedings, Special Session No. 10, Ninth International Conference on Soil Mechanics & Foundation Engineering*, Tokyo, pp. 87–92.

Kitajima, S. & Uwabe, T. (1979) Analysis on seismic damage in anchored sheet-piling bulkheads. *Report of the Japanese Port & Harbour Research Institute*, **18**(1): 67–130 (in Japanese).

Kjellman, W. (1952) Consolidation of clay soil by means of atmospheric pressure. In *Proceedings, Conference on Soil Stabilization*, Massachusetts Institute of Technology, MA, pp. 258–263.

Klaassen, C.D. & Eaton, D.L. (1991) Principles of toxicology. In Amdur, M.O., Doull, J. & Klaassen, C.D. (Eds) *Casarett & Doull's Toxicology: The Basic Science of Poisons*, 4th Edn, Pergamon Press, New York, pp. 12–49.

Klecka, C.M., Wilson, J.T., Klien, N., West, R., Davis, Weaver, J., Kampbell, D. & Wilson, B. (1996) Intrinsic remediation of chlorinated solvents in groundwater. In *Proceedings, Two Day Conference on Intrinsic Bioremediation*, March 18–19. The Cafe Royal, London.

Klotz, J. (1963) Small sample power and efficiency for the one sample Wilcoxon and normal scores tests. *Annals of Mathematics Statistics*, **34**: 624–632.

Klute, A. & Letey, J. (1958) The dependence of ionic diffusion on the moisture content of non-adsorbing porous media. *Soil Science Society of America Proceedings*, **22**: 213–15.

Kmet, P. (1982) *EPA Water Balance Method–Its Use and Limitations*, Wisconsin Department of Natural Resources, Madison, WI.

Kobayashi, S. & Nishimura, N. (1983) Analysis of dynamic soil–structure interactions by boundary integral equation method. In *Proceedings, Third International Symposium on Numerical Methods in Engineering*, Paris, pp. 353–362.

Kobayashi, K., Yao, S. & Yoshiada, N. (1991) Dynamic compliance of pile group considering nonlinear behaviour around piles. In *Proceedings, Second International Conference on Recent Advances in Geotechnical Earthquake Engineering & Soil Dynamics*, University of Missouri at Rolla, Rolla, MO, pp. 785–792.

Kobori, T., Minai, R. & Suzuki, T. (1971) The dynamical ground compliance of a rectangular foundation on a viscoelastic stratum. *Bulletin Disaster Prevention Research Institute*, **20**: 289–329.

Kobori, T., Minai, R. & Baba, K. (1977) Dynamic behaviour of a laterally loaded pile. In *Ninth International Conference of Soil Mechanics*, Tokyo, pp. 175–180.

Kobori, T., Nakazawa, M., Hijikata, K., Kobayashi, Y., Miura, K., Miyamoto, Y. & Moroi, T. (1991) Study on dynamic characteristics of a pile group foundation. In *Proceedings, Second International Conference on Recent Advances in Geotechnical Earthquake Engineering & Soil Dynamics*, University of Missouri at Rolla, Rolla, MO, pp. 853–860.

Koerner, R.M. (1984) Slope stabilization using anchored geotextiles: anchored spider netting. In *Proceedings, Special Geotechnical Engineering for Roads and Bridges Conference*, PennDot, pp. 1–11.

Koerner, R.M. (1998) *Designing with Geosynthetics*, 4th Edn, Prentice-Hall, Englewood Cliffs, NJ.

Koerner, R.M. & Bove, J. (1983) In-plane hydraulic properties of geotextiles. *Geotechnical Testing Journal*, **6**(4): 190–195.

Koerner, R.M. & Daniel, D.E. (1995) A suggested methodology for assessing the technical equivalency of GCLs to CCLs. In Koerner, R.M., Gartung, E. & Zanzinger H. (Eds) *Geosynthetic Clay Liners*, A.A. Balkema, Rotterdam, pp. 73–100.

Koerner, R.M. & Daniel, D.E. (1997) *Final Covers for Solid Waste Landfills and Abandoned Dumps*, American Society of Civil Engineers, Reston, VA.

Koerner, R.M. & Gartung, E. (1995) *Geosynthetic Clay Liners*, Rotterdam, A.A. Balkema.

Koerner, R.M. & Hwu, B.L. (1991) Stability and tension considerations regarding cover soils on geomembrane-lined slopes. *Geotextiles & Geomembranes*, **10**(4): 335–355.

Koerner, G.R. & Koerner, R.M. (1994) Design of landfill leachate-collection filters. *Journal of Geotechnical Engineering, ASCE*, Reston, VA, **120**(10): 1792–1803.

Koerner, R.M. & Koerner, G.R. (1995) *Leachate Clogging Assessment of Geotextile (and Soil) Landfill Filters*, Report CR-819371, US EPA, Cincinnati, OH.

Koerner, R.M. & Robbins, J.C. (1986) *In situ* stabilization of soil slopes using nailed geosynthetics. In *Proceedings, Third International Conference on Geotextiles*, Vienna, Vol. 2, pp. 395–400.

Koerner, R.M. & Soong, T.Y. (1998) Analysis and design of veneer cover soils, (The Giroud Lecture). In Rowe, R.K. (Ed.) *Proceedings, Sixth International Conference on Geosynthetics*, March, Atlanta, GA, Industrial Fabrics Association International, Roseville, MN, Vol. 1, pp. 1–23.

Koerner, R.M., Koerner, G.R. & Eberlé, M.A. (1996) Out-of-plane tensile behavior of geosynthetic-clay liners. *Gesynthetics International*, **3**(2): 277–296.

Koerner, R.M., Wilson-Fahmy, R.F. & Narejo, D. (1996) Puncture protection of geomembranes Part III: examples. *Geosynthetics International*, **3**(5): 655–675.

Koerner, R.M., Carson, D.A., Daniel, D.E. & Bonaparte, R. (1997) Current status of the Cincinnati GCL test plots. In *Proceedings, GRI-10 Conference on Field Performance of Geosynthetics*, Springfield, PA, pp. 153–182.

Kolbasuk, G.M., Lydick, L.D. & Reed, L.S. (1992) Effects of test procedures on geonet transmissivity results. *Journal of Geotextiles & Geomembranes*, **11**(4–6): 153–166.

König, D., Kockel, R. & Jessberger, H.L. (1996) Zur Beutellung der Standsicherhert und zur Prognose der Setzungen von Mischabfalldeponien. In *Proceedings, 12th Nurnberg Deponieseminar*, 75, Eigenverlag LGA, Nurnberg, pp. 95–117.

Konrad, J.-M. (1987) The influence of heat extraction rate in freezing soils. *Cold Reg. Science & Technology*, **14**: 129–137.

Konrad, J.-M. (1989a) Influence of overconsolidation on the freezing characteristics of a clayey silt. *Canadian Geotechnical Journal*, **26**: 9–21.

Konrad, J.-M. (1989b) Effect of freeze–thaw cycles on the freezing characteristics of a clayey silt at various overconsolidation ratios. *Canadian Geotechnical Journal*, **26**: 217–226.

Konrad, J.-M. (1994) Sixteenth Canadian Geotechnical Colloquium. Frost heave in soils: concepts and engineering. *Canadian Geotechnical Journal*, **2**(31): 223–245.

Konrad, J.-M. (1999) Frost susceptibility related to soil index properties. *Canadian Geotechnical Journal*, **36**(3): 403–417.

Konrad, J.-M. and Duquennoi, C. (1993) A model for water transport and ice lensing in freezing soils. *Water Resources Research*, **29**(9): 3109–3024.

Konrad, J.-M. & Morgenstern, N.R. (1980) A mechanistic theory of ice lens formation in fine-grained soils. *Canadian Geotechnical Journal*, **17**: 473–486.

Konrad, J.-M. & Morgenstern, N.R. (1982a) Prediction of frost heave in the laboratory during transient freezing. *Canadian Geotechnical Journal*, **19**(3), 250–259.

Konrad, J.-M. & Morgenstern, N.R. (1982b) Effects of applied pressure on freezing soils. *Canadian Geotechnical Journal*, **19**(4): 494–505.

Konrad, J.-M. & Morgenstern, N.R. (1983) Frost susceptibility of soils in terms of their segregation potential. In *Proceedings, Fourth International Conference on Permafrost*, Fairbanks, AK, pp. 660–665.

Konrad, J.-M. and Seto, J.T.C. (1994) Frost heave characteristics of undistributed sensitive Champlain sea clay. *Canadian Geotechnical Journal*, **31**: 285–298.

Konrad, J.-M. & Shen, M. (1994) Simulation of retaining wall displacement by frost action using the segregation potential approach. In *Ground Freezing 94*, Fremond, M. (Ed.) Balkema, Rotterdam, pp. 265–270.

Kovacs, W.D. & Leo, E. (1981) Cyclic simple shear of large-scale sand samples: Effects of diameter to height ratio. In *Proceedings, International Conference on Recent Advances in Geotechnical Earthquake Engineering & Soil Dynamics*, St Louis, Vol. 3, pp. 897–907.

Kraft, L.M., Ray, R.P. & Kagawa, T. (1981) Theoretical t–z curves. *Journal of Geotechnical Engineering Division, ASCE*, Reston, VA, **107**(11): 1543–1561.

Krahn, J. & Fredlund, D.G. (1972) On total, matric and osmotic suction. *Journal of Soil Science*, **114**(5): 339–348.

Kramer, S.L. (1996) *Geotechnical Earthquake Engineering*, Prentice Hall, Upper Saddle River, NJ.

Kramer, S.L. & Holtz, R.D. (1991) *Soil Improvement and Foundation Remediation with Emphasis on Seismic Hazards*, Report, University of Washington, Seattle.

Kramer, S.R., McDonald, W.J. & Thomson, J.C. (1992) *An Introduction to Trenchless Technology*, Van Nostrand Reinhold, New York.

Kraus, J.F., Benson, C.H., Erickson, A.E. & Chamberlain, E.J. (1997) Freeze–thaw cycling and hydraulic conductivity of bentonite barriers. *Journal of Geotechnical & Geoenvironmental Engineering*, ASCE, Reston, VA, **123**(3): 229–238.

Krinitzsky, E.L., Gould, J.P. & Edinger, P.H. (1993) *Fundamentals of Earthquake-Resistant Construction*, Wiley, New York.

Kruseman, G.P. & de Ridder, N.A. (1990) *Analysis and Evaluation of Pumping Test Data*, International Institute for Land Reclamation and Improvement, The Netherlands.

Kubo, K. (1965) Experimental study of the behaviour of laterally loaded piles. In *Proceedings, Sixth International Conference on Soil Mechanics & Foundation Engineering*, Montreal, Vol. 2, pp. 275–279.

Kueper, B.H. & Frind, E.O. (1991) Two phase flow in heterogeneous porous media: 2. Model development. *Water Resources Research*, **27**(6): 1059–1070.

Kueper, B.H. & McWorter, D.B. (1991) The behavior of dense, nonaqueous phase liquids in fractured clay and rock. *Groundwater*, **29**(5): 716–728.

Kueper, B.H., Pitts, M., Sale, T.C., Simpkin, T. & Wyatt, K. (1997) Technology practices manual for surfactants and cosolvents. Advanced Applied Technology Demonstration Facility, U.S. Department of Defense.

Kuhlemeyer, R.L. (1979) Static and dynamic laterally loaded floating piles. *Journal of the Geotechnical Engineering Division*, ASCE, Reston, VA, **105**(2): 289–304.

Kulhawy, F.H. (1984) Limiting tip and side resistance: fact or fallacy? In Meyer, J.R. (Ed.) *Analyses & Design of Pile Foundations*, American Society of Civil Engineers, Reston, VA, pp. 80–98.

Kulhawy, F.H. (1991) Drilled shaft foundations. In Fang, H.Y. (Ed.) *Foundation Engineering Handbook*, Van Nostrand, Reinhold, New York, NY, pp. 537–552.

Kulhawy, F.H. & Carter, J.P. (1992) Settlement and bearing capacity of foundations on rock masses. In Bell, F.G. (Ed.) *Engineering in Rock Massess*, Butterworth-Heinemann, Oxford, pp. 231–245.

Kulhawy, F.H. & Carter, J.P. (1992) Socketed foundations in rock masses. In Bell, F.G. (Ed.) *Engineering in Rock Masses, Butterworth-Heinemann*, Oxford, pp. 509–529.

Kulhawy, F.H. & Goodman, R.E. (1987) Foundations in rock. In Bell, F.G. (Ed.) *Ground Engineer's Reference Book*, Butterworths, London, Chapter 55.

Kulhawy, F.H. & Mayne P.W. (1990) *Manual on Estimating Soil Properties for Foundation Design*, Report No. EPRI EL-6800, Cornell University.

Kulhawy, F.H. & Phoon, K. (1993) Drilled shaft side resistance in clay soil to rock. In *Proceedings Conference on Design and Performance of Deep Foundations: Piles and Piers in Soil and Soft Rock*, Geotechnical Special Publication No. 38, American Society of Civil Engineers, Reston, VA, pp. 172–183.

Kunze, R.J., Uehara, G. & Graham, K. (1968) Factors important in the calculation of hydraulic conductivity. *Proceedings, Soil Science of America*, **32**: 760–765.

Kushner, E.J. (1976) On determining the statistical parameters for pollution concentration from a truncated data set. *Atmospheric Environment*, **10**: 975–979.

Kuster, E. (1976) Biological degradation of synthetic polymers. In Ranby, B. & Rabek, J.F. (Eds) *long-term properties of polymers and polymeric materials*, J. of Applied Polymer Science 35, Interscience Publication, New York, NY, USA, pp. 395–404.

Kuwabara, F. & Poulos, H.G. (1989) Downdrag forces in a group of piles. *Journal of the Geotechnical Engineering Division*, ASCE, Reston, VA, **115**(6): 806–818.

Lacasse, S. & Nadim, F. (1994) Reliability issues and future challenges in geotechnical engineering for offshore structures (Plenum Paper). In *Seventh International Conference on the Behaviour of Offshore Structures, BOSS'94*, Massachusetts Institute of Technology, MA, Cambridge, pp. 9–38.

Ladanyi, B. (1977) Friction and end bearing tests on bedrock for high capacity socket design: Discussion. *Canadian Geotechnical Journal*, **14**: 153–155.

Ladanyi, B. & Archambault, G. (1970) Simulation

of shear behavior of a jointed rock mass. In *Proceedings, 11th Symposium on Rock Mechanics*, American Institute of Mechanical Engineers, New York, pp. 105–125.

Ladanyi, B. & Roy, A. (1971) Some aspects of bearing capacity of rock mass. In *Proceedings, Seventh Canadian Symposium on Rock Mechanics*, Edmonton, pp. 161–190.

Ladanyi, B., Dufour, R., Larocque, G.S., Samson, L. & Scott, J.S. (1974) *Report of the Subcommittee on Foundations and Near-Surface Structures to the Canadian Advisory Committee on Rock Mechanics.*

Ladd, C.C. (1969) The prediction of *in situ* stress strain behaviour of soft saturated clay during undrained shear. *Proc. Bolkesjo Symp.*, Norway, Norwegian Geotechnical Institute, Oslo, pp. 14–20.

Ladd, C.C. (1991) Stability evaluation during staged construct. *Journal of Geotechnical Engineering, ASCE*, Reston, VA, **117**(4): 540–615.

Ladd, C.C. & Foott, R. (1974) New design procedure for stability of soft clays. *Journal of Geotechnical Engineering Division, ASCE*, Reston, VA, **100**(7): 763–786.

Ladd, C.C., Foott, R., Ishihara, K., Schlosser, F. & Poulos, H.G. (1977) Stress–deformation and strength characteristics, state-of-the-art report. In *Proceedings, Ninth International Conference on Soil Mechanics and Foundation Engineering*, Tokyo, Vol. 2, pp. 421–494.

Lade, P.V. (1977) Elasto-plastic stress–strain theory for cohesionless soil with curved yield surfaces. *International Journal of Solids & Structures*, **13**: 1019–1035.

Lade, P.V. & Kim, M.K. (1995) Single hardening constitutive model for soil, rock and concrete. *International Journal of Solids & Structures*, **32**(14): 1963–1978.

Lade, P.V. & Yamamuro, J.A. (1993) Stability of granular materials in postpeak softening regime. *Journal of Engineering Mechanics, ASCE*, Reston, VA, **119**(1): 128–144.

Laflamme, J.F. & Leroueil, S. (1999) *Analyse des pressions interstitielles mesurées aux sites d'excavation de Saint-Hilaire et de Rivière-Vachon, Québec.* Report GCT-99-10, Prepared for Ministère des Transports du Quèbec, Université Laval, Québec.

Lafleur, J. & Lefebvre, G. (1980) Groundwater regime associated with slope stability in Champlain clay deposits. *Canadian Geotechnical Journal*, **17**(1): 44–53.

Lafleur, J., Silvestri, V., Asselin, R. & Soulié, M. (1988) Behaviour of a test excavation in soft Champlain Sea clay. *Canadian Geotechnical Journal*, **25**(4): 705–715.

LaGatta, M.D., Boardman, B.T., Cooley, B.H. & Daniel, D.E. (1997) Geosynthetic clay liners subjected to differential settlement. *Journal of Geotechnical & Geoenvironmental Engineering, ASCE*, Reston, VA, **123**(5): 402–410.

Laine, D.L. (1991) Analysis of pinhole seam leaks located in geomembrane liners using the electrical leak location method: case histories. In *Proceedings, Geosynthetics '91*, Atlanta, GA, Industrial Fabrics Association International, Roseville, MN, Vol. 1, pp. 239–253.

Lake, C.B. & Rowe, R.K. (1999) The role of contaminant transport in two different geomembrane/geosynthetic clay liner composite liner designs. *Proceedings, Geosynthetics '99*, April, Boston, MA, Industrial Fabrics Association International, Roseville, MN, Vol. 1, pp. 661–670.

Lake, C.B. & Rowe, R.K. (2000a) *Swelling Characteristic of Needlepunched, Thermally Treated GCLs*, Geotextiles and Geomembranes **18**(2): 77–102.

Lake, C.B. & Rowe, R.K. (2000b) *Diffusion of Sodium and Chloride Through Geosynthetic Clay Liners*, Geotextiles and Geomembranes, **18**(2): 103–132.

Lam, T.S.K., Yau, J.H.W. & Permchitt, J. (1991) Side resistance of a rock-socketed caisson. *Hong Kong Engineer*, **19**(2): 17–28.

Lambe, T.W. (1951) *Soil Testing for Engineers*, Wiley, New York.

Lambe, T.W. (1958a) The structure of compacted clay. *Journal of the Soil Mechanics & Foundation Division, ASCE*, Reston, VA, **84**(2): 1654/1–34.

Lambe, T.W. (1958b) The engineering behaviour of compacted clay. *Journal of Soil Mechanics & Foundation Division, ASCE*, Reston, VA, **84**(2): 1655/1–35.

Lambe, T.W. (1964) Methods of estimating settlement. *Journal of the Soil Mechanics & Foundation Division, ASCE*, Reston, VA, **90**(5): 43–67.

Lambe, T.W. (1967) Stress path method. *Journal of the Soil Mechanics & Foundation Division, ASCE*, Reston, VA, **93**(6): 309–331.

Lambe, T.W. & Whitman, R.V. (1979) *Soil mechanics, SI Version*, Wiley, New York.

Landva, A.O. (1980) Vane testing in peat. *Canadian Geotechnical Journal*, **17**(1): 1–19.

Landva, A.O. & Clark, J.I. (1990) Geotechnics of waste fill. In *Geotechnics of Waste Fills–Theory & Practice*, STP 1070, Landua, A. and Knowles, E.D. (Eds.) American Society for Testing & Materials, West Conshohocken, PA, pp. 86–106.

Landva, A.O. & La Rochelle, P. (1983) Compressibility and shear characteristics of Radforth peats. In *Testing of Peats & Organic Soils*, STP 820, P.M. Jarrett (Ed.) ASTM American Society for Testing & Materials, West Conshohocken, PA, pp. 157–191.

Landva, A.O. & Pheeney, P.E. (1980) Peat fabric and structure. *Canadian Geotechnical Journal*, **17**(3): 416–435.

Landva, A.O., Korpijaakko, E.O. & Pheeney, P.E. (1983) Geotechnical classification of peats and organic soils. In *STP 820*, P.M. Jarrett (Ed.) American Society for Testing & Materials, West Conshohocken, PA, pp. 37–51.

Lane, D.T., Benson, C.H. & Bosscher, P.J. (1992) *Hydrologic Observations and Modeling Assessments of Landfill Covers*, Environmental Geotechnics Report No. 92-10, University of Wisconsin, Madison, WI.

Larsson, R. (1980) Undrained shear strength in stability calculations of embankments and foundations on soft clays. *Canadian Geotechnical Journal*, **17**(4): 591–602.

Larsson, R. (1986) *Consolidation of Soft Soils*, Swedish Geotechnical Institute Report No. 29, Linköping, Sweden.

Lav, M.A., Carter, J.P. & Booker, J.R. (1995) The effect of fissures on the bearing capacity of clay. In *Proceedings, 14th Australasian Conference on the Mechanics of Structures & Materials*, Hobart, pp. 38–43.

Law, K.T. (1985) Use of field vane tests under earth structures. *11th International Conference on Soil Mechanics & Foundation Engineering*, San Francisco, Vol. 2, pp. 893–898.

Law, K.T. & Bozozuk, M. (1979) A method of estimating excess pore pressures beneath em bankments on sensitive clays. *Canadian Geotechnical Journal*, **16**(4): 691–702.

Lawson, C.R. (1986) Geosynthetics in soil reinforcement. In *Proceedings, Symposium on Geotextiles in Civil Engineering*, Institution of Engineers Australia, Newcastle, pp. 1–35.

Lea, N.D. & Brawner, C.O. (1963) *Highway Design and Construction Over Peat Deposits in the Lower Mainland of British Columbia*, Highway Research Board Research Record, No. 7, Washington, DC, pp. 1–33.

Leach, G. & Reed, K. (1989) Observations and assessment of the disturbance caused by displacement methods of trenchless construction. In *Proceedings, Fourth International Conference on Trenchless Construction for Utilities, NO-DIG '89* April, London, International Society for Trenchless Technology, London, England, pp. 67–78.

Leblanc, D.R. & Celia, M.A. (1991) *Density-induced downward movement of solutes during a natural-gradient tracer test, Cape Cod, Massachusetts*, Resources Investigation Report 91-4034, US Geological Survey Water, pp. 10–14.

Lebuis, J., Robert, J.M. & Rissmann, P. (1983) Regional mapping of landslides hazard in Québec. In *Symposium on Slopes in Soft Clays*, Linköping, SGI Report No. 17, pp. 205–262.

Lechner, P. (1994) Design criteria for leachate drainage and collection systems. In *Landfilling of Waste: Barriers*. T.H. Christensen, R. Cossu & R. Stegmann (Eds.), E & FN Spon, London, UK, pp. 519–530.

Ledbetter, R.H. & Finn, W.D.L. (1993) *Development and evaluation of remediation strategies by deformation analyses*. In *Proceedings, Special Technical Publication, Geotechnical Practice in Dam Rehabilitation*, April 15–28, American Society of Civil Engineers, Reston, VA.

Lee, C.Y., Poulos, H.G. & Hull, T.H. (1991) Effect of seafloor instability on offshore pile foundations. Canadian Geotechnical Journal, **28**(5): 729, 737.

Lee, I.K. (1993) Analysis and performance of raft and pile-raft systems (keynote lecture). In *Third International Conference on Case Histories in Geotechnical Engineering*, St. Louis.

Lee, K.L. & Singh, A. (1971) Relative density and relative compaction. *Journal of the Soil Mechanics & Foundations Division, ASCE*, Reston, VA, **97**(7): 1049–1052.

Lee, K.L. & Seed, H.B. (1967) Drained strength characteristics of sands. *Journal of the Soil Mechanics & Foundations Division, ASCE*, Reston, VA, **93**(6): 117–141.

Lee, K.L., Jones, C.J.F.P. Sullivan, W.R. & Trolinger, W. (1995) Failure and deformation of four reinforced earth walls in Eastern Tennessee, USA. *Géotechnique*, **45**(4): 749–752.

Lee, M. & Finn, W.D.L. (1978) *DESRA-2: Dynamic Effective Stress Response Analysis of Soil Deposits with Energy Transmitting Boundary Including Assessment of Liquefaction Potential*, Report No. 38, Soil Mechanics Services, University of British Columbia, Vancouver, BC.

Lee, M.K., Lum, K.Y. & Hartford, D.N.D. (1998) Calculation of the seismic risk of an earth dam susceptible to liquefaction. In Dakoulas, P. & Yegian, M. (Eds.), Earthquake Engineering and Soil Dynamics III. Geotechnical Special Publication No. 75. American Society of Civil Engineers, Reston, VA, pp. 1451–1460.

Lee, S.L., Chow, Y.K., Karunaratne, G.P. & Wong, K.Y. (1988) Rational wave equation model for pile driving analysis. *Journal of Geotechnical Engineering, ASCE*, Reston, VA, **114**(3): 306–325.

Leeman, E.R. (1968) The determination of the complete state of stress in rock in a single bore-hole-laboratory and underground measurements. *International Journal of Rock Mechanics & Mineral Science*, **5**: 31–56.

Lefebvre, G. (1981) Fourth Canadian Geotechnical Colloquium: Strength and slope stability in Canadian soft clay deposits. *Canadian Geotechnical Journal*, **18**(3): 420–442.

Lefebvre, G., Langlois, P., Lupien, C. & Lavallée, J.-G. (1984) Laboratory testing and *in situ* behaviour of peat as embankment foundation. *Canadian Geotechnical Journal*, **21**(2): 322–337.

Lefebvre, G., Paré, J.J. & Dascal, O. (1987) Undrained shear strength in the surficial weathered crust. *Canadian Geotechnical Journal*, **24**(1): 23–34.

Leflaivre, E, Khay, M. & Blivet, J.C. (1983) Un nouveau material: le Texsol. *Bulletin de Liason du Laboratoire des Ponts et Chaussees*, **125**: 105–114.

Legget, R.F. (1962) *Geology in Engineering*, McGraw-Hill, 2nd Edn, New York.

Legget, R.F. & Hatheway, A.W. (1988) *Geology & Engineering*, McGraw-Hill, New York.

Lehr, J.H. (1990) Toxicological risk assessment distortions: Part III a different look at environmentalism. *Ground Water*, **28**(3): 330–340.

Lenhard, R.J. & Parker, J.C. (1990) Estimation of free hydrocarbon volume from fluid levels in monitoring wells. *Groundwater*, **28**, 57–67.

Lenhard, R.J., Johnson, T.G. & Parker, J.C. (1993) Experimental observations of non-aqueous-phase liquid subsurface movement. *Journal of Contaminant Hydrology*, **12**: 79–101.

Lentz, R.W., Horst, W.D. & Uppot, J.O. (1985) The permeability of clay to acidic and caustic permeants. In Johnson, A.I., Frobel, R.K., Cavalli, N.J. & Pettersson, C.B. (Eds) *Hydraulic Barriers in Soil and Rock*, STP 874, American Society for Testing & Materials, West Conshohocken, PA, pp. 127–139.

Leonards, G.A. (1962) *Foundation Engineering*, McGraw Hill, New York.

Leonards, G.A. (1972) Settlement of pile foundations in granular soil. *Proceedings, Conference Perf. Earth & Earth-Supp. Structures, ASCE*, Reston, VA, **2**: 1169–1184.

Leonards, G.A. (1979) Stability of slopes in soft clays (special lecture). *Sixth Panamerican Conference on Soil Mechanics & Foundation Engineering*, Lima.

Leonards, G.A. & Narain, J. (1963) Flexibility of clay and cracking of earth dams. *Journal of the Soil Mechanics & Foundations Division, ASCE*, Reston, VA, **89**(2): 47–98.

Leonards, G.A., Cutter, W.A. & Holtz, R.D. (1980) Dynamic compaction of granular soils. *Journal of Geotechnical Engineering, ASCE*, Reston, VA, **106**(1): 35–44.

Leong, E.C. & H. Rahardjo (1997a) A review on soil–water characteristic curve equations. *Journal of Geotechnical Engineering, ASCE*, Reston, VA, **123**(12): 1106–1117.

Leong, E.C. & H. Rahardjo (1997b) Permeability functions for unsaturated soils. *Journal of Geotechnical Engineering, ASCE*, Reston, VA, **123**(12): 1118–1126.

Leroueil, S. (1988) Recent developments in consolidation of natural clays: 10th Canadian Geotechnical Colloquium. *Canadian Geotechnical Journal*, **25**(1): 85–107.

Leroueil, S. (1996) Compressibility of clays: fundamental and practical aspects. *Journal of Geotechnical Engineering, ASCE*, Reston, VA, **122**(7): 534–543.

Leroueil, S. (1997) Critical state soil mechanics and the behaviour of real soils. In Almeida, M. (Ed.) *Proceedings, International Symposium on Recent Developments in Soil & Pavement Mechanics*, Rio de Janeiro, Brazil, pp. 41–80. A.A. Balkema, Brookfield, Vermont.

Leroueil, S. (2000) 39th Rankine Lecture: Cuts and natural slopes: movement and failure mechanisms. *Géotechnique* (in press).

Leroueil, S. & Jamiolkowski, M. (1991) Exploration of soft soil and determination of design parameters, (General Report, Session 1). *Geo-Coast '91*, Yokohama, Vol. 2, pp. 969–998.

Leroueil, S. & Locat, J. (1998) Slope movements: geotechnical characterization, risk assessment and mitigation. In *Proceedings, 11th Danube-European Conference on Soil Mechanics & Geotechnical Engineering*, Poreč, Croatia, pp. 95–106.

Leroueil, S. & Marques, M.E.S. (1996) Importance of strain rate and temperature effects in geotechnical engineering. *ASCE Convention*, Geotechnical Special Publication No. 61 Washington, DC, pp. 1–60.

Leroueil, S. & Tavenas, F. (1986) Discussion on 'Effective stress paths and yielding in soft clays below embankments' by D.J. Folkes & J.H.A. Crooks. *Canadian Geotechnical Journal*, **23**(3): 410–413.

Leroueil, S. & Vaughan, P.R. (1990) The general and congruent effects of structure in natural soils and weak rocks. *Géotechnique*, **40**(3): 467–488.

Leroueil, S., Tavenas, F., Trak, B., La Rochelle, P. & Roy, M. (1978a) Construction pore pressures in clay foundations under embankments, Part I: the Saint-Alban test fills. *Canadian Geotechnical Journal*, **15**(1): 54–65.

Leroueil, S., Tavenas, F., Mieussens, C. & Peignaud, M. (1978b) Construction pore pressures in clay foundations under embankments, Part II: generalized behaviour. *Canadian Geotechnical Journal*, **15**(1): 66–82.

Leroueil, S., Tavenas, F. & Le Bihan, J.-P. (1983) Propriétés caractéristiques des argiles de l'est du Canada. *Canadian Geotechnical Journal*, **20**(4): 681–705.

Leroueil, S., Kabbaj, M., Tavenas, F. & Bouchard, R. (1985a) Stress–strain–strain rate relation for the compressibility of sensitive natural clays. *Géotechnique*, **35**(2): 159–180.

Leroueil, S., Magnan, J.-P. & Tavenas, F. (1985b) *Remblais sur argiles molles*. Lavoisier-Technique et Documentation, France. English version *Embankment on Soft Clays*, Ellis Horwood, Chichester, UK.

Leroueil, S., La Rochelle, P., Tavenas, F. & Roy, M. (1990) Remarks on the stability of temporary cuts. *Canadian Geotechnical Journal*, **27**(5): 687–692.

Leroueil, S., Tardif, J., Roy, M., LaRochelle, P. & Konrad, J.-M. (1991) Effects of frost on the mechanical behaviour of Champlain Sea clays. *Canadian Geotechnical Journal* **28**: 690–697.

Leroueil, S., Le Bihan, J.P. & Bouchard, R. (1992a) Remarks on the design of clay liners in lagoons as hydraulic barriers. *Canadian Geotechnical Journal*, **29**: 512–515.

Leroueil, S., Le Bihan, J.P. & Bouchard, R. (1992b) Discussion of water content–density criteria for compacted soil liners by D. Daniel, & C. Benson. *Journal of Geotechnical Engineering, ASCE*, Reston, VA, **118**(2): 963–965.

Leroueil, S., Demers, D., La Rochelle, P., Martel,

G. & Virely, D. (1995) Practical applications of the piezocone in Champlain sea clays. In *Proceedings, International Symposium on Cone Penetration Testing*, CPT-95, Linköping, pp. 515–522.

Leroueil, S., Perret, D. & Locat, J. (1996a) Strain rate and structuring effects on the compressibility of a young clay. Session on Measuring and Modeling Time Dependent Soil Behavior, *ASCE Convention*, Geological Special Publication No. 61, Washington, DC, pp. 137–150.

Leroueil, S., Vaunat, J., Picarelli, L., Locat, J., Faure, R.M. & Lee, H. (1996b) A geotechnical characterization of slope movements. In *Proceedings, Seventh International Symposium on Landslides*, Trondheim, Vol. 1, pp. 53–74.

Lesage, S., Brown, S. & Millar, K. (1996) Vitamin B12-catalyzed dechlorination of perchloroethylene present as residual DNAPL. *Ground Water Monitoring & Remediation*, **16**(4): 76–85.

Leshchinsky, D., Ling, H.I. & Hanks, G.A. (1995) Unified design approach to geosynthetic reinforced slopes and segmental walls. *Geosynthetics International*, **2**(5): 845–881.

Leussink, H. & Wenz, K.P. (1969) Storage yard foundations on soft cohesive soils. In *Proceedings, Seventh International Conference on Soil Mechanics & Foundation Engineering*, Mexico City, Vol. 2, pp. 149–155.

Levy, J.F. (1970) Sonic pulse method of testing cast-*in-situ* concrete piles. *Ground* Engineering, **3**(3): 17–19.

Li, A.L. & Rowe, R.K. (1999) Reinforced embankments and the effect of consolidation on soft cohesive soil deposits. In *Geosynthetics '99*, Boston. Industrial Fabrics Association International, Roseville, MN, pp. 477–490.

Li, J., Cameron, D.A. & Mills, K.G. (1996) Numerical modelling of covers and slabs subject to seasonal surface suction changes. In *Seventh Australia–New Zealand Conference on Geomechanics*, 1–5 July, Adelaide, M.B. Jaksa, W.S. Kaggwa & D.A. Cameron (Eds.), pp. 424–429.

Liao, S.S.C. & Whitman, R.V. (1986) Overburden correction factors for SPT in sand. *Journal of Geotechnical Engineering*, ASCE, Reston, VA, **112**(3): 373–377.

Liao, S.S.C., Veneziano, D. & Whitman, R.V. (1988) Regression models for evaluating liquefaction probability. *Journal of Geotechnical Engineering, ASCE*, Reston, VA, **114**(4): 389–411.

Liao, S.T. & Roesset, J.M. (1997) Identification of defects in piles through dynamic testing. *In*-

ternational *Journal of Numerical Analytical Methods in Geomechanics*, **21**(4): 277–291.

Little, A.D. Inc. (1985) *Resistance of Flexible Membrane Liners to Chemicals and Wastes*, US EPA Report PB86–119955, Cincinnati, OH.

Littlejohn, G.S. (1993) Chemical grouting. In Moseley, M.P. (Ed.) *Ground Improvement*, Blackie Academic & Professional, Blackwood, NJ, pp. 100–130.

Liu, C.N., Gilbert, R.B., Thiel, R.S. & Wright, S.G. (1997) What is an appropriate factor of safety for landfill cover slopes. In *Proceedings, Geosynthetics '97 Conference*, Long Beach, Industrial Fabrics Association International, Roseville, MN, Vol. 1, pp. 481–496.

Lo, K.Y. (1962) Shear strength properties of a sample of volcanic material of the Valley of Mexico. *Géotechnique*, **22**(4): 303–318.

Lo, K.Y. (1970) The operational strength of fissured clays. *Géotechnique*, **20**: 57–74.

Lo, K.Y. (1978) Regional distribution of *in-situ* horizontal stresses in rocks of Southern Ontario. *Canadian Geotechnical Journal*, **15**(3): 371–381.

Lo, K.Y. (1986) Recent advances in design and evaluation of performance of underground structures in rocks. *Tunnelling & Underground Space Technology*, **4**(2): 171–183.

Lo, K.Y. & Cooke, B.H. (1989) Foundation design for the Skydome Stadium, Toronto. *Canadian Geotechnical Journal*, **26**: 22–33.

Lo, K.Y. & Grass, J.D. (1994) Recent experience with safety assessment of concrete dams on rock foundations. In *Proceedings, Canadian Dam Safety Conference*, October, Winnipeg, pp. 231–250.

Lo, K.Y. & Hefny, A.M. (1993) The evaluation of *in-situ* stresses by hydraulic fracturing tests in anisotropic rocks with mixed-mode fractures. *Canadian Tunnelling Journal*: 59–73.

Lo, K.Y. & Hefny, A.M. (1996) Design of tunnels in rock with long-term time-dependent and nonlinearly stress-dependent deformation. *Canadian Tunnelling Journal*: 179–214.

Lo, K.Y. & Hefny, A.M. (1998) Statistical analysis of the strength of the contact between concrete dams and rock foundations. In *First Annual Conference System Stewardship for Dams & Reservoirs*, September 27–October 1, The Canadian Dam Association, Halifax, NS, pp. 18–33.

Lo, K.Y. & Hori, M. (1979) Deformation and strength properties of some rocks in Southern Ontario. *Canadian Geotechnical Journal*, **16**: 108–120.

Lo, K.Y. & Lee, F. (1974) An evaluation of the stability of natural slopes in plastic Champlain clays. *Canadian Geotechnical Journal*, **11**(1): 165–181.

Lo, K.Y. & Lee, Y.N. (1990) Time dependent deformation behavior of Queenston Shale. *Canadian Geotechnical Journal*, **27**: 461–471.

Lo, K.Y. & Lukajic, B. (1984) Predicted and measured stresses and displacements around the Darlington intake tunnel. *Canadian Geotechnical Journal*, **21**: 147–165.

Lo, K.Y. & Morin, J.P. (1972) Strength anisotropy and time effects of two sensitive clays *Canadian Geotechnical Journal*, **9**(3): 261–277.

Lo, K.Y. & Stermac, A.G. (1965) Failure of an embankment founded on varved clay. *Canadian Geotechnical Journal*, **2**(3): 234–253.

Lo, K.Y. & Yuen, C.M.K. (1981) Design of tunnel lining in rock for long term effects. *Canadian Geotechnical Journal*, **18**: 24–39.

Lo, K.Y., Wai, R.S.C., Palmer, J.H.L. & Quigley, R.M. (1978) Time-dependent deformation of shaly rocks in Southern Ontario. *Canadian Geotechnical Journal*, **15**: 537–547.

Lo, K.Y., Devata, M. & Yuen, C.M.K. (1979a) Performance of a shallow tunnel in a shaly rock with high horizontal stresses. London, ON. Canada. *Tunnelling*, 1–12.

Lo, K.Y., Lukajic, B., Yuen, C.M.K. & Hori, M. (1979b) *In situ* stress measurement in rock overhang at the Ontario Power Generating Station, Niagara Falls. In *Fourth International Congress on Rock Mechanics*, Montreux, Switzerland, Vol. 2, pp. 343–351.

Lo, K.Y., Lukajic, B. & Ogawa, T. (1984) Predicting settlement due to tunneling in clays. *Proceedings, Tunnelling in Soil & Rock, ASCE*, Reston, VA, May 14–18, Atlanta, GA, pp. 128–155.

Lo, K.Y., Cooke, B.H. & Dunbar, D.D. (1987a) Design of buried structures in squeezing rock in Toronto, Canada. *Canadian Geotechnical Journal*, **24**: 232–241.

Lo, K.Y., Yung, T.C.B. & Lukajic, B. (1987b) A field method for the determination of rock mass modulus. *Canadian Geotechnical Journal*, **24**: 406–413.

Lo, K.Y., Lukajic, B., Wang, S., Ogawa, T. & Tsui, K.K. (1990) Evaluation of strength parameters of concrete–rock interface for dam safety assessment. In *Proceedings, Canadian Dam Safety Conference*, September, Toronto, pp. 71–94.

Lo, K.Y., Ogawa, T., Lukajic, B., Smith, G.F. &

Tang, J.H.K. (1991a) The evaluation of stability of existing concrete dams on rock foundations and remedial measures. In *Proceedings, 17th International Congress, International Commission on Large Dams*, Vienna, Austria, pp. 963–990.

Lo, K.Y., Ogawa, T., Lukajic, B. & Dupak, D. (1991b) Measurements of strength parameters of concrete–rock contact at the dam–foundation interface. *Geotechnical Testing Journal, ASTM*, West Conshohocken, PA, **14**(4): 383–394.

Lo, K.Y., Inculet, I.I. & Ho, K.S. (1991c) Field test of electroosmotic strengthening of soft sensitive clay. *Canadian Geotechnical Journal*, **28**(1): 74–83.

Lo, K.Y., Ho, M.S.F., Hefny, A.M. & Adeghe, L. (1997) Measurements of laboratory rates of concrete expansion in Saunders Dam. *The Canadian Geotechnical Conference*, October 20–22, Golden Jubilee, Ottawa, Ontario, pp. 582–592.

Lo, M.C. (1992) *PhD thesis*, University of Texas.

Locat, J. & Leroueil, S. (1997) Landslides risk assessment in sensitive clays: pre-failure, failure, and post-failure issues. *IUGS Workshop on Landslide Risk Assessment*, Honolulu, pp. 261–270.

Loi, J., Fredlund, D.G., Gan, J.K-M. & Widger, R.A. (1992) Monitoring soil suction in an indoor test track facility. *Transportation Research Record*, **1362**: 101–110.

Long, J.H., Daly, J.J. & Gilbert, R.B. (1993) *Structural Integrity of Geosynthetic Lining and Cover Systems for Solid Waste Landfills*, Department of Civil Engineering, University of Illinois, Project No. OSWR 06-005.

Long, J.H., Gilbert, R.B. & Daly, J.J. (1994) Geosynthetic loads in landfill slopes: displacement compatibility. *Journal of Geotechnical Engineering, ASCE*, Reston, VA, **120**(11): 2009–2025.

Lopes, R.F., DuBois, D., Barone, F.S., Matchett, D. & Quigley, R.M. (1992) An evaluation of compacted Question Shale as clay liner material for a landfill. In *Proceedings, 45th Canadian Geotechnical Conferences*, October, Toronto, pp. 71-1–71-10.

Loudon, P.A. (1967) *PhD thesis*, University of Cambridge, England.

Loughney R.W. (1954) Electricity Stiffens Clay Fivefold for Electric Power Plant Excavation. Construction Methods & Equipment **36**(8): 70–82.

Louis, C. (1969) *PhD thesis*, University of Karlsruhe, Germany. Translated by the Rock Mechanics Department, Imperial College, Report 10, London, England.

Lovell, C. (1957) Temperature effects on phase composition and strength of partially frozen soil. *Hwy. Research Board Bulletin* **168**: 74–95.

Low, P.F. (1987) Structural component of the swelling pressure of clay. *Langmuir*, **3**: 18–25.

Low, P.F. (1992) Interparticle forces in clay suspensions: flocculation, viscous flow and swelling. In Güven, N. & Pollastro, R.M. (Eds) *CMS Workshop Lectures, 4: Clay Water Interface and its Rheological Implications*, Clay Minerals Society, pp. 157–190.

Lowe, J. & Zaccheo, P.F. (1991) Subsurface exploration and sampling. In Fang, H.Y. (Ed.) *Foundation Engineering Handbook*, 2nd Edn. Van Nostrand Reinhold, New York, pp. 1–71.

Lozano, N. & Aughenbaugh, N.R. (1995) Flexibility of fine-grained soils. In *Geoenvironment 2000*, Geotechnical Special Publication No. 46, Acar, Y.B. and Daniel, D.C. (Eds.) American Society of Civil Engineers, Reston, VA, Vol. 1, pp. 844–858.

Luco, J.E. & Hadjian, A.H. (1974) Two-dimensional approximations to the three-dimensional soil–structure interaction problem. *Nuclear Engineering & Design*, **31**(2): 195–203.

Luco, J.E. & Westmann, R.A. (1971) Dynamic response of circular footings. *Journal of Engineering Mechanics Division, ASCE*, Reston, VA, **97**(6): 1381–1395.

Luettich, S.M. & Williams, N.D. (1989) Design of vertical drains using the hydraulic conductivity ratio analysis. In *Proceedings Geosynthetics '89*, IFAI, Roseville, MN, pp. 95–103.

Luettich, S.M., Giroud, J.P. & Bachus, R.C. (1992) Geotextile filter design guide. *Geotextiles & Geomembranes*, **11**(4–6): 355–370.

Lukas, R.G. (1986) *Dynamic Compaction for Highway Construction Volume I: Design and Construction Guidelines*, Report, No. FHWA/RD-86/133, Federal Highway Administration, Springfield, VA.

Lukas, R.G. (1995) *Dynamic Compaction*, Geotechnical Engineering Circular No. 1, Report FHWA-SA-95-037, Federal Highway Administration, Springfield, VA.

Lukas, R.G. (1997) Delayed soil improvement after dynamic compaction. In *Ground Improvement, Ground Reinforcement, and*

Ground Treatment: Developments 1987–1997, V.R. Schaefer (Ed.) *ASCE* Press, Reston, VA.

Lundegard, P.D. & Labrecque, D. (1995) Air sparging in a sandy aquifer (Florence, Oregon): Actual and apparent radius of influence. *Journal of Contaminant Hydrology*, **19**(1): 1–28.

Lunne, T., Eide, O. & de Ruiter, J. (1976) Correlations between cone resistance and vane shear strength in some Scandinavian soft to medium stiff clays. *Canadian Geotechnical Journal*, **13**(4): 430–441.

Lunne, T., Lacasse, S. & Rad, N.S. (1990) SPT, CPT, pressuremeter testing and recent developments on *in-situ* testing of soils. Part 1: all tests except SPT (state-of-the-art report). In *12th International Conference on Soil Mechanics & Foundation Engineering*, August 13–18, Rio de Janeiro, Brazil.

Lunne, T., Powell, J.J.M. & Robertson, P.K. (1996) Use of piezocone tests in non-textbook soils. In Craig, C. (Ed.) *Proceedings, International Conference on Advances in Site Investigation Practice*, March 30–31, London, Thomas Telford, London, England, pp. 438–451.

Lunne, T., Robertson, P.K. & Powell, J.J.M. (1997) *Cone Penetration Testing*, Blackie, London, England.

Lupien, C., Lefebvre, G., Rosenberg, P., Pare, J.J. & Lavallée, J.G. (1983) The use of fabrics for improving the placement of till on peat foundation. *62nd Annual Meeting of The Transportation Research Board*, January 17–21. Washington, DC.

Lupini, J.F., Skinner, A.E. & Vaughan, P.-R. (1981) The drained residual strength of cohesive soils. *Géotechnique*, **31**(2): 181–213.

Luster, J. (1986) Physical loss of stabilizers from polymers. In Scott, G. (Ed.) *Developments in Polymer Stabilization – 2*, Applied Science, London, pp. 185–240.

Lutenegger, A.J., DeGroot, D.J., Mirza, C. & Bozozuk, M. (1995) *Recommended Guidelines for Sealing Geotechnical Exploratory Holes*, Transportation Research Board, National Research Council, National Cooperative Highway Research Program, NCHRP Report 378.

Lyman, W.J. & Nomaw, D.C. (1990) *Assessing UST Corrective Action Technologies: Site Assessment and Selection of Unsaturated Zone Treatment Technologies*, Report EPA/600/2-90/011, US Environmental Protection Agency, Cincinnati, OH, 107 p.

Lyman, W.J., Reehl, W.F. & Rosenblatt, D.H.

(1982) *Handbook of Chemical Property Estimation Methods: Environmental Behavior of Organic Compounds*, McGraw-Hill, New York.

Lysmer, J. & Kuhlemeyer, R.L. (1969) Finite dynamic model for infinite media. *Journal of Engineering Mechanics Division, ASCE*, Reston, VA, **95**(4): 859–877.

Lysmer, J., Udaka, T., Tsai, C.F. & Seed, H.B. (1975) *FLUSH–A computer program for approximate 3-D analysis of soil structure interaction problems*, Report No. EERC 75–30, Earthquake Engineering Research Centre, University of California, Berkeley.

MacFarlane, I.C. (Ed.) (1969) *Muskeg Engineering Handbook*, University of Toronto Press, Toronto, ON.

MacFarlane, I.C. & Rutka, A. (1959) Evaluation of road performance over muskeg in Ontario. In *Proceedings, 14th Convention of the Canadian Good Roads Association*, Vancouver, BC, pp. 396–405.

Mackay, D.M. & Vogel T.M. (1985) Groundwater contamination by organic chemicals: uncertainties in assessing impact, Hitchon, B. & Trudell, M. (Eds), *Second Canadian/American Conference on Hydrogeology*. National Water Well Association, Dublin, OH, pp. 50–59.

Mackay, D.M., Freyberg, D.L. & Roberts, P.V. (1986) A natural gradient experiment on solute transport in a sand aquifer. 1. Approach and overview of plume movement. *Water Resources Research*, **22**(13): 2017–2029.

Mackay, D., Shiu, W.Y. & Ma, K.C. (1992–1997) *Illustrated Handbook of Physical–Chemical Properties of Environmental Fate for Organic Chemicals*, Lewis, Boca Raton, FL.

Mackay, D.M. & Cherry, J.A. (1989) Groundwater contamination: limits of pump-and-treat remediation. *Environmental Science & Technology*, **23**(6): 630–636.

MacKay, P. (1997) *MESc thesis*, University of Western Ontario, London, Canada.

Madsen, F.T. & Mitchell, J.K. (1988) Chemical effects on clay fabric and hydraulic conductivity. In Baccini, P. (Ed.) *The Landfill, Reactor and Final Storage*. Lecture Notes in Earth Sciences 20, Berlin, Springer, pp. 201–251.

Mage, D.T. (1982) An objective graphical method for testing normal distributional assumptions using probability plots. *The American Statistician*, **36**(2): 116–120.

Magnan, J.P. (1987) Prediction and behavior of embankments – Settlements and improvements (State-of-the-art report). *International*

Symposium on Geotechnical Engineering of Soft Soils, Mexico City, Vol. 2, pp. 87–123.

Magnan, J.P. (1992) Le rôle du fluage dans les calculs de consolidation et de tassement des sols compressibles. *Bulletin de Liaison des Laboratoires des Ponts et Chaussées*, **180**: 19–24.

Magnan, J.-P. & Deroy, J.-M. (1980) Analyse graphique des tassements observés sous les ouvrages. *Bulletin de Liaison des Laboratoires des Ponts et Chaussées*, **109**: 45–52.

Maidment, D.R. (1993) *Handbook of Hydrology*, McGraw-Hill, New York.

Maine Bureau of Remediation & Waste Management (1997) *An Assessment of Landfill Cover System Barrier Layers Hydraulic Performance*, Augusta, ME.

Mair, R.J. (1987) Discussion on tunnel face pressures in soft clay. In *Proceedings, Eighth Asian Regional Conference on Soil Mechanics & Foundation Engineering*, Kyoto, Japan, Vol. 2, p. 290.

Mair, R.J. & Wood, D.M. (1987) *Pressuremeter Testing Methods and Interpretation*, Construction Industry Research and Information Association (CIRIA), Ground Engineering Report: *In-Situ* Testing. Butterworths, London, England.

Mair, R.J., Taylor, R.N. & Bracegirdle, A. (1993) Subsurface settlement profiles above tunnels in clays. *Géotechnique*, **43**(2): 315–320.

Major, A. (1962) *Dynamics in Civil Engineering*, Akademical Kiado, Budapest, Vols. I–IV.

Makris, N. & Gazetas, G. (1992) Dynamic pile–soil–pile interaction. Part II: Lateral and seismic response. *Earthquake Engineering & Structural Dynamics*, **21**: 145–162.

Mandel, J. & Salençon, J. (1972) Force portante d'un sol sur une assise rigide (étude théorique). *Géotechnique*, **22**(1): 79–93.

Mansur, C.I. & Kaufman, R.I. (1962) Dewatering. In Leonards, G.A. (Ed.) *Foundation Engineering*, McGraw-Hill, New York, pp. 241–350.

Marchetti, S. (1980) *In situ* tests by flat dilatometer. *Journal of Geotechnical Engineering, ASCE*, Reston, VA, **106**(3): 299–321.

Marchetti, S. & Crapps, D.K. (1981) *Flat Dilatometer Manual*, Gainesville, FL.

Marcuson, W.F., Hynes, M.E. & Franklin, A.G. (1992) Seismic stability and permanent deformation analyses: the last twenty five years. In Seed, R.B. & Boulanger, R.W. (Eds) *Proceedings, ASCE Specialty Conference on Stability and Performance of Slopes and Embankments –*

II, Geotechnical Special Publication No. 31, American Society of Civil Engineers, Reston, VA, Vol. I, pp. 552–592.

Marks, P.J., Wujcik, W.J. & Loncar, A.F. (1994) *Remediation Technologies Screening Matrix and Reference Guide*, 2nd Edn, NTIS PB95-104782, *Department of Defense Environmental Technology Transfer Committee*, Washington, DC, 461 p.

Marle, C.M. (1981) *Multiphase Flow in Porous Media*, Gulf Publishing, Houston, TX.

Marley, M.C. & Hoag, G.E. (1984) Induced soil venting for recovery/restoration of gasoline hydrocarbons in the vadose zone. In *Proceedings, Petroleum Hydrocarbons & Organic Chemicals in Ground Water*, November 5–7, Houston, TX, National Ground Water Association, pp. 473–503.

Marley, M.C., Hazebrouck, D.J. & Walsh, M.T. (1992) The application of *in situ* air sparging as an innovative soils and ground water remediation technology. *Ground Water Monitoring Review*, **12**(2): 137–145.

Marshall, T.J. (1958) A relation between permeability and size distribution of pores. *Journal of Soil Science*, **9**: 1–8.

Marsland, A. (1986) The choice of test methods in site investigation. In Hawkins, A.B. (Ed.) *Site Investigation Practice: Assessing BS 5930*. Engineering Geology Special Publication No. 2, Geological Society, London, England, pp. 289–297.

Marston, A. & Anderson, A.O. (1913) The theory of loads on pipes in ditches and tests on cement and clay drain, tile and sewer pipe. Bulletin 31, Engineering Experimental Station, Iowa State College.

Martian, P. (1994) *Calibration of HELP Version 2.0 and Performance Assessment of Three Infiltration Barrier Designs for Hanford Site Remediation*, Draft A, DOE/RL-93-33, Idaho National Engineering Laboratory, Idaho Falls.

Martin, G.R. & Dobry, R. (1994) Earthquake site response and seismic code provisions. *NCEER Bulletin*, **8**(4): 1–6, Buffalo, NY.

Martin, G.R., Finn, W.D.L. and Seed, H.B. (1975) Fundamentals of liquefaction under cyclic loading. *Journal of the Geotechnical Engineering Division, ASCE*, Reston, VA, **101**(GT5): 423–438.

Martin, J.P., Koerner, R.M. & Whitty, J.E. (1984) Experimental friction evaluation of slippage between geomembranes, geotextiles, & soils. In *Proceedings, International Conference on*

Geomembranes, Denver, CO, Industrial Fabrics Association International, Roseville, MN, pp. 191–196.

Massarsch, K.R. (1975) New method for measurement of lateral earth pressure in cohesive soils. *Canadian Geotechnical Journal*, **12**(1): 142–146.

Massarsch, K.R. (1986) Lateral earth pressure at rest in clay. In *Proceedings, Fourth International Geotechnical Seminar*, Singapore, pp. 203–209.

Massarsch, K.R., Holtz, R.D., Holm, B.G. & Fredriksson, A. (1975) Measurement of horizontal *in-situ* stresses. In *Proceedings, ASCE Geotechnical Engineering Division Specialty Conference on In-Situ Measurement of Soil Properties*, June 1–4, North Carolina State University, Raleigh, NC, Vol. 1, pp. 266–286.

Masuda, K., Saseki, F., Urao, K. Veno, K. & Miyamoto, Y. (1986) Simulation analysis of forced vibration test of actual pile foundation by thin layer method. In *Proceedings, Annual Meeting of Architectural Institute of Japan*, Architectural Institute of Japan.

Matar, M. & Salençon, J. (1977) Capacité portante d'une semelle filante sur sol purement cohérent d'épaisseur limitée et de cohésion variable avec la profondeur. *Annales de l'Institut Technique du Bâtiment et des Traveaux Publics, Sols et Fondations*, **352**: 95–107.

Matasovic, N. (1993) *PhD dissertation*, Department of Civil Engineering, University of California, Los Angeles, NV.

Matasovic, N. & Kavazanjian, E., Jr (1998) Cyclic characterization of OII landfill waste. *Journal of Geotechnical & Geoenvironmental Engineering, ASCE*, Reston, VA, **124**(3): 186–196.

Matasovic, N. & Vucetic, M. (1995) Seismic response of soil deposits composed of fully saturated clay and sand layers. In *Proceedings, First International Conference on Earthquake Geotechnical Engineering*, Tokyo, Japan, Vol. 1, pp. 611–616.

Matasovic, N., Kavazanjian, E., Jr & Yan, L. (1997) Newmark deformation analysis with degrading yield acceleration. In *Proceedings, Geosynthetics '97 Conference*, Long Beach, Industrial Fabrics Association International, Roseville, MN, Vol. 2, pp. 989–1000.

Matasovic, N., Kavazanjian, E., Jr. & Anderson, R.L. (1998) Performance of solid waste landfills in earthquakes. *Earthquake Spectra*, **14**(2): 319–334.

Matheron, G. (1973) The intrinsic random functions and their applications. *Advances Appl. Prob.*, **5**: 439–468.

Matheson, L.J. & Tratnyek, P.G. (1994) Processes affecting remediation of contaminated groundwater by dehalogenation with iron. *Environmental Science & Technology*, **28**(11): 2045–2053.

Mathis, H. & Munson, W.E. (1987) *Retaining Wall Failure at Northern Ditch*, Report, Geotechnical Branch, Kentucky Department of Highways, Frankfort.

Matlock, H. (1970) Correlations for design of laterally loaded piles in soft clay. Proceedings, Second *Offshore Technology Conference*, Houston, TX, Vol. 1, pp. 577–594.

Matlock, H. & Foo, S.C. (1980) Axial analysis of piles using a hysteretic and degrading soil model. *Proceedings, Conference on Numerical Methods in Offshore Engineering*, Institution of Civil Engineers, London, pp. 165–185.

Matlock, H., Foo, H.C. & Bryant, L.M. (1978) Simulation of lateral pile behaviour under earthquake motion. *Proceedings, ASCE Speciality Conference on Earthquake Engineering & Soil Dynamics*, Pasadena, CA, Vol. 2, pp. 1065–1084.

Matsui, T. (1993) Case studies on cast-in-place bored piles and some considerations for design. *Proceedings BAP II*, Ghent, Balkema, Rotterdam, pp. 77–102.

Matsuo, H. & Ohara, S. (1960) Dynamic porewater pressure acting on quay walls during earthquakes. *Proceedings, Third World Conference on Earthquake Engineering*, New Zealand, Vol. 1, pp. 130–140.

Matsuzawa, H., Ishibashi, I. & Kawamura, M. (1985) Dynamic soil and water pressures of submerged soils. *Journal of Geotechnical Engineering*, ASCE, Reston, VA, **111**(10): 1161–1176.

Maugeri, M. & Motta, E. (1991) Stresses on piles used to stabilize landslides. In D. Bell, (Ed.) *Landslides*, Balkema, Rotterdam, pp. 785–790.

Mayhew, H.C. & Harding, H.M. (1986) *The Design of Concrete Roads*, TRRL RR87, Transport & Road Research Laboratory, Crowthorne, England.

Mayne, P.W. & Kulhawy, F.H. (1982) K_0–OCR relationships in soil. *Journal of Geotechnical Engineering Division, ASCE*, Reston, VA, **108**(6): 851–872.

Mayne, P.W. & Rix, G.J. (1993) G_{max}–q_c relationships for clays. *Geotechnical Testing Journal, ASTM*, West Conshohocken, PA, **16**(1): 54–60.

Mayne, P.W., Robertson, P.K. & Lunne, T. (1998) Clay stress history evaluated from seismic piezocone tests. In Robertson, P.K. & Mayne, P.W. (Eds) *Proceedings, First International Conference on Site Characterization–ISC'98*, April 19–22, Atlanta, GA, Vols. 1 & 2, Balkema, Rotterdam, pp. 1113–1118.

McBean, E.A. & Rovers, F.A. (1984) Alternatives for handling detection limit data in impact assessments. *Ground Water Monitoring Review*, Spring **4**: 42–44.

McBean, E.A. & Rovers, F.A. (1998) *Statistical Procedures for Analysis of Environmental Monitoring Data and Risk Assessment*. Prentice Hall, NJ.

McBean, E., Kompter, M. & Rovers F. (1988) A critical examination of approximations implicit in Cochran's procedure. *Ground Water Monitoring Review*, Winter, **8**: 83–87.

McBean, E., Rovers, F. & Schmidtke, K. (1990) Risk assessments using relatively simple mathematical models. In McTernan, W.F. & Kaplan, E. (Eds.) *Risk Assessment for Groundwater Pollution Control*, American Society of Civil Engineers, Reston, VA.

McBean, E.A., Mosher, F.R. & Rovers, F.A. (1993) Reliability-based design for leachate collection systems. In Christensen, T.H., Cossu, R. & Stegmann, R. (Eds), *Proceedings, Fourth International Landfill Symposium*, S. Margherita di Pula, Sardinia, pp. 433–441.

McBean, E.A., Rovers, F.A. & Farquhar, G.J. (1995) *Solid Waste Landfill Engineering & Design*, Prentice Hall, Englewood Cliffs, NJ.

McCaulou, D.R., Jewett, D.G. & Huling, S.G. (1995) *Ground Water Issue, Non-Aqueous Phase Liquids Compatibility with Materials Used in Well Construction, Sampling and Remediation*, EPA/540/5-95/503. Office of Solid Waste and Emergency Response, Washington, DC.

McCarthy, J.F. & Zachara, J.M. (1989) Subsurface transport of contaminants. *Environmental Science & Technology*, **23**(5): 496–502.

McCarty, P.L., Reinhard, M. & Rittmann, B.E. (1981) Trace organics in groundwater. Environmental Science & Technology, **15**(1): 40–51.

McCarty, P.L. (1996) Biotic and abiotic transformation of chlorinated solvents in groundwater. In *Symposium on Natural Attenuation of Chlorinated Organics in Groundwater*, September 11–13, Dallas, TX, 540/R-96-509, US Environmental Protection Agency, pp. 5–9.

McClelland, B. (1972) Techniques used in soil sampling at sea. *Offshore*, **32**(3): 51–57.

McDonald, L.A. (1997) Status of risk assessment of dams in Australia. In *Proceedings, International Workshop on Risk Based Dam Safety Evaluations*, Trondheim, Norway, Norwegian National Committee of the International Committee on Large Dams.

McEnroe, B.M. (1993) Maximum saturated depth over landfill liner. *Journal of Environmental Engineering*, **119**(2): 262–270.

McGrath, T.J. (1998) "Pipe-soil interactions during backfill placement" *PhD thesis*, The University of Massachusetts, Amherst.

McGrath, T.J. & Kurdziel, J.B. (1991) SPIDA method for reinforced concrete pipe design. *Journal of Transportation Engineering, ASCE*, Reston, VA, **117**(4): 371–381.

McKay, L.D. & Trudell, M.R. (1989) The sorption of trichloroethylene in clayey till. *Symposium of Groundwater Contamination*, June 14–15, Saskatoon, Saskatchewan, Paper E-40.

McKelvey, J.A. & Deutsch, W.L. (1991) The effect of equipment loading and tapered soil layers on geosynthetic lined landfill slopes. In *Proceedings, 14th Annual Madison Waste Conference*, University of Wisconsin, Madison, pp. 395–411.

McNichols, R.J. & Davis, C.B. (1988) Statistical issues and problems in ground water detection monitoring at hazardous waste facilities, *Ground Water Monitoring Review*, Fall, **8**: 135–150.

McTernan, W.F. & Kaplan, E. (Eds) (1990) *Risk Assessment for Groundwater Pollution Control*, American Society of Civil Engineers, Reston, VA.

McTrans, *McTrans ELSYM5*, University of Florida, Transportation Research Center, Gainesville, FL (*http://www-mcrans.ce.ufl.edu/*).

McVay, M.C., Townsend, F.C. & Williams, R.C. (1992) Design of socketed drilled shafts in limestone. *Journal of Geotechnical Engineering, ASCE*, Reston, VA, **118**(10): 1626–1637.

Meek, J.W. & Veletsos, A.S. (1974) Simple models for foundations in lateral and rocking motions. *Proceedings, Fifth World Conference on Earthquake Engineering*, Rome, Vol. 2, pp. 2610–2613.

Meek, J.W. & Wolf, J.P. (1992a) Cone models for homogeneous soil. *Journal of Geotechnical Engineering, ASCE*, Reston, VA, **118**(4): 667–685.

Meek, J.W. & Wolf, J.P. (1992b) Cone models for soil layer on rigid rock. *Journal of Geotechnical Engineering, ASCE*, Reston, VA, **118**(4): 686–703.

Meek, J.W. & Wolf, J.P. (1994) Cone models for an embedded foundation. *Journal of Geotechnical Engineering, ASCE*, Reston, VA, **120**(1): 60–80.

Mejia, L.H. & Seed, H.B. (1983) Comparison of 2-D and 3-D dynamic analysis of earth dams. *Journal Geotechnical Engineering, ASCE*, Reston, VA, **109**(11): 1383–1398.

Melchior, S. (1997) *In-situ* studies on the performance of landfill caps. In *Proceedings, International Containment Technology Conference*, February, St Petersburg, FL, pp. 365–373.

Mellor, M. & Hawkes, I. (1971) Measurement of tensile strength by diametral compression of disks and annuli. *Engineering Geology*, **5**: 173–225.

MELT (1993) *Regles techniques de conception et de calcul des fondations des ouvrages de genie civil*. CCTG, Fascicule No. 62, Titre V, Min. de L'Equipement du Lodgement et des Transport, Paris.

MEND (1993) *Soilcover User's Manual for an Evaporative Flux Model*, University of Saskatchewan, Saskatoon, Canada.

Mendoza, C.A. & McAlary, T.A. (1989) Modeling of groundwater contamination caused by organic solvent vapors. *Ground Water*, **28**(2): 199–206.

Mercer, F.B., Andrawes, K.Z., McGown, A. & Hytiris, N. (1984) A new method of soil stabilisation. In *Polymer Grid Reinforcement*, Thomas Telford, London, pp. 224–249.

Mercer, J.W. & Cohen, R.M. (1990) A review of immiscible fluids in the subsurface: properties, models, characterization, and remediation, *Journal of Contaminant Hydrology*, **6**: 107–163.

Mesri, G. (1975) Discussion on 'New design procedure for stability of soft clays'. *Journal of Geotechnical Engineering Division, ASCE*, Reston, VA, **101**(4): 409–412.

Mesri, G. (1987) The fourth law of soil mechanics: the law of compressibility. *International Symposium on Geotechnical Engineering of Soft Soils*, Mexico City, Vol. 2, pp. 179–187.

Mesri, G. & Choi, Y.K. (1985) Settlement analysis of embankments on soft clays. *Journal of Geotechnical Engineering Division, ASCE*, Reston, VA, **111**(4): 441–464.

Mesri, G. & Feng, T.-W. (1991) Surcharging to reduce secondary settlements. In *Proceedings, Geo-Coast '91*, Yokohama, pp. 359–364.

Mesri, G. & Godlewski, R.M. (1977) Time and stress compressibility interrelationship. *Journal of Geotechnical Engineering Division, ASCE*, Reston, VA, **103**(5): 417–430.

Mesri, G., Lo, D.O.D. & Feng, T.W. (1994) Settlement of embankments on soft clays. In *Proceedings, Conference on Vertical & Horizontal Deformations of Foundations & Embankments: Settlement '94*, Geotechnical Special Publication No. 40, American Society of Civil Engineers, Reston, VA, Vol. 1, pp. 8–56.

Mesri, G., Shahien, M. & Feng, T.W. (1995) Compressibility parameters during primary consolidation. *International Symposium on Compression & Consolidation of Clayey Soils IS-Hiroshima '95*, Hiroshima, Vol. 2, pp. 1021–1037.

Mesri, G., Stark, T.D., Ajlouni, M.A. & Chen, C.S. (1997) Secondary compression of peat with or without surcharging. *J of Geotechnical Geoenvironmental Engineering, ASCE*, Reston, VA, **123**(5): 411–421.

Mesri, G., Stark, T.D., Ajlouni, M.A. & Chen, C.S. (1999) Closure to discussion of Secondary compression of peat with or without surcharging. *J of Geotechnical Geoenvironmental Engineering, ASCE*, Reston, VA, **125**(2): 164–165.

Meyerhof, G.G. (1953) The bearing capacity of foundations under eccentric and inclined loads. In *Proceedings, Third International Conference on Soil Mechanics & Foundation Engineering*, Zürich, Vol. 1, pp. 440–445.

Meyerhof, G.G. (1956) Penetration tests and bearing capacity of cohesionless soils. *Journal of Soil Mechanics & Foundations Division, ASCE*, Reston, VA, **82**(1): 1–19.

Meyerhof, G.G. (1963) Some recent research on the bearing capacity of foundations. *Canadian Geotechnical Journal*, **1**(1): 16–26.

Meyerhof, G.G. (1974) Ultimate bearing capacity of footings on a sand layer overlying clay. *Canadian Geotechnical Journal*, **11**: 223–229

Meyerhof, G.G. (1976) Bearing capacity and settlement of pile foundations. *Journal of Geotechnical Engineering Division, ASCE*, Reston, VA, **102**(3): 195–228.

Meyerhof, G.G. (1995) Behaviour of pile foundations under special loading consitions. 1994 R.M. Hardy keynote address. *Canadian Geotechnical Journal*, **32**(2): 204–222.

Meyerhof, G.G. & Adams, J.I. (1968) The ultimate uplift capacity of foundations. *Canadian Geotechnical Journal*, **5**(4): 225–244.

Meyerhof, G.G. & Sastry, V.V.R.N. (1978) Bearing capacity of piles in layered soils: Part I & Part II. Canadian Geotechnical Journal, **15**(2): 171–189.

Michalowski, R.L. (1998) Limit analysis in stability calculations of reinforced soil structures. *Geotextiles* & *Geomembranes*, **16**(6): 311–331.

Michalowski, R.L. & Shi, L. (1995) Bearing capacity of footings over two-layer foundation soils. *Journal Geotechnical Engineering Division, ASCE*, Reston, VA, **121**(5): 421–428.

Middendorp, P., Bermingham, P. & Kuiper, B. (1992) Statnamic load testing of foundation piles. In *Proceedings, Fourth International Conference on Application. of Stress Wave Theory to Piles*, The Hague, Balkema, Rotterdam, pp. 581–588.

Middleton, T. & Cherry, J.A. (1996) The effects of chlorinated solvents on the permeability of clays. In Pankow, J.F. & Cherry J.A. (Eds) *Dense Chlorinated Solvents and Other DNAPLs in Groundwater*, Waterloo Educational Services Inc, Rockwood, Ontario, pp. 313–336.

Mikkelsen, P.E. (1996) Field instrumentation. In Turner, A.K. and Schuster, R.L. (Eds) *Landslides: Investigation and Mitigation*, Special Report 247, Transportation Research Board, Washington, DC, 278–316.

Miller, C.T., Poirier-McNeill, M.M. & Mayer, A.S. (1990) Dissolution of trapped nonaqueous phase liquids: mass transfer characteristics. *Water Resources Research*, **26**(11): 2783–2796.

Miller, R.D. (1972) Freezing and heaving of saturated and unsaturated soils. *Highway Research Record*, **393**: 1–11.

Milligan, G.W.E. & Ripley, K.J. (1989) Packing materials in jacked pipe joints. *Proceedings, Fourth International Conference on Trenchless Construction for Utilities, NO-DIG '89*, April, London, International Society for Trenchless Technology, London, UK, pp. 39–48.

Milligan, V. (1975) Field measurement of permeability in soil and rock. In *Proceedings, ASCE Geotechnical Engineering Division Specialty Conference on* In-Situ *Measurement of Soil Properties*, June 1–4, North Carolina State University, Raleigh, NC Vol. 2, pp. 3–36.

Ministères des Transports (1979) *Direction générale des transports interieurs-les ouvrages en Terre Armée–recommendations et règles de l'art*, Paris.

Mitchell, J.K. (1978) *In-situ* techniques for site characterization. In *Proceedings, ASCE Geotechnical Engineering Division Specialty Conference on Site Characterization and Exploration*, Northwestern University, Evanston, IL, pp. 107–129.

Mitchell, J.K. (1981) Soil improvement — State-of-the-art report. In *Proceedings, Tenth International Conference on Soil Mechanics* & *Foundation Engineering*, Stockholm, Vol. 4, pp. 506–565.

Mitchell, J.K. (1986a) Practical problems form surprising soil behavior, 20th Terzaghi lecture. *Journal of Geotechnical Engineering, ASCE*, Reston, VA, **112**(3): 259–289.

Mitchell, J.K. (1986b) Ground improvement evaluation by *in-situ* tests. In *Proceedings, In Situ '86*, Blacksburg, VA, Geotechnical Special Publication No. 6, American Society of Civil Engineers, ASCE Press, Reston, VA, pp. 221–236.

Mitchell, J.K. (1988) New developments in penetration tests and equipment, In De Ruiter, J. (Ed.) *Proceedings, First International Symposium on Penetration Testing, ISOPT-1*, March 20–24, Orlando, FL, A.A. Balkema, Rotterdam, Vol. 1, pp. 245–261.

Mitchell, J.K. (1991) Conduction phenomena: from theory to geotechnical practice. *Géotechnique*, **41**(3): 299–340.

Mitchell, J.K. (1993) *Fundamentals of Soil Behavior*, 2nd Edn, Wiley, New York.

Mitchell, J.K. & Christopher, B.R. (1990) North American practice in reinforced soil systems. In *Design and Performance of Earth Retaining Structures*, Lambe, P.C. and Hansen, L.A. (Eds.) American Society of Civil Engineers, Reston, VA, pp. 322–346.

Mitchell, J.K. & Gardner, W.S. (1975) *In-situ* measurement of volume change characteristics. In *Proceedings, ASCE Geotechnical Engineering Division Specialty Conference on* In-Situ *Measurement of Soil Properties*, June 1–4, North Carolina State University, Raleigh, NC, Vol. 2, pp. 279–345.

Mitchell, J.K. & Solymar, Z.V. (1984) Time-dependent strength gain in freshly deposited or densified sand. *Journal of Geotechnical Engineering, ASCE*, Reston, VA, **110**(11): 1559–1576.

Mitchell, J.K. & van Court, W.A.N. (1997) Barrier design and installation: walls and covers. In Ward, C.H., Cherry, J.A. & Scalf, M.R. (Eds.) *Subsurface Restoration*, Ann Arbor Press, Chelsea, MI, pp. 175–197.

Mitchell, J.K. & Villet, W.C.B. (1987) *Reinforcement of Earth Slopes and Embankments*, National Cooperative Highway Research Program Report 290, Transportation Research Board, National Research Council, Washington, DC.

Mitchell, J.K., Hooper, D.R. & Campanella, R.G. (1965) Permeability of compacted clay. *Journal of the Soil Mechanics & Foundation Division, ASCE*, Reston, VA, **91**(4): 41–65.

Mitchell, J.K., Baxter, C.D.P. & Munson, T.C. (1995) Performance of improved ground during earthquakes. In *Soil Improvement for Earthquake Hazard Mitigation*, Geotechnical Special Publication No. 49, R.D. Hryciw (Ed.) American Society of Civil Engineers, Reston, VA, pp. 1–36.

Mitchell, R.J. & Eden, W.J. (1972) Measured movements of clay slopes in the Ottawa area. *Canadian Journal of Earth Sciences*, **9**: 1001–1013.

Mizuno, H. & Iiba, M. (1992) Dynamic effects of backfill and piles on foundation impedance. In *Proceedings, 10th World Conference on Earthquake Engineering*, Madrid, Spain, Vol. 3, pp. 1823–1828.

MoEE (1993) *Interim Guidelines for the Assessment and Management of Petroleum Contaminated Sites in Ontario*, Ministry of Environment & Energy, Toronto, Ontario.

MoEE (1994) *Proposed Guideline for the Cleanup of Contaminated Sites in Ontario*, Ministry of Environment & Energy, Toronto, Ontario.

MoEE (1996a) The challenge of contaminated sites: remediation approaches in North America. In *Guidance on Sampling & Analytical Methods for Use at Contaminated Sites in Ontario*, Ontario Ministry of Environment & Energy, Standards Development Branch, Toronto, Ontario.

MoEE (1996b) *Guidance on Sampling and Analytical Methods for Use at Contaminated Sites in Ontario*, Ontario Ministry of Environment and Energy Standards Development Branch, Toronto, ON.

MoE (1997) *Guideline for Use at Contaminated Sites in Ontario*. Revised February 1997, Ontario Ministry of Environment, and supporting documents, Toronto, Ontario.

MoE (1998) *Landfill Standards: A Guideline on the Regulatory and Approval Requirements for New or Expanding Landfilling Sites*, and Ontario Regulation 232/98, *Ontario Gazette*, pp. 131–22, 983–996.

Molenaar, A.A.A. (1983) *PhD dissertation*, Delft University of Technology, Delft, Holland.

Mononobe, N. & Matsuo, H. (1929) On the determination of earth pressures during earthquakes. *Proceedings, World Engineering Congress*, 9 p.

Montgomery, J.H. & Welkom, L.M. (1990) *Groundwater Chemicals Desk Reference*, Lewis, Chelsea, MI.

Montgomery, R.J. & Parsons, L.J. (1989) The Omega Hills final cover test plot study: three-year data summary. *Annual Meeting of the National Solid Waste Management Association*, Washington, DC.

Moore, C.A. (1983) *Landfill and Surface Impoundment Performance Evaluation*, SW-869, US EPA, Cincinnati, OH.

Moore, I.D. (1987) The elastic stability of shallow buried tubes. *Géotechnique*, **37**(2): 151–161.

Moore, I.D. (1988) Static response of deeply buried elliptical tubes. *Journal of Geotechnical Engineering, ASCE*, Reston, VA, **114**(6): 672–687.

Moore, I.D. (1989) Elastic buckling of buried flexible tubes – a review of theory and experiment. *Journal of Geotechnical Engineering, ASCE*, Reston, VA, **115**(3): 340–358.

Moore, I.D. (1990) Three dimensional response of elastic tubes. *Solids & Structures*, **26**(4): 391–400.

Moore, I.D. (1994) *Profiled HDPE Pipe Response Under Parallel Plate Loading*, Special Technical Publication 1222, Buried Plastic Pipe Technology, American Society for Testing & Materials, West Conshohocken, PA, **2**: 25–40.

Moore, I.D. & Brachman, R.W.I. (1994) Three dimensional analysis of flexible circular cylinders. *Journal of Geotechnical Engineering, ASCE*, Reston, VA, **120**(10): 1829–1844.

Moore, I.D. & Laidlaw, T.C. (1997) Corrugation buckling in HDPE pipes – measurements and analysis. Paper No. 97-0565, *TRB Annual Meeting*, Washington, DC.

Moore, I.D., Haggag, A. & Selig, E.T. (1994) Buckling strength of flexible cylinders with nonuniform elastic support. *Solids & Structures*, **31**(22): 3041–3058.

Morel, F. (1983) *Principles of Aquatic Chemistry*, Wiley, New York.

Moretrench American Corporation (1967) *Dewatering Systems Applicable to Different Soils*, Moretrench American Corporation, Rockaway, NJ.

Morgan, G.C. (1997) A regulatory perspective on slope hazards and associated risks to life. In *Proceedings, International Workshop on Landslide Risk Assessment*, Honolulu, HI, pp. 285–295.

Morgan, T.A. & Panek, L.A. (1963) *A method for determining stress in rock*. Report of Investigations No. 6312, Bureau of Mines, United States Department of the Interior, Denver, Colorado.

Morgenstern, N.R. (1992) Keynote Paper. The role of analysis in the evaluation of slope stability. In *Proceedings, Sixth International Symposium on Landslides*, Christchurch, Vol. 3, pp. 1615–1629.

Morgenstern, N.R. (1995) Managing risk in geotechnical engineering. In *Proceedings, Tenth Panamerican Conference on Soil Mechanics & Foundation Engineering*, Guadalajara, Vol. 4.

Morgenstern, N.R. & Nixon, J.F. (1971) One dimensional consolidation of thawing soils. *Canadian Geotechnical Journal*, **8**(4): 558–565.

Morgenstern, N.R. & Price, V.E. (1965) The analysis of the stability of general slip surfaces. *Géotechnique*, **15**(1): 79–93.

Morgenstern, N.R. & Sego, D.C. (1981) Performance of temporary tie-backs under winter conditions. *Canadian Geotechnical Journal*, **18**: 566–572.

Morin, P., Leroueil, S. & Samson, L. (1983) Preconsolidation pressure of Champlain clays, Part I—*in situ* determination. *Canadian Geotechnical Journal*, **20**(4): 782–802.

Morin, W.J. & Todor, P.C. (1977) *Laterites and Lateritic Soils and the Problem Soils of the Tropics*, US Agency for International Development, A10/csd 3682.

Moriwaki, Y., Beikae, M. & Idriss, I.M. (1988) Nonlinear seismic analysis of the upper San Fernando Dam under the 1971 San Fernando Earthquake. In *Proceedings, Ninth World Conference on Earthquake Engineering*, Tokyo & Kyoto, Japan, Vol. III, pp. 237–241.

Morrison, R.T. & Boyd, R.N. (1992) *Organic Chemistry*, 6th Edn, Prentice Hall, Englewood Cliffs, NJ, USA.

Moseley, M.P. (Ed.) (1993) *Ground Improvement*, Blackie Academic & Professional, Blackwood, NJ.

Moseley, M.P. & Priebe, H.J. (1993) Vibro techniques. In Moseley, M.P. (Ed.) *Ground Improvement*, Blackie Academic & Professional, Blackwood, NJ, pp. 1–19.

Mroz, Z., Norris, V.A. & Zienkiewicz, O.C. (1981) An anisotropic, critical state model for soils subject to cyclic loading. *Géotechnique*, **31**(4): 451–469.

MTO (1990) *Pavement Design & Rehabilitation Manual*, Publication No. 500-90-01, Ministry of Transportation of Ontario, Downsview, Ontario, Canada.

Mualem, Y. (1976a) A new model for predicting the hydraulic conductivity of unsaturated porous media. *Water Resources Research*, **12**: 513–522.

Mualem, Y. (1976b) Hysteretical models for prediction of the hydraulic conductivity of unsaturated porous media. *Water Resources Research*, **12**: 1248–1254.

Mualem, Y. (1986) Hydraulic conductivity of unsaturated soils: prediction and formulas. In Klute, A. (Ed.) Methods of Soil Analysis, No. 9, Part 1, American Society of Agronomy, Madison, WI, pp. 799–823.

Munfakh, G.A. (1997a) Ground improvement engineering–the state of the US practice: Part 1. methods. *Ground Improvement*, **1**(4): 193–214.

Munfakh, G.A. (1997b) Ground improvement engineering–the state of the US practice: Part 2. applications. *Ground Improvement*, **1**(4): 215–222.

Munfakh, G.A., Abramson, L.W., Barksdale, R.D. & Juran, I. (1987) *In-situ* ground reinforcement. In *Soil Improvement–A Ten Year Update*. Geotechnical Special Publication No. 12, American Society of Civil Engineers, Reston, VA, pp. 1–67.

Murchison, J.M. & O'Neill, M.W. (1984) Evaluation of *P–Y* relationships in cohesionless soils. In Meyer, J.R. (Ed.) *Analysis & Design of Pile Foundations*, American Society of Civil Engineers, Reston, VA, pp. 174–191.

Murdoch, L., Patterson, B. Lasonsky, G. & Harrar, W. (1990) *Technologies of Delivery or Recovery for the Remediation of Hazardous Waste Sites*, Report EPA/600/2-89/066 US Environmental Protection Agency, Cincinnati, OH, 96 p.

Murphy, J.R. (1996) *Understanding and Evaluating Natural Attenuation, Application at Metamora Landfill Site*, Poster Presentation at NGWA/API Petroleum Hydrocarbons & Organic Chemicals in Groundwater Conference, November 13–15, Houston, TX.

Murphy, W.L. & Gilbert, P.A. (1985) *Settlement and Cover Subsidence of Hazardous Waste Landfill*, EPA/600/2-85/035, US Environmental Protection Agency Hazardous Waste Engineering Research Laboratory, Cincinnati.

Murray, R.T. & Irwin, M.J. (1981) *A preliminary Study of TRRL Anchored Earth*, TRRL Supplementary Report 674, Transport & Road Research Laboratory, Crowthorne, UK.

Musser, S.C. (1996) Utah DOT's testing program to determine the soil–structure interaction of pile groups under lateral loads. In *Proceedings, Fourth Caltrans Seismic Research Workshop*, California Department of Transportation, Sacramento.

Mutch, R.D., Ash, R.E. & Caputi, J.R. (1997) New technologies for subsurface barrier wall construction. In *Barrier Technologies for Environmental Management* (Summary of a Workshop), National Academy Press, Washington, DC, D-23-34.

Myles, B. (1993) Practical Aspects of Soil Nailing, Amerad Jord, Svenska Geotechniska Foreningen, Uppsala, September.

Mylleville, B.L.J. & Rowe, R.K. (1988) Some considerations in the design of geosynthetic reinforced embankments on clayey foundations. In *Proceedings, Third Canadian Symposium on Geosynthetics*, October, 29–33, Kitchener, Ontario.

Mylleville, B.L.J. & Rowe, R.K. (1991) On the design of reinforced embankments on soft brittle clays. In *Proceedings, Geosynthetics 91*, February, Atlanta, GA, Industrial Fabrics Association International, Roseville, MN, Vol. 1, pp. 395–408.

Mylonakis, G., Nikolaou, A. & Gazetas, G. (1997) Soil–pile–bridge seismic interaction: kinematic and inertial effects. Part I soft soil. *Journal of Earthquake Engineering & Structural Dynamics*, **26**(3): 337–359.

Myrand, D., Gillham, R.W., Cherry, J.A. & Johnson, R.L. (1987). *Diffusion of Volatile Organic Compounds in Natural Clay Deposits*, University of Waterloo, Waterloo, Ontario, Canada.

Naeim, F. & Lew, M. (1995) On the use of design spectrum compatible motions. *Earthquake spectra*, **III**(1): 111–128.

Nagase, A. (1967) The $\phi = 0$ analysis of stability and unconfined compression strength. *Soils & Foundations*, **11**(4): 35–45.

Narejo, D.B. (1995a) A comparison of puncture behavior of smooth and textured HDPE geomembranes. *Geosynthetics International*, **2**(4): 699–705.

Narejo, D.B. (1995b) Three levels of geomembrane puncture protection. *Geosynthetics International*, **2**(4): 765–769.

Narejo, D.B. (1996) Closure of discussion of 'A comparison of puncture behavior of smooth and textured HDPE geomembranes' and 'Three levels of geomembrane puncture protection.' *Geosynthetics International*, **3**(3): 443–444.

Narejo, D., Koerner, R.M. & Wilson-Fahmy, R.F. (1996) Puncture protection of geomembranes. Part II: experimental. *Geosynthetics International*, **3**(5): 629–653.

Narin Van Court, W.A. & Mitchell, J.K. (1995) New insights into explosive compaction of loose, saturated cohesionless soils. In *Soil Improvement for Earthquake Hazard Mitigation*, Geotechnical Special Publication No. 49, American Society of Civil Engineers, R.D. Hryciw (Ed.), Reston, VA, pp. 51–65.

Nataraj, M.S., Magauti, R.S. & McManis, K.L. (1995). Interface frictional characteristics of Geosynthetics. In *Proceedings, Geosynthetics '95 Conference*, Nashville, TN, Industrial Fabrics Association International, Roseville, MN, Vol. 3, pp. 1057–1069.

Nauroy, J.F., Bruey, F., Le Tirant, P. & Kervadec, J.-P. (1986) Design and installation of piles in calcareous formations. In *Proceedings, Third International Conference on Numerical Methods in Offshore Piling*, Nantes, pp. 461–480.

NavFac (1982) *Soil Mechanics. Design Manual 7.1.* Department of the Navy Naval Facilities Engineering Command, Alexandria, VA, US Government Printing Office, Washington, DC.

NavFac (1983) *Soil Dynamics, Deep Stabilization, & Special Geotechnical Construction, Design Manual 7.3*, US Navy Facilities Command, Alexandria, VA, pp. 77–78.

Nayak, N.V. & Christensen, R.W. (1971). Swelling characteristics of compacted, expansive soils. *Clays & Clay Minerals*, **19**(4): 251–261.

Naylor, D.J. & Hooper, J.A. (1974) An effective stress finite element analysis to predict the short and long term behaviour of a piled raft foundation on London clay. In *Proceedings, Conference On Settlement of Structures*, Cambridge, pp. 394–402.

NCEER (1997) Proceedings of the NCEER workshop on evolution of liquefaction resistance of soils. T.L. Voud & I. Idriss (Eds.), Technical Report, NCEER-7-0022.

Neely, W.B. (1985) Hydrolysis. In Neely, W.B. & Blau, G.E. (Eds) *Environmental Exposure from Chemicals*, Vol. 1, CRC Press, Boca Raton, FL, pp. 157–173.

Nelson, J.D. & Miller, D.J. (1992) *Expansive Soils. Problems and Practice in Foundation and Pavement Engineering*, Wiley Interscience, New York.

Nelson, W. & Hahn, G.J. (1973) Linear estimation of a regression relationship from censored data–Part II: best linear unbiased estimation and theory. *Technometrics*, **15**(1): 133–150.

Nersesova, Z. & Tsytovich, N.A. (1963) Unfrozen water in frozen soils. In *Proceedings, First International Conference on Permafrost*, Purdue University, Lafayette, IN, pp. 230–234.

Newell, C., Acree, S.D. Ross, R.R. & Hulling, S.G. (1995) *Ground Water Issue: Light Nonaqueous Phase Liquids*, EPA/540/S-95/500. Office of Solid Waste and Emergency Response, Washington, DC.

Newmark, N. (1965) Effects of earthquakes on Dams and Embankments, Geotechnique, 15(2): 139–160.

Newmark, N.M. (1965) Effects of earthquakes on dams and embankments. *Géotechnique*, **15**(2): 139–160.

New South Wales (1996) *Design of Reinforced Soil Walls*, RTA QA Specification R57, New South Wales Road and Traffic Authority, Surry Hills, NSW, Australia.

Nichols, W.E. (1991) *Comparative Simulations of a Two-Layer Landfill Barrier Using the HELP Version 2.0 & UNSAT-H Version 2.0 Computer Codes*, PNL-7583, Pacific Northwest Laboratory, Richland, Washington, DC.

Nicholson, R.V., Cherry, J.A. & Reardon, E.J. (1983) Migration of contaminants at a landfill: a case study, 6. Hydrogeochemistry. *Journal of Hydrology* 63(1/2): 131–176.

Nicholson, R.V., Gillham, R.W., Cherry, J.A. & Reardon, E.J. (1989) Reduction of acid generation in mine tailings through the use of moisture-retaining cover layers as oxygen barriers. *Canadian Geotechnical Journal*, **26**(1): 1–8.

Nielsen, D.M. (1991) *Practical Handbook of Ground-Water Monitoring*, Lewis, Chelsea, MI.

Nixon, J.F. (1991) Discrete ice lense theory for frost heave in soils. *Canadian Geotechnical Journal*, **28**(6): 843–859.

Noether, G.E. (1981) Comment. *The American Statistician*, **35**(3): 129.

Nogami, T. (1980) Dynamic stiffness and damping of pile groups in inhomogeneous soil. In *Proceedings Session on Dynamic Response of Pile Foundations: Analytical Aspects, ASCE National Convention*, October, FL, pp. 31–52.

Nogami, T. & Novak, M. (1976) Soil–pile interaction in vertical vibration. *International Journal of Earthquake Engineering & Structural Dynamics*, **4**(3): 277–293.

Noorany, I. (1972) Underwater soil sampling and testing–a state-of-the-art review. In *Underwater Soil Sampling, Testing & Construction Con-*

trol, Special Publication 501, American Society for Testing & Materials, West Conshohocken, PA, pp. 3–41.

Nordal, R.S. & Refsdal, G. (1989) Frost protection in design and construction. *VTT Symposium '95, Frost in Geotechnical Engineering*, Saariselka, Finland, Vol. 1, pp. 127–164.

Nordlund, R.L. (1963) Bearing capacity of piles in cohesionless soils. *Journal of Soil Mechanics & Foundational Division, ASCE*, Reston, VA, **89**(3): 1–35.

Norris, P. & Milligan, G.W.E. (1992) Frictional resistance of jacked concrete pipes at full scale. *Proceedings International Conference on Trenchless Technology, NO-DIG 92*, Paris, Balkema, Rotterdam, pp. 121–128.

Norrish, N.I. & Wyllie, D.C. (1996) Rock slope stability analysis. *Landslides – Investigation & Mitigation*, A.K. Turner & R.L. Schuster (Eds.) Special Report 247, Transportation Research Board, Washington, DC, pp. 391–425.

Norton, D. & Knapp, R. (1977) Transport phenomena in hydrothermal systems: Nature of porosity. *American Journal of Science*, **27**: 913–936.

Novak, M. (1974) Dynamic stiffness and damping of piles. *Canadian Geotechnical Journal*, **11**: 574–598.

Novak, M. (1979) Soil–pile interaction under dynamic loads. In *Proceedings, International Symposium on Numerical Methods in Off-Shore Pilling*, London, England, pp. 41–50.

Novak, M. & Aboul-Ella, F. (1978) Impedance functions of piles in layered media. *Journal of the Engineering Mechanics Division, ASCE*, Reston, VA, **104**(3): 643–661.

Novak, M. & Beredugo, Y.O. (1972) Vertical vibration of embedded footings. *Journal of Soil Mechanics & Foundations Division, ASCE*, Reston, VA, **98**(SM12): 1291–1310.

Novak, M. & El-Sharnouby, B. (1983) Stiffness and damping constants of single piles. *Journal of Geotechnical Engineering Division, ASCE*, Reston, VA, **109**(7): 961–974.

Novak, M. & El-Sharnouby, B. (1984) Evaluation of dynamic experiments on pile group. *Journal of Geotechnical Engineering, ASCE*, Reston, VA, **110**(6): 738–756.

Novak, M. & Grigg, R.F. (1976) Dynamic experiments with small pile foundations. *Canadian Geotechnical Journal*, **13**(4): 372–385.

Novak, M. & Kim, T.C. (1981) Resonant column technique for dynamic testing of cohesive soils. *Canadian Geotechnical Journal*, **18**(3): 448–455.

Novak, M. & Mitwally, H. (1987) Random response of offshore towers with pile–soil–pile interaction. In *Proceeding, Sixth International Symposium on Offshore Mechanics & Arctic Engineering*, Houston, TX, Vol. 1, pp. 329–336.

Novak, M. & Nogami, T. (1977) Soil-pile interaction in horizontal vibration. *International Journal of Earthquake Engineering & Structure Dynamics*, 5(3): 263–282.

Novak, M. & Sachs, K. (1973) Torsional and coupled vibrations of embedded footings. *International Journal of Earthquake Engineering & Structure Dynamics*, 2(1): 11–33.

Novak, M. & Sheta, M. (1980) Approximate approach to contact problems of piles. In *Proceedings, Geotechnical Engineering Division, ASCE National Convention, Dynamic Response of Pile Foundations: Analytical Aspects*, October, FL, pp. 53–79.

Novak, M., Nogami, T. & Aboul-Ella, F. (1978) Dynamic soil reactions for plane stain case. *Journal of the Engineering Mechanics Division, ASCE*, Reston, VA, 104(4): 953–959.

Novak, M., El Naggar, M.H., Sheta, M., El-Hifnawy, L., El-Marsafawi, H. & Ramadan, O. (1999) *DYNA5 a Computer Program for Calculation of Foundation Response to Dynamic Loads*, Geotechnical Research Centre, The University of Western Ontario, London, Ontario.

NRC (1994) National Research Council *Alternatives for Ground Water Cleanup*, National Academy Press, Washington, DC.

NRC (1985) National Research Council, *Liquefaction of Soils During Earthquakes*, National Academy Press, Washington, DC.

NRC (1996) National Research Council, *Rock Fractures & Fluid Flow: Contemporary Understanding & Applications*, National Academy Press, Washington, DC.

NRCC (1995a) *National Building Code of Canada*, National Research Council of Canada, Ottawa.

NRCC (1995b) *User's Guide – NBC 1995, Structural Commentaries (Part 4)*, National Research Council of Canada, Ottawa.

Nunn, M.E. & Merrill, D.B. (1997) *Review of Flexible and Composite Design Methods*, TRL Paper PA/3298/97, Transport Research Laboratory, Crowthorne, Berkshire, UK.

Nyhan, J.W. (1989) *Hydrologic Modeling to Predict Performance of Shallow Land Burial Cover Designs at the Los Alamos National Laboratory*, LA-11533-MS, Los Alamos National Laboratory, Los Alamos, NM.

Oda, M. (1972) Initial fabrics and their relations to mechanical properties of granular materials. *Soils & Foundations*, 12(1): 17–37.

O'Brien, K. & Keyes, G. (1997) Implementation of a funnel-and-gate remediation system. *Proceedings, International Containment Technology Conference* (US Department of Energy/Dupont Company/US Environmental Protection Agency), February 9–12, St Petersburg, FL, pp. 895–901.

O'Connor, M.J., Agar, J.G. & King, R.D. (1984) Practical experience in the management of hydrocarbon vapours in the subsurface. In *Proceedings, Petroleum Hydrocarbons & Organic Chemicals in Ground Water*, November 5–7, Houston, TX, National Ground Water Association, pp. 519–533.

O'Donnell, S.A., Rumar, R.R. & Mitchell, J.K. (1995) *Assessment of Barrier Containment Technologies. A Comprehensive Treatment for Environmental Remedial Applications*, Report PB96-180583.

O'Hannesin, S.F. & Gillham, R.W. (1998) Long-term performance of an *in situ* iron wall for remediation of VOCs. *Ground Water*, 36(1): 164–170.

Ohsaki, Y. & Iwasaki, R. (1973) On dynamic shear moduli and Poisson's ratio of soil deposits. *Soils & Foundations*, 13(4): 61–73.

Oliveira, R. & Graca, J.C. (1987) *In situ* testing of rocks. In Bell, F.G. (Ed.) *Ground Engineer's Reference Book*, Butterworths, London, England, Chapter 26.

Okabe, S. (1926) General theory of earth pressures. *Journal Japan Society of Civil Engineering*, Tokyo, 12(1).

Okamoto, S. (1973) *Introduction to Earthquake Engineering*, University of Tokyo Press, Tokyo, Japan.

O'Kane, M., Wilson, G.W., Barbour, S.L. & Swanson, D.A. (1995) Aspects on the performance of the till cover system at Equity Silver Mines Ltd. *Proceedings, Sudbury 95 – Mining & the Environment*, May 28–June 1, Sudbury, Ontario, Canada, Vol. 2, pp. 565–573.

Olson, R.E. (1974) Shearing strength of kaolinite, illite, & montmorillonite. *Journal of Geotechnical Engineering Division, ASCE*, Reston, VA, 100(11): 1215–1229.

O'Neil, D.A., Baldi, G. & Della Torre, A. (1996) The multi-functional Envirocone® test system. In Craig C. (Ed.) *Proceedings, International Conference on Advances in Site Investigation*

Practice, March 30–31, London, Thomas Telford, London, England, pp. 421–437.

O'Neill, M.W. (1983) Group action in offshore piles. In *Proceedings, ASCE Conference of Geotechnical Practice in* Offshore Engineering, Austin, TX, pp. 25–64.

O'Neill, M.W. & Ha, H.B. (1982) Comparative modelling of vertical pile groups. In *Proceedings, Second International Conference on Numerical Methods in Offshore Piling*, Austin, TX, pp. 399–418.

O'Neill, M.W., Ghazzaly, O.I. & Ha, H.B. (1977) Analysis of three-dimensional pile groups with nonlinear soil response and pile–soil–pile interaction. In *Proceedings, Ninth OTC*, Houston, TX, Paper OTC 2838, pp. 245–256.

Onodera, T.F. (1963) Dynamic investigation of foundation rocks *in situ*. In *Proceedings Fifth US Rock Mechanics Symposium*, University of Minnesota, pp. 517–533.

Ontario MBS (1993) *Contaminant Recognition and Management*, Environmental Advisory Services Unit, Realty Group, Toronto, ON.

OPSD (1994) *Ontario Provincial Standards for Roads and Municipal Services*, Vol. 3, *Drawings for Roads, Barriers, Drainage, Sanitary Sewers, Watermains and Structures*, OPSD 509 and 510 Series, Ontario Provincial Standards, St Catharines, Ontario.

O'Reilly, M.P. & New, B.M. (1982) Settlements above tunnels in the United Kingdom – their magnitude and prediction. In *Proceedings, Tunnelling 82 Symposium*, Institute of Mining and Metallurgy, London, England, pp. 173–181.

O'Reilly, M.P. & Rogers, C.D.F. (1987) Pipe jacking forces. In *Proceedings, International Conference on Foundations & Tunnels*, Edinburgh, Engineering Technics Press, Edinburgh, Scotland, pp. 201–208.

O'Rourke, T.D. (1987) Lateral stability of compressible walls. *Géotechnique*, **37**(2): 145–149.

O'Rourke, T.D. & Jones, C.J.F.P. (1990) Overview of earth retention systems 1970–1990. In *Design & Performance of Earth Retaining Structures*, Lambe, P.C. and Hansen, L.A. (Eds.) American Society of Civil Engineers, Reston, VA, pp. 22–51.

Orth, W.S. & Gillham, R.W. (1996) Dechlorination of trichloroethene in aqueous solution using Fe°. *Environmental Science & Technology*, **30**: 66–71.

Ortigao, J.A.R. (1980) *DSc, thesis*, Federal University of Rio de Janeiro.

Ostendorf, D.W., Moyer, E.E., Richards, R.J., Hinlein, E.S., Xie, Y. & Rajan, R.V. (1992) *LNAPL Distribution and Hydrocarbon Vapor Transport in the Capillary Fringe*, EPA/600/R-92/247.

Osterberg, J.O. (1957) Influence values for vertical stresses in a semi-infinite mass due to an embankment loading. In *Proceedings, Fourth International Conference on Soil Mechanics and Foundation Engineering*, London, Vol. 1, pp. 393–394.

Osterberg, J.O. (1978) Failures in exploration programs. In *Proceedings, ASCE Geotechnical Engineering Division Specialty Conference on Site Characterization & Exploration*, Northwestern University, Evanston, IL, pp. 3–9.

Osterberg, J. (1989) New device for load testing driven and drilled shafts separates friction and end bearing. In *Proceedings, International Conference on Piling & Deep Foundations*, London, pp. 421–427.

Osterberg, J.O. & Gill, S.A. (1973) Load transfer mechanism for piers socketed in hard soils or rock. In *Proceedings, Ninth Canadian Rock Mechanics Symposium*, Montreal, PQ, pp. 235–262.

Othman, M.A. & Benson, C.H. (1993) Effect of freeze–thaw on the hydraulic conductivity and morphology of compacted clay. *Canadian Geotechnical Journal*, **30**(2): 236–246.

Othman, M.A., Bonaparte, R., Gross, B.A. & Schmertmann, G.R. (1995) Design of MSW landfill final cover systems. In *Landfill Closures: Environmental Protection & Land Recovery*, Geotechnical Special Publication No. 53, Dunn, R.J. and Singh U.P. (Eds.) American Society of Civil Engineers, Reston, VA, pp. 218–257.

Ottaviani, M. (1975) Three-dimensional finite element analysis of vertically loaded pile groups. *Géotechnique*, **25**(2): 159–174.

Owen, W.J. & DeRouen, T.A. (1980) Estimation of the mean for lognormal data containing zeroes and left-censored values, with application to the measurement of worker exposure to air contaminants. *Biometrics*, **36**: 707–719.

Palmeira, E.M., Pereira, J.H.F. & da Silva, A.R.L. (1998) Backanalyses of geosynthetic reinforced embankments on soft soils. *Geotextiles & Geomembranes*, **16**(5): 273–292.

Palmer, J.H.L. & Lo, K.Y. (1976) *In-situ* measurements in some near-surface rock formations— Thorold, Ontario. *Canadian Geotechnical Journal*, **13**(1): 1–7.

Pande, G.N. & Pietruszczak, S. (1982) Reflecting surface model for soils. In *Proceedings, International Symposium on Numerical Models in Geomechanics*, Balkema, Rotterdam, The Netherlands, pp. 50–64.

Pankow, J.F., Johnson, R.L. & Cherry, J.A. (1993) Air sparging in gate wells in cutoff walls and trenches for control of plumes of volatile organic compounds (VOCs). *Ground Water*, **31**(4): 654–663.

Pankow, J.F., Feenstra, S., Cherry, J.A. & Ryan, C. (1996) Dense chlorinated solvents in groundwater: background and history of the problem. In Pankow, J.F. & Cherry, J.A. (Eds) *Dense Chlorinated Solvents & Other DNAPLs in Groundwater*, Waterloo Press, Rockwood, Ontario, pp. 1–52.

Paolucci, R. & Pecker, A. (1997) Seismic bearing capacity of shallow strip foundations on dry soils. *Soils & Foundations*, **37**(3): 95–105.

Papagiannakis, A.T. & Schwartz, C.W. (Eds) (1998) *Application of Geotechnical Principles in Pavement Engineering*, Geotechnical Special Publication No. 85, American Society of Civil Engineers, Reston, VA.

Park, J.K., Sakti, J.P. & Hooper J.A. (1995) Effectiveness of geomembranes as barriers of organic compounds. In *Proceedings, Geosynthetics 95*, February, Industrial Fabrics Association International, Roseville, MN, IFAI, Vol. 3, pp. 879–892.

Parry, R.H.G. (1968) Field and laboratory behaviour of a lightly overconsolidated clay. *Géotechnique*, **18**(2): 151–171.

Parry, R.H.G. (1977) Estimating bearing capacity in sand from SPT values. *Journal of Geotechnical Engineering Division*, ASCE, Reston, VA, **103**(9): 1014–1019.

Pasqualini, E., Roccato, M. and Sani, D. (1993) Shear resistance of the interfaces of composite liners. In *Proceedings, Sardinia '93 Fourth International Landfill Symposium*, Cagliari, Italy, Vo. 2, pp. 1457–1471.

Paterson, W.D.O. (1987) *Road Deterioration and Maintenance Effects: Models for Planning & Management*, Johns Hopkins University Press, Baltimore, Maryland, MD.

Patrick, G.C. & Burgess, A.S. (1992) Field studies of non-aqueous phase liquids at two sites in California's Santa Clara Valley. In Weyer, K.U. (Ed.) *Subsurface Contamination by Immiscible Fluids*, Balkema, Rotterdam, pp. 489–501.

Patton, F.D. (1966) Multiple mode of shear failure in rock. In *Proceedings, First Congress of the International Society for Rock Mechanics*, Lisbon, Vol. 1, pp. 509–513.

Paulin, M.J., Phillips, R., Clark, J.I., Trigg, A. & Konuk, I. (1998) A full scale investigation into pipeline/soil interaction. *51st Canadian Geotechnical Conference*, Edmonton, AB, Vol. 1, pp. 241–248.

PCA (1984) *Thickness Design for Concrete Highway and Street Pavements*, Portland Cement Association, Skokie, IL.

PCA (1992) *Design of Concrete Pavements for City Streets*, Information Series 184, Portland Cement Association, Skokie, IL.

Peck, R.B. (1969a) Advantages and limitations of the observational method in applied soil mechanics. Ninth Rankine Lecture. *Géotechnique*, **19**(2): 171–187.

Peck, R.B. (1969b) Deep excavations and tunnelling in soft ground. In *Proceedings, Seventh International Conference, International Society for Soil Mechanics and Foundation Engineering*, Mexico.

Peck, R.B. (1973) Influence of non-technical factors on the quality of embankment dams. In *Embankment-Dam Engineering, Casagrande Volume*, R.C. Hirschfield & S.S. Poulos (Eds.), Wiley, New York, pp. 201–208.

Peck, R.B. (1976) Rock foundations for structures. In *Proceedings, Rock Engineering for foundations & Slopes*, August 15–18, University of Colorado, Boulder, CO, Vol. 2, pp. 1–21.

Peck, R.B., Hanson, W.E. & Thornburn, T.H. (1974) *Foundation Engineering*, 2nd Edn, Wiley, New York,

Pecker, A. (1996) Seismic bearing capacity of shallow foundations. In *Proceedings, 11th World Conference Earthquake Engineering*, Paper No. 2076, Acapulco, Mexico.

Pells, P.J.N. & Turner, R.M. (1979) Elastic solutions for the design and analysis of rock-socketed piles. *Canadian Geotechnical Journal*, **16**: 481–487.

Pender, M.J. (1996) *Aseismic Foundation Design Analysis*, University of Auckland, New Zealand.

Penner, E. (1974) Uplift forces on foundations in frost heaving soils. *Canadian Geotechnical Journal*, **11**(3): 323–338.

Penner, E. & Goodrich, L.E. (1980) Location of segregated ice in frost susceptible soil. In *Proceedings, Second International Symposium on Ground Freezing*, Trondheim, Norway, pp. 626–639.

Penrose, W.R., Polzer, W.L., Essington, E.H.,

Nelson, D.M. & Orlandini, K.A. (1990) Mobility of plutonium and americium through a shallow aquifer in a semiarid region. *Environmental Science & Technology*, **24**: 228–234.

Penzien, J., Scheffey, C.F. & Parmelee, R.A. (1964) Seismic analysis of bridges on long piles. *Journal of the Engineering Mechanics Division, ASCE*, Reston, VA, **90**(3): 223–254.

Pereira, A.T. (1977) *Procedures for Development of CBR Design Curves*, Instruction Report 5-77-1, US Army Corps of Engineers, Vicksburg, MS.

Pessaran, A. (1995) *Hydraulic Characterization of Sand*, M. Eng. Report, University of Saskatchewan, Saskatoon, Canada.

Peters, N., Warner, R.S., Coates, A.L., Logsdon, D.S. & Grube, W.E. (1986) Applicability of the HELP Model in Multilayer Cover Design: A Field Verification and Modeling Assessment. In *Land Disposal of Hazardous Waste–Proceedings, 1986 Research Symposium*, US Environmental Protection Agency, Cincinnati.

Peterson, M.E. & Landis, R.C. (1995) Artificially emplaced floors and bottom barriers. In Rumer, R.R. & Mitchell, J.K. (Eds) *Assessment of Barrier Containment Technologies: A Comprehensive Treatment for Environmental Remediation Applications*, US Department of Energy/US Environmental Protection Agency/Dupont Company, pp. 185–209.

Petrov, R.J. & Rowe, R.K. (1997) Geosynthetic clay liner compatibility by hydraulic conductivity testing: Factors impacting performance. *Canadian Geotechnical Journal*, **34**(6): 863–885.

Petrov, R.J., Rowe, R.K. & Quigley, R.M. (1997a) Comparison of laboratory measured GCL hydraulic conductivity based on three permeameter types. *Geotechnical Testing Journal*, **20**(1): 49–62.

Petrov, R.J., Rowe, R.K. & Quigley, R.M. (1997b) Selected factors influencing GCL hydraulic conductivity. *Journal of Geotechnical & Geoenvironmental Engineering, ASCE*, Reston, VA, **123**(8): 683–695.

Peyton, R.L. & Schroeder, P.R. (1988) Field verification of HELP model for landfills. *Journal of Environmental Engineering, ASCE*, Reston, VA, **114**(2): 247–269.

Peyton, R.L. & Schroeder, P.R. (1993) Water balance for landfills. In Daniel, D.E. (Ed.) *Geotechnical Practice for Waste Disposal*, Chapman & Hall, London, pp. 214–243.

PFRA (1951) *South Saskatchewan River Project: Report on Geological Test Drift-Dam Site 10*, Prairie Farm Rehabilitation Administration, Department of Regional Economic Expansion, Canada.

Pickens, J.F., Grisak, G.E., Avis, J.D., Belanger, D.W. & Thury, M. (1987) Analysis and interpretation of borehole hydraulic tests in deep boreholes: principles, model development and applications. *Water Resources Research*, **23**: 1341–1376.

Pillai, V.S. & Byrne, P.M. (1994) Effect of overburden pressure on liquefaction resistance of sand. *Canadian Geotechnical Journal*, **31**(1): 53–60.

Pilot, G. (1972) Study of five embankments on soft soils. In *Proceeding, ASCE Specialty Conference on Performance of Earth & Earth-Supported Structures*, Purdue University, Lafayette, IN, Vol. 1, 81–99.

Pilot, G. & Moreau, M. (1973) *La stabilité des remblais sur sols mous*, Abaques de calcul, Eyrolles, Paris.

Pilot, G., Trak, B. & La Rochelle, P. (1982) Effective stress analysis of the stability of embankments on soft soils. *Canadian Geotechnical Journal*, **19**(4): 433–450.

PJA (1995) *Guide to Best Practice for the Installation of Pipe Jacks and Microtunnels*, Pipe Jacking Association, London, England.

Piwoni, M.D., & Keeley, J.W. (1990) *Ground Water Issue: Basic Concepts of Contaminant Sorption at Hazardous Waste Sites*, EPA/540/4-90/053. Office of Solid Waste and Emergency Response, Washington, DC.

Pohlman, K.F., Icopini, G.A., McArthur, R.D. & Rosal, C.G. (1994) *Evaluation of sampling and field-filtration methods for the analysis of trace metals in ground water*, EPA/600/R-94/119.

Polo, J.M. & Clemente, J. (1988) Pile group settlement using independent shaft and point loads. *Journal of Geotechnical Engineering, ASCE*, Reston, VA, **114**(4): 469–487.

Polshin, D.E. & Tokar, R.A. (1957) Maximum allowable non-uniform settlement of structures. In *Proceedings, Fourth International Conference on Soil Mechanics & Foundation Engineering*, Vol. 1, pp. 402–405.

Poorooshasb, H.B. (1991) Load settlement response of a compacted fill layer supported by a geosynthetic overlying a void. *Geotextiles & Geomembranes*, **10**(3): 179–201.

Poorooshasb, H.B. & Pietruszczak, S. (1985) On yielding and flow of sand; a generalized two-surface model. *Computers & Geotechnics*, **1**(1): 33–58.

Porter, L.K., Kemper, W.D., Jackson, R.D. & Stewart, B.A. (1960) Chloride diffusion in soils as influenced by moisture content. *Soil Science Society of America Proceedings*, **24**: 460–463.

Potts, D.M., Kovacevic, N. & Vaughan, P.R. (1997) Delayed collapse of cut slopes in stiff clay. *Géotechnique*, **47**(5): 953–982.

Poulos, H.G. (1968) Analysis of the settlement of pile groups. *Géotechnique*, **18**: 449–471.

Poulos, H.G. (1971a) Behaviour of laterally loaded piles: I-single piles. *Journal of Soil Mechanics & Foundation Division, ASCE*, Reston, VA, **97**(5): 711–731.

Poulos, H.G. (1971b) Behaviour of laterally loaded piles: II-pile groups. *Journal of Soil Mechanics & Foundation Division, ASCE*, Reston, VA, **97**(5): 733–751.

Poulos, H.G. (1972) Difficulties in prediction of horizontal deformations of foundations. Journal of Soil Mechanics & Foundation Division, *ASCE*, Reston, VA, **98**(8): 843–848.

Poulos, H.G. (1973) Load–deflection prediction for laterally loaded piles. *Australian Geomechanics Journal*, **3**(1): 1–8.

Poulos, H.G. (1977) Estimation of pile group settlement. *Ground Engineering*, March: **10**: 40–50.

Poulos, H.G. (1979a) Settlement of single piles in non-homogeneous soil. *Journal of Geotechnical Engineering Division, ASCE*, Reston, VA, **105**(5): 627–641.

Poulos, H.G. (1979b) Group factors for pile deflection estimation. *Journal of Geotechnical Engineering Division, ASCE*, Reston, VA, **105**(5): 627–641.

Poulos, H.G. (1982a) Developments in the analysis of static and cyclic lateral response of piles. In *Proceedings, Fourth International Conference on Numerical Methods in Geomechanics*, Edmonton, Vol. 3, pp. 1117–1135.

Poulos, H.G. (1982b) Single pile response to cyclic lateral load. *Journal of Geotechnical Engineering, ASCE*, Reston, VA, **108**(3): 355–375.

Poulos, H.G. (1985) Ultimate lateral pile capacity in two-layer soil. *Geotechnical Engineering*, **16**(1): 25–37.

Poulos, H.G. (1987) Piles and piling. In Bell, F.G. (Ed.) *Ground Engineer's Reference Book*, Butterworths, London, England.

Poulos, H.G. (1988a) Cyclic stability diagram for axially loaded piles. *Journal of Geotechnical Engineering, ASCE*, Reston, VA, **114**(8): 877–895.

Poulos, H.G. (1988b) *Marine geotechnics*, Unwin Hyman, London.

Poulos, H.G. (1989) Pile behaviour-theory and application. *Géotechnique*, **39**(3): 365–415.

Poulos, H.G. (1993) Settlement of bored pile groups. In *Proceedings BAP II*, Ghent, Balkema, Rotterdam, W.F. van Impe (Ed.) pp. 103–117.

Poulos, H.G. (1994a) Settlement prediction for driven piles and pile groups. In *Vertical & Horizontal Deformations of Foundations & Embankments*, Geotechnical Special Publication No. 40, American Society of Civil Engineers, Reston, VA, Vol. 2, pp. 1629–1649.

Poulos, H.G. (1994b) Analysis and design of piles through embankments. In *Proceedings International Conference on Design & Construction of Deep Foundations*, Orlando, FL, Vol. 3, pp. 1403–1421.

Poulos, H.G. (1996) A comparison of methods for the design of piles through embankments. In *Proceedings 12th SE Asian Geotechnical Conference*, Kuala Lumpur, Vol. 2, pp. 157–167.

Poulos, H.G. (1997a) Failure of a building supported on piles. In *Proceedings International Conference On Foundation Failures*, Singapore, Institute of Engineers Singapore, pp. 53–66.

Poulos, H.G. (1997b) Piles subjected to negative friction: a procedure for design. *Geotechnical Engineering*, **28**(1): 23–44.

Poulos, H.G. & Chen L. (1996) Pile response due to unsupported excavation-induced lateral soil movements. Canadian Geotechnical Journal, **33**: 670–677.

Poulos, H.G. & Chen, L. (1997) Pile response due to excavation-induced lateral soil movement. *Journal of Geotechnical & Geoenvironmental Engineering, ASCE*, Reston, VA, **123**(2): 94–99.

Poulos, H.G. & Davis, E.H. (1974) *Elastic Solutions for Soils and Rocks*, Wiley, New York. Reprinted 1991 by University of Sydney.

Poulos, H.G. & Davis, E.H. (1980) *Pile Foundation Analysis and Design*, Wiley, New York.

Poulos, H.G. & Hewitt, C.M. (1986) Axial interaction between dissimilar piles in a group. In *Proceedings, Third International Conference on Numerical Methods in Offshore Piling*, Nantes, pp. 253–270.

Poulos, H.G. & Hull, T.S. (1989) The role of analytical geomechanics in foundation engineering. *Foundation Engineering: Current Principles and Practices, ASCE*, Reston, VA, **2**: 1578–1606.

Poulos, H.G. & Mattes, N. (1969) The behaviour

of axially loaded end bearing piles. *Géotechnique*, **19**(2): 185–300.

Poulos, H.G. & Randolph, M.F. (1983) A study of two methods for pile group analysis. *J. Geotechnical Engineering Division, ASCE*, Reston, VA, **109**(3): pp. 355–372.

Powell, J.J.M. & Shields, H.S. (1995) Field studies of the full displacement pressuremeter in clays. In *Proceedings, Fourth International Symposium on Pressuremeters (ISP4)*, May, Sherbrooke, Quebec, Canada, pp. 239–248.

Powell, J.J.M. & Uglow, I.M. (1985) A comparison of Menard, self-boring and push-in pressuremeter tests in a stiff clay till. IN *Proceedings Conference Offshore Site Investigation, Advances in Underwater Technology & Offshore Engineering*, London, UK, Vol. 3, pp. 201–217.

Powell, J.J.M. & Uglow, I.M. (1988) The interpretation of the Marchetti dilatometer test in UK clays. In *Proceedings, Conference on Penetration Testing in the UK*, July, Birmingham, Thomas Telford, London, UK, pp. 269–273.

Powell, R.M., Puls, R.W., Blowes, D.W., Vogan, J.L., Gillham, R.W., Powell, P.D., Schultz, D., Sivavec, T. and Landis, R. (1998) *Permeable Reactive Barrier Technologies for Contaminant Remediation*, US Environmental Protection Agency, US Office of Research & Development, US Office of Solid Waste & Emergency Response, EPA/600/R-98/125. Washington, DC, 94 p.

Powell, W.D., Potter, J.F., Mayhew, H.C. & Nunn, M.E. (1984) *The Structural Design of Bituminous Roads*, Department of Transport, TRRL Report LR1132, Crowthorne, Berkshire, UK.

Powers, J.P. (1991) Dewatering and groundwater control. In Fang, H.Y. (Ed.) *Foundation Engineering Handbook*, 2nd Edn, Van Nostrand Reinhold, New York, NY, pp. 236–248.

Powers, J.P. (1992) *Construction Dewatering–New Methods and Applications*, 2nd Edn, Wiley, New York, NY.

Powrie, W. (1997) *Soil mechanics–Concepts and Applications*, E. & F.N. Spon, London, UK.

Prakash, S. & Puri, V.K. (1988) *Foundations for Machines: Analysis and Design*, Wiley, New York.

Prandtl, L.C. (1920) Uber die harte plastischen korper. *Nachr. K. Ges. Wiss. Gott., Math.-Phys.* **K1**: 74–85.

Pressley, J.S. & Poulos, H.G. (1986) Finite element analysis of mechanisms of pile group be-

haviour. *International Journal of Numerical Analysis Methods in Geomechanics*, **10**: 213–221.

Preston, J.N. (1996) *Analytical Pavement Design*, SPDM-PC, Highways & Transportation-Shell, Delft.

Prevost, J.-H. (1978) Plasticity theory for soil stress–strain behavior. *Journal of Engineering Mechanics Division, ASCE*, Reston, VA, **104**(5): 1177–1194.

Prevost, J.H. (1981) *DYNAFLOW: A Nonlinear Transient Finite Element Analysis Program*, Princeton University, Princeton, NJ.

Prevost, J.H. & Scanlan, R.H. (1983) Dynamic soil–structure interaction: centrifugal modelling. *Journal of Soil Dynamics & Earthquake Engineering*, **2**(4): 212–221.

Pritchard, M.A. & Savigny, K.W. (1990) Numerical modeling of toppling. *Canadian Geotechnical Journal*, **27**(6): 823–834.

Puls, R.W. & Barcelona, M.J. (1989) Filtration of groundwater samples for metals analyses. *Hazardous Waste & Hazardous Materials*, **6**(4): 385–393.

Puls, R.W. & Barcelona, M.J. (1996) *Ground Water Issue: Low-Flow (Minimal Drawdown) Ground Water Sampling Procedures*, EPA/540/5-95/504. Office of Solid Waste and Emergency Response, Washington, DC.

Puls, R.W. & Paul, C.J. (1995) Low-flow purging and sampling of groundwater monitoring wells with dedicated systems. *Groundwater Monitoring & Remediation*, **XV**(1): 116–123.

Puls, R.W. & Powell, R.M. (1992) Transport of inorganic colloids through natural aquifer material: Implications for contaminant transport. *Environmental Science & Technology*, **26**(3): 614–621.

Puls, R.W., Clark, D.A., Bledsoe, B., Powell, R.M. & Paul, C.J. (1992) Metals in groundwater: Sampling artifacts and reproducibility. *Hazardous Waste & Hazardous Materials*, **9**(2): 149–162.

PWRI (1992) *Design and Construction Manual for Reinforced Soil Structures Using Geotextiles*, Internal Report No. 3117, Public Works Research Institute, Ministry of Construction, Tsukuba, Japan (in Japanese).

Pyke, R., Seed, H.B. & Chan, C.K. (1975) Settlement of sands under multi-directional loading. *Journal of Geotechnical Engineering Division, ASCE*, Reston, VA, **101**(4): 379–398.

Qian, J.H., Zhao, W.B., Cheung, W.B. Cheung, Y.K. & Lee, P.K.K. (1992) The theory and practice of vacuum preloading. *Computers & Geotechnics*, **13**: 103–118.

Quigley, R.M., Yanful, E.K. & Fernandez, F. (1987) Ion transfer by diffusion through clay barriers. *Geotechnical Practice for Waste Disposal*, ASCE Press, Reston, VA, 137–158.

Raabe, A.W. & Esters, K. (1993) Soilfracturing techniques for terminating settlements and restoring levels of buildings and structures, In Moseley, M.P. (Ed.) *Ground Improvement*, Blackie Academic & Professional, Blackwood, NJ, pp. 175–192.

Radhakrishnan, R. & Leung, C.F. (1989) Load transfer behavior of rock-socketed piles. *Journal of Geotechnical Engineering*, **115**(6): 755–768.

Randolph, M.F. (1981) Response of flexible piles to lateral loading. *Géotechnique*, **31**(2): 247–259.

Randolph, M.F. (1983) Design considerations for offshore piles. In *ASCE Special Conference on Geotechnical Practice in Offshore Engineering*, Austin, TX, pp. 422–439.

Randolph, M.F. (1992) *Settlement of Pile Groups*, Lecture No. 4, Modern Methods of Pile Design, University of Sydney, Sydney, Aust.

Randolph, M.F. (1994a) *RATZ. Load transfer analysis of axially loaded piles, Users Manual*, University of Western Australia.

Randolph, M.F. (1994b) Design methods for pile groups and piled rafts. In *Proceedings, 13th International Conference Soil Mechanics & Foundations Engineering*, **5**: 61–82.

Randolph, M.F. & Simons, H. (1986) An improved soil model for one-dimensional pile driving analysis. In *Proceedings, Third International Conference on Numerical Methods in Offshore Piling*, Nantes, pp. 3–17.

Randolph, M.F. & Wroth, C.P. (1978) Analysis of deformation of vertically loaded piles. *Journal of Geotechnical Engineering Division*, ASCE, Reston, VA, **104**(12): 1465–1488.

Randolph, M.F. & Wroth, C.P. (1979) An analysis of the vertical deformation of pile groups. *Géotechnique*, **29**(4): 423–439.

Ranganatham, B.V. & Satyanarayana, B. (1965) A rational method of predicting swelling potential for compacted expansive clays. In *Proceedings, Sixth International Conference on Soil Mechanics & Foundation Engineering*, Montreal, Canada, Vol. 1, pp. 92–96.

Rao, P.S.C., Lee, L.S. & Wood, A.L. (1991) *Solubility, Sorption, and Transport of Hydrophobic Organic Chemicals in Complex Mixtures*, Environmental Research Brief, EPA/600/M-91/009. Robert S. Kerr, Environmental Research Laboratories, Ada, OK.

Rapoport, N.Ya., Livanova, N.M. & Miller, V.B. (1977) On the influence of internal stress on the kinetics of oxidation of oriented polypropylene. *Vysokomol. Soyed.* **A18**(9): 2045–2049. (Translated in *Poly. Sci. USSR*, **18**: 2336–2341).

Rasmuson, A. & Eriksson, J.C. (1986) *Capillary Barriers in Covers for Mine Tailings*, National Swedish Environmental Protection Board, Report 3307.

Raulin, P., Rouques, G. & Toubol, A. (1972) *Calcul de la stabilité des pentes en rupture non circulaire*, Research Report No. 36, Laboratoire Central des Ponts et Chaussées, Paris.

Rausch, E. (1950) *Maschinen Fundamente*, VDI-Verlag, Dusseldorf, pp. 107–232.

Rausche, F., Goble, G.G. & Likins, G. (1985) Dynamic determination of pile capacity. *Journal of Geotechnical Engineering*, ASCE, Reston, VA, **111**(3): 367–387.

Raven, K.G. & Beck, P. (1993) Coal tar and creosote contamination in Ontario. In Wayer, K.U. (Ed.) *Subsurface Contamination by Immiscible Fluids*, Balkema, Rotterdam, pp. 401–440.

Raymond, G.P. (1969) Construction method and stability of embankments on muskeg. *Canadian Geotechnical Journal*, **6**: 81–96.

Raymond, G.P. & Giroud, J.P. (1993) *Geosynthetics Case Histories*, ISSMFE Technical Committee TC9 Publication, Bitech, Richmond, BC, Canada.

Reardon, E.J. & Moddle, P.M. (1985) Gas diffusion co-efficient measurements on uranium mill tailings: implications to cover layer design. *Uranium* **2**: 111–131.

Reddy, K.R., Bandi, S.R., Rohr, J.J., Finy, M. & Siebken, J. (1996) Field evaluation of protective covers for landfill geomembrane liners under construction loading. *Geosynthetics International*, **3**(6): 679–700.

Reddy, K.R., Bandi, S.R., Rohr, J.J., Finy, M. & Siebken, J. (1997) Closure of discussions of 'Field evaluation of protective covers for landfill geomembrane liners under construction loading' by Reddy *et al. Geosynthetics, International*, **4**(5): 543–546.

Reese, L.C., O'Neill, M.W. & Smith, R.E. (1970) Generalized analysis of pile foundations. *Journal of Soil Mechanics & Foundational Division*, ASCE, Reston, VA, **96**(1): 235.

Reese, L.C., Cox, W.R. & Koop, F.D. (1974) Analysis of laterally loaded piles in sand. In *Pro-*

ceedings, Sixth Offshore Technology Conference, Houston, TX, paper OTC 2080, pp. 473–483.

Reid, W.M. & Buchanan, N.W. (1984) Bridge approach support piling. In Proceedings, Conference on Piling & Ground Treatment, Thomas Telford, London, England, pp. 267–274.

Reissner, E. (1936) Stationare, axialsymmetrische durch eine schuttelnde masseerregte schwigungen eines homogenen elastischen halbraumes. (in German) Ingenieur-Archiv, 7(6): 381–396.

Reiter, L. (1990) Earthquake Hazard Analysis: Issues and Insights, Columbia University Press, New York.

Richards, B.G. (1965) Measurement of the free energy of soil moisture by the psychometric technique using thermisters. In Moisture Equilibria & Moisture Changes in Soils Beneath Covered Areas, A Symposium in Print, Australia, Butterworths, pp. 35–46.

Richards, L.A. (1931) Capillary conduction of liquids in porous mediums. Physics, 1: 318–333.

Richards, R. & Elms. D. (1979) Seismic behaviour of gravity retaining walls. Journal of the Geotechnical Engineering Division, ASCE, Reston, VA, 105(4): 449–464.

Richardson, G.N., Kavazanjian, E., Jr. & Matasovic, N. (1995) RCRA Subtitle D (258) Seismic Design Guidance for Municipal Solid Waste Landfill Facilities, Report No. EPA/600/R-95/051, US Environmental Protection Agency, Office of Research & Development, Washington, DC.

Richardson, H.W. (1953) Electric curtain stabilizes wet ground for deep excavation. Construction Methods & Equipment, 35(4):52–58.

Richart, F.E. (1975) Foundation vibrations. In Fang, H.Y. (Ed.) Foundation Engineering Handbook, Van Nostrand Reinhold, New York, pp. 673–699.

Richart, F.E., Hall, J.R. & Woods, R.D. (1970) Vibrations of Soils and Foundations, Pentice-Hall, Englewood Cliffs, NJ.

Richter, C.F. (1958) Elementary Seismology, W.H. Freeman, San Francisco, CA.

Rickards, I. (1994) APSDS: a structural design system for airport and industrial pavements. In Proceedings, Ninth Australian Asphalt Pavement Association Conference, Surfers Paradise, Queensland, Australia.

Ridley, A.M. (1993) PhD thesis, Imperial College, London, UK.

Ridley, A.M. & Burland, J.B. (1993) A new instrument for the measurement of soil moisture suction. Géotechnique, 43: 321–324.

Rieke, R., Vinson, T.S., Mageau, D.W. (1983) The role of specific surface area and related index properties in the frost heave susceptibility of soils. In Proceedings, Fourth International Conference on Permafrost, Fairbanks, Alaska, pp. 1066–1071.

Rifai, H.S., Borden, R.C., Wilson, J.T. & Ward, C.H. (1995) Intrinsic bioattenuation for subsurface restoration. In Hinchee, R.E., Wilson, J.R. & Downey, D.C. (Eds) Intrinsic Remediation, Third International In Situ & On-Site Bioreclamation Symposium, Battelle Press, Columbus, OH, Vol. 3(1), pp. 1–30.

Risk Engineering Inc. (1997) EZ-FRISK Computer Program 4.0, Boulder, CO.

Ripley, C.F. & Leonoff, C.E. (1961) Embankment settlement behaviour on deep peat. In Proceedings, Seventh Muskeg Research Conference, Technical Memorandum No. 71, National Research Council of Canada, Ottawa, ON, pp. 185–204.

Ritchie, J.T. (1972) Model for predicting evaporation from a row crop with incomplete cover. Water Resources Research, 8(5): 1204–1213.

Rittmann, B.E., Fleming, I. & Rowe, R.K. (1996) Leachate chemistry: its implications for clogging. In North American Water & Environment Congress '96, June, Anaheim, CA, Paper 4 (CD Rom).

Rivard, P.J. & Lu, Y. (1978) Shear strength of soft fissured clays. Canadian Geotechnical Journal, 15(3): 382–390.

Rivett, M.O. (1993) A field evaluation of pump-and-treat remediation. In Joint CSCE–ASCE National Conference on Environmental Engineering, July 12–14, Montreal, pp. 1171–1178.

Rixner, J.J., Kraemer, S.R. & Smith, A.D. (1986a) Prefabricated Vertical Drains, 1: Engineering Guidelines, Report No. FHWA-RD-86/168. Federal Highway Administration, Springfield, VA.

Rixner, J.J., Kraemer, S.R. & Smith, A.D. (1986b) Prefabricated Vertical Drains, 2: Summary of Research Effort, Report No. FHWA-RD-86/169. Federal Highway Administration, Springfield, VA.

Robert, Y. (1997) A few comments on pile design. Canadian Geotechnical Journal, 34: 560–567.

Roberts, F.L., Kandhal, P.S., Brown, E.R., Lee, D-Y, and Kennedy, T.W. (1996) Hot Mix As-

phalt Materials, Mixture Design & Construction, National Asphalt Pavement Association, Lanham, MD.

Robertson, P.K. (1986) *In-situ* testing and its application to foundation engineering. *Canadian Geotechnical Journal*, **23**(4): 573–594.

Robertson, P.K. (1990) Soil classification using the cone penetration test. *Canadian Geotechnical Journal*, **27**(1): 151–158.

Robertson, P.K. & Campanella, R.G. (1983) Interpretation of cone penetration tests. Part I: Sand; Part II: Clay. *Canadian Geotechnical Journal*, **20**(4): 718–733.

Robertson, P.K. & Campanella, R.G. (1985) Liquefaction potential of sands using the cone penetration test. *Journal of the Geotechnical Division, ASCE*, Reston, VA, **111**(3): 298–307.

Robertson, P.K., Sully, J.P., Woeller, D.J., Lunne, T. & Gillespie, D.G. (1992) Estimating coefficient of consolidation from piezocone tests. *Canadian Geotechnical Journal*, **29**(4): 551–557.

Robertson, P.K., Lunne, T. & Powell, J.J.M. (1996) Application of penetration tests for geo-environmental purposes. In Craig, C. (Ed.) *Proceedings, International Conference on Advances in Site Investigation Practice*, March 30–31, London, Thomas Telford, London, pp. 407–420.

Robertson, P.K., Lunne, T. & Powell, J.J.M. (1998) Geo-environmental applications of penetration testing. In Robertson, P.K. & Mayne, P.W. (Eds) *Proceedings, First International Conference on Site Characterization–ISC'98*, April 19–22, Atlanta GA, Vols. 1 & 2, A.A. Balkema, Rotterdam, pp. 35–48.

Robertson, W.D. & Cherry, J.A. (1995) *In situ* denitrification of septic system nitrate using reactive porous media barriers: field trials. *Ground Water*, **33**(1): 99–111.

Robertson, W.D. & Cherry, J.A. (1997) Long term performance of the Waterloo denitrification barrier. In *Proceedings, International-Containment Technology Conference*, (US Department of Energy/Dupont Company/US Environmental Protection Agency), February 9–12, St Petersburg, FL pp. 691–696.

Robinsky, E.I. & Bespflug, K.E. (1973) Design of insulated foundations. *Journal of Geotechnical Engineering, ASCE*, Reston, VA, **99**(9): 649–667.

Robinson, H.D. & Gronow, J.R. (1993). A review of landfill leachate composition in the UK. In *Proceedings, Fourth International Landfill Symposium*, Sardinia, pp. 821–832.

Robinson, R.A. & Stokes, R.H. (1970) *Electrolyte Solutions*, Butterworths, London, UK.

Rocha, M. & Silverio, A. (1969) A new method for the complete determination of the state of stress in rock masses. *Géotechnique*, **19**: 116–132.

Rogers, C.D.F. & Chapman, D.N. (1995a) Ground movements caused by trenchless pipe installation techniques. In *Transportation Research Record* 1514, Transportation Research Board, Washington, DC pp. 37–48.

Rogers, C.D.F. & Chapman, D.N. (1995b) An experimental study of pipebursting in sand. *Geotechnical Engineering, Proceedings Institution of Civil Engineers*, **113**(1): 38–50.

Rogers, C.D.F. & Chapman, D.N. (1995c) Ground displacements caused by pipebursting. In *Proceedings of Trenchless Asia 95*, February, Singapore, International Society for Trenchless Technology, London, UK.

Rogers, C.D.F. & Chapman, D.N. (1998) Analytical modelling of ground movements associated with trenchless pipelaying operations. *Geotechnical Engineering, Institution of Civil Engineers*, **131**: 210–222.

Rollins, K.M. (1994) *In-Situ* deep soil improvement. In *Proceedings, Sessions at the ASCE National Convention*, Atlanta, GA, Geotechnical Special Publication No. 45, American Society of Civil Engineers, ASCE Press, Reston, VA.

Rollins, K.M. & Kim, J.H. (1994) US experience with dynamic compaction of collapsible soils. In *In-Situ Deep Soil Improvement*, Geotechnical Special Publication No. 45, American Society of Civil Engineers, ASCE Press, Reston, VA, pp. 26–43.

Roscoe, K.H. & Burland, J.B. (1968) On the generalized stress–strain behaviour of 'wet' clay. In Heymann, J. & Leckie, F.A. (Eds) *Engineering Plasticity*, Cambridge University Press, Cambridge, England, pp. 535–609.

Roscoe, K.H., Schofield, A.N. & Wroth, C.P. (1958) On the yielding of soils. *Géotechnique*, **8**(1): 22–52.

Rosenberg, P. & Journeaux, N.L. (1976) Friction and end bearing tests on bedrock for high capacity socket design. *Canadian Geotechnical Journal*, **13**: 114–124.

Roth, W.H. (1985) *Evaluation of Earthquake Induced Deformations of Pleasant Valley Dam*, Report for the City of Los Angeles, Dames & Moore, Los Angeles.

Rouhani, S., Srivastava, R.M., Desbarats, A.J.,

Cromer, M.V. & Johnson, A.I. (1996) *Geostatistics for Environmental & Geotechnical Applications*, Special Technical Publication 1283, American Society for Testing & Materials, West Conshohocken, PA.

Rowe, P.W. (1952) Anchored sheet pile walls. *Proceedings, Institution of Civil Engineers*, London, England, 1(1): 27–70.

Rowe, P.W. (1957) Sheet pile walls in clay. *Proceedings, Institution of Civil Engineers*, London, England, **7**: 629–654.

Rowe, R.K. (1982) The analysis of an embankment constructed on a geotextile. In *Proceedings, Second International Conference on Geotextiles*, Las Vegas, NA, Industrial Fabrics Association International, Roseville, MN, Vol. 2, pp. 677–682.

Rowe, R.K. (1984) *Recommendations for the Use of Geotextile Reinforcement in the Design of Low Embankments on Very Soft/Weak Soils*, Research Report GEOT-1-84, Faculty of Engineering Science, The University of Western Ontario.

Rowe, R.K. (1988) Contaminant migration through groundwater: The role of modeling in the design of barriers. *Canadian Geotechnical Journal*, **25**(4): 778–798.

Rowe, R.K. (1992) Some challenging applications of geotextiles in filtration and drainage. In *Geotextiles in Filtration & Drainage*, Corbet, S. and King, J. (Eds.) Thomas Telford, London, England, pp. 1–12.

Rowe, R.K. (1994) *Leachate Characterization*, Report to Interim Waste Authority, October 1994. see also Rowe, R.K. (1995) Leachate characterization for MSW landfills. In *Proceedings Fifth International Landfill Symposium*, Sardinia, Italy, Vol. 2, pp. 327–344.

Rowe, R.K. (1995) Considerations in the design of hydraulic control layers. In *Proceedings, Fifth International Landfill Symposium*, Cagliari, Sardinia, Italy, October. Vol. 1, pp. 103–114.

Rowe, R.K. (1996) The role of diffusion and modelling and its impact on groundwater quality. In Areal, M. (Ed.) *Advances in Groundwater Pollution Control & Remediation*, Kluwer Academic Publishers, Netherlands, pp. 371–403.

Rowe, R.K. (1997a) Reinforced embankment behaviour: lessons from a number of case histories. In Almeida, M. (Ed.) *Proceedings, Symposium on Recent Developments in Soil & Pavement Mechanics*, June, Rio de Janeiro, A.A. Balkema, Rotterdam, pp. 147–160.

Rowe, R.K. (1997b) The design of landfill barrier systems: should there be a choice? *Ground Engineering*, **August**: 36–39.

Rowe, R.K. (1998a) Geosynthetics and the minimization of contaminant migration through barrier systems beneath solid waste. In *Proceedings, Sixth International Conference on Geosynthetics*, March, Atlanta, GA, pp. 27–102.

Rowe, R.K. (1998b) From the past to the future of landfill engineering through case histories. In *Proceedings, Fourth International Conference on Case Histories in Geotechnical Engineering*, St Louis, pp. 145–166.

Rowe, R.K. & Armitage, H.H. (1984) *Design of Piles Socketed into Weak Rock*, Research Report GEOT-11-84, University of Western Ontario, London, Ontario.

Rowe, R.K. & Armitage, H.H. (1987) Theoretical solutions for axial deformation of drilled shafts in rock. *Canadian Geotechnical Journal*, **24**: 114–125.

Rowe, R.K. & Armitage, H.H. (1987a) A design method for drilled piers in soft rock. *Canadian Geotechnical Journal*, **24**: 126–142.

Rowe, R.K. & Armitage, H.H. (1987b) A new design method for drilled piers in soft rock: implications relating to three published case histories. In *Proceedings, International Congress on Rock Mechanics*, Montreal, Canada, Vol. 2, pp. 497–502.

Rowe, R.K. & Badv, K. (1996a) Chloride migration through clayey silt underlain by fine sand or silt. *Journal of Geotechnical Engineering, ASCE*, Reston, VA, **122**(1): 60–68.

Rowe, R.K. & Badv, K. (1996b) Advective-diffusive contaminant migration in unsaturated sand and gravel. *Journal of Geotechnical Engineering, ASCE*, Reston, VA, **122**(12): 965–975.

Rowe, R.K. & Barone, F.S. (1991) *Diffusion Tests for Chloride and Dichloromethane in Halton Till*, Report of the Geotechnical Research Centre, University of Western Ontario, London, Ontario, Canada.

Rowe, R.K. & Booker, J.R. (1981a) The behaviour of footings on a non-homogeneous soil mass with a crust. Part I: strip footings. *Canadian Geotechnical Journal*, **18**(2): 250–264.

Rowe, R.K. & Booker, J.R. (1981b) The behaviour of footings on a non-homogeneous soil mass with a crust. Part II: circular footings. *Canadian Geotechnical Journal*, **18**(2): 265–279.

Rowe, R.K. & Booker, J.R. (1989) Analysis of contaminant transport through fractured rock at an Ontario landfill. In *Proceedings, Third Interna-*

tional Symposium on Numerical Methods in Geomechanics, Niagara Falls, pp. 383–390.

Rowe, R.K. & Booker, J.R. (1997a) *POLLUTE v.6.3—ID Pollutant Migration Through a Non-homogeneous soil*, © 1983, 1990, 1994, 1997, GAEA Environmental Engineering, Whitby, ON, Canada.

Rowe, R.K. & Booker, J.R. (1997b) Recent advances in modelling contaminant impact due to clogging (keynote paper). In *Ninth International Conference of the Association for Computer Methods & Advances in Geomechanics*, November, Wuhan, China, Vol. 1, pp. 43–56.

Rowe, R.K. & Davis, E.H. (1982a) The behaviour of anchor plates in clay. *Géotechnique*, **32**(1): 9–23.

Rowe, R.K. & Davis, E.H. (1982b) The behaviour of anchor plates in sand. *Géotechnique*, **32**(1): 25–41.

Rowe, R.K. & Fleming, I.R. (1998) Estimating the time for clogging of leachate collection systems. In *Proceedings, Third International Congress on Environmental Geotechnics*, September, Lisbon, vol. 1, pp. 23–28.

Rowe, R.K. & Lake, C.B. (1999) Geosynthetic clay liner research and design applications. In *Proceedings, Seventh International Landfill Symposium*, October, S. Margherita di Pula, Cagliari, Sardinia, Vol 3, pp. 181–188.

Rowe, R.K. & Li, A.L. (1999) Reinforced embankments over soft foundations under undrained and partially drained conditions. *Geotextiles & Geomembranes*, **17**(3): 129–146.

Rowe, R.K. & Mylleville, B.L.J. (1989) Consideration of strain in the design of reinforced embankments. In *Proceedings, Geosynthetics '89*, San Diego, Industrial Fabrics Association International, Roseville, MN, pp. 124–135.

Rowe, R.K. & Mylleville, B.L.J. (1990) Implications of adopting an allowable geosynthetic strain in estimating stability. In *Fourth International Conference on Geotextiles, Geomembranes & Related Products*, The Hague, Vol. 1, pp. 131–136.

Rowe, R.K. & Mylleville, B.L.J. (1993) The stability of embankments reinforced with steel. *Canadian Geotechnical Journal*, **30**(5): 768–780.

Rowe, R.K. & Mylleville, B.L.J. (1996) A geogrid reinforced embankment on peat over organic silt: a case history. *Canadian Geotechnical Journal*, **33**(1): 106–122.

Rowe, R.K. & Seychuk, J.L. (1995) Alleged non-performance of a geotextile filter. In *Proceedings, Proc Pan-American Conference on Soil Mechanics & Foundation Engineering*, October, Guadalajara, Mexico, Vol. 3, pp. 1654–1667.

Rowe, R.K. & Soderman, K.L. (1984) Comparison of predicted and observed behaviour of two test embankments. *Geotextiles & Geomembranes*, **1**(1): 143.

Rowe, R.K. & Soderman, K.L. (1985a) An approximate method for estimating the stability of geotextile reinforced embankments. *Canadian Geotechnical Journal*, **22**(3): 392–398.

Rowe, R.K. & Soderman, K.L. (1985b) Geotextile reinforcement of embankments on peat. *Geotextiles & Geomembranes*, **2**(4): 277–297.

Rowe, R.K. & Soderman, K.L. (1986) Reinforced embankments on very poor foundations. *Geotextiles & Geomembranes*, **4**(1): 65–81.

Rowe, R.K. & Soderman, K.L. (1987) Stabilization of very soft soils using high strength geosynthetics: the role of finite element analyses. *Geotextiles & Geomembranes*, **6**(1): 53.

Rowe, R.K. & Weaver, T.R. (1997) Contaminant transport in groundwater. In *Proceedings, First Australian–New Zealand Conference on Environmental Geotechnics*, Bouzza, A., Kodikara, J. and Parker, R. (Eds.) November, Melbourne, pp. 97–114.

Rowe, R.K., MacLean, M.D. & Barsvary, A.K. (1984a) The observed behaviour of a geotextile reinforced embankment constructed on peat. *Canadian Geotechnical Journal*, **21**(2): 289–304.

Rowe, R.K., MacLean, M.D. & Soderman, K.L. (1984b) Analyses of a geotextile reinforced embankment constructed on peat. *Canadian Geotechnical Journal*, **21**(3): 563–576.

Rowe, R.K., Caers, C.J. & Barone, F. (1988) Laboratory determination of diffusion and distribution coefficients of contaminants using undisturbed clayey soil. *Canadian Geotechnical Journal*, **25**(1): 108–118.

Rowe, R.K., Caers, C.J. & Chan, C. (1993) Evaluation of a compacted till liner test pad constructed over a granular subliner contingency layer. *Canadian Geotechnical Journal*, **30**(4): 667–689.

Rowe, R.K., Hrapovic, L. & Kosaric, N. (1995a) Diffusion of chloride and dichloro-methane through an HDPE geomembrane. *Geosynthetics International*, **2**(3): 507–536.

Rowe, R.K., Quigley, R.M. & Booker, J.R. (1995b) *Clayey Barrier Systems for Waste Disposal and Facilities*, E. & F.N. Spon. London, U.K.

Rowe, R.K., Fleming, I., Cullimore, R., Kosaric, N. & Quigley, R.M. (1995c) *A Research Study of Clogging and Encrustation in Leachate Collection Systems in Municipal Solid Waste Landfills*, Report to Interim Waste Authority Ltd, Geotechnical Research Centre, University of Western Ontario.

Rowe, R.K., Gnanendran, C.T., Landva, A.O. & Valsangkar, A.J. (1995d) Construction and performance of a full-scale geotextile reinforced test embankment, Sackville, New Brunswick. *Canadian Geotechnical Journal*, **32**: 512–534; and (1996) *Canadian Geotechnical Journal*, Erratum. **33**: 208.

Rowe, R.K., Caers, C.J. & Chan, C. (1996a) The design and operation of a state-of-the-art landfill facility. In *Fourth Canadian Society for Civil Engineering Environmental Engineering Specialty Conference*, May 4, Edmonton, pp. 179–190.

Rowe, R.K., Hrapovic, L. & Armstrong, M.D. (1996b) Diffusion of organic pollutants through HDPE geomembrane and composite liners and its influence on groundwater quality. In *Proceedings, First European Geosynthetics Conference*, October, Maastricht, pp. 737–742.

Rowe, R.K., Petrov, R.J. & Lake, C. (1997a) Compatibility testing and diffusion through GCLs. In *Proceedings, Sixth International Landfill Symposium*, October, S. Margherita di Pula, Cagliari, Italy, Vol. 3, pp. 301–310.

Rowe, R.K., Cooke, A.J., Rittmann, B.E. & Fleming, I.R. (1997b) Some considerations in numerical modelling of leachate collection system clogging, (lead lecture). In *Sixth International Symposium on Numerical Methods in Geomechanics*, July, Montreal, pp. 277–282.

Rowe, R.K., Fleming, I.R., Armstrong, M.D., Millward, S.C., VanGulck, J. & Cullimore, R.D. (1997c) Clogging of leachate collection systems: some preliminary experimental findings. In *Proceedings, 50th Canadian Geotechnical Conference*, October, Ottawa, Vol. 1, pp. 153–160.

Rowe, R.K., Fleming, I.R., Armstrong, M.D., Cooke, A.J., Cullimore, R.D., Rittmann, B.E., Bennett, P. & Longstaffe, F.J. (1997d) Recent advances in understanding the clogging of leachate collection systems. In *Proceedings, Sixth International Landfill Symposium*, October, S. Margherita di Pula, Cagliari, Italy, Vol. 3, pp. 383–392.

Rowe, R.K., Lake, C.B. & Petrov, R.J. (1998a)

Test Apparatus and Procedures for Assessing Inorganic Diffusion Coefficients for Geosynthetic Clay Liners, Geotechnical Research Centre Report, University of Western Ontario.

Rowe, R.K., Caers, C.J., Reynolds, G. & Chan, C. (1998b) *Design of the Halton Landfill*, GEOT-4-98, Faculty of Engineering Science, University of Western Ontario.

Rowe, R.K., Caers, C.J., Reynolds, G. & Chan, C. (2000) Design and construction of the barrier system for the Halton Landfill, *Canadian Geotechnical Journal*, **37**(3): 662–675.

Rowe, R.K., Armstrong, M.D. & Cullimore, D.R. (2000a) *Mass Loading and the Rate of Clogging due to Municipal Solid Waste Leachate*, *Canadian Geotechnical Journal* **37**(2): 355–370.

Rowe, R.K., Armstrong, M.D. & Cullimore, D.R. (2000b) *Particle Size and Clogging of Granular Media Permeated with Leachate. Journal of Geotechnical & Geoenvironmental Engineering, ASCE*, Reston, VA (In press).

Rubright, R. & Welsh, J. (1993) Compaction grouting. In Moseley, M.P. (Ed.) *Ground Improvement*, Blackie Academic & Professional, Blackwood, NJ, pp. 131–148.

Rudolph, D.L., Kachanoski, G.R., Celia, M.A., Leblanc, D.R. & Stevens, J.H. (1996) Infiltration and solute transport experiments in unsaturated sand and gravel, Cape Cod, Massachusetts. Experimental design and overview of results. *Water Resources Research*, **32**(2): 519–532.

Ruhl, J.L. & Daniel, D.E. (1997) Geosynthetic clay liners permeated with chemical solutions and leachates. *Journal of Geotechnical & Geoenvironmental Engineering, ASCE*, Reston, VA, **123**(4): 369–380.

Rumer, R.R. & Mitchell, J.K. (1995) Assessment of barrier containment technologies—a comprehensive treatment for environmental applications. National Technical Information Services, Springfield, VA.

Russam, K. (1958) An investigation into the soil moisture conditions under roads in Trinidad, B.W.I. *Géotechnique*, **8**: 55–71.

Saada, A. & Bianchini, G.S. (Eds) (1987) *Proceedings, International Workshop on Constitutive Equations for Granular Non-Cohesive Soils*, July 22–24, Case Western Reserve University, Cleveland, Ohio, A.A. Balkema.

Saathoff, F. & Sehrbrock, U. (1994) Indicators for selection of protection layers for geomembranes. In *Proceedings, Fifth International*

Conference on Geotextiles, Geomembranes & Related Products, September, Singapore, Vol. 3, pp. 1019–1022.

Sabatini, P.J., Schmertmann, G.R. & Swan, R.H. (1998) Issues in clay/textured geomembrane interface testing. In Rowe, R.K. (Ed.) *Proceedings, Sixth International Conference on Geomembranes*, Atlanta, GA, Industrial Fabrics Association International, Roseville, MN, Vol. 1, pp. 423–426.

Sadigh, K., Chang, C.-Y., Egan, J.A., Makdisi, F. & Youngs, R.R. (1997) SEA96—A new predictive relation for earthquake ground motions in extensional tectonic Regimes. *Seismological Research Letters*, **68**(1): 190–198.

Sagaseta, C. (1987) Analysis of undrained soil deformation due to ground loss. *Géotechnique* **37**(3): 301–320.

Sai, J.O. & Anderson, D.C. (1991) Barrier wall materials for containment of dense nonaqueous phase liquid (DNAPL). *Hazardous Waste & Hazardous Materials*, **9**(4): 317–330.

Saintot, J., Isambert, F., Simon, B. & Potié, G. (1993) Barrage de Borfloc'h Belle-Ile-en-Mer—Borfloc'h Dam Belle-Ile-en-Mer. In *Proceedings, Rencontres 93*, CFGG, Joué-les-Tours, France, Vol. 2, pp. 315–324 (in French & English).

Salanitro, J.P. (1993) The role of bioattenuation in the management of aromatic hydrocarbon plumes in aquifers. *Ground Water Monitoring & Remediation*, **13**(4): 150–161.

Sale, T.C. & Applegate, D. (1997) Mobile NAPL recovery: conceptual, field and mathematical considerations. *Ground Water*, **35**(3): 418–426.

Sällfors, G., Larsson, R. & Ottosson, E. (1996) New Swedish national rules for slope stability analysis. In *Proceedings, Seventh International Symposium on Landslides*, Trondheim, Vol. 1, pp. 377–380.

Salman, A., Elias, V. Juan, I. Lu, S. & Pearce, E. (1997) Durability of geosynthetics based on accelerated laboratory testing. In *Geosynthetics '97*, Long Breach, CA, IFAI, Roseville, MN, pp. 217–234.

Salman, A., Elias, V. & DiMillio, A. (1998) The effect of oxygen pressure, temperature and manufacturing processes on laboratory degradation of polypropylene geosynthetics. In *Proceedings, Sixth International Conference on Geosynthetics*, R.K. Rowe (Ed), Atlanta, GA, IFAI, Roseville, MN, pp. 683–690.

Sara, M.N. (1991) Ground-water monitoring system design. In Nielsen, D.M. (Ed.) *Practical Handbook of Ground-Water Monitoring* Lewis, Chelsea, MI, pp. 17–68.

Sarma, S.K. (1973) Stability analysis of embankments and slopes. *Géotechnique*, **23**(3): 423–433.

Saul, W.E. (1968) Static and dynamic analysis of pile foundations, *Journal of Structural Division, ASCE*, Reston, VA, **94**(ST5): 1077–1100.

Schaefer, V.R. (Ed.) (1997) *Ground Improvement, Ground Reinforcement, and Ground Treatment: Developments 1987–1997*, Geotechnical Special Publication No. 69, American Society of Civil Engineers, Reston, VA.

Schincariol, R.A., Schwartz, F.W. & Mendoza, C.A. (1994) On the generation of instabilities in variable density flow. *Water Resources Research*, **30**(4): 913–927.

Schincariol, R.A., Schwartz, F.W. & Mendoza, C.A. (1997) Instabilities in variable density flows: Stability and sensitivity analyses for homogeneous and heterogeneous media. *Water Resources Research*, **33**(1): 31–41.

Schlosser, F. (1982). Behaviour and design of soil nailing. In *Proceedings, Symposium on Recent Developments in Ground Improvement Techniques*, Bangkok, pp. 299–413.

Schmee, J. & Hahn, G.J. (1979) A simple method for regression analysis with censored data. *Technometrics*, **21**: 417–432.

Schmertmann, J.H. (1970) Static cone to compute static settlement over sand. *Journal of Soil Mechanics & Foundations Division, ASCE*, Reston, VA, **96**(3): 1011–1043.

Schmertmann, J.H. (1975) Measurement of in-situ shear strength. In *Proceedings, ASCE Geotechnical Engineering Division Specialty Conference on In-Situ Measurement of Soil Properties*, June 1–4, North Carolina State University, Raleigh, NC, Vol. II, pp. 57–138.

Schmertmann, J. (1978) *Guidelines for Cone Penetration test–in Performance and Design*, US Department of Transportation, Federal Highways Administration, Washington, DC.

Schmertmann, J.H. (1986) Suggested method for performing the flat dilatometer test. ASTM Subcommittee D.18.02. *ASTM Geotechnical Testing Journal*, **9**(2): 99–102.

Schmertmann, J.H. (1991) The mechanical aging of soils. 25th Terzaghi lecture. *Journal of Geotechnical Engineering, ASCE*, Reston, VA, **117**(9): 1288–1330.

Schmertmann, J.H., Hartman, J.P. & Brown, P.R.

(1978) Improved strain factor diagrams. *Journal of Geotechnical Engineering Division, ASCE*, Reston, VA, **104**(8): 1131–1135.

Schmertmann, J.H., Baker, W., Gupta, R. & Kessler, K. (1986) CPT/DMT QC of ground modification at a power plant. In *Use of In Situ Tests in Geotechnical Engineering*, Geotechnical Special Publication No. 6, American Society of Civil Engineers, ASCE Press, Reston, VA, pp. 985–1001.

Schmidt, B. (1966) Discussion of Earth pressures at rest related to stress history. *Canadian Geotechnical Journal*, **3**(4): 239–242.

Schmidt, H.M., Pas F.W.T., Risseeuw, P. & Voskamp W. (1994) The hydrolytic stability of PET yarns under medium alkaline conditions. In *Proceedings, Fifth International Conference*, September, Singapore, pp. 1153–1158.

Schnabel, H. (1982) *Tiebacks in Foundation Engineering and Construction*, McGraw-Hill, New York, NY.

Schnabel, P., Lysmer, J. & Seed, H.B. (1972) *SHAKE: A Computer Program for Earthquake Response Analysis of Horizontally Layered Sites*, Report No. EERC 72-12, Earthquake Engineering Research Center, University of California, Berkeley, CA.

Schnarr, M.J., Truax, C.T., Farquhar, G.J., Hood, E.D., Gonullu, T. & Stickney, B, (1998) Experiments using potassium permanganate to remediate trichloroethylene and perchloroethylene DNAPLs. *Journal of Contaminant Hydrology*, **29**: 205–224.

Schofield, A.N. & Wroth, C.P. (1968) *Critical State Soil Mechanics*, McGraw-Hill, London.

Schroeder, P.R., Lloyd, C.M. & Zappi, P.A. (1994) *The Hydrologic Evaluation of Landfill Performance (HELP) Model: User's Guide for Version 3*, EPA/600/R-94/168a, US Environmental Protection Agency, Risk Reduction Engineering Laboratory, Cinncinnati, OH.

Schroeder, P.R., Dozier, T.S., Zappi, P.A., McEnroe, B.M., Sjostrom, J.W. & Peyton, R.L. (1994) *The Hydrologic Evaluation of Landfill Performance (HELP) Model: Engineering Documentation for Version 3*, EPA/600/R-94/168b, US Environmental Protection Agency Risk Reduction Engineering Laboratory, Cincinnati, OH.

Schuster, R.L. (1995) Keynote paper: Recent advances in slope stabilization. In *Proceedings, Sixth International Symposium on Landslides*, Christchurch, Vol. 3, pp. 1715–1725.

Schuster, R.L. (1996) Socioeconomic significance

of landslides. *Landslides – Investigation and Mitigation*, Special Report 247, A.K. Turner & R.L. Schuster (Eds.) Transportation Research Board, Washington, DC, pp. 12–35.

Schwartz, D.P. & Coppersmith, K.J. (1984) Fault behaviour and characteristic earthquakes from the Wasatch and San Andreas faults. *Journal of Geophysical Research*, **89**: 5681–5698.

Schwarzenbach, R.P. & Giger, W. (1985) Behavior and fate of halogenated hydrocarbons in groundwater. In Ward, C.H. Giger, W. & McCarty, P.L. (Eds) *Groundwater Quality*, Wiley-Interscience, New York, pp. 446–471.

Schwarzenbach, R.P. & Westall, J. (1981) Transport of nonpolar organic compounds from surface water to groundwater–laboratory studies. *Environmental Science & Technology*, **15**: 1300–1367.

Schwille, F. (1988) *Dense Chlorinated Solvents in Porous and Fractured Media: Model Experiments*, translated by J.F. Pankow, Lewis, Chelsea, MI.

Scott, R.F., Liu, H.P. & Ting, J. (1977) Dynamic pile tests by centrifuge modelling. In *Proceedings, Sixth World Conference on Earthquake Engineering*, New Delhi, India, Vol. 4, pp. 199–203.

Scott, R.F., Ting, J. & Lee, J (1982) Comparison of centrifuge and full-scale dynamic pile tests. In *Proceedings, International Conference on Soil Dynamics & Earthquake Engineering*, Southampton, Vol. 1, pp. 299–309.

SCW (1981) *Stability of Slopes Constructed with Polyester Reinforcing Fabric*, Study Centre for Road Construction, Arnhem, The Netherlands.

SEBJ (1983) *Experts Committee on Soft Clays from the NBR Complex. Final report*, Société d'Énergie de la Baie James Montréal, Québec, Canada.

Seed, H.B. (1979a) Soil liquefaction and cyclic mobility evaluation for level ground during earthquakes. *Journal of the Geotechnical Engineering Division, ASCE*, Reston, VA, **105**(2): 201–255.

Seed, H.B. (1979b) Considerations in the earthquake-resistant design of earth and rockfill dams. *Géotechnique*, **29**(3): 215–263.

Seed, H.B. (1987) Design problems in soil liquefaction. *Journal of Geotechnical Engineering, ASCE*, Reston, VA, **113**(7): 827–845.

Seed, H.B. & Idriss, I.M. (1970) *Soil Moduli and Damping Factors for Dynamic Response Analyses*, Report EERC 70-10, College of Engineering, University of California, Berkeley, CA.

Seed, H.B. & Idriss, I.M. (1971) Simplified procedures for evaluating soil liquefaction potential. *Journal of Soil Mechanics & Foundations Division, ASCE*, Reston, VA, **97**(9): 1249–1273.

Seed, H.B. & Idriss, I.M. (1982) *Ground Motions and Soil Liquefaction During Earthquakes*, Monograph No. 5, Earthquake Engineering Research Institute, Berkeley.

Seed, H.B. & Lee, K.L. (1967) Undrained strength characteristics of cohesionless soils. *Journal of Soil Mechanics & Foundations Division, ASCE*, Reston, VA, **93**(6): 333–360.

Seed, R.B. & Raines, J.R. (1988) *Failure of Flexible Long-span Culverts under Exceptional Live Load*, Transportation Research Record No. 1191, pp. 22–20.

Seed, H.B. & Wilson, S.D. (1966) The Turnagain Heights Landslides, Anchorage, Alaska. In *Proceedings, ASCE Specialty Conference on Stability & Performance of Slopes & Embankments*, Reston, VA, pp. 357–385.

Seed, H.B., Mitchell, J.K. & Chan, C.K. (1960) The strength of compacted cohesive soils. In *ASCE Research Conference on the Strength of Cohesive Soils*, Boulder, CO, pp. 877–964.

Seed, H.B., Mitchell, J.K. & Chan, C.K. (1962a) Swell and swell pressure characteristics of compacted clays. *Highway Research Board Bulletin* **313**: 12–39.

Seed, H.B., Woodward Jr, R.J. & Lundgren, R. (1962b) Prediction of swelling potential for compacted clays. *Journal of Soil Mechanics & Foundations Division, ASCE*, Reston, VA, **88**(3): 53–97.

Seed, H.B., Lee, K.L., Idriss, I.M. & Makdisi, F.F. (1975) The slides in the San Fernando Dams during the earthquake of February 9, 1971. *Journal of Earthquake Engineering Division, ASCE*, Reston, VA, **101**(7): 651–688.

Seed, H.B., Tokimatsu, K., Harder, L.F. & Chung, R.M. (1985) The influence of SPT procedures in soil liquefaction resistance evaluations. *Journal of Geotechnical Engineering, ASCE*, Reston, VA, **111**(12): 1425–1445.

Seed, H.B., Seed, R.B., Harder, L.F. Jr & Jong, H.-L. (1989) *Re-evaluation of the Lower San Fernando Dam, Report 2, Examination of the Post-earthquake Slide of February 9, 1971*, Department of the Army, US Army Corps of Engineers, Washington, DC.

Seed, H.B., Tokimatsu, K., Harder, L.F. & Chung, R.M. (1985) Influence of SPT procedures in soil liquefaction resistance evaluations. *Journal of Geotechnical Engineering, ASCE*, Reston, VA, **111**(12): 1425–1445.

Seed, R.B. & Boulanger, R.W. (1991) Smooth HDPE–clay linear interface shear strengths: compaction effects. *Journal of Geotechnical Engineering, ASCE*, Reston, VA, **117**(4): 686–693.

Seed, R.B. & Harder, L.F. Jr. (1990) SPT-based analysis of cyclic pore pressure generation and undrained residual strength. In *Proceedings, H. Bolton Seed Memorial Symposium*, May, BiTech, pp. 351–376.

Seed, R.B., Mitchell, J.K. & Seed, H.B. (1988) *Slope Stability Failure Investigation: Landfill Unit B-19, Phase I-A, Kettleman Hills, California*, Report No. UCB/GT/88-01, University of California at Berkeley, Berkeley.

Seed, R.B., Dickenson, S. & Mok, C.M. (1991) Seismic response analyses of soft and deep cohesive sites: a brief summary of recent lessons. In *Proceedings, First Annual CALTRANS Seismic Research Seminar*, CALTRANS, Sacramento.

Seeger, S. & Müller, W. (1996) Requirements and testing of protective layer systems for geomembranes. *Geotextiles & Geomembranes*, **14**: 365–376.

Sego, D.C., Robertson, P.K., Sasitharan, S., Kilpatrick, B.L. & Pillai, V.S. (1994) Ground freezing and sampling of foundation soils at Duncan Dam. *Canadian Geotechnical Journal*, **31**(6): 939–950.

Selig, E.T. (1990) Soil properties for plastic pipe installations. In Buczala, G.S. & Cassidy, M.J. (Eds) *Buried Plastic Pipe Technology*, Special Technical Publication 1093, American Society for Testing & Materials, West Conshohocken, PA, pp. 141–158.

Selvadurai, A.P.S. (1979) *Elastic Analysis of Soil–Foundation Interaction*, Elsevier, New York.

Selvadurai, A.P.S. (1985) Numerical simulation of soil–pipeline interaction in a ground subsidence zone. In Jeyapalan, J.K. (Ed.) *International Conference On Advances in Underground Pipeline Engineering*, Madison, WI, pp. 311–319.

Semple, R.M. & Rigden, W.J. (1984) Shaft capacity of driven piles in clay. *Analysis & Design of Pile Foundations*, ASCE Press, Reston, VA, pp. 59–79.

Semprini, L., Kitandis, P.K., Kampbell, D.H. & Wilson, J.T. (1995) Anaerobic transformation of chlorinated aliphatic hydrocarbons in a sand aquifer based on spatial chemical distributions. *Water Resources Research*, **31**(4): 1051–1062.

Serrano, A. & Olalla, C. (1994) Ultimate bearing

capacity of rock masses. *International Journal of Rock Mechanics Mineral Science & Geomechanics Abstracts*, **31**(2): 93–106.

Serrano, A. & Olalla, C. (1996) Ultimate bearing capacity of rock masses. *International Journal of Rock Mechanical Mineral Science & Geomechanics Abstracts*, **33**(4): 327–345.

Sève, G. (1998) *PhD thesis*, École Nationale des Ponts et Chaussées, Paris.

Sève, G. & Durville, J.-L. (1995) Surveillance des glissements de terrain. In *Proceedings, 11th European Conference on Soil Mechanics & Foundation Engineering*, Copenhagen, Vol. 4, pp. 115–120.

Sève, G. & Leca, E. (1995) Stabilization of landslide using large diameter piles. In *Proceedings, 11th European Conference on Soil Mechanics & Foundation Engineering*, Copenhagen, Vol. 6, pp. 179–184.

Sève, G. & Pouget, P. (1998) *Guide technique sur la stabilisation des glissements de terrain*. Techniques et Méthodes des Laboratoires des Ponts et Chausseées, Paris.

Shackelford, C.D. & Daniel, D.E. (1991a) Diffusion in saturated soil. I: Background. *Journal of Geotechnical Engineering*, ASCE, Reston, VA, **117**(3): 467–506.

Shackelford, C.D. & Daniel, D.E. (1991b) Diffusion in saturated soil. II: Results for compacted clay. *Journal of Geotechnical Engineering*, ASCE, Reston, VA, **117**(3): 485–506.

Shahin, M.Y. & Crovetti, J.A. (1987) Determining the effects of utility cut patching on the service life prediction of asphalt concrete pavements. In *Proceedings, Second North American Conference on Managing Pavements*, Toronto, Canada, Vol. 1, pp. 225–236.

Shallenberger, W.C. & Filz, G.M. (1996) Interface strength determination using a large displacement shear box. In *Proceedings, Second International Conference on Environmental Geotechnics*, Osaka Japan, Vol. 1, pp. 147–152.

Shan, H.-Y. (1993) *PhD dissertation*, University of Texas, Austin, TX.

Shan, H.-Y. & Daniel, D.E. (1991) Results of laboratory tests on a geotextile/bentonite linear material. In *Proceedings, Geosynthetics '91*, Industrial Fabrics Association International, Roseville, MN, Vol. 2, pp. 517–535.

Shang, J.Q. (1998) Electroosmotic enhanced preloading consolidation via vertical drains. *Canadian Geotechnical Journal*, **35**(4): 491–499.

Shang, J.Q., Lo, K.Y. & Huang, K.M. (1996) On influencing factors in electro-osmotic consoli-

dation. *Journal of Geotechnical Engineering*, South East Asia Geotechnical Society, **27**(2): 23–26.

Shang, J.Q., Tang, M. & Z. Miao, (1998) Vacuum preloading consolidation of reclaimed land: a case study. *Canadian Geotechnical Journal*, **35**(5): 740–749.

Sharma, S. (1994) *XSTABL, Reference Manual, Version 5*, Interactive Software Designs, Inc., Moscow, ID.

Sharma, H.D. & Lewis, S.P. (1994) *Waste Containment Systems, Waste Stabilization, & Landfills, Design & Evaluation*, Wiley, New York.

Sharma, H.D., Dukes, M.T. & Olsen, D.M. (1990) Field measurements of dynamic moduli and Poisson's ratios of refuse and underlying soils at a landfill site. In *Geotechnics of Waste Landfills – Theory & Practice*, STP 1070, Landva, A. & Knowles, E.D. (Eds.) American Society for Testing & Materials, West Conshohocken, PA, pp. 57–70.

Sharma, H.D., Hullings, D.E. & Greguras, F.D. (1997) Interface strength tests and application to landfill design. In *Proceedings, Geosynthetics '87 Conference*, Long Beach, CA, Vol. 2, pp. 913–926.

Sharma, M., Thomson, N.R. & McBean, E.A. (1995) Linear regression analyses with censored data: estimation of PAH washout ratios and dry deposition velocities to a snow surface. *Canadian Journal of Civil Engineering*: **22**(4): 819–833.

Sharma, M., McBean, E.A. & Thomson, N.R. (1995) Maximum likelihood method for parameter estimation with below-detection data. *Journal of the Environmental Engineering Division*, ASCE, Reston, VA, **11**: 776–784.

Sherard, J.L. (1985) The upstream zone in concrete–face rockfill dams. In Cooke, J.B. & Sherard, J.L. (Eds) *Proceedings, Symposium on Concrete Face Rockfill Dams–Design, Construction, and Performance*, American Society of Civil Engineers, Reston, VA, pp. 618–641.

Sherif, M., Ishibashi, I. & Lee, C. (1982) Earth pressure against rigid retaining Walls. *Journal of the Geotechnical Engineering Division*, ASCE, Reston, VA, **108**(5): 679–695.

Sheta, M. & Novak, M. (1982) Vertical vibration of pile groups. *Journal of Geotechnical Engineering Division*, ASCE, Reston, VA, **108**(4): 570–590.

Shi, X. & Richards, Jr. R. (1995) Seismic bearing capacity with variable shear transfer. In

Kramer, S. & Siddharthan, R. (Eds) *Earthquake Induced Movements and Seismic Remediation of Existing Foundations and Abutments*, Geotechnical Special Publication No. 55, American Society of Civil Engineers, Reston, VA, pp. 17–32.

Shikaze, S., Austrins, C.D., Smyth, D.J.A., Cherry, J.A., Barker, J. & Sudicky, E. (1995) The hydraulics of a funnel-and-gate system. A three-dimensional numerical analysis. *Solutions '95: Proceedings, IAH International Congress*, June, Edmonton, AB.

Shields, D., Chandler, N. & Garnier, J. (1990) Bearing capacity of foundations in slopes. *Journal of Geotechnical Division, ASCE*, Reston, VA, **116**(3): 528–537.

Shoemaker, S.H., Greiner, J.F. & Gillham, R.W. (1995) Permeable reactive barriers. In Rumer, R.R. & Mitchell, J.K. (Eds), *Assessment of Barrier Containment Technologies: A Comprehensive Treatment for Environmental Remediation Applications*. US Department of Energy/US Environmental Protection Agency/Dupont Company, pp. 301–353.

Shoukry, S.N. (Ed.) (1998) Finite element for pavement analysis and design. *Proc., First National Symposium on 3D Finite Element Modeling for Pavement Analysis and Design*, November, West Virginia University, Morgantown, WV-ISBN 093 7058-49-1, pp. 1–305.

SHRP (1994) *The Superpave Mix Design Manual for New Construction and Overlays*, Report A-407, Strategic Highway Research Program, Washington, DC.

Siddle, H.J., Jones, D.B. & Warren, C.D. (1986) The use of tracer techniques to assess groundwater flows in site investigations. In Hawkins, A.B. (Ed.) *Site Investigation Practice: Assessing BS 5930*. Geological Society Engineering Geology Special Publication No. 2, London, England, pp. 375–384.

Sieh, K.E. & Jahns, R.H. (1984) Holocene activity of the San Andreas Fault at Wallace Creek, California. *Bulletin Geological Society of America*, **90**: 883–896.

Silva, W.J. (1988) *Soil Response to Earthquake Ground Motion*, EERI Report NP-5747, Electric Power Research Institute, Palo Alto, CA.

Silva, W.J. & Lee, K. (1987) *WES RASCAL Code for Synthesizing Earthquake Ground Motions*, Department of the Army, US Army Corps of Engineers, Washington, DC.

Simac, M.R., Bathurst, R.J., Berg, R.R. & Lothspeich, S.E. (1993) *National Concrete Masonry Association Segmental Retaining Wall Design Manual*, National Concrete Masonry Association, Herndon, VA.

Simac, M.R., Bathurst, R.J. & Fennessey, T.W. (1997a) Case study of a hybrid gabion basket geosynthetic reinforced soil wall. *Ground Improvement*, **1**(1): 9–17.

Simac, M.R., Bathurst, R.J. & Fennessey, T.W. (1997b) Design of gabion–geosynthetic retaining walls on the Tellico Plains to Robbinsville Highway. In *Geosynthetics'97*, Industrial Fabrics Association International, Roseville, MN, March, Vol. 1, pp. 11–13.

Simms, P.H. & Yanful, E.K. (1997) Monitoring, analysis, and performance of an engineered soil cover near London, Ontario. In *Proceedings, 50th Canadian Geotechnical Conference*, Ottawa, Vol. 1, pp. 42–49.

Simkin, T.J., Sale, T.C., Kueper, B.H., Pitts, M.J., and Wyatt, K. (1999). *Surfactants and Cosolvents for NAPL Remediation: A Technology Practices Manual*. (Eds.) Lowe, D.F., Oubre, C.L., and Ward, C.H. Lewis Publishers, Boca Raton, Floida, 448 p.

Sims, R.C. (1990) Soil remediation techniques at uncontrolled hazardous waste sites: a critical review. *Journal of the Air & Waste Management Association*, **40**(5): 704–732.

Singh, S. & Sun, J.I. (1995) Seismic evaluation of municipal solid waste landfills. In *Geoenvironment 2000*, Acar, Y.B. and Daniel, D.E. (Eds.) Geotechnical Special Publication No. 46, American Society of Civil Engineers, Reston, VA, Vol. 2, pp. 1081–1096.

Sinha, J. & Poulos, H.G. (1996) Behaviour of stiffened raft foundations. In *Seventh Aust–NZ Conference on Geomechanics*, 1–5 July, Adelaide. M.B. Jaksa, W.S. Kaggwa & D.A. Cameron (Eds.), pp. 704–709.

Site Investigation Steering Group, ICE (1993) *Site Investigation in Construction. Part 1: Without Site Investigation Ground is a Hazard; Part 2: Planning, Procurement & Quality Management; Part 3: Specification for Ground Investigation; & Part 4: Guidelines for the Safe Investigation by Drilling of Landfills & Contaminated Land*, Thomas Telford, London, England.

Skempton, A.W. (1957) Discussion. *Proceedings, Institution of Civil Engineers*, **7**: 305–307.

Skempton, A.W. (1964) Long-term stability of clay slopes. *Géotechnique*, **14**(2): 77–101.

Skempton, A.W. (1977) Slope stability of cuttings in brown London clay. *Proceedings, Ninth In-*

ternational Conference on Soil Mechanics & Foundation Engineering, Tokyo, Vol. 3, pp. 261–270.

Skempton, A.W. (1985) Residual strength of clays in landslides, folded strata and the laboratory. Géotechnique, 35(1): 3–18.

Skempton, A.W. (1986) Standard penetration test procedures and the effects in sands of overburden pressure, relative density, particle size, aging and overconsolidation. Géotechnique, 36(3): 425–447.

Skempton, A.W. & MacDonald, D.H. (1956) The allowable settlement of buildings. Proceedings, Institution of Civil Engineers, 5: 727–768.

Sleep, B. (1996) Steam injection for in-situ remediation of DNAPLs in low permeability media. In In Situ Remediation of DNAPL Compounds in Low Permeability Media:Fate/Transport, In Situ Control Technologies & Risk, ORNL/TM-13305, Oak Ridge National Laboratory, Oak Ridges, TN, 13: 1–13.

Slemmons, D.B., Bodin, P. & Zhang, X. (1989) Determination of earthquake size from surface faulting events. In Proceedings, International Seminar on Seismic Zonation, Guangshou, China.

Sloan, S.W. & Randolph, M.F. (1982) Numerical prediction of collapse loads using finite element methods. International Journal for Numerical & Analytical Methods in Geomechanics, 6: 47–76.

Sloan, S.W. & Yu, H.S. (1996) Rigorous plasticity solutions for the bearing capacity factor N_γ. In Proceedings, Seventh Aust–NZ. Conference on Geomechanics, 1–5 July, Adelaide, M.B. Jaksa, W.S. Kaggwa & D.A. Cameron (Eds.), pp. 544–550.

Slocombe, B.C. (1993) Dynamic Compaction. In Moseley, M.P. (Ed.) Ground Improvement, Blackie Academic & Professional, Blackwood, NJ, pp. 20–39.

Small, J.C. & Booker, J.R. (1986) Finite layer analysis of layered elastic materials using a flexibility approach, Part 2. circular and rectangular loadings. International Journal for Numerical Methods in Engineering, 23: 959–978.

Small, J.C. & Booker, J.R. (1996) User's manuals for FLEA (Finite Layer Elastic Analysis) and FLAC (Finite Layer Analysis of Consolidation), Centre for Geotechnical Research, The University of Sydney.

Smith, D.W. (Ed.) (1986) Cold climate utilities manual, Canadian Society for Civil Engineering, Montreal, Quebec.

Smith, E.A.L. (1960) Pile driving analysis by the wave equation. Journal of Soil Mechanics & Foundations Division, ASCE, Reston, VA, 86(4): 35–61.

Smith, L.A., Means, J.L. Chen, A., Alleman, B., Chapman, C.C., Tixier, J.S., Brauning, S.E., Gavaskar, A. & Royer, M.D. Remedial Options for Metals-Contaminated Sites, Battelle Memorial Institute, Columbus, OH.

Smyth, D.J.A., Shikaze, S.G. & Cherry, J.A. (1997) Hydraulic performance of permeable barriers for in situ treatment of contaminated groundwater. Land Contamination & Reclamation, 5(3): 131–137.

Snow, M., Mansour, R., Swan, R.H., Jr & Kavazanjian, E., Jr (1998) Variability of interface shear strengths. In Rowe, R.K. (Ed.) Proceedings, Sixth International Conference on Geomembranes, Atlanta, GA, Industrial Fabrics Association International, Roseville, MN, Vol. 1, pp. 439–442.

Soderman, L. & Milligan, V. (1961) Capacity of friction piles in varved clay increased by electro osmosis. In Fifth International Conference of Soil Mechanics & Foundation Engineering, Paris, Vol. 2, pp. 143–147.

Sokolovskii, V.V. (1960) Statics of Soil Media, Butterworth, London, England.

Sokolovskii, V.V. (1965) Statics of Granular Media, Pergamon Press, London, England.

Solymar, Z.V. (1984) Compaction of alluvial sands by deep blasting. Canadian Geotechnical Journal, 21(2): 305–321.

Solymar, Z.V. & Mitchell, J.K. (1986) Blasting densifies sand. Civil Engineering, ASCE, Reston, VA, 56(3): 46–48.

Solymar, Z.V. & Reed, D.J. (1986) A comparison of foundation compaction techniques. Canadian Geotechnical Journal, 23(3): 271–280.

Solymar, Z.V., Iloabachie, B.C., Gupta, R.C. & Williams, L.R. (1984) Earth foundation treatment at Jebba Dam site. Journal of Geotechnical Engineering, ASCE, Reston, VA, 110(10): 1415–1430.

Somerville, P.G. (1998) Emerging art: earthquake ground motion. In Dakoulas, P., Yegian, M. & Holtz, R.D. (Eds.) Proceedings, Geotechnical Earthquake Engineering in Soil Dynamics III, Geotechnical Special Publication No. 75, American Society of Civil Engineers, Reston, VA, Vol. 1, pp. 1–38.

Somerville, P.G. & Graves, R.W. (1996) Strong ground motions of the Kobe, Japan earthquake of January 17, 1995, and development of forward rupture directivity applicable in California. In *Proceedings, Western Regional Technical* Seminar of Earthquake *Engineering for Dams*, April 11–12, Sacramento, CA, Association of State Dam Safety Officials.

Soong, T.Y. & Koerner, R.M. (1997) *The Design of Drainage Systems Over Geosynthetically Lined Slopes*, GRI Report No. 19, Geosynthetic Research Institute, Drexel University, Philadelphia, PA.

Soong, T.Y. & Koerner, R.M. (1998) Laboratory study of high density polyethylene geomembrane waves. In Rowe, R.K. (Ed.) *Proceedings, Sixth International Conference on Geosynthetics*, March, Atlanta, GA, Vol. 1, Industrial Fabrics Association International, Roseville, MN, pp. 301–306.

Sowers, G.F. (1973) Settlement of waste disposal fills. In *Proceedings, Eighth International Conference on Soil Mechanics & Foundation Engineering*, Moscow, pp. 207–210.

Sowers, G.F. (1979) *Introductory Soil Mechanics and Foundations: Geotechnical Engineering*, 4th Edn, MacMillan, New York.

Sowers, G.B. & Sowers, G.F. (1970) *Introductory Soil Mechanics and Foundations*, 3rd Edn, MacMillan, New York.

Spain, J. (1996) Future vision: compounds with potential for natural attenuation. In *Symposium on Natural Attenuation of Chlorinated Organics in Groundwater*, September 11–13, Dallas, Tx, 540/R-96-509, US Environmental Protection Agency pp. 128–132.

Spangler, M.J. (1956) Stresses in pressure pipelines and protective casting pipes. *Journal of Structural Engineering, ASCE*, Reston, VA, **82**: 1–33.

Spangler, M.J. & Handy, R.L. (1973) *Soil Engineering*, Intext Educational, New York.

Spencer, E. (1967) A method of analysis of the stability of embankments assuming parallel inter-slice forces. *Géotechnique*, **17**(1): 11–26.

Spikula, D. (1996) Subsidence performance of landfills: a seven-year review. In *Proceedings, GRI-10 Conference on Field Performance of Geosynthetics & Geosynthetic Related Systems*, Geosynthetic Research Institute, Philadelphia, PA, pp. 237–244.

Sposito, G. (1984) *The Surface Chemistry of Soils*, Oxford University Press, New York.

Sprague, C.J. (1990) Leachate compatibility of polyester needlepunched nonwoven geotextiles. In Koerner, R.M. (Ed.) *Geosynthetic Testing for Waste Containment Applications, ASTM STP 1081*, American Society for Testing & Materials, West Conshohocken, PA, USA, pp. 212–224.

Sprute, R.H. & Kelsh, R.H. (1980) Dewatering fine-particle suspensions with direct current. In *Proceedings, International Symposium of Fine Particle Process*, Las Vegas, NV, Vol. 2, pp. 1828–1844.

Squillace, P.J., Pankow, J.F., Korte, N.E. & Zogorski, J.S. (1997) Review of the environmental behavior and fate of methyl tert-bretyl ethel. *Environmental Toxicology & Chemistry*, **16**(9): 1836–1844.

SSA (1997) Special issue. *Seismological Research Letters*, **68**(1): 1–255.

Stagg, K.G. (1968) *In situ* tests on the rock mass. In Stagg, K.G. & Zienkiewicz, O.C. (Eds), *Rock Mechanics in Engineering Practice*, Wiley, Chichester, pp. 125–156.

Stain, R.T. & Williams, H.T. (1991) Interpretation of sonic coring results–a research project. In *Proceedings, Fourth International Conference on Piling & Deep Foundations*, Stresa, Balkema, Rotterdam.

Stark, T.D. & Eid, H.T. (1993) Modified Bromhead ring shear apparatus. *Geotechnical Testing Journal, ASTM*, West Conshohocken, PA, **16**(1): 100–107.

Stark, T.D. & Eid, H.T. (1996) Shear behavior of reinforced geosynthetic clay liners. *Geosynthetics International*, **3**(6): 771–786.

Stark, T.D. & Olson (1995) Liquefaction Resistance Using CPT and Field Case Histories. ASCE Journal of Geotechnical Engineering, Reston, VA, **121**(12): 856–869.

Stark, T.D. & Poeppel, A.R. (1994) Landfill liner interface strengths from torsional-ring-shear tests. *Journal of Geotechnical Engineering, ASCE*, Reston, VA, **120**,(3): 597–615.

Stark, T.D., Williamson, T.A. & Eid, H.T. (1996) HDPE geomembrane–geotextile interface shear strength. *Journal of Geotechnical Engineering, ASCE*, Reston, VA, **122**(3): 197–203.

St-Arnaud, G., Morel, R. & Lavallée, J.-G. (1992) *Comportement de la fondation argileuse traitée avec des drains synthétiques sous le remblai d'essai Olga-C*, Internal Report, Hydro-Québec, Service géologie et structures, Montréal, Canada.

Starr, R.C. & Cherry, J.A. (1990) *In situ* barriers for groundwater pollution control. In *Conference on Prevention & Treatment of Soil & Groundwater Contamination in the Petroleum*

Refining & *Distribution Industry*, October 16–17, Montreal, PQ.

Starr, R.C., & Cherry, J.A. (1992) Applications of low permeability cutoff walls for groundwater pollution control. In *Proceedings, 45th* Canadian Geotechnical Conference, October 26–28, Toronto, ON.

Starr, R.C. & Gillham, R.W. (1993) Denitrification and organic carbon availability in two aquifers. *Ground Water*, **31**(6): 934–947.

Starr, R.C. & Cherry, J.A. (1994) Passive *in situ* remediation of contaminated groundwater: the funnel and gate system. *Ground Water*, **32**(3): 465–476.

Stas, C.V. & Kulhawy F.H. (1984) *Critical Evaluation of Design Methods for Foundations Under Axial Uplift and Compression Loading*, EPRI Report EL-3771, Cornell University.

Stefan, J. (1891) Uber die Theorie des Eisbildung, insbesondere uber die Eisbildung im Polameere. Ann. Phys. u. Chem., Neue Folge, **42**(2): 269–286.

Stein, D., Möllers, K. & Bielecki, R., (1989) Microtunnelling, Ernst & Sohn, Berlin.

Stephens, D.B. (1996) *Vadose Zone Hydrology*, CRC Lewis, Boca Raton, FL.

Stermac, A.G., Lo, K.Y., & Barsvary, A. (1967) The performance of an embankment founded on a deep deposit of carved clay. *Canadian Geotechnical Journal*, **4**(1):45–61.

Stewart, D.P., Jewell, R.J. & Randolph, M.F. (1994) Design of piled bridge abutments on soft clay for loading from lateral soil movements. *Geotechnique*, **14**(2): 277–296.

Stocker, M.F., Korber, G.W., Gassler, G. & Gudehus, G. (1979) Soil nailing. In *C.R. Coll. Int. Reinforcement des Sols*, Paris, pp. 105–118.

Stollenwerk, K.G. (1991) *Simulation of Molybdate Sorption with Diffuse Layer Surface-Complexation Model*, Water Resources Investigations Report 91–4034, US Geological Survey, pp. 47–52.

Stroud, M.A. (1974) The Standard Penetration Test in insensitive clays and soft rocks. Proceedings European Symposium on Penetration Testing, Stockholm **2**: 367–375.

Stulgis, R.P., Soydemir, C. & Telgener, R.J. (1995) Predicting landfill settlement. In *Geoenvironment 2000*, Geotechnical Special Publication No. 46, Acar, Y.B. and Daniel D.E. (Eds.) American Society of Civil Engineers, Reston, VA, Vol. 2, pp. 980–993.

STUDENT (W. Gossen) (1908) The probable error of a mean. *Biometrika*, **54**: 1–25.

Stumm, W. & Morgan, J.J. (1996) *Aquatic Chemistry: Chemical Equilibria and Rates in Natural Waters*, 3rd Edn, Wiley, New York.

Sui, T.L., Na, W.J. & Li, D.Z. (1988) Distribution of horizontal frost heaving forces on test retaining wall. In *Proceedings, Fourth Chinese Conference on Glaciology & Geocryology* pp. 108–113. (in Chinese).

Sullivan, W.R., Reese, L.C. & Fenske, C.W. (1979) Unified method for analysis of laterally loaded piles in clay. In *Conference on Numerical Methods in Offshore Piling*, Institution of Civil Engineers, London, pp. 135–146.

Suzuki, Y., Taye, Y., Tokimatsu, K. & Kubota, Y. (1995). Correlation between CPT Data and Dynamic Properties of In-Situ Frozen Samples. Proceedings of the Third International Conference on Recent Advances in Geotechnical Earthquake Engineering and Soil Dynamics, St. Louis, Missouri, April 2–7, 1995. Prakash, Shamsher (Ed.). University of Missouri Rolla, 1995, Vol. 1, pp. 249–252.

Swan, R.H., Bonaparte, R., Bachus, R.C., Rivette, C.A. & Spikula, D.R. (1991) Effect of soil compaction conditions on geomembrane–soil interface strength. *Geotextiles & Geomembranes*, **10**(5): 523–529.

Swane, I.C. & Poulos, H.G. (1984) Shakedown analysis of laterally loaded pile tested in stiff clay. In *Proceedings, Fourth Aust–NZ Conference* on Geomechanics, Perth, Vol. 1, pp. 165–169.

Swedish Geotechnical Society (1995) *CPT'95: Proceedings, International Symposium on Cone Penetration Testing*, October 4–5, Linkoping, Sweden, Vols 1–3, Swedish Geotechnical Institute.

Swedish Road Board (1974) *Bankpälming, särtryck ur verksamhetshan doken, (Embankment Piles)*, Report TV121, Au 110:1 kap 4.2.8.06 och del ii kap 4.3.1 Sweden (in Swedish).

Sy, A. (1996) Pile Dynamics and impact *in-situ* penetration tests. *Canadian Geotechnical Journal*, **34**(6): 952–973.

Sy, A. & Campanella, R.G. (1991a) Wave equation modelling of the SPT. In *Proceedings, ASCE Geotechnical Engineering Congress 1991*, Boulder, CO, Geotechnical Special Publication No. 27, Vol. 1, pp. 225–240.

Sy, A. & Campanella, R.G. (1991b) An alternative method of measuring SPT energy. *Proceedings, Second International Conference on Recent Advances in Geotechnical Engineering & Soil Dynamics*, St Louis, MO, Vol 1, pp. 499–505.

Sy, A & Campanella, R.G. (1992a) Dynamic performance of the Becker hammer and penetration test. In *Proceedings, 45th Canadian Geotechnical Conference*, Toronto, Ontario, Paper 24.

Sy, A. & Campanella, R.G. (1992b) Dynamic measurements of the Becker penetration test with implications for pile driving analysis. In *Proceedings, Fourth International Conference on the Application of Stress-Wave Theory to Piles*. The Hague, The Netherlands, A.A. Balkema Publishers, Rotterdam, pp. 471–478.

Sy, A. & Campanella, R.G. (1994) Becker and standard penetration tests (BPT–SPT) correlations with consideration of casing friction. *Canadian Geotechnical Journal*, **31**: 343–356.

TAC (1997) *Pavement Design and Management Guide*, Transportation Association of Canada, Ottawa, Ontario.

TAI (1983) *Asphalt Overlays for Highways and Streets*, Manual Series No. 17, The Asphalt Institute, Lexington, KY.

TAI (1987a) *Soil Manual*, Asphalt Institute Manual Series No. 10, The Asphalt Institute, Lexington, KY.

TAI (1987b) *Asphalt Pavements for Air Carrier Airports*, Manual Series No. 11, The Asphalt Institute, Lexington, KY.

TAI (1988) *The Asphalt Handbook*, Manual Series No. 4, The Asphalt Institute, Lexington, KY.

TAI (1991a) *Thickness Design – Asphalt Pavements for Highways and Streets*, Manual Series No. 1, The Asphalt Institute, Lexington, KY.

TAI (1991b) *Computer Program CP-4 Thickness Design of Highways and Overlays (HWY)*, The Asphalt Institute, Lexington, KY.

Tajimi, H (1969) Dynamic analysis of a structure embedded in an elastic stratum. In *Proceedings Fourth World Conference on Earthquake Engineering*, Chile, Vol. III, 53–69.

Takahashi, T. (1991) *Debris flow*, Balkema, Rotterdam.

Tarbuck, E.J. & Lutgens, F.K. (1996) *Earth–An Introduction to Physical Geology*, 5th Edn, Prentice Hall, Upper Saddle River, NJ.

TA Siedlungsabfall (1993) Technische anleitung zur verwertung, behandlung und sonstigen entsorgung von Siedlungsabfallen, bundesanzeiger.

Task Force No. 27 (1991) *Guidelines for the Design of Mechanically Stabilized Earth Walls*, AASHTO–AGC–ARTBA Joint Committee, Washington, DC.

Tatsuoka, F. & Kohata, Y. (1994) Stiffness of hard soils and soft rock in engineering applications. In Shibuya, S., Mitachi, T. & Miura S. (Eds) *Proceedings, First International Conference on Pre-Failure Deformation Characteristics of Geomaterials*, Sapporo, Japan, Balkema, Rotterdam, Vol. 2, pp. 947–1063.

Tavenas, F. & Leroueil, S. (1977) Effects of stresses and time on yielding of clays. In *Proceedings, Ninth International Conference on Soil Mechanics & Foundation Engineering*, Tokyo, Vol. 1, pp. 319–326.

Tavenas, F. & Leroueil, S. (1980) The behaviour of embankments on clay foundations. *Canadian Geotechnical Journal*, **17**(2): 236–260.

Tavenas, F. & Leroueil, S. (1981) Creep and failure of slopes in clays. *Canadian Geotechnical Journal*, **18**(1): 106–120.

Tavenas, F. & Leroueil, S. (1987) State-of-the-art on laboratory and *in-situ* stress–strain–time behavior of soft clays. In *Proceedings, International Symposium on Geotechnical Engineering of Soft Soils*, Mexico City, pp. 1–46.

Tavenas, F., Blanchette, G., Leroueil, S., Roy, M. & LaRochelle, P. (1975) Difficulties in the *in-situ* determination of K_0 in soft sensitive clays. In *Proceedings, ASCE Geotechnical Engineering Division Specialty Conference on*, In-Situ *Measurement of Soil Properties*, June 1–4, North Carolina State University, Raleigh, NC, Vol. 1 pp. 450–476.

Tavenas, F., Blanchet, R., Garneau, R. & Leroueil, S. (1978) The stability of stage-constructed embankments on soft clays. *Canadian Geotechnical Journal*, **15**(2): 283–305.

Tavenas, F., Mieussens, C. & Bourges, F. (1979) Lateral displacements in clay foundations under embankments. *Canadian Geotechnical Journal*, **16**(3): 532–550.

Tavenas, F., Trak, B. & Leroueil, S. (1980) Remarks on the validity of stability analyses. *Canadian Geotechnical Journal*, **17**(1): 61–73.

Tavenas, F., Flon, P., Leroueil, S. & Lebuis, J. (1983a) Remolding energy and risk of slide retrogression in sensitive clays. *Symposium on Slope Stability*, Linköping, Sweden, SGI Report No. 17, pp. 423–454.

Tavenas, F., Jean, P., Leblond, P. & Leroueil, S. (1983b) The permeability of natural clays, Part II: permeability characteristics. *Canadian Geotechnical Journal*, **20**(4): 645–660.

Tawfig, K.S., Aggour, M.S. & Al-Sanad, H. (1988) Dynamic properties of cohesive soils from impulse testing. In *Proceedings, Ninth World*

Conference on Earthquake Engineering, Tokyo, Vol. 3, pp. 11–16.

Taylor, D.W. (1948) *Fundamentals of Soil Behavior*, Wiley, New York.

Taylor, M.B. & Erikson, J.S. (1996) Accelerated site characterization. *ASTM Standardization News*, **June**: 34–39.

Tchepak, S. (1997) Private communication.

Tchobanoglous, G., Theisen, H. & Vigil, S. (1993) *Integrated Solid Waste Management*, McGraw-Hill, New York.

Tedd, P., Powell, J.J.M., Charles, J.A. & Uglow, I.M. (1990) *In-situ* measurement of earth pressure using push-in-spade-shaped pressure cells–10 years experience. In *Proceedings, Conference on Geotechnical Instrumentation in Practice*, April 1989, Nottingham, Thomas Telford, London, pp. 701–715.

Terzaghi, K. (1923) Die Berechnung der Durchlassigkeitziffer des Tones aus dem Verlauf der hydrodynamischen Spannungserscheinungen. Original paper published in 1923 and reprinted in *From Theory to Practice in Soil Mechanics*, Wiley, New York, pp. 133–146.

Terzaghi, K. (1925) *Erdbaumechanik auf bodenphysikalischer*, Grundlag, Franz Deuticke, Vienna.

Terzaghi, K. (1943) *Theoretical Soil Mechanics*, Wiley, New York.

Terzaghi, K. & Peck, R.B. (1967) *Soil Mechanics in Engineering Practice*, 2nd Edn, Wiley, New York.

Terzaghi, K., Peck, R.B. & Mesri, G. (1996) *Soil Mechanics in Engineering Practice*, 3rd Edn, Wiley, New York.

Thevanayagam, S. & Nesarajah, S. (Eds) (1996) *International Workshop on Technology Transfer for Vacuum-Induced Consolidation: Engineering and Practice*, Report of a workshop sponsored by the National Science Foundation, Port of Los Angeles, and ISSMFE TC-17, Los Angeles, CA.

Thevanayagam, S., Kavananjian, E., Jacob, A. & Juran, I. (1994) Prospects of vacuum-assisted consolidation for ground improvement of coastal and offshore fills. In Rollins, K.M. (Ed.) In Situ *Deep Soil Improvement*, Special Publication 45, American Society of Civil Engineers, ASCE Press, Reston, VA, pp. 90–105.

Thibodeaux, L.J. (1996) *Environmental Chemodynamics: Movement of Chemicals in Air, Water & Soil*, 2nd Edn, Wiley, New York.

Thiel, R.S. (1997) Discussion of 'Field evaluation of protective covers for landfill geomembrane liners under construction loading' by Reddy *et al. Geosynthetics International*, **4**(5): 543.

Thomas, R.W. & Ancelet, C.R. (1993) The effect of temperature, pressure and oven aging on the high pressure oxidative induction time of different types of stabilizers. In *Geosynthetics '93*, Nashville, TN, IFAI, Roseville, MN, pp. 915–924.

Thombre, M.S., Thomson, B.M. & Barton, L.L. (1997) Use of a permeable biological reaction barrier for groundwater remediation at Uranium Mill Tailings Remedial Action (UMTRA) site. In *Proceedings, International Containment Technology Conference*, February 9–21, St. Petersburg, FL, US Department of Energy/Dupont Company/US Environmental Protection Agency, pp. 744–750.

Thompson, F.L. & Tyler, S.W. (1984) *Comparison of Two Groundwater Flow Models–UNSATID & HELP*, EPRI CS-3695, Electric Power Research Institute, Palo Alto, CA.

Thompson, R.W. (1997) Evaluation protocol for repair for residences damaged by expansive soils. In Houston, S.L. & Fredlund, D.G. (Eds) *Unsaturated Soil Engineering Practice*, Geotechnical Special Publication No. 68, American Society of Civil Engineers, Reston, VA, pp. 255–276.

Thomson, J. (1993) *Pipejacking and Microtunnelling*, Blackie Academic Publishers, Glasgow.

Thornthwaite, C.W. & Mather, J.R. (1955) The water balance. *Publications in Climatology*, **8**(1): 104.

Thornthwaite, C.W. & Mather, J.R. (1957) Instructions and tables for computing potential evapotranspiration and the water balance. *Publications in Climatology*, **10**(3): 185–311.

Ting, J.M. & Scott, R.F. (1984) Static and dynamic lateral pile group action. In *Proceedings, Eighth World Conference on Earthquake Engineering*, San Francisco, Vol. III, pp. 641–648.

Tognon, A.R., Rowe, R.K. & Moore, I.D. (1999) *Large Scale Testing of Geomembrane Protection Layers*, GEOT-16-99, Faculty of Engineering Science, University of Western Ontario, London, Canada.

Tokimatsu, K. & Seed, H.B. (1987) Evaluation of settlements in sands due to earthquake shaking. *Journal of Geotechnical Engineering, ASCE*, Reston, VA, **113**(8):861–879.

Tomlinson, D.W., Thomson, N.R. & Johnson, R.L. (1997) Performance assessment of *in situ* air sparging for the removal of tetrachloroethylene from a mildly heterogeneous sand aquifer.

In *Proceedings, 1997 Petroleum Hydrocarbons & Organic Chemicals in Ground Water: Prevention, Detection & Remediation*, November 12–14, Houston, TX, API/AGWA.

Tomlinson, M.J. (1957) The adhesion of piles driven into clay soils. In *Proceedings, Fourth International Conference on Soil Mechanics & Foundation Engineering*, London, England, Vol. 2, pp. 66–71.

Tomlinson, M.J. (1977) *Pile Design and Construction*, Viewpoint London, England.

Tomlinson, M.J. (1986) *Foundation Design and Construction*, 5th Edn., Longman, Harlow, England.

Tomlinson, M.J. (1995) *Foundation Design and Construction*, 6th Edn., Longman, Harlow, England.

Tòth, P.S. (1993) *In situ* soil mixing. In Moseley, M.P. (Ed.) *Ground Improvement*, Blackie Academic & Professional, Blackwood, NJ, pp. 193–204.

Trak, B., La Rochelle, P., Tavenas, F., Leroueil, S. & Roy, M. (1980) A new approach to the stability analysis of embankments on sensitive clays. *Canadian Geotechnical Journal*, 17(4): 526–544.

Trast, J. & Benson, C.H. (1995) Estimating field hydraulic conductivity at various effective stresses. *Journal of Geotechnical Engineering, ASCE*, Reston, VA, 121(10): 736–740.

Trautmann, C.H. & O'Rourke, T.D. (1985) Lateral force–displacement response of buried pipe. *Journal of Geotechnical Engineering, ASCE*, Reston, VA, 111(9): 1077–1092.

Trautmann, C.H., O'Rourke, T.D. & Kulhawy, F.H. (1985) Uplift force–displacement response of buried pipe. *Journal of Geotechnical Engineering, ASCE*, Reston, VA, 111(9): 1061–1076.

TRB (1986) *Shared Experience in Geotechnical Engineering: Wick Drains*, Transportation Research Circular No. 309, Transportation Research Board, Washington, DC.

TRB (1998) Transportation Research Board, *Catalog of Current State Pavement Design Features*, NCHRP Project 1–32 Report, Transportation Research Board, Washington, DC, (CD-ROM only).

Trochanis, A.M., Bielak J. & Christiano, P. (1988) *A Three-Dimensional Nonlinear Study of Piles Leading to the Development of a Simplified Model*, Technical Report, Grant No. ECE-86/1060, Carnegie Mellon University, Pittsburg, PA, USA.

Trochanis, A.M., Bielak, J. & Christiano, P. (1991) Three-dimensional nonlinear study of piles. *Journal of Geotechnical Engineering, ASCE*, Reston, VA, 117(3): 429–447.

Trow, W.A. & Lo, K.Y. (1989) Horizontal displacements induced by rock excavation: Scotia Plaza, Toronto. *Canadian Geotechnical Journal*, 26: 114–121.

Turner, A.K. & Schuster, R.L. (1996) *Landslides: Investigation and Mitigation*, Special Report 247, Transportation Research Board, Washington, DC.

Turner, M.J. (1997) *Integrity Testing in Piling Practice*, Report 144, Construction Industry Research and Information Association (CIRIA), London, England.

Tuzuki, M., Inada, O. & Yamagishi, M. (1992) Field testing and analysis of dynamic loaded pile group. In *Proceedings, 10th World Conference on Earthquake Engineering*, Madrid, Spain, Vol. 3, pp. 1787–1790.

Udell, K.S. (1996) Thermal treatment of low permeability soils using electrical resistance heating. In *In situ Remediation of DNAPL Compounds in Low Permeability Media: Fate/Transport, In Situ Control Technologies & Risk Reduction*, Siegrist, R.L. & Walden, T.L., (Eds.) ORNL/TM-13305, Oak Ridge National Laboratory, Oak Ridges, TN, pp. 15: 1–18.

Udell, K.S. (1997) Thermally enhanced removal of liquid hydrocarbons contaminants from soils & groundwater. In. Ward, C.H., Cherry, J.A. & Scalf, M.R. (Eds.) *Subsurface Restoration*, Ann Arbor Press, Chelsea, MI pp. 251–270.

Udoh, F.D. (1991) *Minimization of Infiltration into Mining Stockpiles Using Low Permeability Covers*, Dissertation Proposal, Department of Materials Science & Engineering, University of Wisconsin, Madison, WI.

Ullidtz, P. (1987) *Pavement Analysis*, Elsevier, Amsterdam.

Ullidtz, P. (1998) *Modelling Flexible Pavement Response and Performance*, Polyteknisk Forlag, Anker Engelunds Vej 1, DK-2800, Lyngby, Denmark.

Ulrich, C.M. & Kuhlemeyer, R.L. (1973) Coupled rocking and lateral vibrations of embedded footings. *Canadian Geotechnical Journal*, 10(2): 145–160.

United States Office of the Federal Register (1994) *Code of Federal Regulations, 40, Protection of the Environment*, National Archives and Records Administration, pp. 260–268.

Urciuoli, G. (1990) *Contributo alla caratterizzazione geotecnica delle frane dell'Appennino*, Report of the Istituto di Tecnica delle Fondazioni e Costruzioni in Terra, Universita di Napoli Federico II.

US Army (1966) *Arctic and Subarctic Construction: Calculation Methods for Determination of Depths of Freeze and Thaw in Soils*, Technical Manual TM5-852-6/AFM 88-19.

US Army (1983) *Dewatering and Groundwater Control*, Army Joint Departments of the Army, the Air Force & Navy, Technical Manual TM-5-818-5/AFM88-5, Chapter 6/NAVFAC P-418, Washington, DC.

USDA-SCS (1985) *National Engineering Handbook, Section 4, Hydrology*, US Government Printing Office, Washington, DC.

USEPA (1984) 40 CFR Part 136 *Guidelines Establishing Test Procedures for the Analysis of Pollutants Under the Clean Water Act*, October 26, and all subsequent revisions (Federal Register). United States Government Printing Office, Pittsburg, PA.

USEPA (1985) *Modelling Remedial Actions at Uncontrolled Hazardous Waste Sites*, EPA/540/2-85/001.

USEPA (1986a) *RCRA Ground-Water Monitoring Technical Enforcement Guidance Document*, OSWER-9950.1.

USEPA (1986b) *Superfund Health Evaluation Manual*, EPA/540-1-86.

USEPA (1986c) *Test Methods for Evaluating Solid Waste, Physical/Chemical Methods*, 3rd Edn, November (with all subsequent revisions).

USEPA (1987) *A Compendium of Superfund Field Operations Methods*, EPA/540/P-87/001.

USEPA (1989a) *Seminar Publication: Transport and Fate of Contaminants in Subsurface*, EPA/625/4-89/019.

USEPA (1989b) *Integrated Risk Information System (IRIS) Database*, Office of Research & Development.

USEPA (1989c) *RCRA Facility Guidance*, Vols I–IV, EPA/530/SW-89/031.

USEPA (1989d) *Methods for Evaluating Attainment of Cleanup Standards*, Vol. 1, *Soils and Solids Media*, EPA/230/2-89/042.

USEPA (1989e) *Statistical Analysis of Groundwater Monitoring Data at RCRA Facilities*, Interim Final Guidance, Office, of Solid Waste, EPA/530/SW-89/026.

USEPA (1989f) *Risk Assessment Guidance for Superfund*, Vol. 1, *Human Health Evaluation Manual*, Part A, Environmental Protection Agency, Washington, DC.

USEPA (1990) *Guidance for Data Usability in Risk Assessment–Interim Final*, EPA/540/G-90/008.

USEPA (1991a) *Design and Construction of RCRA/CERCLA Final Covers*, EPA/625/4-91/025, US. Environmental Protection Agency, Office of Research & Development, Washington, DC.

USEPA (1991b). *Site Characterization for Subsurface Remediation*, Seminar Publication, EPA/625/4-91/026, Office of Emergency & Remedial Response, Washington, DC.

USEPA (1991c) *Compendium of ERT Groundwater Sampling Procedures*, EPA/540-91.007.

USEPA (1992a) *RCRA Ground-Water Monitoring: Draft Technical Guidance*, EPA/530-R-93-001, Office of Solid Waste, US Environmental Protection Agency, Washington, DC.

USEPA (1992b) *Bioremediation of Hazardous Wastes*, EPA/600/R-92/126, Office of Research & Development, Washington, DC.

USEPA (1992c) *Statistical Analysis of Ground-Water Monitoring Data at RCRA Facilities: Addendum to Interim Final Guidance*, Office of Solid Waste, Washington, DC. Draft.

USEPA (1992d) *Dense Non-Aqueous Phase Liquids–A Workshop Summary*, April 16–18, 1991, Dallas, TX, EPA/600/R-92/030, US Environmental Protection Agency, Washington, DC.

USEPA (1993a) *Solidification/Stabilization and its Application to Waste Materials*, Technical Resource Document, EPA/530-R-93/012, office of Research & Development, Washington, DC.

USEPA (1993b) *Remediation Technologies Screening Matrix and Reference Guide. Solid Waste & Emergency Response* (OS-110W), EPA-542-B-93-005.

USEPA (1994) *Evaluation of Technologies for In situ Cleanup of DNAPL Contaminated Sites*, EPA/600/R-94/120, Office of Research & Development, Washington, DC.

USEPA (1996) *Enabling Document for the New Source Performance Standards and Emissions Guidelines for Municipal Solid Waste Landfills*, EPA-453/R-96-004, US Environmental Protection Agency, Atmospheric Pollution Prevention Division, Washington, DC.

USEPA (1997) *Framework for Environmental Health Risk Management*, Presidential/Congressional Commission on Risk Assessment & Risk Management, Final Report, Vol. 1.

USEPA (1998) *Evaluation of Subsurface Engi-*

neered Barriers at Waste Sites, EPA 542-R-98-005, Office of Solid Waste & Emergency Response (5102G).

USEPA (2000) Technical Guidance for RCRA/CERCLA Final Covers, US Environmental Protection Agency Superfund Office, Washington, DC (in preparation).

USEPA ILM04.0 USEPA Contract Laboratory Program (CLP) Statement of Work (SOW) for Inorganic Analysis.

USEPA, OLM03.2. USEPA Contract Laboratory Program (CLP) Statement of Work (SOW) for Organic Analysis.

US Navy (1983) Soil Dynamics, Deep Stabilization, and Special Geotechnical Construction, Design Manual 7.3, Naval Facilities Engineering Command, Alexandria, VA.

Uthayakumar, M. & Vaid, Y.P. (1998) Static liquefaction of sand under multiaxial loading. Canadian Geotechnical Journal, 35(2): 273–283.

Vaid, Y.P. & Chern, J.C. (1985) Cyclic and monotonic undrained response of saturated sands. Advances in the Art of Testing Soils Under Cyclic Conditions, American Society of Civil Engineers, Reston, VA, 120–147.

Vaid, Y.P. & Negussey, D. (1988) Preparation of Reconstituted Sand Specimens, Advanced Triaxial Testing of Soils & Rock, ASTM STP 977, American Society for Testing & Materials, West Conshohocken, PA, pp. 405–417.

Vaid, Y.P. & Sivathayalan, S. (1996) Static and cyclic liquefaction potential of Fraser Delta sand in simple shear and triaxial tests. Canadian Geotechnical Journal, 33(2): 281–289.

Vaid, Y.P. & Thomas, J. (1994) Post liquefaction behaviour of sand. In Proceedings, 13th International Conference on Soil Mechanics & Foundation Engineering, New Delhi, India.

Vaid, Y.P., Sivathayalan, S. & Stedman, D. (1999) Influence of specimen reconstituting method on the undrained response of sand. ASTM Geotechnical Testing Journal, West Conshohocken, PA, 22(3): 187–195.

Vaid, Y.P., Stedman, D. & Sivathayalan, S. (2000) K_o and K_α factors in cyclic liquefaction. Canadian Geotechnical Journal, submitted.

van Bavel, C.H.M. (1952) Gaseous diffusion and porosity in porous media. Soil Sciences, 73: 91–104.

Van der Molen, W.H. & Van Ommen, H.C. (1988) Transport of solutes in soils and aquifers. Journal of Hydrology, 100: 433–451.

Vanapalli, S.K. (1994) PhD thesis, University of Saskatchewan, Saskatoon, Canada.

Vanapalli, S., Fredlund, D.G., Pufahl, D.E. & Clifton, A.W. (1996) Model for the prediction of shear strength with respect to soil suction. Canadian Geotechnical Journal, 33(3): 379–392.

van der Heijde, P.K.M. & Elnawawy, O.A. (1993) Compilation of Ground-water Models, EPA/600/2-93/118. Robert S. Kerr, Environmental Research Laboratories, Ada, OK.

Vandevivere, P. & Baveye, P. (1992) Effect of bacterial extracellular polymers on the saturated hydraulic conductivity of sand columns. Applied & Environmental Microbiology, 58: 1690–1698.

van Genuchten, M.Th. (1980) A closed-form equation of predicting the hydraulic conductivity of unsaturated soils. Soil Science of America Journal, 44: 892–898.

Van Impe, W.F. (1989) Soil Improvement Techniques and their Evolution, Rotterdam, Balkema, Netherlands.

Van Impe, W.F. (1991) Deformations of deep foundations. In Proceedings, 10th European Conference on Soil Mechanics & Foundation Engineering, Florence, Vol. 3, pp. 1031–1062.

Van Olphen, H. (1977) An Introduction to Clay Colloid Chemistry, 2nd Edn., Wiley Interscience, New York, 318 p.

Van Weele, A.F. (1957) A method of separating the bearing capacity of a test pile into skin friction and point resistance. In Proceedings, Fourth International Conference on Soil Mechanics & Foundation Engineering, London, Vol. 2, pp. 76.

Varnes, D.J. (1978) Slope movement types and processes. In Landslides: Analysis & Control, R.L. Schuster & R. Krizek (Eds.) Special Report 176, Transportation Research Board, Washington, DC, pp. 11–33.

Varnes, D.J. & the IAGE Commission on Landslides & Other Mass Movements on Slopes (1984) Landslide Hazard Zonation–A Review of the Principles and Practice, UNESCO, Paris.

Veletsos, A.S. & Nair, V.V.D. (1974) Torsional vibration of viscoelastic foundation. Journal of the Geotechnical Division, ASCE, Reston, VA, 100(3): 225–246.

Veletsos, A.S. & Verbic, B. (1973) Vibration of viscoelastic foundations. Earthquake Engineering & Structural Dynamics, 2: 87–102.

Veletsos, A.S. & Wei, Y.T. (1971) Lateral and rocking vibrations of footings. Journal of Soil Mechanics & Foundations Division, ASCE, Reston, VA, 97(9): 1227–1248.

Veletsos, A.S. & Younan, A.H. (1994) Dynamic

soil pressures on rigid vertical walls. *Earthquake Engineering and Structural Dynamics*, **23**: 275–301.

Veletsos, A.S., Parikh, V.H. & Younan, A.H. (1995) Dynamic response of a pair of walls retaining a visco-elastic solid. *Earthquake Engineering and Structural Dynamics*, **24**: 1567–1589.

Vesic, A.S. (1969) *Experiments with Instrumented Pile Groups in Sand*, American Society for Testing & Materials, West Conshohocken, PA, STP 444 pp. 177–222.

Vesic, A.S. (1973) Analysis of ultimate loads of shallow foundations. *Journal of Soil Mechanics & Foundations Division, ASCE*, Reston, VA, **99** (SM1): 45–73.

Vesic, A.S. (1975) Bearing capacity of shallow foundations. In Winterkorn, H.F. & Fang, H.-Y. (Eds) *Foundation Engineering Handbook*, Van Nostrand Reinhold, New York.

Viggiani, C. (1981) Ultimate lateral load on piles used to stabilize landslides. In *Proceedings, 10th International Conference on Soil Mechanics & Foundation Engineering*, Stockholm, Vol. 3, pp. 555–560.

Vink, P. (1986) Loss of UV stabilizers from polyolefins during photo-oxidation. In Scott, G. (Ed.) *Developments in Polymer Stabilization–3*, Applied Science, London, England, pp. 117–138.

Vogel, T.M., Criddle, C.S. & McCarty, P.L. (1987) Transformation of halogenated aliphatic compounds. *Environmental Science & Technology*, **21**: 722–736.

Vrymoed, J. (1975) *Dynamic Analysis of Oroville Dam*, Office Report, Department of Water Resources, State of California.

Vucetic, M. & Dobry, R. (1991) Effect of soil plasticity on cyclic response. *Journal of Geotechnical Engineering, ASCE*, Reston, VA, **117**(1): 89–107.

Waas, G. & Hartmann, H.G. (1981) Pile foundations subjected to dynamic horizontal loads. In *European Simulation Meeting Modelling & Simulation of Large Scale Structural Systems*, Capri, Italy.

Wager, O. & Holtz, R.D. (1976) Reinforcing embankments by short sheet piling and tie rods. In *Proceedings International Symposium on New Horizons in Construction Materials*, Lehigh University, Vol. I, pp. 177–185.

Wagner, A.A. (1957) The use of the Unified Soil Classification System by the Bureau of Reclamation. In *Proceedings, Fourth International Conference on Soil Mechanics & Foundation Engineering*, London, England, Vol. I, pp. 125–134.

Walsh, J.B. (1965) The effect of cracks on the compressibility of rock. *Journal of Geophysical Research*, **70**: 381–389.

Walton, W.C. (1962) *Selected Analytical Methods for Well and Aquifer Evaluation*, State of Illinois, Bulletin No. 49, Urbana, IL.

Ward, C.H. (1996) Introductory talk: Where are we now? Moving to a risk-based approach. In *Symposium on Natural Attenuation of Chlorinated Organics in Groundwater*, September 11–13, Dallas, TX, 540/R-96-509, US Environmental Protection Agency, pp. 1–3.

Ward, W.H., Coates, D.J. & Tedd, P. (1976) Performance of Tunnel Support Systems in the Four Fathom Mudstone Tunnelling, '76 Institution Mining & Metallurgy, London, pp. 329–340.

Wardle, L.J. (1996) *CIRLY User's Manual, Version 3.0*, MINCAD Systems, Richmond, Australia.

Wardle, L.J. & Rodway, B. (1995) Development and application of an improved airport pavement design method. In *1995 Transportation Congress*, San Diego, American Society of Civil Engineers, ASCE Press, Reston VA.

Wardle, L.J. & Rodway, B. (1998a) Layered elastic pavement design–recent developments. In *Proceedings, Transport 98*, ARRB Transport Research, Vermont South, Victoria, Australia.

Wardle, L.J. & Rodway, B. (1998b) Recent developments in flexible aircraft pavement design using the layered elastic method. In *Proceedings, Third International Conference on Road & Airfield Pavement Technology*, Beijing, Vol. 2, Huajie International. pp. 882–890.

Warner, J. (1982) Compaction grouting—the first thirty years. In Baker, W.H. (Ed.) *Proceedings, Grouting in Geotechnical Engineering*, New Orleans, American Society of Civil Engineers, ASCE Press, Reston, VA, pp. 694–707.

Warner, J., Schmidt, N., Reed, J., Shepardson, D., Lamb, R. & Wong, W. (1992) Recent advances in compaction grouting technology. In Borden, R.H. Holtz, R.D. & Juran, I. (Ed.) *Grouting, Soil Improvement & Geosynthetics, Proceedings*, New Orleans, Geotechnical Special Publication No. 30, American Society of Civil Engineers, ASCE Press, Reston, VA, pp. 252–264.

Waterloo Centre For Ground Water Research (1989) Short Course, University of Waterloo,

Dense Immiscible Phase Liquid Contaminants in Porous and Fractured Media, Kitchener, Ontario, Canada.

Watts, K.S. & Charles, J.A. (1988) *In-situ* measurement of vertical and horizontal stress from a vertical borehole. *Géotechnique*, **38**(4): 619–626.

Wayne, M.H. & Koerner, R.M. (1993) Correlation between long term flow testing and current geotextile filtration design practice. In *Proceedings, Geosynthetics '93*, IFAI, Roseville, MN, pp. 501–517.

Weatherby, D.E. (1982) *Tiebacks*, Report No. FHW A/RD-81/047, Federal Highway Administration, Springfield, VA.

Weaver, K. (1991) *Dam Foundation Grouting*, American Society of Civil Engineers, ASCE Press, Reston, VA.

Webb, M.C., McGrath, T.J., Zoladz, G.V. & Selig, E.T. (1995) *Field Test Data from Pipe Installation Study*, Geotechnical Report No. NSF95-438I, Department of Civil & Environmental Engineering, University of Massachusetts, Amherst, MA.

Webb, M.C., Sussmann, J. & Selig, E.T. (1999) *Large Span Culvert Field Test Results*, NCHRP Project 12–45, Department of Civil & Environmental Engineering, University of Massachusetts, Amherst, MA.

Webb, S.W., McCord, J.T. & Dwyer, S.F. (1997) Prediction of tilted capillary barrier performance. In *Proceedings, International Containment Technology Conference*, February, St Petersburg, FL, pp. 296–302.

Weber, Jr, W.G. (1969) Performance of embankments constructed over peat. In *Proceedings, Journal of the Soil Mechanics & Foundations Division*, ASCE, Reston, VA, **95**(1): 53–76.

Well, L.W. (1997) *Testing and Acceptance Criteria for Geosynthetic Clay Liners*, ASTM STP 1308, American Society for Testing Materials, West Conshohocken, PA.

Welsh, J.P. (1986) *In situ* testing for ground modification techniques. In *Proceedings*, In Situ '86, Blacksburg, VA, Geotechnical Special Publication No. 6, American Society of Civil Engineers, ASCE Press, Reston, VA, pp. 322–335.

Welsh, J.P. (Ed.) (1987) *Soil Improvement—A Ten Year Update*, Geotechnical Special Publication No. 12, American Society of Civil Engineers, ASCE Press, Reston, VA.

Welsh, J.P., Anderson, R.D., Barksdale, R.P., Satyapriya, C.K., Tumay, M.T. & Wahls, H.E. (1987) Densification. In Welsh, J.P. (Ed.) *Soil Improvement–A Ten Year Update*, Geotechnical Special Publication No. 12, American Society of Civil Engineers, Reston, VA, pp. 67–97.

West, G. (1986) Desk studies, air photograph interpretation and reconnaissance for site investigation. In Hawkins, A.B. (Ed.) *Site Investigation Practice: Assessing BS 5930*, Engineering Geology Special Publication No. 2, Geological Society, London, England, pp. 9–16.

Westergaard, H. (1931) Water pressure on dams during earthquakes. *Transactions of ASCE*, Reston, VA, No. 98 Paper No. 1835: 418–433.

Wheeler, S.J. (1991) Alternative framework for unsaturated soil behavior. *Géotechnique*, **41**(2): 257–261.

Wheeler, S.J. & Sivakumar, V. (1992) Critical state concepts for unsaturated soil. *Proceedings, Seventh International Conference on Expansive Soils*, August 3–5, Dallas, TX, Vol. 1, pp. 167–172.

White, H.L. & Layer, J.P. (1960) The corrugated metal culvert as a compression ring. *Highway Research Board Proceedings*, **39**: 389–397.

White, N.F., Duke, H.R., Sundae, D.K. & Corey, A.T. (1970) Physics of desaturation in porous materials. *Irrigation & Drainage Division*, ASCE, Reston, VA, **96**: 165–191.

Whitman, J. & Christian, J. (1990) Seismic response of retaining structures. In *Symposium Seismic Design for World Port 2020*, Port of Los Angeles, Los Angeles, CA.

Whitman, R.V. (1984) Evaluating calculated risk in geotechnical engineering (17th Terzaghi Lecture). *Journal of Geotechnical Engineering*, **110**(2): 145–188.

Whitman, R. & Liao, S. (1985) *Seismic Design of Retaining Walls*, Miscellaneous Paper GL-85-1, US Army Engineer Waterways Experiment Station, Vicksburg, MS.

Whitman, R.V. & Richart, F.E. Jr (1967) Design procedures for dynamically loaded foundations. *Journal of the Soil Mechanics & Foundations Division*, ASCE, Reston, VA, **93**(6): 169–193.

Whittaker, M., Pollard, S.J.T. & Fallick, A.E. (1995) Characterisation of refractory wastes at heavy oil-contaminated sites: a review of conventional and novel analytical methods. *Environmental Technology*, **16**: 1009–1033.

Whittaker, M., Sprenger, J.G.S. & DuBois, D.D. (1998) Assessing the key components of credible risk assessment at contaminated sites. *Environmental Science & Engineering*, **10**(6): 102–103.

Whittle, R.W. (1996) Recent developments in the cone pressuremeter. In Craig, C. (Ed.) *Proceedings, International Conference on Advances in Site Investigation Practice*, March 30–31, 1995, Thomas Telford, London, England, pp. 533–546.

Whittlestone, A.P. & Ljunggren, C. (1996) A new technique for *in-situ* stress measurement by overcoring. In Craig C. (Ed.) *Proceedings, International Conference Advances in Site Investigation Practice*, March 30–31, 1995, London, Thomas Telford, London, England, pp. 499–509.

Wieczorek, G.F., Lips, E.W. & Ellen, S.D. (1989) Debris flows and hyperconcentrated floods along the Wasatch front, Utah, 1983 & 1984. *Bulletin of the International Association of Engineering Geologists*, **26**(2): 191–208.

Wiedemeier, T.H., Wilson, J.T. Kampbell, D.H., Miller, R.N. & Hansen, J.E. (1995) *Technical Principals of Implementing Intrinsic Remediation with Long-Term Monitoring for Natural Attenuation of Fuel Contamination Dissolved in Groundwater*, US Air Force Center for Environmental Excellence, Technology Transfer Division, Brooks Air Force Base, San Antonio, TX.

Wiedemeir, T.H., Swanson, M.A., Moutoux, D.E., Wilson, J.T., Kampbell, D.H., Hansen, J.E. & Haas, P. (1996a) Overview of technical protocol for natural attenuation of chlorinated aliphatic hydrocarbons in ground water under development for the US Air Force Center of Excellence. In *Symposium on Natural Attenuation of Chlorinated Organics in Groundwater*, September 11–13, Dallas, TX, 540/R-96-509, US Environmental Protection Agency pp. 35–59.

Wiedemeir, T.H., Wilson, J.T. & Kampbell, D.H. (1996b) Natural attenuation of chlorinated aliphatic hydrocarbons at Plattsburgh Air Force Base, New York. In *Symposium on Natural Attenuation of Chlorinated Organics in Groundwater*, September 11–13, Dallas, TX, 540/R-96-509, US Environmental Protection Agency, pp. 65–68.

Wiedemeier, T.H., Swanson, M.A., Moutoux, D.E., Gordon, E.K., Wilson, J.T., Wilson, B.H., Kampbell, D.H., Hansen, J.E., Haas, P. & Chapelle, F.H. (1996) *Technical Protocol for Evaluating Natural Attenuation of Chlorinated Solvents in Groundwater*, Draft Revision 1, Air Force Center for Environmental Excellence, Technology Transfer Division, Brooks Air Force Base, San Antonio, TX.

Wildavsky, A. (1988) *Searching for Safety*, Transaction Publishers, New Brunswick, NJ.

Wilkinson, W.B. (1967) A note on the constant head test to measure soil permeability *in-situ*. *Géotechnique*, **17**(2): 68–71.

Wilkinson, W.B. (1968) Constant head *in-situ* permeability tests in clay strata. *Géotechnique*, **18**(2): 177–194.

Williams, A.F. & Pells, P.J.N. (1981) Side resistance rock sockets in sandstone, mudstone and shale. *Canadian Geotechnical Journal*, **18**(4): 502–513.

Williams, J.G. (1984) *Fracture Mechanics of Polymers*, Ellis Horwood, Ltd, Chichester, England.

Williams, N.D. & Abouzakhm, M.A. (1989) Evaluation of geotextile–soil filtration characteristics using the hydraulic conductivity ratio analysis. *Journal of Geotextiles & Geomembranes*, **8**(1): 1–26.

Williams, N.D. & Houlihan, M.F. (1987) Evaluation of interface friction properties between geosynthetics and soils. In *Proceedings, Geosynthetics '87 Conference*, New Orleans, Industrial Fabrics Association International, Roseville, MN, Vol. 2, pp. 616–627.

Williams, N., Giroud, J.P. & Bonaparte, R. (1984) Properties of plastic nets for liquid and gas drainage associated with geomembranes. In *Proceedings, International Conference on Geomembranes*, June Denver, CO, Vol. 2, pp. 399–404.

Wilson, G.W. (1990) *PhD thesis*, University of Saskatchewan, Saskatoon, Canada.

Wilson, G.W., Fredlund, D.G. & Barbour, S.L. (1994) Coupled soil–atmosphere modeling for evaporation. *Canadian Geotechnical Journal*, **31**(2): 151–161.

Wilson, G.W., Barbour, S.L. & Fredlund D.G. (1997) The effect of soil suction on evaporative fluxes from soil surfaces. *Canadian Geotechnical Journal*, **33**(4): 145–155.

Wilson, J.L. & Conrad, S.H. (1984) Is physical displacement of residual hydrocarbons a realistic possibility in aquifer restoration? In *Proceedings, NWWA/API Conference on Petroleum Hydrocarbons & Organic Chemicals in Groundwater-Prevention, Detection, & Restoration*, National Water Well Association, Dublin, OH, pp. 274–298.

Wilson, R.D. & Mackay, D.M. (1995) A method for passive release of solutes from an unpumped well. *Ground Water*, **33**(6): 936–945.

Wilson, R.D., Mackay, D.M. & Cherry, J.A.

(1997) Arrays of unpumped wells for plume migration control by semi-passive *in situ* remediation. *Ground Water Monitoring & Remediation*, **17**(3): 185–193.

Wilson-Fahmy, R.F. & Koerner, R.M. (1992) *Stability Analysis of Multi-Lined Slopes in Landfill Applications*, GRI Report No. 8, Geosynthetic Research Institute, Drexel University.

Wilson-Fahmy, R.F. & Koerner, R.M. (1993) Finite element analysis of stability of cover soil on geomembrane-lined slopes. In *Proceedings, Geosynthetics '93 Conference*, February Vancouver, BC, Industrial Fabrics Association International, Roseville, MN, Vol. 3, pp. 1425–1438.

Wilson-Fahmy, R.G., Koerner, R.M. & Fleck, J.A. (1993) Unconfined and confined wide width testing of geosynthetics. In Cheng, S.J. (Ed.) *ASTM STP 1190*, American Society for Testing & Materials, West Conshohocken, PA, pp. 44–63.

Wilson-Fahmy, R.F., Narejo, D. & Koerner, R.M. (1996) Puncture protection of geomembranes. Part I: theory. *Geosynthetics International*, **3**(5): 605–628.

Winterkorn, H.F. & Pamukcu, S. (1991) Soil stabilization and grouting. In Fang, H.Y. (Ed.) *Foundation Engineering Handbook*, Van Nostrand Reinhold, New York, NY, pp. 317–378.

Wolf, J.P. (1995) *Foundation Vibration Analysis Using Simple Physical Models*, Prentice-Hall, Englewood Cliffs, NJ.

Wolf, J.P. & Darbre, G.R. (1984) Dynamic-stiffness matrix of soil by the boundary-element method: embedded foundations. *Earthquake Engineering & Structural Dynamics*, **12**: 401–416.

Wolf, J.P. & Darbre, G.R. (1986) Nonlinear soil–structure interaction analysis based on the boundary-element method in time domain with application to embedded foundation. *Earthquake Engineering & Structural Dynamics*, **14**: 83–101.

Wolf, J.P. & von Arx, G.A. (1978) Impedance functions of a group of vertical piles. In *Proceedings, ASCE Specialty Conference on Earthquake Engineering & Soil Dynamics*, Pasadena, CA, Vol. II, pp. 1024–1041.

Wolf, J.P., von Arx, G.A., de Barros, F.C.P. & Kakubo, M. (1981) Seismic analysis of the pile foundation of the reactor building on the NPP Angra 2. *Nuclear Engineering & Design*, **65**(3): 329–341.

Wolf, J.P., Meek, J.W. & Song, Ch. (1992) Cone models for a pile foundation. In Prakash, S. (Ed.) *Piles Under Dynamic Loads*, Geotechnical Special Publication No. 34, American Society of Civil Engineers, Reston, VA, pp. 94–113.

Wong, L.C. & Haug, M.D. (1991) Cyclical closed-system freeze–thaw permeability testing of soil liner & cover materials. *Canadian Geotechnical Journal*, **28**(6): 784–793.

Wood, D.M. (1990) *Soil Behaviour and Critical State Soil Mechanics*, Cambridge University Press, Cambridge, England.

Wood, D.M., Jendele, L., Chan, A.H.C. & Cooper, M.R. (1995) Slope failure by pore pressure recharge: numerical analysis. In *Proceedings, 11th European Conference on Soil Mechanics & Foundation Engineering*, Copenhagen, Vol. 6, pp. 1–8.

Wood, J. (1973) *Earthquake-induced Soil Pressures on Structures*, Report No. EERL 73-05, California Institute of Technology, Pasadena, CA.

Wood, W.W. (1996) Diffusion: the source of confusion. *Ground Water*, **34**(2): 193.

Woods, R.I. & Jewell, R.A. (1990) A computer design method for reinforced soil structures. *Geotextiles & Geomembranes*, **9**(3): 233–259.

Woodward, R., Gardner, W.S. & Greer, D.M. (1972) *Drilled Pier Foundations*, McGraw Hill, New York.

Worotnicki, G. & Walton, R.J. (1976) Triaxial 'hollow inclusion' gauges for the determination of rock stress *in situ*. *Proceedings, ISRM Symposium on Investigation of Stress in Rock & Advances in Shear Measurement*, Sydney supplement, pp. 1–8.

Woyshner, M.R. & Yanful, E.K. (1995) Modelling and field measurements of water percolation through an experimental soil cover on mine tailings. *Canadian Geotechnical Journal*, **32**(4): 601–609.

Wright, S.G. (1991) *UTEXAS3—A Computer Program for Slope Stability Calculations*, Department of Civil Engineering, University of Texas, Austin, TX.

Wroth, C.P. (1984) The interpretation of *in-situ* soil tests, 24th Rankine Lecture. *Géotechnique*, **34**: 449–489.

Wroth, C.P. (1988) Penetration testing – a more rigorous approach to interpretation. In De-Ruiter, J. (Ed.) *Proceedings First International Symposium on Penetration Testing*, ISOPT-1, Orlando, FL, Vol. 1, A.A. Balkema, Rotterdam, pp. 303–311.

Wroth, C.P. & Wood, D.M. (1978) The correlation of index properties with some basic engineering properties of soils. *Canadian Geotechnical Journal*, **15**(2): 137–145.

Wu, G. & Finn, W.D.L. (1996) Seismic pressure against rigid walls. In *Analysis and Design of Retaining Structures Against Earthquakes*, Geotechnical Special Publication No. 60, American Society of Civil Engineers, Reston, VA, pp. 1–19.

Wu, G. & Finn, W.D.L. (1999) Seismic lateral pressures for the design of rigid walls. *Canadian Geotechnical Journal*, **36**(3): 509–522.

Wu, T.H. (1976) *Soil Mechanics*, Allyn & Bacon, Boston, MA.

Wyllie, D.C. (1992) *Foundations on Rock*, Chapman & Hall, London, England.

Wyllie, D.C. & Norrish, N.I. (1996a) Rock strength properties and their measurement. In *Landslides—Investigation and Mitigation*, Special Report 247, Turner, A.K. & Schuster, R.L. (Eds.) Transportation Research Board, Washington, DC, pp. 372–390.

Wyllie, D.C. & Norrish, N.I. (1996b) Stabilization of rock slopes. In *Landslides – Investigation & Mitigation*, Special Report 247, Turner, Transportation Research Board, Washington, DC, Turner, A.K. & Schuster, R.L. (Eds.) pp. 473–504.

Wyss, M. (1979) Estimating Maximum Expectable Magnitude of Earthquakes from Fault Dimensions, *Geology* 7: 336–340.

Yan, L., Matasovic, N. & Kavazanjian, E., Jr. (1996) *YSLIPPM_A Computer Program for Simulation of Dynamic Behaviour of a Rigid Block on an Inclined Plane and Calculation of Permanent Displacements of the Block, User's Manual*, GeoSyntec Consultants, Huntington Beach, CA.

Yanful, E.K. (1993) Oxygen diffusion through soil covers on sulphidic mill tailings. *Journal of Geotechnical Engineering, ASCE*, Reston, VA, **119**(8): 1207–1228.

Yanful, E.K. & Aubé, B. (1993) Modelling moisture-retaining soil covers. In *Proceedings, Joint Canadian Society of Civil Engineers—American Society of Civil Engineers National Conference on Environmental Engineering*, Montreal, Canada, pp. 273–280.

Yanful, E.K., Nesbitt, H.W. & Quigley, R.M. (1988) Heavy metal migration at a landfill site, Sarnia, Ontario, Canada – I. Thermodynamic assessment and chemical interpretations. *Applied Geochemistry*, 3: 522–533.

Yanful, E.K., Bell, A.V. & Woyshner, M.R. (1993a) Design of a composite soil cover for an experimental waste rock pile near Newcastle, New Brunswick, Canada. *Canadian Geotechnical Journal*, **30**(4): 578–587.

Yanful, E.K., Riley, M.D., Woyshner, M.R. & Duncan, J. (1993b) Construction and monitoring of a composite soil cover on an experimental waste rock pile near Newcastle, New Brunswick, Canada. *Canadian Geotechnical Journal*, **30**(4): 588–599.

Yang, D.S. (1994) The applications of soil mix walls in the United States. *Geotechnical News*, **12**(4): 44–47.

Yang, D.S. & Takeshima, S. (1994) Soil mix walls in difficult ground. In In-Situ *Deep Soil Improvement*, Geotechnical Special Publication No. 45. K.M. Rollins (Ed.) American Society of Civil Engineers, ASCE Press, Reston, VA, 106–120.

Yang, D.S., Yagihashi, J.N. & Yoshizawa, S.S. (1998) Swing method for deep mixing. In *Soil Improvement for Big Digs*, Geotechnical Special Publication No. 81, A. Maher & D.S. Yang (Eds.) American Society of Civil Engineers, pp. 111–121, ASCE Press, Reston, VA.

Yasuda, S., Ishihara, K., Harada, K. & Shinkawa, N. (1996) Effectiveness of the ground improvement on the susceptibility of liquefaction observed during the 1995 Hyogoken Nanbu (Kobe) earthquake. In *Proceedings, 11th World Conference on Earthquake Engineering*, Mexico.

Yaws, C.L. (1995) *Handbook of Transport Property Data*, Gulf Publishing Company, Houston, TX.

Yegian, M.K. & Lahalf, A.M. (1992) Dynamic interface shear strength properties of geomembranes & geotextiles. *Journal of Geotechnical Engineering, ASCE*, Reston, VA, **118**(5): 760–779.

Yegian, M.K., Yee, Z.Y. & Harb, J.N. (1995) Seismic response of geosynthetic–soil systems. In *Geoenvironment 2000*, Geotechnical Special Publication No. 46, Acar, Y.B. and Daniel, D.E. (Eds.) American Society of Civil Engineers, Reston, VA, Vol. 2, pp. 1113–1125.

Yeung, A.T.-C. & Mitchell, J.K. (1992) Coupled fluid, chemical, and electrical flows in soil. *Géotechnique*, **43**(4).

Yim, G. & Godin, M. (1993) Long-term heat aging stabilization study of polyethylene and its relationship with oxidative induction time (OIT). In *Geosynthetics '93*, Nashville, TN, IFAI, Roseville, MN, pp. 803–816.

Yong, R.N. (1985) Interaction of clay and industrial waste: a summary review. In Hitchon B. & Trudell, M. (Eds) *Second Canadian/American Conference on Hydrogeology*, National Water Well Association, Westerville, OH, pp. 13–25.

Yood, T.L. & Idris, I. (Eds) (1997) *Proceedings, NCEER Workshop on Evaluation of Liquefaction Resistance of Soils*, Technical Report, NCEER-7-0022.

Yoshida, R., Fredlund, D.G. & Hamilton, J.J. (1983) The prediction of heave of a slab-on-grade floor on Regina clay. *Geotechnical Journal*, **20**(1): 69–81.

Yoshimi, Y., Hatanaka, M. & Oh-oka, H. (1978) Undisturbed sampling of saturated sands by freezing. *Soils & Foundations*, **18**(3): 59–71.

Yoshimine, M., Ishihara, K. & Vargas, W. (1998) Effect of principal stress direction and intermediate principal stress on undrained shear behaviour of sand, *Soils & Foundations*, **38**(3): 179–188.

Youd, T.L. (1973) Factors controlling maximum and minimum densities of sands. In *Evaluation of Relative Density and Its Role in Geotechnical Projects Involving Cohesionless Soils*, ASTM STP 523, American Society for Testing & Materials, West Conshohocken, PA, (Eds.) E.T. Selig and R.S. Ladd, pp. 98–112.

Youd, T.L. & Noble, S.K. (1998) Magnitude scaling factors. In Youd, T.L. (Ed.) *Draft Report, NCEER Workshop on Evaluation of Liquefaction Resistance*, Brigham Young University, Provo, UT.

Young, O.C. & Trott, J.J. (1984) *Buried Rigid Pipes: Structural Design of Pipelines*, Elsevier, London, England.

Youngs, R.R., Chiou, S.-J., Silva, W.J. & Humphrey, J.R. (1997) Strong motion attenuation relationships for subduction zone earthquakes. *Seismological Research Letters*, **68**(1): 58–73.

Yu, P. & Richart, F.E., Jr (1984) Stress ratio effects on shear modulus of dry sands. *Journal of Geotechnical Engineering, ASCE*, Reston, VA, **110**(3): 331–345.

Zanzinger, H. & Gartung, E. (1998) Efficiency of puncture protection layers. In Rowe, R.K. (Ed.) *Proceedings, Sixth International Conference on Geosynthetics*, Vol. 1, March, Atlanta, GA, Industrial Fabrics Association International, Roseville, MN, pp. 301–306.

Zar, J.H. (1982) Power and statistical significance impact evaluation. *Ground Water Monitoring Review*, Summer **2**: 33–35.

Zhang, B.Q. & Small, J.C. (1994) Finite layer analysis of soil-raft-structure interaction. In *Proceedings, XIII International Conference on Soil Mechanics & Foundation Engineering*, January 5–10, New Delhi, India, Vol. 2, pp. 587–590.

Zhang, C. & Moore, I.D. (1998) *Finite Element Analysis of Thermoplastic Pipes*, Transportation Research Record No. 1624, pp. 225–230.

Zhang, H., Schwartz, F.W., Wood, W.W., Garabedian, S.P. & LeBlanc, D.R. (1998) Simulation of variable-density flow and transport of reactive and nonreactive solutes during a tracer test at Cape Cod, Massachusetts. *Water Resources Research*, **34**(1): 67–82.

Zienkiewicz, O.C. (1979) *The Finite Element Method*, 3rd Edn, McGraw-Hill, London, England.

Zornberg, J.G. & Caldwell, J.A. (1998) Design of monocovers for landfills in arid locations. In *Third International Conference on Environmental Geotechnics*, ISSMFE, September, Lisbon, (in press).

Zornberg, J.G. & Giroud, J.P. (1997) Uplift of geomembranes by wind—Extension of equations. *Geosynthetics International*, **4**(2): 187–207, Erratum 6(6): 521–522.

INDEX

Page numbers in *italics* indicate figures. Page numbers followed by "t" indicate tables.

CONVERSION FACTORS US TO SI UNITS

	Multiply	by	to obtain
Length	ft	3.048×10^{-1}	m
	mile	1.609	km
Area	ft^{-2}	9.290×10^{-2}	m^2
	acre	4.047×10^3	m^2
	mi^2	2.590	km^2
Volume	ft^3	2.832×10^{-2}	m^3
	U.S. gal	3.785×10^{-3}	m^3
	U.K. gal	4.546×10^{-3}	m^3
	ft^3	2.832×10	ℓ
	U.S. gal	3.785	ℓ
	U.K. gal	4.546	ℓ
Velocity	$ft\ s^{-1}$	3.048×10^{-1}	$m\ s^{-1}\ (m/s)$
Acceleration	$ft\ s^{-2}\ (ft/s^2)$	3.048×10^{-1}	$m\ s^{-2}\ (m/s^2)$
Mass	lb_m	4.536×10^{-1}	kg
	ton	1.016×10^3	kg
Force and Weight	lb_f	4.448	N
Pressure and Stress	psf	4.788×10	Pa or N $m^{-2}\ (N/m^2)$
	psi	6.895×10^3	Pa or N $m^{-2}\ (N/m^2)$
	atm	1.013×10^5	Pa or N $m^{-2}\ (N/m^2)$
	atm	1.013	bar
Work and Energy	$ft\text{-}lb_f$	1.356	J
	calorie	4.187	J
Mass Density	$lb_m\ ft^{-3}\ (lb_m/ft^3)$	1.602×10	kg $m^{-3}\ (kg/m^3)$
Weight Density	$lb_f\ ft^{-3}\ (lb_f/ft^3)$	1.571×10^2	N $m^{-3}\ (N/m^3)$
Discharge	$ft^3\ s^{-1}\ (ft^3/s)$	2.832×10^{-2}	$m^3\ s^{-1}\ (m^3/s)$
	$ft^3\ s^{-1}\ (ft^3/s)$	2.832×10	$\ell\ s^{-1}\ (\ell/s)$
	U.S. gal min^{-1} (gal/min)	6.309×10^{-5}	$m^3\ s^{-1}\ (m^3/s)$
	U.S. gal min^{-1} (gal/min)	6.309×10^{-2}	$\ell\ s^{-1}\ (\ell/s)$
	U.S. gal min^{-1} (gal/min)	7.576×10^{-5}	$m^3\ m^3\ s^{-1}\ (m^3/s)$
	U.S. gal min^{-1} (gal/min)	7.576×10^{-2}	$\ell\ s^{-1}\ (\ell/s)$
Hydraulic	$ft\ s^{-1}\ (ft/s)$	3.048×10^{-1}	$m\ s^{-1}\ (m/s)$
Conductivity	U.S. gal $day^{-1}\ ft^{-2}$ (gal/day/ft^2)	4.720×10^{-7}	$m\ s^{-1}\ (m/s)$
Transmissivity	$ft^2\ s^{-1}\ (ft^2/s)$	9.290×10^{-2}	$m^2\ s^{-1}\ (m^2/s)$
	U.S. gal $day^{-1}\ ft^{-1}$ (gal/day/ft)	1.438×10^{-7}	$m^2\ s^{-1}\ (m^2/s)$